//

Plant Mitochondria

Plant Mitochondria

Special Issue Editor

Nicolas L. Taylor

MDPI • Basel • Beijing • Wuhan • Barcelona • Belgrade

Special Issue Editor
Nicolas L. Taylor
The University of Western Australia
Australia

Editorial Office
MDPI
St. Alban-Anlage 66
4052 Basel, Switzerland

This is a reprint of articles from the Special Issue published online in the open access journal *International Journal of Molecular Sciences* (ISSN 1422-0067) from 2017 to 2018 (available at: https://www.mdpi.com/journal/ijms/special_issues/plant_mitochondria)

For citation purposes, cite each article independently as indicated on the article page online and as indicated below:

LastName, A.A.; LastName, B.B.; LastName, C.C. Article Title. *Journal Name* **Year**, *Article Number*, Page Range.

ISBN 978-3-03897-550-2 (Pbk)
ISBN 978-3-03897-551-9 (PDF)

© 2019 by the authors. Articles in this book are Open Access and distributed under the Creative Commons Attribution (CC BY) license, which allows users to download, copy and build upon published articles, as long as the author and publisher are properly credited, which ensures maximum dissemination and a wider impact of our publications.

The book as a whole is distributed by MDPI under the terms and conditions of the Creative Commons license CC BY-NC-ND.

Contents

About the Special Issue Editor vii

Preface to "Plant Mitochondria" ix

Nicolas L. Taylor
Editorial for Special Issue "Plant Mitochondria"
Reprinted from: *Int. J. Mol. Sci.* **2018**, *19*, 3849, doi:10.3390/ijms19123849 1

Shin-ichi Arimura, Rina Kurisu, Hajime Sugaya, Naoki Kadoya and Nobuhiro Tsutsumi
Cold Treatment Induces Transient Mitochondrial Fragmentation in *Arabidopsis thaliana* in a Way that Requires DRP3A but not ELM1 or an ELM1-Like Homologue, ELM2
Reprinted from: *Int. J. Mol. Sci.* **2017**, *18*, 2161, doi:10.3390/ijms18102161 7

Pedro Robles, Sergio Navarro-Cartagena, Almudena Ferrández-Ayela, Eva Núñez-Delegido and Víctor Quesada
The Characterization of Arabidopsis *mterf6* Mutants Reveals a New Role for mTERF6 in Tolerance to Abiotic Stress
Reprinted from: *Int. J. Mol. Sci.* **2018**, *19*, 2388, doi:10.3390/ijms19082388 20

Michał Rurek, Magdalena Czołpińska, Tomasz Andrzej Pawłowski, Włodzimierz Krzesiński and Tomasz Spiżewski
Cold and Heat Stress Diversely Alter Both Cauliflower Respiration and Distinct Mitochondrial Proteins Including OXPHOS Components and Matrix Enzymes
Reprinted from: *Int. J. Mol. Sci.* **2018**, *19*, 877, doi:10.3390/ijms19030877 31

Michał Rurek, Magdalena Czołpińska, Tomasz Andrzej Pawłowski, Aleksandra Maria Staszak, Witold Nowak, Włodzimierz Krzesiński and Tomasz Spiżewski
Mitochondrial Biogenesis in Diverse Cauliflower Cultivars under Mild and Severe Drought. Impaired Coordination of Selected Transcript and Proteomic Responses, and Regulation of Various Multifunctional Proteins
Reprinted from: *Int. J. Mol. Sci.* **2018**, *19*, 1130, doi:10.3390/ijms19041130 65

Antje Reddemann and Renate Horn
Recombination Events Involving the *atp9* Gene Are Associated with Male Sterility of CMS PET2 in Sunflower
Reprinted from: *Int. J. Mol. Sci.* **2018**, *19*, 806, doi:10.3390/ijms19030806 94

Helena Štorchová
The Role of Non-Coding RNAs in Cytoplasmic Male Sterility in Flowering Plants
Reprinted from: *Int. J. Mol. Sci.* **2017**, *18*, 2429, doi:10.3390/ijms18112429 111

Natanael Mansilla, Sofia Racca, Diana E. Gras, Daniel H. Gonzalez and Elina Welchen
The Complexity of Mitochondrial Complex IV: An Update of Cytochrome *c* Oxidase Biogenesis in Plants
Reprinted from: *Int. J. Mol. Sci.* **2018**, *19*, 662, doi:10.3390/ijms19030662 124

Anna Podgórska, Monika Ostaszewska-Bugajska, Agata Tarnowska, Maria Burian, Klaudia Borysiuk, Per Gardeström and Bożena Szal
Nitrogen Source Dependent Changes in Central Sugar Metabolism Maintain Cell Wall Assembly in Mitochondrial Complex I-Defective *frostbite1* and Secondarily Affect Programmed Cell Death
Reprinted from: *Int. J. Mol. Sci.* **2018**, *19*, 2206, doi:10.3390/ijms19082206 158

Isabel Velada, Dariusz Grzebelus, Diana Lousa, Cláudio M. Soares, Elisete Santos Macedo, Augusto Peixe, Birgit Arnholdt-Schmitt and Hélia G. Cardoso
AOX1-Subfamily Gene Members in *Olea europaea* cv. "Galega Vulgar"—Gene Characterization and Expression of Transcripts during IBA-Induced In Vitro Adventitious Rooting
Reprinted from: *Int. J. Mol. Sci.* **2018**, *19*, 597, doi:10.3390/ijms19020597 183

Vajira R. Wanniarachchi, Lettee Dametto, Crystal Sweetman, Yuri Shavrukov, David A. Day, Colin L. D. Jenkins and Kathleen L. Soole
Alternative Respiratory Pathway Component Genes (*AOX* and *ND*) in Rice and Barley and Their Response to Stress
Reprinted from: *Int. J. Mol. Sci.* **2018**, *19*, 915, doi:10.3390/ijms19030915 208

Anna Podgórska, Monika Ostaszewska-Bugajska, Klaudia Borysiuk, Agata Tarnowska, Monika Jakubiak, Maria Burian, Allan G. Rasmusson and Bożena Szal
Suppression of External NADPH Dehydrogenase—NDB1 in *Arabidopsis thaliana* Confers Improved Tolerance to Ammonium Toxicity via Efficient Glutathione/Redox Metabolism
Reprinted from: *Int. J. Mol. Sci.* **2018**, *19*, 1412, doi:10.3390/ijms19051412 230

Marie-Hélène Avelange-Macherel, Adrien Candat, Martine Neveu, Dimitri Tolleter and David Macherel
Decoding the Divergent Subcellular Location of Two Highly Similar Paralogous LEA Proteins
Reprinted from: *Int. J. Mol. Sci.* **2018**, *19*, 1620, doi:10.3390/ijms19061620 255

Magdalena Opalińska, Katarzyna Parys and Hanna Jańska
Identification of Physiological Substrates and Binding Partners of the Plant Mitochondrial Protease FTSH4 by the Trapping Approach
Reprinted from: *Int. J. Mol. Sci.* **2017**, *18*, 2455, doi:10.3390/ijms18112455 271

Nan Zhao, Yumei Wang and Jinping Hua
The Roles of Mitochondrion in Intergenomic Gene Transfer in Plants: A Source and a Pool
Reprinted from: *Int. J. Mol. Sci.* **2018**, *19*, 547, doi:10.3390/ijms19020547 282

Alicja Dolzblasz, Edyta M. Gola, Katarzyna Sokołowska, Elwira Smakowska-Luzan, Adriana Twardawska and Hanna Janska
Impairment of Meristem Proliferation in Plants Lacking the Mitochondrial Protease AtFTSH4
Reprinted from: *Int. J. Mol. Sci.* **2018**, *19*, 853, doi:10.3390/ijms19030853 298

Renuka Kolli, Jürgen Soll and Chris Carrie
Plant Mitochondrial Inner Membrane Protein Insertion
Reprinted from: *Int. J. Mol. Sci.* **2018**, *19*, 641, doi:10.3390/ijms19020641 312

Pedro Robles and Víctor Quesada
Emerging Roles of Mitochondrial Ribosomal Proteins in Plant Development
Reprinted from: *Int. J. Mol. Sci.* **2017**, *18*, 2595, doi:10.3390/ijms18122595 333

Michal Zmudjak, Sofia Shevtsov, Laure D. Sultan, Ido Keren and Oren Ostersetzer-Biran
Analysis of the Roles of the Arabidopsis nMAT2 and PMH2 Proteins Provided with New Insights into the Regulation of Group II Intron Splicing in Land-Plant Mitochondria
Reprinted from: *Int. J. Mol. Sci.* **2017**, *18*, 2428, doi:10.3390/ijms18112428 345

Chunli Mao, Yanqiao Zhu, Hang Cheng, Huifang Yan, Liyuan Zhao, Jia Tang, Xiqing Ma and Peisheng Mao
Nitric Oxide Regulates Seedling Growth and Mitochondrial Responses in Aged Oat Seeds
Reprinted from: *Int. J. Mol. Sci.* **2018**, *19*, 1052, doi:10.3390/ijms19041052 370

About the Special Issue Editor

Nicolas Taylor is a Senior Lecturer in the ARC Centre of Excellence in Plant Energy Biology, School of Molecular Sciences and The Institute of Agriculture at The University of Western Australia. He completed his undergraduate studies and MSc at Massey University, New Zealand and in 2000 moved to The University of Western Australia (UWA) to undertake his PhD. After his PhD, he was awarded a European Molecular Biology Organization Long Term Fellowship to study at the Department of Plant Sciences at the University of Oxford, UK. He was recruited back to UWA in 2006 to the newly established ARC Centre of Excellence in Plant Energy Biology. Here, he has applied and developed a wide range of quantitative proteomics approaches and used these in a number of research projects. He is particularly well known for his pioneering work in the development of peptide selective reaction monitoring (SRM) mass spectrometry approaches in plants and the development of tools and resources to enable this analysis. He has been awarded Australian Research Council Post-Doctoral and Future Fellowships at UWA and in 2015, he was awarded the Robson Medal for Research Excellence in Agriculture and Related Areas.

Preface to "Plant Mitochondria"

The primary function of mitochondria is respiration, where the catabolism of substrates is coupled to ATP synthesis via oxidative phosphorylation. In plants, mitochondrial composition is relatively complex and flexible and has specific pathways to support photosynthetic processes in illuminated leaves. Plant mitochondria also play important roles in a variety of cellular processes associated with carbon, nitrogen, phosphorus, and sulfur metabolism. Research on plant mitochondria has rapidly developed in the last few decades with the availability of the genome sequences for a wide range of model and crop plants. Recent prominent themes in plant mitochondrial research include linking mitochondrial composition to environmental stress responses, and how this oxidative stress impacts on the plant mitochondrial function. Similarly, interest in the signaling capacity of mitochondria, the role of reactive oxygen species, and retrograde and anterograde signaling has revealed the transcriptional changes of stress responsive genes as a framework to define specific signals emanating to and from the mitochondrion. There has also been considerable interest in the unique RNA metabolic processes in plant mitochondria, including RNA transcription, RNA editing, the splicing of group I and group II introns and RNA degradation and translation. Despite their identification more than 100 years ago, plant mitochondria remain a significant area of research in the plant sciences.

Nicolas L. Taylor
Special Issue Editor

Editorial

Editorial for Special Issue "Plant Mitochondria"

Nicolas L. Taylor

ARC Centre of Excellence in Plant Energy Biology, School of Molecular Sciences and Institute of Agriculture, The University of Western Australia, Crawley, WA 6009, Australia; nicolas.taylor@uwa.edu.au;
Tel.: +61-8-6488-1107; Fax: +61-8-6488-4401

Received: 28 November 2018; Accepted: 30 November 2018; Published: 3 December 2018

The primary function of mitochondria is respiration, where catabolism of substrates is coupled to adenosine triphosphate (ATP) synthesis via oxidative phosphorylation (OxPhos). Organic acids such as pyruvate and malate produced in the cytosol are oxidised in mitochondria by the tricarboxylic acid (TCA) cycle and subsequently by the electron transport chain (ETC). Energy released by this oxidation is used to synthesise ATP, which is then exported to the cytosol for use in biosynthesis and growth. In plants, mitochondrial composition is relatively complex and flexible and has specific pathways to enable continuous survival during abiotic stress exposure and to support photosynthetic processes in illuminated leaves.

Plant mitochondria are double-membrane organelles where the inner membrane is invaginated to form folds known as cristae to increase the surface area of the membrane. The outer membrane contains relatively few proteins (<100) and is permeable to most small compounds (<Mr = 5 kDa) due to the presence of the pore-forming protein VDAC (voltage dependent anion channel), which is a member of the porin family of ion channels. The inner membrane is the main permeability barrier of the organelle and controls the movement of molecules by means of a series of carrier proteins, many of which are members of mitochondrial substrate carrier family (MSCF). The inner membrane also houses the large complexes that carry out electron transfer in two inter-connected pathways that finish with two terminal oxidases. It is also the site of oxidative phosphorylation (OxPhos) and contains a non-phosphorylating bypass of the classical ETC. The inner membrane also encloses the soluble matrix which contains the enzymes of the TCA cycle and many other soluble proteins involved in a myriad of mitochondrial functions.

Mitochondria are semi-autonomous organelles with their own DNA, protein synthesis, and degradation machinery. The proteins encoded by the mitochondrial genome undergo a range of post-transcriptional and post-translational processing during their synthesis. The mitochondrial genome also encodes a number of pollen abortion related genes involved in controlling plant fertility in a process known as cytoplasmic male sterility (CMS). These CMS plants are used to produce hybrids that benefit from hybrid vigor or heterosis, producing greater biomass and yield. However, the mitochondrial genome encodes only a small portion of the proteins which make up the mitochondrion; the rest are encoded by nuclear genes and synthesised in the cytosol. These proteins are then transported into the mitochondrion by the protein import machinery and assembled with the mitochondrially synthesised subunits to form the large respiratory complexes and other proteins.

Stress tolerance is a very complex trait, involving a multitude of developmental, physiological, and biochemical processes. Compared to other organelles, plant mitochondria are disproportionately involved in stress tolerance, probably because they are a convergence point between metabolism, signaling, and cell fate [1]. Mitochondria are also the site of production of reactive oxygen species (ROS), with the ubiquinone pool and components in Complex I and Complex III the main sites of production. Recently, Complex II has also been shown to produce significant superoxide [2]. Under normal steady state conditions, ROS production is controlled by a complex array of antioxidant enzymes and small molecules that scavenge ROS and limit mitochondrial and cellular damage. However, under some conditions these defences can become overwhelmed and ROS accumulate, leading to damage of proteins, lipids, and DNA.

The number of mitochondria per cell varies with tissue type, with more active cells with high energy demands, such as those in growing meristems, generally equipped with larger numbers of mitochondria per unit cell volume and typically these show faster respiration rates. Research on plant mitochondria has rapidly developed in the last few decades with the availability of genome sequences for a wide range of model and crop plants. Recent prominent themes in the plant mitochondrial research include linking mitochondrial composition to environmental stress responses and how this oxidative stress impacts upon mitochondrial function. Similarly, interest in the signaling capacity of mitochondria (the role reactive oxygen species, retrograde, and anterograde signaling) has revealed the transcriptional changes of stress responsive genes as a framework to define specific signals emanating to and from the mitochondrion. There has also been considerable interest in RNA metabolic processes in plant mitochondria including RNA transcription, RNA editing, the splicing of group I and group II introns, and RNA degradation and translation. Despite their identification more than 100 years ago plant mitochondria remain a significant area of research in the plant sciences.

In this Special Issue, "Plant Mitochondria", a total of 19 articles were accepted with 15 original research articles and 4 review articles broadly covering the field of plant mitochondrial research (Table 1). Manuscripts focused on protein synthesis and degradation [3–6], abiotic stress [7–10], OxPhos [11–14], protein import [15–17], ROS and antioxidants [18], and CMS [19,20].

Table 1. Contributors to the Special Issue "Plant Mitochondria".

Authors	Title	Topics	Type
Arimura et al. [7]	Cold Treatment Induces Transient Mitochondrial Fragmentation in Arabidopsis thaliana in a Way that Requires DRP3A but not ELM1 or an ELM1-Like Homologue, ELM2	Abiotic stress	Original Research
Robles et al. [8]	The Characterization of Arabidopsis mterf6 Mutants Reveals a New Role for mTERF6 in Tolerance to Abiotic Stress	Abiotic stress	Original Research
Rurek et al. [9]	Cold and Heat Stress Diversely Alter Both Cauliflower Respiration and Distinct Mitochondrial Proteins Including OXPHOS Components and Matrix Enzymes	Abiotic stress	Original Research
Rurek et al. [10]	Mitochondrial Biogenesis in Diverse Cauliflower Cultivars under Mild and Severe Drought. Impaired Coordination of Selected Transcript and Proteomic Responses, and Regulation of Various Multifunctional Proteins	Abiotic stress	Original Research
Reddemann et al. [19]	Recombination Events Involving the atp9 Gene Are Associated with Male Sterility of CMS PET2 in Sunflower	cytoplasmic Male Sterility	Original Research
Štorchová et al. [20]	The Role of Non-Coding RNAs in cytoplasmic Male Sterility in Flowering Plants	cytoplasmic Male Sterility	Review
Mansilla et al. [21]	The Complexity of Mitochondrial Complex IV: An Update of Cytochrome c Oxidase Biogenesis in Plants	Oxidative Phosphorylation	Review
Podgórska et al. [12]	Nitrogen Source Dependent Changes in Central Sugar Metabolism Maintain Cell Wall Assembly in Mitochondrial Complex I-Defective frostbite1 and Secondarily Affect Programmed Cell Death	OxPhos	Original Research
Velada et al. [13]	AOX1-Subfamily Gene Members in Olea europaea cv. "Galega Vulgar"—Gene Characterization and Expression of Transcripts during IBA-Induced In Vitro Adventitious Rooting	OxPhos	Original Research
Wanniarachchi et al. [14]	Alternative Respiratory Pathway Component Genes (AOX and ND) in Rice and Barley and Their Response to Stress	OxPhos	Original Research
Podgórska et al. [11]	Suppression of External NADPH Dehydrogenase—NDB1 in Arabidopsis thaliana Confers Improved Tolerance to Ammonium Toxicity via Efficient Glutathione/Redox Metabolism	OxPhos	Original Research
Avelange-Macherel et al. [15]	Decoding the Divergent Subcellular Location of Two Highly Similar Paralogous LEA Proteins	Protein Import	Original Research
Kolli et al. [16]	Plant Mitochondrial Inner Membrane Protein Insertion	Protein Import	Review

Table 1. *Cont.*

Authors	Title	Topics	Type
Zhao et al. [17]	The Roles of Mitochondrion in Intergenomic Gene Transfer in Plants: A Source and a Pool	Protein Import	Original Research
Dolzblasz et al. [3]	Impairment of Meristem Proliferation in Plants Lacking the Mitochondrial Protease AtFTSH4	Protein Synthesis and Degradation	Original Research
Opalińska et al. [4]	Identification of Physiological Substrates and Binding Partners of the Plant Mitochondrial Protease FTSH4 by the Trapping Approach	Protein Synthesis and Degradation	Original Research
Robles et al. [5]	Emerging Roles of Mitochondrial Ribosomal Proteins in Plant Development	Protein Synthesis and Degradation	Review
Zmudjak et al. [6]	Analysis of the Roles of the Arabidopsis nMAT2 and PMH2 Proteins Provided with New Insights into the Regulation of Group II Intron Splicing in Land-Plant Mitochondria	Protein Synthesis and Degradation	Original Research
Mao et al. [18]	Nitric Oxide Regulates Seedling Growth and Mitochondrial Responses in Aged Oat Seeds	ROS & Antioxidants	Original Research

A number of research articles in this Special Issue focused on the responses of mitochondria to abiotic stress, with studies that examined thermal stress (both hot and cold), salinity, and drought. Arimura et al. [7] demonstrated that cold induced mitochondrial fission (which was previously thought to involve the action of both a dynamin-related protein) DRP3A and another plant specific factor ELM1, only requires DRP3A in Arabidopsis. At the same time, they showed that an *ELM1* paralogue (*ELM2*) seemed to have only a limited role in mitochondrial fission in an *elm1* mutant, suggesting that Arabidopsis has a unique, cold induced mitochondrial fission that involves only DRP3A to control the size and shape of mitochondria. The mitochondrial transcription termination factors (mTERFs) which are involved in the control of organellar gene expression (OGE) with mutations in some characterized mTERFs (resulting in plants that have altered responses to salt, high light, heat, or osmotic stress) suggesting a role for these proteins in abiotic stress tolerance. Here Robles et al. [8] showed that strong loss of function mutant *mterf6-2* was hypersensitive to NaCl and mannitol during seedling establishment, while *mterf6-5* showed a greater sensitivity to heat later in development. Rurek et al. presented a pair of research papers that used physiological, proteomic, and transcript analysis approaches to examine the thermal (hot and cold) and drought responses of cauliflower mitochondria [9,10]. In the thermal studies they identified a number of proteins that were temperature responsive including components of OxPhos, photorespiration, porin isoforms, and the TCA cycle. Similarly, in the drought analysis, which examines three different cauliflower cultivars, both OxPhos components and porin isoforms were seen to change in abundance, indicating a significant differential impact on mitochondrial biogenesis between the three cultivars, giving us new insights into the abiotic stress responses of the *Brassica* genus.

Male sterility refers to the inability of a plant to make viable pollen. It can be mediated through nuclear genes leading to genic male sterility (GMS) or through mitochondrial proteins interacting with nuclear genes, leading to cytoplasmic male sterility (CMS). Both GMS and CMS are widely used in agricultural production for the production of hybrid crops that benefit from heterosis. In this Special Issue Štorchová [20] presents a comprehensive review of the role of non-coding RNA in the CMS of flowering plants, while Reddemann and Horn [19] presented research examining the role of *atp9* in the male sterility of CMS PET2 in sunflower. Here they showed that CMS PET2, which has the potential to become an alternative CMS source for commercial breeding, has a duplicated *atp9* with a 271-bp-insertion in the 5' region of one of the *atp9* genes which results in two unique open reading frames (*orf288* and *orf231*). The reduced anther-specific co-transcription of these open reading frames in fertility-restored hybrids supports their involvement in male sterility in CMS PET2.

A total of five papers we submitted examining OxPhos, with two of these focused on identifying non-phosphorylating bypasses of the classical ETC. Wanniarachchi et al. [14] identified and characterised the alternative oxidase (AOX) and the type II NAD(P)H dehydrogenases (NDs) of rice and barley, while Velada et al. [13] characterized the AOX1 subfamily in *Olea europaea* cv. Galega Vulgar

(European olive). Podgórska et al. [12] examined the Complex 1 mutant *fro1* (*frostbite 1*) that has a point mutation in the 8 kDa Fe-S subunit NDUFS4 grown on different nitrogen sources. When these plants were grown on NO_3^- they showed a carbon flux towards nitrogen assimilation and energy production, whereas cellulose integration into the cell wall was restricted. In contrast they showed improved growth on NH_4^+ and not the expected ammonium toxicity syndrome. Similarly, Podgórska et al. [11] showed that plants with external NADPH-dehydrogenase (NDB1) knockdown were resistant to NH_4^+ treatment and had milder oxidative stress symptoms with lower ROS accumulation and induction of glutathione peroxidase-like enzymes and peroxiredoxins antoxidants. Mansilla et al. provided a comprehensive review of the composition and biogenesis of the terminal oxygen acceptor cytochome *c* oxidase (Complex IV) in yeast, mammals, and plants. This revealed that while plants retain many biogenesis features common to other organisms, they have also developed plant specific features.

As the majority of proteins that function in mitochondria are imported from nuclear encoded cytosolic synthesized proteins, studies understanding the process of how mitochondrial protein import is controlled and regulated is vital to alter mitochondrial functions. Here Zhao et al. [17] examined the intergenomic transfer (IGT) from a broad evolutionary perspective by accessing data from nuclear, mitochondrial, and chloroplast genomes in 24 plants, and showed that mitochondrial transfer occurs in all plants examined. Additionally, Avelange-Macherel et al. [15] used two paralogues of late embryogenesis abundant proteins (LEA) (LEA38 (mitochondrial) and LEA2 (cytosolic)) to examine the influence of amino acid sequence of mitochondrial targeting sequences (MTS) on subcellular localisation. They showed that by combining substitution, charge invasion, and segment replacement, they were able to redirect LEA2 to mitochondria, providing an explanation for the loss of mitochondrial localistion after duplication of the ancestral gene. Kolli et al. [16] provided a complete review of unique aspects of plant mitochondrial inner membrane protein insertion using Complex IV as a case study, which revealed the use of Tat machinery for membrane insertion of the Rieske Fe/S protein.

Two papers examined the mitochondrial protease FTSH4, one looking at the impact of a *ftsh4* mutant on meristem proliferation [3], and another identifying physiological substrates and interaction partners using a trapping approach and mass spectrometry [4]. Dolzblasz et al. showed that plants lacking AtFTSH4 show a cessation of growth at both the shoot and root apical meristems when grown at 30 °C, and that this arrest is caused by cell cycle dysregulation and the loss of cell identity. Opalińska et al. revealed a number of novel putative targets for FTSH4 including the mitochondrial pyruvate carrier 4 (MPC4), presequence translocase-associated motor 18 (PAM18), and succinate dehydrogenase (SDH) subunits. Additionally, they showed that FTSH4 is responsible for the degradation of oxidatively damaged proteins in mitochondria. Plant mitochondria contain numerous group II introns which reside in genes. Here Zmudjak et al. [6] showed that the nMAT2 maturase and the RNA helicase PMH2 associate with their intron-RNA targets in large ribonucleoprotein particle in vivo and the splicing efficiencies of the joint intron targets of nMAT2 and PMH2 are more strongly affected in a double *nmat2/pmh2* mutant-line. Together this suggests that these proteins serve as components of a proto-spliceosomal complex in plant mitochondria. Robles et al. [5] provides a thorough review of the phenotypic effects on plant development displayed by mutants of mitoribosomal proteins (mitoRPs) and how they contribute to the elucidation of plant mitoRPs function, the mechanisms that control organelle gene expression, and their contribution to plant growth and morphogenesis.

Mao et al. [18] examined the application of 0.05 mM NO in aged oat seeds and saw an improvement in seed vigor and increased H_2O_2 scavenging ability in mitochondria. Accompanying this were higher activities of CAT, GR, MDHAR, and DHAR in the AsA-GSH scavenging system, enhanced TCA cycle-related enzymes (malate dehydrogenase, succinate-CoA ligase, fumarate hydratase), and activated alternative pathways.

Overall, the 19 contributions published in this special issue illustrate the advances in the field of plant mitochondria and I look forward to catching up with the plant mitochondrial community at the next biannual meeting in Ein Gedi, Israel (https://www.icpmb2019.com/).

Acknowledgments: N.L.T. was funded as an Australian Research Council ARC Future Fellow (FT13010123).

Conflicts of Interest: The author declares no conflict of interest.

Abbreviations

AOX	Alternative Oxidase
CMS	Cytoplasmic Male Sterility
ETC	Electron Transfer Chain
GMS	Genic Male Sterility
IGT	InterGenomic Transfer
LEA	Late Embryogenesis Abundant proteins
mitoRPs	mitochondrial Ribosomal Proteins
MPC4	Mitochondrial Pyruvate Carrier 4
MSCF	Mitochondrial Substrate Carrier Family
mTERFs	mitochondrial Transcription TERmination Factors
NDs	Type II NAD(P)H dehydrogenases
NDB1	external NAD(P)H dehydrogenase
OGE	Organellar Gene Expression
OxPhos	Oxidative Phosphorylation
PAM18	Presequence translocase-Associated Motor 18
ROS	Reactive Oxygen Species
SDH	Succinate DeHydrogenase
TCA	Tricarboxylic Acid Cycle
VDAC	Voltage Dependent Anion Channel

References

1. Taylor, N.L.; Tan, Y.F.; Jacoby, R.P.; Millar, A.H. Abiotic environmental stress induced changes in the Arabidopsis thaliana chloroplast, mitochondria and peroxisome proteomes. *J. Proteomics* **2009**, *72*, 367–378. [CrossRef] [PubMed]
2. Quinlan, C.L.; Orr, A.L.; Perevoshchikova, I.V.; Treberg, J.R.; Ackrell, B.A.; Brand, M.D. Mitochondrial complex II can generate reactive oxygen species at high rates in both the forward and reverse reactions. *J. Biol. Chem.* **2012**. [CrossRef]
3. Dolzblasz, A.; Gola, E.; Sokołowska, K.; Smakowska-Luzan, E.; Twardawska, A.; Janska, H. Impairment of Meristem Proliferation in Plants Lacking the Mitochondrial Protease AtFTSH4. *Int. J. Mol. Sci.* **2018**, *19*, 853. [CrossRef] [PubMed]
4. Opalińska, M.; Parys, K.; Jańska, H. Identification of Physiological Substrates and Binding Partners of the Plant Mitochondrial Protease FTSH4 by the Trapping Approach. *Int. J. Mol. Sci.* **2017**, *18*, 2455. [CrossRef]
5. Robles, P.; Quesada, V. Emerging Roles of Mitochondrial Ribosomal Proteins in Plant Development. *Int. J. Mol. Sci.* **2017**, *18*, 2595. [CrossRef]
6. Zmudjak, M.; Shevtsov, S.; Sultan, L.; Keren, I.; Ostersetzer-Biran, O. Analysis of the Roles of the Arabidopsis nMAT2 and PMH2 Proteins Provided with New Insights into the Regulation of Group II Intron Splicing in Land-Plant Mitochondria. *Int. J. Mol. Sci.* **2017**, *18*, 2428. [CrossRef] [PubMed]
7. Arimura, S.-I.; Kurisu, R.; Sugaya, H.; Kadoya, N.; Tsutsumi, N. Cold Treatment Induces Transient Mitochondrial Fragmentation in Arabidopsis thaliana in a Way that Requires DRP3A but not ELM1 or an ELM1-Like Homologue, ELM2. *Int. J. Mol. Sci.* **2017**, *18*, 2161. [CrossRef] [PubMed]
8. Robles, P.; Navarro-Cartagena, S.; Ferrández-Ayela, A.; Núñez-Delegido, E.; Quesada, V. The Characterization of Arabidopsis mterf6 Mutants Reveals a New Role for mTERF6 in Tolerance to Abiotic Stress. *Int. J. Mol. Sci.* **2018**, *19*, 2388. [CrossRef] [PubMed]
9. Rurek, M.; Czołpińska, M.; Pawłowski, T.; Krzesiński, W.; Spiżewski, T. Cold and Heat Stress Diversely Alter Both Cauliflower Respiration and Distinct Mitochondrial Proteins Including OXPHOS Components and Matrix Enzymes. *Int. J. Mol. Sci.* **2018**, *19*, 877. [CrossRef]

10. Rurek, M.; Czołpińska, M.; Pawłowski, T.; Staszak, A.; Nowak, W.; Krzesiński, W.; Spiżewski, T. Mitochondrial Biogenesis in Diverse Cauliflower Cultivars under Mild and Severe Drought. Impaired Coordination of Selected Transcript and Proteomic Responses, and Regulation of Various Multifunctional Proteins. *Int. J. Mol. Sci.* **2018**, *19*, 1130. [CrossRef] [PubMed]
11. Podgórska, A.; Ostaszewska-Bugajska, M.; Borysiuk, K.; Tarnowska, A.; Jakubiak, M.; Burian, M.; Rasmusson, A.; Szal, B. Suppression of External NADPH Dehydrogenase—NDB1 in Arabidopsis thaliana Confers Improved Tolerance to Ammonium Toxicity via Efficient Glutathione/Redox Metabolism. *Int. J. Mol. Sci.* **2018**, *19*, 1412. [CrossRef] [PubMed]
12. Podgórska, A.; Ostaszewska-Bugajska, M.; Tarnowska, A.; Burian, M.; Borysiuk, K.; Gardeström, P.; Szal, B. Nitrogen Source Dependent Changes in Central Sugar Metabolism Maintain Cell Wall Assembly in Mitochondrial Complex I-Defective frostbite1 and Secondarily Affect Programmed Cell Death. *Int. J. Mol. Sci.* **2018**, *19*, 2206. [CrossRef] [PubMed]
13. Velada, I.; Grzebelus, D.; Lousa, D.; Soares, C.M.; Santos Macedo, E.; Peixe, A.; Arnholdt-Schmitt, B.; Cardoso, H.G. AOX1-Subfamily Gene Members in Olea europaea cv. "Galega Vulgar"—Gene Characterization and Expression of Transcripts during IBA-Induced in Vitro Adventitious Rooting. *Int. J. Mol. Sci.* **2018**, *19*, 597. [CrossRef] [PubMed]
14. Wanniarachchi, V.; Dametto, L.; Sweetman, C.; Shavrukov, Y.; Day, D.; Jenkins, C.; Soole, K. Alternative Respiratory Pathway Component Genes (AOX and ND) in Rice and Barley and Their Response to Stress. *Int. J. Mol. Sci.* **2018**, *19*, 915. [CrossRef] [PubMed]
15. Avelange-Macherel, M.-H.; Candat, A.; Neveu, M.; Tolleter, D.; Macherel, D. Decoding the Divergent Subcellular Location of Two Highly Similar Paralogous LEA Proteins. *Int. J. Mol. Sci.* **2018**, *19*, 1620. [CrossRef] [PubMed]
16. Kolli, R.; Soll, J.; Carrie, C. Plant Mitochondrial Inner Membrane Protein Insertion. *Int. J. Mol. Sci.* **2018**, *19*, 641. [CrossRef]
17. Zhao, N.; Wang, Y.; Hua, J. The Roles of Mitochondrion in Intergenomic Gene Transfer in Plants: A Source and a Pool. *Int. J. Mol. Sci.* **2018**, *19*, 547. [CrossRef]
18. Mao, C.; Zhu, Y.; Cheng, H.; Yan, H.; Zhao, L.; Tang, J.; Ma, X.; Mao, P. Nitric Oxide Regulates Seedling Growth and Mitochondrial Responses in Aged Oat Seeds. *Int. J. Mol. Sci.* **2018**, *19*, 1052. [CrossRef]
19. Reddemann, A.; Horn, R. Recombination Events Involving the atp9 Gene Are Associated with Male Sterility of CMS PET2 in Sunflower. *Int. J. Mol. Sci.* **2018**, *19*, 806. [CrossRef]
20. Štorchová, H. The Role of Non-Coding RNAs in Cytoplasmic Male Sterility in Flowering Plants. *Int. J. Mol. Sci.* **2017**, *18*, 2429. [CrossRef]
21. Mansilla, N.; Racca, S.; Gras, D.; Gonzalez, D.; Welchen, E. The Complexity of Mitochondrial Complex IV: An Update of Cytochrome c Oxidase Biogenesis in Plants. *Int. J. Mol. Sci.* **2018**, *19*, 662. [CrossRef] [PubMed]

© 2018 by the author. Licensee MDPI, Basel, Switzerland. This article is an open access article distributed under the terms and conditions of the Creative Commons Attribution (CC BY) license (http://creativecommons.org/licenses/by/4.0/).

Article

Cold Treatment Induces Transient Mitochondrial Fragmentation in *Arabidopsis thaliana* in a Way that Requires DRP3A but not ELM1 or an ELM1-Like Homologue, ELM2

Shin-ichi Arimura [1,2,*], Rina Kurisu [1], Hajime Sugaya [1], Naoki Kadoya [1] and Nobuhiro Tsutsumi [1]

[1] Graduate School of Agricultural and Life Sciences, The University of Tokyo, 1-1-1 Yayoi, Bunkyo-ku, Tokyo 113-8657, Japan; dandelion3409@yahoo.co.jp (R.K.); hjsugaya64@gmail.com (H.S.); kdynaoki@gmail.com (N.K.); atsutsu@mail.ecc.u-tokyo.ac.jp (N.T.)
[2] Precursory Research for Embryonic Science and Technology (PRESTO), Japan Science and Technology Agency, 4-1-8, Honcho, Kawaguchi, Saitama 332-0012, Japan
* Correspondence: arimura@mail.ecc.u-tokyo.ac.jp; Tel.: +81-3-5841-8158

Received: 6 September 2017; Accepted: 13 October 2017; Published: 17 October 2017

Abstract: The number, size and shape of polymorphic plant mitochondria are determined at least partially by mitochondrial fission. Arabidopsis mitochondria divide through the actions of a dynamin-related protein, DRP3A. Another plant-specific factor, ELM1, was previously shown to localize DRP3A to mitochondrial fission sites. Here, we report that mitochondrial fission is not completely blocked in the Arabidopsis *elm1* mutant and that it is strongly manifested in response to cold treatment. Arabidopsis has an *ELM1* paralogue (*ELM2*) that seems to have only a limited role in mitochondrial fission in the *elm1* mutant. Interestingly, cold-induced mitochondrial fragmentation was also observed in the wild-type, but not in a *drp3a* mutant, suggesting that cold-induced transient mitochondrial fragmentation requires DRP3A but not ELM1 or ELM2. DRP3A: GFP localized from the cytosol to mitochondrial fission sites without ELM1 after cold treatment. Together, these results suggest that Arabidopsis has a novel, cold-induced type of mitochondrial fission in which DRP3A localizes to mitochondrial fission sites without the involvement of ELM1 or ELM2.

Keywords: mitochondrial fission; dynamin; plant mitochondria; mitochondrial division

---

1. Introduction

Mitochondria are not made *de novo* but are created by fission of existing mitochondria [1]. The shape and number of higher plant mitochondria change drastically in response to changing environmental stimuli and changing developmental stages [2–4]. The shape and number of mitochondria are determined at least partially by the balance between mitochondrial fission and mitochondrial fusion. Frequent fission and fusion make it possible to share mitochondrial internal proteins and small molecules in each cell [5].

Mitochondrial fission is mediated by a type of GTPase called dynamin-related proteins (DRPs), which are well conserved in eukaryotes [6–9]. DRPs polymerize into a ring-like spiral structure surrounding mitochondrial fission sites from the outer surface of mitochondria, and then constrict to cleave the mitochondria by their GTPase activity [10–13]. Arabidopsis has 16 *DRP* genes. Two of them, *DRP3A* and *DRP3B* (formerly known as *ADL2a* and *ADL2b*), are most similar to mitochondrial fission-related DRPs in other eukaryotes [14,15]. DRP3A and DRP3B function redundantly and cooperatively in mitochondrial fission [15–20]. DRP3A seems to have a bigger role in mitochondrial fission than DRP3B. In T-DNA insertion mutants of *DRP3A* and *DRP3B* (*drp3a* and *drp3b*), mitochondria are longer and fewer in number than those in the wild type. Moreover, in *drp3a drp3b* double mutants, mitochondria are far more elongated, forming an interconnected network in each cell, because of the severe disruption of mitochondrial fission [19].

Arabidopsis has a plant-specific factor (ELM1) that localizes to the outer surface of mitochondria, where it interacts with DRP3A (and probably DRP3B) to localize them to mitochondrial fission sites [21]. In ethyl methanesulfonate (EMS)-induced and T-DNA insertion-induced *elm1* mutants, the mitochondria are elongated and fewer in number, suggesting that ELM1 is involved in mitochondrial fission [21].

Because the mitochondrial phenotype of *elm1* mutants is not as strong as that of *drp3a drp3b* double mutants, we hypothesized that residual mitochondrial fission occurs in the absence of ELM1. Arabidopsis has an ELM1 homologue of unknown function, ELM2, that is 54% identical (70% similar) to ELM1 at the amino acid sequence level. Here, we tested whether ELM2 is responsible for the residual mitochondrial fission in the absence of ELM1. During the course of this study, we also noticed that mitochondrial fission without ELM1 was transiently manifested by cold treatment. Therefore, we also examined whether transient cold-induced mitochondrial fragmentation needs ELM2 and DRP3A.

2. Results

2.1. Residual Mitochondrial Fission in the Mitochondrial Fission Mutant Elm1

The *drp3a drp3b* double mutant has a single interconnected mitochondrion in each cell because of the malfunction of mitochondrial fission without interruption of mitochondrial fusion [19]. Figure 1 shows representative micrographs of mitochondria in the wild type and *elm1-1* mutant. The latter has a point mutation that puts a termination codon in the middle of the ORF (open reading frame) [21]. Mitochondria in the *elm1-1* and other *elm1* allele mutants are longer and fewer in number than those in the wild type. Even in the mutants with the strongest phenotypes (*elm1-1* and *elm1-6*), each cell still has more than one mitochondrion and some of the mitochondria have particle shapes like those of the wild type (mitochondria indicated by arrows in Figure 1). These results suggest that mitochondrial fission is not completely blocked in the *elm1* mutants.

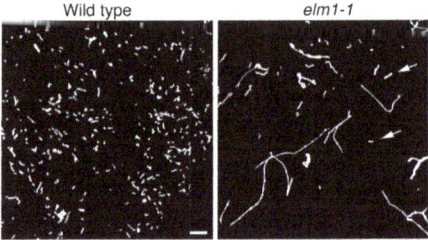

Figure 1. Mitochondrial morphologies in wild-type Arabidopsis and the *elm1-1* mutant. The images show GFP-labeled mitochondria in leaf epidermal cells. Mitochondria in the *elm1-1* mutant are longer and fewer than those in the wild type, because of the disturbance of mitochondrial fission in the mutant. However, *elm1-1* cells still have many short mitochondria (arrows), suggesting that mitochondrial fission can occur without ELM1. Scale bar, 10 µm, is applicable to the both figures.

2.2. Is Mitochondrial Fission without ELM1 due to ELM2?

The Arabidopsis genome has a single paralogue of *ELM1*, called *ELM2* (At5g06180). Its amino acid sequence is 54.0% identical (70% similarity, e-value 4.7×10^{-117}) to that of *ELM1* (Figure 2a). The Arabidopsis genome had no other matches to *ELM1* (the next closest match had an e-value >0.1). When GFP: ELM2 was expressed under the CaMV35S promoter, the green signals seemed to surround the mitochondria (Figure 2b), as was the case with ELM1:GFP in our previous report [21]. This suggests that ELM2, like ELM1, localizes on the outer surface of the outer membrane of mitochondria.

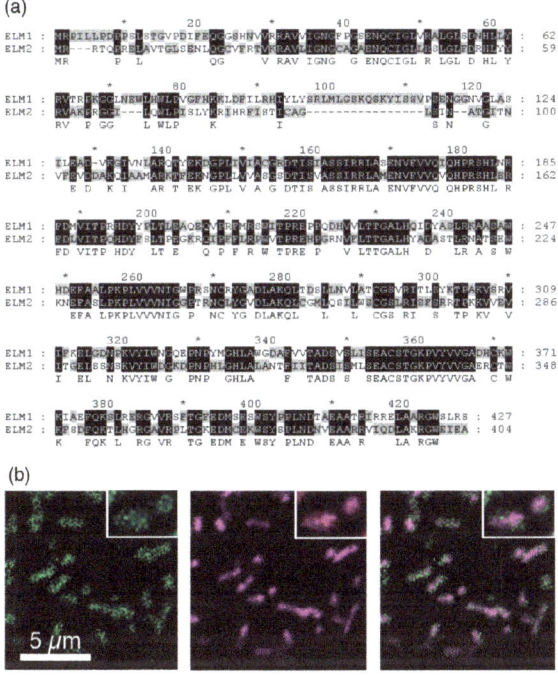

Figure 2. *ELM2* encodes an ELM1-like protein and GFP-tagged ELM2, like ELM1, localizes to the mitochondrial surface. (**a**) Clustal W alignment of ELM1 and ELM2 amino acids sequences. * depicts the positions of numbers in every ten amino acids. (**b**) Localization of GFP-ELM2 surrounding mitochondria. Arabidopsis cultured cells transiently expressing GFP-ELM2 with a mitochondrial marker MitoTracker were examined by confocal laser scanning microscopy (CLSM). A part of a single cell is shown. *Left* and *Center* are separate images obtained with the GFP and MitoTracker, respectively. *Right* is the merged image. Scale bar, 5 μm, is applicable to the other two figures. Upper right insets are X2 enlarged images.

To test the possibility that the ELM2 functions in mitochondrial fission in the same manner as ELM1, the homozygous T-DNA insertion mutant *elm2* (Figure 3a) and the *elm1-1 elm2* double mutant were analyzed. An RT-PCR analysis (Figure 3b) shows that the *elm2* mutants did not accumulate full-length *ELM2* transcripts. The *elm1-1* mutants grew slightly more slowly than the wild type, as reported previously [21], but *elm2* grew almost as well as the wild type and the *elm1-1 elm2* double mutant grew almost as well as the *elm1-1* mutant. Similarly, the mitochondria in *elm2* were as small and numerous as those in the wild type, and the mitochondria in *elm1-1 elm2* double mutant were as long as those in the *elm1-1* mutant (Figure 3d). However, the average planar areas of mitochondria in the *elm2* and *elm1-1 elm2* double mutants were slightly but significantly larger than those in the wild type and *elm1-1* mutants, respectively (Figure 3e). These results suggest that ELM2 has a small effect on mitochondrial fission in the wild type and a small effect in the absence of ELM1. To test whether ELM2 complements ELM1, a chimeric sequence consisting of the *ELM2* ORF with the *ELM1* promoter (Figure 4a) was introduced into the *elm1-1* mutant. The *ELM1* promoter dramatically increased the expression of *ELM2* transcripts (Figure 4b) but did not rescue the mitochondrial fission defect in the *elm1-1* mutant (Figure 4c). This suggests that expression activity of the *ELM2* promoter is much weaker than that of the *ELM1* promoter and that *ELM2*, although a paralogue of *ELM1*, has much weaker activity than ELM1. Furthermore, cells of the *elm1-1 elm2* double mutant still had more than one mitochondrion and some of the mitochondria had particulate shapes (Figure 3d), suggesting that the mitochondria could divide without the involvement of either ELM1 or ELM2.

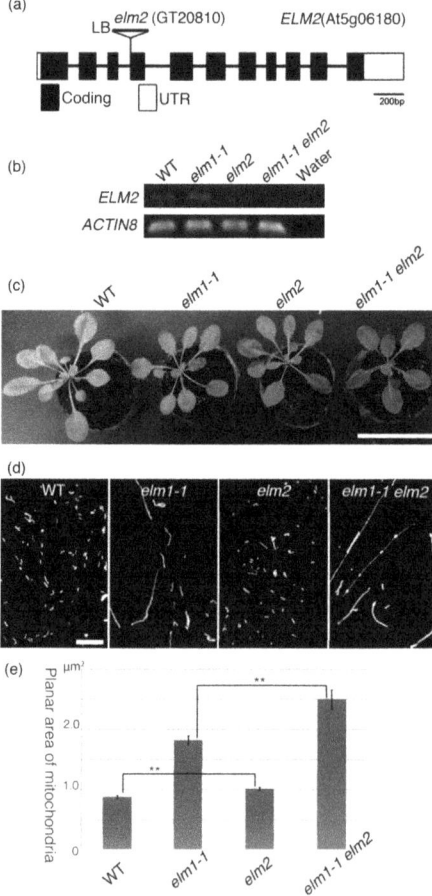

Figure 3. Disruption of *ELM2* does not appear to affect mitochondrial morphology much. (**a**) A T-DNA insertion in the end of the 3rd intron in the *elm2* mutant. (**b**) RT-PCR of full length of *ELM2* ORF (open reading frame) in the wild-type, *elm1-1*, *elm2* and *elm1-1 elm2* double mutants. (**c**) Comparison of growing phenotypes of wild-type, *elm1-1*, *elm2* and *elm1-1 elm2* double mutants. 30-day-old plants. Scale bar, 5 cm. (**d**) Mitochondrial morphologies in the wild-type, *elm1-1*, *elm2* and *elm1-1 elm2* double mutants. Leaf epidermal cells in 14-day-old plants were observed by confocal laser scanning microscopy. Scale bar, 10 μm, is applicable to the four images. (**e**) Average planar areas of mitochondria of wild type and mutants. ($n > 218$ in each of three replications) in each mutant. Error bars show S.E. ** indicates statistical significance at $p < 0.01$.

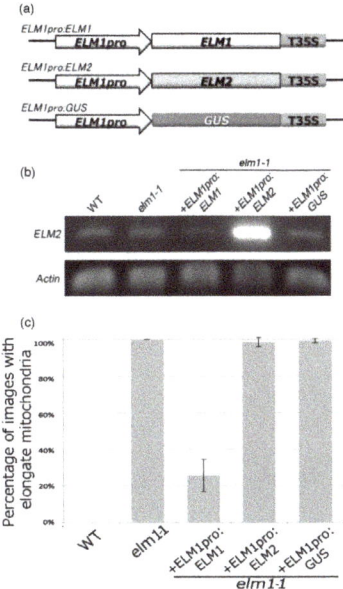

Figure 4. Heterologous complementation test of mitochondrial morphology in the *elm1* mutant by expression of *ELM2*. (**a**) Schematic drawing of DNA constructs used in this study. *ELM1*, *ELM2* and *GUS* coding sequences are attached between the probable promoter, the 950bp upstream region of *ELM1* and the sequence of CaMV35S terminator. (**b**) RT-PCR of the full length of the *ELM2* ORF in the wild-type, *elm1-1*, and three *elm1-1* mutants transformed with *ELM1pro:ELM1*, *ELM1pro:ELM2* and *ELM1pro: GUS* respectively. (**c**) Occurrence of elongate mitochondria in leaf epidermal cells in 14-day-old cotyledons from five Arabidopsis lines (wild-type, *elm1-1*, and three *elm1-1* mutants transformed with *ELM1pro:ELM1*, *ELM1pro:ELM2* and *ELM1pro: GUS*). Occurrence is expressed as the percentage of 40 confocal laser scanning microscopic images obtained from 8 leaves from each line that were judged to have elongated mitochondria (as in the *elm1-1* image in Figure 1). The experiments were repeated three times independently and the results were averaged. Error bars show S.E.

2.3. Transient Mitochondrial Fragmentation by Cold Treatment

During the course of our observations, we noticed that cold treatment induced mitochondrial fragmentation in *elm1*. One hour at 4 °C increased the number and reduced the size of mitochondria, so that they became more like those of the wild type (Figure 5a). Such mitochondrial fragmentation was observed not only in the epidermal cells of leaves, but also in the epidermal cells of stems and roots (Figure S1). The light conditions in the cold treatment did not affect the mitochondrial fragmentation (data not shown). A similar morphological change was observed in *elm1-6*, a T-DNA insertion mutant (data not shown). Cold treatment decreased the area (Figure 5c) and increased the number (Figure 5e) of mitochondria, indicating that mitochondrial fission without ELM1 is manifested by cold in the *elm1-1* mutant. Because mitochondrial morphology is determined by the balance between fission and fusion, cold-induced mitochondrial fragmentation might be increased by down-regulation of mitochondrial fusion. However, whether or not fusion activity changes, mitochondrial fragmentation of the longer mitochondria requires mitochondrial fission. Interestingly, cold treatment also induced mitochondrial fragmentation in the wild-type but not in the *drp3a-1* mutant (Figure 5c,e). Cold also induced mitochondrial fragmentation in the *elm2* and *elm1-1 elm2* double mutants (Figure 5b,d,f). Together, these results suggest that cold-induced mitochondrial fragmentation in the wild type depends on DRP3A but not on ELM1 or ELM2.

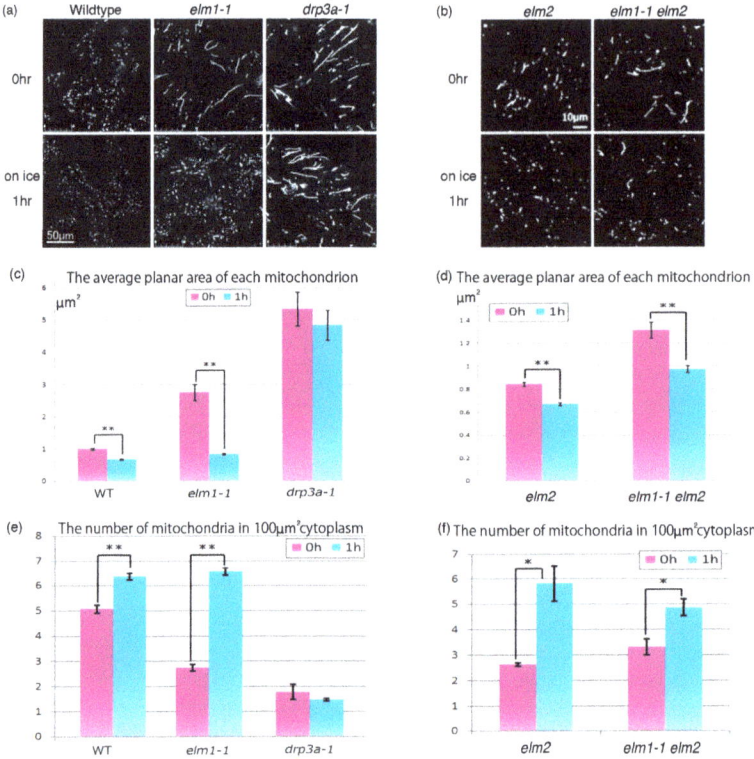

Figure 5. Mitochondrial fragmentation was induced by cold treatment in the wild-type and *elm* mutants but not in the *drp3a-1* mutant. (**a,b**), Representative mitochondrial morphologies in the wild type and mutants at room temperature before and 1 h after 4 °C treatment. Each scale bar is applicable to the all images in (**a,b**), respectively. (**c,d**) Average planar areas of mitochondria in epidermal cells of wild type and mutants before (red bars) and 1 h after (blue bars) cold-temperature treatment ($n > 218$ in each of three replications). (**e,f**) Average number of mitochondria per 100 μm^2 in leaf epidermal cells of wild-type and mutants before (red bars) and 1 h after (blue bars) cold treatment. $n = 3$ Error bars show S.E. ** indicates statistical significance at $p < 0.01$ and * at $p < 0.05$. Because data sets (**a,c,e**) and (**b,d,f**) were collected independently in different conditions (e.g., laser strength, detector gain, etc.), they could not be compared with each other directly.

When the cold treatment was extended to 24 h, the number and shape of the mitochondria in the mutants reverted to their room temperature (22 °C) states (Figure 6), indicating that the cold-induced mitochondrial fragmentation is a transient phenomenon.

2.4. DRP3A Could Localize to Mitochondria without ELM1 at the Cold Treatment

We previously reported that the localization of cytosolic DRP3A to the mitochondrial fission sites required a functional ELM1 [21]. To confirm the present finding that ELM1 was not required for mitochondrial fission following cold treatment, we examined the behavior of DRP3A following cold treatment of *elm1-6* transformed with *DRP3A: GFP* driven by the DRP3A promoter. Before treatment (0 min in Figure 7), the mitochondria had an elongated network shape and the DRP3A: GFP signal was distributed in the cytosol, in agreement with our previous study [21]. DRP3A gradually appeared as small green particles in the cytosol and some of them localized on mitochondria at about 40 min after treatment (Figure 7). Subsequently, the intensity of the green dots increased on the mitochondria, and the intensity of cytosolic green signals decreased. At some locations (an example is shown by the arrows in Figure 7 at 40 and 50 min), the mitochondrial network divided at the green dots, suggesting that DRP3A served to divide the mitochondria at these sites. This result clearly shows that in response to cold treatment, DRP3A localizes from the cytosol to mitochondrial fission sites without ELM1.

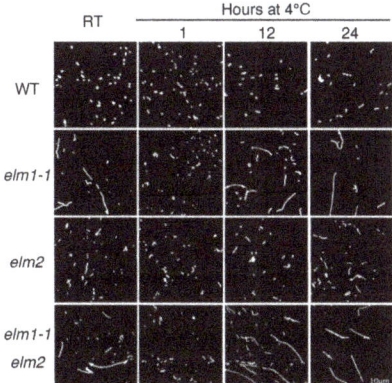

Figure 6. Mitochondrial morphology in the wild type and *elm* mutants after different durations of cold treatment. Mitochondria were observed in leaf epidermal cells of 28-day-old plants grown at 22 °C before and after different durations of 4 °C treatment. Scale bar, 10 μm, is applicable to the all images.

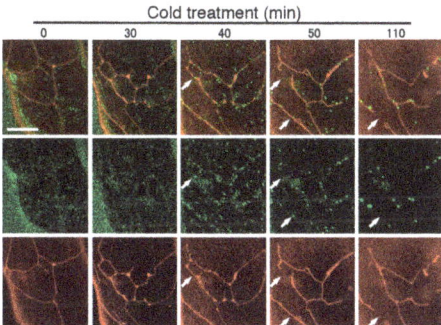

Figure 7. Time course observations of mitochondria and DRP3A in the *elm1* mutant. Images show a double-stained leaf epidermal cell of a 30-day-old *elm1-6* Arabidopsis plant transformed with DRP3Apro:DRP3A: GFP at different times after cold treatment. Bottom panels, mitochondrial network stained with MitoTracker; middle panels, DRP3AGFP; top panels, merged MitoTracker and GFP images. Cytosolic DRP3A: GFP first appeared as a hazy signal (0 and 30 min) and gradually localized and concentrated on mitochondria (40, 50 and 110 min). Arrows indicate sites of mitochondrial fission. Scale bar, 10 μm, is applicable to the all images.

3. Discussion

Mitochondrial fission occurs frequently to counterbalance the opposite event, mitochondrial fusion. In addition, mitochondria divide in accordance with cell division so that they are maintained in each of the daughter cells. In this study, we found that mitochondrial fragmentation can also be induced by cold treatment. However, we cannot rule out the possibility that downregulation of fusion was a contributing factor. Further studies are needed to test this possibility.

Because the mitochondria in *elm1* mutants at room temperature (~22 °C) are usually very elongated (as in Figure 1), ELM1 is apparently important for mitochondrial fission in the wild type at room temperature, in which it localizes DRP3A to mitochondrial fission sites [21]. However, cold-induced fission does not involve either ELM1 or ELM2. ELM2 has only a limited role in mitochondrial fission (Figures 3 and 5). This is further illustrated in the model shown in Figure 8. The finding that the *drp3a* mutant has similar elongated mitochondria at room temperature and 4 °C indicates that DRP3A is required for both types of mitochondrial fission. Mitochondrial fragmentation could be achieved by increasing fission or reducing fusion (or by both). Although we did not examine the effect of cold treatment on mitochondrial fusion, the present results confirm that cold-induced mitochondrial fission required the localization of DRP3A to the mitochondria.

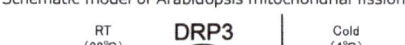
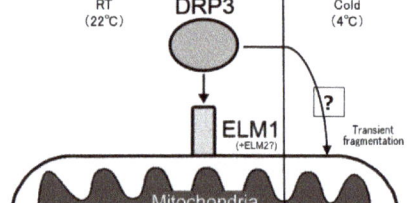

Figure 8. Schematic model of Arabidopsis mitochondrial fission. Two types of mitochondrial fission are drawn. In the normal condition (left, shown as RT (room temperature) 22 °C), the division executor, DRP3 localizes to mitochondria via interaction with ELM1. In the case of mitochondrial fission transiently induced by cold treatment, DRP3 could localize to mitochondria by skipping the help of ELM1 (and ELM2) and underwent fission.

The mechanism by which cold-treatment induced mitochondrial fission is unclear. The simplest idea is that the affinity between DRP3A and the mitochondrial outer membrane is transiently increased by the cold treatment. Purified DRPs in yeast and mammals bind to liposomes without any other proteins [10,13,22], although in vivo DRPs need other proteins to localize to mitochondrial fission sites from the cytosol. If cold treatment increases the affinity between DRP3A and the mitochondrial outer membrane, it would suggest that ELM1 is needed to support the binding of DRP3A to the outer membrane at room temperature but not at cold temperature due to the increased affinity of DRP3A at cold temperature. Further studies are needed to examine the effect of temperature on the affinity between purified DRP3A and the mitochondrial outer membrane.

Another possibility is that cold-induced mitochondrial fission involves other proteins. Tail-anchored proteins FIS1a (BIGYIN), FIS1b, PMD1 and PMD2 were also reported to be involved in mitochondrial division in *A. thaliana* [23–25]. However, none of them have been directly shown to have roles in the localization of DRP3A or DRP3B to the fission sites. PMD1 and PMD2 were shown to contribute to mitochondrial fission independent of DRP3/FIS1 [25]. In addition to protein components, a mitochondrial phospholipid (cardiolipin) was recently shown to stabilize the DRP3 complex on mitochondria [26]. These and unknown other proteins and lipid factors might contribute individually or together to the DRP3A localization and function in cold treatment or other types of mitochondrial fission. Furthermore, in *A. thaliana*, mitochondrial fission is reported to involve dynamin-related proteins other than DRP3A. These include DRP3B as well as the more distantly related DRP5B [20,27]. The relationships between the factors involved in mitochondrial fission, the different types of mitochondrial fission and how they are regulated appear to be more complicated than previously thought.

DRP3A was found to be phosphorylated and dephosphorylated at different stages of the cell cycle [28]. Proteomic analyses have predicted that DRP3A has multiple phosphorylation sites [29,30], but the effects of phosphorylation/dephosphorylation at these sites are unknown. Mammalian Drp1s, which are involved in mitochondrial fission, are also regulated by post-translational modifications other than phosphorylation, such as ubiquitination, SUMOylation and S-nitrosylation reviewed in [31]. Such modifications might also occur in plant DRPs. The overexpression of Arabidopsis UBP27, a mitochondrial outer membrane-bound ubiquitin protease, was recently reported to change mitochondrial morphology by inhibiting the binding of DRP3A and DRP3B to mitochondria, although it is unknown whether DRP3A and DRP3B are direct targets of UBP27 [32].

Plant mitochondria constantly undergo fission and fusion [5]. Such alterations appear to be involved in several activities that are crucial to the health of cells [33,34]. It is unclear what processes may be involved in cold-induced mitochondrial fragmentation in Arabidopsis, although because cold adaptation affects the expression of over 2000 genes in Arabidopsis [35], there are many candidates. Many of these genes are expressed days and weeks after cold treatment, whereas mitochondrial fragmentation occurs within an hour, indicating that it is one of the early responses to cold treatment. In mammalian brown adipose tissue, cold exposure induces thermogenesis, which has been linked to mitochondrial fragmentation through activation of a DRP3A homologue [36]. However, cold stress does not seem to induce mitochondrial thermogenesis through uncoupling proteins in Arabidopsis [37,38]. Further studies are needed to see which of the many metabolic changes in cold stress are responsible for mitochondrial fragmentation in Arabidopsis.

The balance between mitochondrial fission and fusion appears to vary in different tissues and in different environmental conditions in order to change mitochondrial morphology to meet the cells' physiological needs. The shape, distribution and number of mitochondria change in accordance with organ development [2,4,39] and in response to environmental stimuli [3,40]. Changes of mitochondrial morphology, numbers and distribution would affect the three-dimensional distances and attachments between mitochondria and other organelles metabolically related to mitochondria, causing indirect effects on cell metabolisms and physiological states [41]. Thus, a better understanding of the mechanisms underlying the changes in mitochondrial morphology should help to clarify a number of cellular processes in plants.

4. Materials and Methods

4.1. Plant Materials and Growth Conditions

Arabidopsis thaliana ecotype Columbia (Col-0) and its transformant with mitochondrial-targeted GFP [42] were used as wild-type plants in this paper. The EMS mutants *elm1-1* and *drp3a-1* were described previously [21]. All *Arabidopsis* plants were grown in growth chamber at 22 °C under a 14 h photoperiod at 50~100 µmol/m^{2}s. The T-DNA insertion line GT20810 was provided by the Cold Spring Harbor Laboratory (http://www.cshl.edu/). GT20810 was consecutively crossed with Col-0 5 times to obtain a background similar to that of Col-0. The homo-T-DNA insertion line of the BC5F2 was used as *elm2*. The T-DNA insertion was checked by PCR with primers 1 and 2 to detect WT DNA and primers 1 and 3 to detect the T-DNA insertion. The homozygous and heterozygous *elm1-1* point mutations were checked by sequencing and PCR with primers 4 and 6 to detect the mutated DNA and primers 5 and 6 to detect the wild-type DNA. The primers are shown in Table S1.

4.2. Construction of Plasmids

ELM2 ORF was obtained by RT-PCR from *A. thaliana* col-0 RNA with primers 3 and 15 and cloned into pENTR^{TM}/d-TOPO entry vector (Invitrogen). The Ti plasmid expressing GFP: ELM2 fusion protein was constructed by LR reaction of Gateway cloning technology (Invitrogen) with pH7WGF2 destination vector, which was kindly provided from VIB [43]. An In-Fusion HD cloning kit (TaKaRa) was used to make Ti plasmids for expressing *ELM1*, *ELM2* and *GUS* under the *ELM1* promoter. The promoters consisted of 950 bp of the region upstream of the ATG initiation codon of *ELM1*. The promoter was amplified from genomic DNA and the ORFs were amplified by RT-PCR.

The basal Ti-plasmid was pBGWFS7 [43]. Oligonucleotide primers are presented in Table S1 and the combinations of primers to make constructs are presented in Table S2. All PCR for DNA construction was carried out with high-fidelity DNA polymerases. All constructs made in this study were confirmed by sequencing. The T-DNA insertion line *elm1-6* transformed with *DRP3Apro:DRP3A: GFP* used in Figure 7 was previously described [21].

4.3. Agrobacterium Mediated Transformation of Arabidopsis Plants and Cultured Cells

The Ti plasmids described above were transformed into *Agrobacterium tumefaciens* strain C58C1. The Arabidopsis plants in Figure 4 were transformed with *A. tumefaciens* via floral dipping [44]. Transgenic T1 plants were selected on the MS-Agar medium containing 35 mg L^{-1} glufosinate-ammonium (Sigma-Aldrich). The Arabidopsis transgenic cultured cells used in Figure 2b were made as follows. The transformed *Agrobacterium* was cultured in LB medium containing 50 mg L^{-1} hygromycin and 100 mg L^{-1} spectinomycin (O.D. = 0.5 at 600 nm), pelleted and re-suspended in modified MS medium and used to inoculate 10 ml culture of 2-day-old Arabidopsis Col-0 suspension-cultured cells, called Alex. To remove *Agrobacterium*, 30 µL of 250 mg mL^{-1} claforan was added to the culture medium at 1 day after inoculation. The Arabidopsis cells were transferred to fresh media 5 days after the inoculation, and they were observed by microscopes 8 days after the *Agrobacterium* inoculation.

4.4. MitoTracker Orange Staining

The suspension-cultured transformed cells in Figure 2 were stained with 50 µM MitoTracker Orange (Molecular Probes) for 30 min and washed with medium three times. In the experiment in Figure 7, small sections (10~50 mm^2) were cut out from the Arabidopsis leaves with new razor blades and stained with 50 µM MitoTracker Orange (Molecular Probes) for about 60 min.

4.5. Microscopic Observations and Image Analysis

A confocal laser scanning microscope (CLSM) (Nikon TE2000-U and C1Si) was used for all microscopic observations of Arabidopsis leaves and cultured cells with fluorescent fusion proteins or stained with fluorescent dyes. Fluorophores of GFP and MitoTracker Orange were excited by A 488 nm and A 561 nm laser, respectively. Emission signals were detected through a 515/30 nm filter for GFP and a 590/70 nm filter for MitoTracker Orange. All CLSM images were acquired in single focal planes. The acquired images were prepared with Photoshop CS5 (Adobe Systems) and analyzed with Image pro plus 4.0 (Media Cybernetics). The averaged mitochondrial number in every 100 µm^2 microscopic observation area and the averaged area of each mitochondrion were measured and calculated with Image-Pro Plus ver.6.2J (Media Cybernetics) from the CLSM images before and after the cold treatment (Figure 5c–f).

4.6. RT-PCR Analysis

Total RNA for RT-PCR analysis was extracted from about one-month-old Arabidopsis leaves by using an RNeasy plant mini kit (Qiagen) according to the manufacturer's instructions; 400 ng of total RNA were used for RT-PCR analysis. Reverse transcription was carried out with Oligo-dT primer and the Super Script III reverse transcriptase (Invitrogen), and amplified with KOD FX Neo polymerase (TOYOBO). PCR was done with the specific primers 3 and 15 for *ELM2* and 16 and 17 for *ACTIN8* presented in Table S1.

4.7. Cold Treatment

The plantlets and samples on glass slide were incubated in 4 °C incubators. The plantlets were illuminated with a desk-top light with similar strength. To obtain the successive images of single cells under cold treatment in Figure 7, a small petri dish containing cold water and ice was placed on the slide glass on an inverted microscope (Nikon TE2000-U).

Supplementary Materials: Supplementary materials can be found at www.mdpi.com/1422-0067/18/10/2161/s1.

Acknowledgments: We thank M. Karimi (Ghent University, Gent, Belgium) for kindly donating Gateway destination vectors and the Cold Spring Harbor Laboratory for T-DNA insertion line *elm2* mutant. This work was supported by grants partly from the Japanese Science and Technology Agency (PRESTO to S.A.) and partly from the Japan Society for the Promotion of Science and the Ministry of Education, Culture, Sports, Science, and Technology of JAPAN (grant numbers 24248001 to N.T. and 23120507, 24380814 to S.A.).

Author Contributions: Shin-ichi Arimura and Nobuhiro Tsutsumi conceived and designed the experiments; Shin-ichi Arimura, Rina Kurisu, Hajime Sugaya and Naoki Kadoya performed the experiments and analyzed the data; Shin-ichi Arimura and Rina Kurisu wrote the paper.

Conflicts of Interest: The authors declare no conflict of interest.

Abbreviations

DRP   Dynamin-related protein
ELM1   Elongate mitochondria

References

1. Kuroiwa, T.; Kuroiwa, H.; Sakai, A.; Takahashi, H.; Toda, K.; Itoh, R. The division apparatus of plastids and mitochondria. *Int. Rev. Cytol.* **1998**, *181*, 1–41. [PubMed]
2. Logan, D.C.; Leaver, C.J. Mitochondria-targeted GFP highlights the heterogeneity of mitochondrial shape, size and movement within living plant cells. *J. Exp. Bot.* **2000**, *51*, 865–871. [CrossRef] [PubMed]
3. Van Gestel, K.; Verbelen, J.P. Giant mitochondria are a response to low oxygen pressure in cells of tobacco (*Nicotiana tabacum* L.). *J. Exp. Bot.* **2002**, *53*, 1215–1218. [CrossRef] [PubMed]
4. Ito-Inaba, Y.; Sato, M.; Masuko, H.; Hida, Y.; Toyooka, K.; Watanabe, M.; Inaba, T. Developmental changes and organelle biogenesis in the reproductive organs of thermogenic skunk cabbage (*Symplocarpus renifolius*). *J. Exp. Bot.* **2009**, *60*, 3909–3922. [CrossRef] [PubMed]
5. Arimura, S.; Yamamoto, J.; Aida, G.P.; Nakazono, M.; Tsutsumi, N. Frequent fusion and fission of plant mitochondria with unequal nucleoid distribution. *Proc. Natl. Acad. Sci. USA* **2004**, *101*, 7805–7808. [CrossRef] [PubMed]
6. Bleazard, W.; McCaffery, J.M.; King, E.J.; Bale, S.; Mozdy, A.; Tieu, Q.; Nunnari, J.; Shaw, J.M. The dynamin-related GTPase Dnm1 regulates mitochondrial fission in yeast. *Nat. Cell Biol.* **1999**, *1*, 298–304. [PubMed]
7. Sesaki, H.; Jensen, R.E. Division versus fusion: Dnm1p and Fzo1p antagonistically regulate mitochondrial shape. *J. Cell Biol.* **1999**, *147*, 699–706. [CrossRef] [PubMed]
8. Westermann, B. Mitochondrial fusion and fission in cell life and death. *Nat. Rev. Mol. Cell Biol.* **2010**, *11*, 872–884. [CrossRef] [PubMed]
9. Purkanti, R.; Thattai, M. Ancient dynamin segments capture early stages of host-mitochondrial integration. *Proc. Natl. Acad. Sci. USA* **2015**, *112*, 2800–2805. [CrossRef] [PubMed]
10. Ingerman, E.; Perkins, E.M.; Marino, M.; Mears, J.A.; McCaffery, J.M.; Hinshaw, J.E.; Nunnari, J. Dnm1 forms spirals that are structurally tailored to fit mitochondria. *J. Cell Biol.* **2005**, *170*, 1021–1027. [CrossRef] [PubMed]
11. Faelber, K.; Posor, Y.; Gao, S.; Held, M.; Roske, Y.; Schulze, D.; Haucke, V.; Noé, F.; Daumke, O. Crystal structure of nucleotide-free dynamin. *Nature* **2011**, *477*, 556–560. [CrossRef] [PubMed]
12. Ford, M.G.; Jenni, S.; Nunnari, J. The crystal structure of dynamin. *Nature* **2011**, *477*, 561–566. [CrossRef] [PubMed]
13. Mears, J.A.; Lackner, L.L.; Fang, S.; Ingerman, E.; Nunnari, J.; Hinshaw, J.E. Conformational changes in Dnm1 support a contractile mechanism for mitochondrial fission. *Nat. Struct. Mol. Biol.* **2011**, *18*, 20–26. [CrossRef] [PubMed]
14. Hong, Z.; Bednarek, S.Y.; Blumwald, E.; Hwang, I.; Jurgens, G.; Menzel, D.; Osteryoung, K.W.; Raikhel, N.V.; Shinozaki, K.; Tsutsumi, N.; et al. A unified nomenclature for Arabidopsis dynamin-related large GTPases based on homology and possible functions. *Plant Mol. Biol.* **2003**, *53*, 261–265. [CrossRef] [PubMed]

15. Arimura, S.; Tsutsumi, N. A dynamin-like protein (ADL2b), rather than FtsZ, is involved in Arabidopsis mitochondrial division. *Proc. Natl. Acad. Sci. USA* **2002**, *99*, 5727–5731. [CrossRef] [PubMed]
16. Arimura, S.; Aida, G.P.; Fujimoto, M.; Nakazono, M.; Tsutsumi, N. Arabidopsis dynamin-like protein 2a (ADL2a), like ADL2b, is involved in plant mitochondrial division. *Plant Cell Physiol.* **2004**, *45*, 236–242. [CrossRef] [PubMed]
17. Logan, D.C.; Scott, I.; Tobin, A.K. ADL2a, like ADL2b, is involved in the control of higher plant mitochondrial morphology. *J. Exp. Bot.* **2004**, *55*, 783–785. [CrossRef] [PubMed]
18. Mano, S.; Nakamori, C.; Kondo, M.; Hayashi, M.; Nishimura, M. An Arabidopsis dynamin-related protein, DRP3A, controls both peroxisomal and mitochondrial division. *Plant J.* **2004**, *38*, 487–498. [CrossRef] [PubMed]
19. Fujimoto, M.; Arimura, S.; Mano, S.; Kondo, M.; Saito, C.; Ueda, T.; Nakazono, M.; Nakano, A.; Nishimura, M.; Tsutsumi, N. Arabidopsis dynamin-related proteins DRP3A and DRP3B are functionally redundant in mitochondrial fission, but have distinct roles in peroxisomal fission. *Plant J.* **2009**, *58*, 388–400. [CrossRef] [PubMed]
20. Aung, K.; Hu, J. Differential roles of Arabidopsis dynamin-related proteins DRP3A, DRP3B, and DRP5B in organelle division. *J. Integr. Plant Biol.* **2012**, *54*, 921–931. [PubMed]
21. Arimura, S.; Fujimoto, M.; Doniwa, Y.; Kadoya, N.; Nakazono, M.; Sakamoto, W.; Tsutsumi, N. Arabidopsis ELONGATED MITOCHONDRIA1 is required for localization of DYNAMIN-RELATED PROTEIN3A to mitochondrial fission sites. *Plant Cell* **2008**, *20*, 1555–1566. [CrossRef] [PubMed]
22. Francy, C.A.; Alvarez, F.J.; Zhou, L.; Ramachandran, R.; Mears, J.A. The mechanoenzymatic core of dynamin-related protein 1 comprises the minimal machinery required for membrane constriction. *J. Biol. Chem.* **2015**, *290*, 11692–11703. [CrossRef] [PubMed]
23. Scott, I.; Tobin, A.K.; Logan, D.C. BIGYIN, an orthologue of human and yeast FIS1 genes functions in the control of mitochondrial size and number in *Arabidopsis thaliana*. *J. Exp. Bot.* **2006**, *57*, 1275–1280. [CrossRef] [PubMed]
24. Zhang, X.C.; Hu, J.P. FISSION1A and FISSION1B Proteins Mediate the Fission of Peroxisomes and Mitochondria in Arabidopsis. *Mol. Plant* **2008**, *1*, 1036–1047. [CrossRef] [PubMed]
25. Aung, K.; Hu, J. The arabidopsis tail-anchored protein peroxisomal and mitochondrial division factor1 is involved in the morphogenesis and proliferation of peroxisomes and mitochondria. *Plant Cell* **2011**, *23*, 4446–4461. [CrossRef] [PubMed]
26. Pan, R.; Jones, A.D.; Hu, J. Cardiolipin-mediated mitochondrial dynamics and stress response in Arabidopsis. *Plant Cell* **2014**, *26*, 391–409. [CrossRef] [PubMed]
27. Zhang, X.; Hu, J. The Arabidopsis chloroplast division protein DYNAMIN-RELATED PROTEIN5B also mediates peroxisome division. *Plant Cell* **2010**, *22*, 431–442. [CrossRef] [PubMed]
28. Wang, F.; Liu, P.; Zhang, Q.; Zhu, J.; Chen, T.; Arimura, S.; Tsutsumi, N.; Lin, J. Phosphorylation and ubiquitination of dynamin-related proteins (AtDRP3A/3B) synergically regulate mitochondrial proliferation during mitosis. *Plant J.* **2012**, *72*, 43–56. [CrossRef] [PubMed]
29. Heazlewood, J.L.; Durek, P.; Hummel, J.; Selbig, J.; Weckwerth, W.; Walther, D.; Schulze, W.X. PhosPhAt: A database of phosphorylation sites in *Arabidopsis thaliana* and a plant-specific phosphorylation site predictor. *Nucleic Acids Res.* **2008**, *36*, D1015–D1021. [CrossRef] [PubMed]
30. Sugiyama, N.; Nakagami, H.; Mochida, K.; Daudi, A.; Tomita, M.; Shirasu, K.; Ishihama, Y. Large-scale phosphorylation mapping reveals the extent of tyrosine phosphorylation in Arabidopsis. *Mol. Syst. Biol.* **2008**, *4*, 193. [CrossRef] [PubMed]
31. Willems, P.H.; Rossignol, R.; Dieteren, C.E.; Murphy, M.P.; Koopman, W.J. Redox Homeostasis and Mitochondrial Dynamics. *Cell Metab.* **2015**, *22*, 207–218. [CrossRef] [PubMed]
32. Pan, R.; Kaur, N.; Hu, J. The Arabidopsis mitochondrial membrane-bound ubiquitin protease UBP27 contributes to mitochondrial morphogenesis. *Plant J.* **2014**, *78*, 1047–1059. [CrossRef] [PubMed]
33. Scott, I.; Youle, R.J. Mitochondrial fission and fusion. *Essays Biochem.* **2010**, *47*, 85–98. [CrossRef] [PubMed]
34. Mishra, P.; Chan, D.C. Metabolic regulation of mitochondrial dynamics. *J. Cell Biol.* **2016**, *212*, 379–387. [CrossRef] [PubMed]
35. Zhao, C.; Lang, Z.; Zhu, J.K. Cold responsive gene transcription becomes more complex. *Trends Plant Sci.* **2015**, *20*, 466–468. [CrossRef] [PubMed]

36. Wikstrom, J.D.; Mahdaviani, K.; Liesa, M.; Sereda, S.B.; Si, Y.; Las, G.; Twig, G.; Petrovic, N.; Zingaretti, C.; Graham, A.; et al. Hormone-induced mitochondrial fission is utilized by brown adipocytes as an amplification pathway for energy expenditure. *EMBO J.* **2014**, *33*, 418–436. [CrossRef] [PubMed]
37. Watanabe, A.; Nakazono, M.; Tsutsumi, N.; Hirai, A. AtUCP2: A novel isoform of the mitochondrial uncoupling protein of *Arabidopsis thaliana*. *Plant Cell Physiol.* **1999**, *40*, 1160–1166. [CrossRef] [PubMed]
38. Vercesi, A.E.; Borecky, J.; Maia Ide, G.; Arruda, P.; Cuccovia, I.M.; Chaimovich, H. Plant uncoupling mitochondrial proteins. *Annu. Rev. Plant Biol.* **2006**, *57*, 383–404. [CrossRef] [PubMed]
39. Logan, D.C.; Millar, A.H.; Sweetlove, L.J.; Hill, S.A.; Leaver, C.J. Mitochondrial biogenesis during germination in maize embryos. *Plant Physiol.* **2001**, *125*, 662–672. [CrossRef] [PubMed]
40. Scott, I.; Logan, D.C. Mitochondrial morphology transition is an early indicator of subsequent cell death in Arabidopsis. *New Phytol.* **2008**, *177*, 90–101. [CrossRef] [PubMed]
41. El Zawily, A.M.; Schwarzländer, M.; Finkemeier, I.; Johnston, I.G.; Benamar, A.; Cao, Y.; Gissot, C.; Meyer, A.J.; Wilson, K.; Datla, R.; et al. FRIENDLY regulates mitochondrial distribution, fusion, and quality control in Arabidopsis. *Plant Physiol.* **2014**, *166*, 808–828. [CrossRef] [PubMed]
42. Feng, X.G.; Arimura, S.; Hirano, H.Y.; Sakamoto, W.; Tsutsumi, N. Isolation of mutants with aberrant mitochondrial morphology from *Arabidopsis thaliana*. *Genes Genet. Syst.* **2004**, *79*, 301–305. [CrossRef] [PubMed]
43. Karimi, M.; Inze, D.; Depicker, A. GATEWAY ((TM)) vectors for Agrobacterium-mediated plant transformation. *Trends Plant Sci.* **2002**, *7*, 193–195. [CrossRef]
44. Clough, S.J.; Bent, A.F. Floral dip: A simplified method for Agrobacterium-mediated transformation of *Arabidopsis thaliana*. *Plant J.* **1998**, *16*, 735–743. [CrossRef] [PubMed]

© 2017 by the authors. Licensee MDPI, Basel, Switzerland. This article is an open access article distributed under the terms and conditions of the Creative Commons Attribution (CC BY) license (http://creativecommons.org/licenses/by/4.0/).

Communication

The Characterization of Arabidopsis *mterf6* Mutants Reveals a New Role for mTERF6 in Tolerance to Abiotic Stress

Pedro Robles, Sergio Navarro-Cartagena, Almudena Ferrández-Ayela, Eva Núñez-Delegido and Víctor Quesada *

Instituto de Bioingeniería, Universidad Miguel Hernández, Campus de Elche, 03202 Elche, Spain; probles@umh.es (P.R.); s.navarro@umh.es (S.N.-C.); sikalea@hotmail.com (A.F.-A.); eva.nunez@goumh.umh.es (E.N.-D.)
* Correspondence: vquesada@umh.es; Tel.: +34-96-665-88-12; Fax: +34-96-665-85-11

Received: 18 July 2018; Accepted: 11 August 2018; Published: 14 August 2018

Abstract: Exposure of plants to abiotic stresses, such as salinity, cold, heat, or drought, affects their growth and development, and can significantly reduce their productivity. Plants have developed adaptive strategies to deal with situations of abiotic stresses with guarantees of success, which have favoured the expansion and functional diversification of different gene families. The family of mitochondrial transcription termination factors (mTERFs), first identified in animals and more recently in plants, is likely a good example of this. In plants, mTERFs are located in chloroplasts and/or mitochondria, participate in the control of organellar gene expression (OGE), and, compared with animals, the mTERF family is expanded. Furthermore, the mutations in some of the hitherto characterised plant mTERFs result in altered responses to salt, high light, heat, or osmotic stress, which suggests a role for these genes in plant adaptation and tolerance to adverse environmental conditions. In this work, we investigated the effect of impaired mTERF6 function on the tolerance of Arabidopsis to salt, osmotic and moderate heat stresses, and on the response to the abscisic acid (ABA) hormone, required for plants to adapt to abiotic stresses. We found that the strong loss-of-function *mterf6-2* and *mterf6-5* mutants, mainly the former, were hypersensitive to NaCl, mannitol, and ABA during germination and seedling establishment. Additionally, *mterf6-5* exhibited a higher sensitivity to moderate heat stress and a lower response to NaCl and ABA later in development. Our computational analysis revealed considerable changes in the *mTERF6* transcript levels in plants exposed to different abiotic stresses. Together, our results pinpoint a function for Arabidopsis mTERF6 in the tolerance to adverse environmental conditions, and highlight the importance of plant mTERFs, and hence of OGE homeostasis, for proper acclimation to abiotic stress.

Keywords: Arabidopsis; mitochondrial transcription termination factor (mTERF); salt stress; abiotic stresses; abscisic acid (ABA); organellar gene expression (OGE)

1. Introduction

The increased salt content in arable soils severely compromises plant growth and productivity. This is due to osmotic stress, which promotes water loss and hinders its uptake by plant roots, and to ionic stress (Na^+ and Cl^- in most cases), which generates toxicity and hinders the recruitment of other ions [1]. The development of new varieties of more halotolerant crop plants requires unravelling the genetic and molecular mechanisms that underlie tolerance to salinity. It has been proposed that chloroplasts could act as sensors capable of sensing environmental stress, and, by retrograde signalling (from the chloroplast to the nucleus), could coordinate the expression of nuclear genes that allow plants to adapt to stress [2]. In line with this, Leister et al. [3] have reported that perturbed organellar gene expression (OGE) homeostasis activates the acclimation and tolerance responses of plants, likely through retrograde communication. Notwithstanding, information about chloroplasts involvement in the response to abiotic stress in general, and to salinity in particular, is still scarce. We initiated a bioinformatics and reverse genetics approach in the plant model system *Arabidopsis thaliana* to identify novel functions involved in the control of gene expression in chloroplasts. We previously identified and characterised two genes, *MDA1* [4] and *mTERF9* [5], not previously described, which belong to the family of mitochondrial transcription termination factors (mTERFs) [6]. The analysis of the *mda1* and *mterf9* mutants revealed a connection between chloroplast function and the response to salt stress and ABA in Arabidopsis [4,5]. For other Arabidopsis *mTERF* genes besides *MDA1* and *mTERF9*, a role in acquiring tolerance to salinity (*mTERF10* and *mTERF11*) [7], heat (*SHOT1* (*SUPPRESSOR OF HOT1-4 1*)) [8] or high light (*SOLDAT10* (*SINGLET OXYGEN-LINKED DEATH ACTIVATOR10*) [9]) has also been reported. Accordingly, *mda1* and *mterf9* are less sensitive to NaCl than the wild type [4,5], *mterf10* and *mterf11* are salt-hypersensitive [7], whereas *shot1* [8] and *soldat10* [9] show enhanced heat tolerance and constitutive acclimation to light, respectively. In addition to Arabidopsis, the *stm6* mutant (*state transition mutant6*) of the green algae *Chlamydomonas reindhardtii* affected in the *MOC1* (*mterf-like gene of Chlamydomonas1*) gene is light sensitive [10]. Along this line, an emerging role for some *mTERF* genes in the response, tolerance, and/or acclimation of plants to different abiotic stress conditions has been recently proposed, which might, at least in part, explain the expansion and diversification of the plant mTERF family compared with that of animals [11].

mTERF proteins share the presence of a variable number of repeats of a motif called mTERF of about 30 amino acids. In vertebrates, four subfamilies have been identified (MTERF1-4), in which the MTERF1 protein is the first to be characterised [12]. However, plant genomes, especially those from higher plants, contain a larger number of *mTERF* genes than animal genomes [13]. In metazoans, mTERF proteins participate in the control of mitochondrial transcription, and are required for both its initiation and termination [14]. In plants, several molecular functions have been proposed for some of the *mTERF* genes hitherto characterised, all of which are related to the posttranscriptional regulation of chloroplasts and/or mitochondria gene expression. Accordingly, Arabidopsis mTERF15 [15] and maize Zm-mTERF4 [16] are involved in intron splicing in mitochondria, Arabidopsis BELAYA SMERT/RUGOSA2 [17,18] is required for intron splicing in plastids, and *Chlamydomonas reindhardtii* MOC1 promotes the termination of antisense mitochondrial transcription [19]. The Arabidopsis mTERF6 protein, dually targeted to chloroplasts and mitochondria, is involved in the maturation of the chloroplast isoleucine tRNA (*trnI.2*) gene and the aminoacylation of tRNA for isoleucine [20,21].

We previously identified and morphologically characterized a new mutant allele of the Arabidopsis AT4G38160 (*mTERF6*) gene which we dubbed *mterf6-5* after finding it to be allelic of the previously described *mterf6-2* mutant [20,22]. *mterf6-2* and *mterf6-5* are insertional alleles of the SAIL and SALK collection of T-DNA lines (SAIL_360_H09 and SALK_116335 respectively). *mTERF6* transcripts were undetectable in *mterf6-2* plants [20], and significantly reduced in the *mterf6-5* mutant [22]. This caused a substantial delay in plant growth, smaller size than the wild type, and loss of pigmentation in cotyledons, leaves, stems, sepals, and fruits in both mutants. In our growth conditions, these phenotypic traits were much more marked in *mterf6-2* than in *mterf6-5* [22]. Altogether, the data suggest that *mterf6-2* and *mterf6-5* are null and strong hypomorphic alleles respectively, of the *mTERF6* gene [20,22]. Furthermore, the *mterf6-5* mutation enhanced the leaf polarity defects of the *asymmetric leaves1* mutant, and revealed a role for the *mTERF6* gene in adaxial-abaxial leaf patterning [22]. Nevertheless, whether this gene plays a role in tolerance to abiotic stress as reported for other *mTERF* genes remains to be evaluated. To investigate this, we report herein the study of the response of the wild-type Col-0 and the strong loss-of-function *mterf6-2* and *mterf6-5* alleles to the ionic and osmotic stresses caused by the presence of high concentrations of NaCl or mannitol in culture media, respectively. We also evaluated the sensitivity of *mterf6-2* and *mterf6-5* to the abscisic acid (ABA) hormone, involved in plant adaptations to environmental stress. Our results revealed an altered response of the *mterf6* mutants to the stress conditions assayed, which is consistent with the substantial changes in *mTERF6* expression we found in silico after exposing the wild-type to different abiotic stresses.

2. Results

2.1. The mterf6-2 and mterf6-5 Mutants Are Hypersensitive to NaCl and Mannitol

In order to assess whether the *mterf6-5* mutant that we previously identified [22] exhibited altered sensitivity to abiotic stresses, we first analysed its sensitivity, and that of the wild-type Col-0, to the ionic stress produced by NaCl and osmotic stress due to mannitol. We also included the *mterf6-2* mutant, allelic of *mterf6-5*, in the analysis (see above). For this purpose, we first examined the ability of *mterf6-2*, *mterf6-5* and Col-0 seeds to germinate and to form fully expanded green cotyledons (seedling establishment) in the first 2 weeks after seed stratification in the presence of 0, 150, or 200 mM of NaCl or 350 mM of mannitol. In the non-supplemented culture medium, mutants *mterf6-2* and *mterf6-5* respectively yielded, to some extent, lower and similar seed germination ratios than Col-0 (we considered germinated those seeds in which radicle emergence through the seed coat was observed) (Figure 1a). The supplementation of growth medium with NaCl (150 mM) or mannitol (350 mM) did not affect wild-type seed germination, but lowered the *mterf6-2* and *mterf6-5* germination rates, especially those of the former, and the effect was more pronounced from 1 to 5 DAS (days after stratification; Figure 1c,e).

Consistent with the stunted growth of the *mterf6-2* and *mterf6-5* individuals [20,22], the seedling establishment of both mutants was delayed compared with Col-0 (e.g., at 6 DAS in the MS control medium, 10%, 37%, and 99% of the *mterf6-2*, *mterf6-5* and Col-0 seeds yielded seedlings with fully expanded green cotyledons, respectively (Figure 1b)). However, at 10 DAS, 93% and 100% of seedling establishments were achieved for the *mterf6-5* and the wild type, respectively, whereas *mterf6-2* reached a maximum value of 77% at 11 DAS (Figure 1b). The *mterf6-2* and *mterf6-5* seeds yielded substantially lower seedling establishment rates than those of Col-0 in the presence of 150 mM of NaCl or 350 mM of mannitol (Figure 1d,f). Accordingly, the presence of the *mterf6-2* and *mterf6-5* seedlings with green expanded cotyledons could be scored only from 10 DAS in the presence of NaCl or mannitol, while the seedling establishment for Col-0 was observed from 4–5 DAS under the same conditions (Figure 1d,f). Notwithstanding, the *mterf6-2* mutant was more sensitive than *mterf6-5* to NaCl. In line with this, the maximum seedling establishment values for *mterf6-2* and *mterf6-5* in 150 mM NaCl were 14% and 62%, respectively, which were reached at 13 DAS, whereas Col-0 yielded ~100% (Figure 1d). However, a similar strong hypersensitive response to mannitol was found for both mutants throughout the study period (Figure 1f).

We also investigated the response of Col-0 and *mterf6-5* to a higher salt concentration by supplementing the culture medium with 200 mM of NaCl. We found that this condition significantly delayed mutant germination (e.g., at 5 and 10 DAS, 98% and 99% of the wild type and 12% and 80% of the *mterf6-5* seeds germinated, respectively, in 200 mM of NaCl; Table S1), and completely abolished the Col-0 and *mterf6-5* seedling establishments, as we were unable to identify any individual that displayed green expanded cotyledons.

Taken together, our results revealed enhanced sensitivity to salt and osmotic stress during germination, and mainly in the cotyledon greening stage, for the studied *mterf6* mutants.

We evaluated the response of *mterf6-5* to salinity by exposing plants to stress after germination and seedling establishment. To this end, 5 DAS wild-type and mutant seedlings were transferred from the non-supplemented medium to the media supplemented with NaCl (125 or 150 mM), and root length was determined 8 days after transfer (13 DAS; see Materials and Methods; Table S2). The *mterf6-5* plants were significantly less sensitive than the wild-type ones to the inhibition of root growth caused by the presence of either 125 mM of NaCl or 150 mM of NaCl (Table 1; Table S2).

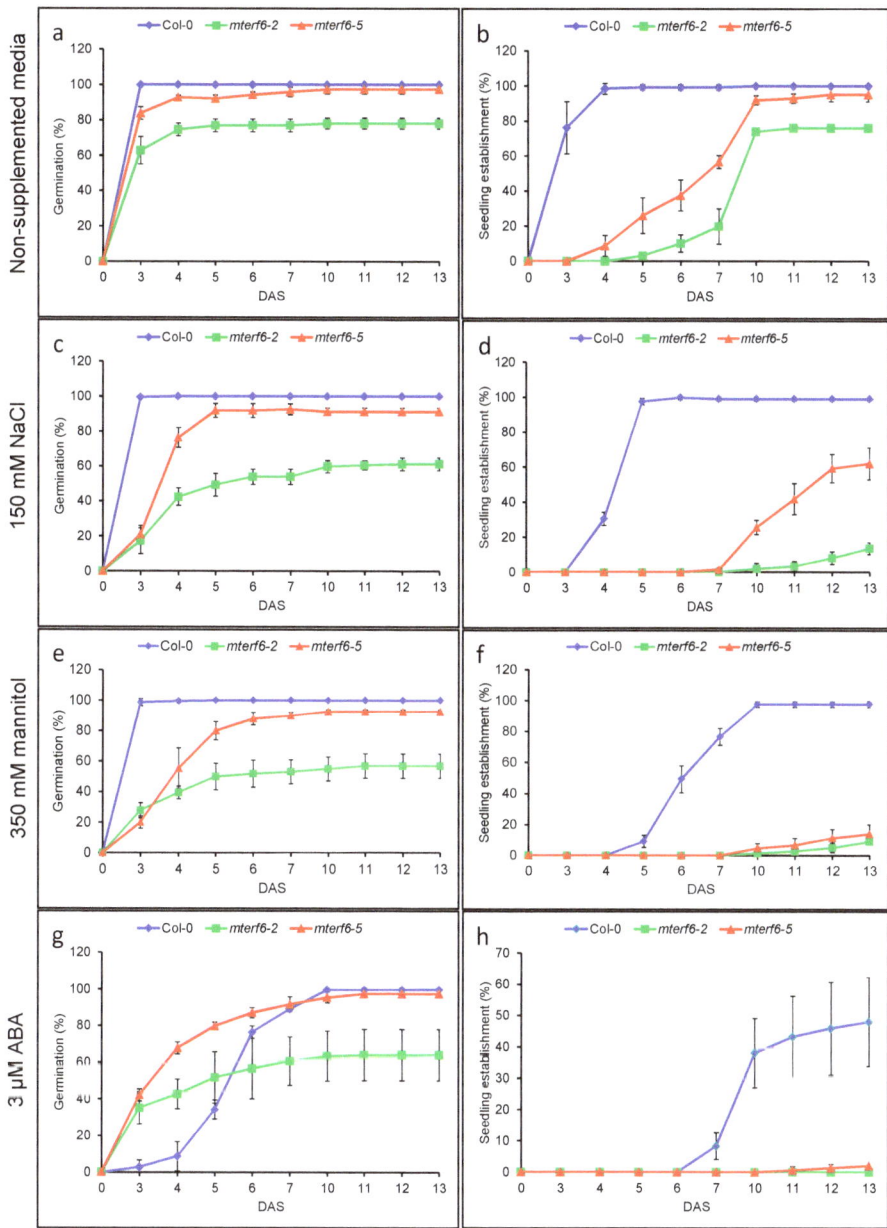

Figure 1. Effects of NaCl, mannitol and ABA on germination and seedling establishment in the wild-type Col-0 and the *mterf6-2* and *mterf6-5* mutants. Each value corresponds to the mean ± standard deviation (SD) of the percentage of germination (**a,c,e,g**) and seedling establishment (**b,d,f,h**) in the growth media either without supplementation (**a,b**) or supplemented with 150 mM of NaCl (**c,d**), 350 mM of mannitol (**e,f**) or 3 μM of ABA (**g,h**) of four replicates of at least 50 seeds each per genotype. DAS: days after stratification.

Table 1. Tolerance of the *mterf6-5* mutant to NaCl and abscisic acid (ABA).

Genotype	Inhibition of Root Length (%)			
	NaCl (mM)		ABA (µM)	
	125	150	5	10
Col-0	64.6 ± 7.2	77.2 ± 4.3	19.4 ± 7.2	29.0 ± 4.5
mterf6-5	55.8 ± 6.0 **	63.5 ± 4.6 **	9.4 ± 13.1 **	23.2 ± 14.7

The values correspond to the root length inhibition percentages of the plants transferred 5 DAS to the media supplemented with either 125 or 150 mM of NaCl or 5 or 10 µM of ABA, which refers to those of plants of the same genotype, which were transferred to the non-supplemented media. Eight days after transfer (13 DAS), the main root length was determined per plant to evaluate their tolerance to these stress conditions (see Materials and Methods). Each value is the mean ± SD of the main root length of at least 20 plants per genotype and condition. The values significantly differed from the Col-0 at ** $p < 0.01$ according to a Student's *t*-test.

To study whether a low *mTERF6* expression altered tolerance to moderate heat stress, the wild-type Col-0 and *mterf6-5* mutant seedlings were exposed 13 days to a higher (28 °C) than normal culture temperature (20 °C). We also compared the response of *mterf6-5* with that of *mterf* mutants *mda1-1* and *mterf9*. The *mterf6-5* mutant was hypersensitive to heat stress because paleness markedly increased and seedling growth was severely impaired, and even arrested, when grown at 28 °C (Figure S1). In contrast, the growth of *mda1-1* and *mterf9* was enhanced at 28 °C, but to a lesser extent than in Col-0 (Figure S1).

2.2. Knock-Down of mTERF6 Alters the Response to ABA

The abscisic acid (ABA) hormone plays a fundamental role in seed germination and in the responses of plants to abiotic stresses [23]. The Arabidopsis mutants deficient in ABA signalling or biosynthesis also exhibited enhanced tolerance to salt stress [24–26]. Therefore, given the enhanced sensitivity of *mterf6-2* and *mterf6-5* to salt and osmotic stress, we investigated whether they also exhibited an altered response to ABA by growing the *mterf6* mutant and Col-0 seedlings in the presence of ABA. As shown in Figure 1a,g, 3 µM of ABA substantially delayed *mterf6-2*, *mterf6-5* and Col-0 germination, but from 3 to 5 DAS both mutant seeds exhibited higher levels of radicle emergence through the seed coat than those of the wild-type. However, when the *mterf6-5* and Col-0 individuals were grown on 6 µM of ABA, seed germination was greater in *mterf6-5* than in Col-0 only at 5 DAS, but both genotypes yielded very low germination values (6% and 3%, respectively; Table S3). In contrast, we found that *mterf6-5* was hypersensitive to ABA from 6–13 DAS. Accordingly at 6, 7, 10, and 13 DAS, 44%, 62%, 99%, and 100% of the Col-0 seeds, and 22%, 36%, 60%, and 82% of the *mterf6-5* seeds germinated, respectively (Table S3).

As regards seedling establishment, exposure to 3 µM ABA considerably reduced it in Col-0 (e.g., up to 48% of the wild-type seedlings under the control condition at 13 DAS), and completely abolished it in *mterf6-2*, while only 2% was found for *mterf6-5* (Figure 1h). When grown on 6 µM ABA, 18% and 42% of the Col-0 seeds yielded seedlings with green expanded cotyledons at 10 and 13 DAS, respectively. As expected, no *mterf6-5* seedlings showing green expanded cotyledons were found from 3 to 13 DAS (Table S3).

We allowed the Col-0 and *mterf6-5* seedlings to grow on the ABA-supplemented medium. At 17 DAS, 6.4% and 1.4% of the mutant seeds (n = 150) yielded individuals that displayed two very tiny leaves in 3 and 6 µM of ABA, respectively. In contrast, 27.7% and 16.2% of the Col-0 seedlings (n = 150) displayed two small leaves in 3 and 6 µM of ABA, respectively.

Taken together, these results indicate that the *mterf6-2* and *mterf6-5* mutants are hypersensitive to ABA principally during seedling establishment.

As we did for NaCl (see Section 2.1), we also investigated the sensitivity of *mterf6-5* to ABA after germination and seedling establishment. To this end, 5 DAS wild-type and mutant plants were transferred from the non-supplemented medium to the media supplemented with ABA (5 or 10 µM). Root length was determined 8 days after transfer (13 DAS; Table S2). As well as for NaCl, the root growth of the *mterf6-5* individuals was significantly more tolerant than that of the Col-0 plants to 5 µM of ABA, whereas inhibition of root length only slightly decreased in 10 µM of ABA (Table 1).

2.3. The Expression of the mTERF6 Gene Changes in Response to Abiotic Stresses

Given the altered sensitivity of the *mterf6* mutants to NaCl, mannitol and ABA, we decided to perform an in silico analysis of the expression of the *mTERF6* gene in response to different abiotic stress conditions. Hence we studied the stress-induced changes in the transcript levels of *mTERF6* with the Arabidopsis AtGenExpress Visualization Tool ([27]; available online: http://weigelworld.org/resources.html) in the roots and aerial parts of the Col-0 seedlings under NaCl, osmotic and drought stresses. The expression values were plotted over time (0, 0.5, 1, 3, 6, 12, and 24 h after treatment started) to obtain a graphical representation of the response of *mTERF6* to these conditions (Figure S2). Compared with the untreated plants, *mTERF6* expression was down-regulated in the green parts of seedlings after 3 h of NaCl (150 mM), mannitol (300 mM) and drought treatments, and mostly in the presence of NaCl and mannitol from 6 to 24 h. This repression peaked 24 h after treatment when the *mTERF6* transcript levels lowered to 16% and 42% of the control plants in response to mannitol and NaCl, respectively (Figure S2a). As regards roots, *mTERF6* expression was down-regulated by salt stress from 1 to 24 h after treatment started. The difference to the control plants was maximum at 6 h (38.6% of the control plants), whereas mannitol slightly increased the *mTERF6* transcript levels at 3 h (28% more than in the control plants), but lowered them from 6 to 24 h, especially at 6 h (63% of the control plants) (Figure S2b). Drought reduced *mTERF6* expression at 1 and 6 h after exposure (77% and 74% of the control plants), but no appreciable differences were found for the remaining time points. As regards the effect of ABA, *mTERF6* expression was down-regulated to 53.4% of the control plants by 10 µM ABA after 3 h, but no noticeable differences were found after 0.5 and 1 h. We also investigated the transcript levels of *mTERF6* using the online data from the At-TAX Arabidopsis whole genome tilling array [28]. Consistently with the AtGenExpress results, we found that the 12-h exposure of the 10-day-old Col-0 seedlings to 200 mM of NaCl, 300 mM of mannitol or 100 µM of ABA markedly reduced *mTERF6* transcript abundance to 30.4%, 45.5% and 45.2% of those of the untreated seedlings, respectively. Slighter differences between the treated and untreated plants were detected after 1 h of exposure under the same conditions.

We experimentally tested by qRT-PCR whether *mTERF6* expression may change in response to NaCl. To this end, RNA was extracted from Col-0 seedlings collected 10 DAS and grown in GM medium supplemented with 100 mM NaCl or in non-supplemented medium. The RNA was retro-transcribed and the cDNAs analyzed by qPCR. Though this condition was different from those used by the Arabidopsis AtGenExpress consortium (see above; [27]), we previously found that it delayed Col-0 growth [4]. We included as a positive control the *RD29A* gene which is induced by salinity [29]. In response to this moderate salt stress, *RD29A* was significantly upregulated (1.7 ± 0.3; $p = 10^{-3}$) whereas *mTERF6* was slightly downregulated (0.8 ± 0.4; $p = 0.2$).

3. Discussion

In this work, we analysed the response of the *mterf6-2* and *mterf6-5* mutants to different abiotic stresses during germination, seedling establishment and for *mterf6-5* later in development. We found that the *mterf6* mutants displayed altered sensitivity to salt, osmotic stress, ABA, and moderate heat stress. Unlike the results obtained with other *mterf*-deficient mutants, such as *mda1* and *mterf9*, which are more insensitive than the wild type to such stresses [4,5], *mterf6-2* and *mterf6-5* were hypersensitive to the inhibition exerted on germination and seedling establishment by high concentrations of NaCl, mannitol, or ABA. *mterf6-2* was always more sensitive than *mterf6-5* to the different abiotic stress conditions studied, which is consistent with its more severe morphological phenotype [22]. The susceptibility of *mterf6* mutants to NaCl was similar to that of *mterf10* and *mterf11* [7], but unlike these mutants, which were as sensitive as the wild type was to ABA, *mterf6-2* and *mterf6-5* were also hypersensitive to this hormone, mainly during seedling establishment.

In line with this, the knock-down of *mTERF6* also reduced seedling tolerance to moderate heat stress and led to impaired growth and development, whereas *mda1-1* and *mterf9* (this work), and *mterf10* and *mterf11* [7], did not show a significantly different response from that of the wild type under this condition. mTERF6 seemed to play a different role further in vegetative development because the deficient *mTERF6* function significantly reduced the sensitivity of roots to the presence of NaCl or ABA in the growth medium. A different susceptibility to salt and ABA during germination and vegetative growth has been previously reported for mutants *mda1* and *mterf9* [4,5].

We extracted the *mTERF6* transcript levels from AtGenExpress [27] by selecting "AtGE Abiostress" as a data source. Consistent with altered tolerance to abiotic stresses, we found that *mTERF6* expression was markedly down-regulated in response to salt, osmotic stress (mannitol) and drought, especially after prolonged exposure (12–24 h) to 150 mM NaCl and 300 mM mannitol. Interestingly, ABA treatment also repressed *mTERF6* expression. We experimentally tested by qRT-PCR *mTERF6* expression in 10 DAS plants grown in mild salt stress conditions (100 mM NaCl), and found that it was slightly but not significantly downregulated, which is likely due to the different stress conditions used to study *mTERF6* expression. Together, our results suggest that the altered tolerance of *mterf6-2* and *mterf6-5* to the tested abiotic stresses could be attributed to its different sensitivity to ABA compared with the wild type, because this hormone plays a fundamental role in plants' response and adaptation to abiotic stress conditions. The involvement of *mTERF6*, *MDA1* (affected in the *mTERF5* gene), and *mTERF9* (the *mda1* and *mterf9* mutants are less sensitive to ABA than the wild type) [4,5], and possibly of *mTERF10* (since a modest overexpression of this gene leads to enhanced germination and growth in the presence of ABA) [7] in abiotic stress tolerance could take place through ABA signalling. Accordingly, several pieces of experimental evidence indicate a role for ABA in plastid-to-nucleus signalling (reviewed in [3]). Therefore, the impaired plastid gene expression may be due to a defective mTERF function perturbing the retrograde communication (from plastids to the nucleus) mediated by ABA under salt or other abiotic stress conditions. As a result, this would alter nuclear gene expression, and hence, tolerance to these environmental conditions. Similarly, Leister and Kleine [21] found that levels of the nuclear transcripts, which encode the chloroplast proteins involved in organellar gene expression (OGE), were affected in the weak *mterf6-1* mutant. Notwithstanding, while some mTERF proteins (e.g., mTERF5, mTERF9 and mTERF10) would negatively modulate Arabidopsis salt tolerance as their down-regulation diminishes sensitivity to ABA and abiotic stresses, mTERF6 would play the opposite role by promoting such tolerance, at least during germination and seedling establishment. Consequently, it could be hypothesised that the outcome of the activity of different mTERF proteins, which act during germination and early vegetative development, might contribute to responses to abiotic stress in these developmental stages. The mTERF6 function in abiotic responses might be conserved in other plant species because the expression of the maize *mTERF12* gene, the orthologue of Arabidopsis *mTERF6*, is substantially altered after NaCl or ABA treatments [30]. Interestingly, the transcript levels of other maize *mTERF* genes also change after exposing maize plants to light/dark treatments, salt, ABA or 1-Naphthaleneacetic acid exposure [30]. The altered levels of the *mTERF6* transcripts after abiotic stress treatments found in silico might be interpreted as being necessary for plants to adapt to adverse environmental conditions. Nevertheless given currently available molecular information, we cannot rule out the notion that changes in the expression of *mTERF6* and other *mTERF* genes under different abiotic stress conditions might result from the perturbation of certain biological processes. Chloroplast homeostasis is likely to be one of these processes altered in *mterf*-deficient mutants, because all the mTERFs involved in the response to salt stresses described to date are targeted to chloroplasts; they also belong to the "chloroplast cluster" (mTERF5, mTERF6 and mTERF9) or to the "chloroplast associated-cluster" (mTERF10 and mTERF11) of proteins by functioning in organelle gene expression, embryogenesis, gene expression, and/or protein catabolism in plants [13]. The altered OGE, and hence chloroplast homeostasis, would account for the delayed growth and greening of the cotyledons of the *mterf6* individuals in relation to Col-0, even in the absence of abiotic stress. However, differences with Col-0 considerably increased when the *mterf6* mutants were exposed to salt, mannitol, or ABA, which indicates that *mterf6-2* and *mterf6-5* sensitivity to abiotic stresses cannot be attributed solely to its defective growth.

The involvement of the mTERF family of genes in the acclimation and tolerance of plants to different abiotic stresses conditions [11,14] is further supported by recent findings in cotton (*Gossypium barbadense*). Accordingly, multiple stress responsive genes have been identified in *G. barbadense* using a normalised cDNA library, constructed after exposure to various abiotic (heat, cold, salt, drought, potassium, and phosphorous deficit) and biotic (*Verticillium dahlia* infection) stress conditions [31]. Remarkably, the mRNAs of 464 transcription factors (TF) have been enriched in this library, and mTERFs are one of the most abundant TF families to have been identified (3.7%) [31].

4. Materials and Methods

4.1. Plant Material and Growth Conditions

Plant cultures and crosses were performed as previously described [4]. The seeds of the *Arabidopsis thaliana* (L.) Heynh. wild-type (WT) accession Columbia-0 (Col-0) were obtained from the Nottingham Arabidopsis Stock Centre (NASC). Seeds of the transferred DNA (T-DNA) insertion lines SAIL_360_H09 (*mterf6-2*), SALK_116335 (*mterf6-5*), SALK_597243 (*mda1-1*) and WiscDsLox474E07 (*mterf9*) were provided by the NASC and are described on the SIGnAL website (available online: http://signal.salk.edu).

4.2. Germination and Growth Sensitivity Assays

For the germination assays, sowings were carried out as described in [4] on Petri dishes filled with GM agar medium (Murashige and Skoog (MS) medium containing 1% sucrose), supplemented with NaCl (150 and 200 mM), mannitol (350 mM) or ABA (3 and 6 µM). The seeds in which radicle emergence was observed were considered to be germinated, whereas seedling establishment was determined as those seedlings that exhibited green and fully expanded cotyledons. Seed germination and seedling establishment were scored from 1 to 13 DAS or from 1 to 24 DAS on Petri dishes, kept at $20 \pm 1\,^\circ\text{C}$ with 72 $\mu\text{mol}\cdot\text{m}^{-2}\cdot\text{s}^{-1}$ of continuous light.

To determine the salt and ABA responses during vegetative growth after seedling establishment, seeds were sown on non-supplemented GM agar medium, and seedlings were transferred on 5 DAS to new Petri dishes supplemented with NaCl (125 or 150 mM) or ABA (5 or 10 µM), and vertically grown. Plant root length was determined after 8 days of NaCl or ABA treatment to evaluate their tolerance to these stress conditions by referring the values to those of the individuals transferred to the control (non-supplemented) media.

For the heat-sensitivity assays, the Col-0, *mda1-1*, *mterf9*, and *mterf6-5* plants were grown on Petri dishes at $28 \pm 1\,^\circ\text{C}$ and $20 \pm 1\,^\circ\text{C}$ for 14 DAS.

4.3. Quantitative RT-PCR (qRT-PCR)

Total RNA was extracted from 80 mg 10 DAS wild-type Col-0 plants grown in the presence or absence of 100 mM NaCl in the GM agar medium. The RNA was resuspended in 40 µL of RNase-free water and DNA removed using the TURBO DNAfree kit (Invitrogen, Waltham, MA, USA) following the manufacturer's instructions. The cDNA preparations and qPCR amplifications were carried out in an ABI PRISM 7000 Sequence Detection System (Applied Biosystems, Waltham, MA, USA) as described in [4] using the oligonucleotides listed on Table S4. Each reaction mix of 20-µL contained 7.5 µL of the SYBR-Green/ROX qPCR Master Kit (Fermentas, Waltham, MA, USA), 0.4 µM of primers, and 1 µL of the cDNA solution. Relative quantification of gene expression data was performed by the $2^{-\Delta\Delta Ct}$ method as described in [4]. Each reaction was done in three replicates, and three different biological replicates were used. The expression levels were normalised to the CT values obtained for the housekeeping *ACTIN2* gene [32], and a Mann–Whitney U-test was applied to the relative expression data obtained.

4.4. Computational Analyses

The expression responses of the *mTERF6* gene under ABA, salt, osmotic, and drought stress were obtained from the AtGenExpress Visualization Tool (available online: http://jsp.weigelworld.org/expviz/expviz.jsp) [27] by selecting the "AtGE Abiostress" as the data source and mean-normalised values. The *mTERF6* expression in response to ABA was also visualised by extracting the tilling array data from TileViz (available online: http://jsp.weigelworld.org/tileviz/tileviz.jsp) [28] by selecting the "Abiotic Stress Dataset" and the mean-normalised values.

5. Conclusions

In summary, the results reported herein reveal a new function for the *mTERF6* gene related to the emerging roles that have been recently proposed for the mTERF family in plants' response and adaptation to different environmental stress conditions. In the plant *mterf* mutants characterised to date which have exhibited altered sensitivity to abiotic stresses, the affected mTERF proteins are involved in OGE [11,13,14]. Hence, this pinpoints an important function for OGE and plastid homeostasis, likely by acting throughout retrograde signalling, in tolerance to adverse environmental conditions, as recently proposed [3]. Further molecular research on the effect of abiotic stresses on the mTERF6 function, and by extension on the remaining mTERFs, is required to shed more light on the contribution of this scarcely known family of genes for plants to cope with abiotic stresses.

Supplementary Materials: Supplementary materials can be found at http://www.mdpi.com/1422-0067/19/8/2388/s1..

Author Contributions: V.Q. and P.R. conceived and designed the experiments. S.N.-C., A.F.-A. and E.N.-D. performed the experiments. V.Q. and P.R. analysed the data. V.Q. and P.R. contributed reagents/materials/analysis tools. V.Q. wrote the manuscript. V.Q., P.R. and E.N.-D. edited the manuscript.

Funding: This research was funded by the Conselleria de Educació of the Generalitat Valenciana (Spain) grant numbers GV/2009/058 and AICO/2015 to VQ.

Conflicts of Interest: The authors declare no conflict of interest.

Abbreviations

mTERF	mitochondrial transcription termination factor
SHOT1	SUPPRESSOR OF HOT1-4 1
OGE	organellar gene expression
SOLDAT10	SINGLET OXYGEN-LINKED DEATH ACTIVATOR10
MOC1	*mterf-like gene of Chlamydomonas1*
ABA	abscisic acid
DAS	days after stratification
mda1	*mterf defective in Arabidopsis1*
RD29A	RESPONSIVE TO DESICCATION29A

References

1. Munns, R.; Tester, M. Mechanisms of salinity tolerance. *Annu. Rev. Plant Biol.* **2008**, *59*, 651–681. [CrossRef] [PubMed]
2. Chan, K.X.; Crisp, P.A.; Estavillo, G.M.; Pogson, B.J. Chloroplast-to-nucleus communication: Current knowledge, experimental strategies and relationship to drought stress signaling. *Plant Signal. Behav.* **2010**, *5*, 1575–1582. [CrossRef] [PubMed]
3. Leister, D.; Liangsheng, W.; Tatjana, K. Organellar Gene Expression and Acclimation of Plants to Environmental Stress. *Front. Plant Sci.* **2017**, *8*, 387. [CrossRef] [PubMed]
4. Robles, P.; Micol, J.L.; Quesada, V. Arabidopsis MDA1, a nuclear-encoded protein, functions in chloroplast development and abiotic stress responses. *PLoS ONE* **2012**, *7*, e42924. [CrossRef] [PubMed]

5. Robles, P.; Micol, J.L.; Quesada, V. Mutations in the plant-conserved MTERF9 alter chloroplast gene expression, development and tolerance to abiotic stress in *Arabidopsis thaliana*. *Physiol. Plant.* **2015**, *154*, 297–313. [CrossRef] [PubMed]
6. Linder, T.; Park, C.B.; Asin-Cayuela, J.; Pellegrini, M.; Larsson, N.G.; Falkenberg, M.; Samuelsson, T.; Gustafsson, C.M. A family of putative transcription termination factors shared amongst metazoans and plants. *Curr. Genet.* **2005**, *48*, 265–269. [CrossRef] [PubMed]
7. Xu, D.; Leister, D.; Kleine, T. *Arabidopsis thaliana* mTERF10 and mTERF11, but not mTERF12, are involved in the response to salt stress. *Front. Plant Sci.* **2017**, *8*, 1213. [CrossRef] [PubMed]
8. Kim, M.; Lee, U.; Small, I.; des Francs-Small, C.C.; Vierling, E. Mutations in an Arabidopsis mitochondrial transcription termination factor-related protein enhance thermotolerance in the absence of the major molecular chaperone HSP101. *Plant Cell* **2012**, *24*, 3349–3365. [CrossRef] [PubMed]
9. Meskauskiene, R.; Würsch, M.; Laloi, C.; Vidi, S.; Coll, N.; Kessler, F.; Baruah, A.; Kim, C.; Apel, K. A mutation in the Arabidopsis mTERF-related plastid protein SOLDAT10 activates retrograde signaling and suppresses $^1O_2$-induced cell death. *Plant J.* **2009**, *60*, 399–410. [CrossRef] [PubMed]
10. Schonfeld, C.; Wobbe, L.; Borgstadt, R.; Kienast, A.; Nixon, P.J.; Kruse, O. The nucleus-encoded protein MOC1 is essential for mitochondrial light acclimation in *Chlamydomonas reinhardtii*. *J. Biol. Chem.* **2004**, *279*, 50366–50374. [CrossRef] [PubMed]
11. Quesada, V. The roles of mitochondrial transcription termination factors (MTERFs) in plants. *Physiol. Plant.* **2016**, *157*, 389–399. [CrossRef] [PubMed]
12. Roberti, M.; Polosa, P.L.; Bruni, F.; Manzari, C.; Deceglie, S.; Gadaleta, M.N.; Cantatore, P. The MTERF family proteins: Mitochondrial transcription regulators and beyond. *Biochim. Biophys. Acta* **2009**, *1787*, 303–311. [CrossRef] [PubMed]
13. Kleine, T. *Arabidopsis thaliana* mTERF proteins: Evolution and functional classification. *Front. Plant Sci.* **2012**, *3*, 233. [CrossRef] [PubMed]
14. Kleine, T.; Leister, D. Emerging functions of mammalian and plant mTERFs. *Biochim. Biophys. Acta* **2015**, *1847*, 786–797. [CrossRef] [PubMed]
15. Hsu, Y.W.; Wang, H.J.; Hsieh, M.H.; Hsieh, H.L.; Jauh, G.Y. Arabidopsis mTERF15 is required for mitochondrial *nad2* intron 3 splicing and functional complex I activity. *PLoS ONE* **2014**, *9*, e112360. [CrossRef] [PubMed]
16. Hammani, K.; Barkan, A. An mTERF domain protein functions in group II intron splicing in maize chloroplasts. *Nucleic Acids Res.* **2014**, *42*, 5033–5042. [CrossRef] [PubMed]
17. Babiychuk, E.; Vandepoele, K.; Wissing, J.; Garcia-Diaz, M.; De Rycke, R.; Akbari, H.; Joubès, J.; Beeckman, T.; Jänsch, L.; Frentzen, M.; et al. Plastid gene expression and plant development require a plastidic protein of the mitochondrial transcription termination factor family. *Proc. Natl. Acad. Sci. USA* **2011**, *108*, 6674–6679. [CrossRef] [PubMed]
18. Quesada, V.; Sarmiento-Mañús, R.; González-Bayón, R.; Hricová, A.; Pérez-Marcos, R.; Graciá-Martínez, E.; Medina-Ruiz, L.; Leyva-Díaz, E.; Ponce, M.R.; Micol, J.L. Arabidopsis *RUGOSA2* encodes an mTERF family member required for mitochondrion, chloroplast and leaf development. *Plant J.* **2011**, *68*, 738–753. [CrossRef] [PubMed]
19. Wobbe, L.; Nixon, P.J. The mTERF protein MOC1 terminates mitochondrial DNA transcription in the unicellular green alga *Chlamydomonas reinhardtii*. *Nucleic Acids Res.* **2013**, *41*, 6553–6567. [CrossRef] [PubMed]
20. Romani, I.; Manavski, N.; Morosetti, A.; Tadini, L.; Maier, S.; Kühn, K.; Ruwe, H.; Schmitz-Linneweber, C.; Wanner, G.; Leister, D.; et al. A Member of the Arabidopsis Mitochondrial Transcription Termination Factor Family Is Required for Maturation of Chloroplast Transfer RNAIle(GAU). *Plant Physiol.* **2015**, *169*, 627–646. [CrossRef] [PubMed]
21. Leister, D.; Kleine, T. Definition of a core module for the nuclear retrograde response to altered organellar gene expression identifies GLK overexpressors as *gun* mutants. *Physiol. Plant.* **2016**, *157*, 297–309. [CrossRef] [PubMed]
22. Robles, P.; Núñez-Delegido, E.; Ferrández-Ayela, A.; Sarmiento-Mañús, R.; Micol, J.L.; Quesada, V. Arabidopsis mTERF6 is required for leaf patterning. *Plant Sci.* **2018**, *266*, 117–129. [CrossRef] [PubMed]
23. Christmann, A.; Moes, D.; Himmelbach, A.; Yang, Y.; Tang, Y.; Grill, E. Integration of abscisic acid signalling into plant responses. *Plant Biol.* **2006**, *8*, 314–325. [CrossRef] [PubMed]

24. Xiong, L.; Ishitani, M.; Lee, H.; Zhu, J.K. The Arabidopsis *LOS5/ABA3* locus encodes a molybdenum cofactor sulfurase and modulates cold stress. *Plant Cell* **2001**, *13*, 2063–2083. [CrossRef] [PubMed]
25. González-Guzmán, M.; Apostolova, N.; Bellés, J.M.; Barrero, J.M.; Piqueras, P.; Ponce, M.R.; Micol, J.L.; Serrano, R.; Rodríguez, P. The short-chain alcohol dehydrogenase ABA2 catalyzes the conversion of xanthoxin to abscisic aldehyde. *Plant Cell* **2002**, *8*, 1833–1846. [CrossRef]
26. Shkolnik-Inbar, D.; Adler, G.; Bar-Zvi, D. ABI4 downregulates expression of the sodium transporter HKT1;1 in Arabidopsis roots and affects salt tolerance. *Plant J.* **2013**, *73*, 993–1005. [CrossRef] [PubMed]
27. Kilian, J.; Whitehead, D.; Horak, J.; Wanke, D.; Weinl, S.; Batistic, O.; D'Angelo, C.; Bornberg-Bauer, E.; Kudla, J.; Harter, K. The AtGenExpress global stress expression data set: Protocols, evaluation and model data analysis of UV-B light, drought and cold stress responses. *Plant J.* **2007**, *50*, 347–363. [CrossRef] [PubMed]
28. Zeller, G.; Henz, S.R.; Widmer, C.K.; Sachsenberg, T.; Rätsch, G.; Weigel, D.; Laubinger, S. Stress-induced changes in the *Arabidopsis thaliana* transcriptome analyzed using whole genome tiling arrays. *Plant J.* **2009**, *58*, 1068–1082. [CrossRef] [PubMed]
29. Yamaguchi-Shinozaki, K.; Shinozaki, K. Characterization of the expression of a desiccation-responsive *RD29* gene of *Arabidopsis thaliana* and analysis of its promoter in transgenic plants. *Mol. Gen. Genet.* **1993**, *236*, 331–340. [CrossRef] [PubMed]
30. Zhao, Y.; Cai, M.; Zhang, X.; Li, Y.; Zhang, J.; Zhao, H.; Kong, F.; Zheng, Y.; Qiu, F. Genome-Wide identification, evolution and expression analysis of mTERF gene family in maize. *PLoS ONE* **2014**, *9*, e94126. [CrossRef] [PubMed]
31. Zhou, B.; Zhang, L.; Ullah, A.; Jin, X.; Yang, X.; Zhang, X. Identification of multiple stress responsive genes by sequencing a normalized cDNA library from sea-land cotton (*Gossypium barbadense* L.). *PLoS ONE* **2016**, *11*, e0152927. [CrossRef] [PubMed]
32. Moschopoulos, A.; Derbyshire, P.; Byrne, M.E. The Arabidopsis organelle-localized glycyl-tRNA synthetase encoded by *EMBRYO DEFECTIVE DEVELOPMENT1* is required for organ patterning. *J. Exp. Bot.* **2012**, *63*, 5233–5243. [CrossRef] [PubMed]

© 2018 by the authors. Licensee MDPI, Basel, Switzerland. This article is an open access article distributed under the terms and conditions of the Creative Commons Attribution (CC BY) license (http://creativecommons.org/licenses/by/4.0/).

Article

Cold and Heat Stress Diversely Alter Both Cauliflower Respiration and Distinct Mitochondrial Proteins Including OXPHOS Components and Matrix Enzymes

Michał Rurek [1,*], Magdalena Czołpińska [1], Tomasz Andrzej Pawłowski [2], Włodzimierz Krzesiński [3] and Tomasz Spiżewski [3]

1. Department of Molecular and Cellular Biology, Institute of Molecular Biology and Biotechnology, Adam Mickiewicz University, Poznań, Umultowska 89, 61-614 Poznań, Poland; magczo@amu.edu.pl
2. Institute of Dendrology, Polish Academy of Sciences, Parkowa 5, 62-035 Kórnik, Poland; tapawlow@man.poznan.pl
3. Department of Vegetable Crops, Poznan University of Life Sciences, Dąbrowskiego 159, 60-594 Poznań, Poland; wlodzimierz.krzesinski@up.poznan.pl (W.K.); tomasz.spizewski@up.poznan.pl (T.S.)
* Correspondence: rurek@amu.edu.pl; Tel.: +48-61-829-5973

Received: 29 January 2018; Accepted: 9 March 2018; Published: 16 March 2018

Abstract: Complex proteomic and physiological approaches for studying cold and heat stress responses in plant mitochondria are still limited. Variations in the mitochondrial proteome of cauliflower (*Brassica oleracea* var. *botrytis*) curds after cold and heat and after stress recovery were assayed by two-dimensional polyacrylamide gel electrophoresis (2D PAGE) in relation to mRNA abundance and respiratory parameters. Quantitative analysis of the mitochondrial proteome revealed numerous stress-affected protein spots. In cold, major downregulations in the level of photorespiratory enzymes, porine isoforms, oxidative phosphorylation (OXPHOS) and some low-abundant proteins were observed. In contrast, carbohydrate metabolism enzymes, heat-shock proteins, translation, protein import, and OXPHOS components were involved in heat response and recovery. Several transcriptomic and metabolic regulation mechanisms are also suggested. Cauliflower plants appeared less susceptible to heat; closed stomata in heat stress resulted in moderate photosynthetic, but only minor respiratory impairments, however, photosystem II performance was unaffected. Decreased photorespiration corresponded with proteomic alterations in cold. Our results show that cold and heat stress not only operate in diverse modes (exemplified by cold-specific accumulation of some heat shock proteins), but exert some associations at molecular and physiological levels. This implies a more complex model of action of investigated stresses on plant mitochondria.

Keywords: cold stress; heat stress; stress recovery; mitochondria; proteomics; respiration; *Brassica*; angiosperms

1. Introduction

Abiotic stress, including excessive cold or heat, cause failure in the cultivation of many plant species. Such conditions may significantly reduce the yield of most major crops. Plants have various physiological and metabolic response mechanisms, which act within the complex network to avoid harm due to unfavorable environmental stimuli [1–3]. Understanding these mechanisms improves our knowledge of stress resistance and will allow the breeding of more appropriate plant varieties.

Numerous aspects of plant responses to cold and heat have been studied. They may differ between plant species [4–6]. Both low and high temperatures can decrease chlorophyll biosynthesis, significantly impeding chloroplast development and potentially resulting in photosystem II (PSII) damage [7–11].

Cold-grown plants generate a vast number of reactive oxygen species (ROS) [12]. Armstrong et al. [13] analyzed temperature-dependent sensitivity of leaf respiration in Arabidopsis during cold acclimation and suggested the importance of an alternative oxidation pathway in this process. Moreover, Talts et al. [14] observed that cold-treated plants often display higher rates of respiration. However, heat stress (depending on its intensity and duration) can exert particularly diverse effects on the photosynthetic apparatus [15], including increased cyclic electron flow around PSI [9,16–19].

Despite reports concerning evident alterations in plant physiological parameters during stress response, data on the correlation of those changes with mitochondrial proteomes are quite limited. Organellar proteomic analyses, including mitochondrial ones, may help to reveal the intrinsic mechanisms of stress response by elucidating the relationship between protein variations and general plant tolerance to environmental factors [20]. Nowadays, characterization of total proteomes or sub-proteomes of important crop and vegetable plants, including cauliflower (Brassica oleracea var. botrytis), appears to be very important [21–26].

The plant mitochondrial proteome is a very dynamic entity which can be remodeled in a plethora of environmental conditions and developmental signals [27,28]. It is known that dozens of nuclear genes encoding mitochondrial proteins respond to stress conditions form a functional network [29]. Using an integrative approach, Cui et al. [30] found 503 Arabidopsis mitochondrial proteins participating in a stress protein interaction network. This suggests the general dependence of plant mitochondria on other plant cell compartments during stress response. Furthermore, Taylor et al. [31] estimated that only 22% of total Arabidopsis organellar proteins that are stress-responsive comprise mitochondrial proteins. It seems that the number of mitochondrial proteins involved in stress response is still underestimated, due to limited complexity of some reports and the fact that a significant number of results came from analyses of total plant proteomes and main metabolic pathways only [25,32–34]. It should be mentioned that the number of low-abundant mitochondrial proteins responsive to temperature stress is still far from being understood [31]. Recently, these issues were improved by the application of isobaric tags for the absolute quantification (iTRAQ) or label-free peptide counting coupled with liquid chromatography-tandem mass spectrometry (LC-MS/MS) [35–38]. Using a gel-free approach, Tan et al. [39] found that cold stress led to a concerted decrease in respiratory protein level, accompanied by an increase in abundance of some import/export protein machinery components. However, the overall amount of cold-responsive proteins was smaller, when compared to other suboptimal stimuli.

Although it is known that temperature stress modulates mitochondrial protein activity, level, biogenesis and interactions [40–42], crucial steps of achieving appropriate coordination during mitochondrial biogenesis in stress need to be further investigated. For instance, Giegé et al. [43] showed that regulation of mitochondrial biogenesis in Arabidopsis cell cultures during sugar starvation seems to be rather coordinated at the complex assembly. Approaches linking molecular and physiological data dealing with temperature stress impact on mitochondria are still welcomed. Some mitochondrial proteins (e.g., alternative oxidase [AOX]) are 'classical' modulators of stress response among plants [44–46]. Regulation of diverse AOX genes varies between monocots and dicots. In a number of plant species, the alterations of AOX protein are less pronounced [47,48]. In addition, AOX may not to be increased in abundance by certain stress treatments, for example chilling [49,50]. The latter phenomenon was also confirmed in our previous study [41]; we reported a significant decline in AOX level in cauliflower mitochondria under cold stress and recovery. Overall AOX gene family responses on proteomic and transcriptomic levels were only partially associated and AOX was a suggested target of translational regulation in diverse temperature treatments. In tobacco (Nicotiana tabacum) leaves, abundance of this protein reached a maximum after 48 h of cold stress and slowly decreased afterwards [51]. This highlights the importance of the length of cold stress treatment for the plant to gain acclimation, presumably by the induction of regular changes in the transcriptome first [52].

Assuming limitations of the deposited data, this work was undertaken to gain a comprehensive view about the influence of cold and heat treatment (as well as cold and heat recovery) on the cauliflower mitochondrial proteome in relation to leaf transpiration and respiration rate, stomatal conductance, rate of leaf photosynthesis, photorespiration as well as chlorophyll content and fluorescence. The current study extends our previous complexomic and functional data [41]. To determine mitochondrial proteome response in relation to plant respiration, we aimed to (1) investigate the dynamic nature of the mitochondrial proteome under cold and heat treatment and stress recovery; (2) identify the most variable proteins in cauliflower inflorescence mitochondrial extracts; and (3) link proteomic and discussed metabolic/functional aspects with alterations of analyzed physiological parameters. On the whole, the broader set of identified proteins responding to cold/heat stress and after stress recovery, which correlate with alterations in plant respiration and some general metabolic demands, was able to be characterized in cauliflower mitochondria.

2. Results

2.1. Proteome Maps of Cauliflower Mitochondria under Stress Conditions

Mitochondrial proteins isolated from curds of control plants and from plants submitted to cold or heat treatment were resolved by two-dimensional gel electrophoresis (2D PAGE). We also examined the mitochondrial proteomes from curds of stress-recovered cauliflower plants with the idea to study the impact of stress on the mitochondrial proteome under stress recovery (Figure S1). 2D gels for investigated variants, including control, were run in triplicate. Individual gel replicates are shown in Figure S2. In order to create master gel, we chose the image of control variant as the reference and then we added the specific spots detected on the gels of remaining variants. The number of spots on silver-stained 2D gels varied from 347 to 511 between all analyzed variants, including the control one. Thus, 694 different spots were taken into account for the building of a synthetic silver-stained master gel (Figure 1). Contrary to silver-stained gels, the number of protein spots on colloidal Coomassie Brilliant Blue (CBB)-stained 2D gels was lower. Finally, for the analysis of spot variation, only 413 spots representing highly abundant proteins from silver-stained gels were taken into account.

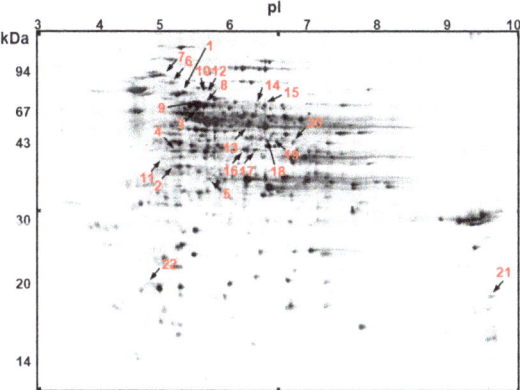

Figure 1. Position of varying spots on 2D silver-stained master gel of cauliflower curd mitochondrial proteome (in total 100 µg of mitochondrial proteins pooled from all experimental variants), including 694 repeatable spots. 24-cm immobilised pH gradient strips (linear pH 3–10) for the first dimension and precast Ettan DALT 12.5% SDS-polyacrylamide gels for the second dimension were used. It shows the position of the 22 variable spots that were mapped and identified (they appear also in Figure 1, Figure S1 and S2 and Table 1). Protein molecular mass standards (Thermo Scientific, Gdańsk, Polska) sizes are given in kilodaltons (kDa); pI-isoelectric point. Further experimental details in Materials and Methods.

Table 1. List of cauliflower mitochondrial proteins the level of which varied during stress treatments.

Spot No. [a]	Mean % Volume [b] K	C; H	CA; HA	Assignment; Species; FunCat [c]	Protein Record Version	UniProt Accession No.	AGI Identifier	Nominal M	Nominal pI	Observed M	Observed pI	Mascot Score; emPAI [d]	Coverage (%)	Uniq. Peps; Tot. Peps [e]
1	0.53 ± 0.02	1.06 ± 0.08 (+2.00)	0.68 ± 0.19 [+1.28]	Mitochondrial heat shock protein 70-1; Arabidopsis; PrF	CAB37531.1	Q9SZJ3	At4g37910	71.4	5.31	79	5.35	4732; 3.61	33	53; 129
2	0.11 ± 0.01	0.21 ± 0.06 [+1.91]	0.22 ± 0.03 (+2.00)	Pyruvate dehydrogenase E1 beta subunit; Arabidopsis; CM	NP_199898.1	Q38799	At5g50850	39.4	5.67	39	5.16	5484; 2.94	29	73; 146
3	0.18 ± 0.01	0.40 ± 0.09 (+2.22)	0.21 ± 0.01 (+1.17)	3-phosphoglycerate dehydrogenase-like protein; Arabidopsis; AM	NP_195146.1	O49485	At4g34200	63.6	6.16	72	5.43	7707; 1.48	31	112; 266
4	0.11 ± 0.01	0.15 ± 0.01 (+1.36)	0.21 ± 0.03 (+1.91)	Phosphoglycerate kinase 1; Arabidopsis; CM	NP_187884.1	Q9LD57	At3g12780	50.1	5.91	43	5.20	630; 1.44	28	9; 26
5	0.15 ± 0.01	0.19 ± 0.04 (+1.27)	0.29 ± 0.03 (+1.93)	Malate dehydrogenase (NAD), mitochondrial; Arabidopsis; CM	NP_564625.1	Q9ZP06	At1g53240	36.0	8.54	37	5.58	409; 0.98	22	10; 18
6	0.19 ± 0.02	0.37 ± 0.05 (+1.95)	0.16 ± 0.03 (−1.18)	Heat shock protein 81-2 (HSP 90 related); Arabidopsis; PrF	NP_200414.1	P55737	At5g56030	80.2	4.95	84	5.20	277; 0.43	14	0; 12
7	0.16 ± 0.05	0.39 ± 0.09 (+2.44)	0.13 ± 0.07 (−1.23)	Heat shock protein 90; Arabidopsis; PrF	BAR00175.1	Q0WRS4	At3g07770	90.8	5.26	89	5.06	1996; 1.11	19	15; 75
8	0.17 ± 0.02	0.64 ± 0.19 (−3.76)	0.45 ± 0.07 (+2.62)	3-phosphoglycerate dehydrogenase-like protein; Arabidopsis; AM	NP_195146.1	O49485	At4g34200	63.6	6.16	73	5.56	6778; 1.74	30	88; 237
9	0.25 ± 0.03	0.72 ± 0.31 (+2.88)	0.72 ± 0.16 (+2.88)	3-phosphoglycerate dehydrogenase-like protein; Arabidopsis; AM	NP_195146.1	O49485	At4g34200	63.6	6.16	74	5.49	7997; 1.36	27	111; 301
10	0.05 ± 0.03	0.06 ± 0.01 (+1.20)	0.19 ± 0.04 (+3.80)	Phosphoglycerate kinase 1; Arabidopsis; CM	NP_187884.1	Q9LD57	At3g12780	50.2	5.91	81	5.50	1376; 2.57	42	22; 53
11	0.14 ± 0.08	0.12 ± 0.11 (−1.17)	1.29 ± 0.25 (+9.21)	Putative succinyl-l-CoA ligase (GDP-forming) beta subunit, mitochondrial; Arabidopsis; CM	NP_179632.1	O82662	At2g20420	45.6	6.30	41	5.15	265; 0.88	17	11; 11
12	0.56 ± 0.15	0.32 ± 0.17 (−1.75)	0.11 ± 0.02 (−5.09)	Putative mitochondrial processing peptidase; Arabidopsis; PrF	BAE98412.1	Q42290	At3g02090	51.5	5.71	80	5.79	833; 1.38	26	17; 42
13	0.36 ± 0.14	0.47 ± 0.09 (+1.30)	0.19 ± 0.04 (−1.59)	ATPase subunit 1; Brassica napus; RC	YP_717155.1	Q6YSN4	AtMg01190	55.4	6.01	58	6.07	3195; 1.12	23	76; 168
14	0.35 ± 0.14	0.22 ± 0.04 (−1.59)	0.08 ± 0.03 (−4.37)	Δ-1-pyrroline-5-carboxylate dehydrogenase precursor; Arabidopsis; AM	AAK73756.1	Q8VZC3	At5g62530	62.2	6.26	70	6.24	784; 0.67	18	3; 41
15	0.54 ± 0.06	0.33 ± 0.06 (−1.64)	0.09 ± 0.01 (−6.00)	Δ-1-pyrroline-5-carboxylate dehydrogenase precursor; Arabidopsis; AM	AAK73756.1	Q8VZC3	At5g62530	62.2	6.26	69	6.33	1622; 0.95	21	0; 78
16	0.60 ± 0.14	0.77 ± 0.11 (+1.28)	0.31 ± 0.05 (−1.93)	Mitochondrial elongation factor Tu; Arabidopsis; PrS	CAA61511.1	Q9ZT91	At4g02930	51.6	5.53	42	6.00	3123; 2.45	37	24; 100

Table 1. Cont.

Spot No.[a]	Mean % Volume[b]			Assignment; Species; FunCat[c]	Protein Record Version	UniProt Accession No.	AGI Identifier	Nominal		Observed		Mascot Score; emPAI[d]	Coverage (%)	Uniq. Peps; Tot. Peps[e]
	K	C; H	CA; HA					M	pI	M	pI			
17	0.31 ± 0.08	0.42 ± 0.07 (**+1.35**)	0.18 ± 0.04 (−1.72)	Mitochondrial elongation factor Tu; Arabidopsis; **PrS**	CAA61511.1	Q9ZT91	At4g02930	51.6	5.53	42	6.19	2502; 2.05	33	35; 77
18	0.58 ± 0.15	0.52 ± 0.07 (−1.11)	0.14 ± 0.02 (**−4.14**)	Isocitrate dehydrogenase-like protein; Arabidopsis; **CM**	CAB87626.1	Q9LYK1	At5g14590	52.3	7.11	45	6.37	697; 0.96	23	14; 23
19	0.26 ± 0.06	0.17 ± 0.03 (**−1.53**)	0.11 ± 0.01 (**−2.36**)	Citrate synthase (SI); Arabidopsis; **CM**	NP_850415.1	P20115	At2g44350	53.1	6.41	48	6.57	999; 0.62	18	34; 40
20	0.40 ± 0.09	0.23 ± 0.07 (**−1.74**)	0.15 ± 0.01 (**−2.67**)	Citrate synthase (SI); Arabidopsis; **CM**	NP_850415.1	P20115	At2g44350	53.1	6.41	47	6.83	2126; 0.83	21	51; 84
21	0.08 ± 0.06	0.16 ± 0.08 (+2.00)	0.61 ± 0.08 (**+7.62**)	10 kDa chaperonin; Arabidopsis; **PrF**	NP_563961.1	P34893	At1g14980	10.8	6.74	19	9.38	107; 1.30	31	8; 8
22	LA[f]	0.80 ± 0.20	0.53 ± 0.01	ATP synthase, d chain, mitochondrial; Arabidopsis; **RC**	NP_190798.1	Q9FT52	At3g52300	19.6	5.09	20	4.59	979; 1.60	25	18; 39

For each spot, values for nominal (computed) and observed isoelectric point (**pI**) and molecular mass (**M**, in kDa) and the mean normalized volume at each of analysed stress and control variants were indicated; [a] Spot number as index in the reference gel; [b] Mean value with the standard deviation of three spot volumes at each analysed stage: control (**K**), cold stress (**C**; **spots 1–7**), heat stress (**H**; **spots 8–22**), cold recovery (**CA**; **spots no. 1–7**), heat recovery (**HA**; **spots 8–22**). In parentheses, the fold change of mean value for stress variants, compared to control. Up-regulations are indicated by + and down-regulations by −. Significant regulations (according to HSD test) are **bolded and underlined**; the ones for cold, cold recovery, heat and heat recovery are shown in blue, green, red and brown highlights, respectively; [c] In **bold**: functional categorisation (FunCat) using data from FunCat scheme (Available online: http://ibis.helmholtz-muenchen.de/funcatDB/). Two-letter legend to all categories: **CM**—carbohydrate metabolism, **PrF**—protein fate, **AM**—amino-acid metabolism, **RC**—respiration (respiratory chain components), **PrS**—protein synthesis; [d] Exponentially modified protein abundance index; [e] Uniq. peps; tot. peps-number of unique peptides; number of total peptides for each protein spot; [f] **LA**—low-abundant protein (with the % volume of circa 0.01); in this case the calculation of the fold change was not applicable. Further experimental details in Materials and Methods.

2.2. Identification of Variable Protein Spots

Twenty two spots (3.2% on the silver-stained master gel) were significantly variable (verified by the analysis of variance [ANOVA] and Tukey's honest significant difference [HSD] test) as detected using Image Master 7 Platinum software. Spot positions depicted in Figure 1, Figure S1 and S2 were calculated from three biological replicates. All spots were successfully identified by LC-MS/MS. The obtained data were used for searching Mascot against the National Centre for Biotechnology Information (NCBI) database (version 20100203). Cauliflower stress-responsive mitochondrial proteins were identified by using the *Viridiplantae* section of the database (with the aid of Arabidopsis and *B. oleracea* nuclear and mitochondrial [53] genomes). To avoid possible misidentifications resulting from large datasets, as pointed out by Schmidt et al. [24], we were able to set the false positive rate threshold to 5%. Identifications of protein spots are presented in Table 1 and properties of individual peptides for each protein spot are given in Table S1. As illustrated, all 22 spots represented 16 non-redundant stress-responsive proteins. We did not exclude spots with multiple protein identification from the analyses, however, we focused on protein identifications (for all stress-responsive spots) confirmed by sufficient parameter quality (highest MOWSE score, emPAI, peptide number and coverage). The disproportion between the quality of mentioned parameters for the initial and remaining records in the MASCOT search list allowed us to do so. The percentage of sequence coverage ranged from 14% to 42% and the total number of identified peptides varied from 12 to 301. Among all spots, mitochondrial proteins in all but one spot (spot No. 13 corresponding to *Brassica napus* protein sequence) were identified based on their high similarity to Arabidopsis sequences. In addition, 17 Arabidopsis proteins also showed a very high or 100% sequence identity with *B. oleracea* var. *oleracea* records (Table S2). The experimental molecular mass corresponded roughly to the theoretical value for the majority of spots. Some proteins including phosphoglycerate kinase isoform 1 (PGK1; spots No. 4, 10; Table 1), mitochondrial elongation factor Tu (mtEF-Tu; spots No. 16, 17), isocitrate dehydrogenase (IDH; spot No. 18) and citrate synthase (CS; spots No. 19, 20) showed a few kDa decrease in molecular mass between theoretical and gel values. We are rather convinced that this did not result from the excessive proteolysis in cauliflower mitochondria.

Seven spots (approximately 1% of all) displayed significant variations in their abundance after cold stress and cold recovery. According to Tukey's HSD test, four proteins, including three members of the heat shock protein (HSP) family (spots No. 1, 6, 7) and 3-phosphoglycerate dehydrogenase (PGDH)–like protein (spot No. 3) were increased in abundance during cold treatment. After cold recovery another three proteins were significantly increased in abundance, namely pyruvate dehydrogenase subunit β (PDHβ; spot No. 2), PGK1 (spot No. 4) as well as NAD^+-dependent malate dehydrogenase (MDH; spot No. 5; Table 1).

After heat stress and heat recovery, 15 responsive spots (representing 11 non-redundant proteins) were identified (approximately 2.2% of all spots; Table 1). Four proteins PGDH-like protein, Δ-1-pyrroline-5-carboxylate dehydrogenase (P5CDH), mtEF-Tu as well as CS were represented by double spots (No. 8/9, 14/15, 16/17 and 19/20, respectively) displaying slightly different molecular mass and pI values. According to Tukey's HSD test, it appeared that three proteins significantly raised their level during heat treatment: PGDH–like protein (spots No. 8), mtEF-Tu (spots No. 17), as well as mitochondrial ATP synthase subunit d (ATPQ; spot No. 22), but two proteins: P5CDH (spot No. 15) and CS (spots No. 19/20) were decreased in abundance.

Some variations were also observed after heat recovery. Here, we detected a more intense accumulation of a broad set of proteins, namely PGDH-like protein (spots No. 8, 9), PGK1 (spot No. 10, succinyl-CoA ligase subunit β (SCLβ; spot No. 11), chaperonin 10 (CPN10; spot No. 21), and ATPQ (spot No. 22). Notably, CPN10 extensively increased in abundance. In those conditions, we also noticed a significant decline of the ATP synthase subunit α (ATP1; spot No. 13), mitochondrial processing peptidase subunit β (MPPβ; spot No. 12), P5CDH (spots No. 14, 15), mtEF-Tu (spots No. 16/17), IDH (spot No. 18) and CS level (spots No. 19/20; Table 1). In addition, with the help of polyclonal antibodies, we verified the abundance of ATP1 (the only protein from our data encoded by the mitochondrial genome) after heat recovery on 2D immunoblots. As illustrated in Figure 2 (all 2D immunoblot replicates are shown in Figure S2), the respective variations of this protein assayed by immunoblotting roughly followed the protein variations on the silver stained 2D gels. Notably, four proteins that were identified as double spots showed very similar response after heat and heat recovery, which is in favour for the correctness of their assignments (Table 1).

Full peptide data from the error-tolerant MASCOT search (Table S3a) allow to get a general view on the extent posttranslational protein modifications (PTMs) within proteins (listed in Table 1) from double spots (No. 8/9, 14/15, 16/17 and 19/20). As was shown, different proteins were identified in each spot pair. Protein multi-spotting frequently accompanies 2D gel data, however, we only investigated stress-responsive double spots (Table 1). We wanted to check whether the presence of unknown PTMs could be associated with protein multi-spotting (Table S3b). We estimated a number of modified residues from spectra of tryptic peptides as well as expected molecular mass difference from the extent of given PTM between compared spots. We focused on phosphorylations, deamidations, methylations, formylations and ethylations. Because no phosphoprotein enrichment was performed, we were unable to further characterize the phosphoproteome in cauliflower mitochondria, despite phosphorylated residues accounting mostly for total theoretical molecular mass difference between double spots. However, only limited correlation was found between this value and the experimental molecular mass difference for double spots. Therefore, multi-spotting of some analysed proteins came rather from non-investigated modifications and/or from the expression of gene family members.

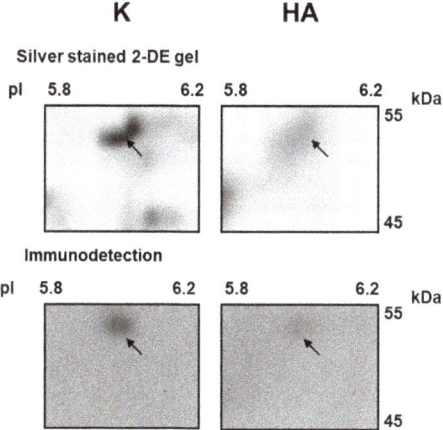

Figure 2. Immunoblotting of ATP1 on 2D blots containing cauliflower curd mitochondrial proteins from control grown plants (**K**) and from heat-recovered plants (**HA**). 100 μg of mitochondrial proteins were loaded onto all gels. 24-cm immobilized pH gradient strips (linear pH 3–10) for the first dimension and precast Ettan DALT 12.5% SDS-polyacrylamide gels for the second dimension were used. For protein transfer onto Immobilone membrane, a semidry system was applied. Blots were probed with polyclonal antibodies raised against mitochondrial ATP synthase subunit α (ATP1). Detection was carried with chemiluminescence assays after incubation with HRP-conjugated secondary antibody. Representative results (from triplicates) are shown. For the comparison, panels showing fragments of silver-stained 2D gels that contains spots for ATP1 are displayed. Arrows (indicated in each blot) show the position of ATP1 on immunoblots and 2D gels. Protein molecular mass standard (Thermo Scientific, Gdańsk, Polska) sizes are given in kilodaltons (kDa); pI-isoelectric point. Further experimental details in the text.

2.3. Functional Categorization of Identified Proteins

Based on Arabidopsis protein orthologues, we used a functional categorization (FunCat) scheme at the Munich Information Center for Protein Sequences database (Available online: http://ibis.helmholtz-muenchen.de/funcatDB/) for clustering of stress-responsive proteins resolved on 2D gels into five functional categories (Figure 3 and Table 1).

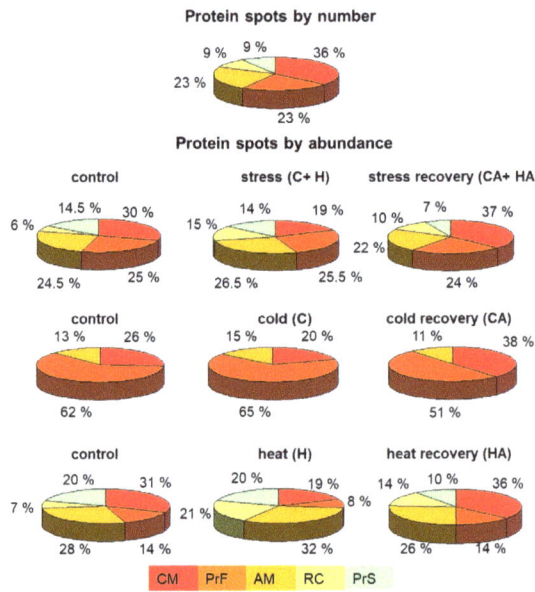

Figure 3. Functional categorization of cauliflower stress-responsive protein spots, analysed by 2D PAGE. Spots were analysed both according to their number (**upper**) as well as densitometrical volume (abundance; **below**) calculated for each of functional groups in control grown plants, cold- or heat-stressed plants (**C** or **H**, respectively) and cold- or heat-recovered plants (**CA** or **HA**, respectively). Bar legend of categories: **CM**—carbohydrate metabolism, **PrF**—protein fate, **AM**—amino acid metabolism, **RC**—respiration (respiratory chain components), **PrS**—protein synthesis.

Counting the number of spots within each category (Figure 3, panel: protein spots by number), it appeared that the majority of cauliflower mitochondrial protein spots responsive to cold and heat stress belonged to the class participating in carbohydrate metabolism, including tricarboxylic acid (TCA) cycle components (about 36% spots) as well as amino acid metabolism and protein fate (each of approximately 23%). The next ones were represented by respiratory chain (RC) components and protein synthesis apparatus (each of approximately 9%). Interestingly, eight spots (36%) representing six proteins were already annotated as stress responsive in the MIPS database.

Spots linked to RC components increased in abundance after heat stress, as well as after heat recovery, however, the ones linked to amino acid metabolism were upregulated after cold and heat stress. In contrast, spots linked with carbohydrate metabolism decreased in abundance after cold and heat, but markedly upregulated in cold- and heat-recovered plants (Figure 3, panel: protein spots by abundance), Interestingly, the total abundance of spots related to the protein fate showed some increase after cold, but neither after heat stress (where it was decreased), nor after recovery phase. It seems that the majority of identified protein spots that belonged to the protein fate class appeared responsive in cold stress and cold recovery, which indicates its overall importance in low temperature response in cauliflower mitochondria. Many protein functional classes, however, were regulated by heat and heat recovery (Figure 3, panel: protein spots by abundance, at the bottom).

2.4. Effect of Cold Stress on Abundance of Additional Mitochondrial Proteins

Due to the fact that the number of cold-regulated proteins was lower than those regulated by heat stress in cauliflower mitochondria, we decided to verify our analyses by additional immunoblotting assays (Figure 4). All immunoblot replicates are shown in Figure S2. The level of selected proteins was monitored in mitochondria isolated from curds of cauliflower plants grown either in control conditions or submitted to cold, or from cold-recovered plants. To verify protein loading, blots were Coomassie-stained.

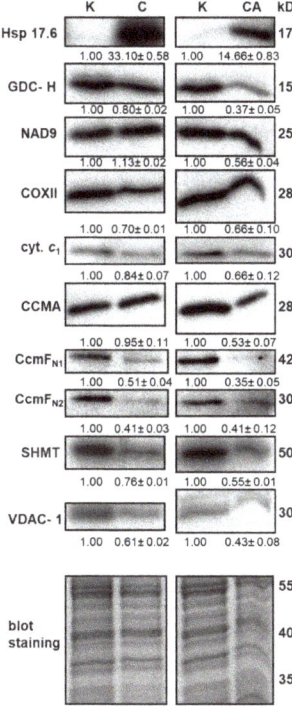

Figure 4. Immunoblotting of proteins from control grown plants (**K**), cold-stressed plants (**C**) and from cold-recovered plants (**CA**). About 10 µg of mitochondrial proteins from cauliflower curds were loaded onto SDS-polyacrylamide gels. Proteins were transferred onto an Immobilone membrane using a semidry system. All assays were performed using specific primary antibodies raised against heat shock protein 17.6C class I (Hsp17.6), glycine decarboxylase subunit H (GDC-H), NADH dehydrogenase (CI) subunit 9 (NAD9), cyt. c oxidase (CIV) subunit 2 (COXII), cyt. c_1, cyt. c maturation proteins (CCMA, CcmF$_{N1}$, CcmF$_{N2}$), serine hydroxymethyltransferase 1 (SHMT) and voltage-dependent anion channel 1 (VDAC-1). Detection was carried out with chemiluminescence assays after incubation with HRP-conjugated secondary antibody. Representative results from triplicates are shown. The relative abundance of bands is given below each panel. The abundance in stress conditions (value ± SD) is standardized to 1.00 in control variants. For the loading control, blot staining with Coomassie Brilliant Blue is additionally shown. Protein molecular mass standard (Thermo Scientific, Gdańsk, Polska) sizes are given in kilodaltons (kDa). Further experimental details in Materials and Methods.

We assayed the level of glycine decarboxylase subunit-H (GDC-H), serine hydroxymethyltransferase 1 (SHMT), mitochondrial porine isoform 1 (VDAC-1), some OXPHOS proteins, including complex I (CI) subunit 9 (NAD9), cytochrome c_1, complex IV (CIV) subunit 2 (COXII) as well as proteins engaged in cytochrome c (cyt. c) maturation in plant mitochondria, particularly ABC transporter I family member 1 (CCMA) and CcmF N-terminal-like mitochondrial proteins 1 and 2 (CcmF$_{N1}$ and CcmF$_{N2}$, respectively). With the application of specific antibodies, we also investigated the level of cytoplasmic small Hsp17.6 of class I (sHsp17.6C-CI), that interacts with mitochondria under temperature stress [54]. It appeared, that the level of Hsp17.6C-CI associated with mitochondrial membranes increased extensively after cold stress and remained quite high after cold recovery (Figure 4). The abundance of GDC-H showed almost a three-fold change decrease after cold recovery, but only slightly after cold stress. A similar decrease in the level of CcmF$_{N1}$ and CcmF$_{N2}$ proteins was observed in cold and cold recovery conditions (up to two- and three-fold change, respectively). In contrast, the accumulation of the CCMA transporter protein was not affected by cold; however, it was decreased (by almost 50%) after cold recovery. In the tested conditions, the relative abundance of SHMT and VDAC-1 was also decreased. Regarding RC proteins, we detected a small upregulation of NAD9 subunit of CI after cold stress and a subsequent major decline under cold recovery as well as a small downregulation of COXII and cyt. c_1 in stress (Figure 4).

2.5. Association between Protein and Transcript Level

Besides analyses of cauliflower mitochondrial proteome, we studied how proteomic response is accompanied by transcript alterations. For the rapid assessment of both patterns, we employed RT-semiqPCR and the level of five mitochondrial and 11 nuclear messengers was assayed (Figure S3) and compared with the abundance of some OXPHOS proteins encoded by mitochondrial genome and proteins coded by nuclear genome investigated in this study (Sections 2.2 and 2.4).

None of the mitochondrial mRNAs showed alterations in their abundance associated with the protein level. *nad9* and *coxII* messengers (coding for CI subunit 9 and CIV subunit 2) were regulated inversely compared with the respective proteomic data in cold and cold recovery. *atp1* transcripts (coding for ATP synthase subunit 1) responded only in cold recovery. Furthermore, expression profiles of genes coding for subunits of the same protein complexes (e.g., *nad* genes for CI and *atp*/*ATP* genes for ATP synthase) were largely un-associated. This is true both for selected mitochondrial as well as nuclear genes. The level of *ATP2* nuclear transcripts was slightly decreased in cold and heat, and decreased furthermore in stress recovery.

Parallel RNA/protein accumulation patterns were noted in case of the only three nuclear genes coding for HSP70 isoform 1 (*HSP70-1*) in cold and cold recovery, mitochondrial processing peptidase subunit (*MPPβ*) as well as for Δ-1-pyrroline-5-carboxylate dehydrogenase (*P5CDH*) in heat recovery. However, those variations were quite minor. In case of the remaining regulations, we can see that the decreased mRNA level (particularly for some transcripts, e.g., *Hsp17.6* in cold and cold recovery and *SCLβ* together with *CPN10* in heat recovery) did not correlate with the increased protein abundance in stress and recovery. Our results indicate also for the down-regulation of transcripts coding for two enzymes of Pro catabolism: proline dehydrogenase (*PRODH*) and Δ-1-pyrroline-5-carboxylate dehydrogenase (*PRODH* and *P5CDH*, respectively) in cold and heat recovery (Figure S3).

2.6. Cauliflower Physiological Responses to Cold and Heat Stress, and after Stress Recovery

We also studied how leaf respiration was affected after cessation of cold and heat treatment as well as after post-stress plant recovery. By using an appropriate assay [55], we determined mitochondrial respiration in the light (non-photorespiratory intracellular decarboxylation; R_d) in gas phase as the rate of CO_2 release. In addition, we also measured the rate of respiration of darkened leaves (R_n). It appeared, that the respiratory production of CO_2 in illuminated leaves was lowered in cold-stressed plants; however, under cold recovery, a significant burst of R_d was observed. R_n rate was also lowered after cold treatment and remained so after cold recovery. In contrast, both R_d and R_n rates significantly increased in heat stress and decreased almost to control stage values after heat recovery (Figure 5).

To gain a more complete view of cauliflower plant physiological status, we also assayed the impact of stress conditions and stress recovery on leaf transpiration (E) rate, stomatal conductance (g_s) as well as essential photosynthetic parameters. We detected a decrease in E rate as well as lower g_s value under cold and heat stress, but not after cold recovery. However, after heat recovery, leaf transpiration was slightly elevated (Figure 6).

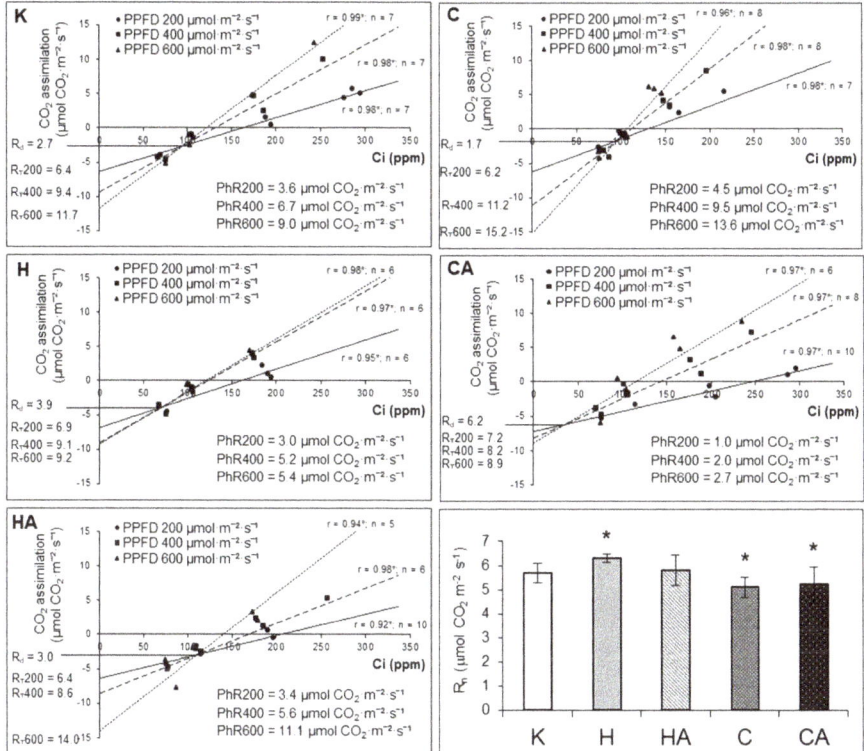

Figure 5. Changes in cauliflower leaf light (R_d), dark (R_n) respiration as well as total light (R_T) respiration and (PhR) photorespiration (all expressed in $\mu mol \cdot CO_2 \cdot m^{-2} \cdot s^{-1}$) at 200 ($R_T$200 and PhR200), 400 ($R_T$400 and PhR400) and 600 (R_T600 and PhR600) $\mu mol \cdot m^{-2} \cdot s^{-1}$ illumination rate in control grown (**K**), heat-stressed (**H**), heat-recovered (**HA**), cold-stressed (**C**) and cold-recovered (**CA**) plants. All parameters were measured on 3-month-old plants with fully developed leaves with the application of an infrared gas analyser. Data were recorded after at least 2 h of illumination. During the experiment, each of the analysed leaves were placed into a 6-cm^2 chamber of the analyser. Results were recorded after initial leaf acclimation to the desired light and CO_2 concentration, relative humidity and temperature. The R_d rate was determined according to the Laisk [55] method. The photorespiration rate for each PPFD value was determined as the difference between R_T and R_d values. Error bars denote ± S.D. Asterisks indicate significantly different curves at $p = 0.05$ (Student's t-test). Further experimental details in Materials and Methods.

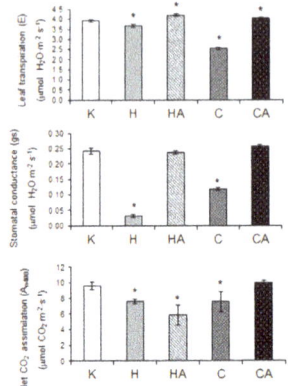

Figure 6. Changes in cauliflower leaf transpiration, stomatal conductance to water vapour and net CO_2 assimilation rate at 400 $\mu mol \cdot m^{-2} \cdot s^{-1}$ illumination in control grown (**K**), cold-stressed (**C**), heat-stressed (**H**), cold-recovered (**CA**) and heat-recovered (**HA**) plants. All parameters were measured on 3-month-old plants with fully developed leaves with the application of an infrared gas analyser. Data were recorded after at least 2 h of illumination. During the experiment, each of the analysed leaves were placed into a 6-cm^2 chamber of the analyser. Results were recorded after initial leaf acclimation to the desired light and CO_2 concentration, relative humidity and temperature. Bars are means ± SD ($n > 3$) and asterisks indicate significant differences ($p < 0.05$; Student's t-test) from the control (**K**). Further experimental details in Materials and Methods.

To investigate whether all those responses were also accompanied by impaired photosynthetic performance, we also measured the rate of net CO_2 assimilation at three photosynthetic photon flux densities (PPFDs)—200, 400 and 600 $\mu mol \cdot m^{-2} \cdot s^{-1}$. Here, the net photosynthesis intensity was presented only for 400 $\mu mol \cdot m^{-2} \cdot s^{-1}$ (A_{n400}), which appeared the most optimal PPFD; the respective net CO_2 assimilation rate values at the remaining photon flux densities (A_{n200}, A_{n600}) followed similar to A_{n400} trends in stress response. The rate of A_{n400} was markedly decreased after cold, heat and also after heat recovery and, generally, it accompanied similar variations in stomatal closure and leaf transpiration (Figure 6).

Notably, all those parameters did not correlate with alterations in variable (Fv) to maximal (Fm) chlorophyll fluorescence ratio, which appeared relatively constant for all investigated stress conditions. However, Fv and Fm significantly decreased both after heat and cold stress as well as after heat recovery. The relative chlorophyll content (assayed by chlorophyll meter) was affected only after cold stress and cold recovery (Figure 7). Due to the fact that in cauliflower curds, which are not involved in CO_2 assimilation, the decrease in abundance of two main photorespiratory enzymes (GDC and SHMT) was noticed (Figure 4), we also aimed to investigate photorespiration (PhR) in photosynthetically active organs in fully expanded leaves. Using Laisk's [55] method, we determined the ratio of photosynthetic rate under three investigated PPFDs between ambient and low CO_2 concentration. It appeared that PhR at all PPFDs markedly increased in cold-stressed plants; however, after cold recovery it was severely impaired. In contrast, heat stress and heat recovery resulted only in the slight decline of PhR200 and PhR400 values, whereas PhR600 was more affected at heat stress, but it was recovered after heat recovery (Figure 5).

Overall, we showed that cauliflower plants, besides mitochondrial proteome plasticity at the physiological level, display only partial but diverse alterations in various photosynthetic and respiratory parameters.

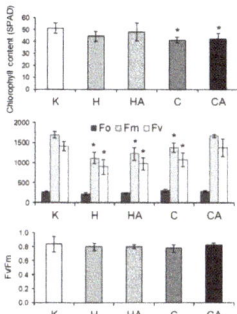

Figure 7. Changes in cauliflower leaf chlorophyll content minimal (Fo), maximal (Fm) and variable (Fv) fluorescence and Fv/Fm ratio in control grown (**K**), cold-stressed (**C**), heat-stressed (**H**), cold-recovered (**CA**) and heat-recovered (**HA**) plants. Chlorophyll content measured with a chlorophyll meter was expressed in relative units. Chlorophyll fluorescence was measured using a portable fluorometer. Before measurement, leaves were dark adapted for 30 min. Photochemical efficiency of PSII could be estimated from the Fv/Fm ratio, where Fv is the difference between Fm and Fo. Bars are means ± SD ($n > 3$) and asterisks indicate significant differences ($p < 0.05$; Student's t-test) from the control (**K**). Further experimental details in 'Materials and Methods'.

3. Discussion

3.1. Identification of Cauliflower Stress-Responsive Proteins by MS Analysis

In order to obtain a more general view of the impact of cold and heat stress on the functioning of cauliflower mitochondria, we began our study by their proteome analysis. Using 2D PAGE, 22 stress-responsive spots representing 19 non-redundant proteins were selected. Although some proteins belong to the general components of the abiotic stress response [40], in this study the list of cauliflower mitochondrial proteins responsive to temperature stress was broadened by stress recovery data showing new candidates (Table 1). Our previous studies suggested that stress recovery is associated with the possible acquiring of stress tolerance by cauliflower displaying some alterations within the mitochondrial OXPHOS and dehydrin-like proteins [41,56]. We would like to complement the study of mitochondrial complexome [57] by extended physiological and proteomic analyses and to follow the importance of stress recovery conditions in such assays.

Cauliflower is closely genetically related with other *Brassica* species and the identification of mitochondrial proteins was conducted based on protein sequence similarity between *Brassicaceae* members. Schmidt et al. [24] and Zhu et al. [36] have identified some proteins (e.g., ATPQ, CPN10, MDI I, PDHβ and HSP81-1) that appeared to be stress-responsive in our study. The presence of protein spots containing glycolytic enzymes (for instance PGK1) (Table 1) was not curious, because this enzyme was reported to be associated with outer mitochondrial membrane [58,59]. Such a finding was concluded mainly from the measurements of its enzymatic activity in mitochondrial extracts, however, the cytosolic member of this enzyme family (At1g79550), distinct to the Arabidopsis homolog (At3g12780) of cauliflower protein, was also identified in a large protein complex associated with mitochondrial membranes [58,60]. Interestingly, Arabidopsis PGK1 ortholog from plastid proteome showed cold response [61], whereas cauliflower mitochondrial protein was affected after heat recovery.

Four cauliflower mitochondrial proteins (3-PGDH, P5CDH, mtEF-Tu and CS) were represented as double spots. The presence of multiple spots on 2D gels was reported in numerous proteomic analyses, also including proteins analysed in this study (Figure S1) [21,22,32,50,62–67]. Consequently, we determined the extent to which posttranslational modifications might be responsible for the presence of multiple spots for the investigated proteins. Due to the lack of quantitative analysis including laborious enrichment of protein extracts in modified proteins and technical limitations of our protein separation methods, we were not able to accurately analyse majority of PTMs. Instead, we focused on a few selected modifications only (Tables S3a and S3b). However, various algorithms used for the PTM prediction among Arabidopsis emphasize that our data is largely novel and also significantly broadens deposited records. Among investigated modifications, many phosphorylated and methylated peptides were detected. Phosphorylation, together with oxidation, belong to the most important PTMs, regulating the activity of many stress-responsive proteins; in plant mitochondria, phosphorylation has particularly been studied in detail [34,68–71]. Energy and transport proteins, HSPs and even RC components were identified as potent phosphorylation targets [71]. Among proteins that were present in multiple spots in our study, it was shown that rice (*Oryza sativa*) CS can be phosphorylated [72] and mtEF-Tu was subjected to oxidation [73]. Overall, multi-spotting of cauliflower mitochondrial protein may depend not only on the presence of different PTMs, but largely on multigenic families coding novel protein isoforms, which resulted from the complex evolution of *Brassica* nuclear genomes as they underwent numerous chromosomal doublings, hybridizations and rearrangements [74]. More sensitive and quantitative proteomic assays should be implemented in the future for the better characterization of PTMs in cauliflower mitochondrial proteome.

We also noticed minor differences in molecular mass between nominal and observed values of some cauliflower mitochondrial proteins (Table 1). However, such discrepancies may be even more evident due to protein degradation [75]. Taylor et al. [50] and Imin et al. [76] have shown that abiotic stress could induce accumulation of protein degradation products. We routinely used protease inhibitors for the preparation of mitochondria, therefore, we think that extensive proteolysis could not account for major molecular mass discrepancies. Overall, despite the general similarity of 2D maps, it seems that numerous mitochondrial proteins may slightly differ in some physicochemical properties between Arabidopsis and cauliflower. We expected this from our previous analyses [23].

3.2. Variations in Pattern of Cauliflower Mitochondrial Proteome in Stress and Stress Recovery, and Their Metabolic Relevance

From the identified 16 stress-responsive proteins, at least an ca. two-fold change in variations in protein abundance were shown for most of them (Table 1). Under heat stress and heat recovery more proteins which varied in abundance were identified, compared to cold/cold recovery. From our data, only four stress-responsive proteins (ATP1, NAD9, COXII, CcmF$_{NI}$, CcmF$_{NII}$) are encoded in the plant mitochondrial genome (Table 1, Figure 4). Such a discrepancy may be due to the fact that the proteomic data allow the estimation of only a limited amount of mitochondrial proteins participating in various stress responses [31], therefore, literature inventories of those proteins are still far from complete. Rurek [40] lists almost 82 cold- and 52 heat-responsive plant mitochondrial proteins and only five proteins encoded by mitochondrial genome within them. In the up-dated Heidarvand et al. [42] review, only four proteins encoded in mitochondria contrast with the 44 nuclear-encoded cold responsive proteins. However, the modulation of plant mitochondrial biogenesis may rather depend on the regulation of the level of nuclear-encoded proteins governing assembly of macromolecular complexes, as has been speculated for sucrose-starved Arabidopsis cell cultures [43]. Overall, temperature stress response seems to involve no more than ca. 5% mitochondrial proteins encoded by mtDNA.

Some proteins detected in our study were previously shown to vary under diverse abiotic stress conditions. PDH participates in regulation of carbon flux from glycolysis to TCA cycle. In the published data, upregulations of PDH subunits prevail; in pea (*Pisum sativum*) mitochondria, PDHβ proteolytic products accumulate [50,77,78]. In rice leaves, however, contrasting PDHα responses (similarly to HSP90, see below) were noted under diverse cold conditions [37]. It is known that also other components of PDH complex including dehydrolipoamide dehydrogenase, may be upregulated during heat stress [79]. In contrast, we showed extensive accumulation of PDHβ after cold recovery, but not after heat treatment. In rice, PDHβ was downregulated during hypoxia [80], however, subunit-α of this enzyme increased during heat treatment and decreased in abundance after stress cessation [81]. Our results suggest that despite the overall number of major cold-responsive mitochondrial proteins being lower than those regulated by heat, it seems that carbon transfer from glycolysis to TCA cycle is increased in cauliflower cold response. Nonetheless, stress can regulate plant energetic and metabolic demands, including ATP/ADP intracellular and intramitochondrial ratio and the need for carbon skeletons [82].

Mitochondrial NAD^+-dependent MDH, which was increased in abundance after cold recovery in cauliflower mitochondria, in Arabidopsis was accumulated in response to different environmental stimuli including cold de-acclimation (but not cold acclimation) [32,33,83,84]. Arabidopsis MDH1 was suggested to belong to translational regulation targets [85]. The level of this enzyme (together with CS) was diversely modulated by various chilling conditions; generally, MDH abundance increased in cold-sensitive plant species [26,37,50,78,86,87]. Dumont et al. [88] investigating alterations in MDH abundance in diverse pea genotypes submitted to the combined cold and frost action and obtained contrasting results depending on the stress treatment and duration, similarly to the Yin et al. [89] and Cheng et al. [90] studies on MDH1 level in soybean (*Glycine max*) embryonic axes. Interestingly, CS responses depend on the severity of the temperature treatment (e.g., in the severe chilling the abundance of this enzyme declined), whereas under moderate treatment it increased [91]. During 2-day-long heat stress, MDH was also diversely downregulated in two *Agrostis* species depending on their thermotolerance [92]. Such a decrease in abundance was also reported for soybean MDH [89]. The significant up-regulation of cauliflower MDH only to cold recovery suggests that it may be the cold recovery marker [93]. However, heat recovery appeared detrimental for the level of this enzyme in cauliflower mitochondria [41]; cauliflower IDH and CS markedly declined after a 2-day-long heat recovery. Similar changes were reported for CS in heat adapted *Populus euphratica* [79]. In general, heat (which may lead to intramitochondrial oxidative damage) results in TCA enzymes, mitochondrial NADH pool and ATP synthesis impairments [94] and cold stress results in general stimulation of respiratory metabolism.

It appeared that cold stress causes an increase in the level of cauliflower HSPs; interestingly, in our study HSP70 and HSP90 increased more than in pea (*Pea sativum*) and rice leaves and peach (*Prunus persica*) barks [38,50,95]. A similar trend was observed in rice during salinity [96] and heat action in Arabidopsis [97]. However, mitochondrial HSP70 declined in abundance in stored or detached peach fruits submitted to prolonged cold [77,98]. Van Aken et al. [29] reported that Arabidopsis mitochondrial heat shock proteins responded only slightly to some forms of abiotic stress, for example HSP70 in the case of Cd treatment [83]. Another protein, HSP81-2, appeared to be cold-responsive in cauliflower mitochondria, contrary to the Arabidopsis ortholog, which was regulated by heat [31,97]. Notably, the regulation of HSP90 level in rice leaves depended on cold duration [38]. CPN10 remained unaffected after cold stress in pea [50], but in cauliflower mitochondria this protein accumulated very extensively under heat recovery. Also, mitochondrial sHsp22 was induced preferentially by heat (but not by cold) in soybean seedlings [25]. Together with FunCat data, all those findings suggest that the accumulation of some HSPs in cauliflower mitochondria may be specific for the preferential temperature stress conditions. Some HSPs can also diversely participate in various stress conditions, leading to distinct stress responses. It should also be noted, that the expression of two proteins (HSP70 and MDH1) regulated by low-temperature treatment, as well as additional proteins (mtEF-Tu, CS) responded to heat/heat recovery in our study and is known to be modulated by the specific glycine-rich protein (displaying RNA chaperone activity) under cold adaptation in Arabidopsis plants [99].

Cauliflower mtEF-Tu increased in abundance mainly after heat stress and did not last after heat recovery; overall, this may imply that the mitochondrial translation apparatus is impaired after heat cessation and rapid shift to control growth conditions of cauliflower, which was also observed for instance in chilled soybean embryo axes [89]. mtEF-Tu together with β-subunit of succinyl-CoA ligase increased in abundance in drought and partially in flood and MPP and ATP1 by salinity in Arabidopsis [32,84]. Curiously, β-subunit of succinyl-CoA ligase showed heat duration-dependent responses in soybean roots and rice leaves [25,37]. The major downregulation of succinyl-CoA ligase β-subunit in cauliflower mitochondria followed alterations of other TCA cycle components (IDH, CS) after heat stress [37,79], but not after heat recovery. Therefore, we can speculate that succinyl-CoA ligase may be preferentially accumulated in cauliflower during heat recovery in order to adjust the mitochondrial metabolism to control conditions.

ATP1 belongs to the proteins with level alterations dependent on the given species as well as stress intensity and duration [42]. In our study, ATP1 was declined in abundance after heat recovery. Similar trends were noted for pea, Arabidopsis, and *Zea mays* in a course of chilling, prolonged heat, CuCl or H_2O_2 treatment [39,50,100,101]. In contrast, cauliflower ATP1 abundance slightly increased in heat, similarly to the unassembled subunit *b* of ATP synthase [41]. We also found that heat caused a vast increase in abundance of ATP synthase d-subunit, contrary to its major downregulation reported by Gammulla et al. [37] and Tan et al. [39] for cold-stressed Arabidopsis cell cultures and heat-treated rice leaves, respectively. During oat (*Avena sativa*) seed storage, ATP1 level consistently declined as temperature increased from 35 to 50 °C, whereas subunit d of ATP synthase initially increased and then decreased in abundance under the same treatment; notably, subunits d and α were differentially accumulated at 10% and 16% moisture content, respectively [102]. Overall, those findings suggest that demand for ATP synthesis during heat treatment increases and the excess of de novo synthesized diverse ATP synthase subunits (e.g., mitochondrially encoded ATP1 or nuclear-encoded ATP7 proteins) is likely to be assembled into novel ATP synthase holocomplexes, labile in heat recovery [41].

Regarding the decrease in the level of MPPβ after heat recovery, we think that this may reflect the impairment of the import machinery, which may not be fully restored after stress recovery: according to our previous study [41], another subunit—MPPα appeared also to be down-regulated in heat recovery. Gammulla et al. [37] and Neilson et al. [38] noticed contrasting changes in the level of MPP subunits in rice leaves under low temperature and overall downregulations under heat. The level of MPP subunits underwent major changes in flood, indicating mitochondrial damage [84]. The influence of abiotic stresses on the efficiency of protein import into plant mitochondria was investigated, inter alia, by Taylor et al. [103] and Giegé et al. [43]. Taylor et al. [103] observed import inhibition of all tested pre-proteins into pea mitochondria during thermal stress. In turn, Giegé et al. [43] reported that the capacity for in vitro mitochondrial protein import is not affected after sucrose starvation in Arabidopsis cell cultures. Owing to our present and previous results [41], the pattern of protein import into cauliflower mitochondria under temperature stress should be investigated.

Another down-regulated cauliflower mitochondrial protein in heat was P5CDH, an enzyme involved in the proline degradation pathway of the Pro/P5C cycle [104]. Enzymes of this cycle, including P5C synthetase and proline dehydrogenase (ProDH) could be reciprocally expressed under stress. Moreover, ProDH closely associates with the OXPHOS system [42,105]; it was suggested that P5CDH prevents oxidative stress and electron run-off within the mitochondrial respiratory chain during Pro metabolism [106]. Free Pro accumulated in leaves of cold-treated cauliflower of wild type and mutant clones selected on hydroxyproline-containing medium, however, after salinity stress in mutated populations [107,108]. Interestingly, the level of *P5CDH* messengers significantly decreased in Arabidopsis plants expressing ectopically P5C synthetase 1 in response to heat stress. Pro accumulation impeded Arabidopsis seedlings growth in heat stress and may not serve as a protective osmolyte [109]. Therefore, it would be important to determine whether the decrease in abundance of P5CDH in cauliflower curds is associated with the increased Pro level after heat stress and heat recovery.

To extend our knowledge regarding cauliflower cold-responsive proteins, we carried out immunoblotting using antisera against dedicated proteins (Figure 3). We observed accumulation of cytosolic Hsp17.6C-CI after cold stress and recovery, indicating interaction of small HSPs with cauliflower mitochondrial membranes under prolonged cold treatment (as speculated by Rikhvanov et al. [54] for heat-stressed Arabidopsis cell cultures) and the importance of stress recovery phase in gaining stress resistance. Overall, HSPs are known to form oligomeric complexes with stress-affected proteins [110]. Important photorespiratory enzymes, GDC and SHMT, were decreased in abundance after cold recovery, similarly to *Agrostis scabra*, *A. stolonifera*, Arabidopsis, pea, *P. cathayana*, rice and wheat (*Triticum aestivum*) proteins in cold, heat and drought [37,38,50,92,111–113]. This observation is consistent with the reported declined levels of those enzymes in plant mitochondria under unfavourable conditions, leading to photorespiratory impairments [111]. However, under microspore development in rice plants submitted to cold, GDC-H was upregulated [114]. Interestingly, such up-regulation of GDC-H was also reported in pea leaves under frost and independently to cold tolerance and in the case of SHMT- in cold and salinity [37,88,115]. GDC-H slightly increases in abundance also in the early stages of low temperature action [116]. In accordance with that observation, as a part of protective mechanisms, significant accumulation of GDC-H transcripts in Arabidopsis leaves in response to short cold treatment was also reported [117].

We also determined variations in the level of some proteins engaged in maturation of cyt. c. Interestingly, major level downregulations of $CcmF_{N1}$ and $CcmF_{N2}$ proteins suggest that components of cyt. c maturation apparatus, including putative heme lyase components, may be sensitive to temperature stress. Generally, evidence for alterations of the level of those proteins in plant mitochondria during stress conditions are quite scarce. However, Naydenov et al. [118] found that $CcmF_N$ messengers responded during three-day-long cold exposure in maize embryos.

Regarding the level of other mitochondrial proteins during cold stress and cold recovery, VDAC-1 was downregulated in cauliflower mitochondria, which is generally in accordance with previously published results [37,77,94,119]. Interestingly, according to our previous study [41], we found the affected level of another VDAC isoform (VDAC-2) under heat recovery only. In addition, we detected some level of regulation of selected RC components, such as NAD9, COXII and cyt. c_1. Despite Tan et al. [39] having found cyt. c_1 abundance alterations (roughly followed by our data) among a number of Arabidopsis proteins in terms of their decline in cold, they did not identify COXII among them. However, those authors also reported the increased level of NAD9 protein in chilled Arabidopsis cell cultures. Longer cold acclimation led to the downregulation of this protein in wheat crowns, which is similar to our data [112], but 72 h-long cold stress resulted in NAD9 increase [37]. Selected CIV subunits [e.g., 6b-1 in chickpea (*Cicer arietinum*)] could also decline in abundance in cold, which suggests that overall respiratory activity decreased [120]. In most of the investigated plants, the level of COXII increased under low temperature and the overall changes in NAD9 abundance seem to be species-specific under temperature stress [5,121,122]. OXPHOS components are heat action sites in cauliflower [41]. It should be underlined that cold/cold recovery responses in cauliflower mitochondria also resulted in few protein upregulations, as was evident from 2D PAGE.

Finally, immunoblotting results extended the 2D PAGE data for cold-regulated proteins and the current study has also broadened knowledge on temperature stress responsive mitochondrial proteins, compared to our previous complexomic data [41]. Obtained data indicate for the variations of the same mitochondrial proteins in analysed stress conditions between cauliflower and other plant species. Few novel proteins, representing various pathways of mitochondrial metabolism, were discovered as responsive ones in thermal stress in cauliflower mitochondria. Therefore, one could speculate that numerous signalling pathways may be induced during action of cold or heat stress to alter the pattern of mitochondrial proteome. Various metabolic pathways (e.g., TCA cycle) may diversely participate in a particular stress response, which results in a plethora of various proteomic effects for cold, heat stress conditions and for stress recovery. In addition, the imbalance between proteomic and transcriptomic responses investigated in this study suggest that messengers accumulated at lowered levels have to be more efficiently used for translation, presumably by adaptive alterations in transcript/ribosome associations. Moreover, lack of coordination of expression profiles of diverse genes coding for subunits of the same complexes (CI, ATP synthase) in diverse temperature treatments points to the putative aberrations in the biogenesis of OXPHOS complexes also in cold and cold recovery and extends our previous data [41].

3.3. Cauliflower Leaf Respiratory Responses to Cold and Heat Stress

Temperature belongs to the critical factors controlling plant growth and development. Understanding both molecular, physiological as well as metabolic responses of crop and vegetable species to temperature stress in order to improve their tolerance and sustain high field yields is crucial [123]. However, the data regarding physiological functioning of *Brassica* species, including cauliflower, in cold and heat treatment [15,124] is still insufficient, contrary to some other environmental conditions, such as salinity or cadmium treatment [125–129].

Cauliflower is one of the most agriculturally important vegetable crops worldwide [107]. Notably, cauliflower and kale (*B. oleracea* var. *acephala*) belong to species better cold- and frost-adapted than Arabidopsis [107,130]. In our study, cold or heat stress was applied to cauliflower plants at the early stage of curd development, which enabled us to study the stress response of plants both at the molecular and physiological level. Previously, we used polarographic assays for investigating physiological properties and the activity of alternative pathway under temperature stress and recovery in isolated cauliflower mitochondria [41]. For physiological measurements in the current report, fully developed cauliflower leaves instead of curds were chosen; leaves, contrary to other plant organs, appeared more cold sensitive, which makes them most suitable for physiological assays [131]. We determined leaf respiration rate (by gas-exchange measurements on illuminated and darkened leaves), transpiration, stomatal opening, net CO_2 assimilation rate as well as chlorophyll level and fluorescence parameters which appeared to be affected by the same treatments to various extents (Figures 5–7), complementing our previous data.

In our study, the increase of R_n and R_d after heat stress and their subsequent decrease to the level of control variant during heat recovery suggest that adaptive forces of respiratory metabolism of cauliflower leaves to thermal treatment depends on stress duration. The increase of respiration after heat stress was also assayed in a number of plants [e.g., pepper (*Capsicum annuum*) leaves], which is a thermotolerant species with effective energy dissipation and ROS scavenging systems [132]. Due to the fact that cold stress acted for a longer period and appeared even more detrimental than heat, we did not observe R_n return to the level of control variant after cold recovery. This indicates for some irreversible effects in the cold, contrary to heat response (Figure 5). Such temperature recovery is expected to control energetic needs during acclimation, because of larger maintenance costs due to the increased activity of numerous enzymes [14].

The rate of respiration belongs to the first processes affected in plants also subjected to the low temperature treatments [27]. In the illuminated cauliflower leaves, R_d burst after cold recovery was evident; also for cold-acclimated Arabidopsis plants light respiration increased [14]. In some plants, however, the increase in respiration rate is visible at the early stage of cold treatment and it declines afterwards [42]. Overall, the importance of R_d in thermal adaptation is suggested. In addition, Talts et al. [14] showed that R_n of Arabidopsis leaves was more sensitive to cold stress than R_d. We noticed a more evident decrease in R_d rather than in R_n after cold treatment (Figure 5). However, in various winter and spring wheat and rye (*Secale cereale*) cultivars, chilling also resulted in a small R_n increase [133]. Apart from the known various temperature treatments, species-specific respiratory responses are suggested between various plant species.

In our study, cold, heat and heat recovery resulted in a significant decrease of net photosynthesis rate, which was partially accompanied by decreased stomatal conductance [14,132,134,135]. Similarly, to cauliflower data, the post-cold plant acclimation resulted in recovery of photosynthesis [134]. Copolovici et al. [135] pointed out the relevance of various cold/heat treatments for different photosynthesis and stomatal conductance decrease in tomato (*Solanum lycopersicum*) leaves, which is also important in our case (Figure 6). Dahal et al. [133] showed that in some wheat and rye cultivars, cold resulted in a decrease of both net CO_2 assimilation and as well as leaf transpiration and stomatal conductance. Leaf transpiration, stomatal conductance, chlorophyll content and photosynthesis also responded to cold in *P. cathayana* [136]. The decreased photosynthetic CO_2 assimilation rate in cauliflower leaves also correlated with an apparent decrease in stomatal conductance and transpiration rate after cold stress; however, in heat stress, stomata were closed to an even greater extent (Figure 6). Similarly, heat stress affected photosynthetic parameters and decreased stomatal conductance in grapevine (*Vitis amurensis*) and tobacco leaves [137,138]; in cauliflower leaves after heat recovery, despite rapid stomatal opening, the net photosynthetic rate remained decreased (Figure 6).

Under temperature stress, chlorophyll level and fluorescence as well as PS performance could be affected to various extents [10,124,137,139]. In our case, the decrease of Fm and Fv was accompanied by lower amounts of chlorophyll in cauliflower leaves only after cold treatment, and chlorophyll fluorescence parameters were not restored only after heat recovery. The lower chlorophyll content in leaves of cold stressed plants may suggest some damage in photosynthetic apparatus because photosynthetic rate was decreased. Generally, heat stress may result in the decrease of Fv/Fm [132]. However, in our study, PSII performance was largely unaffected due to the overall stable Fv/Fm ratio in all stress conditions investigated (Figure 7). It is known that the heat damage of PSII, accompanied by a decrease of CO_2 assimilation rate, occurs when severe stress conditions (exceeded 42 °C) were applied on illuminated leaves; however, this damage could be restored either in cases when a 'point of no return' is not exceeded or when exogenous Ca is applied for stomatal opening [138,140,141]. We conclude that despite our heat treatment conditions bordered with this threshold between mild and severe conditions, closed stomata in heat stress resulted in overall photosynthetic, but not respiratory decrease (Figures 5 and 6) and overall less susceptibility of cauliflower to heat than cold treatment.

We also noticed an association between decreased photorespiration rate and GDC-H and SHMT levels in cold recovery (Figures 4 and 5). Photorespiratory decline under temperature stress may result from GDC and SHMT downregulations in abundance and/or activity [50,92,111,117]. In our study, photorespiratory impairment was observed after heat treatment and heat recovery. Here, decreased photorespiration corresponded with decreased photosynthetic activity (due to over-reduction of the photosynthetic chain) and appeared irreparable after heat recovery. Also, in pepper leaves, heat treatment decreased both net photosynthetic as well as photorespiratory rate [132]. Interestingly, in cauliflower mitochondrial proteome, the increased level of MDH was associated with GDC-H and SHMT downregulation in cold stress and cold recovery. Mitochondrial MDH, which assists in metabolic flux through the TCA cycle, could operate in a reverse manner, by reducing oxaloacetate to malate, providing NAD^+ for photorespiratory glycine decarboxylation [142]. Regarding our study, mitochondrial MDH did not respond to heat treatment and heat recovery. Despite distinct tissues (curds and leaves) being chosen for proteomic and physiological experiments, still we may speculate whether cauliflower NAD^+-dependent MDH is engaged rather in the increase of the NADH pool inside mitochondria by acting within the TCA cycle, and not in NAD^+ regeneration necessary for photorespiration in cold stress, because the level of MDH and GDC-H were regulated conversely. Further experimental attempts are necessary to elucidate this issue.

Figure 8 summarizes results of our study. In general, we suggest that distinct cold and heat stress responses act in various way not only on the cauliflower mitochondrial proteome, but also on investigated transcript alterations and physiological parameters related with respiration with limited association. As it was pointed out above, associations between photorespiration rate and the dedicated enzymes were clearly seen. Contrasting conditions of temperature stress and recovery may result in diversely affected pre protein import to cauliflower mitochondria, impaired metabolite exchange and altered chaperoning activity together with TCA and OXPHOS functioning. In addition, transcript accumulation is proposed to compensate affected protein pool. Nevertheless, specificity of studied physiological and molecular responses to cold and heat stress between cauliflower and other plant species were easily observed and should be investigated in the future in more detail.

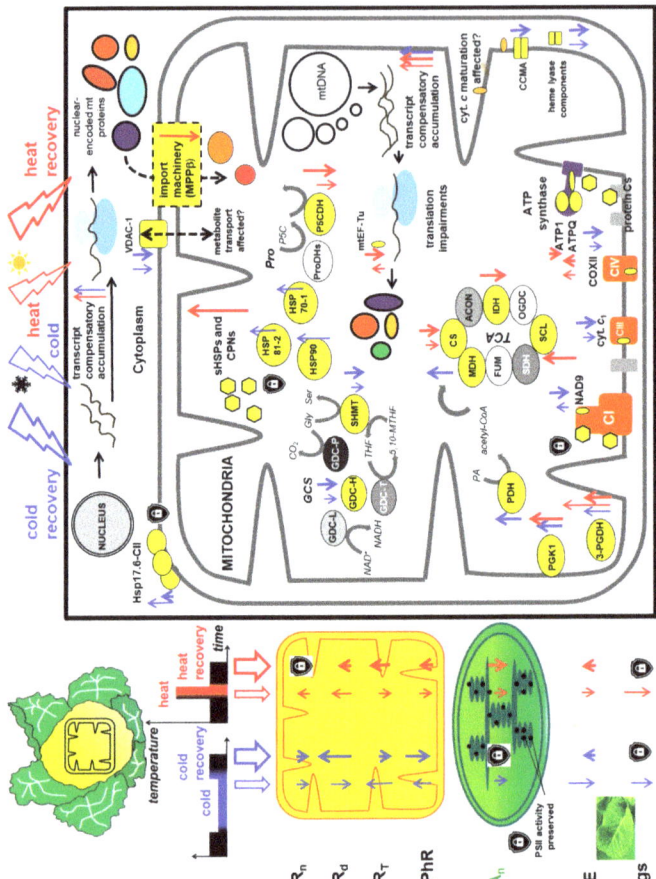

Figure 8. Proposed model of the impact of temperature stress and recovery on mitochondrial biogenesis in cauliflower curds. Investigated phenomena are depicted. Regulations in the cold and heat are shown by thin blue and red arrows; regulations in stress recovery by the respective thick blue and red arrows. Up-regulations and down-regulations are denoted by arrow heads raised up and down, respectively. Icons with a shield inside depict sustained physiological parameters or phenomena (**left panel**) or protective function of selected proteins in mitochondrial biogenesis (**right panel**). **Left panel:** stress dosage scheme and regulations in key physiological parameters. **Right Panel:** Proteins and complex subunits regulated by abundance are yellow-marked. Some regulatory steps affected by stress conditions are highlighted by discontinuous lines and arrows. Important abbreviations: A, net photosynthetic rate; ATP1, ATPQ, ATP synthase subunits 1 and d; CCM, cytochrome *c* maturation; CPN, chaperonin; C(s), complex(es); COX, cytochrome *c* oxidase; CS, citrate synthase; cyt. *c*, cytochrome *c*; E, leaf transpiration; GCS, glycine cleavage system; GDC, glycine decarboxylase; gs, stomatal conductance; HSP(s), heat shock protein(s); IDH, isocitrate dehydrogenase; MDH, malate dehydrogenase; MPP, mitochondrial processing peptidase; mt, mitochondrial; NAD9, complex I subunit 9; P5CDH, 1-Δ-pyrroline-5-carboxylate dehydrogenase; PDH, pyruvate dehydrogenase; 3-PGDH, 3-phosphoglycerate dehydrogenase; PGK1, phosphoglycerate kinase isoform 1; PhR, photorespiration; ProDH(s), proline dehydrogenase(s); R_d, light respiration; R_n, dark respiration; R_T, total light respiration; SCL, succinyl-CoA ligase; SHMT, serine hydroxymethyltransferase; TCA, tricarboxylic acid; VDAC, voltage-dependent anion channel. Further data in the text.

4. Materials and Methods

4.1. Plant Material, Growth Conditions and Stress Treatment

Cauliflower (*Brassica oleracea* var. *botrytis* subvar. *cauliflora* DC cv. 'Diadom') seeds were purchased from Bejo Zaden (Warmenhuizen, Holland). Cauliflower seedlings were grown in 0.09 dm^3 pots filled with peat substrate for growing cruciferous vegetables (Kronen–Clasmann, Gryfice, Poland). Seedlings with 3–4 leaves were transferred to larger containers (5 dm^3 in volume). Plants were grown for three months in cultivation chambers at a local breeding station (Poznan University of Life Sciences, Poland) at 23/19 °C (D/N) and 70% relative humidity under photon flux density 200 µmol·m^{-2}·s^{-1} (16 h of light/8 h of dark). After three months of growth corresponding to the young inflorescence (10 cm in diameter) stage, plants were divided into a few sets for stress treatment and the parallel control variants (plants grown in conditions described above).

Two stress variants were tested in this study: the direct stress treatment-heat or cold and post-stress plant cultivation (stress recovery). For the application of cold stress, plants before the isolation of mitochondria were transferred for ten days to 8 °C. Heat treatment (40 °C) was applied to growing plants for 4 h before the isolation of mitochondria. After stopping the stress treatment, part of cauliflower plants were transferred to the standard growth conditions for 48 h for the stress recovery. Curds (5 mm topmost layer) were directly harvested either after stopping the stress treatment or after stress recovery.

4.2. Gas Exchange Measurements

All analyses were carried out on at least three fully developed leaves from three 3-month-old plants. Leaves were taken from each plant representing all experimental variants (control versus stress-treated or control versus stress recovered plants). At least three biological replicates were analyzed. All parameters (the rate of total CO_2 assimilation [A_g], A_n, total respiration rate [R_T], R_d, R_n, E, and g_s) were measured using an LI-6400 XT infrared gas analyzer (LI-COR, Lincoln, NE, USA) and adjusted to the enclosed leaf area determined by an LI-300 leaf meter (LI-COR, Lincoln, NE, USA). Data were recorded after at least 2 h-long illumination. During the experiment, each of the analyzed leaves were placed into a 6-cm^2 chamber of the analyzer. Results were recorded after initial leaf acclimation to the desired light and CO_2 concentration, relative humidity and temperature. Gas-exchange parameters were recorded after leaf acclimated in the gas exchange chamber under the following conditions: PPFD of 400 µmol·m^{-2}·s^{-1}, 50% of the relative humidity (RH), 22 °C, 350 ppm of CO_2. CO_2 assimilation rate was also determined at two additional PPFD values (200 and 600 µmol·m^{-2}·s^{-1}).

R_d rate was determined according to Laisk [55]. For each leaf, CO_2 assimilation rate representing a given R_T was recorded during decreasing intercellular CO_2 concentration (C_i) to 0 ppm at 22 °C and 50% RH, and for each of the three different PPFD values (200, 400 and 600 µmol·m^{-2}·s^{-1}). For each PPFD, the linear regression of CO_2 assimilation (A) versus C_i was calculated (A/C_i curve) and the photorespiration rate (PhR for each PPFD value, denoted as PhR200, PhR400 and PhR600) was determined as the difference between R_T and R_d values (the last one expressed as a given CO_2 evolution rate at the point of crossing of all A/C_i curves). The R_n rate was extrapolated from the A value during decreased PPFD to 0 µmol·m^{-2}·s^{-1} from A/PPFD curve.

4.3. Chlorophyll Content and Fluorescence Measurements

Chlorophyll content was measured with a SPAD-502 chlorophyll meter (Konica Minolta, Wrocław, Poland) and expressed in relative units. Chlorophyll fluorescence was determined using a portable fluorometer (PAM-2000; Heinz Walz GmbH, Effeltrich, Germany) in a dark room with stable conditions. Before measurement, leaves were dark adapted for 30 min. Minimal fluorescence (Fo) was measured under 650 nm wavelength at a very low intensity (0.8 µmol·m^{-2}·s^{-1}). Fm was estimated after 1 s application of the saturating pulse of white light (3000 µmol·m^{-2}·s^{-1}). PSII photochemical efficiency was estimated from the Fv/Fm ratio, where Fv stands for the difference between Fm and Fo.

4.4. Preparation of Mitochondria

Mitochondria from 100 to 500 g of 5 mm-thick apical layer of cauliflower curds were isolated using a modified protocol of Boutry et al. [143], as described by Pawlowski et al. [23]. During isolation, the Complete Mini EDTA-free Protease Inhibitor Cocktail (Merck Poland, Warsaw, Poland) was added. Protein concentration was determined by the Bradford [144] method, using BSA as a calibrator.

4.5. Control Assays

Purity assays of isolated mitochondria (measurement of activities of mitochondrial cyt. *c* oxidase, peroxisomal catalase, plastid alkaline pyrophosphatase and cytoplasmic alcohol dehydrogenase) were conducted according to Pawlowski et al. [23]. Additionally, the purity of isolated mitochondria was verified by transmission electron microscopy (JEOL 1200EXII, Jeol, Peabody, MA, USA; [56]).

4.6. Preparation of Samples for Two-Dimensional Electrophoresis (2D SDS-PAGE)

Freshly isolated samples of cauliflower mitochondria were precipitated with trichloroacetic acid at −20 °C overnight [145,146]. After centrifugation for 5 min (16,000× *g*, 4 °C), pellets were washed once with 1 mL of acetone supplemented with 20 mM dithiothreitol (DTT) and re-centrifuged as described above. After vacuum drying, pellets were resuspended in the lysis buffer (7 M urea, 2 M thiourea, 0.5% (*w*/*v*) 3-((3-cholamidopropyl)dimethylammonio)-1-propanesulfonate (CHAPS), 1.5% (*w*/*v*) DTT, 0.5% (*v*/*v*) pharmalyte, pH 3–10) and protein concentration was determined either with the modified Bradford assay [147] or using a 2D Quant Kit (GE Healthcare, Warsaw, Poland).

4.7. 2D SDS-PAGE

All analyses were conducted at 15 °C; at least three biological replicates were analysed. Mitochondrial proteins (100 µg for silver nitrate staining or 500 µg for colloidal CBB) were first separated according to their charge on rehydrated Immobiline dry strips (24 cm, containing linear gradient of pH 3–10) with the rehydration buffer (8 M urea, 2% (*w*/*v*) CHAPS, 0.3% (*w*/*v*) DTT, 2% (*v*/*v*) pharmalyte, pH 3 to10) on an IPGphor apparatus (GE Healthcare, Uppsala, Sweden). Conditions for isoelectrofocusing (IEF) were as follows: 1 h at 500 V (step), 1 h at 1000 V (gradient), 3 h at 8000 V (gradient) and, finally, 5.5 h at 8000 V (step). The strips were either stored at −80 °C or they were directly treated for 10 min with solution A (6 M urea, 50 mM Tris-HCl, pH 6.8, 30% (*v*/*v*) glycerol, 2% (*w*/*v*) SDS, 0.25% [*w*/*v*] DTT) and for the same time with solution B (solution A supplemented with 4.5% (*w*/*v*) iodoacetamide without DTT) and subjected for the second dimension run (SDS-PAGE).

For SDS-PAGE precast Ettan DALT 12.5% (*w*/*v*) polyacrylamide gels (GE Healthcare) and an Ettan Dalt Six electrophoretical chamber (for six gels) were used. Conditions for the run were as follows: 45 min at 80 V and 15 h at 120 V. After electrophoresis, proteins on gel triplicates were either silver stained [148] for protein variation analysis or stained with colloidal CBB, according to Neuhoff et al. [149] for MS analyses. 2D gels were scanned, analysed using 2D Image Master 7 Platinum software (GE Healthcare) and the normalized quantitative volume of protein spots was determined.

4.8. Statistical Analysis of 2D Protein Pattern Variations

Protein spots showing variations in abundance were submitted to ANOVA to select spots for which stress treatment of post-stress plant cultivation had a significant effect ($p < 0.05$) on their volume. Additionally, the most variable proteins were also checked using Tukey's HSD test (JMP Software v8, SAS Institute, Cary, NC, USA). These variable proteins were further identified by MS.

4.9. Protein Identification by MS

For MS analysis, gel spots were subjected to a standard 'in-gel' digestion procedure during which proteins were reduced with 100 mM (w/v) DTT (for 30 min at 56 °C), alkylated with iodoacetamide (45 min at room temperature in the dark) and digested overnight with trypsin (sequencing Grade Modified Trypsin—Promega V5111). Resulting peptides were eluted from the gel with 0.1% (v/v) trifluoroacetic acid, 2% (v/v) acetonitrile.

Peptide mixtures were separated by liquid chromatography prior to molecular mass measurements (LC coupled to a linear ion trap-Fourier transform ion cyclotron resonance (LTQ-FTICR) mass spectrometer) on Orbitrap Velos mass spectrometer (Thermo Electron Corporation, San Jose, CA, USA) at the Mass Spectrometry Laboratory (Institute of Biochemistry and Biophysics, Polish Academy of Sciences, Warsaw, Poland). Peptide mixture was applied to an RP-18 precolumn (nanoACQUITY Symmetry® C18—Waters 186003514, Waters, Warsaw, Poland) using water containing 0.1% (v/v) trifluoroacetic acid as mobile phase and then transferred to a nano-HPLC RP-18 column (nanoACQUITY BEH C18—Waters 186003545) using an acetonitrile gradient (0 to 60% (v/v) acetonitrile for 120 min) in the presence of 0.05% (v/v) formic acid with a flow rate of 150 nL·min^{-1}. The column outlet was directly coupled to the ion source of the spectrometer working in the regime of data dependent MS to MS/MS switch. A blank run ensuring lack of cross contamination from previous samples preceded each analysis.

Acquired raw data were processed by Mascot Distiller followed by Mascot search (Matrix Science, London, UK, 8-processor on-site license) against NCBInr (version 20100203) with taxonomy restricted to *Viridiplantae*. Search parameters for precursor and product ions mass tolerance were 40 ppm and 0.8 Da, respectively, with allowance made for one missed trypsin cleavage, and the following fixed modifications: cysteine carbamidomethylation and allowed variable modifications: lysine carbamidomethylation and methionine oxidation, serine, threonine and tyrosine phosphorylation as well as deamidations, methylations, formylations and ethylations. Peptides with Mascot Score exceeding the threshold value corresponding to <5% False Positive Rate, calculated by Mascot procedure, were considered to be positively identified. Phosphorylation sites were predicted by PhosPhAt v4.0 (Available online: http://phosphat.uni-hohenheim.de/; [150]), NetPhos v2.0, available online: (http://www.cbs.dtu.dk/services/NetPhos-2.0/; [151]) and MUsite v1.0 (Available online: www.musite.net; [152]). Methylation sites were predicted by PMes (Available online: http://bioinfo.ncu.edu.cn/inquiries_PMeS.aspx; [153]). The data were compared with Arabidopsis data at PPDB (Available online: http://ppdb.tc.cornell.edu/dbsearch/searchmod.aspx). PPDB experimental sources concerned Zybailov et al. [154] as well as Kim et al. [155] data. Additional modified residues were predicted by FindMod (Available online: https://web.expasy.org/findmod/; [156]).

4.10. SDS-PAGE and Immunoblotting

Aliquots containing 20 µg of mitochondrial proteins were separated by 12% (w/v) SDS-PAGE [157]. For immunoassays, proteins were electroblotted from 1D (SDS PAGE) or 2D (IEF/SDS-PAGE) gels onto polyvinylidene difluoride Immobilon-P membranes (Merck, Warsaw, Poland), using a Sedryt semidry blotting apparatus (Kucharczyk, Warsaw, Poland). Membranes were CBB-stained to ensure that equal amounts of proteins were transferred. After destaining and subsequent blocking of the membrane, they were incubated overnight with antibodies. Indicated antibodies (Table S4) were purchased from Agrisera (Vännäs, Sweden). Antibodies against cyt. c_1 and ATP1 were kindly donated by Prof. Gottfried Schatz (University of Basel). Hsp17.6 antisera were a generous gift from Prof. Elisabeth Vierling (University of Massachusetts, Amherst, MA, USA). Antibodies against NAD9 and CCMA were produced by [158] and [159], respectively. CcmF$_{N1}$ and CcmF$_{N2}$ antisera were generated by [160]. Bound sera were detected using an anti-rabbit immunoglobulin G horseradish peroxidase or alkaline phosphatase conjugate diluted to 1/10,000 (BioRad Polska, Warsaw, Poland) and visualized with enhanced chemiluminescent reagents (GE Healthcare, Warsaw, Poland) or with Lumi-Phos WB Chemiluminescent Substrate (Life Technologies Poland, Warsaw, Poland). Immunoblotting images in triplicates were analyzed by Multi Gauge (v2.2, Fujifilm, Tokio, Japan) software and the representative pattern was presented. Band intensities were calibrated to the protein loading in the linear relationship (the control denoted as 1.00); the other bands were calculated relative to this value.

4.11. RNA Isolation and RT-semiqPCR

Total RNA from cauliflower curds was extracted using Trizol reagent or an EZ-10 Spin Column Plant RNA Mini-Preps Kit (BioBasic, Markham, ON, Canada) according to the manufacturer's protocol. Genomic DNA contaminants were removed by RQ1 DNase I free of RNase (Promega Poland, Warsaw, Poland). cDNA was synthesized using 1 µg of RNA, 0.2 µg of random hexamers mixture from HexaLabel DNA Labeling Kit (Thermo Scientific, Gdańsk, Poland) and 200 units of M-MLV reverse transcriptase (Promega Poland, Warsaw, Poland) in a 20 µL total volume for 1 h at 37 °C. After first strand synthesis, the reaction mixture was diluted with 10 mM Tris-HCl, pH 8.0 three or six times; after normalization, aliquots of 1–2 µL were subjected to RT-semiquantitative multiplex PCR (RT-semiqPCR) in a 15 µL total volume.

RT-semiqPCR was performed in an Applied Biosystems 2720 thermal cycler (Applied Biosystems Poland, Warsaw, Poland) with the following profile: 3 min at 95 °C followed by 25–26 cycles depending on amplicon of 20 s at 95 °C, 30 s at 55 °C (except 58 °C and 50 °C for *coxII* and *CPN10*, respectively) and 30 s at 72 °C, and with a final incubation for 5 min at 72 °C. PCR products were separated on a 1.5% agarose gel and stained with ethidium bromide. The gels were documented using a GBOX XL1.4 (TK Biotech, Warsaw, Poland) imaging system and quantified with Multi Gauge (v.2.2, Fujifilm, Tokio, Japan). For RT-PCR assays, two biological and at least three technical replicates were included.

Cauliflower cDNA fragments for selected mitochondrial proteins were amplified using specific primers (Table S5); a 239-bp fragment of cauliflower actin1 (ACT1) cDNA was used as an internal standard. The amplicons were directly sequenced bi-directionally (Big Dye Terminator v.3.1 Cycle Sequencing kit, Applied Biosystems Poland, Warsaw, Poland) on an ABI Prism 31–30 XL system (Applied Biosystems Poland, Warsaw, Poland) for sequence identity verification.

5. Conclusions

Our approach comprises general data about variations regarding cold and heat stress responses in the mitochondrial proteome of cauliflower and in physiological parameters, related particularly to plant respiration. It appeared that the set of cauliflower mitochondrial proteins responded to temperature stress conditions as well as to the stress recovery varied from the previously described ones. These results significantly extend the deposited data also by means of investigated quantitative alterations. However, investigated proteomic, transcriptomic and respiratory physiological responses related to the functioning of cauliflower mitochondria in stress were largely not associated. For instance, the rates of respiration in illuminated leaves together with leaf transpiration and photorespiration were significantly affected by cold and/or cold recovery, despite more proteins of various functional classes being involved in heat/heat recovery. Studied transcripts and protein alterations in temperature stress and recovery involve contrasting responses. Interestingly, the expression patterns of genes coding for various CI and ATP synthase subunits also differ. According to our previous data [41], this may suggest perturbations in the biogenesis of OXPHOS complexes. Owing to the scarce representation of cold- and heat-affected proteins encoded in the mitochondrial genome during mitochondrial response to temperature stress and recovery, modulation of cauliflower mitochondrial biogenesis under the investigated stimuli may depend rather on the massive regulation of nuclear-encoded proteins.

We would like to emphasize that heat-regulated proteins were distinct (with minor exceptions) from the ones regulated by cold/cold recovery. Overall, we (1) noticed the impaired photorespiration rate which was followed by alterations in photorespiratory enzymes after cold recovery; (2) suggested possible metabolic impairments in various TCA components and Pro catabolism (downregulations of *PRODH* and *P5CDH* transcripts in cold and heat recovery were also notable), and in protein import apparatus; (3) observed elevated demand for ATP synthesis after heat/heat recovery (e.g., ATP1 and ATPQ level); (4) noticed evident downregulation of some RC subunits (e.g., ATP1, NAD9, COXII) and the sensitivity of c-type cytochrome biogenesis apparatus to cold stress and cold recovery; and (5) compared selected proteomic and transcriptomic responses providing additional data on their participation in temperature stress and recovery. Our data show that selected regulations cannot be fully restored after temperature recovery. All these results imply the necessity (1) to go deeper in the quantitative analysis of protein posttranslational modifications and (2) to study further tissue-specific proteomic and physiological alterations.

Supplementary Materials: Supplementary materials can be found at http://www.mdpi.com/1422-0067/19/3/877/s1.

Acknowledgments: The work was supported by the Ministry of Science and Higher Education, Poland, grant number: N N303 338835. We gratefully thank Mikołaj Knaflewski, Alina Kałużewicz, and Anna Zaworska (Poznan University of Life Sciences) for the valuable help during cauliflower plant cultivation and the gas exchange measurements. We would like to thank Michał Dadlez, Janusz Dębski and the staff from Mass Spectrometry Laboratory (Institute of Biochemistry and Biophysics, Polish Academy of Sciences, Warsaw) for MS analyses. We thank Ludmiła Bladocha (Institute of Dendrology, Polish Academy of Sciences, Kórnik) as well as Grzegorz Pietkiewicz (Adam Mickiewicz University, Poznań) for valuable technical assistance.

Author Contributions: Michał Rurek, Tomasz Pawłowski and Włodzimierz Krzesiński conceived and designed the experiments; Michał Rurek was the principle investigator, who designed this study, performed extraction of mitochondria, control assays, carried out SDS-PAGE and immunoblotting, analysed proteomic results (together with FunCat), performed RT-semiqPCR assays, analysed data from those assays and wrote the manuscript; Magdalena Czołpińska isolated mitochondria, prepared protein samples for SDS-PAGE and immunoblotting, analyzed MS data and wrote the manuscript; Tomasz Pawłowski prepared protein samples for 2D PAGE, performed 2D PAGE and the statistical analysis of spot variations, selected stress-responsive protein spots, submitted protein spots for MS analyses and participated in writing the paper; Włodzimierz Krzesiński cultivated plant material, and conducted all physiological analyses (gas exchange measurements and fluorescence assays), analysed their results; Tomasz Spiżewski assisted in cultivation and maintenance of the plant material in control conditions and after stress dosage, prepared nutrient media, subjected plants to stress conditions and participated in physiological analyses.

Conflicts of Interest: The authors declare no conflicts of interest.

Abbreviations

A_n	net CO_2 assimilation rate
A_g	total CO_2 assimilation rate
ANOVA	analysis of variance
AOX	alternative oxidase
ATP1	ATP synthase subunit α
BSA	bovine serum albumin
CBB	Coomassie Brilliant Blue
Ccm/CCM	cytochrome c maturation
CAPS	3-[(3-cholamidopropyl)dimethylammonio]-1-propanesulfonate
C_i	intercellular CO_2 concentration
COX	cytochrome c oxidase
CPN	chaperonin
CS	citrate synthase
2D PAGE	two-dimensional gel electrophoresis
DTT	dithiothreitol
E	transpiration
EDTA	ethylenediaminetetraacetic acid
EGTA	ethylene glycol-bis(β-aminoethyl ether)-N,N,N',N'-tetraacetic acid
EF	elongation factor
Fm	maximal fluorescence
Fo	minimal fluorescence
FunCat	functional categorization
Fv	variable fluorescence
GDC	glycine decarboxylase
g_s	stomatal conductance

HSD	honest significant difference
HSP	heat shock protein
IDH	isocitrate dehydrogenase
IEF	isoelectrofocusing
iTRAQ	isobaric tags for the absolute quantification
LC-MS/MS	liquid chromatography-tandem mass spectrometry
MDH	malate dehydrogenase
MIPS	Munich Information Center for Protein Sequences
MPP	mitochondrial processing peptidase
NAD	complex I subunit (mitochondrially-encoded)
NCBI	National Center for Biotechnology Information
OXPHOS	oxidative phosphorylation
P5CDH	Δ-1-pyrroline-5-carboxylate dehydrogenase
PDH	pyruvate dehydrogenase
PGDH	3-phosphoglycerate dehydrogenase
PGK	phosphoglycerate kinase
PhR	photorespiration rate
PPFD	photosynthetic photon flux density
PRODH	proline dehydrogenase
PS	photosystem
PTM	posttranslational protein modification
RC	respiratory chain
R_d	respiration in the light (day respiration) rate
RH	relative humidity
R_n	respiration in the dark (night respiration) rate
ROS	reactive oxygen species
R_T	total respiration rate
SCL	succinyl-CoA ligase
SHMT	serine hydroxy-methyl aminotransferase
TCA	tricarboxylic acid
VDAC	voltage-dependent anion channel

References

1. Bray, E.A.; Bailey-Serres, J.; Weretilnyk, E. Responses to abiotic stresses. In *Biochemistry and Molecular Biology of Plants*, 1st ed.; Buchanan, B., Gruissem, W., Jones, R., Eds.; American Society of Plant Physiologists: Rockville, MD, USA, 2000; pp. 158–1249, ISBN1 13 978-0943088372, ISBN2 10 0943088372.
2. Bohnert, H.J.; Nelson, D.E.; Jensen, R.G. Adaptations to Environmental Stresses. *Plant Cell* **1995**, *7*, 1099–1111. [CrossRef] [PubMed]
3. Krishnan, A.; Pereira, A. Integrative approaches for mining transcriptional regulatory programs in Arabidopsis. *Brief. Funct. Genom. Proteom.* **2008**, *7*, 264–274. [CrossRef] [PubMed]
4. Ribas-Carbo, M.; Aroca, R.; Gonzàlez-Meler, M.A.; Irigoyen, J.J.; Sánchez-Díaz, M. The Electron Partitioning between the Cytochrome and Alternative Respiratory Pathways during Chilling Recovery in Two Cultivars of Maize Differing in Chilling Sensitivity. *Plant Physiol.* **2000**, *122*, 199–204. [CrossRef] [PubMed]
5. Kurimoto, K.; Millar, A.H.; Lambers, H.; Day, D.A.; Noguchi, K. Maintenance of Growth Rate at Low Temperature in Rice and Wheat Cultivars with a High Degree of Respiratory Homeostasis is Associated with a High Efficiency of Respiratory ATP Production. *Plant Cell Physiol.* **2004**, *45*, 1015–1022. [CrossRef] [PubMed]
6. Armstrong, A.F.; Logan, D.C.; Tobin, A.K.; O'Toole, P.; Atkin, O.K. Heterogeneity of plant mitochondrial responses underpinning respiratory acclimation to the cold in *Arabidopsis thaliana* leaves. *Plant Cell Environ.* **2006**, *29*, 940–949. [CrossRef] [PubMed]
7. Yang, M.-T.; Chen, S.-L.; Lin, C.-Y.; Chen, Y.-M. Chilling stress suppresses chloroplast development and nuclear gene expression in leaves of mung bean seedlings. *Planta* **2005**, *221*, 374–385. [CrossRef] [PubMed]
8. Mohanty, S.; Grimm, B.; Tripathy, B.C. Light and dark modulation of chlorophyll biosynthetic genes in response to temperature. *Planta* **2006**, *224*, 692–699. [CrossRef] [PubMed]

9. Allakhverdiev, S.I.; Kreslavski, V.D.; Klimov, V.V.; Los, D.A.; Carpentier, R.; Mohanty, P. Heat stress: An overview of molecular responses in photosynthesis. *Photosynth. Res.* **2008**, *98*, 541–550. [CrossRef] [PubMed]
10. Dutta, S.; Mohanty, S.; Tripathy, B.C. Role of Temperature Stress on Chloroplast Biogenesis and Protein Import in Pea. *Plant Physiol.* **2009**, *150*, 1050–1061. [CrossRef] [PubMed]
11. Lütz, C. Cell physiology of plants growing in cold environments. *Protoplasma* **2010**, *244*, 53–73. [CrossRef] [PubMed]
12. Wise, R.R. Chilling-enhanced photooxidation: The production, action and study of reactive oxygen species produced during chilling in the light. *Photosynth. Res.* **1995**, *45*, 79–97. [CrossRef] [PubMed]
13. Armstrong, A.F.; Badger, M.R.; Day, D.A.; Barthet, M.M.; Smith, P.M.C.; Millar, A.H.; Whelan, J.; Atkin, O.K. Dynamic changes in the mitochondrial electron transport chain underpinning cold acclimation of leaf respiration. *Plant Cell Environ.* **2008**, *31*, 1156–1169. [CrossRef] [PubMed]
14. Talts, P.; Pärnik, T.; Gardeström, P.; Keerberg, O. Respiratory acclimation in *Arabidopsis thaliana* leaves at low temperature. *J. Plant Physiol.* **2004**, *161*, 573–579. [CrossRef] [PubMed]
15. Díaz, M.; de Haro, V.; Muñoz, R.; Quiles, M.J. Chlororespiration is involved in the adaptation of *Brassica* plants to heat and high light intensity. *Plant Cell Environ.* **2007**, *30*, 1578–1585. [CrossRef] [PubMed]
16. Pastenes, C.; Horton, P. Effect of High Temperature on Photosynthesis in Beans (I. Oxygen Evolution and Chlorophyll Fluorescence). *Plant Physiol.* **1996**, *112*, 1245–1251. [CrossRef] [PubMed]
17. Havaux, M. Short-term responses of Photosystem I to heat stress. *Photosynth. Res.* **1996**, *47*, 85–97. [CrossRef] [PubMed]
18. Bukhov, N.G.; Samson, G.; Carpentier, R. Nonphotosynthetic Reduction of the Intersystem Electron Transport Chain of Chloroplasts Following Heat Stress. Steady-State Rate. *Photochem. Photobiol.* **2000**, *72*, 351–357. [CrossRef]
19. Bukhov, N.G.; Wiese, C.; Neimanis, S.; Heber, U. Heat sensitivity of chloroplasts and leaves: Leakage of protons from thylakoids and reversible activation of cyclic electron transport. *Photosynth. Res.* **1999**, *59*, 81–93. [CrossRef]
20. Hossain, Z.; Nouri, M.-Z.; Komatsu, S. Plant cell organelle proteomics in response to abiotic stress. *J. Proteome Res.* **2012**, *11*, 37–48. [CrossRef] [PubMed]
21. Bardel, J.; Louwagie, M.; Jaquinod, M.; Jourdain, A.; Luche, S.; Rabilloud, T.; Macherel, D.; Garin, J.; Bourguignon, J. A survey of the plant mitochondrial proteome in relation to development. *Proteomics* **2002**, *2*, 880–898. [CrossRef]
22. Heazlewood, J.L.; Howell, K.A.; Whelan, J.; Millar, A.H. Towards an Analysis of the Rice Mitochondrial Proteome. *Plant Physiol.* **2003**, *132*, 230–242. [CrossRef] [PubMed]
23. Pawlowski, T.; Rurek, M.; Janicka, S.; Raczynska, K.D.; Augustyniak, H. Preliminary analysis of the cauliflower mitochondrial proteome. *Acta Physiol. Plant.* **2005**, *27*, 275–281. [CrossRef]
24. Schmidt, U.G.; Endler, A.; Schelbert, S.; Brunner, A.; Schnell, M.; Neuhaus, H.E.; Marty-Mazars, D.; Marty, F.; Baginsky, S.; Martinoia, E. Novel Tonoplast Transporters Identified Using a Proteomic Approach with Vacuoles Isolated from Cauliflower Buds. *Plant Physiol.* **2007**, *145*, 216–229. [CrossRef] [PubMed]
25. Ahsan, N.; Donnart, T.; Nouri, M.-Z.; Komatsu, S. Tissue-Specific Defense and Thermo-Adaptive Mechanisms of Soybean Seedlings under Heat Stress Revealed by Proteomic Approach. *J. Proteome Res.* **2010**, *9*, 4189–4204. [CrossRef] [PubMed]
26. Koehler, G.; Wilson, R.C.; Goodpaster, J.V.; Sønsteby, A.; Lai, X.; Witzmann, F.A.; You, J.-S.; Rohloff, J.; Randall, S.K.; Alsheikh, M. Proteomic Study of Low-Temperature Responses in Strawberry Cultivars (*Fragaria* × *ananassa*) That Differ in Cold tolerance. *Plant Physiol.* **2012**, *159*, 1787–1805. [CrossRef] [PubMed]
27. Lee, B.; Lee, H.; Xiong, L.; Zhu, J.-K. A Mitochondrial Complex I Defect Impairs Cold-Regulated Nuclear Gene Expression. *Plant Cell* **2002**, *14*, 1235–1251. [CrossRef] [PubMed]
28. Millar, A.H.; Heazlewood, J.L.; Kristensen, B.K.; Braun, H.-P.; Møller, I.M. The plant mitochondrial proteome. *Trends Plant Sci.* **2005**, *10*, 36–43. [CrossRef] [PubMed]
29. Van Aken, O.; Zhang, B.; Carrie, C.; Uggalla, V.; Paynter, E.; Giraud, E.; Whelan, J. Defining the Mitochondrial Stress Response in *Arabidopsis thaliana*. *Mol. Plant* **2009**, *2*, 1310–1324. [CrossRef] [PubMed]
30. Cui, J.; Liu, J.; Li, Y.; Shi, T. Integrative Identification of Arabidopsis Mitochondrial Proteome and Its Function Exploitation through Protein Interaction Network. *PLoS ONE* **2011**, *6*, e16022. [CrossRef] [PubMed]

31. Taylor, N.L.; Tan, Y.-F.; Jacoby, R.P.; Millar, A.H. Abiotic environmental stress induced changes in the *Arabidopsis thaliana* chloroplast, mitochondria and peroxisome proteomes. *J. Proteom.* **2009**, *72*, 367–378. [CrossRef] [PubMed]
32. Ndimba, B.K.; Chivasa, S.; Simon, W.J.; Slabas, A.R. Identification of *Arabidopsis* salt and osmotic stress responsive proteins using two-dimensional difference gel electrophoresis and mass spectrometry. *Proteomics* **2005**, *5*, 4185–4196. [CrossRef] [PubMed]
33. Jiang, Y.; Yang, B.; Harris, N.S.; Deyholos, M.K. Comparative proteomic analysis of NaCl stress-responsive proteins in *Arabidopsis* roots. *J. Exp. Bot.* **2007**, *58*, 3591–3607. [CrossRef] [PubMed]
34. Kosová, K.; Vítámvás, P.; Prášil, I.T.; Renaut, J. Plant proteome changes under abiotic stress—Contribution of proteomics studies to understanding plant stress response. *J. Proteom.* **2011**, *74*, 1301–1322. [CrossRef] [PubMed]
35. Dunkley, T.P.J.; Hester, S.; Shadforth, I.P.; Runions, J.; Weimar, T.; Hanton, S.L.; Griffin, J.L.; Bessant, C.; Brandizzi, F.; Hawes, C.; et al. Mapping the *Arabidopsis* organelle proteome. *Proc. Natl. Acad. Sci. USA* **2006**, *103*, 6518–6523. [CrossRef] [PubMed]
36. Zhu, M.; Dai, S.; McClung, S.; Yan, X.; Chen, S. Functional Differentiation of *Brassica napus* Guard Cells and Mesophyll Cells Revealed by Comparative Proteomics. *Mol. Cell. Proteom.* **2009**, *8*, 752–766. [CrossRef] [PubMed]
37. Gammulla, C.G.; Pascovici, D.; Atwell, B.J.; Haynes, P.A. Differential proteomic response of rice (*Oryza sativa*) leaves exposed to high- and low-temperature stress. *Proteomics* **2011**, *11*, 2839–2850. [CrossRef] [PubMed]
38. Neilson, K.A.; Mariani, M.; Haynes, P.A. Quantitative proteomic analysis of cold-responsive proteins in rice. *Proteomics* **2011**, *11*, 1696–1706. [CrossRef] [PubMed]
39. Tan, Y.-F.; Millar, A.H.; Taylor, N.L. Components of Mitochondrial Oxidative Phosphorylation Vary in Abundance Following Exposure to Cold and Chemical Stresses. *J. Proteome Res.* **2012**, *11*, 3860–3879. [CrossRef] [PubMed]
40. Rurek, M. Plant mitochondria under a variety of temperature stress conditions. *Mitochondrion* **2014**, *19*, 289–294. [CrossRef] [PubMed]
41. Rurek, M.; Woyda-Ploszczyca, A.M.; Jarmuszkiewicz, W. Biogenesis of mitochondria in cauliflower (*Brassica oleracea* var. *botrytis*) curds subjected to temperature stress and recovery involves regulation of the complexome, respiratory chain activity, organellar translation and ultrastructure. *Biochi. Biophys. Acta-Bioenerg.* **2015**, *1847*, 399–417. [CrossRef] [PubMed]
42. Heidarvand, L.; Millar, A.H.; Taylor, N.L. Responses of the Mitochondrial Respiratory System to Low Temperature in Plants. *Crit. Rev. Plant Sci.* **2017**, *36*, 217–240. [CrossRef]
43. Giegé, P.; Sweetlove, L.J.; Cognat, V.; Leaver, C.J. Coordination of Nuclear and Mitochondrial Genome Expression during Mitochondrial Biogenesis in Arabidopsis. *Plant Cell* **2005**, *17*, 1497–1512. [CrossRef] [PubMed]
44. Millenaar, F.F.; Lambers, H. The Alternative Oxidase: In vivo Regulation and Function. *Plant Biol.* **2003**, *5*, 2–15. [CrossRef]
45. Lambers, H.; Robinson, S.A.; Ribas-Carbo, M. Plant Respiration: From Cell to Ecosystem. In *Advances in Photosynthesis and Respiration Series*, 1st ed.; Lambers, H., Ribas-Carbo, M., Eds.; Springer: Dordrecht, The Netherlands, 2005; Volume 18, pp. 1–15, ISBN 9781402035883.
46. Vanlerberghe, G.C.; Cvetkovska, M.; Wang, J. Is the maintenance of homeostatic mitochondrial signaling during stress a physiological role for alternative oxidase? *Physiol. Plant.* **2009**, *137*, 392–406. [CrossRef] [PubMed]
47. Grabelnych, O.I.; Sumina, O.N.; Funderat, S.P.; Pobezhimova, T.P.; Voinikov, V.K.; Kolesnichenko, A.V. The distribution of electron transport between the main cytochrome and alternative pathways in plant mitochondria during short-term cold stress and cold hardening. *J. Therm. Biol.* **2004**, *29*, 165–175. [CrossRef]
48. Sugie, A.; Naydenov, N.; Mizuno, N.; Nakamura, C.; Takumi, S. Overexpression of wheat alternative oxidase gene *Waox1a* alters respiration capacity and response to reactive oxygen species under low temperature in transgenic *Arabidopsis*. *Genes Genet. Syst.* **2006**, *81*, 349–354. [CrossRef] [PubMed]
49. Popov, V.N.; Purvis, A.C.; Skulachev, V.P.; Wagner, A.M. Stress-induced changes in ubiquinone concentration and alternative oxidase in plant mitochondria. *Biosci. Rep.* **2001**, *21*, 369–379. [CrossRef] [PubMed]
50. Taylor, N.L.; Heazlewood, J.L.; Day, D.A.; Millar, A.H. Differential Impact of Environmental Stresses on the Pea Mitochondrial Proteome. *Mol. Cell. Proteom.* **2005**, *4*, 1122–1133. [CrossRef] [PubMed]

51. Wang, J.; Rajakulendran, N.; Amirsadeghi, S.; Vanlerberghe, G.C. Impact of mitochondrial alternative oxidase expression on the response of *Nicotiana tabacum* to cold temperature. *Physiol. Plant.* **2011**, *142*, 339–351. [CrossRef] [PubMed]
52. Fowler, S.; Thomashow, M.F. Arabidopsis Transcriptome Profiling Indicates That Multiple Regulatory Pathways Are Activated during Cold Acclimation in Addition to the CBF Cold Response Pathway. *Plant Cell* **2002**, *14*, 1675–1690. [CrossRef] [PubMed]
53. Chang, S.; Yang, T.; Du, T.; Huang, Y.; Chen, J.; Yan, J.; He, J.; Guan, R. Mitochondrial genome sequencing helps show the evolutionary mechanism of mitochondrial genome formation in *Brassica*. *BMC Genom.* **2011**, *12*, 497. [CrossRef] [PubMed]
54. Rikhvanov, E.G.; Gamburg, K.Z.; Varakina, N.N.; Rusaleva, T.M.; Fedoseeva, I.V.; Tauson, E.L.; Stupnikova, I.V.; Stepanov, A.V.; Borovskii, G.B.; Voinikov, V.K. Nuclear-mitochondrial cross-talk during heat shock in Arabidopsis cell culture. *Plant J.* **2007**, *52*, 763–778. [CrossRef] [PubMed]
55. Laisk, A. *Kinetics of Photosynthesis and Photorespiration in C_3-Plants*; Nauka: Moscow, Russia, 1977.
56. Rurek, M. Diverse accumulation of several dehydrin-like proteins in cauliflower (*Brassica oleracea* var. *botrytis*), *Arabidopsis thaliana* and yellow lupin (*Lupinus luteus*) mitochondria under cold and heat stress. *BMC Plant Biol.* **2010**, *10*, 181. [CrossRef] [PubMed]
57. Yan, S.-P.; Zhang, Q.-Y.; Tang, Z.-C.; Su, W.-A.; Sun, W.-N. Comparative Proteomic Analysis Provides New Insights into Chilling Stress Responses in Rice. *Mol. Cell. Proteom.* **2006**, *5*, 484–496. [CrossRef] [PubMed]
58. Giegé, P.; Heazlewood, J.L.; Roessner-Tunali, U.; Millar, A.H.; Fernie, A.R.; Leaver, C.J.; Sweetlove, L.J. Enzymes of Glycolysis Are Functionally Associated with the Mitochondrion in Arabidopsis Cells. *Plant Cell* **2003**, *15*, 2140–2151. [CrossRef] [PubMed]
59. Heazlewood, J.L.; Tonti-Filippini, J.S.; Gout, A.M.; Day, D.A.; Whelan, J.; Millar, A.H. Experimental Analysis of the Arabidopsis Mitochondrial Proteome Highlights Signaling and Regulatory Components, Provides Assessment of Targeting Prediction Programs, and Indicates Plant-Specific Mitochondrial Proteins. *Plant Cell* **2004**, *16*, 241–256. [CrossRef] [PubMed]
60. Graham, J.W.A.; Williams, T.C.R.; Morgan, M.; Fernie, A.R.; Ratcliffe, R.G.; Sweetlove, L.J. Glycolytic Enzymes Associate Dynamically with Mitochondria in Response to Respiratory Demand and Support Substrate Channeling. *Plant Cell* **2007**, *19*, 3723–3738. [CrossRef] [PubMed]
61. Goulas, E.; Schubert, M.; Kieselbach, T.; Kleczkowski, L.A.; Gardeström, P.; Schröder, W.; Hurry, V. The chloroplast lumen and stromal proteomes of *Arabidopsis thaliana* show differential sensitivity to short- and long-term exposure to low temperature. *Plant J.* **2006**, *47*, 720–734. [CrossRef] [PubMed]
62. Millar, A.H.; Sweetlove, L.J.; Giegé, P.; Leaver, C.J. Analysis of the Arabidopsis Mitochondrial Proteome. *Plant Physiol.* **2001**, *127*, 1711–1727. [CrossRef] [PubMed]
63. Kruft, V.; Eubel, H.; Jänsch, L.; Werhahn, W.; Braun, H.P. Proteomic Approach to Identify Novel Mitochondrial Proteins in Arabidopsis. *Plant Physiol.* **2001**, *127*, 1694–1710. [CrossRef] [PubMed]
64. Giegé, P.; Sweetlove, L.J.; Leaver, C.J. Identification of mitochondrial protein complexes in *Arabidopsis* using two-dimensional blue-native polyacrylamide gel electrophoresis. *Plant Mol. Biol. Rep.* **2003**, *21*, 133–144. [CrossRef]
65. Huang, S.; Taylor, N.L.; Narsai, R.; Eubel, H.; Whelan, J.; Millar, A.H. Experimental Analysis of the Rice Mitochondrial Proteome, Its Biogenesis, and Heterogeneity. *Plant Physiol.* **2009**, *149*, 719–734. [CrossRef] [PubMed]
66. Dubinin, J.; Braun, H.-P.; Schmitz, U.; Colditz, F. The mitochondrial proteome of the model legume *Medicago truncatula*. *Biochim. Biophys. Acta* **2011**, *1814*, 1658–1668. [CrossRef] [PubMed]
67. Taylor, N.L.; Heazlewood, J.L.; Millar, A.H. The *Arabidopsis thaliana* 2-D gel mitochondrial proteome: Refining the value of reference maps for assessing protein abundance, contaminants and post-translational modifications. *Proteomics* **2011**, *11*, 1720–1733. [CrossRef] [PubMed]
68. Ito, J.; Heazlewood, J.L.; Millar, A.H. The plant mitochondrial proteome and the challenge of defining the posttranslational modifications responsible for signalling and stress effects on respiratory functions. *Physiol. Plant.* **2007**, *129*, 207–224. [CrossRef]
69. Ito, J.; Taylor, N.L.; Castleden, I.; Weckwerth, W.; Millar, A.H.; Heazlewood, J.L. A survey of the *Arabidopsis thaliana* mitochondrial phosphoproteome. *Proteomics* **2009**, *9*, 4229–4240. [CrossRef] [PubMed]

70. Huang, C.; Verrillo, F.; Renzone, G.; Arena, S.; Rocco, M.; Scaloni, A.; Marra, M. Response to biotic and oxidative stress in *Arabidopsis thaliana*: Analysis of variably phosphorylated proteins. *J. Proteom.* **2011**, *74*, 1934–1949. [CrossRef] [PubMed]
71. Havelund, J.F.; Thelen, J.J.; Møller, I.M. Biochemistry, proteomics, and phosphoproteomics of plant mitochondria from non-photosynthetic cells. *Front. Plant Sci.* **2013**, *4*, 51. [CrossRef] [PubMed]
72. Khan, M.; Takasaki, H.; Komatsu, S. Comprehensive Phosphoproteome Analysis in Rice and Identification of Phosphoproteins Responsive to Different Hormones/Stresses. *J. Proteome Res.* **2005**, *4*, 1592–1599. [CrossRef] [PubMed]
73. Solheim, C.; Li, L.; Hatzopoulos, P.; Millar, A.H. Loss of Lon1 in Arabidopsis Changes the Mitochondrial Proteome Leading to Altered Metabolite Profiles and Growth Retardation without an Accumulation of Oxidative Damage. *Plant Physiol.* **2012**, *160*, 1187–1203. [CrossRef] [PubMed]
74. Parkin, I.A.P.; Robinson, S.J. Exploring the Paradoxes of the Brassica Genome Architecture. In *Genetics, Genomics and Breeding of Vegetable Brassicas*, 1st ed.; Sadowski, J., Kole, C., Eds.; Science Publishers: Enfield, NH, USA; Boca Raton, FL, USA, 2011; pp. 328–348; ISBN 9781578087068.
75. Faurobert, M.; Mihr, C.; Bertin, N.; Pawlowski, T.; Negroni, L.; Sommerer, N.; Causse, M. Major Proteome Variations Associated with Cherry Tomato Pericarp Development and Ripening. *Plant Physiol.* **2007**, *143*, 1327–1346. [CrossRef] [PubMed]
76. Imin, N.; Kerim, T.; Rolfe, B.G.; Weinman, J.J. Effect of early cold stress on the maturation of rice anthers. *Proteomics* **2004**, *4*, 1873–1882. [CrossRef] [PubMed]
77. Qin, G.; Meng, X.; Wang, Q.; Tian, S. Oxidative Damage of Mitochondrial Proteins Contributes to Fruit Senescence: A Redox Proteomics Analysis. *J. Proteome Res.* **2009**, *8*, 2449–2462. [CrossRef] [PubMed]
78. Yun, Z.; Jin, S.; Ding, Y.; Wang, Z.; Gao, H.; Pan, Z.; Xu, J.; Cheng, Y.; Deng, X. Comparative transcriptomics and proteomics analysis of citrus fruit, to improve understanding of the effect of low temperature on maintaining fruit quality during lengthy post-harvest storage. *J. Exp. Bot.* **2012**, *63*, 2873–2893. [CrossRef] [PubMed]
79. Ferreira, S.; Hjernø, K.; Larsen, M.; Wingsle, G.; Larsen, P.; Fey, S.; Roepstorff, P.; Salomé Pais, M. Proteome Profiling of *Populus euphratica* Oliv. Upon Heat Stress. *Ann. Bot.* **2006**, *98*, 361–377. [CrossRef] [PubMed]
80. Howell, K.A.; Cheng, K.; Murcha, M.W.; Jenkin, L.E.; Millar, A.H.; Whelan, J. Oxygen Initiation of Respiration and Mitochondrial Biogenesis in Rice. *J. Biol. Chem.* **2007**, *282*, 15619–15631. [CrossRef] [PubMed]
81. Lee, D.-G.; Ahsan, N.; Lee, S.-H.; Kang, K.Y.; Bahk, J.D.; Lee, I.-J.; Lee, B.-H. A proteomic approach in analyzing heat-responsive proteins in rice leaves. *Proteomics* **2007**, *7*, 3369–3383. [CrossRef] [PubMed]
82. Millar, A.H.; Whelan, J.; Soole, K.L.; Day, D.A. Organization and regulation of mitochondrial respiration in plants. *Annu. Rev. Plant Biol.* **2011**, *62*, 79–104. [CrossRef] [PubMed]
83. Sarry, J.-E.; Kuhn, L.; Ducruix, C.; Lafaye, A.; Junot, C.; Hugouvieux, V.; Jourdain, A.; Bastien, O.; Fievet, J.B.; Vailhen, D.; et al. The early responses of *Arabidopsis thaliana* cells to cadmium exposure explored by protein and metabolite profiling analyses. *Proteomics* **2006**, *6*, 2180–2198. [CrossRef] [PubMed]
84. Komatsu, S.; Yamamoto, A.; Nakamura, T.; Nouri, M.-Z.; Nanjo, Y.; Nishizawa, K.; Furukawa, K. Comprehensive Analysis of Mitochondria in Roots and Hypocotyls of Soybean under Flooding Stress using Proteomics and Metabolomics Techniques. *J. Proteome Res.* **2011**, *10*, 3993–4004. [CrossRef] [PubMed]
85. Nakaminami, K.; Matsui, A.; Nakagami, H.; Minami, A.; Nomura, Y.; Tanaka, M.; Morosawa, T.; Ishida, J.; Takahashi, S.; Uemura, M.; et al. Analysis of Differential Expression Patterns of mRNA and Protein During Cold-Acclimation and De-Acclimation in *Arabidopsis*. *Mol. Cell. Proteom.* **2014**, *13*, 3602–3611. [CrossRef] [PubMed]
86. Li, T.; Xu, S.L.; Oses-Prieto, J.A.; Putil, S.; Xu, P.; Wang, R.J.; Li, K.H.; Maltby, D.A.; An, L.H.; Burlingame, A.L.; et al. Proteomics Analysis Reveals Post-Translational Mechanisms for Cold-Induced Metabolic Changes in *Arabidopsis*. *Mol. Plant* **2011**, *4*, 361–374. [CrossRef] [PubMed]
87. Rinalducci, S.; Egidi, M.G.; Karimzadeh, G.; Jazii, F.R.; Zolla, L. Proteomic analysis of a spring wheat cultivar in response to prolonged cold stress. *Electrophoresis* **2011**, *32*, 1807–1818. [CrossRef] [PubMed]
88. Dumont, E.; Bahrman, N.; Goulas, E.; Valot, B.; Sellier, H.; Hilbert, J.-L.; Vuylsteker, C.; Lejeune-Hénaut, I.; Delbreil, B. A proteomic approach to decipher chilling response from cold acclimation in pea (*Pisum sativum* L.). *Plant Sci.* **2011**, *180*, 86–98. [CrossRef] [PubMed]
89. Yin, G.; Sun, H.; Xin, X.; Qin, G.; Liang, Z.; Jing, X. Mitochondrial Damage in the Soybean Seed Axis during Imbibition at Chilling Temperatures. *Plant Cell. Physiol.* **2009**, *50*, 1305–1318. [CrossRef] [PubMed]

90. Cheng, L.; Gao, X.; Li, S.; Shi, M.; Javeed, H.; Jing, X.; Yang, G.; He, G. Proteomic analysis of soybean [*Glycine max* (L.) Meer.] seeds during imbibition at chilling temperature. *Mol. Breed.* **2010**, *26*, 1–17. [CrossRef]
91. Sánchez-Bel, P.; Egea, I.; Sanchez-Ballesta, M.T.; Martinez-Madrid, C.; Fernandez-Garcia, N.; Romojaro, F.; Olmos, E.; Estrella, E.; Bolarin, M.C.; Flores, F.B. Understanding the mechanisms of chilling injury in bell pepper fruits using the proteomic approach. *J. Proteom.* **2012**, *75*, 5463–5478. [CrossRef] [PubMed]
92. Xu, C.; Huang, B. Differential proteomic response to heat stress in thermal *Agrostis scabra* and heat-sensitive *Agrostis stolonifera*. *Physiol. Plant.* **2010**, *139*, 192–204. [CrossRef] [PubMed]
93. Vítámvás, P.; Prášil, I.T.; Kosová, K.; Planchon, S.; Renaut, J. Analysis of proteome and frost tolerance in chromosome 5A and 5B reciprocal substitution lines between two winter wheats during long-term cold acclimation. *Proteomics* **2012**, *12*, 68–85. [CrossRef] [PubMed]
94. Sweetlove, L.J.; Heazlewood, J.L.; Herald, V.; Holtzapffel, R.; Day, D.A.; Leaver, C.J.; Millar, A.H. The impact of oxidative stress on *Arabidopsis* mitochondria. *Plant J.* **2002**, *32*, 891–904. [CrossRef] [PubMed]
95. Renaut, J.; Hausman, J.-F.; Bassett, C.; Artlip, T.; Cauchie, H.-M.; Witters, E.; Wisniewski, M. Quantitative proteomic analysis of short photoperiod and low-temperature responses in bark tissues of peach (*Prunus persica* L. Batsch). *Tree Genet. Genomes* **2008**, *4*, 589–600. [CrossRef]
96. Chen, X.; Wang, Y.; Li, J.; Jiang, A.; Cheng, Y.; Zhang, W. Mitochondrial proteome during salt stress-induced programmed cell death in rice. *Plant Physiol. Biochem.* **2009**, *47*, 407–415. [CrossRef] [PubMed]
97. Palmblad, M.; Mills, D.J.; Bindschedler, L.V. Heat-Shock Response in *Arabidopsis thaliana* Explored by Multiplexed Quantitative Proteomics Using Differential Metabolic Labeling. *J. Proteome Res.* **2008**, *7*, 780–785. [CrossRef] [PubMed]
98. Wu, X.; Jiang, L.; Yu, M.; An, X.; Ma, R.; Yu, Z. Proteomic analysis of changes in mitochondrial protein expression during peach fruit ripening and senescence. *J. Proteom.* **2016**, *147*, 197–211. [CrossRef] [PubMed]
99. Kim, J.Y.; Park, S.J.; Jang, B.; Jung, C.-H.; Ahn, S.J.; Goh, C.-H.; Cho, K.; Han, O.; Kang, H. Functional characterization of a glycine-rich RNA-binding protein 2 in *Arabidopsis thaliana* under abiotic stress conditions. *Plant J.* **2007**, *50*, 439–451. [CrossRef] [PubMed]
100. Cui, S.; Huang, F.; Wang, J.; Ma, X.; Cheng, Y.; Liu, J. A proteomic analysis of cold stress responses in rice seedlings. *Proteomics* **2005**, *5*, 3162–3172. [CrossRef] [PubMed]
101. Xin, X.; Lin, X.-H.; Zhou, Y.-C.; Chen, X.-L.; Liu, X.; Lu, X.-X. Proteome analysis of maize seeds: The effect of artificial ageing. *Physiol. Plant.* **2011**, *143*, 126–138. [CrossRef] [PubMed]
102. Chen, L.; Chen, Q.; Kong, L.; Xia, F.; Yan, H.; Zhu, Y.; Mao, P. Proteomic and Physiological Analysis of the Response of Oat (*Avena sativa*) Seeds to Heat Stress under Different Moisture Conditions. *Front. Plant Sci.* **2016**, *7*, 896. [CrossRef] [PubMed]
103. Taylor, N.L.; Rudhe, C.; Hulett, J.M.; Lithgow, T.; Glaser, E.; Day, D.A.; Millar, A.H.; Whelan, J. Environmental stresses inhibit and stimulate different protein import pathways in plant mitochondria. *FEBS Lett.* **2003**, *547*, 125–130. [CrossRef]
104. Funck, D.; Eckard, S.; Müller, G. Non-redundant functions of two proline dehydrogenase isoforms in Arabidopsis. *BMC Plant Biol.* **2010**, *10*, 70. [CrossRef] [PubMed]
105. Peng, Z.; Lu, Q.; Verma, D.P. Reciprocal regulation of delta 1-pyrroline-5-carboxylate synthetase and proline dehydrogenase genes controls proline levels during and after osmotic stress in plants. *Mol. Gen. Genet.* **1996**, *253*, 334–341. [CrossRef] [PubMed]
106. Miller, G.; Honig, A.; Stein, H.; Suzuki, N.; Mittler, R.; Zilberstein, A. Unraveling Δ^1-Pyrroline-5-Carboxylate-Proline Cycle in Plants by Uncoupled Expression of Proline Oxidation Enzymes. *J. Biol. Chem.* **2009**, *284*, 26482–26492. [CrossRef] [PubMed]
107. Fuller, M.P.; Metwali, E.M.R.; Eed, M.H.; Jellings, A.J. Evaluation of Abiotic Stress Resistance in Mutated Populations of Cauliflower (*Brassica oleracea* var. botrytis). *Plant Cell Tissue Organ Cult.* **2006**, *86*, 239. [CrossRef]
108. Hadi, F.; Gilpin, M.; Fuller, M.P. Identification and expression analysis of CBF/DREB1 and COR15 genes in mutants of *Brassica oleracea* var. *botrytis* with enhanced proline production and frost resistance. *Plant Physiol. Biochem.* **2011**, *49*, 1323–1332. [CrossRef] [PubMed]
109. Lv, W.-T.; Lin, B.; Zhang, M.; Hua, X.-J. Proline Accumulation is Inhibitory to Arabidopsis Seedlings during Heat Stress. *Plant Physiol.* **2011**, *156*, 1921–1933. [CrossRef] [PubMed]

110. Lee, U.; Wie, C.; Escobar, M.; Williams, B.; Hong, S.-W.; Vierling, E. Genetic Analysis Reveals Domain Interactions of Arabidopsis Hsp100/Clpb and Cooperation with the Small Heat Shock Protein Chaperone System. *Plant Cell* **2005**, *17*, 559–571. [CrossRef] [PubMed]
111. Taylor, N.L.; Day, D.A.; Millar, A.H. Environmental Stress Causes Oxidative Damage to Plant Mitochondria Leading to Inhibition of Glycine Decarboxylase. *J. Biol. Chem.* **2002**, *277*, 42663–42668. [CrossRef] [PubMed]
112. Herman, E.M.; Rotter, K.; Premakumar, R.; Elwinger, G.; Bae, H.; Ehler-King, L.; Chen, S.; Livingston, D.P. Additional freeze hardiness in wheat acquired by exposure to −3 °C is associated with extensive physiological, morphological, and molecular changes. *J. Exp. Bot.* **2006**, *57*, 3601–3618. [CrossRef] [PubMed]
113. Zhang, S.; Chen, F.; Peng, S.; Ma, W.; Korpelainen, H.; Li, C. Comparative physiological, ultrastructural and proteomic analyses reveal sexual differences in the responses of *Populus cathayana* under drought stress. *Proteomics* **2010**, *10*, 2661–2677. [CrossRef] [PubMed]
114. Imin, N.; Kerim, T.; Weinman, J.J.; Rolfe, B.G. Low Temperature Treatment at the Young Microspore Stage Induces Protein Changes in Rice Anthers. *Mol. Cell. Proteom.* **2006**, *5*, 274–292. [CrossRef] [PubMed]
115. Kim, D.-W.; Rakwal, R.; Agrawal, G.K.; Jung, Y.-H.; Shibato, J.; Jwa, N.-S.; Iwahashi, Y.; Iwahashi, H.; Kim, D.H.; Shim, I.-S.; et al. A hydroponic rice seedling culture model system for investigating proteome of salt stress in rice leaf. *Electrophoresis* **2005**, *26*, 4521–4539. [CrossRef] [PubMed]
116. Wang, X.Y.; Shan, X.H.; Wu, Y.; Su, S.Z.; Li, S.P.; Liu, H.K.; Han, J.Y.; Xue, C.M.; Yuan, Y.P. iTRAQ-based quantitative proteomic analysis reveals new metabolic pathways responding to chilling stress in maize seedlings. *J. Proteom.* **2016**, *146*, 14–24. [CrossRef] [PubMed]
117. Byun, Y.-J.; Kim, H.-J.; Lee, D.-H. LongSAGE analysis of the early response to cold stress in Arabidopsis leaf. *Planta* **2009**, *229*, 1181–1200. [CrossRef] [PubMed]
118. Naydenov, N.G.; Khanam, S.; Siniauskaya, M.; Nakamura, C. Profiling of mitochondrial transcriptome in germinating wheat embryos and seedlings subjected to cold, salinity and osmotic stresses. *Genes Genet. Syst.* **2010**, *85*, 31–42. [CrossRef] [PubMed]
119. Jacoby, R.P.; Millar, A.H.; Taylor, N.L. Wheat Mitochondrial Proteomes Provide New Links between Antioxidant Defense and Plant Salinity Tolerance. *J. Proteome Res.* **2010**, *9*, 6595–6604. [CrossRef] [PubMed]
120. Heidarvand, L.; Maali-Amiri, R. Physio-biochemical and proteome analysis of chickpea in early phases of cold stress. *J. Plant Physiol.* **2013**, *170*, 459–469. [CrossRef] [PubMed]
121. Holtzapffel, R.C.; Finnegan, P.M.; Millar, A.H.; Badger, M.R.; Day, D.A. Mitochondrial protein expression in tomato fruit during on-vine ripening and cold storage. *Funct. Plant Biol.* **2002**, *29*, 827–834. [CrossRef]
122. Bocian, A.; Kosmala, A.; Rapacz, M.; Jurczyk, B.; Marczak, L.; Zwierzykowski, Z. Differences in leaf proteome response to cold acclimation between *Lolium perenne* plants with distinct levels of frost tolerance. *J. Plant Physiol.* **2011**, *168*, 1271–1279. [CrossRef] [PubMed]
123. Wahid, A.; Gelani, S.; Ashraf, M.; Foolad, M. Heat tolerance in plants: An overview. *Environ. Exp. Bot.* **2007**, *61*, 199–223. [CrossRef]
124. Lin, K.-H.; Huang, H.-C.; Lin, C.-Y. Cloning, expression and physiological analysis of broccoli catalase gene and Chinese cabbage ascorbate peroxidase gene under heat stress. *Plant Cell Rep.* **2010**, *29*, 575–593. [CrossRef] [PubMed]
125. Ashraf, M. Relationships between growth and gas exchange characteristics in some salt-tolerant amphidiploid *Brassica* species in relation to their diploid parents. *Environ. Exp. Bot.* **2001**, *45*, 155–163. [CrossRef]
126. Baryla, A.; Carrier, P.; Franck, F.; Coulomb, C.; Sahut, C.; Havaux, M. Leaf chlorosis in oilseed rape plants (*Brassica napus*) grown on cadmium-polluted soil: Causes and consequences for photosynthesis and growth. *Planta* **2001**, *212*, 696–709. [CrossRef] [PubMed]
127. Gill, S.S.; Khan, N.A.; Tuteja, N. Differential cadmium stress tolerance in five indian mustard (*Brassica juncea* L.) cultivars: An evaluation of the role of antioxidant machinery. *Plant Signal. Behav.* **2011**, *6*, 293–300. [CrossRef] [PubMed]
128. Hayat, S.; Maheshwari, P.; Wani, A.S.; Irfan, M.; Alyemeni, M.N.; Ahmad, A. Comparative effect of 28 homobrassinolide and salicylic acid in the amelioration of NaCl stress in *Brassica juncea* L. *Plant Physiol. Biochem.* **2012**, *53*, 61–68. [CrossRef] [PubMed]
129. Yang, Y.; Zheng, Q.; Liu, M.; Long, X.; Liu, Z.; Shen, Q.; Guo, S. Difference in Sodium Spatial Distribution in the Shoot of Two Canola Cultivars under Saline Stress. *Plant Cell. Physiol.* **2012**, *53*, 1083–1092. [CrossRef] [PubMed]

130. Bunce, J.A. Acclimation of photosynthesis to temperature in *Arabidopsis thaliana* and *Brassica oleracea*. *Photosynthetica* **2008**, *46*, 517–524. [CrossRef]
131. Hu, W.H.; Shi, K.; Song, X.S.; Xia, X.J.; Zhou, Y.H.; Yu, J.Q. Different effects of chilling on respiration in leaves and roots of cucumber (*Cucumis sativus*). *Plant Physiol. Biochem.* **2006**, *44*, 837–843. [CrossRef] [PubMed]
132. Hu, W.H.; Xiao, Y.A.; Zeng, J.J.; Hu, X.H. Photosynthesis, respiration and antioxidant enzymes in pepper leaves under drought and heat stresses. *Biol. Plant.* **2010**, *54*, 761–765. [CrossRef]
133. Dahal, K.; Kane, K.; Gadapati, W.; Webb, E.; Savitch, L.V.; Singh, J.; Sharma, P.; Sarhan, F.; Longstaffe, F.J.; Grodzinski, B.; et al. The effects of phenotypic plasticity on photosynthetic performance in winter rye, winter wheat and *Brassica napus*. *Physiol. Plant.* **2012**, *144*, 169–188. [CrossRef] [PubMed]
134. Savitch, L.V.; Barker-Astrom, J.; Ivanov, A.G.; Hurry, V.; Oquist, G.; Huner, N.P.; Gardeström, P. Cold acclimation of *Arabidopsis thaliana* results in incomplete recovery of photosynthetic capacity, associated with an increased reduction of the chloroplast stroma. *Planta* **2001**, *214*, 295–303. [CrossRef] [PubMed]
135. Copolovici, L.; Kännaste, A.; Pazouki, L.; Niinemets, U. Emissions of green leaf volatiles and terpenoids from *Solanum lycopersicum* are quantitatively related to the severity of cold and heat shock treatments. *J. Plant Physiol.* **2012**, *169*, 664–672. [CrossRef] [PubMed]
136. Zhang, S.; Jiang, H.; Peng, S.; Korpelainen, H.; Li, C. Sex-related differences in morphological, physiological, and ultrastructural responses of *Populus cathayana* to chilling. *J. Exp. Bot.* **2011**, *62*, 675–686. [CrossRef] [PubMed]
137. Luo, H.-B.; Ma, L.; Xi, H.-F.; Duan, W.; Li, S.-H.; Loescher, W.; Wang, J.-F.; Wang, L.-J. Photosynthetic Responses to Heat Treatments at Different Temperatures and Following Recovery in Grapevine (*Vitis amurensis* L.) Leaves. *PLoS ONE* **2011**, *6*, e23033. [CrossRef] [PubMed]
138. Tan, W.; wei Meng, Q.; Brestic, M.; Olsovska, K.; Yang, X. Photosynthesis is improved by exogenous calcium in heat-stressed tobacco plants. *J. Plant Physiol.* **2011**, *168*, 2063–2071. [CrossRef] [PubMed]
139. Efeoğlu, B.; Terzioğlu, S. Photosynthetic responses of two wheat varieties to high temperature. *EurAsian J. BioSci.* **2009**, *3*, 97–106. [CrossRef]
140. Yin, Y.; Li, S.; Liao, W.; Lu, Q.; Wen, X.; Lu, C. Photosystem II photochemistry, photoinhibition, and the xanthophyll cycle in heat-stressed rice leaves. *J. Plant Physiol.* **2010**, *167*, 959–966. [CrossRef] [PubMed]
141. Hüve, K.; Bichele, I.; Rasulov, B.; Niinemets, U. When it is too hot for photosynthesis: Heat-induced instability of photosynthesis in relation to respiratory burst, cell permeability changes and H_2O_2 formation. *Plant Cell Environ.* **2011**, *34*, 113–126. [CrossRef] [PubMed]
142. Tomaz, T.; Bagard, M.; Pracharoenwattana, I.; Lindén, P.; Lee, C.P.; Carroll, A.J.; Ströher, E.; Smith, S.M.; Gardeström, P.; Millar, A.H. Mitochondrial Malate Dehydrogenase Lowers Leaf Respiration and Alters Photorespiration and Plant Growth in Arabidopsis. *Plant Physiol.* **2010**, *154*, 1143–1157. [CrossRef] [PubMed]
143. Boutry, M.; Faber, A.-M.; Charbonnier, M.; Briquet, M. Microanalysis of plant mitochondrial protein synthesis products. *Plant Mol. Biol.* **1984**, *3*, 445–452. [CrossRef] [PubMed]
144. Bradford, M.M. A rapid and sensitive method for the quantitation of microgram quantities of protein utilizing the principle of protein-dye binding. *Anal. Biochem.* **1976**, *72*, 248–254. [CrossRef]
145. Staszak, A.M.; Pawłowski, T.A. Proteomic Analysis of Embryogenesis and the Acquisition of Seed Dormancy in Norway Maple (*Acer platanoides* L.). *Int. J. Mol. Sci.* **2014**, *15*, 10868–10891. [CrossRef] [PubMed]
146. Pawłowski, T.A.; Staszak, A.M. Analysis of the embryo proteome of sycamore (*Acer pseudoplatanus* L.) seeds reveals a distinct class of proteins regulating dormancy release. *J. Plant Physiol.* **2016**, *195*, 9–22. [CrossRef] [PubMed]
147. Ramagli, L.S.; Rodriguez, L.V. Quantitation of microgram amounts of protein in two-dimensional polyacrylamide gel electrophoresis sample buffer. *Electrophoresis* **1985**, *6*, 559–563. [CrossRef]
148. Heukeshoven, J.; Dernick, R. Silver staining of proteins. In *Electrophoresis Forum*, 1st ed.; Radola, B.J., Ed.; Technische Universität München: Munich, Germany, 1986.
149. Neuhoff, V.; Arold, N.; Taube, D.; Ehrhardt, W. Improved staining of proteins in polyacrylamide gels including isoelectric focusing gels with clear background at nanogram sensitivity using Coomassie Brilliant Blue G-250 and R-250. *Electrophoresis* **1988**, *9*, 255–262. [CrossRef] [PubMed]
150. Durek, P.; Schmidt, R.; Heazlewood, J.L.; Jones, A.; MacLean, D.; Nagel, A.; Kersten, B.; Schulze, W.X. PhosPhAt: The *Arabidopsis thaliana* phosphorylation site database. An update. *Nucleic Acids Res.* **2010**, *38*, D828–D834. [CrossRef] [PubMed]

151. Blom, N.; Gammeltoft, S.; Brunak, S. Sequence and structure-based prediction of eukaryotic protein phosphorylation sites. *J. Mol. Biol.* **1999**, *294*, 1351–1362. [CrossRef] [PubMed]
152. Gao, J.; Thelen, J.J.; Dunker, A.K.; Xu, D. Musite, a Tool for Global Prediction of General and Kinase-specific Phosphorylation Sites. *Mol. Cell. Proteom.* **2010**, *9*, 2586–2600. [CrossRef] [PubMed]
153. Shi, S.-P.; Qiu, J.-D.; Sun, X.-Y.; Suo, S.-B.; Huang, S.-Y.; Liang, R.-P. PMeS: Prediction of Methylation Sites Based on Enhanced Feature Encoding Scheme. *PLoS ONE* **2012**, *7*, e38772. [CrossRef] [PubMed]
154. Zybailov, B.; Sun, Q.; van Wijk, K.J. Workflow for Large Scale Detection and Validation of Peptide Modifications by RPLC-LTQ-Orbitrap: Application to the *Arabidopsis thaliana* Leaf Proteome and an Online Modified Peptide Library. *Anal. Chem.* **2009**, *81*, 8015–8024. [CrossRef] [PubMed]
155. Kim, J.; Rudella, A.; Ramirez Rodriguez, V.; Zybailov, B.; Olinares, P.D.B.; van Wijk, K.J. Subunits of the Plastid ClpPR Protease Complex Have Differential Contributions to Embryogenesis, Plastid Biogenesis, and Plant Development in *Arabidopsis*. *Plant Cell* **2009**, *21*, 1669–1692. [CrossRef] [PubMed]
156. Wilkins, M.R.; Gasteiger, E.; Gooley, A.A.; Herbert, B.R.; Molloy, M.P.; Binz, P.A.; Ou, K.; Sanchez, J.C.; Bairoch, A.; Williams, K.L.; et al. High-throughput mass spectrometric discovery of protein post-translational modifications. *J. Mol. Biol.* **1999**, *289*, 645–657. [CrossRef] [PubMed]
157. Laemmli, U.K. Cleavage of Structural Proteins during the Assembly of the Head of Bacteriophage T4. *Nature* **1970**, *227*, 680–685. [CrossRef] [PubMed]
158. Lamattina, L.; Gonzalez, D.; Gualberto, J.; Grienenberger, J.M. Higher plant mitochondria encode an homologue of the nuclear-encoded 30-kDa subunit of bovine mitochondrial complex I. *Eur. J. Biochem.* **1993**, *217*, 831–838. [CrossRef] [PubMed]
159. Rayapuram, N.; Hagenmuller, J.; Grienenberger, J.M.; Bonnard, G.; Giegé, P. The Three Mitochondrial Encoded Ccmf Proteins Form a Complex That Interacts with CCMH and *c*-Type Apocytochromes in *Arabidopsis*. *J. Biol. Chem.* **2008**, *283*, 25200–25208. [CrossRef] [PubMed]
160. Rayapuram, N.; Hagenmuller, J.; Grienenberger, J.-M.; Giegé, P.; Bonnard, G. AtCCMA Interacts with AtCcmB to Form a Novel Mitochondrial ABC Transporter Involved in Cytochrome *c* Maturation in *Arabidopsis*. *J. Biol. Chem.* **2007**, *282*, 21015–21023. [CrossRef] [PubMed]

© 2018 by the authors. Licensee MDPI, Basel, Switzerland. This article is an open access article distributed under the terms and conditions of the Creative Commons Attribution (CC BY) license (http://creativecommons.org/licenses/by/4.0/).

Article

Mitochondrial Biogenesis in Diverse Cauliflower Cultivars under Mild and Severe Drought. Impaired Coordination of Selected Transcript and Proteomic Responses, and Regulation of Various Multifunctional Proteins

Michał Rurek [1,*], Magdalena Czołpińska [1], Tomasz Andrzej Pawłowski [2], Aleksandra Maria Staszak [2,3], Witold Nowak [4], Włodzimierz Krzesiński [5] and Tomasz Spiżewski [5]

1. Department of Molecular and Cellular Biology, Institute of Molecular Biology and Biotechnology, Adam Mickiewicz University, Poznań, Umultowska 89, 61-614 Poznań, Poland; magczo@amu.edu.pl
2. Institute of Dendrology, Polish Academy of Sciences, Parkowa 5, 62-035 Kórnik, Poland; tapawlow@man.poznan.pl (T.A.P.); a.staszak@uwb.edu.pl (A.M.S.)
3. Present address: Department of Plant Physiology, Institute of Biology, Faculty of Biology and Chemistry, University of Białystok, Ciołkowskiego 1J, 15-245 Białystok, Poland
4. Molecular Biology Techniques Laboratory, Faculty of Biology, Adam Mickiewicz University, Poznań, Umultowska 89, 61-614 Poznań, Poland; nowak@amu.edu.pl
5. Department of Vegetable Crops, Poznan University of Life Sciences, Dąbrowskiego 159, 60-594 Poznań, Poland; wlodzimierz.krzesinski@up.poznan.pl (W.K.); tomasz.spizewski@up.poznan.pl (T.S.)
* Correspondence: rurek@amu.edu.pl; Tel.: +48-61-829-5973

Received: 29 January 2018; Accepted: 4 April 2018; Published: 10 April 2018

Abstract: Mitochondrial responses under drought within *Brassica* genus are poorly understood. The main goal of this study was to investigate mitochondrial biogenesis of three cauliflower (*Brassica oleracea* var. *botrytis*) cultivars with varying drought tolerance. Diverse quantitative changes (decreases in abundance mostly) in the mitochondrial proteome were assessed by two-dimensional gel electrophoresis (2D PAGE) coupled with liquid chromatography-tandem mass spectrometry (LC-MS/MS). Respiratory (e.g., complex II, IV (CII, CIV) and ATP synthase subunits), transporter (including diverse porin isoforms) and matrix multifunctional proteins (e.g., components of RNA editing machinery) were diversely affected in their abundance under two drought levels. Western immunoassays showed additional cultivar-specific responses of selected mitochondrial proteins. Dehydrin-related tryptic peptides (found in several 2D spots) immunopositive with dehydrin-specific antisera highlighted the relevance of mitochondrial dehydrin-like proteins for the drought response. The abundance of selected mRNAs participating in drought response was also determined. We conclude that mitochondrial biogenesis was strongly, but diversely affected in various cauliflower cultivars, and associated with drought tolerance at the proteomic and functional levels. However, discussed alternative oxidase (AOX) regulation at the RNA and protein level were largely uncoordinated due to the altered availability of transcripts for translation, mRNA/ribosome interactions, and/or miRNA impact on transcript abundance and translation.

Keywords: dehydrins; 2D PAGE; drought; mitochondrial biogenesis; mitochondrial proteome; plant transcriptome

1. Introduction

Under drought, plants respond through numerous physiological and molecular mechanisms [1]. Each plant species possesses a unique drought resistance response, which is accompanied by the diverse sensitivity of selected growth and metabolic processes to progressing stress conditions [2]. The balance between water uptake and transpiration is controlled by water potential. The leaf surface controls CO_2 assimilation as well as photosynthetic and respiration rates. Excessive transpiration may decrease the water potential in plants, resulting in growth cessation [3]. Initially, drought results in stomatal closure and declined transpiration to prevent further water losses among drought-sensitive species, in osmolyte synthesis, and consequently, leaf cell growth inhibition. Drought regulates leaf respiration in various directions, nevertheless, those alterations may enable prompt stress recovery [4–8].

The mitochondrial proteome is a highly dynamic entity containing at least 1500 diverse proteins (1060 of which have been identified in potato (*Solanum tuberosum*) mitochondria by Salvato et al. [9]) actively responding to environmental conditions. Over the past 15 years, significant progress has been made on the elucidation of key steps of mitochondrial biogenesis, which implies finely coordinated expression of mitochondrial and nuclear genes that can be disrupted under stress action [10–13]. Taylor et al. [14] estimated 22% of the stress-responsive organellar proteins in *Arabidopsis* to be targeted to mitochondria, but the number of mitochondrial proteins involved in diverse stress response still remains underestimated. More complex studies integrating various approaches for better understanding of drought responses are required.

Drought also results in dynamic alterations within the cellular transcriptome and proteome [7,15–17], including the mitochondrial proteome [18,19]. Drought response is often connected with the increase or induction of diverse protective proteins, such as dehydrins [20]. Mitochondrial proteins (including key enzymes) may be directly involved in developing of drought tolerance as well; under drought, some novel protein isoforms may be also induced, although the proteolysis of some mitochondrial proteins was also reported [19,21–27]. Variations in the abundance of numerous mitochondrial proteins, however, may not clearly correspond with the drought intensity.

Brassica genus contains important plant species for worldwide agriculture. Total cellular proteomic and transcriptomic responses of *Brassica* species in drought have already been investigated, although without deepened attention towards elucidation of the particular aspects of mitochondrial biogenesis [28–32]. Interestingly, drought response of close *Brassica* relatives, including *Thellungiella*, are distinct from that observed for *Arabidopsis* [33]. However, reports comparing responses of *Brassica* cultivars with contrasting drought tolerance are still limited [34–38], contrary to other species data [6,39–48]. Search for protein markers in order to develop drought-tolerant plant accessions belongs to the current goals of proteomic analyses [7].

Owing to recent research trends, this work was undertaken to gain a comprehensive view of the influence of middle and severe water deficiency conditions on the mitochondrial biogenesis of three cauliflower (*Brassica oleracea* var. *botrytis*) cultivars displaying diverse drought tolerance. Previously, we studied cauliflower mitochondrial biogenesis under temperature stress and subsequent recovery [13]. Cauliflower belongs to vegetables with the major cultivation yield in Central Europe. The early generative phase of curd ripening belongs to the key developmental stages with some physiological demands. In addition, due to the size of its vegetative organs, it is sensitive to the low water level in the soil. To determine mitochondrial responses in relation to plant respiration, we aimed (1) to investigate the dynamic nature of the mitochondrial proteome; (2) to identify the most variable proteins in cauliflower curd mitochondria; (3) to determine the abundance of selected mitochondrial proteins, including dehydrin-like proteins previously investigated by us [49]; (4) to analyze relevant proteomic and transcriptomic alterations; and (5) to link them with the physiological level (respiratory and photorespiratory alterations) for the discussion of the general responses to mild and severe water deficit. This is the first comprehensive study of the mitochondrial proteome of the *Brassica* genus member which allowed characterization of a broadened set of drought-responsive mitochondrial proteins in the cultivar context. It highlights the participation not only of oxidative phosphorylation (OXPHOS) proteins, but a number of multifunctional mitochondrial proteins (including RNA editing factors and dehydrin-like proteins) in drought response.

2. Results

2.1. Respiration and Photorespiration Pattern in Cauliflower Leaves

In order to study mitochondrial response at physiological and molecular levels in "*Adelanto*" ("*A*") and "*Casper*" ("*C*") drought-sensitive cultivars, as well as in the "*Pionier*" ("*P*") drought-tolerant cultivar, total light (R_T), day (R_d), and night (R_n) respiration, as well as the photorespiration (PhR) rates were determined.

Generally, R_T slightly increased under mild drought, but markedly decreased in severe treatment in "*A*" and especially in "*C*". In "*P*", R_T rate only slightly decreased under severe drought (Figure S1). R_d (the average from all illumination conditions), as well as R_n increased progressively in the drought-sensitive cultivars under all investigated treatments. R_d rate was markedly decreased in "*P*" in mild drought. The highest increase of R_n and R_d rates was noted under severe drought in "*A*" and "*P*"; Figure S1). Interestingly, respiratory effects depended on the cultivar. In "*P*" R_d exceeded R_n, both in control conditions of growth and under severe drought treatment, while in "*A*" cultivar the R_d exceeded R_n rate only under the severe stress. In "*C*" R_d was always lower than R_n, regardless of the drought level.

PhR rate progressively declined along the stress duration in "*A*" and "*C*". In contrast, a slight increase of PhR rate was observed after mild drought in "*P*" (Figure S1). In the course of the drought progression, both in "*A*" and (to a lesser extent) in "*C*" cultivar, the rapid increase of the contribution of R_d and the decrease of the contribution of PhR in the R_T value was noticed. Under mild drought in "*P*" leaves, R_d and PhR rates contributed in the other way to R_T rate (Figure S1).

Next, we investigated the dynamic nature of the cauliflower mitochondrial proteome in three cultivars under water deficit. We studied whether diverse pools of drought-responsive mitochondrial proteins accompany the analyzed respiratory alterations in stress-sensitive and stress-tolerant cultivars.

2.2. Specificity of Mitochondrial Proteome Alterations under Drought in Diverse Cauliflower Cultivars

Mitochondrial proteins isolated from curds of cauliflower plants of three investigated cultivars (Section 2.1) grown in control conditions (0) as well as in mild (1) and severe (2) water deficit were resolved by two-dimensional gel electrophoresis (2D PAGE). Experimental variants were analyzed in pairs as follows: "*A1*" vs. "*A0*", "*A2*" vs. "*A0*", "*C1*" vs. "*C0*", "*C2*" vs. "*C0*", "*P1*" vs. "*P0*", "*P2*" vs. "*P0*". 2D gels for nine different variants, including control, were run in triplicate. The Coomassie Brilliant Blue (CBB)-stained master gel (a fused image) was created based on pooled samples containing equal amounts of mitochondrial proteins from all experimental variants, resulting in 370 different spots. The number of spots on the particular 2D gel varied from 231 to 370 between all analyzed variants (including controls) for each cultivar (Figure S2).

Thirty-two spots (8.65% of all spots from the master gel) from all cultivars significantly varied in abundance and their positions (Figures 1 and S2) were assessed from three biological replicates. Experimental as well as statistical data for responsive protein spots are shown in Tables S1 and S2 and on Figure S3. Mitochondrial proteomes varied among all cultivars. Massively decreased in abundance spots exceeded those ones which increased in abundance. Under mild drought, four (in "*A*"), six (in "*C*"), and two spots (in "*P*") were specifically decreased in abundance. In the same conditions, a specific increase in abundance for five, three, and two spots in those cultivars, respectively, was noted. Under severe drought, six, two, and no spots were specifically decreased in abundance, and only one, two, and two spots were specifically increased in abundance in the mentioned cultivars, respectively (Figures 2 and S3). Thus, increased abundance in spots specific to the given drought level dominated over common spots for both treatments. The number of spots diversely affected in abundance between investigated drought-sensitive and drought-tolerant cultivars exceeded the number of such spots within the drought-sensitive cultivars. Conversely, spots commonly affected in abundance increased in number between the drought-sensitive cultivars (Figure 2). Drought responses for the drought-tolerant cultivar ("*P*") were particularly specific (alterations in abundance related with the distinct set of protein spots when compared to the impact of mild stress).

In addition, most of protein spots displayed the relationship in abundance, evaluated by the correlation analysis (Table S3). Negatively correlated pairs exceeded in number the positively correlated ones in their abundance across all cultivars and treatments. All significant correlations are strong; the correlation pattern exemplifies the diversity of spot abundance alterations in drought.

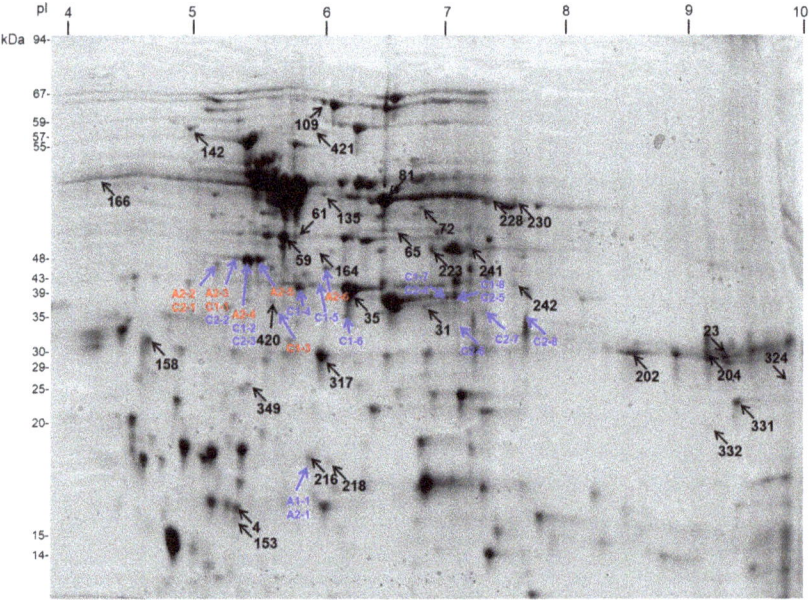

Figure 1. Position of 32 drought-responsive spots (*black arrows*) on Coomassie Brilliant Blue (CBB)-stained two–dimensional (2D) master gel with separated cauliflower mitochondrial proteins (proposed identities of all protein spots are shown in Table S1 and peptide data in Table S4). This map also shows positions of additional protein spots (*blue arrows*) from two-dimensional (2D) gels containing resolved mitochondrial proteins of "*Adelanto*" (A) and "*Casper*" (C), that were cut out and used for the identification of tryptic peptides specific to dehydrin-like proteins (proposed identities for those spots appear in Table S6). Identifiers for spots containing the mentioned peptides are marked *in red* (remaining labels—*in blue*). For molecular mass calibration (kDa) of protein spots, PageRuler Prestained Protein Ladder (Thermo Scientific, Gdańsk, Poland) and Low Molecular Weight (LMW)-SDS Marker Kit (GE Healthcare Poland, Warsaw, Poland) are used. For calibration of spot isoelectric point (pI), Broad pI Kit (GE Healthcare) is used. Further data found in the text.

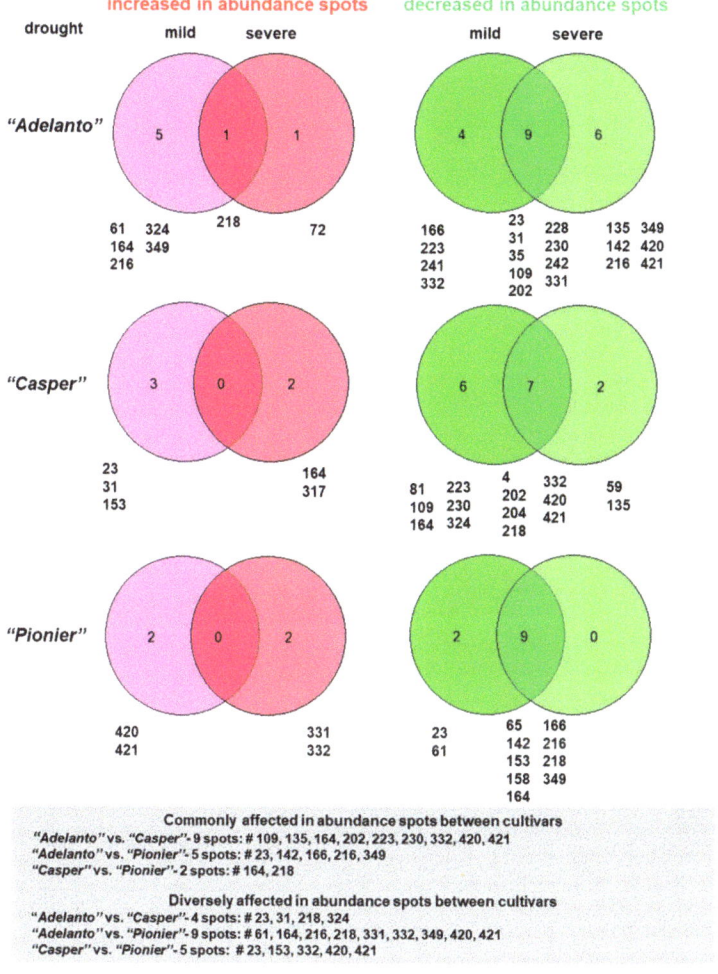

Figure 2. Venn diagrams showing distribution of increased and decreased in abundance specific/common protein spots to mild and severe drought across investigated cultivars. Numbers refer to protein spot identifiers. Increased (magenta and pink diagrams) and decreased (light and dark green diagrams) in abundance spots between cultivars are marked below each diagram; commonly and diversely affected in abundance spots are also listed. Further data in the text.

2.3. Functional Categorization of Drought-Responsive Proteins in Diverse Cultivars of Cauliflower

Proteins from spots collected from the master gel were identified by the tandem mass spectrometry coupled with liquid chromatography (LC-MS/MS). Obtained data were used for searching Mascot against the *Viridiplantae* section of the NCBInr database (version 20160525 containing 88005140 protein sequences). In addition, Gelmap tool (Available online: https://gelmap.de/projects-arabidopsis/) was applied to compare cauliflower and *Arabidopsis* proteomic maps, and to validate MS identifications. Because mitochondrial proteome from the non-green apical part of cauliflower curds was investigated, the use of a 2D PAGE reference map of *Arabidopsis* cell culture mitochondrial proteome was particularly advisable. In some cases, we also used the map representing *Arabidopsis* mitochondrial proteome of green tissues [13].

Identities of protein spots are presented in Table S1; individual peptides for each protein spot are also listed in Table S4. All spots represented 91 non-redundant proteins within records that fit to the experimental data. Of this number, 69 non-redundant proteins with the highest probability of identification were found (bolded records in Table S1). Proteins were identified based mostly on high similarity to sequences of diverse cruciferous species. Various respiratory (e.g., ATP synthase, proteins for respiratory complexes (CII and CIV) biogenesis, mostly decreased in abundance), transporter (e.g., diverse voltage-dependent anion channel (VDAC) isoforms and dicarboxylate antiporters) and matrix proteins (ex. heat shock-proteins (HSPs), DNA-binding proteins, RNA editing and translation factors, mitochondrial thioredoxins, diverse multifunctional enzymes for amino acid, carbohydrate, lipid, and nucleotide metabolism, and some novel proteins) responded to drought. For instance, diversely affected in abundance spots between cauliflower cultivars (Section 2.2; Figure 2) included, inter alia, VDAC isoform 2, α/β hydrolase domain-containing protein 11, RNA editing factor 6, copper ion binding protein, mitochondrial elongation factor EF-Tu, single-stranded DNA-binding protein WHY2 (mitochondrial isoform X1), NADH-cytochrome b_5 reductase-like protein, mitoribosomal protein L21, malonyl-CoA-acyl carrier protein, SWIB/MDM2 domain superfamily protein, HSPs (e.g., HSP70-9) and a few uncharacterized proteins. Selected alterations in protein abundance are further discussed in Section 3.2.

Based on *Arabidopsis* protein orthologs, next we used the functional classification by the Munich Information Center for Protein Sequences (MIPS) at VirtualPlant 1.3 (Available online: http://virtualplant.bio.nyu.edu; [50]) for the clustering of drought-responsive proteins resolved on 2D gels into the functional categories (Table S5). The majority of proteins belonged to the class participating in various metabolic routes (ca. 44%). Next classes were represented by C compounds and carbohydrate metabolism (23.1%), amino acid metabolism (18.7%), cell rescue, defense and virulence (16.5%), energy conversions (12,1%) as well as in N and S metabolism proteins (7.7%). Participation of electron transport (7.7%), complex cofactor binding proteins (6.6%) and folding proteins (4.4%) in drought response were also distinctive (Table S5).

In the case of drought-tolerant "P" cultivar, some differences in the abundance of functional categories across proteins compared to the all-cultivar data were observed. C-compound and carbohydrate metabolism and energy conversion proteins contributed to a lesser degree (18.6 and 9.3%, respectively), however, N- and S-metabolism proteins, folding and stress-responding proteins contributed to a greater extent (9.3%, 7%, and 16.3%, respectively). Stress response proteins within cell rescue, defense, and virulence category were significantly enriched in drought-tolerant cultivar (16.3%) (Table S5).

Accordingly to the data in Tables S1 and S2, proteins within spots No. 23, 135, 164, 204, 331, 332, and 421 seem to be the best candidates for stress tolerance in cauliflower. These are mainly VDAC isoforms, ATP synthase subunit β (ATP2), At1g18480-like protein, WHY2 factor (isoform X1), OB-fold-like protein, and NADH-cytochrome b_5 reductase-like protein, as well as HSP70-9 (which notably increased in abundance in "P"). Thus, they represent quite broad protein classes (oxidative phosphorylation (OXPHOS) proteins, as well as proteins involved in metabolite exchange, mitochondrial DNA-binding proteins, some enzymes and chaperones). Additional candidate proteins for the drought tolerance, according to results of immunoassays (Sections 2.4 and 2.5) are listed in the final Conclusions section.

2.4. Abundance of Key Matrix Proteins, cyt. c and Components of Dissipating Energy Systems is Diversely Affected under Two Drought Levels across Cauliflower Cultivars

To assess the abundance of additional, drought-responsive mitochondrial proteins and to validate abundance of selected proteins identified in 2D spots, we performed Western immunoassays (Figure 3), using specific antisera dedicated against investigated proteins.

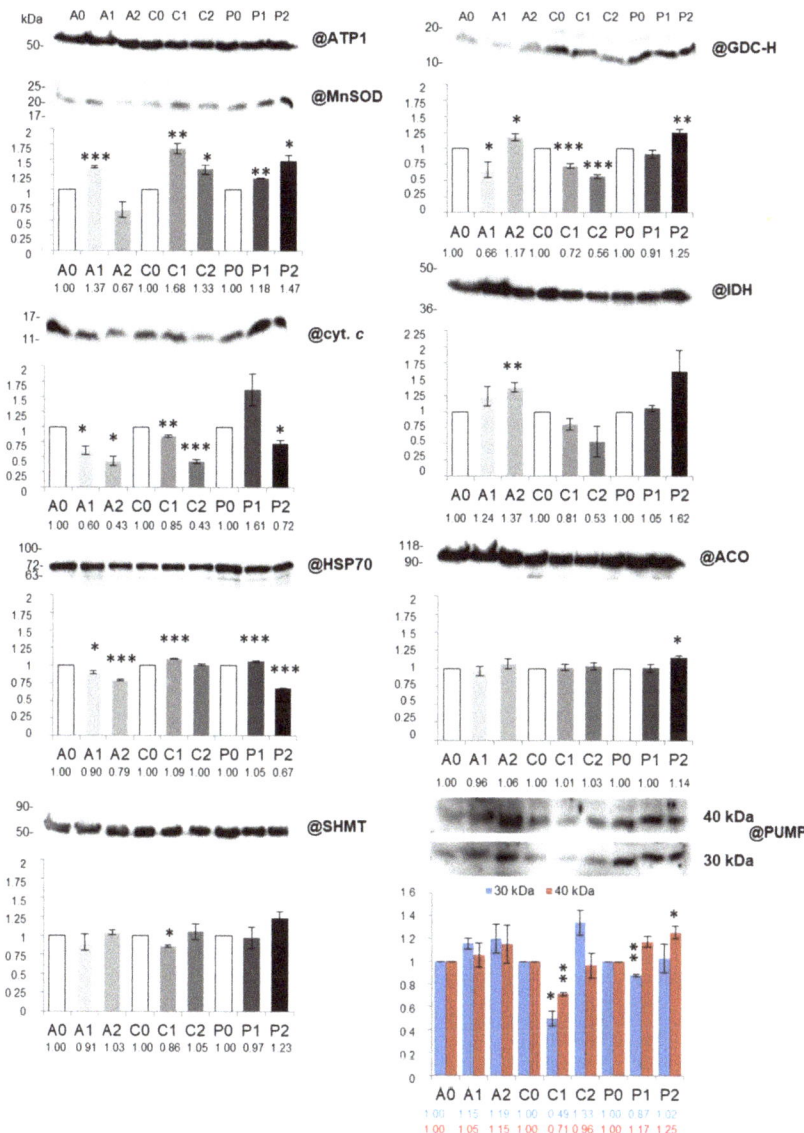

Figure 3. Abundance of Mn-superoxide dismutase (@MnSOD), cytochrome *c* (@cyt. *c*), heat shock protein 70 (@HSP70), serine hydroxymethyltransferase (@SHMT), glycine decarboxylase subunit H (@GDC-H), isocitrate dehydrogenase (@IDH), aconitase (@ACO) and 40 and 30kDa isoforms of uncoupling proteins (@PUMP) in mitochondria isolated from control grown "*Adelanto*", "*Casper*", and "*Pionier*" plants (A0, C0, P0), plants grown in moderate (A1, C1, P1) and severe water deficiency (A2, C2, P2, respectively). Results from representative SDS-polyacrylamide gel blots using listed antibodies (@) are shown. For loading control, antibody against mitochondrial ATP synthase subunit α (@ATP1) is used. For molecular mass calibration, PageRuler Prestained Protein Ladder (Thermo Scientific, Gdańsk, Poland) is applied. Protein molecular mass is indicated in kDa. All results are presented as mean values (±SE) from triplicate detection. Significant alterations are marked with asterisks: ***, $p < 0.001$, **, $p < 0.01$, *, $p < 0.05$ versus control values for each cultivar. Further data in the text.

The accumulation profile of those proteins varied depending on cultivar and stress intensity. Under mild drought, a significant decrease in glycine decarboxylase subunit H (GDC-H) abundance was visible only in "A" and "C" mitochondria, however, in severe stress, such a decrease was detected only in "C"; in other stress variants (especially in "P") GDC-H abundance increased. Strikingly, another important photorespiratory enzyme, serine hydroxymethyltransferase (SHMT), decreased in abundance in "C" under mild drought only, and remained stable in other experimental variants. Changes in GDC-H and SHMT abundance analyzed by immunoassays were not fully associated with accumulation of spots No. 158, 228, 230 and 241 (Figure 3; Table S1).

The abundance of mitochondrial heat shock protein HSP70 decreased in "A" and "P" (particularly in severe drought) and slightly increased in "C" under mild stress (Figure 3). Interestingly, changes in HSP70 abundance roughly agreed with 2D PAGE data (spot No. 421; Table S1). Accumulation of isocitrate dehydrogenase (IDH) increased only in severe drought in "A" (Figure 3); in contrast, spots No. 223 and 241 containing this protein under mild drought were decreased in abundance in the same cultivar (Table S1). Aconitase (ACO) abundance (the applied antibodies can cross-react with both ACO2 and ACO3 isoforms) was relatively stable, and it was increased only in "P" under severe treatment; ACO2 decrease in abundance in drought-sensitive cultivars was detected by 2D PAGE (spot No. 109; Table S1). The abundance of Mn superoxide dismutase (Mn-SOD) increased in almost all investigated stress conditions, except severe drought treatment in "A" (Figure 3). Accordingly, we noticed variations in the abundance of spot No. 4 containing a mitochondrial-like glutaredoxin in this cultivar (Table S1). The intramitochondrial pool of cytochrome c (cyt. c) was decreased in severe drought in all investigated cauliflower cultivars, but in mild stress, among drought-sensitive ones ("A" and "C"; Figure 3) only.

In addition, the abundance of selected components of energy dissipating systems was investigated. Antisera against potato detected two isoforms of plant-uncoupling mitochondrial proteins (PUMPs) of 30 and 45 kDa in cauliflower mitochondria. The abundance of both polypeptides representing isoforms of uncoupling proteins was decreased significantly in "C" and to a lesser extent in "P" under mild drought. However, in severe drought, we noticed visible increase in abundance of PUMP 40 kDa in "P" (Figure 3). We also characterized the alternative oxidase (AOX) abundance. Monoclonal antibodies raised against the *Sauromatum guttatum* enzyme cross-reacted with three polypeptides of 29–36 kDa in cauliflower mitochondria (Figure 4A). Polypeptides of 29 and 33 kDa were significantly decreased in their abundance in all investigated cauliflower cultivars; in "P", such a trend was less pronounced than among drought-sensitive cultivars. In contrast, a massive accumulation of AOX 36 kDa protein after severe drought treatment of "A" was noted. In general, variations in abundance of PUMP 40 kDa and AOX 36 kDa isoforms differentiate drought-sensitive and tolerant cauliflower cultivars.

2.5. Pattern of Dehydrin-Like Proteins (Dlps) in Cauliflower Mitochondria is Affected in Abundance by Drought

We also investigated the pattern of cauliflower mitochondrial dehydrin-like proteins (dlps) under mild and severe drought (Figure 5). According with our previous data [49], three independent dehydrin-specific antisera were used (Section 4.8).

Obtained results indicate the response in abundance of low-molecular weight dlps (18–80 kDa) in "A" and "C". Stressgen antibodies detected the most evident increase in abundance of 18 and 27 kDa dlps (and to a lesser extent, in "A" for 37 kDa protein) under two drought levels in "A" (mild drought) and "C" (severe treatment). Large-sized dlps (ca. 80 kDa) also markedly increased in abundance in drought-sensitive cultivars, although the highest increase in abundance of those proteins was noted under mild stress in "A" (Figure 5A,B).

Antisera against dehydrin SK_3-motif recognized smaller alterations, but there was a significant increase in abundance of ca. 30/35 kDa dlps under both drought conditions in "A" and under severe drought in "C" (Figure 5C). Alterations in the abundance of middle and large-sized dlps were less pronounced; the accumulation of dlps of 55–65 kDa slightly decreased under severe drought in "A". Similar results were observed for 30 and 40 kDa dlps in "P".

Overall, dlps in "P", contrary to "A" and "C" were relatively stable in abundance under the analyzed adverse conditions, and dlps of 18 and 27 kDa differentiate best the drought-sensitive from the drought-tolerant cauliflower cultivars.

Relationships between abundance of immunodetected proteins (Sections 2.4 and 2.5) were evaluated by correlation analysis (Table S3). We found strong positive correlation between IDH and GDC-H abundance, as well as between components of energy-dissipating systems (AOX 29 and 33 kDa isoforms, and in addition, AOX 33 kDa and PUMP 40 kDa isoforms). Positive correlations involved also dlps of various sizes, confirming their participation in drought response. On the contrary, accumulation of relatively stable IDH negatively correlated with drought-affected HSP70. Similarly, accumulation of PUMP 30 kDa isoform negatively correlated with and cyt. c, and dlp 18 kDa accumulation with GDC-H and PUMP 40 kDa isoform abundance. Negative correlations involved also AOX 33 kDa isoform (not induced by drought in our study) and dlp 80 kDa (Table S3).

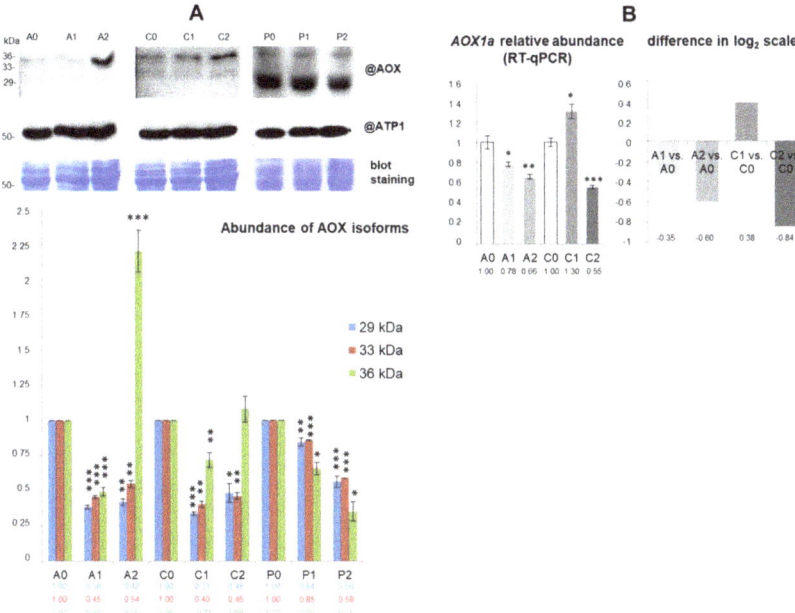

Figure 4. AOX protein and mRNA abundance in mitochondria isolated from control grown "*Adelanto*", "*Casper*", and "*Pionier*" plants (A0, C0, P0), plants grown in moderate (A1, C1, P1) and severe water deficiency (A2, C2, P2, respectively). (**A**) Analysis of detected AOX polypeptides (29, 33, 36 kDa) on representative SDS-polyacrylamide gel blots using respective antibodies (@AOX). For loading control, antibody against mitochondrial ATP synthase subunit α (@ATP1) is used. Equal protein loading is also shown by Coomassie Brilliant Blue (CBB) staining of Western blots. For molecular mass calibration, PageRuler Prestained Protein Ladder (Thermo Scientific, Gdańsk, Poland) is applied. Protein molecular mass is indicated in kDa. (**B**) Relative abundance of *AOX1a* by reverse transcription quantitative PCR (RT-qPCR). Graph *at the left*, relative abundance normalized to average level (mean log expression = 1). Graph *at the right*, differences in $\log_2$ scale between treated and control variants. For normalization, actin1 (*ACT1*) is used. Results of analyses are presented as mean values (±SE) from triplicate detection. Significant alterations in (**A**) and (**B**) are marked with asterisks: ***, $p < 0.001$, **, $p < 0.01$, *, $p < 0.05$ versus control values for each cultivar. Further data found in the text.

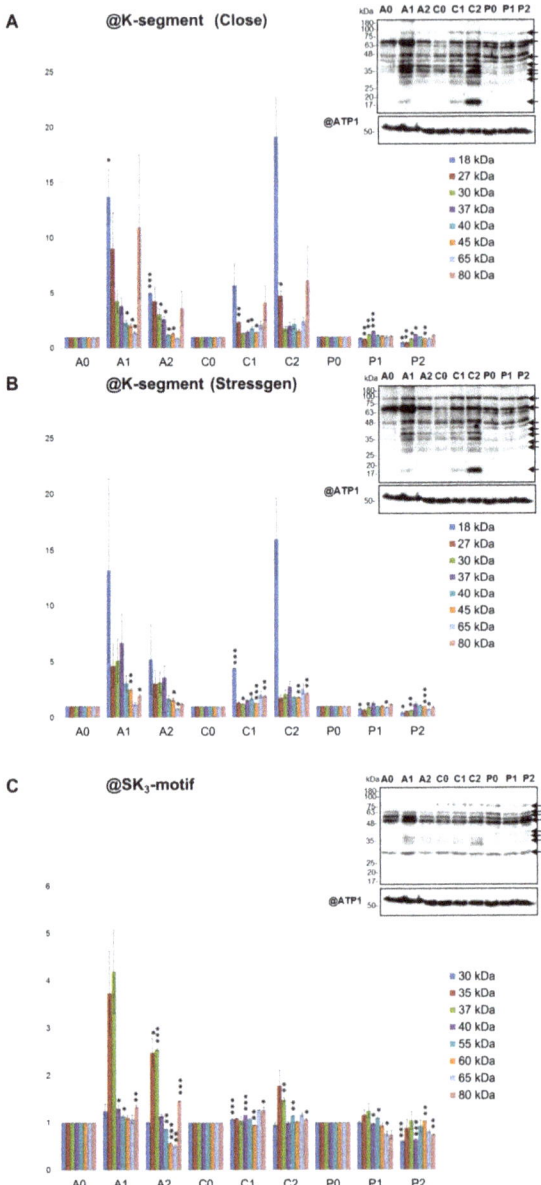

Figure 5. (pp. 11 and 12) Dlps abundance in mitochondria from control grown "*Adelanto*", "*Casper*" and "*Pionier*" plants (A0, C0, P1), plants grown in moderate (A1, C1, P1) and severe water deficiency (A2, C2, P2, respectively). Results of analysis of dlps of various size (in kDa) on representative SDS-polyacrylamide gel blots with antibodies directed against (**A**) dehydrin K-segment from Close [51] (@K-segment (Close)) or from (**B**) Stressgen (@K-segment (Stressgen)), or (**C**) antibodies recognizing dehydrin SK_3 motif (@SK_3) are shown. For loading control, antibody against mitochondrial ATP synthase subunit α (@ATP1) is used. For molecular mass calibration, PageRuler Prestained Protein Ladder (Thermo Scientific, Gdańsk, Poland) is applied. The protein molecular mass is indicated in kDa. Investigated dlps are marked by arrows on gel blots. Significant alterations from triplicate detection in (**A**), (**B**) and (**C**) are marked with asterisks: ***, $p < 0.001$, **, $p < 0.01$, *, $p < 0.05$ versus control values for each cultivar. Further data in the text.

Int. J. Mol. Sci. **2018**, *19*, 1130

2.6. Identification of Drought-Responsive Spots Containing Putative Dehydrin-Like Proteins

The most notable changes in the accumulation of dlps were noticed in "*A*" and "*C*" cultivars under all analyzed drought conditions (Section 2.5). To further characterize the drought-responsive cauliflower dlps, we prepared 2D blots with separated whole mitochondrial proteins from "*A*" and "*C*" cultivars submitted to mild and severe drought, and we used such blots to immunodetect dlps (within the particular protein spots) by antisera against dehydrin K-segment (Section 4.8). Results of the dlp immunodetection on 2D spots are shown in Figure 6.

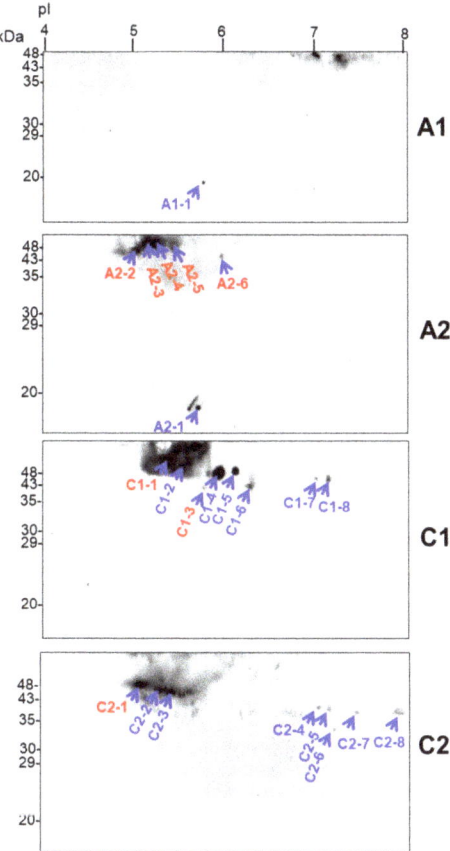

Figure 6. Representative pattern of dlps of "*Adelanto*" and "*Casper*" plants grown in moderate (A1, C1) and severe drought (A2, C2, respectively) immunodetected with antibodies directed against dehydrin K-segment [52] on two dimensional (2D) blots. Spots referring to detected proteins that were cut out from the respective 2D gels for protein identification by liquid chromatography-tandem mass spectrometry (LC-MS/MS, *blue arrows*) also appear in Figure 1 (denoted *in blue* and *red*). Proposed identities for those spots and all tryptic peptide data are indicated in Table S6. Identifiers of spots containing tryptic peptides specific to dehydrins are marked *in red* (remaining labels are shown *in blue*). For molecular mass calibration (kDa) of protein spots, PageRuler Prestained Protein Ladder (Thermo Scientific, Gdańsk, Poland) and Low Molecular Weight (LMW)-SDS Marker Kit (GE Healthcare Poland, Warsaw, Poland) are used. For calibration of spot isoelectric point (pI), Broad pI Kit (GE Healthcare) is applied. Further data found in the text.

Due to the assay sensitivity, we were able to detect several spots per investigated cultivar/treatment within pI range of ca. 5–8 and molecular weight of 18–48 kDa (35–48 kDa mostly) with varying abundance under the water deficit (Section 2.5). However, the only single spot was immunodetected in "A" cultivar under mild drought. Certain spots (C1-7 and C1-8, C2-4 to C2-6) migrated in more neutral pI values, whereas the others (C2-7 and C2-8) represented basic proteins. Based on immunoassay results, we cut out all immunodetected protein spots from 2D gels, and proteins were identified by LC-MS/MS. Position of all those spots from Western blots are superimposed on the spot pattern within the master gel image (Figure 1), and protein identities within individual spots are shown in Table S6.

Notably, only selected spots contained dehydrin-related tryptic peptides that showed high similarity of their protein sequences to the selected *Brassica* dehydrins. These were: five spots (A2-2 to A2-6) in "A" cultivar under severe drought, two spots (C1-1 and C1-3) in "C" under mild stress, and the single spot (C2-1, all indicated in red on Figure 6) in the latter cultivar under severe water deficit. Detected tryptic peptides were highly similar to early response dehydrins (ERD14 and ERD14-like) from various *Brassica* species, particularly *B. oleracea* var. *oleracea* and *B. rapa* (GenBank accession.version identifiers XP_013592580.1 and XP_009128158.1, respectively; Table S6). However, analyzed protein spots also contained highly abundant mitochondrial proteins (listed in Table S6). Finally, we focused on the comparison of selected proteomic and transcriptomic responses important for cauliflower mitochondrial biogenesis in drought, including abundance profiling of mRNA coding for the identified dehydrins.

2.7. mRNA Abundance and Coordination of Mitochondrial Biogenesis in Drought

We examined the abundance of selected nuclear transcripts coding the investigated drought-responsive proteins. RT-qPCR assays were carried out with application of primers specific for the dedicated cDNA fragments (Table S7). As an internal calibrator of gene expression, cauliflower actin1 (*ACT1*) mRNA was applied (the partial sequence was previously cloned and deposited in GenBank under accession.version KC631780.1; [13]). The cauliflower *AOX1a* partial sequence was previously cloned and deposited in GenBank (accession.version KC631778.1; [13]). Accumulation of *AOX1a* mRNAs was significantly increased under mild drought in "C"; on the contrary, they were strongly decreased in abundance under mild and severe drought in "A" and severe drought in "C" (Figure 4B).

We studied the abundance of *PRODH* and *P5CDH* transcripts coding for important Pro catabolism enzymes (proline dehydrogenase and Δ-1-pyrroline-5-carboxylate dehydrogenase, respectively). Generally, the abundance of *PRODH* mRNA decreased in "C" in both drought treatments; the most severe impact was visible under mild drought. In "A" cultivar, a significant decrease of *PRODH* mRNA abundance was noted only in mild stress. Contrasting with this, *P5CDH* transcript accumulation was only markedly elevated in "C" under severe drought (Figure 7).

The abundance of mRNA coding for ERD14 and ERD14-like dehydrins (Section 2.6) was also determined. The highest increase in both transcript abundance was noticed in "A" cultivar in mild and severe drought, which is in line with the induction of *ERD14* gene expression within the short dehydration period, as well as with the presence of dehydrin-related tryptic peptides in analyzed protein spots. Notably, *ERD14* and *ERD14 like* transcripts were significantly decreased in abundance in "C" under severe water deficit (Figure 7).

Finally, we investigated the accumulation of mRNAs coding for five selected transcription factors (CBF1a, CBF1b, CBF2, CBF4) related to stress response. *CBF1a* and *CBF2* transcripts were increased in abundance in "A", but they declined in "C" cultivar, as drought progressed. Strikingly, *CBF1b* mRNA showed substantial increase in abundance only in "C" in mild and severe drought treatments. Increase in abundance of *CBF4* mRNAs was also very well noted under severe water deficiency in "C" (Figure 7).

Correlation analysis revealed no significant relationship between accumulation pattern of investigated transcripts (Table S3).

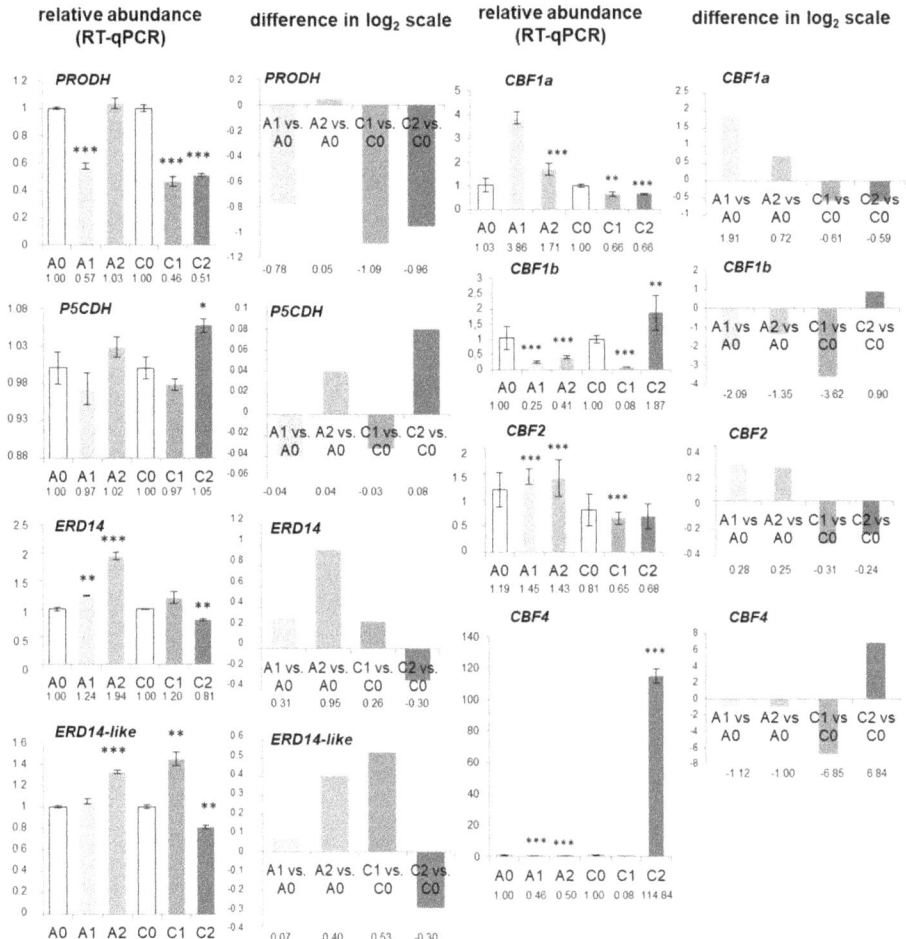

Figure 7. Relative abundance of transcripts (by reverse transcription quantitative PCR (RT-qPCR)) for cauliflower proline dehydrogenase (*PRODH*), Δ-1-pyrroline-5-carboxylate dehydrogenase (*P5CDH*), ERD14 dehydrin (*ERD14*), ERD14-like dehydrin (*ERD14-like*), as well as transcripts for transcription factors (*CBF1a*, *CBF1b*, *CBF2*, *CBF4*) in control grown "*Adelanto*", "*Casper*", and "*Pionier*" plants (A0, C0, P0), plants grown in moderate (A1, C1, P1) and severe drought (A2, C2, P2, respectively). Differences in log$_2$ scale between treated and control variants are also shown. In all cases, for normalization, actin1 (*ACT1*) is used. Significant alterations are marked with asterisks: ***, $p < 0.001$, **, $p < 0.01$, *, $p < 0.05$ versus control values for each cultivar. Further data found in the text.

3. Discussion

3.1. Physiological Response of Cauliflower Cultivars under Mild and Severe Drought

We studied leaf respiratory responses within three distinct cauliflower cultivars with varying drought tolerance. Previously, we showed that the respiratory rate exceeded the photosynthetic one in cauliflower leaves under severe (but not moderate) water deficiency [52]. This emphasizes the importance of adequate respiratory adaptations in this species under the mentioned unfavorable conditions. Respiration and photorespiration become a part of the complex network response under water deficiency, and photorespiration participates in oxidative damage avoidance while optimizing photosynthesis [53]. We noticed enhancement of respiration among drought-sensitive cultivars (contrary to "P") under mild drought; the respiration rate was often decreased in severe treatment. Interestingly, respiratory decline in such conditions is a well-known phenomenon (also for *Brassica* species) and R_n rate is affected much under fast drying [19,30,54]. Mitochondrial R_d can be also inhibited by drought; it markedly increases in prolonged drought but declines under short water deficit [55–57]. In our case, the effect depended on the cultivar. Generally, drought-sensitive crop cultivars exhibit larger R_T decreases, than sensitive ones; nevertheless, both effects could be reversed under drought recovery [58].

The increase of photorespiratory to gross CO_2 assimilation ratio under drought is often required in order to protect the photosynthetic machinery against photoinhibition. In field-grown *Gossypium hirsutum*, drought resulted in affected stomatal conductance and elevated respiratory and PhR rates, while photosynthetic electron transport was not affected [40,55,59]. The progressive alteration of PhR rate in "A" and "C" along the stress duration is in line with Liu et al. [60] data, suggesting that the PhR in drought-sensitive cultivars cannot be a major energy dissipation strategy, as was reported for some Asiatic and Mediterranean-originated plant species. In contrast to that, the visible increase of PhR rate in the mild treatment in "P" is known among drought-tolerant species [61]. Notably, the general trend in PhR response could be reversed at the early vegetative stage in some species with varying drought tolerance [62]. In drought-tolerant and drought-sensitive cultivars of *Malus domestica*, even moderate drought resulted in major PhR decline [63].

Overall, observed alterations in the respiratory parameters coincide with the level of drought tolerance among investigated cauliflower cultivars and suggest distinct regulation of drought physiological responses in "P".

3.2. Mitochondrial Response to Drought Involves Diverse Multifunctional OXPHOS, Transporter and Matrix Proteins in Various Cauliflower Cultivars

We also investigated drought-resulted alterations within the cauliflower mitochondrial proteome. Since the experimental molecular mass of protein spots corresponded roughly to the theoretical one, we are rather convinced that investigated stress conditions do not result in excessive proteolysis [19]. Proteins identified in some double spots (e.g., No. 228, 230) showed similar responses, which is in favor of the correctness of their assignments (Table S1). We applied functional classification for the clustering of drought-responsive proteins (Table S5). Some functional classes are often underrepresented, highlighting the relevance of organelle-specific studies; on the other hand, key stress-related enzymes (for carbohydrate and amino acid metabolism) are often overrepresented in drought response [7,45,64,65], whereas proteins related to protein folding and degradation may decline in abundance [30].

Energy and carbohydrate metabolism proteins play a significant role in drought response [44,45]. Malate dehydrogenase (MDH), succinyl-CoA ligase subunit β and ACO2 (spots No. 35, 59 and 109) were decreased in abundance in drought-sensitive cauliflower cultivars (the present study), in roots of drought-sensitive rapeseed (*Brassica napus*) cultivar [35], as well as in other reports elucidating the impact of the extended water deficit on the plant proteome [46,48], but contrary to the Ndimba et al. [18] study on the sorbitol-induced drought. Variations of MDH accumulation are linked to increased NADH demands [38]. ACO2 isoform is predominantly localized in plant mitochondria, and ACO-containing complexes were shown to be unstable in stress [13,66].

Decrease in abundance of subunits of cauliflower OXPHOS complexes (e.g., CI and CII) was similar to some studies [19,67]. Massive decrease in accumulation of CII subunit 1 and 5 (SDH1 and SDH5; spots No. 4, 23, 204; Table S1) differed from barley (*Hordeum vulgare*) drought response pattern [47]; CII subunits were increased in abundance after drought recovery in *Populus euphratica* [2]. Contrary to other reports [31,44], CIV subunits were unaffected in our study. ATP synthase was one of the expected complexes that appeared decreased in abundance in cauliflower mitochondria [13]. Its assembly may be affected as diverse cellular energy demands rise under stress [68–70]. According to our data (spots No. 135, 166), a decrease in abundance of ATP synthase subunits under water deficit was reported by a number of studies [16,35,39], contrary to other ones [43]. Notably, ATP synthase subunit 24 kDa was increased in abundance in "C" cultivar under severe drought (spot No. 317; Table S1).

Water deficit results in decrease in abundance of pyruvate dehydrogenase (PDH) subunits [19]. However, alterations in abundance of *Hippophae rhamnoides* PDH E1 subunit α in drought [65] contrasts with our data (decreased in abundance spots No. 223, 241 for "A" and "C"; Table S1). To enhance pyruvate import to mitochondria, *NRGA1* coding for a mitochondrial pyruvate carrier is often co-expressed with other carrier genes [71,72]. Genes for mitochondrial dicarboxylate transporters may be induced in water deficit [67]. In our study, PDH decrease in abundance was not associated with co-regulation of any of the specific substrate carriers among proteins affected in abundance under drought. The accumulation of dicarboxylate/tricarboxylate mitochondrial transporters declined in drought-sensitive cultivars instead (spots No. 202, 204; Table S1).

A decrease in abundance of mitochondrial processing peptidase subunit β (MPPβ; spot No. 166) in "A" (mild stress) and "P" (all treatments) was noted, in addition to translocase of the inner mitochondrial membrane subunit (TIM44-2-like; spot No. 230) in "A" (Table S1). Regulation of genes coding for MPP subunits in stress has been already reported [14,18,40,46,67]. It is nonetheless known that drought may alter protein import into plant mitochondria [73]. Results of our study indicate some perturbations in the general import pathway.

Some spots (No. 23 (VDAC2-like and adenosine nucleotide translocator protein) and 31 (hydrolase domain-containing protein and carbonic anhydrase)) showed inconsistent alterations among drought-sensitive cultivars [68]. Distinct VDAC isoforms which increased in abundance in *Brassica rapa* under prolonged drought and in drought-tolerant wheat cultivars [32,41,48] contrasted with the decreased accumulation of VDAC isoforms in drought-sensitive cauliflower cultivars (spots No. 23, 202-204; Table S1).

In general, distinct alterations within the mitochondrial proteome are potentially associated with drought tolerance (Section 2.3; [45]). Such alterations encompass mitochondrial DNA/RNA-binding proteins, chaperonins, heat-shock proteins, as well as a number of enzymes for mitochondrial metabolism. Interestingly, we did not notice any biases towards protein spots increased in abundance in drought-tolerant cultivar, contrary to Mohammadi et al. [35], however, we observed some differences in the distribution of functional categories across responsive proteins (Section 2.3; Table S5). ssDNA-binding proteins, as well as proteins related to RNA metabolism and translation (e.g., RNA editing factors 1 and 6, and mitochondrial EF-Tu), were diversely affected in abundance between drought-sensitive and tolerant cauliflower cultivars (spots No. 59, 61, 218, 331, 332). Also, the abundance of SWIB/MDM2 domain superfamily protein, calcineurin-like metallophosphoesterase superfamily, sucrose/ferredoxin-like proteins, mitoribosomal protein L21, chaperonin and HSP70 isoforms differentiated cultivars with diverse drought tolerance (spots No. 153, 164, 349, 420, and 421). A prevalent decrease in abundance after mild drought treatment and more specific alterations in protein spot accumulation in severe conditions likely represent specific adaptations in the cauliflower mitochondrial proteome to diverse drought conditions. However, functional implications of the observed proteomic alterations require further exploration.

3.3. Diverse Variations in Abundance of Matrix Proteins, cyt. c, Components of Dissipating Energy Systems and Dehydrin-Like Proteins Across Drought Treatments/Cultivars

In order to obtain a more complete view on the abundance of additional, drought-affected proteins we extended results of 2D PAGE analysis by Western immunoassays.

GDC and SHMT belong to important photorespiratory enzymes, often regulated in stress conditions (including drought) [15,19,30,40,45,68]. Ford et al. [41] reported contrasting regulation of GDC, as well as SHMT, which depended on drought tolerance. In our study, changes in GDC-H abundance roughly corresponded to the decline in the photorespiration rate only in drought-sensitive cultivars under both mild and severe drought (Section 2.1; Figure S1). SHMT abundance did not correlate with GDC-H alterations (Figure 3) and remained relatively stable. In contrast, HSP70 abundance alterations were not associated with drought tolerance (Figure 3). In such conditions, an evident HSP70 increase in abundance was noted in pea mitochondria [19]. The HSP70 and IDH abundance pattern is in line with typical variations of those proteins in stress [7,18,41]. Diverse variations in the abundance of ACO isoforms may be a part of the adaptive response of Krebs cycle, due to the altered NADH and carbon skeleton demands in drought. In some cases, however, enzymes of Krebs cycle are massively decreased in abundance in water deficiency [74].

Mn-SOD displays distinct expression pattern between cultivars with variable drought sensitivity, often by the increase and decrease in abundance in drought and drought recovery, respectively [14,30,32,41,44,75]. Cauliflower Mn-SOD increased in abundance in most of all investigated treatments and cultivars; the maximal increase was observed among drought-sensitive cultivars (Figure 3) and was accompanied by variations in abundance of spot No. 4 containing mitochondrial-like glutaredoxin in "C" (Table S1), suggesting that redox regulation likely accompanies drought response [2].

According to our results, cyt. *c* can be released from cauliflower mitochondria even under the first level of water deficiency. Such phenomenon often accompanies programmed cell death (PCD), which indeed was suggested by us in temperature stress response [13]. However, in some plant species (e.g., in pea mitochondria), drought did not result in PCD appearance [19].

Plant mitochondrial energy dissipating systems including AOX, PUMPs, and membrane channels functionally modulate drought response by influencing key signaling pathways [8]. We used antisera that were able to immunodetect respective cauliflower PUMP isoforms. An observed decrease in abundance prevailed the pattern, and only in severe drought was the increase in PUMP 40 kDa abundance in "P" noted (Figure 3). Also, the abundance of selected PUMP isoforms in pea mitochondria was increased in drought [19]. To extend our proteomic data on further components of energy dissipating systems, we also investigated AOX protein accumulation. This enzyme controls respiration rate and photosynthetic efficiency in drought, and often increases in abundance during severe water deficit and re-watering [22,76,77]. Three polypeptides of 29–36 kDa in cauliflower mitochondria [13] likely represent AOX isoforms resulting from the expression of *AOX* gene family and diverse posttranslational modifications of this protein (Figure 4A). Our results contrast with the data of Taylor et al. [19], who showed drought-affected induction of similarly sized 31 kDa pea isoform. However, abundance of AOX polypeptides could be substantially increased among drought-sensitive cultivars.

Under progressing drought, abundance of late embryogenesis abundant (LEA) proteins preserving the stability of membrane proteins and adjusting intracellular osmotic pressure often increases [32,78]. Some LEA proteins, including dlps, were found to be mitochondrially-localized in a number of crop species [79–82]. Our previous report showed that accumulation of mitochondrial dlps was altered in abundance under temperature stress and after stress recovery [49]. We extended those analyses to three cultivars of cauliflower and to mild and severe drought treatments (Figure 5) using three independent antisera. Broadly-sized dlps responded in abundance as a part of mitochondrial adaptations to water deficiency, because such a response was visible only in drought-sensitive cultivars. The stable dlp accumulation in "P" mitochondria agrees with drought tolerance of this cultivar and with the relatively lower increase in dehydrin mRNA abundance (e.g., *DHN8*) under progressed drought among highly adapted plants [36]. Alterations in dlps abundance, as well as the induction pattern of other dlps in drought (Figure 5), substantially extend the data on mitochondrial proteins related to dehydrins in *Brassicaceae*.

Owing to results on the massive increase of dlps abundance under investigated drought treatments in stress-sensitive cultivars, we identified putative dlps involved in response to water deficit in those cultivars by tandem MS. Protein spots immunopositive with dehydrin K-segment-specific antibodies were selected, and several of them contained tryptic peptides with high sequence similarity to selected *Brassica* dehydrins (Figure 6, Table S6). Due to the fact that highly abundant proteins in spots would hamper identification of dlps, we did not employ protein microsequencing for identification of chosen dlps, and subsequent amplification and cloning full-length cDNAs. We were unable to determine the N-terminal sequence for the given dlps. Notably, depletion of any abundant proteins would enhance the risk of sample cross-contamination.

3.4. Some Transcriptomic Responses and the Coordination of the Mitochondrial Biogenesis in Drought

We analyzed transcriptomic responses to drought connected with the profiling of the abundance of some nuclear transcripts. Since observed proteomic alterations (particularly those connected with dlps abundance) were especially distinctive among drought-sensitive cultivars, we focused on such experimental variants only for this part of our study.

The choice of *AOX1a* mRNAs for our study can be justified by the fact that accumulation of *AOX1a* transcripts is often affected by diverse stress stimuli [83]; other members of the AOX family, (e.g., *AOX1d*), can be increased in abundance in drought, as well [31]. In our study, the most severe drought, the most intense decrease of *AOX1a* accumulation in "A" was observed. A major imbalance between protein and mRNA abundance in "C" and "A" cultivars was noticed. In "A", two immunoreactive AOX protein bands decreased in abundance under mild and severe stress (which is in line with *AOX1a* transcript accumulation pattern in the same conditions), whereas the third band (36 kDa) notably increased in abundance in the severe treatment (Section 2.4; Figure 4). In this cultivar, under severe drought, the lowered amount of *AOX1a* mRNA contrasted with enhanced accumulation of 36 kDa AOX isoform. In "C", no significant increase in abundance of AOX protein was visible, contrary to the notable increase in abundance of *AOX1a* transcripts under mild drought. Thus, *AOX1a* mRNA compensatory increase in abundance was accompanied by the decrease in protein abundance and vice versa. Alterations of *AOX1a* mRNAs to drought involved only selected cauliflower cultivars and stress conditions (Figure 4B). Lack of coordination between AOX protein and mRNA may result from synthesis of investigated immunoreactive polypeptides from transcripts coding distinct AOX isoforms (especially 33 kDa protein). Regulation of *AOX1a* mature mRNA abundance may depend on changes in selection of transcripts for translation, diverse mRNA/ribosome interactions and/or affected protein synthesis [7]. The decreased pool of those transcripts in some investigated variants suggests their efficient translation leading to protein excess and thus those mRNAs seems to be fully translatable.

Participation of non-coding RNAs in the regulation of the abundance of investigated transcripts or their translational efficiency should be also considered. Growing evidence supports the presence of miRNAs (nuclear-encoded and processed before import to mitochondria) or components of miRNA biogenesis machinery within mitochondria [84–87]. We focused on in silico miRNA candidates that may putatively target *B. oleracea AOX1a* by psRNATarget (Available online: http://plantgrn.noble.org/psRNATarget/analysis?function=2; [88]) search (Table S8). Arabidopsis and Brassica miRNAs were taken into account, because of the relative high similarity of *AOX1a* nucleotide sequence (GenBank accession.version KC631778.1) between those genera. According to our data, most of the predicted miRNAs resulted rather in messenger degradation, than in affected translation (Table S8). This does not obviously exclude the possibility that multiple non-coding RNAs could influence *AOX1a* mRNA and protein abundance.

Proline, the potent osmoprotectant, over-accumulates under drought and decreases under drought recovery [7,29,30,89]. Its variations may not simply correspond to the level of drought adaptation [36]. Previously, we showed that cauliflower P5CDH protein increased in abundance both in heat and heat recovery [13]. We postulate the reciprocal regulation of the abundance of *ProDH* as well as *P5CDH* transcripts under severe drought in "C". Decrease in abundance of *ProDH* mRNA was not equal between investigated cultivars, indicating diverse osmolyte accumulation in those cultivars [37].

The respective proteomic responses related with dlps profile were very evident for the drought-sensitive cultivars (Section 2.5). Henceforth, we determined the abundance of transcripts coding ERD14 and ERD14-like dehydrins, because tryptic peptides with high sequence similarity to those proteins were found in protein spots on 2D blots that were immunoreactive with dehydrin K-segment antisera (Section 2.6; Figure 6, Table S6). *ERD* transcripts (including *ERD14*) are known to be highly abundant under ABA, salinity, cold, and drought [64,90,91]. In general, our results suggest that positive transcriptomic response progressing with water deficit coincides with strong increase in abundance of dlps in "A" (drought-sensitive) cultivar (Figure 5).

Finally, to obtain a more complex view on cauliflower drought responses, we examined the abundance of transcripts coding for several nuclear transcription factors (TFs), which did not respond in protein abundance in our study, belonging to C-repeat/dehydration-responsive element binding (CBF/DREB) subfamily (from ETHYLENE RESPONSE FACTOR (ERF) family). These proteins contain conserved DNA-binding domains, and are regulated by a number of stressors, including cold and drought [92]. Notably, DREB subfamily contains at least 91 known members in *B. oleracea*; the whole ERF family and DREB A-4 subgroup are particularly enriched comparing with *Arabidopsis* data. *B. oleracea CBF* genes displayed various expression patterns. In our study, *CBF1a* expression showed contrasting responses between two drought-sensitive cultivars; the most visible increase in abundance was noted in severe drought for "A" (*CBF1a*) and "C" (*CBF1b*). Results obtained for *CBF1b*, and particularly, for *CBF4* transcripts, roughly agree with the late induction pattern in *B. oleracea* [93]. *CBF4* and *DREB1* regulons were suggested to participate in drought adaptation or enhance drought tolerance, but participation of some other CBFs in this process is still controversial [94,95]. Therefore, our results extend such findings by showing alterations in cauliflower *CBF1a*, *CBF1b*, and *CBF2* mRNA abundance under drought (the latter displayed similar expression profile to *CBF1a*). Rapid responses of *CBF1a*, *CBF2*, and *CBF4* genes to mild drought suggest that they may participate in positive drought signaling in "A" and "C", respectively.

4. Materials and Methods

4.1. Growth of Plant Material and Stress Application

Seeds of three analyzed cauliflower (*Brassica oleracea* var. *botrytis* subvar. *cauliflora* DC) cultivars ("Adelanto", "Casper" and "Pionier") were obtained from Bejo Zaden (Warmenhuizen, The Netherlands). Cauliflower seedlings were produced in 0.09 dm^3 pots filled with peat substrate (Kronen-Clasmann, Gryfice, Poland). Seedlings with 3–4 leaves were transferred to 5 dm^3 containers. Plants were grown for three months in cultivation chambers at a local breeding station (Poznan University of Life Sciences, Poland) at 23/19 °C (day/night) and 70% relative humidity under photon flux density 200 µmol·m^{-2}s^{-1} (16 h of light/8 h of dark). Stress conditions were applied to plants with developing curds up to a curd diameter of 7–10 cm. The water capacity of the substrate was 40% (v/v) and under drought stress, the water content decreased to 22% (v/v) (mild drought, the first level of drought) and to 15% (v/v) (severe drought, the second level of drought). After the occurrence of the assumed level of drought stress, plants were irrigated to 40% (v/v) and then drought treatment was applied again and repeated for ca. 2–3 weeks. Curds were harvested immediately after stress treatment cessation. Duration of the drought stress was estimated on the basis of relative water content (RWC) in the soil and in plant leaves and curds. RWC in developed, mature cauliflower leaves in mild drought was achieved at RWC of 94%, 92%, and 95% for "A", "C", and "P" cultivars, respectively. Under severe drought, RWC values lasted 69%, 74%, and 73% for the mentioned three cultivars, respectively. RWC of cauliflower leaves under severe drought referred to the third and fourth day of the drought treatment.

4.2. Physiological Analyses

Physiological analyses were conducted on well-developed cauliflower leaves with an LCpro+ infrared gas analyzer (ADC BioScientific Ltd., Hoddesdon, UK). To obtain more reliable results, extra experimental replicas ($n = 8$) were used. R_d rate was determined according to the Laisk [96] method. CO_2 assimilation rate representing R_T rate was recorded during intercellular CO_2 concentration (C_i) decreased to 0 ppm at 22 °C and 50% relative humidity. For each value of photosynthetic photon flux density (PPFD) at 200, 400, and 600 $\mu mol \cdot m^{-2} s^{-1}$, the linear regression of CO_2 assimilation (A) versus C_i was calculated. The intersection of three regression lines was determined by minimizing the sum of squares of errors between the measured values and calculated for each PPFD level, while minimizing the standard deviation for the intersection. PhR rate was determined as the difference between R_T and R_d values at C_i 0 ppm (the latter one expressed as a given CO_2 evolution rate at the crossing point of all A/C_i curves). R_n rate was measured after 30 min of adaptation to the dark. The applied drought treatments were necessary for visible alterations in respiratory and photorespiratory rates (Figure S1). The given drought exposures (Section 3.1) were necessary to observe proteomic and physiological alterations.

4.3. Isolation of Mitochondria, Purity Assays, and Protein Determination

Mitochondria from the topmost 5 mm-thick layer of cauliflower curds were extracted by differential centrifugation and purified in Percoll gradients according to Pawlowski et al. [97]. During isolation, the Complete Mini EDTA-free Protease Inhibitor Cocktail (Merck Poland, Warsaw, Poland) was used. Purity assays of isolated mitochondria were conducted according to previous reports [49,97]. Protein content was determined using a Bio-Rad Protein Assay (Bio-Rad Poland, Warsaw, Poland), with bovine serum albumin as a standard curve calibrator. The efficiency of organellar preparations (proteins per 100 g of cauliflower curds) varied from 0.9–3.5 mg for "C" and "P" and 0.1–1.9 mg for "A" cultivar.

4.4. Sample Preparation for the Two Dimensional Isoelectric Focusing/SDS Polyacrylamide Gel Electrophoresis (2D IEF/SDS-PAGE)

Proteins were extracted and precipitated overnight at −20 °C in a 10% solution of trichloroacetic acid in acetone containing 20 mM dithiothreitol (DTT) by the method of Staszak and Pawłowski [98]. After centrifugation (16,000× g for 5 min at 4 °C), resulting pellets were washed three times with 1 mL of acetone supplemented with 20 mM DTT. Samples were re-centrifuged after each washing, and resulting pellets were vacuum dried and then resuspended in lysis buffer (7 M urea, 2 M thiourea, 2% CHAPS, 1.5% DTT, 0.5% IPG buffer pH 4–7), and supplemented with Protease Inhibitor Cocktail (Roche, Basel, Switzerland) according to the manufacturer's suggestions. Protein concentration in processed samples was determined using the Bradford assay [99].

4.5. 2D IEF/SDS-PAGE

All analyses were conducted at 25 °C with at least three biological replicas. Proteins (500 µg for CBB staining) were first separated according to their charge on rehydrated Immobiline DryStrip Gels (24 cm in length, containing linear gradient of pH 3–10) with rehydration buffer (6 M urea, 2 M thiourea, 2% CHAPS, 20 mM DTT, and 0.5% Pharmalyte, pH 4–7) on an Ettan IPGphor 3 IEF System (GE Healthcare, Uppsala, Sweden). The program for isoelectric focusing was applied according to the manufacturer's suggestions. Strips were either stored at −80°C or they were directly treated for 10 min with equilibration solution I (6 M urea, 1.5 M Tris-HCl, pH 8.8, 30% glycerol, 2% SDS, and 1% DTT) and for the same time with equilibration in solution II (solution I supplemented with 2.5% 2-iodoacetamide, without DTT) and subjected to a second dimension run (SDS-PAGE).

For SDS-PAGE, Ettan DALT 12.5% Precast Polyacrylamide Gels and an Ettan DALTsix Electrophoresis System (both from GE Healthcare) were used. Conditions for the run were as follows:

1 h at 80 V and 5 h at 500 V. Broad pI Kit, pH 3–10 (GE Healthcare) for protein spot pI calibration within 3.5–9.3, as well as PageRuler Prestained Protein Ladder (Thermo Scientific, Gdańsk, Poland) and LMW-SDS Marker Kit (GE Healthcare) for protein spot MW calibration were used. Resolved proteins were stained with colloidal CBB, which, in addition to visualization and quantification, also allowed for downstream MS analysis [100].

4.6. Proteome Analysis

Gels were scanned and evaluated using ImageMaster 2D Platinum v7.0 software (GE Healthcare). After spot detection, 2D gels (three gels from three independent biological samples) were aligned and matched, and normalized spot volumes were determined quantitatively. For each matched spot, the percent volume was calculated as the volume divided by the total volume of matched spots. Spots with variations in abundance were subjected to ANOVA, Tukey–Kramer HSD tests and contrast analysis (JMP software, SAS Institute, Cary, NC, USA) to assign spots that significantly varied ($p < 0.05$) in abundance for two factors: drought and cultivar, and their interactions (Table S2). An unpaired two-tailed Student's *t*-test was used to assign significant variations in abundance (a given drought treatment vs. control) within analyzed cultivars (Section 4.10; Table S1). Proteins were subsequently identified by MS from spots that significantly varied in abundance.

4.7. Protein Identification by Mass Spectrometry (MS)

Gel spots were subjected to a standard *"in-gel digestion"* procedure in which proteins were reduced with 10 mM DTT (for 30 min at 56 °C), alkylated with 55 mM 2-iodoacetamide (45 min in the dark at room temperature) and digested overnight with trypsin (Promega, Madison, WI, USA) in 25 mM ammonium bicarbonate. The resulting peptides were eluted from the gel matrix with 0.1% trifluoroacetic acid in 2% acetonitrile.

Peptide mixtures were analyzed by liquid chromatography coupled to a mass spectrometer in the Laboratory of Mass Spectrometry (Institute of Biochemistry and Biophysics, Polish Academy of Sciences, Warsaw, Poland). Samples were concentrated and desalted on a RP-C18 pre-column (nanoACQUITY Symmetry® C18, Waters, Milford, MA, USA), and further peptide separation was achieved on a nano-Ultra Performance Liquid Chromatography (UPLC) RP-C18 column (BEH130 C18 column, 75 µm id, 250 mm long; Waters, Milford, MA, USA) of a nanoACQUITY UPLC system, using a linear 0–60% CAN gradient for 120 min in the presence of 0.05% formic acid with a flow rate of 150 nL min^{-1}. The column outlet was directly coupled to the electrospray ionization (ESI) ion source of an Orbitrap Velos type mass spectrometer (Thermo Electron Corp., San Jose, CA, USA), working in the regime of data dependent MS to MS/MS switch. An electrospray voltage of 1.5 kV was used. A blank run preventing cross-contamination from previous samples preceded each analysis.

Proteins were identified using the Mascot search algorithm (Available online: www.matrixscience.com) against the NCBInr (Available online: www.ncbi.nig.gov) databases. Protein identification was performed using the Mascot search probability-based molecular weight search (MOWSE) score. The ion score was $-10 \times \log(P)$, where P was the probability that the observed match was a random event. To avoid possible misidentifications resulting from the implementation of large datasets, as pointed out by Schmidt et al. [101], we were able to set the threshold of false positive rate. Peptides with a Mascot score exceeding the threshold value, corresponding to a <5% false positive rate as calculated by the Mascot procedure, were considered to be positively identified.

4.8. SDS-PAGE, Western Blotting, and Immunodetection of Proteins

Proteins resolved by one-dimensional SDS polyacrylamide gel electrophoresis (12% SDS-PAGE; [49]) were electroblotted in semidry conditions onto Immobilon-P membranes (Merck, Warsaw, Poland), using a TE77 PWR ECL Semi-Dry blotting apparatus (GE Healthcare Life Science Poland, Warsaw, Poland) and standard transfer buffer (20% methanol, 48 mM Tris, 39 mM glycine, 0.0375% SDS). Proteins resolved on 2D gels were electroblotted in semidry conditions using the same apparatus and alternative transfer buffer (10% methanol, 10 mM CAPS pH 11.0). Protein immunodetection was carried out with rabbit polyclonal antibodies directed against Mn-SOD (product. No. AS09 524, 1:5000 dilution), cyt. *c* (product No. AS08 343A, 1:5000), mitochondrial HSP70 (product No. AS08 347, 1:4000), SHMT1 (product. No. AS05 075, 1:10,000), GDC-H (product No. AS05 074, 1:4000), IDH (product No. AS06 203A, 1:4000), aconitase (product No. AS09 521, 1:5000; all antisera listed above from Agrisera, Vännäs, Sweden), PUMP (1:1000; [102,103]), dehydrin K-segment (1:1000, a gift of T.J. Close, University of California at Riverside, USA; [51]), dehydrin K-segment with N terminal cysteine on the synthetic peptide (product No. PLA-100, 1:1400; Stressgen, Victoria, BC, Canada), SK$_3$-motif of *Solanum sogarandinum* DHN24 dehydrin (1:500, a gift of T. Rorat, Institute of Plant Genetics, Polish Academy of Sciences, Poznań, Poland; [104]) and mouse monoclonal antibodies directed against AOX (1:1000; [105]), and ATP1 (1:200; [106]; both antisera donated by T. Elthon, University of Lincoln, Lincoln, NE, USA). Immunoassay details were described previously [13,49]. Enhanced chemiluminescence (ECL) signals were quantified with Multi Gauge (v2.2, Fujifilm, Tokio, Japan).

4.9. RNA Isolation and RT-qPCR

Total RNA from cauliflower curds was extracted using an EZ-10 Spin Column Plant RNA Mini-Preps Kit (BioBasic, Markham, ON, Canada) according to the manufacturer's protocol. Genomic DNA contaminants were removed by RQ1 DNase I free of RNase (Promega Poland, Warsaw, Poland). cDNA was synthesized using 1 µg of RNA, 0.2 µg of random hexamers mixture from HexaLabel DNA Labeling Kit (Thermo Scientific, Gdańsk, Poland) and 200 units of M-MLV reverse transcriptase (Promega Poland, Warsaw, Poland) in a 20 µL total volume for 1 h at 37 °C. After first strand synthesis, the reaction mixture was diluted with 10 mM Tris-HCl, pH 8.0 three or six times, and after normalization, aliquots of 1–2 µL were subjected to RT-quantitative PCR (RT-qPCR) using a Thermo Scientific Luminaris Color HiGreen High ROX qPCR Master Mix kit on an Applied Biosystems StepOnePlus Real-Time PCR System. The following profile was used: 5 min at 95 °C followed by 40 cycles of 20 s at 95 °C, 1 min at 60 °C, and finally, a melting step. The quality of qRT-PCR assays was verified by LinRegPCR (v. 2012.3, Heart Failure Research Center, Academic Medical Center, Amsterdam, The Netherlands). Outliers were manually removed. Two biological and at least three technical replicates were included.

Fragments of cauliflower cDNA for selected proteins were amplified using specific primers (Table S7); a 239 bp fragment of cauliflower actin1 (*ACT1*) cDNA was used as an internal standard. Amplicons were directly sequenced bi-directionally (Big Dye Terminator v. 3.1 Cycle Sequencing kit, Applied Biosystems Poland, Warsaw, Poland) on an ABI Prism 31-30 XL system (Applied Biosystems Poland, Warsaw, Poland) for sequence identity verification. In the case of *AOX1a* and *ACT1*, amplicons were additionally cloned to a pGEM T-Easy vector with the pGEM T-Easy Ligation System II (Promega) before sequencing.

4.10. Statistical Analysis

All experiments were conducted in triplicate, unless otherwise indicated. Results of densitometric analyses (Sections 2.4, 2.5, and 2.7), and 2D PAGE spot pattern alterations based on the spot volume (Sections 2.2 and 2.3) are presented as means ± SE. An unpaired two-tailed Student's *t*-test was used to identify significant differences; in particular, differences were considered to be statistically significant if $p < 0.05$ (*), $p < 0.01$ (**), or $p < 0.001$ (***). Significant correlations (Table S3) among variables were calculated using Spearman's correlation coefficient ($p < 0.05$) with the help of STATISTICA 13.1 (StatSoft Poland, Kraków, Poland) software.

5. Conclusions

Results of our study that significantly broaden *Brassica* data suggest that plant mitochondria (across distinct cultivars) are actively engaged in the response to water deficit. Variations within the mitochondrial proteome of investigated cauliflower cultivars encompass major decreases in abundance; increased in abundance spots were specific to the intensity of the water deficit. Investigated proteomic variations coincided roughly with drought tolerance. Mitochondrial porin isoforms, ATP synthase subunit, DNA-binding proteins, heat shock proteins, components of energy-dissipating systems (AOX isoform 36 kDa, PUMP isoform 40 kDa) as well as dehydrin-like proteins (18 and 27 kDa) investigated in our study are among the best candidates for stress tolerance markers, highlighting diversity of drought responses within cauliflower mitochondria. Identification of dehydrin-specific tryptic peptides in several spots from 2D gels additionally indicates for the relevant participation of such proteins in acclimation to water deficit.

The future study of the dynamic pattern of PTMs [30] among cauliflower drought-responsive proteins will use spectral data obtained from MS/MS peptide sequencing. Owing to the relevance of protein phosphorylation in the regulation of protein activity and stress-signaling pathways [107], we will pay special attention to such protein modification.

Alterations of transcripts for the stress-affected AOX isoform were largely unassociated with the proteomic ones, which is contrary to findings from our previous study on plant mitochondrial biogenesis in temperature stress and thermal recovery [13]. We suggest that the enhanced availability of *AOX1a* transcripts for translation may be an important regulatory point for the drought response of cauliflower mitochondria. Such results could be at least partially explained by Nakaminami et al. [108] findings on the imbalanced mRNA/protein pools and the altered pattern of translation initiation through stress acclimation and de-acclimation. Profiling of TF expression highlights valuable variations (additional to the ones at the protein level) in drought responses between stress-sensitive cultivars.

Additional studies are required (1) to elucidate the impact of drought on transcript binding to ribosomes; (2) to investigate protein biosynthesis patterns [13] and mechanisms of increased selection of mRNA for translation; (3) to analyze participation of non-coding RNAs in the mitochondrial biogenesis (with emphasis on the cauliflower microtranscriptome targeting investigated mRNA); and (4) to characterize cauliflower genes with expression pattern regulated by the investigated CBF factors under drought.

Supplementary Materials: Supplementary materials can be found at http://www.mdpi.com/1422-0067/19/4/1130/s1.

Acknowledgments: This work was supported by OPUS grant of National Science Centre, Poland (grant No. 2011/03/B/NZ9/05237) as well as by KNOW Poznan RNA Centre (grant No. 01/KNOW2/2014). We would like also to thank Mikołaj Knaflewski, Alina Kałużewicz and Anna Zaworska (Poznan University of Life Sciences, Poland) for additional help during plant material cultivation and physiological analyses. We gratefully thank Grzegorz Pietkiewicz and Jakub Kosicki (Adam Mickiewicz University, Poznań) for technical assistance and valuable participation in data statistical analysis, respectively. We are grateful to Michał Dadlez and the staff of the Mass Spectrometry Laboratory (Institute of Biochemistry and Biophysics, Polish Academy of Sciences, Warsaw, Poland) for MS analyses and the help with protein identification.

Author Contributions: Michał Rurek was principle investigator, who designed this study, performed extraction of mitochondrial proteins, carried out all immunoassays, analyzed proteomic results, prepared and wrote the paper; Magdalena Czołpińska performed isolation of total and mitochondrial RNA, performed RT-qPCR assays and substantially participated in paper writing; Tomasz Andrzej Pawłowski prepared protein samples for 2D PAGE, participated partially in 2D PAGE and the statistical analysis of spot variations, selected stress-responsive protein spots, submitted protein spots for MS analyses, participated in preparation of 2D blots and substantially in paper writing; Aleksandra Maria Staszak participated partially in 2D PAGE, statistical analysis of spot variations and in paper writing; Witold Nowak assisted in primer design, amplicon sequencing, and participated in writing of the paper; Włodzimierz Krzesiński cultivated plant material, conducted all physiological assays, analyzed their results and participated in paper writing; Tomasz Spiżewski assisted in cultivation and maintenance of the plant material in control conditions and after stress dosage, prepared nutrient media, subjected plants to stress conditions and participated in physiological analyses. All Authors have approved the submitted version and agreed to be personally accountable for their own contributions and for ensuring that questions related to the accuracy or integrity of any part of the work, even those in which the Author was not personally involved, are appropriately investigated, resolved, and documented in the literature.

Conflicts of Interest: The authors declare no conflicts of interest.

Abbreviations

ACO	aconitase
AOX	alternative oxidase
ATP1, ATP2	mitochondrial ATP synthase subunit α or subunit β
CBB	Coomassie Brilliant Blue
CBF/DREB	C-repeat/dehydration-responsive element binding
CI, CII, CIV	respiratory complexes I, II and IV
CAPS	3-[(3-cholamidopropyl)dimethylammonio]-1-propanesulfonate
CHAPS	3-[(3-cholamidopropyl)dimethylammonio]-2-hydroxy-1-propanesulfonate
2D PAGE	two-dimensional gel electrophoresis
DHN	dehydrin
dlp	dehydrin-like protein
DTT	dithiothreitol
EDTA	ethylenediaminetetraacetic acid
EF	elongation factor
ERD	early response to dehydration
ERF	ETHYLENE RESPONSE FACTOR
FDH	formate dehydrogenase
GDC	glycine decarboxylase
HSP	heat shock protein
IDH	isocitrate dehydrogenase
IEF	isoelectrofocusing
LC-MS/MS	liquid chromatography-tandem mass spectrometry
LEA	late embryogenesis abundant
MDH	malate dehydrogenase
miRNA	microRNA
MPP	mitochondrial processing peptidase
OXPHOS	oxidative phosphorylation
P5CDH	Δ-1-pyrroline-5-carboxylate dehydrogenase
PDH	pyruvate dehydrogenase
PhR	photorespiration rate
PPFD	photosynthetic photon flux density
ProDH	proline dehydrogenase
PTMs	posttranslational protein modifications
PUMP	plant-uncoupling mitochondrial protein
R_d	respiration in the light (day respiration) rate
R_n	respiration in the dark (night respiration) rate
ROS	reactive oxygen species
R_T	total respiration rate
RT-qPCR	reverse transcription quantitative PCR
RWC	relative water content
SDH	succinate dehydrogenase (complex II)
SHMT	serine hydroxymethyl aminotransferase
SOD	superoxide dismutase
TF(s)	transcription factor(s)
UPLC	nano-ultra performance liquid chromatography
VDAC	voltage-dependent anion channel

References

1. Zhang, S.; Chen, F.; Peng, S.; Ma, W.; Korpelainen, H.; Li, C. Comparative physiological, ultrastructural and proteomic analyses reveal sexual differences in the responses of *Populus cathayana* under drought stress. *Proteomics* **2010**, *10*, 2661–2677. [CrossRef] [PubMed]

2. Bogeat-Triboulot, M.-B.; Brosché, M.; Renaut, J.; Jouve, L.; Le Thiec, D.; Fayyaz, P.; Vinocur, B.; Witters, E.; Laukens, K.; Teichmann, T.; et al. Gradual Soil Water Depletion Results in Reversible Changes of Gene Expression, Protein Profiles, Ecophysiology, and Growth Performance in *Populus euphratica*, a Poplar Growing in Arid Regions. *Plant Physiol.* **2007**, *143*, 876–892. [CrossRef] [PubMed]
3. Li, Y.L.; Stanghellini, C. Analysis of the effect of EC and potential transpiration on vegetative growth of tomato. *Sci. Hort.* **2001**, *89*, 9–21. [CrossRef]
4. Chaves, M.M. How Plants Cope with Water Stress in the Field? Photosynthesis and Growth. *Ann. Bot.* **2002**, *89*, 907–916. [CrossRef] [PubMed]
5. Flexas, J.; Medrano, H. Drought-inhibition of Photosynthesis in C3 Plants: Stomatal and Non-stomatal Limitations Revisited. *Ann. Bot.* **2002**, *89*, 183–189. [CrossRef] [PubMed]
6. Alvarez, S.; Roy Choudhury, S.; Pandey, S. Comparative Quantitative Proteomics Analysis of the ABA Response of Roots of Drought-Sensitive and Drought-Tolerant Wheat Varieties Identifies Proteomic Signatures of Drought Adaptability. *J. Proteome Res.* **2014**, *13*, 1688–1701. [CrossRef] [PubMed]
7. Johnová, P.; Skalák, J.; Saiz-Fernández, I.; Brzobohatý, B. Plant responses to ambient temperature fluctuations and water-limiting conditions: A proteome-wide perspective. *Biochim. Biophys. Acta* **2016**, *1864*, 916–931. [CrossRef] [PubMed]
8. Atkin, O.K.; Macherel, D. The crucial role of plant mitochondria in orchestrating drought tolerance. *Ann. Bot.* **2008**, *103*, 581–597. [CrossRef] [PubMed]
9. Salvato, F.; Havelund, J.F.; Chen, M.; Rao, R.S.P.; Rogowska-Wrzesinska, A.; Jensen, O.N.; Gang, D.R.; Thelen, J.J.; Møller, I.M. The Potato Tuber Mitochondrial Proteome. *Plant Physiol.* **2014**, *164*, 637–653. [CrossRef] [PubMed]
10. Møller, I.M. What is hot in plant mitochondria? *Physiol. Plant.* **2016**, *157*, 256–263. [CrossRef] [PubMed]
11. Giegé, P.; Sweetlove, L.J.; Cognat, V.; Leaver, C.J. Coordination of Nuclear and Mitochondrial Genome Expression during Mitochondrial Biogenesis in Arabidopsis. *Plant Cell* **2005**, *17*, 1497–1512. [CrossRef] [PubMed]
12. Howell, K.A.; Cheng, K.; Murcha, M.W.; Jenkin, L.E.; Millar, A.H.; Whelan, J. Oxygen Initiation of Respiration and Mitochondrial Biogenesis in Rice. *J. Biol. Chem.* **2007**, *282*, 15619–15631. [CrossRef] [PubMed]
13. Rurek, M.; Woyda-Ploszczyca, A.M.; Jarmuszkiewicz, W. Biogenesis of mitochondria in cauliflower (*Brassica oleracea* var. *botrytis*) curds subjected to temperature stress and recovery involves regulation of the complexome, respiratory chain activity, organellar translation and ultrastructure. *BBA Bioenerg.* **2015**, *1847*, 399–417. [CrossRef]
14. Taylor, N.L.; Tan, Y.-F.; Jacoby, R.P.; Millar, A.H. Abiotic environmental stress induced changes in the *Arabidopsis thaliana* chloroplast, mitochondria and peroxisome proteomes. *J. Proteom.* **2009**, *72*, 367–378. [CrossRef] [PubMed]
15. Ali, G.M.; Komatsu, S. Proteomic Analysis of Rice Leaf Sheath during Drought Stress. *J. Proteome Res.* **2006**, *5*, 396–403. [CrossRef] [PubMed]
16. Aranjuelo, I.; Molero, G.; Erice, G.; Avice, J.C.; Nogués, S. Plant physiology and proteomics reveals the leaf response to drought in alfalfa (*Medicago sativa* L.). *J. Exp. Bot.* **2011**, *62*, 111–123. [CrossRef] [PubMed]
17. Kosová, K.; Vítámvás, P.; Urban, M.O.; Klíma, M.; Roy, A.; Prášil, I.T. Biological Networks Underlying Abiotic Stress Tolerance in Temperate Crops–A Proteomic Perspective. *Int. J. Mol. Sci.* **2015**, *16*, 20913–20942. [CrossRef] [PubMed]
18. Ndimba, B.K.; Chivasa, S.; Simon, W.J.; Slabas, A.R. Identification of *Arabidopsis* salt and osmotic stress responsive proteins using two-dimensional difference gel electrophoresis and mass spectrometry. *Proteomics* **2005**, *5*, 4185–4196. [CrossRef] [PubMed]
19. Taylor, N.L.; Heazlewood, J.L.; Day, D.A.; Millar, A.H. Differential Impact of Environmental Stresses on the Pea Mitochondrial Proteome. *Mol. Cell Proteom.* **2005**, *4*, 1122–1133. [CrossRef] [PubMed]
20. Bies-Ethève, N.; Gaubier-Comella, P.; Debures, A.; Lasserre, E.; Jobet, E.; Raynal, M.; Cooke, R.; Delseny, M. Inventory, evolution and expression profiling diversity of the LEA (late embryogenesis abundant) protein gene family in *Arabidopsis thaliana*. *Plant Mol. Biol.* **2008**, *67*, 107–124. [CrossRef] [PubMed]
21. Taylor, N.L.; Day, D.A.; Millar, A.H. Environmental Stress Causes Oxidative Damage to Plant Mitochondria Leading to Inhibition of Glycine Decarboxylase. *J. Biol. Chem.* **2002**, *277*, 42663–42668. [CrossRef] [PubMed]

22. Vanlerberghe, G.C. Alternative Oxidase: A Mitochondrial Respiratory Pathway to Maintain Metabolic and Signaling Homeostasis During Abiotic and Biotic Stress in Plants. *Int. J. Mol. Sci.* **2013**, *14*, 6805–6847. [CrossRef] [PubMed]
23. Zadražnik, T.; Hollung, K.; Egge-Jacobsen, W.; Meglič, V.; Šuštar-Vozlič, J. Differential proteomic analysis of drought stress response in leaves of common bean (*Phaseolus vulgaris* L.). *J. Proteom.* **2013**, *78*, 254–272. [CrossRef] [PubMed]
24. Yu, C.; Wang, L.; Xu, S.; Zeng, Y.; He, C.; Chen, C.; Huang, W.; Zhu, Y.; Hu, J. Mitochondrial ORFH79 is Essential for Drought and Salt Tolerance in Rice. *Plant Cell Physiol.* **2015**, *56*, 2248–2258. [CrossRef] [PubMed]
25. Reddy, P.S.; Kavi Kishor, P.B.; Seiler, C.; Kuhlmann, M.; Eschen-Lippold, L.; Lee, J.; Reddy, M.K.; Sreenivasulu, N. Unraveling Regulation of the Small Heat Shock Proteins by the Heat Shock Factor HvHsfB2c in Barley: Its Implications in Drought Stress Response and Seed Development. *PLoS ONE* **2014**, *9*, e89125. [CrossRef] [PubMed]
26. Wu, G.; Wilen, R.W.; Robertson, A.J.; Gusta, L.V. Isolation, Chromosomal Localization, and Differential Expression of Mitochondrial Manganese Superoxide Dismutase and Chloroplastic Copper/Zinc Superoxide Dismutase Genes in Wheat. *Plant Physiol.* **1999**, *120*, 513–520. [CrossRef] [PubMed]
27. Huseynova, I.M.; Aliyeva, D.R.; Aliyev, J.A. Subcellular localization and responses of superoxide dismutase isoforms in local wheat varieties subjected to continuous soil drought. *Plant Physiol. Biochem.* **2014**, *81*, 54–60. [CrossRef] [PubMed]
28. Dalal, M.; Tayal, D.; Chinnusamy, V.; Bansal, K.C. Abiotic stress and ABA-inducible Group 4 LEA from *Brassica napus* plays a key role in salt and drought tolerance. *J. Biotechnol.* **2009**, *139*, 137–145. [CrossRef] [PubMed]
29. Efeoğlu, B.; Ekmekçi, Y.; Çiçek, N. Physiological responses of three maize cultivars to drought stress and recovery. *S. Afr. J. Bot.* **2009**, *75*, 34–42. [CrossRef]
30. Koh, J.; Chen, G.; Yoo, M.-J.; Zhu, N.; Dufresne, D.; Erickson, J.E.; Shao, H.; Chen, S. Comparative Proteomic Analysis of *Brassica napus* in Response to Drought Stress. *J. Proteome Res.* **2015**, *14*, 3068–3081. [CrossRef] [PubMed]
31. Zhang, J.; Mason, A.S.; Wu, J.; Liu, S.; Zhang, X.; Luo, T.; Redden, R.; Batley, J.; Hu, L.; Yan, G. Identification of Putative Candidate Genes for Water Stress Tolerance in Canola (*Brassica napus*). *Front. Plant Sci.* **2015**, *6*. [CrossRef] [PubMed]
32. Kwon, S.-W.; Kim, M.; Kim, H.; Lee, J. Shotgun Quantitative Proteomic Analysis of Proteins Responding to Drought Stress in *Brassica rapa* L. (Inbred Line "Chiifu"). *Int. J. Genom.* **2016**, *2016*, 4235808. [CrossRef]
33. Wong, C.E.; Li, Y.; Whitty, B.R.; Díaz-Camino, C.; Akhter, S.R.; Brandle, J.E.; Golding, G.B.; Weretilnyk, E.A.; Moffatt, B.A.; Griffith, M. Expressed sequence tags from the Yukon ecotype of *Thellungiella* reveal that gene expression in response to cold, drought and salinity shows little overlap. *Plant Mol. Biol.* **2005**, *58*, 561–574. [CrossRef] [PubMed]
34. Li, Z.; Zhao, L.; Kai, G.; Yu, S.; Cao, Y.; Pang, Y.; Sun, X.; Tang, K. Cloning and expression analysis of a water stress-induced gene from *Brassica oleracea*. *Plant Physiol. Biochem.* **2004**, *42*, 789–794. [CrossRef] [PubMed]
35. Mohammadi, P.P.; Moieni, A.; Komatsu, S. Comparative proteome analysis of drought sensitive and drought-tolerant rapeseed roots and their hybrid F1 line under drought stress. *Amino Acids* **2012**, *43*, 2137–2152. [CrossRef] [PubMed]
36. De Mezer, M.; Turska-Taraska, A.; Kaczmarek, Z.; Glowacka, K.; Swarcewicz, B.; Rorat, T. Differential physiological and molecular response of barley genotypes to water deficit. *Plant Physiol. Biochem.* **2014**, *80*, 234–248. [CrossRef] [PubMed]
37. Das, A.; Mukhopadhyay, M.; Sarkar, B.; Saha, D.; Mondal, T.K. Influence of drought stress on cellular ultrastructure and antioxidant system in tea cultivars with different drought sensitivities. *J. Environ. Biol.* **2015**, *36*, 875–882. [PubMed]
38. Urban, M.O.; Vašek, J.; Klíma, M.; Krtková, J.; Kosová, K.; Prášil, I.T.; Vítámvás, P. Proteomic and physiological approach reveals drought-induced changes in rapeseeds: Water-saver and water-spender strategy. *J. Proteom.* **2017**, *152*, 188–205. [CrossRef] [PubMed]
39. Vincent, D.; Ergül, A.; Bohlman, M.C.; Tattersall, E.A.R.; Tillett, R.L.; Wheatley, M.D.; Woolsey, R.; Quilici, D.R.; Joets, J.; Schlauch, K.; et al. Proteomic analysis reveals differences between *Vitis vinifera* L. cv. Chardonnay and cv. Cabernet Sauvignon and their responses to water deficit and salinity. *J. Exp. Bot.* **2007**, *58*, 1873–1892. [CrossRef] [PubMed]

40. Bonhomme, L.; Monclus, R.; Vincent, D.; Carpin, S.; Lomenech, A.-M.; Plomion, C.; Brignolas, F.; Morabito, D. Leaf proteome analysis of eight *Populus xeuramericana* genotypes: Genetic variation in drought response and in water-use efficiency involves photosynthesis-related proteins. *Proteomics* **2009**, *9*, 4121–4142. [CrossRef] [PubMed]
41. Ford, K.L.; Cassin, A.; Bacic, A. Quantitative Proteomic Analysis of Wheat Cultivars with Differing Drought Stress Tolerance. *Front. Plant Sci.* **2011**, *2*, 44. [CrossRef] [PubMed]
42. Ge, P.; Ma, C.; Wang, S.; Gao, L.; Li, X.; Guo, G.; Ma, W.; Yan, Y. Comparative proteomic analysis of grain development in two spring wheat varieties under drought stress. *Anal. Bioanal. Chem.* **2012**, *402*, 1297–1313. [CrossRef] [PubMed]
43. Ashoub, A.; Beckhaus, T.; Berberich, T.; Karas, M.; Brüggemann, W. Comparative analysis of barley leaf proteome as affected by drought stress. *Planta* **2013**, *237*, 771–781. [CrossRef] [PubMed]
44. Budak, H.; Akpinar, B.A.; Unver, T.; Turktas, M. Proteome changes in wild and modern wheat leaves upon drought stress by two-dimensional electrophoresis and nanoLC-ESI-MS/MS. *Plant Mol. Biol.* **2013**, *83*, 89–103. [CrossRef] [PubMed]
45. Kausar, R.; Arshad, M.; Shahzad, A.; Komatsu, S. Proteomics analysis of sensitive and tolerant barley genotypes under drought stress. *Amino Acids* **2013**, *44*, 345–359. [CrossRef] [PubMed]
46. Oliveira, T.M.; da Silva, F.R.; Bonatto, D.; Neves, D.M.; Morillon, R.; Maserti, B.E.; Filho, M.A.C.; Costa, M.; Pirovani, C.P.; Gesteira, A.S. Comparative study of the protein profiles of Sunki mandarin and Rangpur lime plants in response to water deficit. *BMC Plant Biol.* **2015**, *15*, 69. [CrossRef] [PubMed]
47. Vítámvás, P.; Urban, M.O.; Škodáček, Z.; Kosová, K.; Pitelková, I.; Vítámvás, J.; Renaut, J.; Prášil, I.T. Quantitative analysis of proteome extracted from barley crowns grown under different drought conditions. *Front. Plant Sci.* **2015**, *6*, 479. [CrossRef] [PubMed]
48. Cheng, L.; Wang, Y.; He, Q.; Li, H.; Zhang, X.; Zhang, F. Comparative proteomics illustrates the complexity of drought resistance mechanisms in two wheat (*Triticum aestivum* L.) cultivars under dehydration and rehydration. *BMC Plant Biol.* **2016**, *16*, 188. [CrossRef] [PubMed]
49. Rurek, M. Diverse accumulation of several dehydrin-like proteins in cauliflower (*Brassica oleracea* var. *botrytis*), *Arabidopsis thaliana* and yellow lupin (*Lupinus luteus*) mitochondria under cold and heat stress. *BMC Plant Biol.* **2010**, *10*, 181. [CrossRef]
50. Katari, M.S.; Nowicki, S.D.; Aceituno, F.F.; Nero, D.; Kelfer, J.; Thompson, L.P.; Cabello, J.M.; Davidson, R.S.; Goldberg, A.P.; Shasha, D.E.; et al. VirtualPlant: A software platform to support systems biology research. *Plant Physiol.* **2010**, *152*, 500–515. [CrossRef] [PubMed]
51. Close, T.J.; Fenton, R.D.; Moonan, F. A view of plant dehydrins using antibodies specific to the carboxy terminal peptide. *Plant Mol. Biol.* **1993**, *23*, 279–286. [CrossRef] [PubMed]
52. Krzesiński, W.; Kałużewicz, A.; Frąszczak, B.; Zaworska, A.; Lisiecka, J. Cauliflower's response to drought stress. *Nauka Przyroda Technol.* **2016**, *10*. [CrossRef]
53. Voss, I.; Sunil, B.; Scheibe, R.; Raghavendra, A.S. Emerging concept for the role of photorespiration as an important part of abiotic stress response. *Plant Biol. (Stuttg.)* **2013**, *15*, 713–722. [CrossRef] [PubMed]
54. Kim, J.; van Iersel, M.W. Slowly developing drought stress increases photosynthetic acclimation of *Catharanthus roseus*. *Physiol. Plant.* **2011**, *143*, 166–177. [CrossRef] [PubMed]
55. Haupt-Herting, S.; Klug, K.; Fock, H.P. A New Approach to Measure Gross CO_2 Fluxes in Leaves. Gross CO_2 Assimilation, Photorespiration, and Mitochondrial Respiration in the Light in Tomato under Drought Stress. *Plant Physiol.* **2001**, *126*, 388–396. [CrossRef] [PubMed]
56. Campos, H.; Trejo, C.; Peña-Valdivia, C.B.; García-Nava, R.; Conde-Martínez, F.V.; Cruz-Ortega, M.R. Stomatal and non-stomatal limitations of bell pepper (*Capsicum annuum* L.) plants under water stress and re-watering: Delayed restoration of photosynthesis during recovery. *Environ. Exp. Bot.* **2014**, *98*, 56–64. [CrossRef]
57. Sperlich, D.; Barbeta, A.; Ogaya, R.; Sabaté, S.; Peñuelas, J. Balance between carbon gain and loss under long-term drought: impacts on foliar respiration and photosynthesis in *Quercus ilex* L. *J. Exp. Bot.* **2016**, *67*, 821–833. [CrossRef] [PubMed]
58. Vassileva, V.; Simova-Stoilova, L.; Demirevska, K.; Feller, U. Variety-specific response of wheat (*Triticum aestivum* L.) leaf mitochondria to drought stress. *J. Plant Res.* **2009**, *122*, 445–454. [CrossRef] [PubMed]

59. Chastain, D.R.; Snider, J.L.; Collins, G.D.; Perry, C.D.; Whitaker, J.; Byrd, S.A. Water deficit in field-grown *Gossypium hirsutum* primarily limits net photosynthesis by decreasing stomatal conductance, increasing photorespiration, and increasing the ratio of dark respiration to gross photosynthesis. *J. Plant Physiol.* **2014**, *171*, 1576–1585. [CrossRef] [PubMed]
60. Liu, C.; Wang, Y.; Pan, K.; Wang, Q.; Liang, J.; Jin, Y.; Tariq, A. The Synergistic Responses of Different Photoprotective Pathways in Dwarf Bamboo (*Fargesia rufa*) to Drought and Subsequent Rewatering. *Front. Plant Sci.* **2017**, *8*, 489. [CrossRef] [PubMed]
61. Abogadallah, G.M. Differential regulation of photorespiratory gene expression by moderate and severe salt and drought stress in relation to oxidative stress. *Plant Sci.* **2011**, *180*, 540–547. [CrossRef] [PubMed]
62. Lima Neto, M.C.; Cerqueira, J.V.A.; da Cunha, J.R.; Ribeiro, R.V.; Silveira, J.A.G. Cyclic electron flow, NPQ and photorespiration are crucial for the establishment of young plants of *Ricinus communis* and *Jatropha curcas* exposed to drought. *Plant Biol. (Stuttg.)* **2017**, *19*, 650–659. [CrossRef] [PubMed]
63. Zhou, S.; Li, M.; Guan, Q.; Liu, F.; Zhang, S.; Chen, W.; Yin, L.; Qin, Y.; Ma, F. Physiological and proteome analysis suggest critical roles for the photosynthetic system for high water-use efficiency under drought stress in *Malus*. *Plant Sci.* **2015**, *236*, 44–60. [CrossRef] [PubMed]
64. Chen, L.; Ren, F.; Zhong, H.; Feng, Y.; Jiang, W.; Li, X. Identification and expression analysis of genes in response to high-salinity and drought stresses in *Brassica napus*. *Acta Biochim. Biophys. Sin. (Shanghai)* **2010**, *42*, 154–164. [CrossRef] [PubMed]
65. He, C.Y.; Zhang, G.Y.; Zhang, J.G.; Duan, A.G.; Luo, H.M. Physiological, biochemical, and proteome profiling reveals key pathways underlying the drought stress responses of *Hippophae rhamnoides*. *Proteomics* **2016**, *16*, 2688–2697. [CrossRef] [PubMed]
66. Bernard, D.G.; Cheng, Y.; Zhao, Y.; Balk, J. An Allelic Mutant Series of *ATM3* Reveals Its Key Role in the Biogenesis of Cytosolic Iron-Sulfur Proteins in Arabidopsis. *Plant Physiol.* **2009**, *151*, 590–602. [CrossRef] [PubMed]
67. Landi, S.; Hausman, J.-F.; Guerriero, G.; Esposito, S. Poaceae vs. Abiotic Stress: Focus on Drought and Salt Stress, Recent Insights and Perspectives. *Front. Plant Sci.* **2017**, *8*, 1214. [CrossRef] [PubMed]
68. Caruso, G.; Cavaliere, C.; Foglia, P.; Gubbiotti, R.; Samperi, R.; Laganà, A. Analysis of drought responsive proteins in wheat (*Triticum durum*) by 2D-PAGE and MALDI-TOF mass spectrometry. *Plant Sci.* **2009**, *177*, 570–576. [CrossRef]
69. Hamilton, C.A.; Good, A.G.; Taylor, G.J. Induction of Vacuolar ATPase and Mitochondrial ATP Synthase By Aluminum in an Aluminum-Resistant Cultivar of Wheat. *Plant Physiol.* **2001**, *125*, 2068–2077. [CrossRef] [PubMed]
70. Moghadam, A.A.; Taghavi, S.M.; Niazi, A.; Djavaheri, M.; Ebrahimie, E. Isolation and in silico functional analysis of *MtATP6*, a 6-kDa subunit of mitochondrial F_1F_0-ATP synthase, in response to abiotic stress. *Genet. Mol. Res.* **2012**, *11*, 3547–3567. [CrossRef] [PubMed]
71. Li, C.-L.; Wang, M.; Ma, X.-Y.; Zhang, W. NRGA1, a Putative Mitochondrial Pyruvate Carrier, Mediates ABA Regulation of Guard Cell Ion Channels and Drought Stress Responses in Arabidopsis. *Mol. Plant* **2014**, *7*, 1508–1521. [CrossRef] [PubMed]
72. Wang, M.; Ma, X.; Shen, J.; Li, C.; Zhang, W. The ongoing story: The mitochondria pyruvate carrier 1 in plant stress response in Arabidopsis. *Plant Signal. Behav.* **2014**, *9*, e973810. [CrossRef] [PubMed]
73. Taylor, N.L.; Rudhe, C.; Hulett, J.M.; Lithgow, T.; Glaser, E.; Day, D.A.; Millar, A.H.; Whelan, J. Environmental stresses inhibit and stimulate different protein import pathways in plant mitochondria. *FEBS Lett.* **2003**, *547*, 125–130. [CrossRef]
74. Riccardi, F.; Gazeau, P.; de Vienne, D.; Zivy, M. Protein Changes in Response to Progressive Water Deficit in Maize: Quantitative Variation and Polypeptide Identification. *Plant Physiol.* **1998**, *117*, 1253–1263. [CrossRef] [PubMed]
75. Kaouthar, F.; Ameny, F.-K.; Yosra, K.; Walid, S.; Ali, G.; Faiçal, B. Responses of transgenic *Arabidopsis* plants and recombinant yeast cells expressing a novel durum wheat manganese superoxide dismutase *TdMnSOD* to various abiotic stresses. *J. Plant Physiol.* **2016**, *198*, 56–68. [CrossRef] [PubMed]
76. Dahal, K.; Wang, J.; Martyn, G.D.; Rahimy, F.; Vanlerberghe, G.C. Mitochondrial Alternative Oxidase Maintains Respiration and Preserves Photosynthetic Capacity during Moderate Drought in *Nicotiana tabacum*. *Plant Physiol.* **2014**, *166*, 1560–1574. [CrossRef] [PubMed]

77. Galle, A.; Florez-Sarasa, I.; Thameur, A.; de Paepe, R.; Flexas, J.; Ribas-Carbo, M. Effects of drought stress and subsequent rewatering on photosynthetic and respiratory pathways in *Nicotiana sylvestris* wild type and the mitochondrial complex I-deficient CMSII mutant. *J. Exp. Bot.* **2010**, *61*, 765–775. [CrossRef] [PubMed]
78. Hincha, D.K.; Thalhammer, A. LEA proteins: IDPs with versatile functions in cellular dehydration tolerance. *Biochem. Soc. Trans.* **2012**, *40*, 1000–1003. [CrossRef] [PubMed]
79. Borovskii, G.B.; Stupnikova, I.V.; Antipina, A.I.; Vladimirova, S.V.; Voinikov, V.K. Accumulation of dehydrin-like proteins in the mitochondria of cereals in response to cold, freezing, drought and ABA treatment. *BMC Plant Biol.* **2002**, *2*, 5. [CrossRef]
80. Grelet, J.; Benamar, A.; Teyssier, E.; Avelange-Macherel, M.-H.; Grunwald, D.; Macherel, D. Identification in Pea Seed Mitochondria of a Late-Embryogenesis Abundant Protein Able to Protect Enzymes from Drying. *Plant Physiol.* **2005**, *137*, 157–167. [CrossRef] [PubMed]
81. Boswell, L.C.; Moore, D.S.; Hand, S.C. Quantification of cellular protein expression and molecular features of group 3 LEA proteins from embryos of *Artemia franciscana*. *Cell Stress Chaperones* **2014**, *19*, 329–341. [CrossRef] [PubMed]
82. Boswell, L.C.; Hand, S.C. Intracellular localization of group 3 LEA proteins in embryos of *Artemia franciscana*. *Tissue Cell* **2014**, *46*, 514–519. [CrossRef] [PubMed]
83. Clifton, R.; Millar, A.H.; Whelan, J. Alternative oxidases in Arabidopsis: A comparative analysis of differential expression in the gene family provides new insights into function of non-phosphorylating bypasses. *Biochim. Biophys. Acta* **2006**, *1757*, 730–741. [CrossRef] [PubMed]
84. Das, S.; Ferlito, M.; Kent, O.A.; Fox-Talbot, K.; Wang, R.; Liu, D.; Raghavachari, N.; Yang, Y.; Wheelan, S.J.; Murphy, E.; et al. Nuclear miRNA Regulates the Mitochondrial Genome in the Heart. *Circ. Res.* **2012**, *110*, 1596–1603. [CrossRef] [PubMed]
85. Leung, A.K.L. The Whereabouts of microRNA Actions: Cytoplasm and Beyond. *Trends Cell Biol.* **2015**, *25*, 601–610. [CrossRef] [PubMed]
86. Ro, S.; Ma, H.-Y.; Park, C.; Ortogero, N.; Song, R.; Hennig, G.W.; Zheng, H.; Lin, Y.-M.; Moro, L.; Hsieh, J.-T.; et al. The mitochondrial genome encodes abundant small noncoding RNAs. *Cell Res.* **2013**, *23*, 759–774. [CrossRef] [PubMed]
87. Rurek, M. Participation of non-coding RNAs in plant organelle biogenesis. *Acta Biochim. Pol.* **2016**, *63*, 653–663. [CrossRef] [PubMed]
88. Dai, X.; Zhao, P.X. psRNATarget: A plant small RNA target analysis server. *Nucleic Acids Res.* **2011**, *39*, W155–W159. [CrossRef] [PubMed]
89. Wu, H.; Wu, X.; Li, Z.; Duan, L.; Zhang, M. Physiological Evaluation of Drought Stress Tolerance and Recovery in Cauliflower (*Brassica oleracea* L.) Seedlings Treated with Methyl Jasmonate and Coronatine. *J. Plant Growth Regul.* **2012**, *31*, 113–123. [CrossRef]
90. Kiyosue, T.; Yamaguchi-Shinozaki, K.; Shinozaki, K. Characterization of two cDNAs (ERD10 and ERD14) corresponding to genes that respond rapidly to dehydration stress in Arabidopsis thaliana. *Plant Cell Physiol.* **1994**, *35*, 225–231. [PubMed]
91. Nylander, M.; Svensson, J.; Palva, E.T.; Welin, B.V. Stress-induced accumulation and tissue-specific localization of dehydrins in Arabidopsis thaliana. *Plant Mol. Biol.* **2001**, *45*, 263–279. [CrossRef] [PubMed]
92. Mizoi, J.; Shinozaki, K.; Yamaguchi-Shinozaki, K. AP2/ERF family transcription factors in plant abiotic stress responses. *Biochim. Biophys. Acta* **2012**, *1819*, 86–96. [CrossRef] [PubMed]
93. Thamilarasan, S.K.; Park, J.-I.; Jung, H.-J.; Nou, I.-S. Genome-wide analysis of the distribution of AP2/ERF transcription factors reveals duplication and CBFs genes elucidate their potential function in *Brassica oleracea*. *BMC Genom.* **2014**, *15*, 422. [CrossRef] [PubMed]
94. Haake, V.; Cook, D.; Riechmann, J.; Pineda, O.; Thomashow, M.F.; Zhang, J.Z. Transcription Factor CBF4 Is a Regulator of Drought Adaptation in Arabidopsis. *Plant Physiol.* **2002**, *130*, 639–648. [CrossRef] [PubMed]
95. Mohavedi, S.; Tabatabaei, B.E.S.; Alizade, H.; Ghobadi, C.; Yamchi, A.; Khaksar, G. Constitutive expression of *Arabidopsis DREB1B* in transgenic potato enhances drought and freezing tolerance. *Biol. Plant.* **2012**, *56*, 37–42. [CrossRef]
96. Laisk, A. *Kinetics of Photosynthesis and Photorespiration in C_3-Plants*; Nauka: Profsouznaya Ulitsa, Moscow, 1977.
97. Pawlowski, T.; Rurek, M.; Janicka, S.; Raczynska, K.D.; Augustyniak, H. Preliminary analysis of the cauliflower mitochondrial proteome. *Acta Physiol. Plant.* **2005**, *27*, 275–281. [CrossRef]

98. Staszak, A.; Pawłowski, T. Proteomic Analysis of Embryogenesis and the Acquisition of Seed Dormancy in Norway Maple (*Acer platanoides* L.). *Int. J. Mol. Sci.* **2014**, *15*, 10868–10891. [CrossRef] [PubMed]
99. Ramagli, L.S.; Rodriguez, L.V. Quantitation of microgram amounts of protein in two-dimensional polyacrylamide gel electrophoresis sample buffer. *Electrophoresis* **1985**, *6*, 559–563. [CrossRef]
100. Neuhoff, V.; Arold, N.; Taube, D.; Ehrhardt, W. Improved staining of proteins in polyacrylamide gels including isoelectric focusing gels with clear background at nanogram sensitivity using Coomassie Brilliant Blue G-250 and R-250. *Electrophoresis* **1988**, *9*, 255–262. [CrossRef] [PubMed]
101. Schmidt, U.G.; Endler, A.; Schelbert, S.; Brunner, A.; Schnell, M.; Neuhaus, H.E.; Marty-Mazars, D.; Marty, F.; Baginsky, S.; Martinoia, E. Novel Tonoplast Transporters Identified Using a Proteomic Approach with Vacuoles Isolated from Cauliflower Buds. *Plant Physiol.* **2007**, *145*, 216–229. [CrossRef] [PubMed]
102. Nantes, I.L.; Fagian, M.M.; Catisti, R.; Arruda, P.; Maia, I.G.; Vercesi, A.E. Low temperature and aging-promoted expression of PUMP in potato tuber mitochondria. *FEBS Lett.* **1999**, *457*, 103–106. [CrossRef]
103. Ježek, P.; Žáčková, M.; Košařová, J.; Rodrigues, E.T.; Madeira, V.M.; Vicente, J.A. Occurrence of plant-uncoupling mitochondrial protein (PUMP) in diverse organs and tissues of several plants. *J. Bioenerg. Biomembr.* **2000**, *32*, 549–561. [CrossRef] [PubMed]
104. Rorat, T.; Szabala, B.M.; Grygorowicz, W.J.; Wojtowicz, B.; Yin, Z.; Rey, P. Expression of SK_3-type dehydrin in transporting organs is associated with cold acclimation in *Solanum* species. *Planta* **2006**, *224*, 205–221. [CrossRef] [PubMed]
105. Elthon, T.E.; Nickels, R.L.; McIntosh, L. Monoclonal Antibodies to the Alternative Oxidase of Higher Plant Mitochondria. *Plant Physiol.* **1989**, *89*, 1311–1317. [CrossRef] [PubMed]
106. Luethy, M.H.; Horak, A.; Elthon, T.E. Monoclonal Antibodies to the α- and β-Subunits of the Plant Mitochondrial F_1-ATPase. *Plant Physiol.* **1993**, *101*, 931–937. [CrossRef] [PubMed]
107. Havelund, J.F.; Thelen, J.J.; Møller, I.M. Biochemistry, proteomics, and phosphoproteomics of plant mitochondria from non-photosynthetic cells. *Front. Plant Sci.* **2013**, *4*, 51. [CrossRef] [PubMed]
108. Nakaminami, K.; Matsui, A.; Nakagami, H.; Minami, A.; Nomura, Y.; Tanaka, M.; Morosawa, T.; Ishida, J.; Takahashi, S.; Uemura, M.; et al. Analysis of Differential Expression Patterns of mRNA and Protein During Cold-Acclimation and De-Acclimation in *Arabidopsis*. *Mol. Cell. Proteom.* **2014**, *13*, 3602–3611. [CrossRef] [PubMed]

© 2018 by the authors. Licensee MDPI, Basel, Switzerland. This article is an open access article distributed under the terms and conditions of the Creative Commons Attribution (CC BY) license (http://creativecommons.org/licenses/by/4.0/).

Article

Recombination Events Involving the *atp9* Gene Are Associated with Male Sterility of CMS PET2 in Sunflower

Antje Reddemann and Renate Horn *

Institut für Biowissenschaften, Abt. Pflanzengenetik, Universität Rostock, Albert-Einstein-Straße 3, D-18059 Rostock, Germany; antje.reddemann@web.de
* Correspondence: renate.horn@uni-rostock.de; Tel.: +49-381-498-6170; Fax: +49-381-498-6112

Received: 6 February 2018; Accepted: 6 March 2018; Published: 11 March 2018

Abstract: Cytoplasmic male sterility (CMS) systems represent ideal mutants to study the role of mitochondria in pollen development. In sunflower, CMS PET2 also has the potential to become an alternative CMS source for commercial sunflower hybrid breeding. CMS PET2 originates from an interspecific cross of *H. petiolaris* and *H. annuus* as CMS PET1, but results in a different CMS mechanism. Southern analyses revealed differences for *atp6*, *atp9* and *cob* between CMS PET2, CMS PET1 and the male-fertile line HA89. A second identical copy of *atp6* was present on an additional CMS PET2-specific fragment. In addition, the *atp9* gene was duplicated. However, this duplication was followed by an insertion of 271 bp of unknown origin in the 5′ coding region of the *atp9* gene in CMS PET2, which led to the creation of two unique open reading frames *orf288* and *orf231*. The first 53 bp of *orf288* are identical to the 5′ end of *atp9*. *Orf231* consists apart from the first 3 bp, being part of the 271-bp-insertion, of the last 228 bp of *atp9*. These CMS PET2-specific orfs are co-transcribed. All 11 editing sites of the *atp9* gene present in *orf231* are fully edited. The anther-specific reduction of the co-transcript in fertility-restored hybrids supports the involvement in male-sterility based on CMS PET2.

Keywords: *atp9*; cytoplasmic male sterility; CMS PET1; CMS PET2; *Helianthus annuus*; plant mitochondria; recombination; RNA-editing; respiration

---

1. Introduction

Cytoplasmic male sterility (CMS) is a maternally inherited incapability of higher plants to produce or shed functional pollen [1]. CMS has been described for more than 150 plant species [2,3], and is often associated with mitochondrial rearrangements and the expression of new open reading frames (orfs) leading to the translation of unique proteins that appear to interfere with mitochondrial functions and pollen development [4]. Exploitation of CMS systems is the most cost-effective way to produce hybrids. Hybrid production is widely used in field crops to gain enhanced yield and yield stability by using heterosis effects [5]. Hybrid breeding based on a CMS-system most frequently consists of a three line system: the CMS line, which is maintained by an isonuclear maintainer line present on a normal fertile cytoplasm, and a restorer line carrying one or two dominant nuclear restorer-of-fertility (*Rf*) genes to restore male fertility in the F1-hybrids. These restorer genes interact with the mitochondrial transcripts to suppress the deleterious effect of CMS by diverse mechanisms and thereby allow the production of male-fertile F1-hybrids [1,6].

In sunflower, CMS PET1 is so far the only cytoplasm worldwide used for the commercial hybrid production [7]. CMS PET1 was originally derived from the interspecific cross of *H. petiolaris* Nutt. × *H. annuus* [8] and has been used the last 50 years in sunflower hybrid breeding [9]. This reduction to one CMS source carries the risk of a high vulnerability of the cytoplasm to pathogen attacks, as the interaction of *Bipolaris maydis* with the T-cytoplasm in maize has demonstrated [10,11]. However, more than 70 CMS sources have been reported in the FAO Technical Consultation of the European Cooperative Research Network on Sunflower [12]. These CMS sources have either occurred spontaneously in wild populations or were derived from interspecific crosses or mutagenesis experiments. Prerequisite of the utilization of one of these alternative CMS sources in hybrid breeding is their molecular characterization and the identification of suitable restorer lines. This has happened so far only for very few of these alternative CMS sources [13–18]. In sunflower, the most comprehensive study of 28 male sterile cytoplasms and the fertile, normal cytoplasm was based on hybridization patterns obtained with different mitochondrial genes, which grouped the CMS sources and the fertile cytoplasms into 10 classes using the UPGMA (Unweighted Pair Group Method with Arithmetic Mean) method [19,20].

In this study, CMS PET2, an alternative CMS source derived from an interspecific cross *Helianthus petiolaris* Nutt. × *Helianthus annuus* L. [21], was analyzed. Whereas CMS PET1 went together with nine PET1-like CMS sources [22] into group MT-θ, CMS PET2 grouped together with CMS GIG1 (*H. giganteus* L. × *H. annuus* L., [23]) into MT-γ [19]. Although CMS PET1 and CMS PET2 have the same parental origin regarding the involved species, the differences in the restriction fragment patterns between the mtDNAs of CMS PET1 and CMS PET2 indicate a different molecular mechanism behind male sterility in these two CMS cytoplasms [19]. In CMS PET1, pollen development is aborted in the tetrad stage of meiosis II due to premature programmed cell death of the tapetum cells initiated by the release of cytochrome C from the mitochondria [24,25]. Only very rudimentary very small anthers are formed in CMS PET1, but sunflower lines carrying the PET2 male-sterile cytoplasm still form medium sized anthers (Figure 1). For CMS PET1, it is known that reorganization of mtDNA generated a new open reading frame *orfH522* (coding for a protein of about 16 kDa), which is co-transcribed with *atpA*, now called atp1, and is responsible for the male-sterile phenotype in sunflower CMS PET1 [26–28]. Anther-specific reduction of the co-transcript of *atp1* and *orfH522* restores male fertility [29,30]. Heterologous expression of *orfH522* in tapetal cell layers of tobacco induced male sterility [31], whereas RNAi-mediated silencing of *orfH522* restored fertility [32]. For CMS PET2 and CMS GIG1, a CMS-specific protein of 12.4 kDa was identified by *in organello* translation [15]. However, the corresponding open reading frame has not yet been identified.

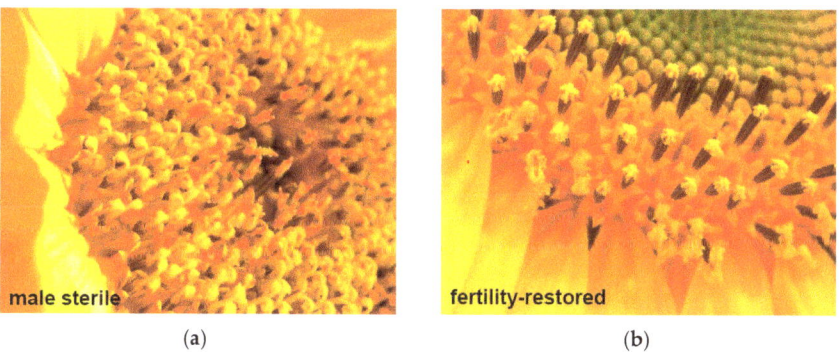

Figure 1. Inflorescence in the present of the CMS PET2 in a maintainer and restorer background. (**a**) CMS PET2 × RHA265, male sterile; (**b**) CMS PET2 × IH-51, fertility-restored hybrid.

In recent years, mitochondrial DNAs have been sequenced to answer several questions [33–37]. Some mitochondrial genome sequences have been successfully used to identify CMS-specific open reading frames [38–40], whereas in other studies no clear answers regarding the CMS mechanism were obtained like in wheat [41] or in pigeonpea [42]. However, only the fertile mitochondrial genome with 300,945 bp [43] and the mtDNAs of CMS PET1 (MG735191.1) have been sequenced and assembled in sunflower, but not CMS PET2.

2. Results

2.1. Identification of Recombination Events in CMS PET2

To investigate the cause of aberrant pollen development in CMS PET2, we analyzed the organization of the mitochondrial DNA, the occurrence of specific new open reading frames, their transcription profiles and their absence in CMS PET1 and the male-fertile line HA89 as reference. The occurrence of two novel open reading frames, *orf288* and *orf231*, inside a second copy of the *atp9* gene and their possible role in causing male-sterility in CMS PET2 are discussed. In addition, diagnostic markers for CMS PET1, CMS PET2 and the fertile cytoplasm in sunflower are presented that can be applied in hybrid breeding.

2. Results

2.1. Identification of Recombination Events in CMS PET2

In Southern hybridizations using *Hind*III as restriction enzyme, CMS PET2 showed identical fragment patterns with CMS PET1 and the male fertile line HA89 for *atp8*, *coxIII* and *nd5* as probes. For the mitochondrial genes *atp6*, *atp9* and *cob*, an additional fragment was present in CMS PET2, apart from the fragments detected for HA89 and CMS PET1 (Figure 2). To establish more precisely the alterations present in CMS PET2, all fragments were cloned and sequenced. In total, 35 open reading frames (>201 bp) were detected, six corresponding to the mitochondrial genes used as probes and 12 were only present in the CMS PET2-specific fragments (Table S1). The organization of all open reading frames is shown in detail in Figures 3–5. For *atp6* as probe, CMS PET2 showed a fragment of 1.2 kb identical to CMS PET1 and the male-fertile line HA89. The additional CMS PET2-specific fragment of 2.5 kb carried a second intact copy of *atp6'* (1056 bp) and two different orfs, *orf321* and *orf255* (Figure 3). For *atp9* as probe, one identical fragment of 3.4-kb was present in CMS PET1, fertile and CMS PET2, and an additional fragment of 4.1-kb was only visible in CMS PET2. The 4.1-kb-fragment contained a split *atp9* gene, which resulted in two new open reading frames of 228 bp and 231 bp, and three additional orfs, *orf285*, *orf267* and *orf627* (Figure 4). Hybridization with *cob* as probe showed the same fragment of 7.3 kb in all cytoplasms carrying the 5' end of *cob* (Figure 5A), as well as a 3.9-kb-fragment for the fertile cytoplasm and CMS PET1, but a 5.5-kb-fragment for CMS PET2 carrying the 3' end of *cob* (Figure 5B). A recombination within *orf843* resulted in a shortened *orf366* and an enlarged *Hind*III fragment in CMS PET2. Seven additional orfs could be identified on this fragment, one of these represents *coxIII*.

Comparison of the fragment sequences with the mitochondrial sequence of HA412 (accession no KF815390) revealed that the fragments present in the male-fertile, normal HA89 and in CMS PET2 were 99–100% homologous to HA412 (Table S2). For the CMS PET2 specific 2.5-kb-fragment (*atp6*) a recombination between two identical areas (326 bp in size) in the mitochondrial DNA seems to be the reason for the larger fragment.

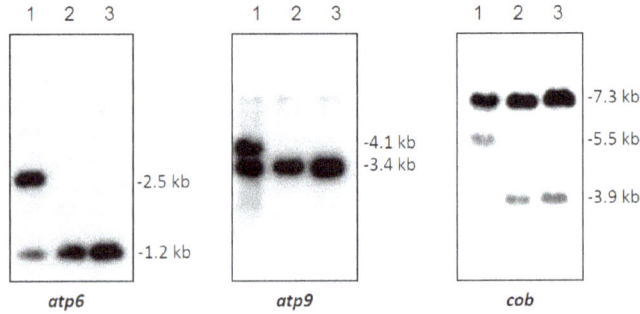

Figure 2. Southern hybridization pattern of mtDNA digested with *Hind*III using the mitochondrial genes *atp6*, *atp9* and *cob* as probes. Lanes: 1, CMS PET2 (male sterile); 2, CMS PET1; 3, HA89 (fertile, normal cytoplasm).

Also, the 5.5-kb-fragment specific for CMS PET2 (*cob*) represents a recombination event between two mtDNA regions, even though there is a small area with no homology between them. However, the 4.1-kb-fragment specific for CMS PET2 (*atp9*) represents a scramble of four small fragments with homology to mitochondrial sequences (>100 bp and <600 bp in size) interrupted by sequences with no homology to the mtDNA.

To answer the question if the detected 12 open reading frames in the additional CMS PET2-specific fragments (Table S1) were unique for CMS PET2 or moved via recombination from other places to these new locations, all orfs present on the CMS PET2-specific fragments were also amplified by PCR with sequence tagged site primers (Table S3) in *H. annuus*, *H. petiolaris*, CMS PET2 and the fertility-restored hybrid CMS PET2 × IH-51. Five open reading frames—*orf288*, *orf231*, *orf285*, *orf267* and *orf627*—proved to be unique to CMS PET2 (Figure S1). All of these were localized in the 4.1-kb-fragment of CMS PET2, which had hybridized to *atp9*. These orfs were also present in the fertility-restored hybrid.

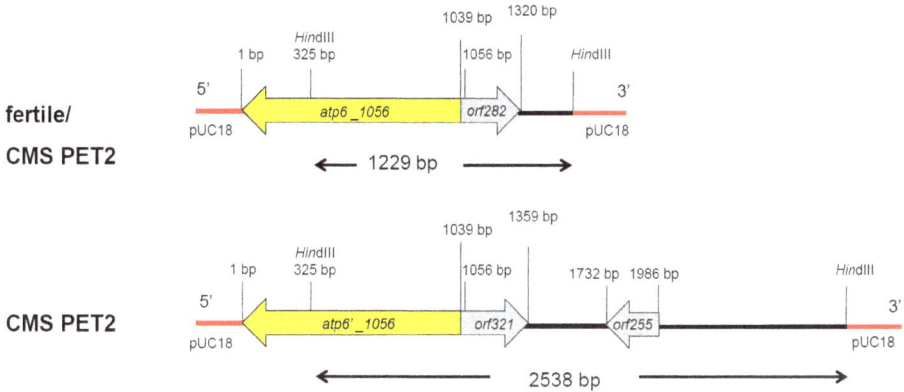

Figure 3. Comparison of the organization of the open reading frames on the hybridization fragments obtained by using *atp6* as probe in the normal, fertile HA89 and CMS PET2.

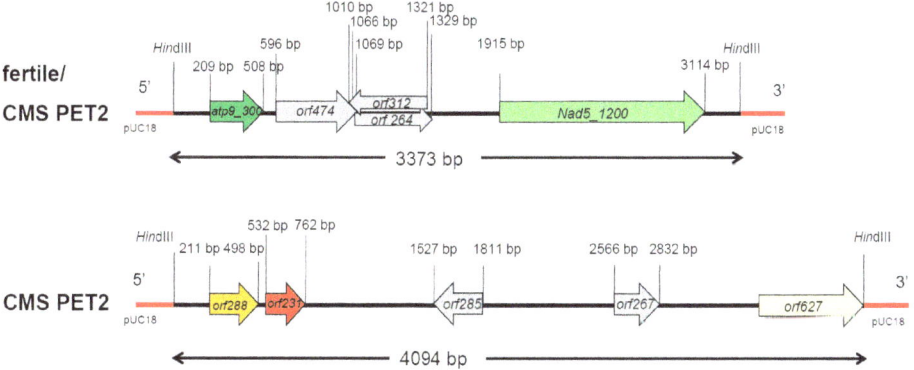

Figure 4. Comparison of the organization of the open reading frames on the hybridization fragments obtained by using *atp9* as probe in the normal, fertile HA89 and CMS PET2.

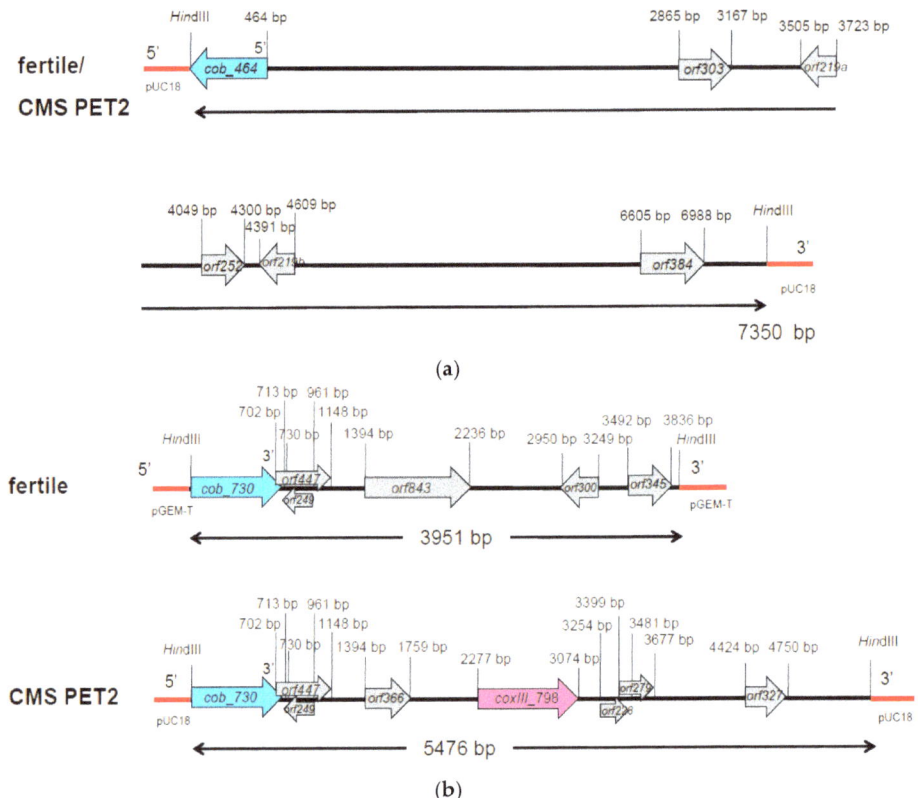

Figure 5. Comparison of the organization of the open reading frames on the hybridization fragments obtained by using *cob* as probe in the normal, fertile HA89 and CMS PET2. (**a**) 5′ cob fragment (7.3 kb) of fertile cytoplasm and CMS PET2, (**b**) 3′ cob fragment for the fertile cytoplasm (3.9 kb) and CMS PET2 (5.5 kb).

2.2. Origin of Orf288 and Orf231 in CMS PET2

After the duplication of the *atp9* gene, an insertion of 271 bp occurred in the 5′ coding region of the *atp9* gene (Figure 6A), creating two new open reading frames, *orf288*, coding for a potential protein of 11.1 kDa, and *orf231*, encoding a 7.9-kDa-protein. Blast analyses indicated that the insertion represents a unique sequence not present elsewhere in genomes. *Orf231* showed 87.4% homology to the *atp9* gene of sunflower. Comparison between the *orf288* and *orfH522*, responsible for male-sterile phenotype in CMS PET1, showed 33.3% homology and *orf288* versus *orfB* 35.3%. These results underline the specificity of *orf288* to CMS PET2. It is interesting to note that the first 53 bp of *orf288* are identical with the 5′ coding region of the *atp9* gene in the CMS PET1 cytoplasm and the male-fertile line HA89. Moreover 19 bp of *atp9* were deleted by the insertion event and the last three base pairs of the 271-bp-insertion act as start codon for *orf231*, which otherwise consists of the 5′ deleted *atp9* (−72 bp/+3 bp) located 33 bp downstream of *orf288*. Furthermore, a small direct repeat of 10 bp (ACTGCTAATC) could be found inside of the 271-bp-insertion (Figure S2). To characterize the protein encoded by *orf288*: it encodes a protein of 95 aa (pI 7.84, Mw 11.1 kDa), of which 16.9% are basic and 10.6% acidic amino acids including a transmembrane domain of 23 aa (Figure 6B).

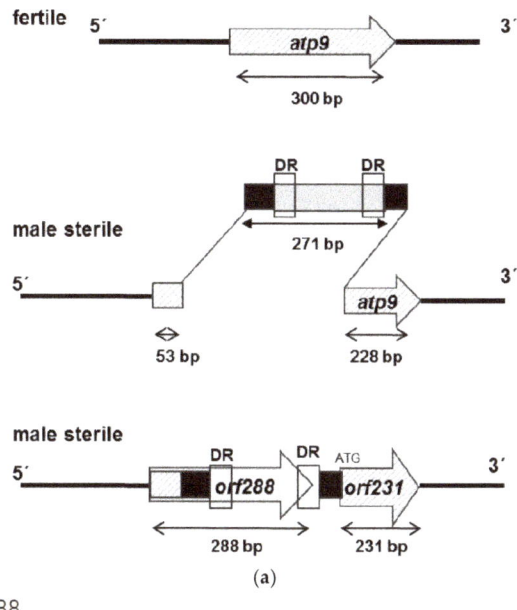

ORF288

MKKKKREENDQLEMLEGALITLIDNIFVKFLLCLLLILVSFLIYTYDRSFRVHQQTLLWAHQ
HNVTPGVSFLYKIRVGHDGTTLDPIPLPVKEHC

ORF231 edited

MAATIALAGAAIGIGNVFSSLIHSVARNPSLAKQLFGYAILGFALTEAIALFALMMAFLILFVF

(b)

Figure 6. Model for the creation of the CMS-PET2 specific *orf288* and *orf231* by an insertion event of 271 bp into the duplicated *atp9* gene. (**a**) Scheme; (**b**) Amino acid sequences of *orf288* and *orf231* (edited). Transmembrane domains as predicted by TMHMM are marked by red bars.

2.3. Expression and RNA Editing of Orf288 and Orf231

In order to characterize *orf288* and *orf231* as potential candidates for male sterility in CMS PET2, semi-quantitative RT-PCRs were performed. Mitochondrial *18S rRNA* was used as internal standard and *atp9* (*orf300*), *atp6* (*orf1056*) and *cob* (*orf1194*) as references. In leaves, disk florets and anthers unique signals for the co-transcript of *orf288* and *orf231* (552 bp) were only detected in CMS PET2 and the fertility-restored hybrid (Figure 7A). On the other hand, expression of *atp9* (*orf300*), *atp6* (*orf1056*) and *cob* (*orf1194*) occurred in CMS PET2, the fertility-restored hybrid as well as the male fertile line HA89 without any observable differences in signal intensity.

In sunflower, 11 editing sites had been described for the *atp9*-mRNA [44], and these were also found in the new *orf231* (Figure S3). All sites are fully edited as in *atp9*, resulting in a protein of 64 amino acids with a molecular weight of 6.7 kDa (pI 8.37, Figure 6B). Inspecting *orf288*, no editing sites could be identified. In contrast to the high signal intensity in CMS PET2, the co-transcript of *orf288* and *orf231* was significantly reduced in the fertility-restored hybrid and not present in the male-fertile line HA89 (Figure 7A). Quantification of the signal intensity revealed a strong down-regulation of the co-transcript (552 bp) in the fertility-restored hybrid by 2.7 in leaves, by 1.9 in disk florets and by 5.4 in anthers in comparison to CMS PET2 (Figure 7B).

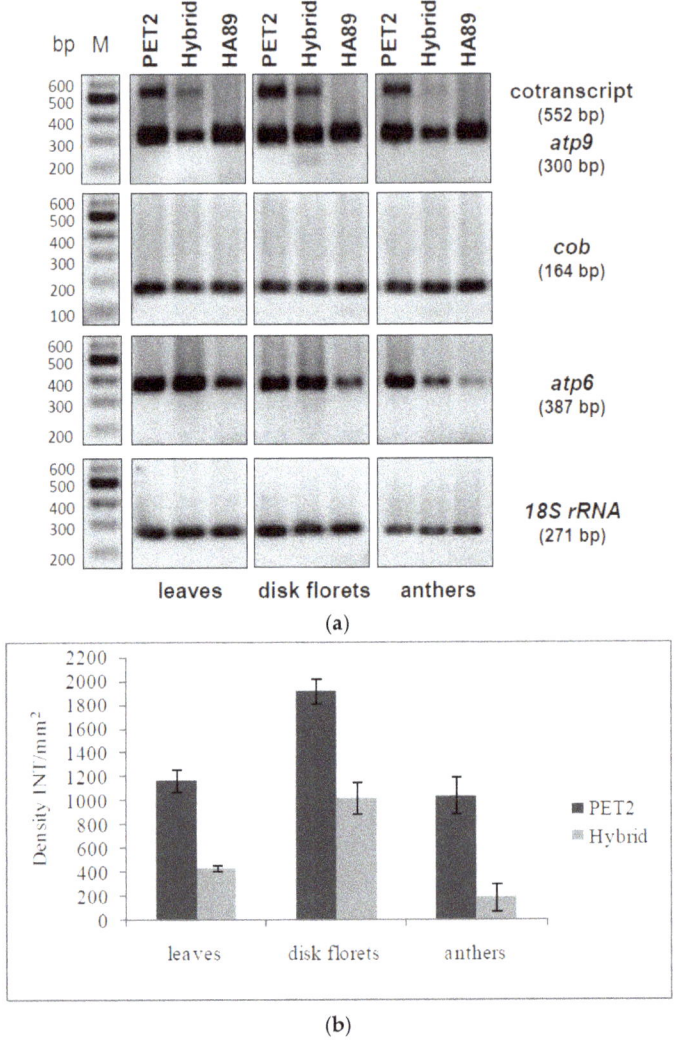

Figure 7. RT-PCR analysis of CMS PET2, the fertility-restored hybrid and the male-fertile line HA89 (**a**) PCR amplification products using primer specific for *atp9* (*orf300*), *cob* (*orf1194*), *atp6* (*orf1056*) and *18S rRNA*. Expected fragment sizes are given in brackets. Lanes: 1, CMS PET2 (male sterile); 2, PET2 × IH-51 (fertile F1-hybrid); 3, HA89 (fertile, normal cytoplasm); M, 100 bp marker; (**b**) Quantification of the co-transcript expression level (552 bp) in CMS PET2 and the fertility-restored hybrid in leaves, disk florets and anthers by densitometry. Dark grey: CMS PET2, light grey: fertility-restored hybrid (CMS PET2 × IH-51).

2.4. Presence of Orf288 and Orf231 in CMS GIG1

Southern hybridization had grouped CMS PET2 and CMS GIG1 together into the MT-γ group [19]. Therefore it was interesting to see if CMS GIG1 also contained *orf288* and *orf231*. Cloning and sequencing of the two PCR products (743 and 491 bp) obtained by the primer combination PET2spec_for/PET2spec_rev identified these two open reading frames also in CMS GIG1 (Figure S4). The second functional copy of the *atp9* was also present. The sequences of *orf288*, *orf231* and *atp9* are

100% identical between CMS PET2 and CMS GIG1. This confirms the relevance of *orf288* and *orf231* for the male sterile phenotype in the presence of these two CMS cytoplasms.

2.5. Protein Analyses of Orf288 and Orf231

For biochemical verification, ORF288 and ORF231 were produced as recombinant proteins after overexpression in *E. coli* using the vector pET28a. Induction of the expression by IPTG reduced the growth of the bacteria heavily indicating a cytotoxic effect of both proteins (data not shown). In both cases, the His-tagged proteins were purified by affinity chromatography on Ni-NTA columns, but this was hampered by the fact that the two membrane proteins ORF288 and ORF231 were overexpressed in form of inclusion bodies. Immuno-blotting analysis using the specific Anti-ORF288 antibody, raised against a peptide derived from ORF288, revealed a strong signal of 57 kDa in CMS PET2 and a much weaker of the same size in HA89 (Figure 8). A specific band of 11 kDa, representing the ORF288 in isolated CMS PET2 mitochondria and protein plant extracts was not observed, but a signal of 16 kDa serving as positive control for antibody reaction against the recombinant protein. It is conceivable that the 57-kDa-signal represents an aggregation of six to eight subunits of ORF288 or that ORF288 was enclosed in another multimeric structure.

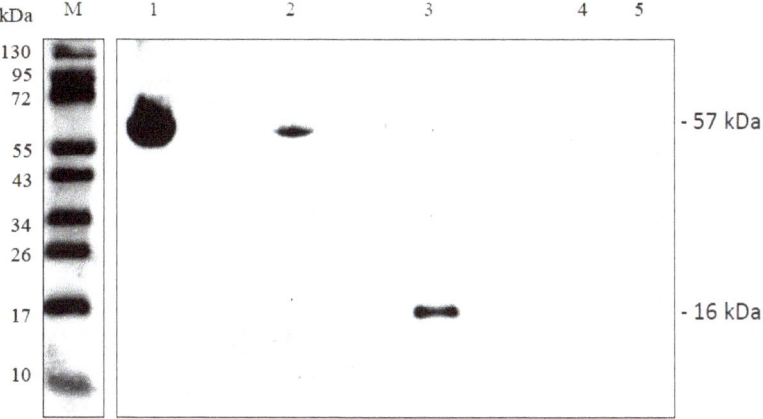

Figure 8. Immuno-blot using complete affinity-purified peptide Anti-ORF288 polyclonal antibody against 10 µg of isolated mitochondria from CMS PET2 and the fertile line HA89, ORF288 recombinant protein and protein plant extracts of CMS PET2 and HA89. Lanes: 1, CMS PET2 mitochondrial extracts; 2, HA89 mitochondrial extracts; 3, recombinant ORF288; 4, CMS PET2 whole plant protein extracts; 5, HA89 whole plant protein extracts; M, prestained protein ladder

2.6. Development of Diagnostic Markers for CMS PET1 and CMS PET2 Cytoplasm

To use different CMS cytoplasms in commercial sunflower hybrid breeding it is essential to distinguish the CMS sources by diagnostic markers. For this purpose four markers were developed (Figure 9): (1) marker HRO_ATP9-PET2 distinguishing the normal, fertile cytoplasm from CMS PET2, (2) marker HRO_PET1 specific for CMS PET1, also showing the absence of *orfH522* in CMS PET2 (3) marker HRO_ATP1-PET1 detecting CMS PET1 in combination with the internal *atp1* control, and (4) marker HRO_ATP1 as internal PCR control present in all cytoplasms. Application of these four diagnostic markers allows a clear differentiation between CMS PET1, CMS PET2 and the fertile cytoplasm in sunflower.

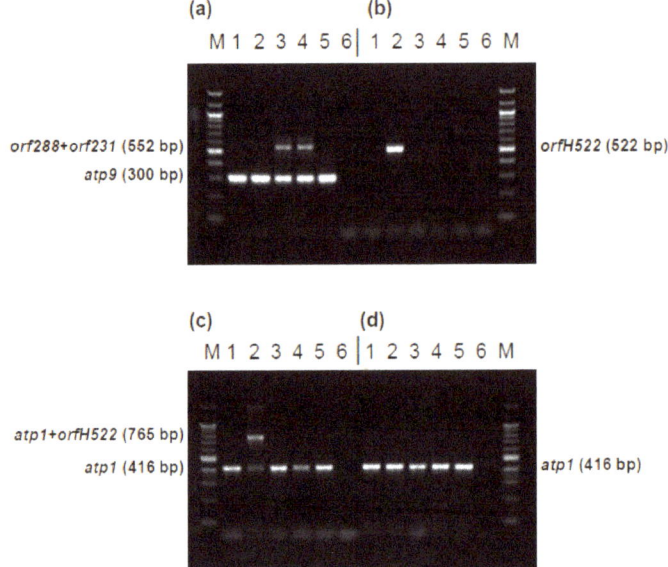

Figure 9. Development of markers specific for CMS PET1, CMS PET2 and the fertile, normal cytoplasm in sunflower. (**a**) Marker HRO_ATP9-PET2 using primer combination orf300_for/orf300_rev; (**b**) Marker HRO_PET1 using primers orfH522_for/orfH522_rev; (**c**) Marker HRO_ATP1-PET1 using primer combination M_atp1_for/M_atp1_rev/orfH522_rev; (**d**) Marker HRO_ATP1 using primer combination M_atp1_for/M_atp1_rev; Lanes: 1, HA342 (male fertile, normal cytoplasm); 2, CMS PET1 × HA342 (male sterile); 3, CMS PET2 × IH51 (fertility restored F1-hybrid); 4, CMS PET2 × RHA265 (male sterile); 5, RHA265 (restorer line, male fertile, normal cytoplasm); 6, PCR negative control; M, 100 bp marker.

3. Discussion

3.1. Molecular Mechanisms behind Male Sterility in CMS PET2

In this study, a correlation between male sterility in CMS PET2 and mitochondrial rearrangements involving the *atp9* gene was identified. In addition, CMS PET2 showed a duplication of *atp6* and a recombination in the second *cob*-specific fragment. However, only the co-transcript of the split second copy of *atp9*, creating two new open reading frames *orf231* and *orf288*, showed a clear reduction of the co-transcript in the fertility-restored hybrids. This reduction of the co-transcript, especially in the anthers of fertility-restored hybrids, indicates an involvement in male sterility. This raised the question, how the proteins encoded by the edited *orf231* (6.7 kDa) or *orf288* (11.1 kDa) and/or their co-transcript could be responsible for male-sterility. In a number of CMS systems, one of the first visible signs of CMS is the premature degeneration of the tapetum layer in the anthers [6]. In the PET1-mediated male-sterility this happens after meiosis II [24]. Release of cytochrome C from the mitochondria in the male-sterile lines leads to a premature programmed cell death [25]. The fact that the transcription rate of the co-transcript of *orf288* and *orf231* is reduced by 5.4 in the anthers of fertility-restored hybrids in comparisons to the male sterile PET2, indicates that reduction of the co-transcript in anther-specific tissues may play an essential role in restoring microspore development. Although the expression of *atp9*, which is highly edited in sunflower [44], is not changed in CMS PET2, the expression of *orf288* and *orf231* might interfere with the correct function of the membrane bound F0-part of the F1F0-ATP synthase due to the partial homology to *atp9* and could thereby be responsible for male sterility in CMS PET2. Lack of ATP would hamper the production of functional pollen, a highly energy demanding process [45]. However, investigations of the respiratory activity and the use of the alternative pathway did not show any weaknesses for CMS PET2 and CMS GIG1 in mitochondria isolated from etiolated seedlings [46]. This indicates that changes in the mitochondrial respiration activity might only be visible in the generative tissue due to interaction with flower-specific factors.

Overexpression of both open reading frames *orf231* and *orf288* leads to a reduction in bacterial growth and indicates that these two proteins might be cytotoxic in E. coli as observed for other CMS-specific proteins [10,31]. The CMS-specific protein of 12.4 kDa observed in the in organello translation products [15] could correspond to the 11.1-kDa-protein encoded by *orf288*. The edited normal *atp9* in sunflower would have a size of 9.2 kDa. It would thereby be smaller than the detected CMS-specific protein, but would have overlapped in the gels with a potential protein of 6.7 kDa encoded by an edited *orf231* mRNA. The antibody produced against the *orf288* peptide showed a strong band of 57 kDa in CMS PET2 and a much weaker in HA89 mitochondrial extracts. The peptide represents the first 14 amino acids of ORF288 and is therefore not only identical to ORF288 but also to ATP9, which explains the signal in HA89. The ATP9 protein tends to form oligomeric structures that are not separated by SDS treatment [47,48] explaining the signal with a larger complex. The signal in CMS PET2 is very strong indicating a higher accumulation of protein complex than in the male-fertile HA89. This complex might represent aggregations of ATP9, ORF288 or ORF231 or combinations of the proteins. In the Owen male sterile cytoplasm in sugar beet, an ATP9 ring structure, not yet assembled into complex V, was also observed [49]. In addition, free ATP9 was accumulated in male-sterile line and reduced upon fertility-restoration. Combining the results of Blue native gel electrophoresis, in-gel activity assays and LC-MS-MS-MS it could be demonstrated for the Owen cytoplasm that preSATP6, the CMS-specific component [50], interacts with the assembly and the activity of F1F0-ATPase, probably via the ATP9 protein [49]. Interestingly, the restorer-of-fertility *Rf1(X)* in sugar beet represents an unusual restorer gene, which encodes a homolog of an OMA1 protein [51]. OMA1 in combination with OXA1 interacts with ATP9 and supports the correct assembly of complex V [52]. Assuming that *Rf1(X)* in sugar beet acts in the same way, fertility restoration leads, by an unknown mechanism, to a reduction in ATP9, but not in preSATP6 [49]. Even accumulation of normal ATP9 may play a role in the CMS mechanism by effects on the mitochondrial proteome [49].

There are still interesting questions for CMS PET2 to be answered to fully understand the CMS mechanism. It would be interesting to see by Blue native gel electrophoresis whether products of *orf288* and *orf231* associate with the F1F0-ATP synthase or other complexes of the respiratory chain and thereby might interfere with the ATP production, especially in the anthers. As proof of concept transgenic tobacco plants expressing either *orf288* or *orf231* under the control of an anther-specific promoter and targeted to the mitochondria should be produced to see if the products of these orfs induce male sterility.

3.2. Recombination Events Leading to Rearrangements at Atp9

Rapid changes in the genome structure via illegitimate recombination activity and emergence of novel split and chimeric gene structures, represent an apparent vulnerability of plant mitochondrial DNA and lead to CMS phenotypes in higher plants [33,53]. A 10-bp-direct repeat was observed within the 271-bp-insertion in the *atp9* gene in CMS PET2. This might be responsible for illegitimate recombination activity [54] and thereby involved in the creation of the male-sterile phenotype in CMS PET2. The use of the flanking regions of mitochondrial respiratory/ATP synthesis-related genes is typical for CMS-specific new orfs [4,6,54]. Especially, parts of *atp6*, *atp8* and *atp9* are most frequently involved in creating CMS-specific new open reading frames [1]. This seems to be also true for CMS PET2, where the insertion event of 271 bp uses the promoter and the first 53 bp of the 5' coding region of the *atp9* gene to create *orf288*. In addition, the insertion provides an ATG start codon, which allows the creation of *orf231* by using the remaining 3' part of the *atp9* and the termination signal. The *orf231* represents the C-terminal transmembrane part of *atp9* as does *orf77* associated with S cytoplasmic male sterility in maize [55]. CMS GIG1, which grouped together with CMS PET2 in the MT-γ group [19], showed the same hybridization pattern as CMS PET2 with *atp9*, *atp6* and *cob* as probes. This was surprising as well as the presence of *orf288* and *orf231* in CMS GIG1 because of the different origin of the CMS sources. CMS GIG1 resulted from an interspecific cross of *H. giganteus* and *H. annuus* [23]). However, also the PET1-like cytoplasms have different origins but the same CMS-mechanism [22]. In addition, for CMS PEF1, which originates from an interspecific cross of *H. petiolaris* ssp. *fallax* with *H. annuus* [56], a modification at the *atp9* gene seems to be responsible for male sterility [13]. Here a 500-bp-insertion in the 3' UTR of the *atp9* gene was identified as cause. Alterations of the *atp9* gene region were also found in *Daucus carota* [57,58], in *Brassica napus* 'Tournefortii-Stiewe' [59] as well as *Boehmeria nivea* [60] and may result in dysfunctions of mitochondria in form of insufficient supply of ATP for pollen development.

3.3. Role of RNA Editing in Creating a Functional ATP9

RNA editing, a posttranscriptional process, which is essential to produce functional proteins as it frequently leads to amino acid exchanges, was observed for *atp9*. In this process the genetic information is typically changed from C-to-U on mRNA level by deamination [61–63]. The *atp9*-mRNA represents one of the best studied RNA editing objects. In most plants like *Oenothera* [64] and potato [65], it is highly edited. Changes in RNA editing can be involved in creating CMS because of amino acid transitions like S to L, P to L and S to F as consequence of the process [65]. In the case of *atp9*, RNA editing is required for creating a functional protein [66]. The fatal role of abnormal *atp9* transcripts for pollen development has been demonstrated in transgenic tobacco and *Arabidopsis thaliana* plants by expressing the unedited *atp9* gene and targeting it to mitochondria [67,68]. Mitochondrial dysfunctions affected normal anther development, especially the fate of the tapetum cell layer and reduced pollen formation [69]. RNA editing tends to increase the proportion of hydrophobic amino acids like leucine, which are crucial for complex formation as well as integrity of proteins [70]. In mitochondria unedited ATP9 protein involved in the proton channel may affect the functioning of F1F0-ATP synthase and thereby reduce the ATP production. However, the CMS PET2-specific *orf231*, which carries all 11 editing sites of *atp9* in sunflower [44], is fully edited.

3.4. CMS PET2 as Alternative to CMS PET1 in Commercial Hybrid Breeding

An alternative CMS source for commercial sunflower hybrid breeding would require that it is based on a different CMS mechanism than CMS PET1. The fact that RHA265, a restorer line of CMS PET1, represents a maintainer line of CMS PET2 [14], has already been a good indicator that CMS PET2 is different from CMS PET1 with regard to the CMS mechanism. In addition, clear differences in the anther morphology between plants carrying CMS PET2 or CMS PET1 also pointed to another mechanism leading to male sterility in CMS PET2. Previous studies on CMS PET1 had shown that a 17-kb-region of the mitochondrial genome, a 12-kb-inversion and a 5-kb-insertion/deletion, flanked by 261-bp inverted-repeats, are involved in PET1 male-sterility [27,71]. The 5-kb-insertion created the new *orfH522* downstream of *atp1* that encodes a 16-kDa-protein, which accumulates in male-sterile and fertility-restored CMS PET1 seedlings [26,28]. The results obtained so far for CMS PET2 indicate that the molecular mechanisms behind the PET2 male-sterility in sunflower depends on recombination events involving the *atp9* gene, which lead to the two new open reading frames *orf288* and *orf231*. In addition, no PCR signal for *orfH522* was visible in CMS PET2. Thereby CMS PET2 clearly differs from the mechanism in CMS PET1.

To use CMS PET2 as alternative male sterility sources in commercial hybrid breeding requires restorer lines with very good fertility restoration capacity and markers to introduce the fertility restorer gene into a breeding pool. In test crosses using five restorer lines of CMS PET1, the line IH-51 produced 100% fertile plants when CMS PET2 was used as mother [14]. Segregation analyses showed that a single restorer gene is responsible for fertility restoration. AFLP markers linked to the restorer gene *Rf_PET2*, which is also located on linkage group 13 as *Rf1*, were identified [72].

Easy differentiation of cytoplasms is essential if different CMS sources are intended to be used in hybrid breeding. This study here, also presents molecular markers distinguishing CMS PET1, CMS

PET2 and the fertile normal cytoplasm in sunflower by easy to use simple PCR-markers. Also in *Raphanus sativa* markers were developed to differentiate the Ogura cytoplasm from other mitochondrial types and to identify novel sub-stoichiometric organizations [73]. In addition, 12 CMS-specific markers as well as SSR-markers were developed to discriminate the mitochondrial genomes present in six *Brassica* species [74]. Also, in *Gossypium hirsutum*, Zhang et al. [75] obtained SCAR and SSR markers to differentiate between CMS and maintainer lines.

Improvements in important agronomical traits can get neglected by restricting hybrid breeding to a single CMS cytoplasm. In potato, cytoplasm specific markers were used to characterize 1217 European potato cultivars with regard to the presence of the six known cytoplasm types (T, D, W, A, M and P) [76]. With regard to agronomic important traits, the W-cytoplasm could be correlated with increased tuber starch content and later maturity whereas the D- and M-type of cytoplasm showed more resistance towards late blight [76]. Diversification on the cytoplasm side can help to improve the agronomic performance of a crop and to reduce the vulnerability to pathogens. The results presented in this study may be the first steps to use more than the CMS PET1 cytoplasm in sunflower commercial hybrid breeding.

4. Materials and Methods

4.1. Plant Material

Three male-fertile sunflower inbred lines, HA89 (maintainer line of CMS PET1 and CMS PET2), RHA265 (maintainer line of CMS PET2, restorer line of CMS PET1), and IH-51 (restorer line of CMS PET2), three CMS-lines PET1 [8], PET2 [21], GIG1 [23], and the fertility-restored hybrid PET2 (RHA265) × IH-51 were used in this study. The CMS lines were obtained from Hervé Serieys within the FAO program [12]. The plants were cultivated in the greenhouse under controlled conditions.

4.2. DNA Isolation and Labeling

Total genomic DNA was extracted from leaves according to the protocol of Doyle and Doyle [77]. Mitochondrial DNA was isolated using the procedure of Horn [78]. The mitochondrial genes *atp6* (subunit 6 of the ATPase gene of sunflower), *atp9* (subunit 9 of the ATPase gene of sunflower) and *cob* (apocytochrome b gene of sunflower) were used as probes after amplification with gene specific primers (Table S3). Primers were designed using web primer (available online: http://www.yeastgenome.org/cgi-bin/web-primer). The probes were labeled, using three different labeling systems: ECL Direct™ Nucleic Acid Labeling and Detection System (Amersham, GE Healthcare, Munich, Germany), ^{32}P radioactively labeled overgo-primer (Hartmann Analytic GmbH, Braunschweig, Germany) and Prime-It II Random Primer Labeling Kit (Agilent Technologies, Santa Clara, CA, USA) according to the supplier's instructions.

4.3. Cloning and Sequencing

The mitochondrial DNA was digested with *Hind*III restriction endonuclease (Fermentas, St. Leon-Rot, Germany), separated on a 0.8% agarose gel and blotted on Hybond N+ membrane (Amersham, GE Healthcare, Munich, Germany) using the procedure of Evans et al. [79] (1994). Hybridization with *atp6*, *atp9* and *cob* as probes using ECL Direct™ Nucleic Acid Labeling and Detection System (Amersham, GE Healthcare, Munich, Germany) according to the manufacturer's recommendations followed. Blots were washed in 0.5 × SSC, 0.4% SDS, 6 M urea at 42 °C for 20 min two times, 2 × SSC, 42 °C for 5 min and exposed to Amersham Hyperfilm ECL (Amersham, GE Healthcare, Munich, Germany) for 0.5–6 h. The *Hind*III digested mtDNA was cloned into pUC18 vectors and the resulting recombinant plasmids were used to prepare a mitochondrial DNA library. Positive clones were selected by *Hind*III hybridization pattern with *atp9*, *atp6* and *cob* as probes (Table S3) and sequenced. The 3.9-kb-*cob* fragment from HA89 was amplified by PCR and cloned into the pGEM-T Easy vector according to the manufacturer's protocol (Promega, Mannheim, Germany). Blast searches against the NCBI database (available online: http://blast.ncbi.nlm.nih.gov/Blast.cgi) and ORF-Finder program (available online: https://www.ncbi.nlm.nih.gov/orffinder/) were used to detect homologies and open reading frames. Molecular weights and isoelectric points were calculated online (available online: http://web.expasy.org/compute_pi/). Transmembrane domains were predicted using the TMHMM Server v.2.0 (available online: http://www.cbs.dtu.dk/services/TMHMM-2.0/).

4.4. RNA Isolation and Reverse Transcriptase (RT)-PCR Analysis

Total RNA was extracted from leaves using TRI Reagent (Sigma-Aldrich Biochemie GmbH, Hamburg, Germany). The RNA extraction from anthers and disk florets was performed by using RNeasy Plant Mini Kit (Qiagen, Hilden, Germany) according to the manufacture's manual. The reverse transcription reaction (RT) was realized with 1 µg RNA and 200 U RevertAid™ H Minus M-MuLV reverse transcriptase (Fermentas, St. Leon-Rot, Germany), routinely carried out at 42 °C followed by incubation at 70 °C for 10 min. Each PCR-experiment was accompanied by the following controls, a reaction in which: (1) no RNA or DNA template was added, (2) DNase treated RNA was added and (3) RNA but no reverse transcriptase was added. All reactions did not yield detectable product. Semiquantitative RT-PCR were performed using the primers given in Table S3. RNA loading was standardized relative to sunflower mitochondrial 18S rRNA.

4.5. Overexpression of Recombinant Orfs in E. coli

The *orf288* and *orf231* (5′ deleted *atp9*) were amplified by PCR using the gene-specific primers with corresponding cleavage sites for *SacI*/*Hind*III (Table S3). PCR fragments were cloned into pGEM-T Easy vector according to the manufacturer's protocol (Promega, Mannheim, Germany) and sequenced with T7 primer. DNA from verified plasmids was cut with *SacI*/*Hind*III and the fragments were cloned into the expression vector pET28a (Novagen, Madison, WI, USA). *E. coli* BL21 strains expressing a fusion protein were grown in LB medium to an OD600 of 0.6. The gene expression was induced by addition of 1 mM IPTG. The fusion proteins carried N-terminal His-tags for purification with the Ni-NTA Fast Start Kit, according to the protocol of Qiagen (Hilden, Germany). The cells were harvested and resuspended in 10 mL digestion buffer pH 8.0 containing 50 mM NaH_2PO_4, 300 mM NaCl and 1 mg/mL lysozyme. The protein was extracted by ultrasonic treatments (4 × 30 s, 90 W) under ice-cooling. Afterwards efficient purification from cleared *E. coli* lysates was done with Ni-NTA columns under denaturing conditions. The eluted proteins were checked regarding purity using SDS-PAGE.

4.6. Immuno-Blotting

A complete affinity-purified peptide Anti-ORF288 polyclonal antibody was produced by GenScript USA (Piscataway, NJ, USA) against the peptide MKKKREENDQLEMC, representing the first 14 amino acids of *orf288* plus a C added for KLH conjugation. For the Western blot, 10 µg of recombinant protein and mitochondrial protein extracts from CMS PET2 and the fertile line were separated in 12% Tris-Tricine gels and blotted onto nitrocellulose membranes (Amersham, GE Healthcare, Munich, Germany). After visualization by Ponceau S staining (0.2% Ponceau S in 0.25% acetic acid) the membrane was blocked in TBS-Triton pH 7.6 (2.42 g Tris, 8 g NaCl, 2.5 mL 20% Triton) with 5% skim milk for 2 h at 25 °C and incubated with the antisera after 10fold dilution in TBS-Triton with 5% skim milk overnight at 4 °C under shaking. After washing with TBS-Triton the secondary peroxidase-labeled anti-rabbit IgG HRP-linked antibody (GE Healthcare, Munich, Germany) diluted 1:2000 was added for 2 h at 25 °C under shaking. The blots were exposed to Amersham Hyperfilm ECL (Amersham, GE Healthcare, Munich, Germany).

4.7. Accession Numbers

The online available accession numbers (https://www.ncbi.nlm.nih.gov/) for *Helianthus annuus* are: *atp6* (X82388), *atp9* (X51895), *nd5* (AF258785.1), *orfB* (X57669.1), *coxIII* (X57669), *cob* (X98362), *orfH522* (X55963), *18S rRNA* (AF107577), complete mitochondrial genome cultivar HA412 (KF815390.1).

Sequences of the *Hind*III fragments obtained with *atp6*, atp9*atp9* and *cob* as probe for CMS PET2 and HA89 were deposited in Genbank under the accession numbers MF828616-MF828625.

Supplementary Materials: Supplementary materials can be found at http://www.mdpi.com/1422-0067/19/3/806/s1.

Acknowledgments: We are grateful to Friedt (University Gießen, Germany) for multiplication of seeds in the field station Groß-Gerau, to the group of Ralf Bastrop (University Rostock, Germany) for part of the sequencing. Sequencing was finalized using the GATC sequencing service. We also like to thank Monja Sundt for the excellent technical assistance (University Rostock, Germany). This work was supported by two grants (HO1593/6-1 and HO1593/6-2) from the Deutsche Forschungsgemeinschaft (DFG, German Research Foundation).

Author Contributions: Antje Reddemann hybridization, cloning and sequencing of most fragments, cloning of *orf288* in CMS GIG1, orf screening, RNA editing analyses, RT-PCR analyses and quantification, immunological studies, draft manuscript; Renate Horn finalizing the results and the manuscript, cloning the 3.9-kb-fragment HA89, final analyses of the orfs and sequence comparison to HA412, verification of the presence of *orf231* and intact *atp9* in CMS GIG1, additional orf screening, contribution of the pictures of CMS PET2 and fertility-restored hybrid, development of the markers for CMS PET2, CMS PET1 and the fertile cytoplasm.

Conflicts of Interest: The authors declare no conflict of interest.

References

1. Horn, R.; Gupta, J.K.; Colombo, N. Mitochondrion role in molecular basis of cytoplasmic male sterility. *Mitochondrion* **2014**, *19*, 198–205. [CrossRef] [PubMed]
2. Kaul, M.L.H. *Male Sterility in Higher Plants. Monographs in Theoretical and Applied Genetics*; Springer: Berlin/Heidelberg, Germany; New York, NY, USA, 1988.
3. Laser, K.D.; Lersten, N.R. Anatomy and cytology of microsporogenesis in cytoplasmic male sterile angiosperms. *Bot. Rev.* **1972**, *33*, 337–346. [CrossRef]
4. Hanson, M.R.; Bentolila, S. Interactions of mitochondrial and nuclear genes that affect male gametophyte development. *Plant Cell* **2004**, *16* (Suppl. 1), S154–S169. [CrossRef] [PubMed]
5. Bohra, A.; Jha, U.C.; Adhimoolam, P.; Bisht, D.; Singh, N.P. Cytoplasmic male sterility (CMS) in hybrid breeding in field crops. *Plant Cell Rep.* **2016**, *35*, 967–993. [CrossRef] [PubMed]
6. Schnable, P.S.; Wise, R.P. The molecular basis of cytoplasmic male sterility and fertility restoration. *Trends Plant Sci.* **1998**, *3*, 175–180. [CrossRef]
7. Reddy, C.V.C.M.; Sinha, R.; Reddy, A.V.V.; Reddy, Y.R. Maintenance of male sterility and fertility restoration in different CMS sources in sunflower (*Helianthus annuus* L.). *Asian J. Plant Sci.* **2008**, *7*, 762–766. [CrossRef]
8. Leclercq, P. Une sterilite male chez le tournesol. *Ann. Amelior. Plants* **1969**, *19*, 99–106.
9. Vear, F. Changes in sunflower breeding over the last fifty years. *OCL* **2016**, *23*, D202. [CrossRef]
10. Miller, R.J.; Koepe, D.E. Southern corn leaf blight: Susceptible and resistant mitochondria. *Science* **1971**, *173*, 67–69. [CrossRef] [PubMed]
11. Levings, C.S. The Texas cytoplasm of maize—Cytoplasmic male-sterility and disease susceptibility. *Science* **1990**, *250*, 942–947. [CrossRef] [PubMed]
12. Serieys, H. Identification, study and utilisation in breeding programs of new CMS sources in the FAO subnetwork. In *Proceedings of the 2005 Sunflower Subnetwork Progress Report, Novi Sad, Serbia and Montenegro, 17–20 July 2005*; FAO: Rome, Italy, 2005; pp. 47–53.
13. De la Canal, L.; Crouzillat, D.; Quetier, F.; Ledoigt, G. A transcriptional alteration on the *atp9* gene is associated with a sunflower male-sterile cytoplasm. *Theor. Appl. Genet.* **2001**, *102*, 1185–1189. [CrossRef]
14. Horn, R.; Friedt, W. Fertility restoration of new CMS sources in sunflower. *Plant Breed.* **1997**, *116*, 317–322. [CrossRef]
15. Horn, R.; Friedt, W. CMS sources in sunflower: Different origin but same mechanism? *Theor. Appl. Genet.* **1999**, *98*, 195–201. [CrossRef]
16. Liu, Z.; Wang, D.; Feng, J.; Seiler, G.J.; Cai, X.; Jan, C.-C. Diversifying sunflower germplasm by integration and mapping of a novel male fertility restoration gene. *Genetics* **2013**, *193*, 727–737. [CrossRef] [PubMed]
17. Schnabel, U.; Engelmann, U.; Horn, R. Development of markers for the use of the PEF1-cytoplasm in sunflower hybrid breeding. *Plant Breed.* **2008**, *127*, 587–591. [CrossRef]
18. Dimitrijevic, A.; Horn, R. Sunflower hybrid breeding: From markers to genomic selection. *Front. Plant Sci.* **2018**, *8*, 2238. [CrossRef] [PubMed]

19. Horn, R. Molecular diversity of male sterility inducing and male-fertile cytoplasms in the genus *Helianthus*. *Theor. Appl. Genet.* **2002**, *104*, 562–570. [CrossRef] [PubMed]
20. Horn, R.; Kusterer, B.; Lazarescu, E.; Prüfe, M.; Özdemir, N.; Friedt, W. Molecular diversity of CMS sources and fertility restoration in the genus *Helianthus*. *Helia* **2002**, *25*, 29–40. [CrossRef]
21. Whelan, E.D.P.; Dedio, W. Registration of sunflower germplasm composite crosses CMG-1, CMG-2, and CMG-3. *Crop Sci.* **1980**, *20*, 832. [CrossRef]
22. Horn, R.; Hustedt, J.E.G.; Horstmeyer, A.; Hahnen, J.; Zetsche, K.; Friedt, W. The CMS-associated 16 kDa protein encoded by *orfH522* is also present in other male sterile cytoplasms of sunflower. *Plant Mol. Biol.* **1996**, *30*, 523–538. [CrossRef] [PubMed]
23. Whelan, E.D.P. Cytoplasmic male sterility in *Helianthus giganteus* L. × *H. annuus* L. interspecific hybrids. *Crop Sci.* **1981**, *21*, 855–858.
24. Horner, H. A comparative light- and electron-microscopic study of microsporogenesis in male-fertile and cytoplasmic male-sterile sunflower (*Helianthus annuus*). *Am. J. Bot.* **1977**, *64*, 745–759. [CrossRef]
25. Balk, J.; Leaver, C.J. The PET1-CMS mitochondrial mutation in sunflower is associated with premature programmed cell death and cytochrome c release. *Plant Cell* **2001**, *13*, 1803–1818. [CrossRef] [PubMed]
26. Horn, R.; Köhler, R.H.; Zetsche, K. A mitochondrial 16 kDa protein is associated with cytoplasmic male sterility in sunflower. *Plant Mol. Biol.* **1991**, *17*, 29–36. [CrossRef] [PubMed]
27. Köhler, R.H.; Horn, R.; Lössl, A.; Zetsche, K. Cytoplasmic male sterility in sunflower is correlated with the co-transcription of a new open reading frame with the *atpA* gene. *Mol. Gen. Genet.* **1991**, *227*, 369–376. [CrossRef] [PubMed]
28. Laver, H.K.; Reynolds, S.J.; Monéger, F.; Leaver, C.J. Mitochondrial genome organization and expression associated with cytoplasmic male sterility in sunflower (*Helianthus annuus*). *Plant J.* **1991**, *1*, 185–193. [CrossRef] [PubMed]
29. Monéger, F.; Smart, C.J.; Leaver, C.J. Nuclear restoration of cytoplasmic male sterility in sunflower is associated with the tissue-specific regulation of a novel mitochondrial gene. *EMBO J.* **1994**, *13*, 8–17. [PubMed]
30. Smart, C.J.; Moneger, F.; Leaver, C.J. Cell-specific regulation of gene expression in mitochondria during anther development in sunflower. *Plant Cell* **1994**, *6*, 811–825. [CrossRef] [PubMed]
31. Nizampatnam, N.R.; Doodhi, H.; Narasimhan, Y.K.; Mulpuri, S.; Viswanathaswamy, D.K. Expression of sunflower cytoplasmic male sterility-associated open reading frame, *orfH522* induces male sterility in transgenic tobacco plants. *Planta* **2009**, *229*, 987–1001. [CrossRef] [PubMed]
32. Nizampatnam, N.R.; Kumar, V.D. Intron hairpin and transitive RNAi mediated silencing of *orfH522* transcripts restores male fertility in transgenic male sterile tobacco plants expressing *orfH522*. *Plant Mol. Biol.* **2011**, *76*, 557–573. [CrossRef] [PubMed]
33. Notsu, Y.; Masood, S.; Nishikawa, T.; Kubo, N.; Akiduki, G.; Nakazono, M.; Hirai, A.; Kadowaki, K. The complete sequence of the rice (*Oryza sativa* L.) mitochondrial genome: Frequent DNA sequence acquisition and loss during the evolution of flowering plants. *Mol. Genet. Genom.* **2002**, *268*, 434–445. [CrossRef] [PubMed]
34. Unseld, M.; Marienfeld, J.R.; Brandt, P.; Brennicke, A. The mitochondrial genome of *Arabidopsis thaliana* contains 57 genes in 366,924 nucleotides. *Nat. Genet.* **1997**, *15*, 57–61. [CrossRef] [PubMed]
35. Tanaka, Y.; Tsuda, M.; Yasumoto, K.; Terachi, T.; Yamagishi, H. The complete mitochondrial genome sequence of *Brassica oleracea* and analysis of coexisting mitotypes. *Curr. Genetics* **2014**, *60*, 277–284. [CrossRef] [PubMed]
36. Bentolila, S.; Stefanov, S. A reevaluation of rice mitochondrial evolution based on the complete sequence of male-fertile and male-sterile mitochondrial genomes. *Plant Physiol.* **2012**, *158*, 996–1017. [CrossRef] [PubMed]
37. Slipiko, M.; Myszczynski, K.; Buczkowska-Chmielewska, K.; Baczkiewicz, A.; Szczecinska, M.; Sawicki, J. Comparative analysis of four *Calypogeia* species revealed unexpected change in evolutionarily-stable liverwort mitogenomes. *Genes* **2017**, *8*, 395. [CrossRef] [PubMed]
38. Igarashi, K.; Kazama, T.; Motomura, K.; Toriyama, K. Whole genomic sequencing of RT98 mitochondria derived from *Oryza rufipogon* and Northern blot analysis to uncover a cytoplasmic male sterility-associated gene. *Plant Cell Physiol.* **2013**, *54*, 237–243. [CrossRef] [PubMed]
39. Okazaki, M.; Kazama, T.; Murata, H.; Motomura, K.; Toriyama, K. Whole mitochondrial genome sequencing and transcriptional analysis to uncover an RT102-type cytoplasmic male sterility-associated candidate gene derived from *Oryza rufipogon*. *Plant Cell Physiol.* **2013**, *54*, 1560–1568. [CrossRef] [PubMed]

40. Tanaka, Y.; Tsuda, M.; Yasumoto, K.; Yamagishi, H.; Terachi, T. A complete mitochondrial genome sequence of Ogura-type male-sterile cytoplasm and its comparative analysis with that of a normal cytoplasm in radish (*Raphanus sativus* L.). *BMC Genom.* **2012**, *13*, 352. [CrossRef] [PubMed]
41. Liu, H.; Cui, P.; Zhan, K.; Lin, Q.; Zhuo, G.; Guo, X.; Ding, F.; Yang, W.; Liu, D.; Hu, S.; et al. Comparative analysis of mitochondrial genomes between a wheat K-type cytoplasmic male sterility (CMS) line and its maintainer line. *BMC Genom.* **2011**, *12*, 163. [CrossRef] [PubMed]
42. Tuteja, R.; Saxena, R.K.; Davila, J.; Shah, T.; Chen, W.; Xiao, Y.L.; Fan, G.; Saxena, K.B.; Alverson, A.J.; Spillane, C.; et al. Cytoplasmic male sterility-associated chimeric open reading frames identified by mitochondrial genome sequencing of four *Cajanus* genotypes. *DNA Res.* **2013**, *20*, 485–495. [CrossRef] [PubMed]
43. Grassa, C.J.; Ebert, D.P.; Kane, N.C.; Rieseberg, L.H. Complete mitochondrial genome sequence of sunflower (*Helianthus annuus* L.). *Genome Announc.* **2016**, *4*, e00981-16. [CrossRef] [PubMed]
44. Recipon, H. The sequence of the sunflower mitochondrial ATPase subunit 9 gene. *Nucleic Acid Res.* **1990**, *18*, 1644. [CrossRef] [PubMed]
45. Luo, D.; Xu, H.; Liu, Z.; Guo, J.; Li, H.; Chen, L.; Fang, C.; Zhang, Q.; Bai, M.; Yao, N.; et al. A detrimental mitochondrial-nuclear interaction causes cytoplasmic male sterility in rice. *Nat. Genet.* **2013**, *45*, 573–577. [CrossRef] [PubMed]
46. Leipner, J.; Horn, R. Nuclear and cytoplasmic differences in the mitochondrial respiration and protein expression of CMS and maintainer lines of sunflower. *Euphytica* **2002**, *123*, 411–419. [CrossRef]
47. Dabbeni-Sala, F.; Rail, A.K.; Lippe, G. proteomics. In *Proteomics—Human Diseases and Protein Functions*; Man, T.K., Flores, R.J., Eds.; INTECH Open Access Publisher: Rijeka, Croatia, 2012; pp. 161–188.
48. Rak, M.; Gokova, S.; Tzagoloff, A. Modular assembly of yeast mitochondrial ATP synthase. *EMBO J.* **2011**, *30*, 920–930. [CrossRef] [PubMed]
49. Wesolowski, W.; Szklarczyk, M.; Szalonek, M.; Slowinska, J. Analysis of the mitochondrial proteome in cytoplasmic male-sterile and male-fertile beets. *J. Proteom.* **2015**, *119*, 61–74. [CrossRef] [PubMed]
50. Yamamoto, M.P.; Kubo, T.; Mikami, T. The 5′-leader sequence of sugar beet mitochondrial *atp6* encodes a novel polypeptide that is characteristic of Owen cytoplasmic male sterility. *Mol. Gen. Genom.* **2005**, *273*, 342–349. [CrossRef] [PubMed]
51. Matsuhira, H.; Kagami, H.; Kurata, M.; Kitazaka, K.; Matsunaga, M.; Hamaguchi, Y.; Hagihara, E.; Ueda, M.; Harada, M.; Muramatsu, A.; et al. Unusual and typical features of a novel restorer-of-fertility gene of sugar beet (*Beta vulgaris* L.). *Genetics* **2012**, *192*, 1347–1358. [CrossRef] [PubMed]
52. Jia, L.; Dienhardt, M.K.; Stuart, R.A. Oxa1 directly interacts with ATP9 and mediates its assembly into the mitochondrial F_1F_0-ATP synthase complex. *Mol. Biol. Cell* **2007**, *18*, 1897–18908. [CrossRef] [PubMed]
53. Kühn, K.; Gualberto, J.M. Recombination in the stability, repair and evolution of the mitochondrial genome. *Adv. Bot. Res.* **2012**, *63*, 215–252.
54. Mackenzie, S.; McIntosh, L. Higher plant mitochondria. *Plant Cell* **1999**, *11*, 571–586. [CrossRef] [PubMed]
55. Gallagher, L.J.; Betz, S.K.; Chase, C.D. Mitochondrial RNA editing truncates a chimeric open reading frame associated with S male-sterility in maize. *Curr. Genet.* **2002**, *42*, 179–184. [CrossRef] [PubMed]
56. Serieys, H.; Vincourt, P. Characterisation of some new CMS sources from *Helianthus* genus. *Helia* **1987**, *10*, 9–13.
57. Szklarczyk, M.; Oczkowski, M.; Augustyniak, H.; Börner, T.; Linke, B.; Michalik, B. Organisation and expression of mitochondrial *atp9* genes from CMS and fertile carrots. *Theor. Appl. Genet.* **2000**, *100*, 263–270. [CrossRef]
58. Szklarczyk, M.; Szymanski, M.; Wójik-Jagla, M.; Simon, P.W.; Weihe, A.; Börner, T. Mitochondrial *atp9* genes from petaloid male-sterile and male fertile carrots differ in their status of heteroplasmy recombination involvement, post-transcriptional processing as well as accumulation of RNA and protein product. *Theor. Appl. Genet.* **2014**, *127*, 1689–1701. [CrossRef] [PubMed]
59. Dieterich, J.H.; Braun, H.P.; Schmitz, U.K. Alloplasmic male sterility in *Brassica napus* (CMS 'Tournefortii-Stiewe') is associated with a special gene arrangement around a novel *atp9* gene. *Mol. Genet. Genom.* **1993**, *269*, 723–731. [CrossRef] [PubMed]
60. Liu, X.-L.; Zhang, S.-W.; Duan, J.-Q.; Du, G.-H.; Liu, F.-H. Mitochondrial genes *atp6* and *atp9* cloned and characterized from ramie (*Boehmeria nivea* (L.) Gaud.) and their relationship with cytoplasmic male sterility. *Mol. Breed.* **2012**, *30*, 23–32. [CrossRef]

61. Takena, M.; Verbitskiy, D.; Zehrmann, A.; Härtel, B.; Bayer-Csaszar, E.; Glass, F.; Brennicke, A. RNA editing in plant mitochondria—Connecting RNA target sequences and acting proteins. *Mitochondrion* **2014**, *19*, 191–197. [CrossRef] [PubMed]
62. Hammani, K.; Giegé, P. RNA metabolism in plant mitochondria. *Cell* **2014**, *19*, 380–389. [CrossRef] [PubMed]
63. Giege, P.; Brennicke, A. RNA editing in *Arabidopsis* mitochondria effects 441 C to U changes in ORFs. *Proc. Natl. Acad. Sci. USA* **1999**, *96*, 15324–15329. [CrossRef] [PubMed]
64. Schuster, W.; Brennicke, A. RNA editing of ATPase subunit 9 transcripts in *Oenothera* mitochondria. *FEBS Lett.* **1990**, *268*, 252–256. [CrossRef]
65. Dellorto, P.; Moenne, A.; Graves, P.V.; Jordana, X. The potato mitochondrial ATP synthase subunit-9—Gene structure, RNA editing and partial protein-sequence. *Plant Sci.* **1993**, *88*, 45–53. [CrossRef]
66. Bégu, D.; Graves, P.-V.; Domec, C.; Arselin, G.; Litvak, S.; Araya, A. RNA editing of wheat mitochondrial ATP synthase subunit 9: Direct protein and cDNA sequencing. *Plant Cell* **1990**, *2*, 1283–1290. [CrossRef] [PubMed]
67. Hernould, M.; Suharsono, S.; Litvak, S.; Araya, A.; Mouras, A. Male-sterility induction in transgenic tobacco plants with an unedited *atp9* mitochondrial gene from wheat. *Proc. Natl. Acad. Sci. USA* **1993**, *90*, 2370–2374. [CrossRef] [PubMed]
68. Gomez-Casati, D.F.; Busi, M.V.; Gonzalez-Schain, N.; Mouras, A.; Zabaleta, E.J.; Araya, A. A mitochondrial dysfunction induces the expression of nuclear-encoded complex I genes in engineered male sterile *Arabidopsis thaliana*. *FEBS Lett.* **2002**, *532*, 70–74. [CrossRef]
69. Zabaleta, E.; Mouras, A.; Hernould, M.; Araya, A. Transgenic male-sterile plant induced by an unedited *atp9* gene is restored to fertility by inhibiting its expression with antisense RNA. *Proc. Natl. Acad. Sci. USA* **1996**, *93*, 11259–11263. [CrossRef] [PubMed]
70. Kalinati, N.Y.; Kumar, D.V.; Reddy, S.S. RNA editing of the *nad3* and *atp9* mitochondrial gene transcripts of safflower (*Carthamus tinctorius*). *Int. J. Integr. Biol.* **2008**, *3*, 143–149.
71. Siculella, L.; Palmer, J.D. Physical and gene organization of mitochondrial DNA in fertile and male sterile sunflower. CMS-associated alterations in structure and transcription of the *atpA* gene. *Nucl. Acids Res.* **1988**, *16*, 3787–3799. [CrossRef] [PubMed]
72. Horn, R.; Reddemann, A.; Drumeva, M. Comparison of cytoplasmic male sterility based on PET1 and PET2 cytoplasm in sunflower (*Helianthus annuus* L.). In Proceedings of the 2016 19th International Sunflower Conference, Edirne, Turkey, 29 May–6 June 2016; pp. 620–629.
73. Kim, S.; Lim, H.; Park, S.; Cho, K.-H.; Sung, S.-K.; Oh, D.-G.; Kim, K.-T. Identification of a novel mitochondrial genome type and development of molecular markers for cytoplasm classification in radish (*Raphanus sativus* L.). *Theor. Appl. Genet.* **2007**, *115*, 1137–1145. [CrossRef] [PubMed]
74. Liu, G.; Zhao, Z.; Xiao, M.; Mason, A.S.; Yan, H.; Zhou, Q.; Fu, D. Repetitive sequence characterization and development of SSR and CMS-specific markers in the *Brassica* mitochondrial genomes. *Mol. Breed.* **2015**, *35*, 219. [CrossRef]
75. Zhang, X.; Meng, Z.; Zhou, T.; Sun, G.; Shi, J.; Yu, Y.; Zhang, R.; Guo, S. Mitochondrial SCAR and SSR markers for distinguishing cytoplasmic male sterile lines from their isogenic maintainer lines in cotton. *Plant Breed.* **2012**, *131*, 563–570. [CrossRef]
76. Sanetomo, R.; Gebhardt, C. Cytoplasmic genome types of European potatoes and their effects on complex agronomic traits. *BMC Plant Biol.* **2015**, *15*, 162. [CrossRef] [PubMed]
77. Doyle, J.J.; Doyle, J.L. A rapid DNA isolation procedure for small quantities of fresh leaf tissue. *Phytochem. Bull.* **1987**, *19*, 11–15.
78. Horn, R. Technical protocol for mitochondria isolation for different studies. In *Alternative Respiratory Pathways in Higher Plants*; Gupta, K.J., Mur, L.A., Neelwarne, B., Eds.; John Wiley & Sons, Ltd.: Chichester, UK, 2015; pp. 347–358. [CrossRef]
79. Evans, M.R.; Bertera, A.L.; Harris, D.W. The Southern blot. *Mol. Biotechnol.* **1994**, *1*, 1–12. [CrossRef] [PubMed]

© 2018 by the authors. Licensee MDPI, Basel, Switzerland. This article is an open access article distributed under the terms and conditions of the Creative Commons Attribution (CC BY) license (http://creativecommons.org/licenses/by/4.0/).

Review

The Role of Non-Coding RNAs in Cytoplasmic Male Sterility in Flowering Plants

Helena Štorchová

Institute of Experimental Botany of the Czech Academy of Sciences, Rozvojová 263, 16502 Prague, Czech Republic; storchova@ueb.cas.cz

Received: 1 November 2017; Accepted: 14 November 2017; Published: 16 November 2017

Abstract: The interactions between mitochondria and nucleus substantially influence plant development, stress response and morphological features. The prominent example of a mitochondrial-nuclear interaction is cytoplasmic male sterility (CMS), when plants produce aborted anthers or inviable pollen. The genes responsible for CMS are located in mitochondrial genome, but their expression is controlled by nuclear genes, called fertility restorers. Recent explosion of high-throughput sequencing methods enabled to study transcriptomic alterations in the level of non-coding RNAs under CMS biogenesis. We summarize current knowledge of the role of nucleus encoded regulatory non-coding RNAs (long non-coding RNA, microRNA as well as small interfering RNA) in CMS. We also focus on the emerging data of non-coding RNAs encoded by mitochondrial genome and their possible involvement in mitochondrial-nuclear interactions and CMS development.

Keywords: cytoplasmic male sterility; non-coding RNA; global transcriptome; gene expression; pollen development

1. Introduction

In plants, male sterility refers to the inability to generate viable pollen. It is encoded by nuclear genes leading to the genic male sterility (GMS) or by mitochondrial genes interacting with nuclear genes resulting in the development of cytoplasmic male sterility (CMS). Both kinds of male sterility are broadly utilized in agriculture for the production of hybrid crops providing higher yield than inbred parents [1]. The existence of male-sterile lines eliminates the need for laborious sterilization in a long array of crops including rice (*Oryza sativa*), maize (*Zea mays*), wheat (*Triticum aestivum*), sorghum (*Sorghum bicolor*), sunflower (*Helianthus annuus*) and sugar beet (*Beta vulgaris*). Despite more than two centuries of research and high economic importance, CMS mechanisms remain poorly understood. CMS represents a special case of mitochondrial-nuclear interaction, which is regulated at multiple levels. New discoveries highlight the contributions of non-coding RNAs—a genomic "dark matter" [2]—to the complex regulatory network controlling CMS. In this review, I discuss recent observations and evidence for the action of various classes of plant non-coding RNAs in CMS biogenesis and pollen development.

2. Mitochondrial CMS Genes and Their Mode of Action

The mitochondrial and nuclear genes involved in CMS biogenesis or associated with CMS are very diverse [3,4]. Mitochondrial CMS genes are often chimeric, comprised of pieces of essential genes or unknown open reading frames (ORF) [5–8]. Chimeric CMS genes can be generated by intramolecular recombination events. For example, maize male-sterile Texas (CMS-T), one of the first CMS lines used in agriculture, possesses *T-urf13* gene [9], derived from at least seven recombination events involving *atp6* and *rrn26* mitochondrial genes. Functional copies of *atp6* and *rrn26* remain in another part of the mitochondrial genome [9]. Another example of a mitochondrial CMS gene is a mutation in *cox2* encoding the cytochrome *c* oxidase subunit 2 truncated protein appearance [10]. CMS genes are very diverse not only across angiosperms, but also within species. For example, numerous CMS systems were described in rice [11,12], maize [9,13], and sugar beet [10,14]. Accordingly, we may expect them to employ similarly diverse modes of action.

The precise details how CMS gene expression impairs mitochondria and the pollen development is not known, but several models have been proposed. Mitochondrial CMS genes may code for cytotoxic proteins like URF13 in maize CMS-T [9], or they may cause energy deficiency during energetically highly energetically demanding male (but not female) reproductive development. Many CMS proteins are hydrophobic and could interfere with oxidative phosphorylation (OXPHOS) complexes within inner mitochondrial membrane [14–16], which may decrease the ATP production. Another mechanism can be a premature or delayed programmed cell death (PCD) of the tapetum, the innermost cell layer of the anther wall [17], crucial for the pollen development [18]. PCD of the tapetum, which provides nutrients for the pollen maturation, must be properly timed. CMS genes are often transcribed both in vegetative tissues and anthers, but the respective proteins are produced only at a specific time and tissue. Rice CMS-WA [19] provides a clear example: the WA352 protein accumulates only in tapetal cells and only at the microspore mother cell stage, although *WA352* transcripts are constitutively present in all tissues. Accurate spatiotemporal patterning of CMS-associated mitochondrial genome expression requires fine-tuned regulation at transcriptional, post-transcriptional, translational and post-translational levels. This is achieved by employing plethora of transcription factors and regulatory non-coding RNAs which affect transcript longevity and translation efficiency [2,20].

3. Restoration of Fertility by Nuclear Genes

The sterility effects of the mitochondrial CMS genes may be inhibited by the nuclear *Restorer of fertility* (*Rf*) genes, which re-enable the development of functional anthers, fertile pollen, and hermaphroditic flowers [21]. The *Rf* genes suppress CMS genes' sterilizing effect by degrading or cleaving their mRNAs [22,23], or by post-transcriptional modification including the RNA editing [24]. Alternatively, either the translation of CMS-associated transcript may be blocked [19,25], or the CMS protein degraded [26].

The majority of *Rf* genes belong to the large family of *Pentatricopeptide Repeat* (*PPR*) genes [27]. This gene family, highly expanded in flowering plants, controls multiple aspects of the organellar gene expression, including RNA editing, RNA stabilization and processing, and translation initiation. All PPR proteins contain P-type 35 amino acid domains, each of which recognizes a single nucleotide of RNA [28]. The *Rf* genes constitute a specific *PPR* subfamily called *Restorer of fertility-like* (*RFL*), which shows an accelerated evolutionary rate [29], and numerous domain-level recombination events [30]. Mitochondrial CMS genes and *RFL* genes may co-evolve similarly to a host-pathogen system [31].

In keeping with their tremendous diversity, not all *Rf* genes code for PPR proteins. The *Rf2* gene in maize CMS-T encodes mitochondrial aldehyde dehydrogenase and restores male fertility at a metabolic level [32,33]. Another example is the restoration factor *Rf17* in rice, bearing protein sequence similarity with acyl-carrier proteins [34], which restores male fertility by retrograde mitochondrial-nuclear signaling pathway.

Whereas CMS has been well studied in agricultural plants, its occurrence in the remaining species has been under the less attention. CMS forms the basis for the widespread plant reproduction system-gynodioecy, characterized by the co-occurrence of male-sterile (female) and hermaphrodite individuals in the same populations [35]. However, only a handful CMS systems from natural populations were studied at the molecular level [36–39]. Domestication is associated with a strong selection for beneficial features which also results in the loss of genetic diversity. We may therefore expect an even more diverse collection of CMS-associated genes in the wild than in agricultural species investigated so far.

4. Non-Coding RNAs in Pollen Development and CMS

CMS as a specific case of mitochondrial-nuclear interaction is a complex phenomenon which has to be tightly regulated at multiple levels. Its intricacy became apparent with the recent onset of high throughput methods which enabled complete genome and global transcriptome sequencing [40–43]. Besides long studied transcription factors [44,45], non-coding RNAs (ncRNAs) became to draw attention.

Non-coding RNAs contain no large open reading frame (ORF) and are therefore presumed not to encode proteins. They comprise well-studied structural RNAs, such as rRNA, tRNA, snoRNA, snRNA etc., and regulatory RNA. The latter are either longer than 200 nt (long non-coding RNA-lncRNA), or shorter (small RNA–sRNA) [20,46]. Among sRNAs, microRNAs (miRNAs) are the known regulators of gene expression at the post-transcriptional level [47,48]. Another subclass of ncRNAs is represented by small-interfering RNAs (siRNAs), which are involved in the defense against viruses and mobile elements [49–51]. Next, *trans*-acting small interfering RNAs (ta-siRNAs) are endogenous regulatory elements participating in complex signaling networks [52]. Of all the classes of regulatory RNAs, only the function of miRNAs in CMS biogenesis has been investigated in detail.

5. miRNAs

miRNAs are small RNAs (about 21 nt) that guide the RNA-induced silencing complex (RISC) to the target transcripts, inducing their cleavage or translational inhibition [47]. They are present in most eukaryotic organisms, but their biogenesis and signaling pathways notably differ between plants and animals [53]. miRNAs are encoded by their own genes at various genomic loci. RNA polymerase II generates a long primary miRNA transcript (pri-miRNA), which contains a hairpin with the miRNA sequence. Plant pri-miRNAs are cut in the nucleus by the complex comprised of RNase DICER LIKE1 (DCL1) and additional proteins e.g., HYPONASTIC LEAVES1 (HYL1) and SERRATE (SE) [54], producing the miRNA duplex. This duplex is transported to the cytoplasm, where it interacts with ARGONAUTE1 (AGO1) to form RISC and to guide it to target genes (Figure 1). Some miRNAs are evolutionary conserved, but many are species-specific [55].

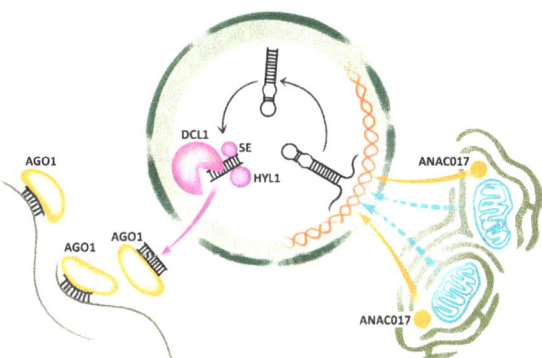

Figure 1. Induction and biogenesis of miRNA during CMS. Mitochondrial biogenesis is altered by the action of cytoplasmic male sterility (CMS)-associated genes sending retrograde signals to the nucleus by means of the NAC transcription factor ANAC017 localized close to endoplasmatic reticulum and/or by other unknown factors. They trigger miRNA gene expression and the production of pri-miRNA, which is subsequently trimmed by the complex containing DICER LIKE1 (DCL1), HYPONASTIC LEAVES1 (HYL), SERRATE (SE) and other proteins in the nucleus. Afterwards, miRNA duplexes are transported to the cytoplasm, where they join ARGONAUTE1 (AGO1), find target mRNAs and initiate its cleavage or translation inhibition by RNA-induced silencing complex (RISC). The figure is based on well-supported model except for blue dashed arrows representing an unknown signal.

Plant miRNAs control multiple aspects of plant development and stress response, including shoot and root apical development, leaf and trichome development, floral transition and fruit size as well as nutrition-, drought-, salinity- and heat-stress responses [48]. They are also prominent regulators of the pollen development [56]. In the last decade, numerous studies have compared the microtranscriptomes of CMS and fertile lines of agricultural species. (Tables 1 and S1).

Hundreds of miRNAs have been identified by these studies, many of them belonging to the novel ones. Some miRNAs were differentially expressed between sterile and fertile lines (e.g., 47 in *Brassica juncea* [57]; 42 in cybrid pummelo (*Citrus grandis*) [58]; 87 in *Brassica rapa* CMS-Ogura [59]). Evolutionarily conserved miR156/7a targeting *SQUAMOSA PROMOTER BINDING PROTEIN-LIKE* (*SPL*), which regulates flowering, leaf shape and also tapetum development, was frequently present among differentially expressed miRNAs. Other examples of miRNAs involved in pollen development are miR166 targeting the transcription factor HD-ZIPIII [60], or miR167 targeting *AUXIN RESPONSE FACTOR* (*ARF*) genes, which control anther dehiscence [57]. A broad array of metabolic processes-catabolism of fatty acids [61], sugar transport [61] or inorganic phosphate homeostasis [58] occurring in anthers were affected by the differentially expressed miRNAs. Differentially expressed genes identified by the comparison between the cytoplasmic mRNA-derived transcriptomes of CMS and fertile lines belonged to the similar functional categories as differentially expressed miRNAs. They were involved in starch and sucrose metabolism, amino acid and sulphur metabolism, flavonoid biosynthesis, or pollen development [62,63].

Whereas the impact of mitochondrial CMS genes on the global transcriptome is well described in many crops, the retrograde signal which communicates the mitochondrial impairments to the nucleus of tapetal or pollen cells and triggers diverse cascades of regulatory elements is still elusive. The recently described NAC transcription factor ANAC017 [64] is not the only factor responsible for the mitochondrial retrograde signaling [65] and the mediators communicating between mitochondria and nucleus in CMS are yet to be discovered. In parallel with animal mitochondria [2], we may assume that not only proteins, but also ncRNAs may convey retrograde signals (Figure 1).

Table 1. The examples of miRNAs and their putative target genes differentially expressed between the CMS lines and their maintainers in various crops.

miRNA	Putative Target Genes	Target Gene Functions	References
Maize CMS C48-2			
Zma-miR397c	Laccase	Oxidation of phenolic substrates	[61]
Zma-miR601	Flavin-containing monooxygenase (FMO)	Auxin biosynthesis	
	Enoyl-CoA hydratase	Catabolism of fatty acids	
Zma-miR604	Monosaccharide transport protein 2 (STP2)	Uptake of glucose from callose degradation	
Brassica juncea hybrid			
miR156a	SPL transcription factors	Floral transition, tapetum development	[57]
miR167a	Auxin response factor (ARF6/ARF8)	Anther dehiscence	
miR319a	TCP transcription factors	Floral induction	
miR395a	ATP sulphurylase (APS)	Sulphur metabolism	
Rice MeixiangA			
osa-miR528-3p	F-box containing protein	Proteolytic turnover through proteasome	[42]
osa-miR1432-5p	Metal cation transporter	Cation homeostasis	
osa-miR2118c	NBS-LRR	Disease-resistance related proteins	
Brassica oleracea Bo01-12A			
bol-miR157a	SPL transcription factors	Floral transition, tapetum development	[40]
bol-miR171a	SCARECROW-like (SCL) transcription factor	GA mediated action	
bol-miR172	APETALA2 (AP2) transcription factor	Floral transition	
bol-miR824	MADS-box transcription factor-like	Plant development	
Brassica rapa CMS-Ogura			
bra-miR157a	SPL transcription factors	Floral transition, tapetum development	[59]
bra-miR158-3p	PPR-RFL	RNA metabolism in organelles	
bra-miR159a	MYB81 transcription factor	Flowering	
bra-miR164a	CUP SHAPED COTYLEDON 1	Meristem development	
bra-miR172a	APETALA2 (AP2) transcription factor	Floral transition	
bra-miR5712	VACUOLAR ATP SYNTHASE SUBUNIT A	Male gametophyte development	
bra-miR5716	Zinc finger transcription factor	Drought stress response	
bra-miR6030	CC-NBS-LRR	Disease-resistance related proteins	

Table 1. Cont.

miRNA	Putative Target Genes	Target Gene Functions	References
Glycine max NJCMS1A			
gma-miR166a-3p	HD-ZIPIII transcription factor	Vascular nad cell wall development	[60]
gma-miR169b	Nuclear factor Y (NF-YA) transcription factor	Flowering	
gma-miR171a	SCARECROW-like (SCL) transcription factor	GA mediated action	
gma-miR394b-5p	F-box protein	Proteolytic turnover through proteasome	
gma-miR395c	Sulphate transporter 2.1-like	Sulphur metabolism	
gma-miR396k-5p	bHLH79 transcription factor	Floral development	
gma-miR397a	Laccase	Oxidation of phenolic substrates	
gma-miR408c-3p	Plastocyanin-like	Copper metabolism	
Raphanus sativus CMS-WA			
miR-158b-3p	PPR-RFL	RNA metabolism in organelles	[43]
miR161	Mechanosensitive channel of small conductance-like 10 (MSL10)	Mechanosensitive ion channel, cell death induction	
miR395a	putative F-box/kelch-repeat (KFB)	Proteolytic turnover through proteasome	
Pummelo cybrid line			
cga-miR156a.1	SPL transcription factors	Floral transition, tapetum development	[58]
cga-miR399a.1	UBC (ubiquitin-conjugating E2 enzyme)	Phosphate (P_i) homeostasis	
cga-miR827	Basic leucine zipper (bZIP)	Pollen and flower development	

6. siRNAs and ta-si RNAs

Unlike miRNAs, siRNAs mediate the silencing of the same genes from which they originate [53]. They are typically 20–24 nt long and are cleaved from a long precursor dsRNA, which may be derived from viruses, transposons or the combination of a sense and antisense transcript. They guide RISC complex to degrade complementary RNA of viral or endogenous origin, or inhibit translation of respective mRNAs. Alternatively, siRNAs mediate de novo modification to form transcriptionally inactive chromatin by recruiting DNA- and histone-modifying enzymes to the specific chromosomal targets [66]. Plant siRNAs may move from cell to cell, but also at longer distance through plasmodesmata or the vascular phloem tissue [53]. They facilitate the communication among individual organs, fine-tuning the response to environmental cues.

The complex interplay between siRNAs and miRNAs is illustrated by the action of ta-si RNAs. They are produced from the transcripts of *TRANS-ACTING SIRNA* (*TAS*) genes which are initially cleaved by the specific miRNAs. Instead being degraded, *TAS* cleavage products are transcribed by the RNA-dependent polymerase and subsequently diced by the DCL4 complex. The resulting siRNAs have a phased pattern, starting at the miRNA cleavage site. They may target the parental or different genes, frequently the members of large gene families [67,68].

The miR173-*TAS1/2-PPR* ta-si pathway has been described in *Arabidopsis thaliana* [68,69]. It is triggered by miR173 which initiates the cleavage of *TAS1/2* transcripts and the subsequent production of ta-si RNAs targeting selected *PPR* genes. This pathway is highly conserved across angiosperms [70]. Considering the pervasive influence of *PPR* genes on mitochondrial metabolism, the participation of ta-si RNA in CMS biogenesis is highly plausible, but has not yet been demonstrated. Another example of ta-si pathway which may play a role in the CMS biogenesis is the miR390-*TAS3* module. It may influence pollen development by modulating *ARF* gene expression [68].

7. lncRNA

The field of lncRNAs research has expanded and accelerated recently [71,72]. lncRNAs represent the most diverse class of regulatory ncRNA. They are capped and polyadenylated [73,74]; some of them, however, do not contain poly(A) tail [75]. Their modes of action are very diverse. They may be produced as antisense transcripts and to inhibit *sense* transcription (e.g., COOLAIR, a cold-induced antisense transcript of the floral inhibitor *FLOWERING LOCUS C* (*FLC*) in *A. thaliana* [76]).

Some lncRNAs influence alternative splicing. They interact with the nuclear speckle RNA-binding protein (NSR), which forms complexes with pre-mRNAs. At least two lncRNAs compete with target pre-mRNAs for NSR and modify their alternative splicing during lateral root formation [77].

The chromatin remodeling caused by lncRNA was also described. The lncRNA *COLDAIR* is transcribed from the intron of the *FLC* gene under cold temperature. It interacts with Polycomb Repressive Complex 2 (PRC2), recruits it to *FLC* and induces *FLC* repression. Silencing the floral inhibitor *FLC* activates flowering in the course of vernalization [78].

lncRNAs often affect the male fertility and the pollen development. Ma and coworkers [79] reported that lnc transcript *zm401* was essential for the tapetum and pollen development in maize. *Long Day Specific Male Fertility Associated RNA* (*LDMAR*) encodes a 1236 nt lncRNA necessary for male fertility under long days. A mutation reducing *LDMAR* transcript levels leads to premature PCD in anthers and male sterility. This example refers to GMS and not to CMS, as no mitochondrial genes are involved in male sterility.

Some lncRNAs harboring miRNA binding sites function as endogenous target mimics (eTMs) to reduce the repression imposed by miRNAs [80] and to affect the reproductive development–e.g., osa-eTM160 and ath-eTM160 in rice and *A. thaliana*, respectively. They attenuate the repression imposed by miR160 on *ARFs*, which leads to the failure of pollen production [80,81].

The crosstalk between lncRNA and siRNA biogenesis has been reported in rice. lncRNAs serve as the source of siRNAs associated with the *MEIOSIS ARRESTED AT LEPTOTENE1* (*MEL1*) transcript [82].

MEL1 protein is necessary for the meiotic progress and its loss of function results in aberrant vacuolation of spore mother cells and impaired male fertility.

Global analyses of long non-coding transcriptomes have greatly expanded our understanding of lncRNA function. The regulatory role of lncRNA in the course of fruit development of hot peppers is particularly known [83]. However, no comprehensive study of lncRNA participation in CMS biogenesis has been published.

8. Non-Coding RNAs Encoded by the Mitochondrial Genome

Whereas knowledge about the regulatory functions of plant nuclear-encoded ncRNAs has been steadily accumulating [47,53,71,72], evidence about the role of ncRNAs encoded by organellar genomes remains sparse. Dietrich and coworkers [84] provided an overview of the functions of organellar ncRNA in plants and animals, including the well-documented regulatory roles of both small and long ncRNAs in animal mitochondria and plant chloroplasts, but they provide only a few candidates of plant mitochondrial ncRNA. Additional examples of mitochondrial ncRNA were described in the comprehensive review on plant organelle biogenesis by Rurek [85]. Small RNAs may be the products of degradation of longer transcripts [86], or may arise due to the relaxed transcription of intergenic regions [75]. As mitochondrial DNA is often transferred to the nucleus [87], a nuclear origin for the already reported mitochondrial ncRNA cannot be excluded.

Ruwe and coworkers [88] described sRNA clusters near the 3′ ends of mitochondrial transcripts in *A. thaliana*. They can stabilize the transcripts or compete for PPR proteins.

A non-coding mitochondrial transcript about 500 nt long was reported by Holec and coworkers [89] in *A. thaliana*. It carried short stretches of sequence homology with 18S rRNA and tRNA, and exhibited editing sites. No conclusion about its possible function was drawn.

lncRNA accumulated and edited preferentially in male sterile plants (but not in their restored siblings) was documented in bladder campion (*Silene vulgaris*) [90]. Its sequence was not similar to any known sequence in GenBank, it was transcribed from its own promoter. Although it cannot be determined whether this lncRNA is the molecular cause or a consequence of male sterility, it becomes a very first example of mitochondrial ncRNA molecule associated with CMS in plants.

9. Future Perspectives

The role of nucleus-encoded miRNAs in CMS has been addressed by numerous studies, as documented in Table 1. However, the function of other classes of regulatory ncRNAs requires additional investigation. In addition to performing new experiments, existing data should be reexamined for the current insights. For example, data sets utilized for the analyses of miRNAs, may be used to study siRNAs in CMS biogenesis. Similarly, CMS-related transcriptomes constructed from polyA-enriched or rRNA-depleted samples may provide information about novel lncRNAs. Given the near-ubiquity of ncRNA involvement in plant development and stress response [48,53,71], investigating their role in CMS should be a priority.

Whereas a plethora of transcription factors and miRNAs induced by CMS and influencing plant metabolism, including mitochondrial functions, have been reported (Table 1), very little is known about retrograde signaling from mitochondria to nucleus (Figure 1). The application of genome editing and the existence loss-of-function mutations in candidate genes in many species with CMS make it possible to reveal the novel candidate genes involved in retrograde signaling during CMS.

Agricultural species went through genetic bottlenecks in the course of domestication which decreased their genetic variation in many genomic regions. As the studies of CMS have been performed primarily in crops, many important aspects or features may have been missed. The investigations of plants from natural populations will likely reveal novel characteristics associated with CMS [36,37,90].

Supplementary Materials: Supplementary materials can be found at www.mdpi.com/1422-0067/18/11/2429/s1.

Acknowledgments: The author thanks James D. Stone for valuable comments and linguistic correction, and Filip Štorch for help with drawing the figure. Financial support was provided by the grant of the Grant Agency of the Czech Republic 16-09220S.

Author Contributions: Helena Štorchová is the only author of this manuscript.

Conflicts of Interest: The author declares no conflict of interest.

References

1. Tester, M.; Langridge, P. Breeding technologies to increase crop production in a changing world. *Science* **2010**, *327*, 818–822. [CrossRef] [PubMed]
2. Vendramin, R.; Marine, J.C.; Leucci, E. Non-coding RNAs: The dark side of nuclear-mitochondrial communication. *EMBO J.* **2017**, *36*, 1123–1133. [CrossRef] [PubMed]
3. Horn, R.; Gupta, K.J.; Colombo, N. Mitochondrion role in molecular basis of cytoplasmic male sterility. *Mitochondrion* **2014**, *19*, 198–205. [CrossRef] [PubMed]
4. Touzet, P.; Meyer, E.H. Cytoplasmic male sterility and mitochondrial metabolism in plants. *Mitochondrion* **2014**, *19*, 166–171. [CrossRef] [PubMed]
5. Hanson, M.R.; Bentolila, S. Interactions of mitochondrial and nuclear genes that affect male gametophyte development. *Plant Cell* **2004**, *16*, S154–S169. [CrossRef] [PubMed]
6. Gillman, J.D.; Bentolila, S.; Hanson, M.R. The petunia restorer of fertility protein is part of a large mitochondrial complex that interacts with transcripts of the CMS-associated locus. *Plant J.* **2007**, *49*, 217–227. [CrossRef] [PubMed]
7. Kim, D.H.; Kang, J.G.; Kim, B.D. Isolation and characterization of the cytoplasmic male sterility-associated *orf456* gene of chili pepper (*Capsicum annuum* L.). *Plant Mol. Biol.* **2007**, *63*, 519–532. [CrossRef] [PubMed]
8. Duroc, Y.; Hiard, S.; Vrielynck, N.; Ragu, S.; Budar, F. The Ogura sterility-inducing protein forms a large complex without interfering with the oxidative phosphorylation components in rapeseed mitochondria. *Plant Mol. Biol.* **2009**, *70*, 123–137. [CrossRef] [PubMed]
9. Dewey, R.E.; Timothy, D.H.; Levings, C.S. A mitochondrial protein associated with cytoplasmic male-sterility in the T-cytoplasm of maize. *Proc. Natl. Acad. Sci. USA* **1987**, *84*, 5374–5378. [CrossRef] [PubMed]
10. Ducos, E.; Touzet, P.; Boutry, M. The male sterile G cytoplasm of wild beet displays modified mitochondrial respiratory complexes. *Plant J.* **2001**, *26*, 171–180. [CrossRef]
11. Itabashi, E.; Kazama, T.; Toriyama, K. Characterization of cytoplasmic male sterility of rice with Lead Rice cytoplasm in comparison with that with Chinsurah Boro II cytoplasm. *Plant Cell Rep.* **2009**, *28*, 233–239. [CrossRef] [PubMed]
12. Wang, K.; Gao, F.; Ji, Y.X.; Liu, Y.; Dan, Z.W.; Yang, P.F.; Zhu, Y.G.; Li, S.Q. ORFH79 impairs mitochondrial function via interaction with a subunit of electron transport chain complex III in Honglian cytoplasmic male sterile rice. *New Phytol.* **2013**, *198*, 408–418. [CrossRef] [PubMed]
13. Zabala, G.; Gabay-Laughnan, S.; Laughnan, J.R. The nuclear gene Rf3 affects the expression of the mitochondrial chimeric sequence R implicated in S-type male sterility in maize. *Genetics* **1997**, *147*, 847–860. [PubMed]
14. Yamamoto, M.P.; Kubo, T.; Mikami, T. The 5'-leader sequence of sugar beet mitochondrial *atp6* encodes a novel polypeptide that is characteristic of Owen cytoplasmic male sterility. *Mol. Genet. Genom.* **2005**, *273*, 342–349. [CrossRef] [PubMed]
15. Grelon, M.; Budar, F.; Bonhomme, S.; Pelletier, G. Ogura cytoplasmic male sterility (CMS)-associated *ORF138* is translated into a mitochondrial-membrane polypeptide in male-sterile *Brassica* cybrids. *Mol. Gen. Genet.* **1994**, *243*, 540–547. [CrossRef] [PubMed]
16. Kazama, T.; Itabashi, E.; Fujii, S.; Nakamura, T.; Toriyama, K. Mitochondrial ORF79 levels determine pollen abortion in cytoplasmic male sterile rice. *Plant J.* **2016**, *85*, 707–716. [CrossRef] [PubMed]
17. Balk, J.; Leaver, C.J. The PET1-CMS mitochondrial mutation in sunflower is associated with premature programmed cell death and cytochrome *c* release. *Plant Cell* **2001**, *13*, 1803–1818. [CrossRef] [PubMed]
18. Twell, D. Male gametogenesis and germline specification in flowering plants. *Sex. Plant Reprod.* **2011**, *24*, 149–160. [CrossRef] [PubMed]

19. Luo, D.P.; Xu, H.; Liu, Z.L.; Guo, J.X.; Li, H.Y.; Chen, L.T.; Fang, C.; Zhang, Q.Y.; Bai, M.; Yao, N.; et al. A detrimental mitochondrial-nuclear interaction causes cytoplasmic male sterility in rice. *Nat. Genet.* **2013**, *45*, 573–577. [CrossRef] [PubMed]
20. Shafiq, S.; Li, J.R.; Sun, Q.W. Functions of plants long non-coding RNAs. *Biochim. Biophys. Acta* **2016**, *1859*, 155–162. [CrossRef] [PubMed]
21. Schnable, P.S.; Wise, R.P. The molecular basis of cytoplasmic male sterility and fertility restoration. *Trends Plant Sci.* **1998**, *3*, 175–180. [CrossRef]
22. Sabar, M.; Gagliardi, D.; Balk, J.; Leaver, C.J. ORFB is a subunit of F1FO-ATP synthase: Insight into the basis of cytoplasmic male sterility in sunflower. *EMBO Rep.* **2003**, *4*, 381–386. [CrossRef] [PubMed]
23. Kazama, T.; Nakamura, T.; Watanabe, M.; Sugita, M.; Toriyama, K. Suppression mechanism of mitochondrial ORF79 accumulation by Rf1 protein in BT-type cytoplasmic male sterile rice. *Plant J.* **2008**, *55*, 619–628. [CrossRef] [PubMed]
24. Chakraborty, A.; Mitra, J.; Bhattacharyya, J.; Pradhan, S.; Sikdar, N.; Das, S.; Chakraborty, S.; Kumar, S.; Lakhanpaul, S.; Sen, S.K. Transgenic expression of an unedited mitochondrial *orfB* gene product from wild abortive (WA) cytoplasm of rice (*Oryza sativa* L.) generates male sterility in fertile rice lines. *Planta* **2015**, *241*, 1463–1479. [CrossRef] [PubMed]
25. Uyttewaal, M.; Arnal, N.; Quadrado, M.; Martin-Canadell, A.; Vrielynck, N.; Hiard, S.; Gherbi, H.; Bendahmane, A.; Budar, F.; Mireau, H. Characterization of *Raphanus sativus* pentatricopeptide repeat proteins encoded by the fertility restorer locus for Ogura cytoplasmic male sterility. *Plant Cell* **2008**, *20*, 3331–3345. [CrossRef] [PubMed]
26. Sarria, R.; Lyznik, A.; Vallejos, C.E.; Mackenzie, S.A. A cytoplasmic male sterility-associated mitochondrial peptide in common bean is post-translationally regulated. *Plant Cell* **1998**, *10*, 1217–1228. [CrossRef] [PubMed]
27. Gaborieau, L.; Brown, G.G.; Mireau, H. The propensity of pentatricopeptide repeat genes to evolve into restorers of cytoplasmic male sterility. *Front. Plant Sci.* **2016**, *7*, 1816. [CrossRef] [PubMed]
28. Takenaka, M.; Zehrmann, A.; Verbitskiy, D.; Haertel, B.; Brennicke, A. RNA Editing in plants and its evolution. *Annu. Rev. Genet.* **2013**, *47*, 335–352. [CrossRef] [PubMed]
29. Fujii, S.; Bond, C.S.; Small, I.D. Selection patterns on restorer-like genes reveal a conflict between nuclear and mitochondrial genomes throughout angiosperm evolution. *Proc. Natl. Acad. Sci. USA* **2011**, *108*, 1723–1728. [CrossRef] [PubMed]
30. Melonek, J.; Stone, J.D.; Small, I. Evolutionary plasticity of restorer-of-fertility-like proteins in rice. *Sci. Rep.* **2016**, *6*, 35152. [CrossRef] [PubMed]
31. Gouyon, P.H.; Couvet, D. A conflict between two sexes, females and hermaphrodites. In *The Evolution of Sex and Its Consequences*; Stearns, S.C., Ed.; Birkhauser Verlag: Basel, Switzerland, 1987; pp. 245–261, ISBN 978-3-0348-6273-8.
32. Cui, X.; Wise, R.P.; Schnable, P.S. The rf2 nuclear restorer gene of male-sterile T-cytoplasm maize. *Science* **1996**, *272*, 1334–1336. [CrossRef] [PubMed]
33. Liu, F.; Cui, X.; Horner, H.T.; Weiner, H.; Schnable, P.S. Mitochondrial aldehyde dehydrogenase activity is required for male fertility in maize. *Plant Cell* **2001**, *13*, 1063–1078. [CrossRef] [PubMed]
34. Fujii, S.; Toriyama, K. Suppressed expression of RETROGRADE-REGULATED MALE STERILITY restores pollen fertility in cytoplasmic male sterile rice plants. *Proc. Natl. Acad. Sci. USA* **2009**, *106*, 9513–9518. [CrossRef] [PubMed]
35. McCauley, D.E.; Olson, M.S. Do recent findings in plant mitochondrial molecular and population genetics have implications for the study of gynodioecy and cytonuclear conflict? *Evolution* **2008**, *62*, 1013–1025. [CrossRef] [PubMed]
36. Case, A.L.; Willis, J.H. Hybrid male sterility in *Mimulus* (Phrymaceae) is associated with a geographically restricted mitochondrial rearrangement. *Evolution* **2008**, *62*, 1026–1039. [CrossRef] [PubMed]
37. Darracq, A.; Varré, J.S.; Marechal-Drouard, L.; Courseaux, A.; Castric, V.; Saumitou-Laprade, P.; Oztas, S.; Lenoble, P.; Vacherie, B.; Barbe, V.; et al. Structural and content diversity of mitochondrial genome in beet: A comparative genomic analysis. *Genome Biol. Evol.* **2011**, *3*, 723–736. [CrossRef] [PubMed]
38. Mower, J.P.; Case, A.L.; Floro, E.R.; Willis, J.H. Evidence against equimolarity of large repeat arrangements and a predominant master circle structure of the mitochondrial genome from a monkeyflower (*Mimulus guttatus*) lineage with cryptic CMS. *Genome Biol. Evol.* **2012**, *4*, 670–686. [CrossRef] [PubMed]

39. Štorchová, H.; Müller, K.; Lau, S.; Olson, M.S. Mosaic origin of a complex chimeric mitochondrial gene in *Silene vulgaris*. *PLoS ONE* **2012**, *7*, e30401. [CrossRef] [PubMed]
40. Song, J.H.; Yang, J.; Pan, F.; Jin, B. Differential expression of microRNAs may regulate pollen development in *Brassica oleracea*. *Genet. Mol. Res.* **2015**, *14*, 15024–15034. [CrossRef] [PubMed]
41. Stone, J.D.; Štorchová, H. The application of RNA-seq to the comprehensive analysis of plant mitochondrial transcriptomes. *Mol. Genet. Genom.* **2015**, *290*, 1–9. [CrossRef] [PubMed]
42. Yan, J.J.; Zhang, H.Y.; Zheng, Y.Z.; Ding, Y. Comparative expression profiling of miRNAs between the cytoplasmic male sterile line MeixiangA and its maintainer line MeixiangB during rice anther development. *Planta* **2015**, *241*, 109–123. [CrossRef] [PubMed]
43. Zhang, W.; Xie, Y.; Xu, L.; Wang, Y.; Zhu, X.W.; Wang, R.H.; Zhang, Y.; Muleke, E.M.; Liu, L.W. Identification of microRNAs and Their Target Genes Explores miRNA-Mediated Regulatory Network of Cytoplasmic Male Sterility Occurrence during Anther Development in Radish (*Raphanus sativus* L.). *Front. Plant Sci.* **2016**, *7*, 1054. [CrossRef] [PubMed]
44. Liu, T.K.; Li, Y.; Zhang, C.W.; Duan, W.K.; Huang, F.Y.; Hou, X.L. Basic helix-loop-helix transcription factor *BcbHLHpol* functions as a positive regulator of pollen development in non-heading Chinese cabbage. *Funct. Integr. Genom.* **2014**, *14*, 731–739. [CrossRef] [PubMed]
45. Li, Y.W.; Ding, X.L.; Wang, X.; He, T.T.; Zhang, H.; Yang, L.S.; Wang, T.L.; Chen, L.F.; Gai, J.Y.; Yang, S.P. Genome-wide comparative analysis of DNA methylation between soybean cytoplasmic male-sterile line NJCMS5A and its maintainer NJCMS5B. *BMC Genom.* **2017**, *18*, 596. [CrossRef] [PubMed]
46. Guttman, M.; Rinn, J.L. Modular regulatory principles of large non-coding RNAs. *Nature* **2012**, *482*, 339–346. [CrossRef] [PubMed]
47. Achkar, N.P.; Cambiagno, D.A.; Manavella, P.A. miRNA biogenesis: A dynamic pathway. *Trends Plant Sci.* **2016**, *21*, 1034–1044. [CrossRef] [PubMed]
48. Li, S.J.; Castillo-Gonzalez, C.; Yu, B.; Zhang, X.R. The functions of plant small RNAs in development and in stress responses. *Plant J.* **2017**, *90*, 654–670. [CrossRef] [PubMed]
49. Hamilton, A.; Voinnet, O.; Chappell, L.; Baulcombe, D. Two classes of short interfering RNA in RNA silencing. *EMBO J.* **2002**, *21*, 4671–4679. [CrossRef] [PubMed]
50. Zilberman, D.; Cao, X.F.; Jacobsen, S.E. *ARGONAUTE4* control of locus-specific siRNA accumulation and DNA and histone methylation. *Science* **2003**, *299*, 716–719. [CrossRef] [PubMed]
51. Burkhart, K.B.; Guang, S.; Buckley, B.A.; Wong, L.; Bochner, A.F.; Kennedy, S. A pre-mRNA-associating factor links endogenous siRNAs to chromatin regulation. *PLoS Genet.* **2011**, *7*, e1002249. [CrossRef] [PubMed]
52. MacLean, D.; Elina, N.; Havecker, E.R.; Heimstaedt, S.B.; Studholme, D.J.; Baulcombe, D.C. Evidence for Large Complex Networks of Plant Short Silencing RNAs. *PLoS ONE* **2010**, *5*, e9901. [CrossRef] [PubMed]
53. Kamthan, A.; Chaudhuri, A.; Kamthan, M.; Datta, A. Small RNAs in plants: Recent development and application for crop improvement. *Front. Plant Sci.* **2015**, *6*, 208. [CrossRef]
54. Rogers, K.; Chen, X.M. Biogenesis, Turnover, and Mode of Action of Plant MicroRNAs. *Plant Cell* **2013**, *25*, 2383–2399. [CrossRef] [PubMed]
55. Cui, J.; You, C.J.; Chen, X.M. The evolution of microRNAs in plants. *Curr. Opin Plant Biol.* **2017**, *35*, 61–67. [CrossRef] [PubMed]
56. Li, H.; Wang, Y.; Wu, M.; Li, L.H.; Jin, C.; Zhang, Q.L.; Chen, C.B.; Song, W.Q.; Wang, C.G. Small RNA Sequencing Reveals Differential miRNA Expression in the Early Development of Broccoli (*Brassica oleracea* var. *italica*) Pollen. *Front. Plant Sci.* **2017**, *8*, 404. [CrossRef] [PubMed]
57. Yang, J.H.; Liu, X.Y.; Xu, B.C.; Zhao, N.; Yang, X.D.; Zhang, M.F. Identification of miRNAs and their targets using high-throughput sequencing and degradome analysis in cytoplasmic male-sterile and its maintainer fertile lines of *Brassica juncea*. *BMC Genom.* **2013**, *14*, 9. [CrossRef] [PubMed]
58. Fang, Y.N.; Zheng, B.B.; Wang, L.; Wu, X.M.; Xu, Q.; Guo, W.W. High-throughput sequencing and degradome analysis reveal altered expression of miRNAs and their targets in a male-sterile cybrid pummelo (*Citrus grandis*). *BMC Genom.* **2016**, *17*, 591. [CrossRef] [PubMed]
59. Wei, X.C.; Zhang, X.H.; Yao, Q.J.; Yuan, Y.X.; Li, X.X.; We, F.; Zhao, Y.Y.; Zhang, Q.; Wang, Z.Y.; Jiang, W.S.; et al. The miRNAs and their regulatory networks responsible for pollen abortion in Ogura-CMS Chinese cabbage revealed by high-throughput sequencing of miRNAs, degradomes, and transcriptomes. *Front. Plant Sci.* **2015**, *6*, 894. [CrossRef] [PubMed]

60. Ding, X.L.; Li, J.J.; Zhang, H.; He, T.T.; Han, S.H.; Li, Y.W.; Yang, S.P.; Gai, J.Y. Identification of miRNAs and their targets by high-throughput sequencing and degradome analysis in cytoplasmic male-sterile line NJCMS1A and its maintainer NJCMS1B of soybean. *BMC Genom.* **2016**, *17*, 24. [CrossRef] [PubMed]
61. Shen, Y.; Zhang, Z.M.; Lin, H.J.; Liu, H.J.; Chen, J.; Peng, H.; Cao, M.J.; Rong, T.Z.; Pan, G.T. Cytoplasmic male sterility-regulated novel microRNAs from maize. *Funct. Integr. Genom.* **2011**, *11*, 179–191. [CrossRef] [PubMed]
62. Li, J.J.; Han, S.H.; Ding, X.L.; He, T.T.; Dai, J.Y.; Yang, S.P.; Gai, J.Y. Comparative Transcriptome Analysis between the Cytoplasmic Male Sterile Line NJCMS1A and Its Maintainer NJCMS1B in Soybean (*Glycine max* (L.) Merr.). *PLoS ONE* **2015**, *10*, e0126771. [CrossRef] [PubMed]
63. Li, J.J.; Yang, S.P.; Gai, J.Y. Transcriptome comparative analysis between the cytoplasmic male sterile line and fertile line in soybean (*Glycine max* (L.) Merr.). *Genes Genom.* **2017**, *39*, 1117–1127. [CrossRef]
64. Ng, S.; Ivanova, A.; Duncan, O.; Law, S.R.; van Aken, O.; de Clercq, I.; Wang, Y.; Carrie, C.; Xu, L.; Kmiec, B.; et al. A membrane-Bound NAC transcription factor, ANAC017, mediates mitochondrial retrograde signaling in *Arabidopsis*. *Plant Cell* **2013**, *25*, 3450–3471. [CrossRef] [PubMed]
65. Van Aken, O.; Ford, E.; Lister, R.; Huang, S.B.; Millar, A.H. Retrograde signalling caused by heritable mitochondrial dysfunction is partially mediated by ANAC017 and improves plant performance. *Plant J.* **2016**, *88*, 542–558. [CrossRef] [PubMed]
66. Castel, S.E.; Martienssen, R.A. RNA interference in the nucleus: Roles for small RNAs in transcription, epigenetics and beyond. *Nat. Rev. Genet.* **2013**, *14*, 100–112. [CrossRef] [PubMed]
67. Allen, E.; Xie, Z.X.; Gustafson, A.M.; Carrington, J.C. microRNA-directed phasing during trans-acting siRNA biogenesis in plants. *Cell* **2005**, *121*, 207–221. [CrossRef] [PubMed]
68. Yoshikawa, M.; Peragine, A.; Park, M.Y.; Poethig, R.S. A pathway for the biogenesis of trans-acting siRNAs in *Arabidopsis*. *Genes Dev.* **2005**, *19*, 2164–2175. [CrossRef] [PubMed]
69. Howell, M.D.; Fahlgren, N.; Chapman, E.J.; Cumbie, J.S.; Sullivan, C.M.; Givan, S.A.; Kasschau, K.D.; Carrington, J.C. Genome-wide analysis of the RNA-DEPENDENT RNA POLYMERASE6/DICER-LIKE4 pathway in *Arabidopsis* reveals dependency on miRNA- and tasiRNA-directed targeting. *Plant Cell* **2007**, *19*, 926–942. [CrossRef] [PubMed]
70. Xia, R.; Meyers, B.C.; Liu, Z.C.; Beers, E.P.; Ye, S.Q.; Liu, Z.R. MicroRNA Superfamilies Descended from miR390 and Their Roles in Secondary Small Interfering RNA Biogenesis in Eudicots. *Plant Cell* **2013**, *25*, 1555–1572. [CrossRef] [PubMed]
71. Heo, J.B.; Lee, Y.S. Molecular functions of long noncoding transcripts in plants. *J. Plant Biol.* **2015**, *58*, 361–365. [CrossRef]
72. Liu, J.; Wang, H.; Chua, N.H. Long noncoding RNA transcriptome of plants. *Plant Biotechnol. J.* **2015**, *13*, 319–328. [CrossRef] [PubMed]
73. Lu, T.; Zhu, C.; Lu, G.; Guo, Y.; Zhou, Y.; Zhang, Z.; Zhao, Y.; Li, W.; Lu, Y.; Tang, W.; et al. Strand-specific RNA-seq reveals widespread occurrence of novel *cis*-natural antisense transcripts in rice. *BMC Genom.* **2012**, *13*, 721. [CrossRef] [PubMed]
74. Wu, X.H.; Liu, M.; Downie, B.; Liang, C.; Ji, G.; Li, Q.Q.; Hunt, A.G. Genome-wide landscape of polyadenylation in *Arabidopsis* provides evidence for extensive alternative polyadenylation. *Proc. Natl. Acad. Sci. USA* **2011**, *108*, 12533–12538. [CrossRef] [PubMed]
75. Wang, Y.; Wang, X.; Deng, W.; Fan, X.; Liu, T.T.; He, G.; Chen, R.; Terzaghi, W.; Zhu, D.; Deng, X.W. Genomic Features and Regulatory Roles of Intermediate-Sized Non-Coding RNAs in *Arabidopsis*. *Mol. Plant* **2014**, *7*, 514–527. [CrossRef] [PubMed]
76. Swiezewski, S.; Liu, F.; Magusin, A.; Dean, C. Cold-induced silencing by long antisense transcripts of an Arabidopsis Polycomb target. *Nature* **2009**, *462*, 799–802. [CrossRef] [PubMed]
77. Bardou, F.; Ariel, F.; Simpson, C.G.; Romero-Barrios, N.; Laporte, P.; Balzergue, S.; Brown, J.W.S.; Crespi, M. Long noncoding RNA modulates alternative splicing regulators in *Arabidopsis*. *Dev. Cell* **2014**, *30*, 166–176. [CrossRef] [PubMed]
78. Heo, J.B.; Sung, S. Vernalization-mediated epigenetic silencing by a long intronic noncoding RNA. *Science* **2011**, *331*, 76–79. [CrossRef] [PubMed]
79. Ma, J.X.; Yan, B.X.; Qu, Y.Y.; Qin, F.F.; Yang, Y.T.; Hao, X.J.; Yu, J.J.; Zhao, Q.; Zhu, D.Y.; Ao, G.M. *Zm401*, a short-open reading-frame mRNA or noncoding RNA, is essential for tapetum and microspore development and can regulate the floret formation in maize. *J. Cell Biochem.* **2008**, *105*, 136–145. [CrossRef] [PubMed]

80. Wang, M.; Wu, H.J.; Fang, J.; Chu, C.C.; Wang, X.J. A long noncoding RNA involved in rice reproductive development by negatively regulating osa-miR160. *Sci. Bull.* **2017**, *62*, 470–475. [CrossRef]
81. Wu, H.J.; Wang, Z.M.; Wang, M.; Wang, X.J. Widespread long noncoding RNAs as endogenous target mimics for microRNAs in plants. *Plant Physiol.* **2013**, *161*, 1875–1884. [CrossRef] [PubMed]
82. Komiya, R.; Nonomura, K.I. Isolation and bioinformatic analyses of small RNAs interacting with germ cell-specific Argonaute in rice. In *PIWI-Interacting RNAs. Methods in Molecular Biology*; Humana Press: Totowa, NJ, USA, 2014; Volume 1093, pp. 1–249. [CrossRef]
83. Ou, L.J.; Liu, Z.B.; Zhang, Z.Q.; Wei, G.; Zhang, Y.P.; Kang, L.Y.; Yang, B.Z.; Yang, S.; Lv, J.H.; Liu, Y.H. Noncoding and coding transcriptome analysis reveals the regulation roles of long noncoding RNAs in fruit development of hot pepper (*Capsicum annuum* L.). *Plant Growth Regul.* **2017**, *83*, 141–156. [CrossRef]
84. Dietrich, A.; Wallet, C.; Iqbal, R.K.; Gualberto, J.M.; Lotfi, F. Organellar non-coding RNAs: Emerging regulation mechanisms. *Biochimie* **2015**, *117*, 48–62. [CrossRef] [PubMed]
85. Rurek, M. Participation of non-coding RNAs in plant organelle biogenesis. *Acta Biochim. Pol.* **2016**, *63*, 653–663. [CrossRef] [PubMed]
86. Wu, Z.Q.; Stone, J.D.; Štorchová, H.; Sloan, D.B. High transcript abundance, RNA editing, and small RNAs originating from intergenic regions in the massive mitochondrial genome of the angiosperm *Silene noctiflora*. *BMC Genom.* **2015**, *16*, 938. [CrossRef] [PubMed]
87. Richly, E.; Leister, D. NUMTs in sequenced eukaryotic genomes. *Mol. Biol. Evol.* **2004**, *21*, 1081–1084. [CrossRef] [PubMed]
88. Ruwe, H.; Wang, G.W.; Gusewski, S.; Schmitz-Linneweber, C. Systematic analysis of plant mitochondrial and chloroplast small RNAs suggests organelle-specific mRNA stabilization mechanisms. *Nucleic Acids Res.* **2016**, *44*, 7406–7417. [CrossRef] [PubMed]
89. Holec, S.; Lange, H.; Kuhn, K.; Alioua, M.; Borner, T.; Gagliardi, D. Relaxed transcription in *Arabidopsis* mitochondria is counterbalanced by RNA stability control mediated by polyadenylation and polynucleotide phosphorylase. *Mol. Cell. Biol.* **2006**, *26*, 2869–2876. [CrossRef] [PubMed]
90. Stone, J.D.; Koloušková, P.; Sloan, D.B.; Štorchová, H. Non-coding RNA may be associated with cytoplasmic male sterility in *Silene vulgaris*. *J. Exp. Bot.* **2017**, *68*, 1599–1612. [CrossRef] [PubMed]

© 2017 by the author. Licensee MDPI, Basel, Switzerland. This article is an open access article distributed under the terms and conditions of the Creative Commons Attribution (CC BY) license (http://creativecommons.org/licenses/by/4.0/).

Review

The Complexity of Mitochondrial Complex IV: An Update of Cytochrome *c* Oxidase Biogenesis in Plants

Natanael Mansilla, Sofia Racca, Diana E. Gras, Daniel H. Gonzalez * and Elina Welchen *

Instituto de Agrobiotecnología del Litoral (CONICET-UNL), Cátedra de Biología Celular y Molecular, Facultad de Bioquímica y Ciencias Biológicas, Universidad Nacional del Litoral, 3000 Santa Fe, Argentina; nmansilla@fbcb.unl.edu.ar (N.M.); sofia.racca@hotmail.com (S.R.); grasdiana@gmail.com (D.E.G.)
* Correspondence: dhgonza@fbcb.unl.edu.ar (D.H.G.); ewelchen@fbcb.unl.edu.ar (E.W.);
 Tel.: +54-0342-4511370 (5022) (E.W.)

Received: 8 January 2018; Accepted: 29 January 2018; Published: 27 February 2018

Abstract: Mitochondrial respiration is an energy producing process that involves the coordinated action of several protein complexes embedded in the inner membrane to finally produce ATP. Complex IV or Cytochrome *c* Oxidase (COX) is the last electron acceptor of the respiratory chain, involved in the reduction of O_2 to H_2O. COX is a multimeric complex formed by multiple structural subunits encoded in two different genomes, prosthetic groups (heme *a* and heme a_3), and metallic centers (Cu_A and Cu_B). Tens of accessory proteins are required for mitochondrial RNA processing, synthesis and delivery of prosthetic groups and metallic centers, and for the final assembly of subunits to build a functional complex. In this review, we perform a comparative analysis of COX composition and biogenesis factors in yeast, mammals and plants. We also describe possible external and internal factors controlling the expression of structural proteins and assembly factors at the transcriptional and post-translational levels, and the effect of deficiencies in different steps of COX biogenesis to infer the role of COX in different aspects of plant development. We conclude that COX assembly in plants has conserved and specific features, probably due to the incorporation of a different set of subunits during evolution.

Keywords: mETC; OXPHOS; COX; plant growth; biogenesis

---

1. Introduction

Mitochondrial respiration is responsible for energy production in most eukaryotic organisms. The respiratory chain is composed of several multiprotein complexes attached to the inner mitochondrial membrane (IMM), which are involved in the transfer of electrons and the translocation of H^+ for ATP synthesis through oxidative phosphorylation (OXPHOS). The respiratory complexes also contain prosthetic groups, which are essential for the electron transfer reactions. To establish a functional respiratory complex, several steps must be orderly followed, starting with the transcription of genes in two different compartments (the nucleus and mitochondria), the editing and processing of transcripts synthesized in the organelle, the synthesis, membrane translocation and assembly of the structural subunits that will finally conform the multiprotein complex and the synthesis and insertion of the prosthetic groups. All these steps must be finely controlled and are critical to ensure the assembly of a functional complex for the successful operation of mitochondrial respiration [1].

One of the respiratory complexes, complex IV or cytochrome *c* oxidase (COX), catalyzes the transfer of electrons from reduced cytochrome *c* (CYT*c*) to the final acceptor of electrons, O_2, in a process that is coupled to H^+ translocation for ATP production. Among the evidence that highlights the importance of the assembly process involved in COX biogenesis and function, one can mention the existence of more than 30 proteins specifically responsible for the synthesis and delivery of redox-active metal centers and prosthetic groups [2], and for the insertion and maintenance of properly assembled subunits into the IMM [3–5], the existence of multiple tissue- and growth condition-specific isoforms, and the existence of a significant number of severe human diseases connected to COX dysfunction [6,7]. In plants, multiple editing proteins and RNA-processing factors (pentatricopeptide repeat protein (PPR), RNA-editing factor interacting/ Multiple organellar RNA editing factor proteins (RIP/MORF), Organelle RNA recognition motif protein (ORRM) and the Organelle zinc finger editing factor family proteins (OZ)) [8,9] are required for proper expression of subunits encoded in mitochondria. In addition, mitochondrial respiratory activity must be connected to the energy requirements of the cells and those imposed by the environment. Thus, the COX assembly process must be regulated by internal and external factors in order to ensure the generation of the necessary energy through the mitochondrial OXPHOS pathway. The lack of mutant plants in COX components, due to embryonic lethality, highlight the importance of COX activity in plants and poses difficulties to the study of the assembly process. In this review, we summarize current knowledge about the COX assembly process and its regulation in plants, with references to other model systems in which the process has been more thoroughly characterized.

2. COX Biogenesis: from Yeast, to Mammals, to Plants

COX is present in mitochondria and also in several groups of prokaryotes, pointing to an endosymbiotic origin of the mitochondrial enzyme [10]. Bacterial COX is mainly composed of three subunits that form the catalytic core: COX1, COX2 and COX3. The eukaryotic enzyme is much more complex, with about 10 to 14 additional subunits that were most likely added after endosymbiosis [11,12]. The prokaryotic origin of COX is also reflected by the fact that the catalytic-core subunits are encoded in the mitochondrial genome in almost all organisms, with only a few exceptions, notably that of COX2, which is encoded in the nuclear genome in some leguminous plants [13]. The additional subunits are universally encoded in the nucleus and their evolutionary origin is not clear [3].

In addition to its polypeptidic components, COX contains redox cofactors involved in electron transfer reactions. A di-copper center (Cu_A), present in COX2, receives electrons from reduced CYTc [14]. These electrons are then transferred to a heme a group present in COX1, to a binuclear heme a_3-Cu_B center located in the same subunit, and finally to O_2. Assembly of these redox cofactors requires the participation of a set of proteins which are conserved in prokaryotic and eukaryotic COX containing organisms. These are COX10 and COX15, involved in heme A synthesis [15–17], COX11 and SCO (Synthesis of Cytochrome c Oxidase) [18–20], involved in copper insertion, and SURF1 (Surfeit1), the role of which is not completely clear but may be related to heme A insertion into COX1 [21,22]. In parallel with the increased complexity of the eukaryotic enzyme, mitochondrial COX assembly requires a multitude of additional factors, about 30 described up to now. COX biogenesis has been widely studied in mammals and yeast (*Saccharomyces cerevisiae*). Because of the advantages of using *S. cerevisiae* as a model and the possibility of isolating respiratory-deficient mutants in this organism, studies in yeast pioneered the identification of proteins required for COX assembly [4,5,11,23]. The assembly process has been also extensively characterized in humans and other mammalian systems, due to the relevance of mitochondrial disorders in human diseases. In this sense, several revisions were recently published [5–7,11,12]. In the next paragraphs, we will briefly describe what is known about the COX assembly process in yeast and mammalian systems.

Current evidence suggests that COX is assembled from different modules, which are preassembled around the catalytic-core subunits [4,24,25]. COX1 is embedded in the IMM and contains 12 membrane-spanning domains. In yeast, it is inserted into the membrane during its synthesis with the help of Oxa1, Cox14 and Coa3. While Oxa1 has a more general role in the insertion of proteins in the IMM, Cox14 and Coa3 are COX1 chaperones that specifically aid in COX assembly and remain bound to COX1 after its insertion into the membrane. This complex also contains Mss51 and the Hsp70 chaperone Ssc1, involved in the feedback regulation of COX1 translation [26]. In humans, CMC1 (C-X9-C motif containing 1), instead of Mss51 and Hsp70, would be added to the complex to provide stability to the assembly intermediate [27]. Once in the membrane, heme A is added to COX1. Mss51 and Hsp70, as well as CMC1 in humans, probably exit the complex during this process, while the assemblies factor Coa1 and the structural subunits Cox5a and Cox6 are incorporated [28]. Heme A is synthesized from heme B in a two-step process catalyzed by the IMM enzymes Cox10 (heme O synthase) and Cox15 (heme A synthase). Due to its reactivity, it is assumed that heme is never released into the medium and is always bound to proteins [4]. However, how heme A is delivered to COX1 is not known. One possibility is that Cox15 directly handles heme A to COX1. Cox15 action requires its oligomerization and this is promoted by Pet117, which may aid in heme delivery to COX1 [29]. Another protein, SURF1, has also been involved in heme A insertion [21]. SURF1, named Shy1 in yeast, incorporates to the assembly complex at the time of heme A addition. The protein from the bacterium *Paracoccus denitrificans* is able to bind heme A, leading to the proposal that it functions as a heme chaperone during heme A insertion. However, mutation of a conserved histidine putatively involved in heme A binding in the yeast protein does not affect COX assembly. Alternatively, SURF1 may participate in heme A insertion as a COX1 chaperone, and it is possible that its function has diverged during evolution. This is also supported by the fact that SURF1 is not essential for COX assembly, but only to optimize the process [30]. Assuming that it is unlikely that heme A insertion occurs spontaneously, the actual factor directly responsible for this process remains unknown.

Assembly of the Cu_B center also takes place in the membrane after Mss51 release. Its relationship with heme A insertion is not known. Copper insertion is affected by Cox11, a membrane bound protein with a globular domain located in the intermembrane space (IMS) [31]. The Cox11 globular domain contains two conserved cysteines involved in copper binding. A third conserved cysteine is required for copper insertion into COX1, but not for copper binding by Cox11. This cysteine, in its reduced state, would participate in the transfer of copper from Cox11 to COX1 [32]. Another protein, Cox19, is required to maintain the reduced state of this conserved cysteine [33]. Cox19 is a copper-binding protein from the IMS. It has a twin CX_9C motif found in several IMS proteins, many of which are related to COX assembly. Even if Cox19 was shown to bind copper, it is not involved in copper delivery to Cox11. This function is performed by another IMS twin CX_9C protein, Cox17 [34,35]. In addition to the twin CX_9C motif, Cox17 contains a CCXC copper-binding site that partially overlaps with the first CX_9C. The source of copper for Cox17 is unknown. After proposals that Cox17 could translocate from the cytoplasm to the IMS with bound copper, the general consensus is currently that copper reaches the IMS bound to a low-molecular-weight ligand of unknown structure, and that the source of copper for Cox17 metalation is the mitochondrial matrix [36]. Accordingly, two IMM transporters, Pic2 and Mrs3, phosphate and iron carriers, respectively, were found to be able to transport copper into the matrix [37].

Cox17 also participates in the delivery of copper for the assembly of the Cu_A center in COX2. In this case, copper is transferred to Sco1, a protein with a similar location to Cox11. Sco1 and other SCO proteins have two conserved cysteines located in a CX_3C motif, involved in copper binding together with a distal histidine [34,38]. With variations in different organisms, SCO proteins were implicated in the reduction of the copper-binding cysteines located in COX2 and/or the direct transfer of copper to COX2 (see [2] for details). Another protein from the IMS, Coa6, also participates in copper delivery to COX2, but its exact role is not clear at present [39].

Before copper insertion, COX2 must be correctly inserted into the IMM. COX2 has two membrane-spanning domains and a globular domain located in the IMS, where the Cu_A center is located. Thus, translocation of this domain from the site of synthesis, the matrix, to the IMS is also required. Membrane insertion is initiated by Oxa1 and the COX2-specific chaperone Cox20 [27,40]. During this process, an N-terminal extension involved in COX2 translocation is cleaved by the IMM protease Imp1 [41]. Then, Cox18, an Oxa1-like protein, together with Mss2 and Pnt1, promote the translocation of the globular domain through the IMM [40]. In humans, the COX2 module would also contain COX5B, COX6C, COX7B, COX7C and COX8A before assembling to COX1 and additional subunits [5,42]. Finally, COX3 would be incorporated to the complex. COX3 has been found in assembly intermediates containing Cox4, Cox7 and Cox13 in yeast, and COX6A, COX6B and COX7A, equivalent to yeast Cox13, Cox12 and Cox7, in humans [4,42].

It is evident that the COX assembly process has species-specific variations. This is reflected by the fact that certain assembly factors are present in some species but not in others. In addition, when putative homologs are present, they do not necessarily have similar roles. OXA1L, the human homolog of yeast Oxa1, for example, does not seem to be involved in COX1 or COX2 membrane translocation, as its counterpart in yeast [43]. In another example, yeast and humans have two SCO proteins, named 1 and 2. However, only Sco1 is essential for COX assembly in yeast, while both SCO1 and SCO2 have specific, non-redundant functions in COX assembly in humans [38,40,44]. In addition, the presence of alternative isoforms of certain subunits indicates that different forms of COX, probably with different properties, may coexist in the same organism or even in the same organelle [6,7]. Finally, COX may be found as a monomer, a dimer, or a component of supercomplexes, in association with other respiratory complexes. COX subunit composition varies in these different assemblies, suggesting that the incorporation of certain subunits modulates the formation of these structures. It has been proposed that the non-essential subunits Cox12 and Cox13 are involved in the formation of COX dimers. Efficient incorporation of Cox12 and Cox13 to COX requires Rcf1, a yeast protein involved in the accumulation of supercomplexes [45], suggesting that Cox12 and/or Cox13 may also participate in the incorporation of COX into supercomplexes [5]. Current views assume that the formation of dimers and supercomplexes changes the properties of the enzyme and may be used to modulate energy production according to cellular needs [46–48]. How these processes are regulated and whether COX assembly dynamically responds to these cellular needs is currently unknown.

3. COX Biogenesis Proteins in Plants

COX biogenesis in plants has not been extensively studied and most proteins associated with its structure and assembly were annotated by sequence homology. Trying to shed light on this scenario, we searched for genes encoding putative homologs of yeast and mammalian COX subunits or biogenesis proteins in the genome of the model plant *Arabidopsis thaliana*. As starting point, we used the information recently reviewed by Timón-Gómez et.al. [5] and Kadenbach [12]. We completed the list of candidates with the yeast proteins compiled by several authors [4,11,49–54] or by using a "keyword search" in the yeast genome database (yeastgenome.org; [55]). As an inclusion criterion, we only considered the yeast proteins for which experimental evidence about its biological connection with respiratory Complex IV was available. As a result, a total of 73 proteins from yeast and 62 candidates from mammals were included in the list. Of these, 26 are specific to yeast, five are unique to mammals, and 44 proteins are shared by the two organisms (see Table 1).

Table 1. COX subunits and assembly factors identified in yeast, humans and Arabidopsis.

Yeast (S. cerevisiae)	Humans	Role described in yeast and/or mammals	Arabidopsis	AGI
Mitochondrial catalytic core subunits				
Cox1	COX1	Cytochrome c oxidase subunit 1	COX1	ATMG01360
Cox2	COX2	Cytochrome c oxidase subunit 2	COX2	ATMG00160
Cox3	COX3	Cytochrome c oxidase subunit 3	COX3	ATMG00730
				AT2G07687
Structural nuclear subunits				
Cox4	COX5b		COX5b-1	AT3G15640
			COX5b-2	AT1G80230
			COX5b-3	AT1G52710
Cox5a	COX4-1, COX4-2 [a]		-	-
Cox5b [b]	-		-	-
Cox6	COX5A		-	-
Cox7	COX7A1, COX7A2 [a]	Required for COX assembly and function	-	-
Cox8	COX7C		-	-
-	COX8-1, COX8-2, COX8-3 [a]		-	-
Cox9	COX6C, COX7B, COX8A, COX8B		-	-

Table 1. Cont.

Yeast (S. cerevisiae)	Humans		Arabidopsis	
COX12/COXVIb	COX6B1, COX6B2 [a]	Not essential for COX assembly or function	COX6b-1	AT1G22450
			COX6b-2	AT5G57815
			COX6b-3	AT4G28060
			COX6b-4	AT1G32710
Cox13/CoxVIa	COX6A1 [a] COX6A2 [a]		COX6a	AT4G37830
-	-		COX5c-1	AT2G47380
-	-		COX5c-2	AT3G62400
-	-		COX5c-3	AT5G61310
-	-		COX-X1-1	AT5G27760
-	-		COX-X2-1	AT4G00860
-	-		COX-X2-2	AT1G01170
-	-		COX-X3	AT1G72020
-	-		COX-X4	AT4G21105
-	-		COX-X5	AT3G43410
-	-		COX-X6	AT2G16460
COX Assembly Factors				
		Role described in yeast and/or mammals		AGI
Membrane insertion and processing of catalytic-core subunits				
Oxa1	OXA1L	Mitochondrial insertase, mediates insertion of COX subunits into the IMM	OXA1 [d]	AT5G62050
			OXA1L [d]	AT2G46470
Cox20	COX20	COX2 chaperone for copper metalation	-	-
Cox18	COX18	Translocation and export of the COX2 C-terminal tail into the IMS	-	-
Mss2	-	Peripherally bound IMM protein of the mitochondrial matrix; involved in membrane insertion of COX2 C-terminus [b]	-	-
Pnt1	-	Export of the COX2 C-terminal tail [b]	-	-
Imp1	IMMP1	Catalytic subunit of the IMM peptidase complex; required for maturation of mitochondrial proteins of the IMS	Peptidase S24/S26A/S26B/S26C	AT1G53530
				AT1G29960
				AT1G23465
Imp2	IMMP2	Required for the stability and activity of Imp1	MYB3R-3	AT3G08980
-	NDUFA4 NDUFA4L2	Assembly factor for COX or supercomplexes in mitochondria of growing cells and cancer tissues [c]	-	-
Heme A Biosynthesis and Insertion				
Cox10	COX10	Farnesylation of heme B	COX10	AT2G44520
Cox15	COX15	Heme A synthase required for the hydroxylation of heme O to form heme A	COX15	AT5G56090
Yah1	FDX2	Collaborates with COX15 in heme O oxidation. Essential for heme A and Fe/S protein biosynthesis	MFDX1	AT4G05450
			MFDX2	AT4G21090
Ahr	ADR	Collaborates with COX15 in heme O oxidation Pyridine nucleotide-disulphide oxidoreductase family protein	MFDR	AT4G32360
Shy1	SURF1	Required for efficient COX assembly in the IMM. Involved in a step of COX1 translation and assembly; proposed to participate in heme A delivery	SURF1-1	AT3G17910
			SURF1-2	AT1G48510
Copper Trafficking and Insertion				
Sco1	SCO1	Copper chaperone, transporting copper to the CuA site on COX2	HCC1	AT3G08950
Sco2	SCO2		HCC2	AT4G39740
Coa6	COA6	Cooperates with SCO2 in the metalation of CuA	COA6-L	AT5G58005
Cox11	COX11	Assembly of CuB in COX1	COX11	AT1G02410
Cox17	COX17	Copper metallochaperone that transfers copper to SCO1 and COX11	COX17-1	AT3G15352
			COX17-2	AT1G53030
Cox19	COX19	Interacts with COX11 as a reductant, critical for COX11 activity	COX19-1	AT1G66590
			COX19-2	AT1G69750
Cox23	COX23	COX assembly factor, unknown function	COX23	AT1G02160
				AT5G09570
Pet191	PET191	Protein required for COX assembly; contains a twin CX9C motif; imported into the IMS via the MIA import machinery	PET191	AT1G10865
Cmc1	CMC1	Stabilizes the COX1-COX14-COA3 complex prior to the incorporation of subunits COX4 and COX5a. Maintains COX1 in a maturation-competent state before insertion of its prosthetic groups	CMC1	AT5G16060

Table 1. *Cont.*

Yeast (*S. cerevisiae*)	Humans		*Arabidopsis*	
Cmc2	CMC2	COX biogenesis protein	CMC2	AT4G21192
Mir1 Pic2	SLC25A3	Mitochondrial copper and phosphate carrier; imports copper and inorganic phosphate into mitochondria.	PHT3-1 PHT3-2 PHT3-3	AT5G14040 AT3G48850 AT2G17270
Mrs3	MITOFERRIN-1 MITOFERRIN-2	Iron transporter; mediates Fe^{2+} and copper transport across the IMM; mitochondrial carrier family member	Mitochondrial substrate carrier protein	AT1G07030 AT2G30160 AT5G42130
Cox Assembly (other)				
Cox14	COX14 (C12orf62)	Involved in translational regulation of COX1, avoiding COX1 aggregation before assembly	-	-
Cox16	COX16	Mitochondrial IMM protein; required for COX assembly	COX assembly protein	AT4G14145
Cox24	-	Mitochondrial IMM protein; required for accumulation of spliced COX1 mRNA [b]	-	-
Cox26	-	Stabilizes the formation of Complex III-IV supercomplexes [b]	-	-
Coi1	-	Interacts with subunits of Complexes III and IV. Essential for supercomplex formation [b]	-	-
Rcf1	HIGD1A	Stabilizes the COX4-COX5A module and promotes its assembly with COX1 [c]	ATL48	AT3G48030
	HIGD2A	Supports the formation of a class of COX-containing supercomplexes [c]		
Rcf2	-	Required for late-stage assembly of the COX12 and COX13 subunits and for COX activity [b]	ATHIGD3	AT3G05550
Coa1	COA1/MITRAC15	Required for assembly of Complex I and Complex IV in mammals. Interacts with Shy1 during the early stages of assembly in yeast	-	-
Coa2	-	Acts downstream of assembly factors Mss51 and Coa1 and interacts with assembly factor Shy1 [b]	-	-
-	COA3/MITRAC12	Required for efficient translation of COX1 [c]	-	-
Coa3/Cox25	-	Required for COX assembly; involved in translational regulation of COX1 and prevention of COX1 aggregation before assembly [b]	-	-
Coa4	COA4	Twin CX9C protein involved in COX assembly and/or stability		
-	MITRAC7	Chaperone-like assembly factor required to stabilize newly synthesized COX1 and to prevent its premature turnover [c]	-	-
-	MR-1S	Short isoform of the myofibrillogenesis regulator 1 (MR-1S). Interacts PET100 and PET117 chaperones		
	TMEM177	TMEM177 associates with newly synthesized COX2 and SCO2 in a COX20-dependent manner		
Mss51	-	*COX1* mRNA-specific translation activator. Influences COX1 assembly into COX [b]		
Mss18	-	Required for efficient splicing of mitochondrial *COX1* [b]	-	-
Mba1	-	Membrane-associated mitochondrial ribosome receptor; possible role in protein export from the matrix to the IMM [b]		
Mne1	-	Involved in *COX1* mRNA intron splicing [b]	-	-
Mrs1	-	Splicing protein; required for splicing of two mitochondrial group I introns [b]		
Mrp1	-		-	-
Mrp17	-	Mitochondrial ribosomal proteins specific for *COX2* and *COX3* mRNA [b,e]		
Mrp21	-			
Mrp51	-			
Mrpl36	-			
Pet54	-	Protein required to activate translation of the *COX3* mRNA, to process the aI5β intron on the *COX1* transcript, and required for Cox1 synthesis [b]	-	-
Pet100	PET100	Chaperone that facilitates COX assembly	-	-
Pet111	-	Mitochondrial translational activator specific for the *COX2* mRNA [b]	-	-

Table 1. Cont.

Yeast (S. cerevisiae)	Humans		Arabidopsis	
Pet117	PET117	Assembly factor that couples heme A synthesis to Complex IV assembly	-	-
Pet122	-	Mitochondrial translational activator specific for the COX3 mRNA [b]	-	-
Pet123	-	Mitochondrial ribosomal protein of the small subunit [b]	-	-
Pet494	-	Mitochondrial translational activator specific for the COX3 mRNA [b]	-	-
Pet309	LRPPRC	Specific translational activator for the COX1 mRNA	PPR superfamily protein [d]	AT2G02150
				AT1G52640
				AT5G16640
Dcp29	TACO1	Translational activator of mitochondria-encoded COX1	-	-
Oma1	OMA1	Metalloendopeptidase that is part of the quality control system in the IMM; important for respiratory supercomplex stability	MIO24.13	AT5G51740
Mam33	C1QBP	Specific translational activator for the mitochondrial COX1 mRNA	MAM33-L Mitochondrial glycoprotein family	AT5G02050
				AT1G80720
				AT1G15870
				AT3G55605
				AT5G05990
				AT2G39795
				AT4G31930
				AT2G41600
Ssc1/HSP70	HSPA9	Facilitates translational regulation of COX biogenesis	mtHSC70-1	AT4G37910
			mtHSC70-2	AT5G09590
AI1/Q0050	-	Intron maturase; type II family protein	MATR	ATMG00520
AI2/Q0055	-			

[a] Mammalian tissue specific variants (COX4-1, COX6A1, COX6B1, COX7A2, COX8-2: Liver/Ubiquitous; COX4-2: Lung; COX6A2, COX7A1, COX8-1: Heart; COX6B2, COX8-3: Testis). [b] Function described only in yeast. [c] Function described only in mammals. [d] Only proteins with the highest sequence identity were included (p-value $< 10^{-8}$ for PPR family proteins and p-value $< 10^{-20}$ for OXA-related proteins). [e] Information from the yeast genome database. AGI: *Arabidopsis* genome initiative.

Using this dataset, we performed a bioinformatics analysis in the Phytozome 12 database (https://phytozome.jgi.doe.gov/pz/portal.html, [56]) to identify putative homologous proteins in the *Arabidopsis thaliana* genome. After this, experimental or predicted mitochondrial localization was corroborated for these candidate proteins (SUBA4, http://suba.live/, [57]; ARAMEMNON, http://aramemnon.uni-koeln.de/, [58]; Complexome, https://complexomemap.de/at_mito_leaves/, [59]). A total of 72 genes encoding putative *Arabidopsis* homologous COX-related proteins were found, which sum to 10 for already reported plant-specific COX subunits [59,60] (Table 1 and Table S1). From these, 4 encode the highly conserved structural catalytic-core subunits, usually encoded in the mitochondrial genome [61]. For COX3, an extra copy in the nuclear chromosome 2 is also present (Table 1), which is 100% identical to the mitochondrial gene. This region of chromosome 2, of around 600 kbp, presents hybridization with probes of mitochondrial DNA [62]. Nuclear *COX3* genes seem to be present also in other *Brassicaceae*, but it is not clear if they arise from the same DNA transfer event. To evaluate this, we performed a synteny analysis for the nuclear *Arabidopsis COX3* gene. Using Genomicus (Version 16.03, http://www.genomicus.biologie.ens.fr/genomicus-plants-16.03/cgi-bin/search.pl; [63]) and Plant Genome Duplication Database (PGDD; http://chibba.agtec.uga.edu/duplication/index/locus; [64]) databases, we observed that the gene order observed in *A. thaliana* is not conserved in other *Brassicaceae*, not even in *A. lyrata*. For comparison, other nuclear genes, such as At*COX10* or At*HCC2*, show gene arrangement conservation in all *Brassicaceae* (Figure S1a). This suggests that the *Arabidopsis* nuclear *COX3* gene is the result of a recent DNA transfer from the mitochondrion to the nucleus in *A. thaliana*. The region of 30 kbp that contains the nuclear *COX3* gene presents 15 additional genes, most of them almost identical at the sequence level with genes present in mitochondria, encoding structural OXPHOS proteins or mitochondrial ribosomal proteins (Figure S1b). Surprisingly, the 30 kbp region in chromosome 2 is transcribed at similar levels as the one found in the mitochondrial genome (Figure S1b). However, we presume that the transcripts of the nuclear version of *COX3* are not edited and the protein, if synthesized, is probably not able to be imported into mitochondria. The gene coding for COX2 is absent in the mitochondrial genome and has migrated to the nucleus in *Glycine max* and other leguminous plants. In this case, Gm*COX2* seems to be fully edited and also acquired a pre-sequence for import into mitochondria, thus allowing the replacement of the mitochondrial *COX2* copy [65].

We also identified eight genes that encode three accessory COX subunits with sequence homology to yeast and human proteins (COX5b, COX6a and COX6b). It is noteworthy that the number of isoforms of COX5b and COX6b is higher in Arabidopsis than in the other organisms (Table 1). For COX5b, one of the genes, *COX5b-3*, encodes a smaller isoform due to a deletion in the N-terminal region. No evidence for the existence of the corresponding protein is available. For COX6b, one long (COX6b-1; 191 amino acids) and two short (COX6b-2 and COX6b-3; 78 amino acids) forms are present. A fourth isoform (COX6b-4) has an intermediate length and has not been identified experimentally. It is remarkable that homologs for several of the subunits conserved in mammals and yeast were not identified in Arabidopsis (Table 1). Several of these are small proteins of about 100 amino acids or less, which may hinder their identification if their sequence diverged. However, putative homologs of these subunits were not identified in proteomic studies of mitochondrial complexes, suggesting that plant COX differs in this respect from its yeast and human counterparts [59,66]. Additionally, proteomic studies identified a set of subunits that seem to be plant-specific. COX5c was identified in early studies of plant COX purification. Putative additional subunits, named COX-X, were also identified [59,60]. Several of them co-migrate with other COX subunits in Blue Native Polyacrylamide Gel Electrophoresis (BN-PAGE), suggesting that they are structural COX subunits (Figure S2). In summary, plant COX retains the catalytic-core subunits, which are universally conserved in eukaryotes and prokaryotes, and a few additional structural subunits, while several subunits are not conserved and seem to be plant-specific. It is possible that this is due to extensive sequence divergence, but also that the evolutionary acquisition of new accessory subunits was different in plants than in fungi and mammals. This implies that some aspects of the COX assembly process may also differ in plants. It is surprising that two of the conserved subunits, COX6a and COX6b, are non-essential subunits, while many yeast and human essential subunits do not seem to be present in plants. COX6a and COX6b have been implicated in the formation of COX oligomers (either dimers or supercomplexes), and their conservation suggests an ancestral role and the importance of oligomerization for COX function. Related to this, putative homologs of Rcf1 and Rcf2, required for supercomplex formation in yeast, are also present in plants [67]. In addition, a putative homolog of the COX assembly factor COX16 is also present in Arabidopsis. Recently, Cox16 was found associated to Cox1 assembly intermediates in yeast, but also as a substoichiometric subunit in mature COX and in supercomplexes, suggesting that it may also be a non-essential subunit that regulates COX properties or oligomerization [68].

Several putative COX assembly factors were also identified by sequence homology in Arabidopsis. The five factors already present in prokaryotes required for co-factor assembly (COX10, COX15, COX11, SCO and SURF1) are present in *Arabidopsis* (Table 1 and Table S1). Single genes encode the first three proteins, and those encoding COX10 and COX11 have been characterized, confirming their function as COX assembly factors [17,69]. SCO proteins are encoded by two genes, named *HCC1* and *HCC2*. These genes were also characterized, and only HCC1 seems to be required for COX assembly [18–20]. HCC2 is present in angiosperms and gymnosperms and has lost the conserved cysteines and histidines required for copper binding. Current evidence suggests that HCC2 function is related to copper and stress responses in plants [18]. For SURF1, also two genes, that seem to be the consequence of a recent duplication within the *Brassicaceae*, can be recognized in Arabidopsis.

Putative homologs of other eukaryotic proteins known or supposed to be involved in copper trafficking and insertion, like COX17, COX19, COX23, COA6, CMC1, CMC2, and PET191, were also identified. Arabidopsis genes encoding COX17 and COX19 (two genes for each protein) have been characterized and are able to complement the corresponding yeast *null* mutants, suggesting that the proteins are indeed COX assembly factors [70–72]. In addition, COX17 was involved in the regulation of plant stress responses [72,73]. Homologs of two proteins required for heme A synthesis, Yah1 and Arh1 in yeast, are also present in *Arabidopsis*. These proteins are mitochondrial ferredoxin and ferredoxin reductase, respectively, and are required for electron transfer reactions related to the conversion of heme O to heme A. It is not known if the Arabidopsis proteins are required for heme A synthesis. It was reported that they may participate in biotin synthesis instead [74], and it is likely that they have general roles in electron transfer reactions in the mitochondrial matrix.

Interestingly, homologs of COX18 and COX20, specifically required for the insertion of core COX subunits into the IMM, were not identified. Instead, two OXA1 insertase homologs are present in Arabidopsis [75]. One or both of these proteins may be functional in COX subunit insertion. As mentioned above, OXA1 participates in this process in yeast but apparently not in humans [76], but COX18 and COX20 are required in both organisms. Arabidopsis also contains two additional mitochondrial proteins, named OXA2, with lower sequence homology to yeast Oxa1 and that may participate in the assembly process [77]. Homologs of the Imp1/2 IMM peptidase are also present in plants. It is not known if they are particularly involved in COX2 biogenesis, since they probably have a general function in IMM protein translocation. As a general rule, assembly factors involved in the biogenesis of the redox-active centers are present in plants, while most COX-specific factors involved in other aspects of COX biogenesis seem to be lacking. Considering the different subunit composition of plant COX, probably different factors, yet to be discovered, are involved in this process.

Among the proteins required for COX biogenesis in yeast, several of them are RNA binding and/or processing proteins involved in RNA maturation or translational regulation within mitochondria. Proteins with sequence homology to some of these proteins are found in *Arabidopsis*, but it is not clear if they are involved in COX biogenesis. Only proteins that present a *p*-value $< 10^{-8}$ about their sequence identity with RNA processing proteins previously identified in yeast or mammals were included in Table 1. Plant mitochondria have a complex RNA metabolism that involves hundreds of proteins. Unlike mammals and yeast, flowering plants must convert hundreds of cytidines to uridines in their chloroplast and mitochondrial transcripts through a post-transcriptional process known as RNA editing. To successfully carry out this process, plants have editosome complexes comprised of pentatricopeptide repeat (PPR) proteins [8], and non-PPR editing factors, including RIPs/MORFs, ORRMs, OZs and protoporphrinogen IX oxidase 1 (PPO1) (recently reviewed in [9]). Several proteins have been experimentally demonstrated as responsible for the editing and maturation of transcripts encoding mitochondrial COX components, and these are listed in Table S1 and described in Table 2 in detail.

Table 2. Mutants that exhibit COX altered composition and/or activity.

Affected Process	Name/AGI	Mutant Name/Code	Description	Phenotype	Complex IV Accumulation/Respiration	Ref.
			COX Assembly Factors			
Copper delivery and insertion	HCC1 AT3G08950	hcc1-1 SALK_057821 hcc1-2 GABI-Kat923A11	Homolog of the yeast Copper Chaperone Sco1	Embryos are arrested at various developmental stages. Altered response of root elongation to copper	Very low levels of COX activity in embryos and rosette leaves	[18,19]
	HCC2 AT4G39740	hcc2-1 GABI-Kat843H01 hcc2-2 GABI-at640A10	Homolog of the yeast Copper Chaperone Sco1	Knockout lines exhibit only mild growth suppression. More sensitive to UV-B treatment	Normal COX activity	[18,20]
	COX17-1 AT3G15352 COX17-2 AT1G53030	cox17-2 SALK_062021C +amiR-COX17-1	Cytochrome c oxidase 17, a soluble protein from the IMS that participates in the transfer of copper to COX	Smaller rosettes and roots. Severity of the phenotype is related with the decrease in the level of COX17	Slight decrease in COX activity in cox17-2 mutants	[73]
	COX11 AT1G02410	KD1 SALK_105793 KD2 SALK_003445C	Cytochrome c oxidase 11, involved in copper delivery to COX	Defects in pollen germination, root growth inhibition, smaller rosettes and leaf curling	Reduced COX activity	[69]
Heme A synthesis	COX10 AT2G44520	cox10 SAIL_1283_D03.V1	Homolog to the yeast farnesyltransferase Cox10 that catalyzes the conversion of heme B to heme O in the heme A biosynthesis pathway	Homozygous mutants are embryo-lethal. Heterozygous mutant plants show early onset and progression of natural and dark-induced senescence	Reduced COX activity. Normal levels of total respiration; lower levels of cyanide-sensitive respiration, increased AOX respiration.	[17]
			RNA Processing Factors [a]			
Affected Process	Name/AGI	Mutant Name and Code	Description	Phenotype	Complex IV Accumulation/Respiration	Ref.
Editing	RIP1/MORF8 AT3G15000	rip1/morf8 FLAG_150D11	Lacks the PPR motif. Interacts with the chloroplast-PPR protein RARE1	Dwarf phenotype	NA	[78]
	MORF1 AT4G20020	morf1 WiscDsLox419C10	Required for RNA editing in plant mitochondria	Abortion of homozygous mutant seeds	NA	[79]
	COD1 AT2G35030	cod1-1 (SALK_000882) cod1-2 (SALK_615308)	Mitochondria-localized PLS-subfamily PPR protein	Germination deficiency; shoot and root growth retardation. No viable pollen	Absence of COX activity	[80]
	MEF13 AT3G02330	mef13-1 EMS mutant mef13-2/SALK_097270C	E-type PPR protein required for editing at 8 sites in Arabidopsis. Dual targeted protein	Growth retardation	NA	[81]
	ORRM4 AT1G74230	orrm4 SALK_023061C	Organelle RNA Recognition Motif-containing protein	Vegetative growth and flowering retardation	NA	[82]
	DEK10 Zea mays	dek10/dek10-N1176A	Defective kernel 10; encodes an E-subgroup PPR protein in maize	Small kernels and vegetative growth delay	Reduced Complex IV activity and COX2 accumulation	[83]

Table 2. Cont.

	COX Assembly Factors					
Affected Process	Name/AGI	Mutant Name/Code	Description	Phenotype	Complex IV Accumulation/Respiration	Ref.
Splicing	ppPPR_78/*Physcomitrella patens*	ΔppPPR_78	PPR protein family member	Slight growth retardation	NA	[84,85]
	ppPPR_77/*Physcomitrella patens*	ΔppPPR_77	PPR protein family member	Severe growth retardation	NA	[86]
	WTF9 AT2G39120	*wtf9-1* SALK_022250 *wtf9-2* SALK_143433	Protein involved in the splicing of group II introns in mitochondria	Severely stunted shoots and roots. Small flowers, small anthers and little pollen. Aborted seeds	Reduced COX2 accumulation	[87]
	mCSF-1 AT4G31010	*mcsf-1* SALK-086774	Mitochondria-localized CRM family member required for the processing of many mitochondrial introns	Germination delayed. Altered growth pattern and delayed development	Reduced Complex I and Complex IV activity. Reduced levels of COX2 and decrease of fully-assembled COX. Reduced respiration	[88]
	mMAT2 AT5G46920 PMH2 AT3G22330	Double mutant *nmat2* x *pmh2* SALK-064659 x SAIL-628C06	Members of the maturase and RNA helicase families; function in the splicing of many introns in Arabidopsis mitochondria	Growth delay and alterations in vegetative and reproductive development	Less sensitive to KCN (COX inhibitor)	[89]
	ppPPR_43/*Physcomitrella patens*	ΔppPPR_43	Involved in the splicing of COX1	Severe growth retardation	NA	[90]

	Other Mutants Affected in Cox Assembly/Activity					
Affected Process	Name/AGI	Mutant Name and Code	Description	Phenotype	Complex IV Accumulation/Respiration	Ref.
Mitochondrial transcription	RPOTmp AT5G15700	*rpoTmp-1* GABI_286E07 *rpoTmp-2* SALK_132842	T3/T7 phage-type dual-targeted RNA polymerase	Delayed plant development, wrinkly rosette leaves	Reduced abundance of the respiratory Complexes I and IV	[91,92]
RNA processing	SLO2 AT2G13600	*slo2-2* SALK_021900 *slo2-3* Tilling line T94087	PPR protein belonging to the E+ subclass of the PLS subfamily	Retarded leaf emergence, delayed development, late flowering and smaller roots	Marked reduction in Complexes I, III, and IV	[93]
Mitochondrial translation	MRPL1 AT2G42710	*mrpl1-1* SALK_206492C *mrpl1-3* SALK_014201	Mitochondrial Ribosomal Protein L1	Delayed plant growth	Reduced levels of Complexes I and IV. COX2 protein is severely reduced. Reduced respiration	[14]
Protein processing	LON1 AT5G26660	*lon1-1* EMS-line *lon1-2* SALK_012797	ATP-dependent protease and chaperone	Retarded growth of both shoots and roots	Lower abundance of Complexes I, IV, and V	[95,96]
Complex IV abundance and stability	CYTC-1 AT1G22840 CYTC-2 AT4G10040	*cytc-1a* GK586B10 *cytc-1b* SALK_143142 *cytc-2a* SALK_037790 *cytc-2b* SALK_029663	Cytochrome c (CYTc) is an IMS electron carrier that transfers electrons between Complexes III and IV	Knock-out of both genes produces embryo-lethality. Plants exhibit smaller rosettes with a pronounced decrease in parenchymatic cell size and a delay in plant development	Decreased levels of Complex IV. Normal levels of total respiration; lower levels of cyanide-sensitive respiration and increased AOX-respiration	[97]

[a] Part of the information contained in this table was extracted from Colas des Francs-Small C. and Small I. [98]. NA: not analyzed.

We analyzed the possible arrangement of COX subunits and assembly factors into protein complexes or supercomplexes using the server Complexome (https://complexomemap.de/at_mito_leaves/ [59]). Forty-three proteins were detected in the database, built with proteins isolated from Arabidopsis leaves [59]. As expected, structural subunits, including several COX-X subunits, were grouped together in 2 bands migrating at about 210 and 300 kDa. The most abundant band (210 kDa) does not contain COX6b-1, the larger form of COX6b, which is present in the 300 kDa complex (Figure S2). The Rcf1 homolog ATL48 also migrates with other COX subunits, suggesting that it may be part of plant COX. In yeast, it was proposed that Rcf1 is not a stoichiometric COX subunit and that it rather acts as a late-maturation factor. The human homolog, HIGD1A, forms a complex with early COX1 assembly intermediates. A group of COX subunits also appears at 700 and 1700 kDa, probably reflecting previously described III_2 + IV and III_2 + IV + I supercomplexes [99]. COX11, HCC1, HCC2 and SURF1-1, all migrated similarly in a region slightly below 100 kDa. The MW calculated for these proteins is between 30 and 40 kDa, which suggests that they are probably dimers or trimers or that they interact with other proteins (Figure S2). The formation of dimers was suggested for SCO proteins from humans. All of these proteins are also present in regions of slower migration, perhaps reflecting their incorporation into assembly complexes. COX15 forms oligomers of about 120 kDa, probably also dimers or trimers. Oligomerization of Cox15 in yeast was suggested to be important for its function. Homologs of Pet117, involved in Cox15 oligomerization, were not found in Arabidopsis. Unfortunately, the migration patterns of COX1 and COX3 are not provided in the Complexome database, not allowing the identification of putative assembly intermediates containing these proteins.

4. Factors Regulating COX Assembly and Activity

4.1. Regulation of Gene Expression

4.1.1. Tissue- and Organ-Specific Expression

To analyze the transcriptional regulation of the expression of genes encoding COX-related proteins, we performed a meta-analysis by exploring publicly available microarray data included in the Genevestigator database (https://genevestigator.com/gv/doc/intro_plant.jsp [100]). We obtained information about the expression profile of 67 genes according to organ/tissue, plant developmental stage, (Figure 1) and responses to several perturbations (Figures 2 and S3). Hierarchical clustering of the expression data across different tissues and cell types obtained from the "Anatomy" and "Development" tools showed that several genes for COX structural components or editing and assembly factors are preferentially expressed in roots and endoreduplicative giant cells, in the shoot apical meristem, and at different stages during embryogenesis (Figure 1). This agrees with previous experimental analysis of the expression patterns of some of these genes [17–19,72,101–107], suggesting that they are expressed in tissues with high energy demands. In agreement with the central role of mitochondria in cellular metabolism and as energy suppliers during developmental processes and cellular differentiation, COX biogenesis seems to be regulated at different stages of development, and according to organ and tissue type.

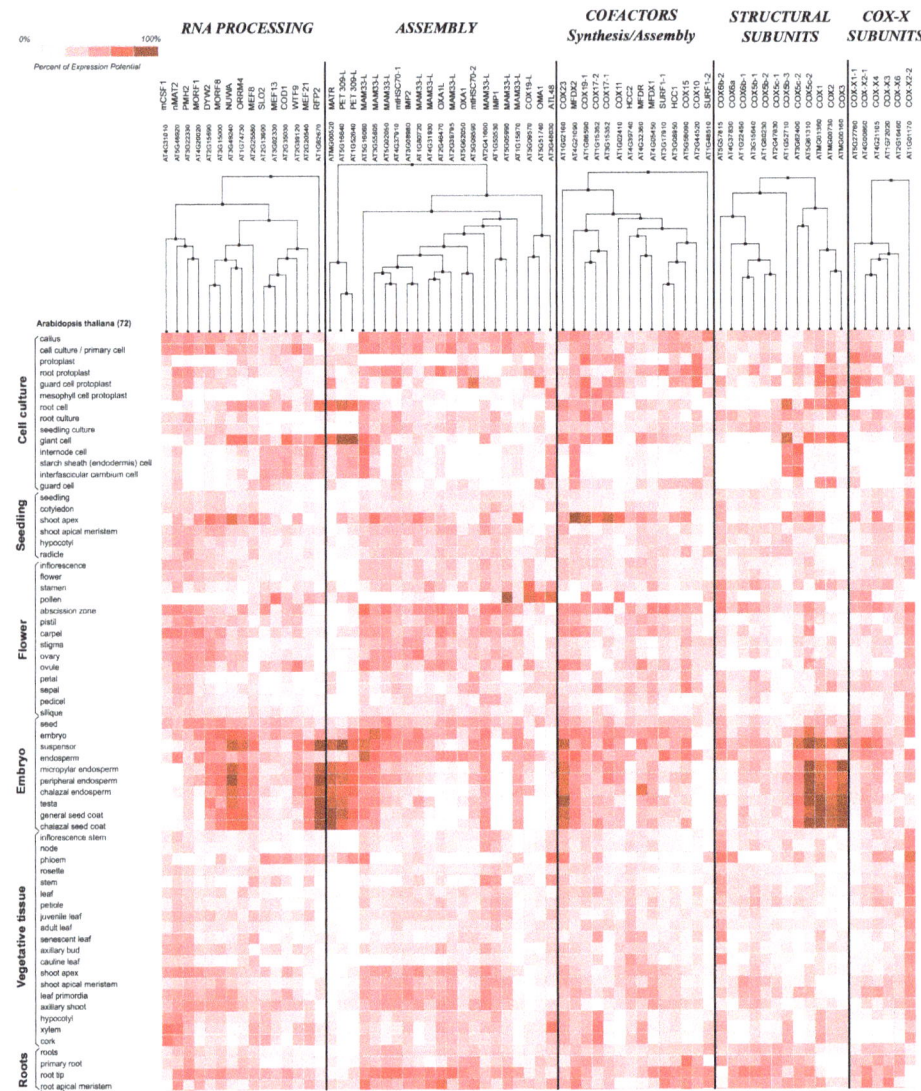

Figure 1. Hierarchical clustering of expression data across different tissues, cell types and developmental stages. Meta-analysis of the expression of genes encoding 68 COX-related proteins according to tissue- and cell-type, and in different developmental stages. Candidate proteins were classified into five different categories according to their putative or demonstrated role in COX biogenesis. Hierarchical clustering was performed within each category. Transcriptional data were collected and analyzed using publicly available microarray data included in the Genevestigator database (https://genevestigator.com/gv/doc/intro_plant.jsp, [100]). The expression level is represented as percent of maximal expression in the dataset analyzed.

4.1.2. Expression of Different Isoforms

In mammals, 6 of the 11 nuclear-encoded subunits possess differential tissue- or environmental condition-specific isoforms [7]. This may result in COX isoforms with different activity or regulatory properties, different affinity for oxygen, or different interactions with other OXPHOS components [6,7]. In Arabidopsis, some genes coding for different COX isoforms that theoretically fulfill the same function show specialized expression characteristics. As examples, *COX5b-1* and *COX5b-2* are preferentially expressed in meristematic regions or in vegetative tissues and are differentially regulated by external stimuli [104,108–110]. This is related to the presence of different *cis* regulatory elements in their promoter regions [111]. Interestingly, the expression patterns and responses of *CYTC-1* and *CYTC-2*, both genes encoding the electron carrier CYTc, which is required for complex IV stability [97], are similar to those of *COX5b-2* and *COX5b-1*, respectively [105,106]. In addition, they contain similar responsive elements in their promoters, suggesting the existence of co-evolution in their expression mechanisms. A model where one of the genes maintained its ancient expression characteristics while the other one incorporated novel elements that allowed a progressive specialization was proposed [109]. Whether the different isoforms confer different properties to plant COX is not known.

As a possible example of functionalization, the AtCOX19-1 gene produces two transcripts by alternative splicing that originate proteins with different N-terminal ends. The smaller AtCOX19-1 isoform is the only one able to restore growth of the yeast *cox19* null mutant [101]. A more extreme case of divergence is represented by genes encoding plant SCO proteins. AtHCC1 is the Arabidopsis homolog of yeast Sco1, involved in copper delivery to COX, while AtHCC2 functions in copper sensing and stress responses, but does not seem to be required for COX assembly [18–20].

4.1.3. Response to Nutrients

We also performed a meta-analysis for transcriptional data related with "Perturbations" available in Genevestigator [100]. As a criterion of selection for posterior analysis, we included all experiments in which at least 30% of genes included in the analysis were differentially expressed with a fold change higher than 2. We applied a clustering analysis within individual gene categories and the images were fused to improve visualization of the results (Figure 2; see Figure S3 for a magnification of Figure 2A).

Differential expression of mitochondrial genes (*COX1*, *COX2*, *COX3* and *MATR*), relative to nuclear genes, was observed, as reported previously [112]. Higher expression of nuclear genes, mainly those related to editing and assembly factors, was observed in plants treated with glucose or shifted to high-light conditions. An opposite behavior was observed in plants in which the target of rapamycin (TOR) kinase, a regulator of plant growth in response to nutrients [113], was silenced (Figure 2B). Strengthening the connection between light, development and mitochondria, a marked increase in expression is observed in mutants of the members of the COP9 signalosome (*csn3-1*, *csn4-1*, and *csn5*), and in the *cop1-4* mutant [114], with altered photomorphogenesis (Figure 2C). It was previously observed that transcript levels for *COX5b*, *COX5c*, *COX6a* and *COX6b* genes are regulated by light, metabolizable sugars and nitrogen sources [102–104,107–110,115,116], and that carbohydrates produce an increase in transcript levels for nuclear genes but not for the mitochondrial ones [102,112]. It was proposed that carbon and nitrogen sources act in concert to regulate the expression of different components of the respiratory pathway [117].

We evaluated the presence of common regulatory elements in the promoter regions of nuclear COX-related genes. To achieve this, the sequences corresponding to the first 500 bp upstream of the transcriptional start site of each gene were obtained and loaded in the MEME server [118]. Three elements were statistically over-represented: wwTGGGCY ($p < 10^{-29}$), rAAgAArA ($p < 10^{-15}$), and TTTTkTTT ($p < 10^{-2}$) (Figure S4). The last two motifs do not match with any previously recognized regulatory element present in the TOMTOM database (http://meme-suite.org/tools/tomtom; [118]). The element wwTGGGCY matches with the previously reported Site II element (TGGGCY) [117,119]. Site II elements are present in the promoter regions of genes encoding COX components [105–108,117,119]. This regulatory motif has been reported as relevant in regulation of gene expression in meristematic regions [105,119], in the response to light and carbohydrates [117,119], in the coordination of gene expression for the biogenesis of the OXPHOS machinery [117], in the connection between the expression of mitochondrial proteins and the circadian clock [120], as well as in coordinating the expression between chloroplast, peroxisome and mitochondrial components [1].

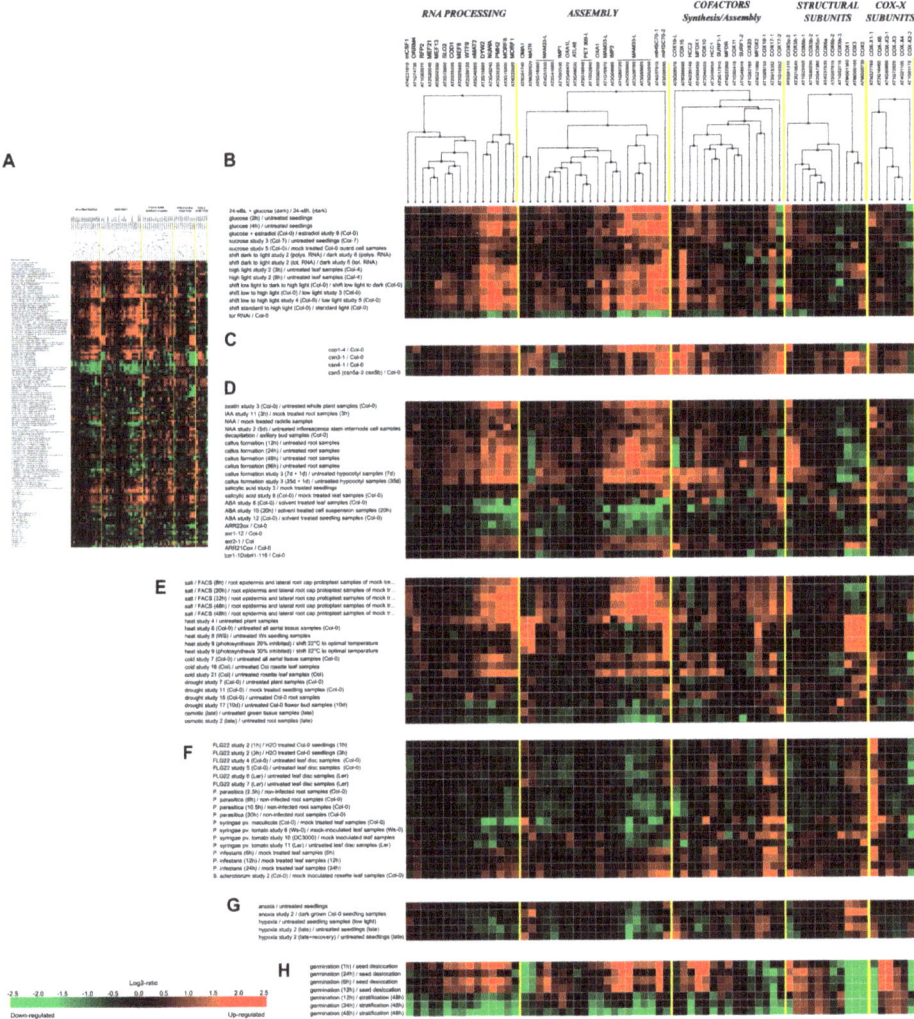

Figure 2. Meta-analysis of transcriptional data for genes encoding COX-related proteins in response to different perturbations or in several mutant backgrounds. (**A**) Complete hierarchical clustering of the expression data. A larger image is available in Figure S3. (**B–H**) Detail of specific parts of the clustering for transcriptional responses to nutrients and light (**B**); in mutants in members of the COP9 signalosome and the *cop1-4* mutant (**C**); in response to hormones (**D**); abiotic stress (**E**); biotic stress (**F**); during oxygen deprivation (**G**); and during germination (**H**). Expression level is represented as log2-ratio of differential expression, in red for up-regulation and in green for down-regulation.

4.1.4. Response to Hormones

We described evidence that COX activity is connected to plant growth and development, and there are several examples that alterations in COX activity seriously impact on plant fitness (Table 2). This is somewhat unsurprising, considering that mitochondrial OXPHOS is probably required for obtaining the energy necessary for these processes. Interestingly, the opposite also seems to happen, since a regulation of plant growth programs over COX exist. Expression of nuclear genes coding for COX proteins is particularly connected to hormonal control of plant growth. Plant hormones are connected to mitochondrial activity through abscisic acid (ABA), auxins, cytokinins (CKs), jasmonic acid (JA), and salicylic acid (SA), regulators of the expression of nuclear genes encoding mitochondrial components [121]. In this sense, *COX6b-1* is induced by gibberelins (GA_3) [107], ABA increases the expression of *COX5b-1* and *CYTC-2*, and *COX6b-3* is induced by CKs [105–107,111]. A recent study demonstrated that CKs can differentially activate the alternative respiratory pathway in roots subjected to oxidative stress, decreasing reactive oxygen species (ROS) production and drought-induced oxidative stress, and thus promoting root growth [122]. Furthermore, the expression of the metallochaperones AtCOX17 and AtCOX19 is induced by SA, a hormone connected to biotic stress [72–77,99–101]. Hormones also seem to have impact on the posttranslational regulation of COX assembly factors. AtCOX17 and the hypoxia regulated protein AtATL8 are differentially phosphorylated in response to auxins [123], while the same effect was seen in AtMATR after treatment with brassinosteroids (BRs) [124].

The analysis presented in Figure 2D allows the discrimination of different responses of COX-related genes to hormones or inhibitors of hormonal pathways. There is an increased expression of several RNA-processing and assembly proteins in aerial plant tissue treated by CK (zeatin), in roots and inflorescence stem internode tissues exposed to the auxins IAA (indole-3-acetic acid) and NAA (1-naphthaleneacetic acid), and during the first 96 h in callus-inducing medium. During the last stages of callus formation, there is an enhancement of the expression of nuclear genes for structural proteins, while mitochondrial genes are down-regulated. Among the biotic stress-related hormones, SA causes a general weak induction that is more pronounced for genes encoding co-factor synthesis or assembly proteins. A clear opposite behavior is observed after ABA treatments, where all genes seem to be down-regulated.

When mutants in hormonal synthesis and signaling pathways are evaluated, opposite expression is observed for RNA processing and assembly proteins in plants with altered response to auxins. While in *axr2-1* mutants, where auxin responses are impaired [125], they are up-regulated, in *axr1-12* auxin-resistant plants their expression is decreased. The same behavior is observed in plants overexpressing the *A. thaliana* response regulators (ARRs) ARR21C or ARR22, being the expression down-regulated in ARR22ox plants, that exhibit a dwarf phenotype previously associated with CK deficiency [126]. Gene expression is also decreased in a *bzr1-1D* mutant background [127], in which the brassinosteroid (BR) signaling pathway and several aspects of plant growth and development are impaired (Figure 2D).

Mitochondrial dysfunction due to alterations in mitochondrial respiratory proteins has been associated to a decrease in auxin responses [128,129]. Future experiments will be required to establish the molecular network connecting hormones and COX biogenesis/function.

4.1.5. Stress Conditions

Mitochondria fulfill a central role during perception, signaling and orchestrated responses to biotic and abiotic stress in plants [130–133]. In relation to this, transcript levels for AtCOX17 and AtCOX19 are increased by exposition to toxic concentrations of Cu^{2+}, Zn^{2+} and Fe^{2+}, to compounds that produce ROS, and after wounding and infection with the pathogen *Pseudomonas syringae* [72,101]. The silencing of AtCOX17 genes originates a decreased response of several stress genetic markers, including those that comprise the mitochondrial retrograde response and are normally induced by mitochondrial dysfunctions [73]. It was proposed that AtCOX17, in addition to its function as a copper metallochaperone for COX assembly, acts as a component of signaling pathways that link plant stress to gene expression responses [73]. As another example, *AtCOX5b-2* is induced in *cyp79B2/cyp79B3* mutants, deficient in aliphatic glucosinolates, important secondary metabolites in the defense against insects and pathogens [134].

Analysis of transcriptional data in response to different stress situations (Figure 2E) shows a coordinate increase in the expression of several RNA processing and assembly factors with decreased expression of mitochondrial genes in root cells exposed to salt. The opposite occurs during heat stress, where COX catalytic-core genes are highly induced, while no changes are observed during cold treatments. Drought and osmotic stress do not generate a marked behavior.

Considering the responses to biotic stress, a decrease in the expression of several nuclear genes is observed when plants are exposed to elicitors (Flg22), bacterial pathogens (*P. syringae*), or the oomycete *P. parasitica*. A contrasting behavior is observed for mitochondrial genes and those nuclear genes coding for the metallochaperones HCC1, COX19, COX17-1, and COX17-2, among a few others, in agreement with previous results [72,101] (Figure 2F). A general increase in expression is observed during infection with necrotrophic pathogens like *Phytophthora infestans* or *Sclerotinia sclerotiorum* (Figure 2F). One possibility is that COX assembly or structural proteins accumulate under certain stress conditions to actively replace damaged or inactive COX subunits to avoid or diminish ROS generation due to a dysfunctional mitochondrial cyanide-sensitive respiration.

4.2. ROS- and Redox-Regulated Mechanisms Involved in COX Biogenesis

4.2.1. Regulation by Oxygen Availability

Cellular respiration consumes oxygen as terminal acceptor of electrons from reduced equivalents produced by the cell during metabolism. In mammals and yeast, different COX isoforms, with different affinity for oxygen, presumably exist. The expression of subunits Cox5a and Cox5b from yeast, and COX4-1 and COX4-2 from mammals, is differentially regulated by oxygen tension, being the isoform expressed under hypoxia responsible for increasing COX catalytic efficiency. This adapts the respiratory electron transfer rate to environmental requirements [135]. As another example, expression of the human *COX10* gene is regulated by the microRNA miR-210, which is induced under hypoxia to reduce COX10 levels and thus regulates the rate of oxygen consumption and mitochondrial metabolism in conditions of low oxygen availability [136]. In plants, transcript levels of genes encoding rice COX5b and COX5c and sunflower COX5c are severely reduced under hypoxia and recovered to initial levels when plants are returned to normal aerobic conditions [112,137]. Transcriptomic data indicate that nuclear genes are down-regulated under anoxia and hypoxia in Arabidopsis, whereas mitochondrial genes appear to be up-regulated (Figure 2G). This suggests that COX biogenesis may also be under the control of oxygen availability in plants.

4.2.2. Regulation of Import and Oxidative Folding

Mitochondria are particularly important in regulating ROS-mediated processes [138]. During respiration, ROS are inevitably produced, mainly in Complexes I, II and III [139]. Paradoxically, COX, that should be one the major oxygen consumers in the cell, does not generate ROS, but its biogenesis involves several ROS and redox regulated steps [135]. While cells have evolved strategies to avoid the generation of pro-oxidant compounds during COX biogenesis, redox biology and ROS exert a control during the assembly and copper delivery to the complex. Mitochondria are rich in thiol-containing proteins that, together with the couple reduced/oxidized glutathione (GSH/GSSG), are specifically compartmentalized in the matrix and the IMS [140–142]. Many IMS proteins that are synthesized in the cytosol are imported through a conserved redox-regulated disulfide relay system established by the receptor MIA40 and the sulfhydryl oxidase ERV1 [143]. Some of the imported proteins, and MIA40 itself, share conserved CX_3C or CX_9C motifs. Thirteen of the twin CX_9C family members are involved in COX biogenesis in yeast and are conserved in animals and plants [144]. In mammalian cells, the kinetics of import through the MIA40/ERV1 system depend on the glutathione pool, thus providing a connection between the redox state of the IMS and the amount of imported CX_9C proteins [145]. In plants, while AtERV1 was found to be critical for mitochondrial biogenesis, AtMIA40 seems to be dispensable for import to the IMS [143,146] and recent evidence suggests that it only improves substrate specificity [147]. A second connection between the MIA40-ERV1 import pathway and respiration is established by CYTc, the final electron acceptor of this pathway that transfers them to COX [148,149]. Thus, the cellular redox state, and more precisely that of the IMS, defines the amount and the proper oxidative folding of CX_9C proteins imported by the MIA40/ERV1 system, which is connected to the respiratory chain and, at the end, to the energetic status of the cell.

4.2.3. Regulation of Copper Metalation

Copper insertion into COX1 and COX2 involves several redox-dependent conformational changes of metallochaperones from the IMS [2,150,151]. COX17 forms disulfide bridges and changes its oligomerization state depending on the redox state, thus adopting different conformations for copper transfer to COX [34,35]. Besides Cox17, other proteins with twin CX_9C motifs reside in the IMS and were demonstrated as essential for copper uptake and COX assembly in yeast and humans (Table 1): CMC1, CMC2 [27,152], COX19 [153,154], COX23 [155], PET191 [156], and COA6 [54,157]. CX_9C proteins could also play a role in copper transfer towards COX17 by modulating the local IMS redox environment [135,158]. In addition, the redox state of human SCO1 modulates the copper content of cells, and COX19, that partitions between the cytosol and the IMS in a copper-dependent manner, was postulated as a transducer of the signal that connects SCO1 with copper transporters of the plasma membrane [159].

COX assembly factors responsible for metalation of the catalytic centers have their counterparts in plants. Two Arabidopsis genes, *AtCOX17-1* and *AtCOX17-2*, encode functional plant homologs able to complement the respiratory deficiency of a yeast *cox17* null mutant [70–72]. While plants in which both *AtCOX17* genes are silenced show severe phenotypic changes, presumably due to decreased COX activity, the silencing of either gene does not alter plant growth [73]. However, these plants show altered ROS levels and a reduced response to stress conditions. Arabidopsis also encodes two proteins with homology to yeast Sco1, named HCC1 and HCC2 (*Homolog to yeast Copper Chaperone*), but HCC2 lacks the canonical copper-binding motif described in Sco1 [18,19]. Several pieces of evidence point to HCC1 as a COX assembly factor in plants: (i) HCC1 is able to complement a yeast *sco1* mutant [19]; and (ii) knockout of *HCC1* produces embryo arrest, while heterozygous mutants show a decrease in COX activity [18–20] (Table 2). In addition, knockout plants in *HCC2* show normal COX activity levels and no obvious phenotype under normal growth conditions [18–20]. Changes in the expression of HCC1 and HCC2 also cause altered redox homeostasis and responses to copper and stress conditions [18–20]. *Arabidopsis* also contains a COX11 homolog that seems to be essential for normal COX activity. Altered levels of this protein affect pollen germination, development, growth, and copper homeostasis [69] (Table 2). Finally, two different genes, *AtCOX19-1* and *AtCOX19-2*, encode putative homologs of yeast Cox19 [101]. Altered levels of AtCOX19 impact on plant phenotypic responses to changes in copper and iron levels in the growth medium and in the expression of metal-responsive genes.

4.2.4. Heme A Synthesis and COX1 mRNA Translation

Within the process of Complex IV formation, Cox1 hemylation is essential and, since free heme is toxic to the cell, its synthesis and insertion must be precisely coordinated. Studies in yeast established that Cox1 is translationally regulated by heme B availability [4], while heme A biosynthesis is regulated by the redox-dependent oligomerization of Cox10 (heme O synthase) [21,51]. After Cox1 synthesis, Mss51, a *Cox1* mRNA-specific processing factor and translational activator, forms a Cox1 preassembly complex that is only disrupted when the prosthetic groups are inserted into Cox1. Formation of this complex limits the availability of Mss51 for *Cox1* translational activation, thus linking Cox1 synthesis to COX assembly. It was recently demonstrated that Mss51 is activated by heme B and that the redox environment modulates the affinity of Mss51 for heme, thus compromising Mss51 function in *Cox1* mRNA translation [11]. In addition, Pet54 is required for the activation of *Cox1* mRNA translation by Mss51, probably influencing its hemylation or conformational state [160]. Coordination of COX biogenesis with heme availability and the redox state are probably important to avoid the accumulation of pro-oxidative reactive assembly intermediates and to minimize the potential damaging effects of ROS. Nothing is known about the regulation of the synthesis of COX1 and other subunits encoded in mitochondria in plants.

4.3. Regulation by Phosphorylation

Phosphoproteomic studies identified several phosphorylated proteins in purified mitochondria from Arabidopsis cell suspensions [161]. Recent studies in mammalian and yeast mitochondria demonstrate that reversible phosphorylation of proteins on serine, threonine and tyrosine residues is an important biological regulatory mechanism across a broad range of important mitochondrial processes. In mammals, there is solid evidence that COX activity is regulated by phosphorylation and, to date, between 14 and 18 in vivo phosphorylation sites were identified by mass spectrometry (MS) in several Complex IV proteins [162,163]. In some cases, connections of the phosphorylation state of COX components with respiratory energy demands, human diseases, and cell destiny mediated by apoptosis were established [162,163]. In plants, phosphorylation of COX components is less studied and there is discrete evidence that this PTMS specifically impacts on COX proteins in plants exposed to nutritional deficiencies or exposed to hormones (Table S1). AtCOX5b-1, AtCOX6a and several PPR proteins are differentially phosphorylated in a *atm/atr* serine/threonine protein kinase double mutant when plant rosettes are exposed to ionizing DNA-damaging radiation [164]. Furthermore, AtCOX5b-1, AtCOX6a, several PPR proteins and members of the glycoprotein family homologous to MAM33 from yeast were identified in a proteome-wide profiling of phosphopeptides in nine-day-old Arabidopsis seedlings [94]. These proteins were proposed as part of a phosphor-regulatory network involved in biological processes like central metabolism and cell signaling, regulating plant growth and development [94,161]. Several MAM33-like proteins, AtHCC1 and AtSURF1-2 were identified as differentially phosphorylated when nitrogen sources are supplied to growing Arabidopsis seedlings in the form of nitrate or ammonium [165]. As mentioned, AtCOX17 and AtATL8 are differentially phosphorylated in the presence of auxin when this hormone is tested in a lateral root-inducing system [123], and AtMATR2 is part of a group of proteins that are rapidly modified by phosphorylation in response to BRs [124]. To conclude that these modifications have real impact on the regulation of COX activity and are connected to mitochondrial energy generation by respiration or to additional functions will require an in-depth future analysis.

5. Mutant Plants with Altered Cox Activity

In this section, we present a survey of plant mutants related to COX assembly or mutants in which COX activity is, either directly or indirectly, altered (Table 2). This may be useful to understand the role of COX in plant physiology and development.

5.1. Mutants in COX Assembly Factors

The silencing of *AtCOX17-1* in a *cox17-2* knock-out background generates plants with small rosettes and severely delayed development [73]. Silencing of *COX17-1* or *COX17-2* independently does not have any phenotypic consequence and does not affect COX activity, but causes a decrease in the response of plants to stress [73]. COX11 knock-down plants show reduced COX activity [69]. In addition, they exhibit a short root phenotype and reduced pollen germination [69]. HCC1 is essential for plant development [18,19]. *AtHCC1/athcc1* heterozygous mutant plants produce 25% defective seeds with embryos that stop their development at the heart stage [18,19]. Rescue of *hcc1* knock-out embryo lethality by complementation with the *HCC1* gene under the control of the embryo-specific *ABSCISIC ACID INSENSITIVE 3* promoter caused growth arrest at early plant developmental stages, supporting the role of HCC1 for plant growth and development [20]. Heterozygous *AtHCC1/athcc1* mutants show a ca. 50% decrease in COX activity and no phenotypic alterations. Homozygous loss-of-function mutants in *AtHHC2* do not affect plant development or COX activity.

Null mutations in *AtCOX10* also lead to embryonic lethality [17]. Heterozygous *AtCOX10/atcox10* plants show reduced COX activity and normal levels of total respiration at the expense of an increase in alternative respiration. In addition, AtCOX10 seems to be implicated in the progression of senescence, since heterozygous mutants show an increased rate of chlorophyll loss and photosynthesis decay during both natural and dark-induced senescence [17].

5.2. Mutants in RNA Processing Factors Influencing Mitochondrial COX-Coding Genes

There is a group of mutant plants where the mutation affects nuclear-coding genes that are necessary for correct mitochondrial RNA processing, including protection of 5′ and 3′ termini, splicing and editing, that constitute a relevant tool to understand the effects of defective COX over plant growth [98]. RNA editing in land plant organelles alters the coding content of transcripts through site-specific exchanges of cytidines into uridines and vice versa. In model plant species such as rice or *Arabidopsis*, there are about 500 editing sites, indicating the importance of this mechanism [166,167]. One of the most important plant-specific protein family involved in RNA editing is the PPR protein family. PPR proteins are sequence-specific RNA binding proteins that identify C residues for editing. All the reported PPR proteins required for RNA editing are members of E or DYW subclasses of the PPR family [84,168,169].

PpPPR_78, identified and characterized by Uchida et al. [84], is a protein involved in *COX1* editing in *Physcomitrella patens*. Disruption of the *PpPPR_78* gene results in editing defects in *rps_14* (*rps14-C137*) and *cox1* (*cox1-C755*), with no changes in other editing sites [85]. At the phenotypic level, mutant plants display slight growth retardation. This suggests that the editing events are not crucial for RPS-14 and COX1 function. PpPPR_77, another member of the PPR protein family from *Physcomitrella patens* was described by Othani et al. [86] as a mitochondria-localized protein involved in RNA editing. PpPPR_77 mutants show editing defects in *cox2* (*cox2-C370*) and *cox3* (*cox3-C733*) sites, but not in other editing sites. The mutant has severe growth retardation.

Arabidopsis Cytochrome c Oxidase Deficient 1 (*COD1*) encodes a mitochondria-localized PLS-subfamily PPR protein involved in the editing of two distant sites in the *COX2* transcript [80]. Lack of COD1 generates embryo lethality, since no homozygous mutant plants could be identified in the progeny. Immature homozygous mutant seeds could be germinated using a high-sugar medium supplemented with nutrients and co-factors. However, their subsequent growth was retarded, and plants showed a proliferation of short leaves with limited stems and a delay in root development. Occasionally, some plants were able to produce flowers which did not produce viable pollen. COX activity could not be detected in mutant lines and no signal of native Complex IV was observed using anti-COX2 antibodies. MEF13 is a E domain PPR protein required for editing at eight sites in *Arabidopsis thaliana*, including *COX3-314* [81]. *mef13-1* mutants exhibit delayed growth at the vegetative phase but no defects were detected at later stages.

DEK10 encodes a E-subgroup PPR protein from maize that is specifically involved in the C to U editing at *nad3-61*, *nad3-62* and *cox2-550* [83]. Defects in these editing events reduces the function of Complex I and Complex IV in the electron transport chain and affect mitochondrial respiration and embryo, endosperm and seedling development. Homozygous *dek10* mutant kernels are small and their weight is 73% reduced compared to wild-type. Vegetative growth is also severely delayed in the mutants. Lack of DEK10 affects COX2 protein accumulation, while Complex I and Complex IV activities are reduced compared to wild-type.

In addition to PPR proteins, other protein families were also identified as RNA editing factors in mitochondria. One of these families is called ORRM (Organelle RNA Recognition Motif-containing protein) and plays a crucial role in mitochondrial RNA editing in *Arabidopsis thaliana* [9,82]. The absence of ORRM4 expression causes defective mitochondrial editing at 264 sites in 31 mitochondrial transcripts, among them *COX2* and *COX3*. *Orrm4* mutant plants show slow growth and the average flowering time is delayed three days in comparison with wild-type plants.

MORF family proteins (multiple organellar RNA editing factor) are required for RNA editing in plant mitochondria as additional components of the RNA editing machinery [79]. The etilmetanosulfonato (EMS)-mutated Arabidopsis *morf1* mutant shows reduced RNA editing at more than 40 mitochondrial sites, without defects in other RNA processing steps or RNA stability. COX3-257 is one of the sites affected by MORF1 mutation, with a 50% decrease in editing events of COX3 transcripts. Homozygous mutant plants with T-DNA insertion in the MORF1 gene are not viable, due to the abortion of seed development. This finding suggests that this protein is essential and that the EMS mutant is hypomorphic and that residual MORF1 activity is sufficient for plant viability.

Plant mitochondrial gene expression involves the splicing of introns from the coding regions. The mitochondrial introns of angiosperms are classified as group II-type. Particularly in Arabidopsis, these include 23 introns found within Complex I, CYT*c* biogenesis factors, Complex IV and ribosomal protein genes. This process relies on many different RNA-binding proteins or co-factors. Two members of the maturase and RNA helicase families, nMAT2 and PMH2, participate in the splicing of several transcripts in *Arabidopsis* mitochondria, including COX2 [89]. *mat2/pmh2* double mutants show a decrease of mRNA levels of multiple mitochondrial genes, and exhibit growth delay and alterations in vegetative and reproductive development. No change in total respiration was detected. However, inhibition of Complex IV by KCN appeared to have a reduced effect on the respiratory activity of *mat2/pmh2* plants.

Besides their role in editing, the PPR protein family is also involved in other steps of RNA metabolism in plant organelles. Particularly, PpPPR_43 has been reported as a splicing regulator of COX1 mRNA in *Physcomitrella patens* [90]. The PpPPR_43 disruptants show normal editing, but mature COX1 transcripts levels are substantially reduced. Instead of the mature COX1 transcript, unexpectedly long transcripts were detected. A defect in PpPPR_43 strongly inhibits the splicing of COX1 intron 3, without affecting the splicing of introns 1, 2 and 4. Mutant plants show severe growth retardation.

Most 5′ termini of plant mitochondrial transcripts are generated post-transcriptionally [170]. RFP2 is a PPR protein involved in the formation of COX3 5′ transcript ends. It has been reported that COX3 mRNA 5′-end formation is affected in *rfp2* mutants, resulting in a substantial reduction of mature COX3 transcripts. However, the impaired 5′ processing does not interfere with protein accumulation.

The plant organellar RNA recognition (PORR) domain was recognized to be an RNA binding domain, and most of the members of this family are involved in mRNA splicing in chloroplasts and mitochondria. WTF9 is involved in the splicing of group II introns in mitochondria and is required for the CYT*c* maturation pathway [87]. Mutant plants lacking WTF9 show severely stunted growth. They survive to flowering but the flowers are small with very small anthers and small amounts of pollen. This results in very small siliques with a few aborted seeds. The levels of CYT*c*, cytochrome c_1, and the COX2 subunit are severely reduced in mutant mitochondria, which is accompanied by a dramatic increase of the alternative oxidase, which is commonly induced when the respiratory chain is compromised.

Not all Mitochondrial Editing Factors (MEFs) are essential for plant growth. This is the case for MEF21 [171] or MEF26 [172], two RNA processing factors belonging to the E- or DYW-groups of PPR mitochondrial proteins, respectively. The *cox3-257* site targeted by MEF21 changes a Ser to a conserved Phe residue at amino acid 86 in COX3 [171]. MEF26 is involved in editing of *cox3-311* and *nad4-166* sites, changing codon identities from Ser to Phe and Arg to Trp, respectively. Mutants *mef26-1* and *mef 26-2* exhibit strong deficits in COX3 edition but only a 20% decrease in the edition of NAD4 [172]. Mutant plants in any of these proteins do not exhibit phenotypic defects under normal growth conditions.

The CRM domain is a recently recognized RNA-binding motif of bacterial origin that is present in several group II intron splicing factors. mCSF-1 is a mitochondria-localized CRM family member required for the processing of many mitochondrial introns, and is involved in the biogenesis of respiratory Complexes I and IV in Arabidopsis [88]. Embryo development is arrested during the early globular stage in knock-out *mcsf-1* mutant plants, showing that mCSF-1 is an essential gene.

Knock-down *mcsf-1* mutants show germination delay and growth alterations. The respiratory activity is lower in *mcsf-1* plants, and the relative accumulation of native Complexes I and IV is reduced.

5.3. Mutant Plants Indirectly Connected to COX Activity

Mutations in the dual-targeted mitochondrial and plastidial RPOTmp RNA polymerase cause defects in transcript abundance of mitochondrial genes and a dramatic decrease of Complex I and Complex IV levels [91]. *rpoTmp* mutant lines show wrinkly rosette leaves with delayed development from germination onwards [91].

SLO2 (SLOW GROWTH 2) is a PPR protein belonging to the E+ subclass of the PLS subfamily. In *slo2* mutants, seven editing changes producing four amino acid changes in NAD4L (S37L), NAD1 (T1M), NAD7 (L247F) and MTTB (P49S) were detected. These mutants show a marked reduction in Complexes I, III and IV of the mETC. The *slo2* mutants are characterized by retarded leaf emergence, delayed development, late flowering and smaller roots. Sugar, ATP and NAD$^+$ contents are lower in *slo2* mutants, and plant phenotype is enhanced in the absence of sucrose, suggesting a defect in energy metabolism [93].

MRPL1 (Mitochondrial Ribosomal Protein L1) is an essential mitochondrial ribosomal protein. Consistent with its role, levels of mitochondrial COX2 are severely reduced in *mrpl1* mutants, whereas no changes are observed in the expression of the nuclear encoded-mitochondrial proteins. *mrpl1-1* and *mrpl1-3* mutants exhibit delayed growth under normal growth conditions, with reduced respiration compared to wild-type plants, whereas the levels of Complex I and Complex IV are strongly reduced [173]. The mito-nuclear protein imbalance induces a proteotoxic stress, thus activating MAPK6 and hormonal (mainly auxin and ethylene) signaling pathways [173].

Mutants in the *Arabidopsis* mitochondrial LON1 protease [95] show a retarded growth phenotype [96]. Mitochondrial proteome characterization of *lon1* mutant plants showed an altered abundance of enzymes of the TCA cycle and OXPHOS [95]. In addition, the *lon1* mutant has lower abundance of Complexes I, IV, and V [95].

Welchen et al. [97] studied the role of CYT*c*, which transfers electrons from Complex III to Complex IV, in plant development and redox homeostasis. In Arabidopsis, CYT*c* is encoded by two genes, designated *CYTC-1* and *CYTC-2*. Knock-out of *CYTC-1* and *CYTC-2* causes embryo developmental arrest indicating that CYT*c* is essential for plant development. Moreover, double T-DNA insertion mutant plants with considerably reduced CYT*c* levels show delayed development, altered expression of stress responsive genes and reduced COX activity [97]. These results suggest that CYT*c* plays an important role in the control of stability of COX in plant mitochondria [97].

6. Requirement of COX Activity for Plant Growth and Development

The analysis of mutants suggests that COX activity is important for several aspects of plant development.

6.1. Embryogenesis

As we mentioned, COX components are preferentially expressed in plant tissues with high energy demands, like shoot and root meristems, anthers and, significantly, during all stages of embryogenesis (Figure 1). These characteristics suggest that an optimal COX activity may be essential during embryogenesis, in the initial phases of the formation of a new plant. In agreement with this, lack of the COX biogenesis factors AtCOX10 and AtHCC1 originate embryo lethality at various developmental stages, mostly at the heart [18,19] and torpedo [17] stages. There are also several examples of mutants in different COX editing and RNA processing factors, or in proteins connected to Complex IV stability, that also exhibit severe embryo lethality [79,80,87,97,98] (Table 2 for details).

6.2. Germination

Mitochondria are dynamic organelles that constantly change their form and function in response to different external and internal factors, such as organ, tissue, developmental stage, environmental stimuli, and cell energy and metabolic demands, among others ([174,175], and reviewed in [1]). Mitochondria play a crucial role during germination, providing cellular energy via oxidative phosphorylation [175]. Studies performed in maize and rice demonstrated that respiration increases rapidly during the first 24 h in seeds exposed to stratification conditions (dark, 4 °C) [176]. This is due to the execution of a program where the components of the protein import machinery are already present in undifferentiated promitochondrial structures to facilitate a rapid rate of mitochondrial biogenesis after imbibition [177]. In rice, transcripts for two nuclear-encoded COX components (Os*Cox5b-1* and Os*Cox5b-2*) reached maximum levels after 24 h of imbibition, being COX activity detected at 12 h post-imbibition [177]. In Arabidopsis, promitochondria are bioenergetically active immediately upon hydration and respiration increases rapidly within the first hour if evaluated at 21 °C [178]. Law et al. [175] showed the existence of changes in mitochondrial number, size, and morphology after 12 h of imbibition in continuous light and established a triphasic progression model based on the expression profile of genes encoding functional proteins for mitochondrial biogenesis during germination. First, there is an early increase of transcripts associated with nucleic acid metabolism (DNA transcription and RNA editing/processing), followed by transcripts encoding proteins associated with protein synthesis and import and, finally, transcripts encoding proteins of the mitochondrial electron transport chain (mETC) [175–177]. Transcripts for components of the mETC reach maximum levels during the transition between promitochondria and mature mitochondria, characterized by a reduced biogenesis and an increase in the bioenergetic and metabolic function of the organelle [175]. This model is in agreement with the data showed in Figure 2H where a clear initial increment in transcripts for RNA processing and cofactor assembly/synthesis proteins can be identified, in comparison with an opposite behavior for the mitochondrial transcripts, when the expression is evaluated in imbibed seeds compared to dry seeds (Figure 2H). Beyond the transcriptional data, there are several examples of Arabidopsis mutants affected in COX composition and activity that exhibit problems and serious delays during the germinative phase [88,97]. In an extreme example, *cod1-1* an *cod1-2* are mutants in an RNA editing factor that lack COX activity and are only able to germinate if embryos from immature seeds are cultivated in a special medium [80] (see below, Table 2)

6.3. Vegetative Growth and Development

There are several examples portraying the connection between COX and plant growth and development. Abnormalities or dysfunctions in COX composition and/or activity severely impact on growth and development in Arabidopsis [79,81,82,87,89,98], maize [83] and *Physcomitrella patens* [84,86,90]. Due to lower COX activity, plants exhibit smaller rosettes [73,92,97] with altered rosette leaf shape [78,91], and root growth impairment [69,97]. The *cod1* mutant, with negligible COX activity, exhibits an anarchical slow-growing phenotype leading to a small bush-like structure. Root development is also severely compromised and only sporadically plants are able to produce flowers, with no viable pollen [80] (Table 2).

6.4. Senescence

Most mutants in proteins connected with COX activity exhibit delayed growth and development, and this is usually connected to an extended life period [98]. As an opposite example, *Arabidopsis* heterozygous mutants in *AtCOX10*, that possess defective respiration with lower levels of COX activity, exhibit an early onset and an accelerated progression of natural senescence and dark-induced senescence [17]. It was postulated that lower COX activity levels may cause a decrease in energy availability, thus accelerating the use of reserves. Alternatively, dysfunctional COX assembly intermediates may have deleterious effects on the natural progression of the plant developmental program [17]. Further studies, in which COX activity is specifically affected during plant senescence, are required to evaluate the actual requirement of COX for this process.

7. Conclusions

COX biogenesis is a modular process controlled by finely regulated steps to assemble a functional respiratory complex for the successful operation of the mETC. Expression of proteins encoded in two genomes must be coordinated, including those responsible for COX catalytic activity, for the synthesis and delivery of redox-active metal centers and prosthetic groups, for the assembly of subunits, and for the multiple RNA processing events. COX has several plant-specific subunits and, while mitochondria-encoded, catalytic-core subunits are extremely conserved, homologs of several yeast and human nuclear-encoded subunits are missing in plants. In a similar way, COX biogenesis proteins related to the assembly of the redox-active centers are conserved in plants, but other COX assembly factors are not. Plant COX assembly factors also seem to have acquired novel functions, not directly related to COX assembly. The presence of alternative isoforms of certain subunits suggests that COX composition is finely regulated according to cell- and tissue-type, and plant developmental stage. This may establish COX complexes with different properties, able to adjust COX function to energy demands. In agreement with the connections between mitochondrial respiratory activity and energy demands for growth, expression of COX components is regulated at the transcriptional level by nutrients and the diurnal cycle. In addition, several energy-demanding processes, such as embryogenesis and germination, also generate an increase in the expression of the majority of COX-related genes. COX activity is also under hormonal control and there are several pieces of evidence for possible regulation at the transcriptional and post-translational levels. A decrease in COX activity produces strong alterations of plant growth and development, often leading to embryonic lethality. This can be thought of as a double-check control mechanism, where hormones regulate COX biogenesis and growth, and COX is directly connected to plant growth and development through its activity. Finally, several COX assembly factors were recognized as stress-related proteins, acting in response to abiotic and biotic stress in plants. As we previously speculated, COX assembly factors or structural proteins may accumulate under stress to avoid or diminish ROS generation due to dysfunctional COX and/or to actively replace damaged COX subunits. Much less is known about COX biogenesis in plants than in yeast and mammals. Further work is needed to clarify the role and significance of plant-specific subunits, to identify putative plant-specific COX assembly factors, to unequivocally establish the interactions between the different proteins involved in COX biogenesis, and to elucidate if the stress-related functions of several accessory proteins are related to COX activity or more widely connected to retrograde signaling and other mitochondrial regulatory pathways.

Supplementary Materials: Supplementary materials can be found at www.mdpi.com/422-0067/19/3/662/s1.

Acknowledgments: This work was supported by grants from ANPCyT (Agencia Nacional de Promoción Científica y Tecnológica, Argentina) by grants PICT-3758 and PICT-2307 and Universidad Nacional del Litoral (UNL-CAI+D 50420150100007LI). Diana E. Gras, Daniel H. Gonzalez and Elina Welchen are members of CONICET. Natanael Mansilla and Sofía Racca are CONICET fellows.

Conflicts of Interest: The authors declare no conflict of interest.

Abbreviations

COX	Cytochrome *c* oxidase
IMM	Inner mitochondrial membrane
IMS	Intermembrane space
mETC	Mitochondrial electron transport chain
MS	Mass spectrometry
OXPHOS	Oxidative phosphorylation
PTMS	Posttranslational modifications

References

1. Welchen, E.; García, L.; Mansilla, N.; Gonzalez, D.H. Coordination of plant mitochondrial biogenesis: Keeping pace with cellular requirements. *Front. Plant Sci.* **2014**, *4*, 551. [CrossRef] [PubMed]
2. Jett, K.A.; Leary, S.C. Building the Cu$_A$ site of cytochrome *c* oxidase: A complicated, redox-dependent process driven by a surprisingly large complement of accessory proteins. *J. Biol. Chem.* **2017**. [CrossRef] [PubMed]
3. Khalimonchuk, O.; Rodel, G. Biogenesis of cytochrome *c* oxidase. *Mitochondrion* **2005**, *5*, 363–388. [CrossRef] [PubMed]
4. Soto, I.C.; Fontanesi, F.; Liu, J.; Barrientos, A. Biogenesis and assembly of eukaryotic cytochrome *c* oxidase catalytic core. *Biochim. Biophys. Acta* **2012**, *1817*, 883–897. [CrossRef] [PubMed]
5. Timón-Gómez, A.; Nývltová, E.; Abriata, L.A.; Vila, A.J.; Hosler, J.; Barrientos, A. Mitochondrial cytochrome *c* oxidase biogenesis: Recent developments. *Semin. Cell Dev. Biol.* **2017**. [CrossRef] [PubMed]
6. Rak, M.; Bénit, P.; Chrétien, D.; Bouchereau, J.; Schiff, M.; El-Khoury, R.; Tzagoloff, A.; Rustin, P. Mitochondrial cytochrome *c* oxidase deficiency. *Clin. Sci.* **2016**, *130*, 393–407. [CrossRef] [PubMed]
7. Sinkler, C.A.; Kalpage, H.; Shay, J.; Lee, I.; Malek, M.H.; Grossman, L.I.; Huttemann, M. Tissue- and Condition-Specific Isoforms of Mammalian Cytochrome *c* Oxidase Subunits: From Function to Human Disease. *Oxid. Med. Cell. Longev.* **2017**, *2017*, 1534056. [CrossRef] [PubMed]
8. Hammani, K.; Gobert, A.; Small, I.; Giegé, P. A PPR protein involved in regulating nuclear genes encoding mitochondrial proteins? *Plant Signal. Behav.* **2011**, *6*, 748–750. [CrossRef] [PubMed]
9. Sun, T.; Bentolila, S.; Hanson, M.R. The Unexpected Diversity of Plant Organelle RNA Editosomes. *Trends Plant Sci.* **2016**, *21*, 962–973. [CrossRef] [PubMed]
10. Gray, M.; Burger, G.; Lang, B. Mitochondrial evolution. *Science* **1991**, *283*, 1476–1481. [CrossRef]
11. Soto, I.; Barrientos, A. Mitochondrial Cytochrome *c* Oxidase Biogenesis Is Regulated by the Redox State of a Heme-Binding Translational Activator. *Antioxid. Redox Signal.* **2016**, *24*, 281–298. [CrossRef] [PubMed]
12. Kadenbach, B. Regulation of Mammalian 13-Subunit Cytochrome *c* Oxidase and Binding of other Proteins: Role of NDUFA4. *Trends Endocrinol. Metab.* **2017**, *28*, 761–770. [CrossRef] [PubMed]
13. Cox, C.J.; Foster, P.G.; Hirt, R.P.; Harris, S.R.; Embley, T.M. The archaebacterial origin of eukaryotes. *Proc. Natl. Acad. Sci. USA* **2008**, *105*, 20356–20361. [CrossRef] [PubMed]
14. Poyton, R.O.; McEwen, J.E. Crosstalk between nuclear and mitochondrial genomes. *Annu. Rev. Biochem.* **1996**, *65*, 563–607. [CrossRef] [PubMed]
15. Barros, M.H.; Tzagoloff, A. Regulation of the heme A biosynthetic pathway in *Saccharomyces cerevisiae*. *FEBS Lett.* **2002**, *516*, 119–123. [CrossRef]
16. Pecina, P.; Houstková, H.; Hansíková, H.; Zeman, J.; Houstek, J. Genetic defects of cytochrome *c* oxidase assembly. *Physiol. Res.* **2004**, *53*, 213–223. [CrossRef]
17. Mansilla, N.; Garcia, L.; Gonzalez, D.H.; Welchen, E. AtCOX10, a protein involved in haem *o* synthesis during cytochrome *c* oxidase biogenesis, is essential for plant embryogenesis and modulates the progression of senescence. *J. Exp. Bot.* **2014**, *66*, 6761–6775. [CrossRef] [PubMed]
18. Attallah, C.V.; Welchen, E.; Martin, A.P.; Spinelli, S.V.; Bonnard, G.; Palatnik, J.F.; Gonzalez, D.H. Plants contain two SCO proteins that are differentially involved in cytochrome *c* oxidase function and copper and redox homeostasis. *J. Exp. Bot.* **2011**, *62*, 4281–4294. [CrossRef] [PubMed]
19. Steinebrunner, I.; Landschreiber, M.; Krause-Buchholz, U.; Teichmann, J.; Rödel, G. HCC1, the Arabidopsis homologue of the yeast mitochondrial copper chaperone SCO1, is essential for embryonic development. *J. Exp. Bot.* **2011**, *62*, 319–330. [CrossRef] [PubMed]
20. Steinebrunner, I.; Gey, U.; Andres, M.; Garcia, L.; Gonzalez, D.H. Divergent functions of the Arabidopsis mitochondrial SCO proteins: HCC1 is essential for COX activity while HCC2 is involved in the UV-B stress response. *Front. Plant Sci.* **2014**, *5*, 87. [CrossRef] [PubMed]
21. Khalimonchuk, O.; Bestwick, M.; Meunier, B.; Watts, T.C.; Winge, D.R. Formation of the redox cofactor centers during Cox1 maturation in yeast cytochrome oxidase. *Mol. Cell. Biol.* **2010**, *30*, 1004–1017. [CrossRef] [PubMed]
22. Lee, I.C.; El-Hattab, A.W.; Wang, J.; Li, F.Y.; Weng, S.W.; Craigen, W.J.; Wong, L.J. SURF1-associated Leigh syndrome: A case series and novel mutations. *Hum. Mutat.* **2012**, *33*, 1192–1200. [CrossRef] [PubMed]

23. Barrientos, A.; Gouget, K.; Horn, D.; Soto, I.C.; Fontanesi, F. Suppression mechanisms of COX assembly defects in yeast and human: Insights into the COX assembly process. *Biochim. Biophys. Acta* **2009**, *1793*, 97–107. [CrossRef] [PubMed]
24. Nijtmans, L.G.; Taanman, J.W.; Muijsers, A.O.; Speijer, D.; van den Bogert, C. Assembly of cytochrome-*c* oxidase in cultured human cells. *Eur. J. Biochem.* **1998**, *254*, 389–394. [CrossRef] [PubMed]
25. McStay, G.P.; Su, C.H.; Tzagoloff, A. Modular assembly of yeast cytochrome oxidase. *Mol. Biol. Cell* **2013**, *24*, 440–452. [CrossRef] [PubMed]
26. Mick, D.U.; Vukotic, M.; Piechura, H.; Meyer, H.E.; Warscheid, B.; Deckers, M.; Rehling, P. Coa3 and Cox14 are essential for negative feedback regulation of COX1 translation in mitochondria. *J. Cell Biol.* **2010**, *191*, 141–154. [CrossRef] [PubMed]
27. Bourens, M.; Barrientos, A. A CMC1-knockout reveals translation-independent control of human mitochondrial complex IV biogenesis. *EMBO Rep.* **2017**, *18*, 477–494. [CrossRef] [PubMed]
28. Fontanesi, F.; Soto, I.C.; Horn, D.; Barrientos, A. Mss51 and Ssc1 facilitate translational regulation of cytochrome *c* oxidase biogenesis. *Mol. Cell. Biol.* **2010**, *30*, 245–259. [CrossRef] [PubMed]
29. Taylor, N.G.; Swenson, S.; Harris, N.J.; Germany, E.M.; Fox, J.L.; Khalimonchuk, O. The assembly factor Pet117 couples heme A synthase activity to cytochrome oxidase assembly. *J. Biol. Chem.* **2016**, *292*, 1815–1825. [CrossRef] [PubMed]
30. Bestwick, M.; Jeong, M.Y.; Khalimonchuk, O.; Kim, H.; Winge, D.R. Analysis of Leigh syndrome mutations in the yeast SURF1 homolog reveals a new member of the cytochrome oxidase assembly factor family. *Mol. Cell. Biol.* **2010**, *30*, 4480–4491. [CrossRef] [PubMed]
31. Banci, L.; Bertini, I.; Cantini, F.; Ciofi-Baffoni, S.; Gonnelli, L.; Mangani, S. Solution structure of Cox11, a novel type of beta-immunoglobulin-like fold involved in CuB site formation of cytochrome *c* oxidase. *J. Biol. Chem.* **2004**, *279*, 34833–34839. [CrossRef] [PubMed]
32. Carr, H.S.; George, G.N.; Winge, D.R. Yeast Cox11, a protein essential for cytochrome *c* oxidase assembly, is a Cu(I)-binding protein. *J. Biol. Chem.* **2002**, *277*, 31237–31242. [CrossRef] [PubMed]
33. Bode, M.; Woellhaf, M.W.; Bohnert, M.; Van der Laan, M.; Sommer, F.; Jung, M.; Zimmermann, R.; Schroda, M.; Herrmann, J.M. Redox-regulated dynamic interplay between Cox19 and the copper-binding protein Cox11 in theintermembrane space of mitochondria facilitates biogenesis of cytochrome *c* oxidase. *Mol. Biol. Cell* **2015**, *26*, 2385–2401. [CrossRef] [PubMed]
34. Glerum, D.M.; Shtanko, A.; Tzagoloff, A. Characterization of COX17, a yeast gene involved in copper metabolism and assembly of cytochrome oxidase. *J. Biol. Chem.* **1996**, *271*, 14504–14509. [CrossRef] [PubMed]
35. Beers, J.; Glerum, D.M.; Tzagoloff, A. Purification, characterization, and localization of yeast Cox17p, a mitochondrial copper shuttle. *J. Biol. Chem.* **1997**, *272*, 33191–33196. [CrossRef] [PubMed]
36. Cobine, P.A.; Pierrel, F.; Bestwick, M.L.; Winge, D.R. Mitochondrial matrix copper complex used in metallation of cytochrome oxidase and superoxide dismutase. *J. Biol. Chem.* **2006**, *281*, 36552–36559. [CrossRef] [PubMed]
37. Vest, K.E.; Wang, J.; Gammon, M.G.; Maynard, M.K.; White, O.L.; Cobine, L.A.; Mahone, W.K.; Cobine, P.A. Overlap of copper and iron uptake systems in mitochondria in *Saccharomyces cerevisiae*. *Open Biol.* **2016**, *6*, 150–223. [CrossRef] [PubMed]
38. Banci, L.; Bertini, I.; Cavallaro, G.; Rosato, A. The functions of Sco proteins from genome-based analysis. *J. Proteome Res.* **2007**, *6*, 1568–1579. [CrossRef] [PubMed]
39. Stroud, D.A.; Maher, M.J.; Lindau, C.; Vögtle, F.N.; Frazier, A.E.; Surgenor, E.; Mountford, H.; Singh, A.P.; Bonas, M.; Oeljeklaus, S.; et al. COA6 is a mitochondrial complex IV assembly factor critical for biogenesis of mtDNA-encoded COX2. *Hum. Mol. Genet.* **2015**, *24*, 5404–5415. [CrossRef] [PubMed]
40. Bourens, M.; Boulet, A.; Leary, S.C.; Barrientos, A. Human COX20 cooperates with SCO1 and SCO2 to mature COX2 and promote the assembly of cytochrome *c* oxidase. *Hum. Mol. Genet.* **2014**, *23*, 2901–2913. [CrossRef] [PubMed]
41. Abriata, L.A. Structural models and considerations on the COA6, COX18 and COX20 factors that assist assembly of human cytochrome *c* oxidase subunit II. *bioRxiv* **2017**. [CrossRef]
42. Vidoni, S.; Harbour, M.E.; Guerrero-Castillo, S.; Signes, A.; Ding, S.; Fearnley, I.M.; Taylor, R.W.; Tiranti, V.; Arnold, S.; Fernandez-Vizarra, E.; et al. MR-1S interacts with PET100 and PET117 in module-based assembly of human cytochrome *c* oxidase. *Cell Rep.* **2017**, *18*, 1727–1738. [CrossRef] [PubMed]

43. Stiburek, L.; Fornuskova, D.; Wenchich, L.; Pejznochova, M.; Hansikova, H.; Zeman, J. Knockdown of human Oxa1l impairs the biogenesis of F1Fo-ATP synthase and NADH:ubiquinone oxidoreductase. *J. Mol. Biol.* **2007**, *34*, 506–516. [CrossRef] [PubMed]
44. Leary, S.C.; Sasarman, F.; Nishimura, T.; Shoubridge, E.A. Human SCO2 is required for the synthesis of CO II and as a thiol-disulphide oxidoreductase for SCO1. *Hum. Mol. Genet.* **2009**, *18*, 2230–2240. [CrossRef] [PubMed]
45. Garlich, J.; Strecker, V.; Wittig, I.; Stuart, R.A. Mutational analysis of the QRRQ motif in the yeast Hig1 type 2 protein Rcf1 reveals a regulatory role for the cytochrome *c* oxidase complex. *J. Biol. Chem.* **2017**, *292*, 5216–5226. [CrossRef] [PubMed]
46. Pierron, D.; Wildman, D.E.; Huttemann, M.; Markondapatnaikuni, G.C.; Aras, S.; Grossman, L.I. Cytochrome *c* oxidase: Evolution of control via nuclear subunit addition. *Biochim. Biophys. Acta* **2012**, *1817*, 590–597. [CrossRef] [PubMed]
47. Wu, M.; Gu, J.; Guo, R.; Huang, Y.; Yang, M. Structure of mammalian respiratory supercomplex $I_1III_2IV_1$. *Cell* **2016**, *167*, 1598–1609. [CrossRef] [PubMed]
48. Gu, J.; Wu, M.; Guo, R.; Yan, K.; Lei, J.; Gao, N.; Yang, M. The architecture of the mammalian respirasome. *Nature* **2016**, *537*, 639–643. [CrossRef] [PubMed]
49. Fontanesi, F.; Diaz, F.; Barrientos, A. Evaluation of the mitochondrial respiratory chain and oxidative phosphorylation system using yeast models of OXPHOS deficiencies. *Curr. Protoc. Hum. Genet.* **2009**. [CrossRef]
50. Fontanesi, F.; Clemente, P.; Barrientos, A. Cox25 teams up with Mss51, Ssc1, and Cox14 to regulate mitochondrial cytochrome *c* oxidase subunit 1 expression and assembly in *Saccharomyces cerevisiae*. *J. Biol. Chem.* **2011**, *286*, 555–566. [CrossRef] [PubMed]
51. Khalimonchuk, O.; Kim, H.; Watts, T.; Perez-Martinez, X.; Winge, D.R. Oligomerization of heme *o* synthase in cytochrome oxidase biogenesis is mediated by cytochrome oxidase assembly factor Coa2. *J. Biol. Chem.* **2012**, *287*, 26715–26726. [CrossRef] [PubMed]
52. Khalimonchuk, O.; Jeong, M.Y.; Watts, T.; Ferris, E.; Winge, D.R. Selective Oma1 protease-mediated proteolysis of Cox1 subunit of cytochrome oxidase in assembly mutants. *J. Biol. Chem.* **2012**, *287*, 7289–7300. [CrossRef] [PubMed]
53. García-Villegas, R.; Camacho-Villasana, Y.; Shingú-Vázquez, M.Á.; Cabrera-Orefice, A.; Uribe-Carvajal, S.; Fox, T.D.; Pérez-Martínez, X. The Cox1 C-terminal domain is a central regulator of cytochrome *c* oxidase biogenesis in yeast mitochondria. *J. Biol. Chem.* **2017**, *292*, 10912–10925. [CrossRef] [PubMed]
54. Pacheu-Grau, D.; Bareth, B.; Dudek, J.; Juris, L.; Vögtle, F.; Wissel, M.; Leary, S.; Dennerlein, S.; Rehling, P.; Deckers, M. Cooperation between COA6 and SCO2 in COX2 maturation during cytochrome *c* oxidase assembly links two mitochondrial cardiomyopathies. *Cell Metab.* **2015**, *21*, 823–833. [CrossRef] [PubMed]
55. Cherry, J.M.; Hong, E.L.; Amundsen, C.; Balakrishnan, R.; Binkley, G.; Chan, E.T.; Christie, K.R.; Costanzo, M.C.; Dwight, S.S.; Engel, S.R.; et al. Saccharomyces Genome Database: The genomics resource of budding yeast. *Nucleic Acids Res.* **2012**, D700–D705. [CrossRef] [PubMed]
56. Goodstein, D.M.; Shu, S.; Howson, R.; Neupane, R.; Hayes, R.D.; Fazo, J.; Mitros, T.; Dirks, W.; Hellsten, U.; Putnam, N.; et al. Phytozome: A comparative platform for green plant genomics. *Nucleic Acids Res.* **2012**, D1178–D1186. [CrossRef] [PubMed]
57. Hooper, C.M.; Castleden, I.R.; Tanz, S.K.; Aryamanesh, N.; Millar, A.H. SUBA4: The interactive data analysis centre for Arabidopsis subcellular protein locations. *Nucleic Acids Res.* **2017**, *45*, D1064–D1074. [CrossRef] [PubMed]
58. Schwacke, R.; Schneider, A.; van der Graaff, E.; Fischer, K.; Catoni, E.; Desimone, M.; Frommer, W.B.; Flügge, U.I.; Kunze, R. ARAMEMNON, a novel database for Arabidopsis integral membrane proteins. *Plant Physiol.* **2003**, *131*, 16–26. [CrossRef] [PubMed]
59. Senkler, J.; Senkler, M.; Eubel, H.; Hildebrandt, T.; Lengwenus, C.; Schertl, P.; Schwarzländer, M.; Wagner, S.; Wittig, I.; Braun, H.P. The mitochondrial complexome of *Arabidopsis thaliana*. *Plant J.* **2017**, *89*, 1079–1092. [CrossRef] [PubMed]
60. Millar, A.; Eubel, H.; Jänsch, L.; Kruft, V.; Heazlewood, J.; Braun, H. Mitochondrial cytochrome *c* oxidase and succinate dehydrogenase complexes contain plant specific subunits. *Plant Mol. Biol.* **2004**, *56*, 77–90. [CrossRef] [PubMed]

61. Unseld, M.; Marienfeld, J.; Brandt, P.; Brennicke, A. The mitochondrial genome of *Arabidopsis thaliana* contains 57 genes in 366,924 nucleotides. *Nat. Genet.* **1997**, *15*, 57–61. [CrossRef] [PubMed]
62. Forner, J.; Weber, B.; Wiethölter, C.; Meyer, R.C.; Binder, S. Distant sequences determine 5′ end formation of cox3 transcripts in *Arabidopsis thaliana* ecotype C24. *Nucleic Acids Res.* **2005**, *33*, 4673–4682. [CrossRef] [PubMed]
63. Muffato, M.; Louis, A.; Poisnel, C.E.; Crollius, H.R. Genomicus: A database and a browser to study gene synteny in modern and ancestral genomes. *Bioinformatics* **2010**, *26*, 1119–1121. [CrossRef] [PubMed]
64. Lee, T.H.; Tang, H.; Wang, X.; Paterson, A.H. PGDD: A database of gene and genome duplication in plants. *Nucleic Acids Res.* **2013**, D1152–D1158. [CrossRef] [PubMed]
65. Daley, D.O.; Adams, K.L.; Clifton, R.; Qualmann, S.; Millar, A.H.; Palmer, J.D.; Pratje, E.; Whelan, J. Gene transfer from mitochondrion to nucleus: Novel mechanisms for gene activation from Cox2. *Plant J.* **2002**, *30*, 11–21. [CrossRef] [PubMed]
66. Klodmann, J.; Senkler, M.; Rode, C.; Braun, H. Defining the Protein Complex Proteome of Plant Mitochondria. *Plant Physiol.* **2011**, *157*, 587–598. [CrossRef] [PubMed]
67. Vukotic, M.; Oeljeklaus, S.; Wiese, S.; Vogtle, F.N.; Meisinger, C.; Meyer, H.E.; Zieseniss, A.; Katschinski, D.M.; Jans, D.C.; Jakobs, S.; et al. Rcf1 mediates cytochrome oxidase assembly and respirasome formation, revealing heterogeneity of the enzyme complex. *Cell Metab.* **2012**, *7*, 336–347. [CrossRef] [PubMed]
68. Su, C.; Tzagoloff, A. Cox16 protein is physically associated with Cox1p assembly intermediates and with cytochrome oxidase. *J. Biol. Chem.* **2017**, *292*, 16277–16283. [CrossRef] [PubMed]
69. Radin, I.; Mansilla, N.; Rödel, G.; Steinebrunner, I. The Arabidopsis COX11 Homolog is Essential for Cytochrome *c* Oxidase Activity. *Front. Plant Sci.* **2015**, *6*, 1091. [CrossRef] [PubMed]
70. Balandin, T.; Castresana, C. AtCOX17, an Arabidopsis homolog of the yeast copper chaperone COX17. *Plant Physiol.* **2002**, *129*, 1852–1857. [CrossRef] [PubMed]
71. Wintz, H.; Vulpe, C. Plant copper chaperones. *Biochem. Soc. Trans.* **2002**, *30*, 732–735. [CrossRef] [PubMed]
72. Attallah, C.V.; Welchen, E.; Gonzalez, D.H. The promoters of *Arabidopsis thaliana* genes AtCOX17-1 and -2, encoding a copper chaperone involved in cytochrome *c* oxidase biogenesis, are preferentially active in roots and anthers and induced by biotic and abiotic stress. *Physiol. Plant* **2007**, *129*, 123–134. [CrossRef]
73. Garcia, L.; Welchen, E.; Gey, U.; Arce, A.L.; Steinebrunner, I.; Gonzalez, D.H. The cytochrome *c* oxidase biogenesis factor AtCOX17 modulates stress responses in Arabidopsis. *Plant Cell Environ.* **2016**, *39*, 628–644. [CrossRef] [PubMed]
74. Picciocchi, A.; Douce, R.; Alban, C. The plant biotin synthase reaction. Identification and characterization of essential mitochondrial accessory protein components. *J. Biol. Chem.* **2003**, *278*, 24966–24975. [CrossRef] [PubMed]
75. Hamel, P.; Sakamoto, W.; Wintz, H.; Dujardin, G. Functional complementation of an oxa1-Yeast mutation identifies an *Arabidopsis thaliana* cDNA involved in the assembly of respiratory complexes. *Plant J.* **1997**, *6*, 1319–1327. [CrossRef]
76. Sakamoto, W.; Spielewoy, N.; Bonnard, G.; Murata, M.; Wintz, H. Mitochondrial localization of AtOXA1, an arabidopsis homologue of yeast Oxa1p involved in the insertion and assembly of protein complexes in mitochondrial inner membrane. *Plant Cell Physiol.* **2000**, *41*, 1157–1163. [CrossRef] [PubMed]
77. Benz, M.; Soll, J.; Ankele, E. *Arabidopsis thaliana* Oxa proteins locate to mitochondria and fulfill essential roles during embryo development. *Planta* **2013**, *237*, 573–588. [CrossRef] [PubMed]
78. Bentolila, S.; Hellera, W.P.; Suna, T.; Babinaa, A.M.; Frisob, G.; Van Wijkb, K.J.; Hanson, M.R. RIP1, a member of an Arabidopsis protein family, interacts with the protein RARE1 and broadly affects RNA editing. *Proc. Natl. Acad. Sci. USA* **2012**, *109*, E1453–E1461. [CrossRef] [PubMed]
79. Takenaka, M.; Zehrmann, A.; Verbitskiy, D.; Kugelmann, M.; Härtel, B.; Brennicke, A. Multiple organellar RNA editing factor (MORF) family proteins are required for RNA editing in mitochondria and plastids of plants. *Proc. Natl. Acad. Sci. USA* **2012**, *109*, 5104–5109. [CrossRef] [PubMed]
80. Dahan, J.; Tcherkez, G.; Macherel, D.; Benamar, A.; Belkram, K.; Quadrado, M.; Arnal, N.; Mireau, H. Disruption of the CYTOCHROME C OXIDASE DEFICIENT 1 gene leads to cytochrome *c* oxidase depletion and reorchestrated respiratory metabolism in Arabidopsis. *Plant Physiol.* **2014**, *166*, 1788–1802. [CrossRef] [PubMed]

81. Glass, F.; Hartel, B.; Zehrmann, A.; Verbitskiy, D.; Takenaka, M. MEF13 Requires MORF3 and MORF8 for RNA Editing at Eight Targets in Mitochondrial mRNAs in *Arabidopsis thaliana*. *Mol. Plant* **2015**, *8*, 1466–1477. [CrossRef] [PubMed]
82. Shi, X.; Germain, A.; Hanson, M.R.; Bentolila, S. RNA Recognition Motif-Containing Protein ORRM4 Broadly Affects Mitochondrial RNA Editing and Impacts Plant Development and Flowering. *Plant Physiol.* **2016**, *170*, 294–309. [CrossRef] [PubMed]
83. Qi, W.; Tian, Z.; Lu, L.; Chen, X.; Chen, X.; Zhang, W.; Song, R. Editing of mitochondrial transcripts nad3 and cox2 by Dek10 is essential for mitochondrial function and maize plant development. *Genetics* **2017**, *205*, 1489–1501. [CrossRef] [PubMed]
84. Uchida, M.; Ohtani, S.; Ichinose, M.; Sugita, C.; Sugita, M. The PPR-DYW proteins are required for RNA editing of rps14, cox1 and nad5 transcripts in *Physcomitrella patens* mitochondria. *FEBS Lett.* **2011**, *585*, 2367–2371. [CrossRef] [PubMed]
85. Rüdinger, M.; Szövényi, P.; Rensing, S.A.; Knoop, V. Assigning DYW-type PPR proteins to RNA editing sites in the funariid mosses *Physcomitrella patens* and *Funaria hygrometrica*. *Plant J.* **2011**, *67*, 370–380. [CrossRef] [PubMed]
86. Ohtani, S.; Ichinose, M.; Tasaki, E.; Aoki, Y.; Komura, Y.; Sugita, M. Targeted gene disruption identifies three PPR-DYW proteins involved in RNA editing for five editing sites of the moss mitochondrial transcripts. *Plant Cell Physiol.* **2010**, *51*, 1942–1949. [CrossRef] [PubMed]
87. Colas des Francs-Small, C.; Kroeger, T.; Zmudjak, M.; Ostersetzer-Biran, O.; Rahimi, N.; Small, I.; Barkan, A. A PORR domain protein required for $rpl2$ and $ccmF_C$ intron splicing and for the biogenesis of c-type cytochromes in Arabidopsis mitochondria. *Plant J.* **2012**, *69*, 996–1005. [CrossRef] [PubMed]
88. Zmudjak, M.; Colas des Francs-Small, C.; Keren, I.; Shaya, F.; Belausov, E.; Small, I.; Ostersetzer-Biran, O. mCSF1, a nucleus-encoded CRM protein required for the processing of many mitochondrial introns, is involved in the biogenesis of respiratory complexes I and IV in Arabidopsis. *New Phytol.* **2013**, *199*, 379–394. [CrossRef] [PubMed]
89. Zmudjak, M.; Shevtsov, S.; Sultan, L.D.; Keren, I. Analysis of the Roles of the Arabidopsis nMAT2 and PMH2 Proteins Provided with New Insights into the Regulation of Group II Intron Splicing in Land-Plant Mitochondria. *Int. J. Mol. Sci.* **2017**, *18*, 2428. [CrossRef] [PubMed]
90. Ichinose, M.; Tasaki, E.; Sugita, C.; Sugita, M. A PPR-DYW protein is required for splicing of a group II intron of cox1 pre-mRNA in *Physcomitrella patens*. *Plant J.* **2012**, *70*, 271–278. [CrossRef] [PubMed]
91. Kühn, K.; Richter, U.; Meyer, E.H.; Delannoy, E.; de Longevialle, A.F.; O'Toole, N.; Börner, T.; Millar, A.H.; Small, I.D.; Whelan, J. Phage-type RNA polymerase RPOTmp performs gene-specific transcription in mitochondria of *Arabidopsis thaliana*. *Plant Cell* **2009**, *21*, 2762–2779. [CrossRef] [PubMed]
92. Kühn, K.; Yin, G.; Duncan, O.; Law, S.R.; Kubiszewski-Jakubiak, S.; Kaur, P.; Meyer, E.; Wang, Y.; Small, C.C.; Giraud, E.; et al. Decreasing electron flux through the cytochrome and/or alternative respiratory pathways triggers common and distinct cellular responses dependent on growth conditions. *Plant Physiol.* **2015**, *167*, 228–250. [CrossRef] [PubMed]
93. Zhu, Q.; Dugardeyn, J.; Zhang, C.; Takenaka, M.; Kühn, K.; Craddock, C.; Smalle, J.; Karampelias, M.; Denecke, J.; Peters, J.; et al. SLO2, a mitochondrial pentatricopeptide repeat protein affecting several RNA editing sites, is required for energy metabolism. *Plant J.* **2012**, *71*, 836–849. [CrossRef] [PubMed]
94. Wang, X.; Bian, Y.; Cheng, K.; Gu, L.F.; Ye, M.; Zou, H.; Sun, S.S.; He, J. A large-scale protein phosphorylation analysis reveals novel phosphorylation motifs and phosphoregulatory networks in Arabidopsis. *J. Proteom.* **2013**, *78*, 486–498. [CrossRef] [PubMed]
95. Solheim, C.; Li, L.; Hatzopoulos, P.; Millar, A.H. Loss of Lon1 in Arabidopsis changes the mitochondrial proteome leading to altered metabolite profiles and growth retardation without an accumulation of oxidative damage. *Plant Physiol.* **2012**, *160*, 1187–1203. [CrossRef] [PubMed]
96. Rigas, S.; Daras, G.; Laxa, M.; Marathias, N.; Fasseas, C.; Sweetlove, L.J.; Hatzopoulos, P. Role of Lon1 protease in post-germinative growth and maintenance of mitochondrial function in *Arabidopsis thaliana*. *New Phytol.* **2009**, *181*, 588–600. [CrossRef] [PubMed]
97. Welchen, E.; Hildebrandt, T.M.; Lewejohann, D.; Gonzalez, D.H.; Braun, H.P. Lack of cytochrome c in Arabidopsis decreases stability of Complex IV and modifies redox metabolism without affecting Complexes I and III. *Biochim. Biophys. Acta* **2012**, *1817*, 990–1001. [CrossRef] [PubMed]

98. Colas des Francs-Small, C.; Small, I. Surrogate mutants for studying mitochondrially encoded functions. *Biochimie* **2014**, *100*, 234–242. [CrossRef] [PubMed]
99. Eubel, H.; Jansch, L.; Braun, H.P. New Insights into the Respiratory Chain of Plant Mitochondria. Supercomplexes and a Unique Composition of Complex II. *Plant Physiol.* **2003**, *133*, 274–286. [CrossRef] [PubMed]
100. Hruz, T.; Laule, O.; Szabo, G.; Wessendorp, F.; Bleuler, S.; Oertle, L.; Widmayer, P.; Gruissem, W.; Zimmermann, P. Genevestigator V3: A reference expression database for the meta-analysis of transcriptomes. *Adv. Bioinform.* **2008**, 420747. [CrossRef] [PubMed]
101. Attallah, C.V.; Welchen, E.; Pujol, C.; Bonnard, G.; Gonzalez, D.H. Characterization of *Arabidopsis thaliana* genes encoding functional homologues of the yeast metal chaperone Cox19p, involved in cytochrome *c* oxidase biogenesis. *Plant Mol. Biol.* **2007**, *65*, 343–355. [CrossRef] [PubMed]
102. Curi, G.C.; Welchen, E.; Chan, R.L.; Gonzalez, D.H. Nuclear and mitochondrial genes encoding cytochrome *c* oxidase subunits respond differently to the same metabolic factors. *Plant Physiol. Biochem.* **2003**, *41*, 689–693. [CrossRef]
103. Welchen, E.; Chan, R.L.; Gonzalez, D.H. Metabolic regulation of genes encoding cytochrome *c* and cytochrome *c* oxidase subunit Vb in *Arabidopsis*. *Plant Cell Environ.* **2002**, *25*, 1605–1615. [CrossRef]
104. Welchen, E.; Chan, R.L.; Gonzalez, D.H. The promoter of the *Arabidopsis* nuclear gene COX5b-1, encoding subunit 5b of the mitochondrial cytochrome *c* oxidase, directs tissue-specific expression by a combination of positive and negative regulatory elements. *J. Exp. Bot.* **2004**, *55*, 1997–2004. [CrossRef] [PubMed]
105. Welchen, E.; Gonzalez, D.H. Differential expression of the *Arabidopsis* cytochrome *c* genes *Cytc-1* and *Cytc-2*. Evidence for the involvement of TCP-domain protein-binding elements in anther and meristem specific expression of the *Cytc-1* gene. *Plant Physiol.* **2005**, *139*, 88–100. [CrossRef] [PubMed]
106. Welchen, E.; Viola, I.L.; Kim, H.J.; Prendes, L.P.; Comelli, R.N.; Hong, J.C.; Gonzalez, D.H. A segment containing a G-box and an ACGT motif confers differential expression characteristics and responses to the *Arabidopsis Cytc-2* gene, encoding an isoform of cytochrome *c*. *J. Exp. Bot.* **2009**, *60*, 829–845. [CrossRef] [PubMed]
107. Mufarrege, E.F.; Curi, G.C.; Gonzalez, D.H. Common sets of promoter elements determine the expression characteristics of three *Arabidopsis* genes encoding isoforms of mitochondrial cytochrome *c* oxidase subunit 6b. *Plant Cell Physiol.* **2009**, *50*, 1393–1399. [CrossRef] [PubMed]
108. Comelli, R.N.; Gonzalez, D.H. Identification of regulatory elements involved in expression and induction by sucrose and UV-B light of the *Arabidopsis thaliana* COX5b-2 gene, encoding an isoform of cytochrome *c* oxidase subunit 5b. *Physiol. Plant* **2009**, *137*, 213–224. [CrossRef] [PubMed]
109. Comelli, R.N.; Gonzalez, D.H. Divergent regulatory mechanisms in the response of respiratory chain component genes to carbohydrates suggests a model for gene evolution after duplication. *Plant Signal. Behav.* **2009**, *4*, 1179–1181. [CrossRef] [PubMed]
110. Comelli, R.N.; Viola, I.L.; Gonzalez, D.H. Characterization of promoter elements required for expression and induction by sucrose of the *Arabidopsis* COX5b-1 nuclear gene, encoding the zinc-binding subunit of cytochrome *c* oxidase. *Plant Mol. Biol.* **2009**, *69*, 729–743. [CrossRef] [PubMed]
111. Comelli, R.N.; Welchen, E.; Kim, H.J.; Hong, J.C.; Gonzalez, D.H. Delta subclass HD-Zip proteins and a B-3 AP$_2$/ERF transcription factor interact with promoter elements required for expression of the Arabidopsis cytochrome *c* oxidase 5b-1 gene. *Plant Mol. Biol.* **2012**, *80*, 157–167. [CrossRef] [PubMed]
112. Giegé, P.; Sweetlove, L.J.; Cognat, V.; Leaver, C.J. Coordination of nuclear and mitochondrial genome expression during mitochondrial biogenesis in Arabidopsis. *Plant Cell* **2005**, *17*, 1497–1512. [CrossRef] [PubMed]
113. Xiong, Y.; Sheen, J. Novel links in the plant TOR kinase signaling network. *Curr. Opin. Plant Biol.* **2015**, *28*, 83–91. [CrossRef] [PubMed]
114. Wei, N.; Deng, X. The COP9 signalosome. *Annu. Rev. Cell Dev. Biol.* **2003**, *19*, 261–286. [CrossRef] [PubMed]
115. Curi, G.C.; Chan, R.L.; Gonzalez, D.H. The leader intron of *Arabidopsis thaliana* genes encoding cytochrome *c* oxidase subunit 5c promotes high-level expression by increasing transcript abundance and translation efficiency. *J. Exp. Bot.* **2005**, *56*, 2563–2571. [CrossRef] [PubMed]
116. Curi, G.C.; Chan, R.L.; Gonzalez, D.H. Genes encoding cytochrome *c* oxidase subunit 5c from sunflower (*Helianthus annuus* L.) are regulated by nitrate and oxygen availability. *Plant Sci.* **2002**, *163*, 897–905. [CrossRef]

117. Gonzalez, D.H.; Welchen, E.; Attallah, C.V.; Comelli, R.N.; Mufarrege, E.F. Transcriptional coordination of the biogenesis of the oxidative phosphorylation machinery in plants. *Plant J.* **2007**, *51*, 105–116. [CrossRef] [PubMed]
118. Bailey, T.L.; Boden, M.; Buske, F.A.; Frith, M.; Grant, C.E.; Clementi, L.; Ren, J.; Li, W.W.; Noble, W.S. MEME SUITE: Tools for motif discovery and searching. *Nucleic Acids Res.* **2009**, *37*, W202–W208. [CrossRef] [PubMed]
119. Welchen, E.; Gonzalez, D.H. Overrepresentation of elements recognized by TCP-domain transcription factors in the upstream regions of nuclear genes encoding components of the mitochondrial oxidative phosphorylation Machinery. *Plant Physiol.* **2006**, *141*, 540–545. [CrossRef] [PubMed]
120. Giraud, E.; Ng, S.; Carrie, C.; Duncan, O.; Low, J.; Lee, C.P.; Van Aken, O.; Millar, A.H.; Murcha, M.; Whelan, J. TCP transcription factors link the regulation of genes encoding mitochondrial proteins with the circadian clock in *Arabidopsis thaliana*. *Plant Cell* **2010**, *22*, 3921–3934. [CrossRef] [PubMed]
121. Berkowitz, O.; De Clercq, I.; Van Breusegem, F.; Whelan, J. Interaction between hormonal and mitochondrial signalling during growth, development and in plant defence responses. *Plant Cell Environ.* **2016**, *39*, 1127–1139. [CrossRef] [PubMed]
122. Xu, X.; Burgess, P.; Zhang, X.; Huang, B. Enhancing cytokinin synthesis by overexpressing *ipt* alleviated drought inhibition of root growth through activating ROS-scavenging systems in *Agrostis stolonifera*. *J. Exp. Bot.* **2016**, *67*, 1979–1992. [CrossRef] [PubMed]
123. Zhang, H.; Zhou, H.; Berke, L.; Heck, A.J.; Mohammed, S.; Scheres, B.; Menke, F.L. Quantitative phosphoproteomics after auxin-stimulated lateral root induction identifies an SNX1 protein phosphorylation site required for growth. *Mol. Cell. Proteom.* **2013**, *12*, 1158–1169. [CrossRef] [PubMed]
124. Lin, L.; Hsu, C.; Hu, C.; Ko, S.; Hsieh, H.; Huang, H.; Juan, H. Integrating Phosphoproteomics and Bioinformatics to Study Brassinosteroid-Regulated Phosphorylation Dynamics in *Arabidopsis*. *BMC Genom.* **2015**, *16*, 533. [CrossRef] [PubMed]
125. Llorente, F.; Muskett, P.; Sánchez-Vallet, A.; López, G.; Ramos, B.; Sánchez-Rodríguez, C.; Jordá, L.; Parker, J.; Molina, A. Repression of the auxin response pathway increases *Arabidopsis* susceptibility to necrotrophic fungi. *Mol. Plant* **2008**, *1*, 496–509. [CrossRef] [PubMed]
126. Kiba, T.; Aoki, K.; Sakakibara, H.; Mizuno, T. Arabidopsis response regulator, ARR22, ectopic expression of which results in phenotypes similar to the wol cytokinin-receptor mutant. *Plant Cell Physiol.* **2004**, *45*, 1063–1077. [CrossRef] [PubMed]
127. Wang, Z.Y.; Nakano, T.; Gendron, J.; He, J.; Chen, M.; Vafeados, D.; Yang, Y.; Fujioka, S.; Yoshida, S.; Asami, T.; et al. Nuclear-localized BZR1 mediates brassinosteroid-induced growth and feedback suppression of brassinosteroid biosynthesis. *Dev. Cell* **2002**, *2*, 505–513. [CrossRef]
128. Tognetti, V.B.; Muhlenbock, P.; Van Breusegem, F. Stress homeostasis—The redox and auxin perspective. *Plant Cell Environ.* **2012**, *35*, 321–333. [CrossRef] [PubMed]
129. Zhang, S.; Wu, J.; Yuan, D.; Zhang, D.; Huang, Z.; Xiao, L.; Yang, C. Perturbation of auxin homeostasis caused by mitochondrial FtSH4 gene-mediated peroxidase accumulation regulates arabidopsis architecture. *Mol. Plant* **2014**, *7*, 856–873. [CrossRef] [PubMed]
130. Zhang, B.; Van Aken, O.; Thatcher, L.; De Clercq, I.; Duncan, O.; Law, S.R.; Murcha, M.W.; van der Merwe, M.; Seifi, H.S.; Carrie, C.; et al. The mitochondrial outer membrane AAA ATPase AtOM66 affects cell death and pathogen resistance in *Arabidopsis thaliana*. *Plant J.* **2014**, *80*, 709–727. [CrossRef] [PubMed]
131. Colombatti, F.; Gonzalez, D.H.; Welchen, E. Plant mitochondria under pathogen attack: A sigh of relief or a last breath? *Mitochondrion* **2014**, *19*, 238–244. [CrossRef] [PubMed]
132. Schwarzländer, M.; König, A.C.; Sweetlove, L.J.; Finkemeier, I. The impact of impaired mitochondrial function on retrograde signalling: A meta-analysis of transcriptomic responses. *J. Exp. Bot.* **2012**, *63*, 1735–1750. [CrossRef] [PubMed]
133. Van Aken, O.; Zhang, B.; Carrie, C.; Uggalla, V.; Paynter, E.; Giraud, E.; Whelan, J. Defining the mitochondrial stress response in *Arabidopsis thaliana*. *Mol. Plant* **2009**, *2*, 1310–1324. [CrossRef] [PubMed]
134. Mostafa, I.; Zhu, N.; Yoo, M.; Balmant, K.; Misra, B.; Dufresne, C.; Abou-Hashem, M.; Chen, S.; El-Domiaty, M. New nodes and edges in the glucosinolate molecular network revealed by proteomics and metabolomics of Arabidopsis myb28/29 and cyp79B2/B3 glucosinolate mutants. *J. Proteom.* **2016**, *138*, 1–19. [CrossRef] [PubMed]

135. Bourens, M.; Fontanesi, F.; Soto, I.C.; Liu, J.; Barrientos, A. Redox and reactive oxygen species regulation of mitochondrial cytochrome c oxidase biogenesis. *Antioxid. Redox Signal.* **2013**, *19*, 1940–1952. [CrossRef] [PubMed]
136. Chan, S.Y.; Zhang, Y.Y.; Hemann, C.; Mahoney, C.E.; Zweier, J.L.; Loscalzo, J. MicroRNA-210 controls mitochondrial metabolism during hypoxia by repressing the iron-sulfur cluster assembly proteins ISCU1/2. *Cell Metab.* **2009**, *10*, 273–284. [CrossRef] [PubMed]
137. Tsuji, H.; Nakazono, M.; Saisho, D.; Tsutsumi, N.; Hirai, A. Transcript levels of the nuclear-encoded respiratory genes in rice decrease by oxygen deprivation: Evidence for involvement of calcium in expression of the alternative oxidase 1a gene. *FEBS Lett.* **2000**, *471*, 201–204. [CrossRef]
138. Considine, M.J.; Diaz-Vivancos, P.; Kerchev, P.; Signorelli, S.; Agudelo-Romero, P.; Gibbs, D.; Foyer, C.H. Learning To Breathe: Developmental Phase Transitions in Oxygen Status. *Trends Plant Sci.* **2017**, *22*, 140–153. [CrossRef] [PubMed]
139. Murphy, M.P. How mitochondria produce reactive oxygen species. *Biochem. J.* **2009**, *417*, 1–13. [CrossRef] [PubMed]
140. Kojer, K.; Bien, M.; Gangel, H.; Morgan, B.; Dick, T.P.; Riemer, J. Glutathione redox potential in the mitochondrial intermembrane space redox potential in the mitochondrial intermembrane space is linked to the cytosol and impacts the Mia40 redox state. *EMBO J.* **2012**, *31*, 3169–3182. [CrossRef] [PubMed]
141. Riemer, J.; Schwarzländer, M.; Conrad, M.; Herrmann, J.M. Thiol switches in mitochondria: Operation and physiological relevance. *Biol. Chem.* **2015**, *39*, 465–482. [CrossRef] [PubMed]
142. Nietzel, T.; Mostertz, J.; Hochgräfe, F.; Schwarzländer, M. Redox regulation of mitochondrial proteins and proteomes by cysteine thiol switches. *Mitochondrion* **2017**, *33*, 72–83. [CrossRef] [PubMed]
143. Carrie, C.; Giraud, E.; Duncan, O.; Xu, L.; Wang, Y.; Huang, S.; Clifton, R.; Murcha, M.; Filipovska, A.; Rackham, O.; et al. Conserved and novel functions for *Arabidopsis thaliana* MIA40 in assembly of proteins in mitochondria and peroxisomes. *J. Biol. Chem.* **2010**, *285*, 36138–36148. [CrossRef] [PubMed]
144. Longen, S.; Bien, M.; Bihlmaier, K.; Kloeppel, C.; Kauff, F.; Hammermeister, M.; Westermann, B.; Herrmann, J.M.; Riemer, J. Systematic analysis of the twin Cx(9)C protein family. *J. Mol. Biol.* **2009**, *393*, 356–368. [CrossRef] [PubMed]
145. Fischer, M.; Horn, S.; Belkacemi, A.; Kojer, K.; Petrungaro, C.; Habich, M.; Ali, M.; Küttner, V.; Bien, M.; Kauff, F.; et al. Protein import and oxidative folding in the mitochondrial intermembrane space of intact mammalian cells. *Mol. Biol. Cell* **2013**, *24*, 2160–2170. [CrossRef] [PubMed]
146. Carrie, C.; Soll, J. To Mia or not to Mia: Stepwise evolution of the mitochondrial intermembrane space disulfide relay. *BMC Biol.* **2017**, *15*, 119. [CrossRef] [PubMed]
147. Peleh, V.; Zannini, F.; Backes, S.; Rouhier, N.; Herrmann, J.M. Erv1 of *Arabidopsis thaliana* can directly oxidize mitochondrial intermembrane space proteins in the absence of redox-active Mia40. *BMC Biol.* **2017**, *15*, 106. [CrossRef] [PubMed]
148. Bihlmaier, K.; Mesecke, N.; Terzyiska, N.; Bien, M.; Hell, K.; Herrmann, J.M. The disulfide relay system of mitochondria is connected to the respiratory chain. *J. Cell Biol.* **2007**, *179*, 389–395. [CrossRef] [PubMed]
149. Banci, L.; Bertini, I.; Cefaro, C.; Ciofi-Baffoni, S.; Gallo, A.; Martinelli, M.; Sideris, D.P.; Katrakili, N.; Tokatlidis, K. MIA40 is an oxidoreductase that catalyzes oxidative protein folding in mitochondria. *Nat. Struct. Mol. Biol.* **2009**, *16*, 198–206. [CrossRef] [PubMed]
150. Horn, D.; Barrientos, A. Mitochondrial copper metabolism and delivery to cytochrome c oxidase. *IUBMB Life* **2008**, *60*, 421–429. [CrossRef] [PubMed]
151. Morgada, M.N.; Abriata, L.A.; Cefaro, C.; Gajda, K.; Banci, L.; Vila, A.J. Loop recognition and copper-mediated disulfide reduction underpin metal site assembly of CuA in human cytochrome oxidase. *Proc. Natl. Acad. Sci. USA* **2015**, *112*, 11771–11776. [CrossRef] [PubMed]
152. Horn, D.; Al-Ali, H.; Barrientos, A. Cmc1p is a conserved mitochondrial twin CX9C protein involved in cytochrome c oxidase biogenesis. *Mol. Cell. Biol.* **2008**, *28*, 4354–4364. [CrossRef] [PubMed]
153. Nobrega, M.P.; Bandeira, S.C.; Beers, J.; Tzagoloff, A. Characterization of COX19, a widely distributed gene required for expression of mitochondrial cytochrome oxidase. *J. Biol. Chem.* **2002**, *277*, 40206–40211. [CrossRef] [PubMed]

154. Rigby, K.; Zhang, L.; Cobine, P.A.; George, G.N. Winge, D.R. Characterization of the cytochrome *c* oxidase assembly factor Cox19 of *Saccharomyces cerevisiae*. *J. Biol. Chem.* **2007**, *282*, 10233–10242. [CrossRef] [PubMed]
155. Barros, M.H.; Johnson, A.; Tzagoloff, A. COX23, a homologue of COX17, is required for cytochrome oxidase assembly. *J. Biol. Chem.* **2004**, *279*, 31943–31947. [CrossRef] [PubMed]
156. McEwen, J.E.; Hong, K.H.; Park, S.; Preciado, G.T. Sequence and chromosomal localization of two PET genes required for cytochrome *c* oxidase assembly in *Saccharomyces cerevisiae*. *Curr. Genet.* **1993**, *23*, 9–14. [CrossRef] [PubMed]
157. Ghosh, A.; Trivedi, P.P.; Timbalia, S.A.; Griffin, A.T.; Rahn, J.J.; Chan, S.S.; Gohil, V.M. Copper supplementation restores cytochrome c oxidase assembly defect in a mitochondrial disease model of COA6 deficiency. *Hum. Mol. Genet.* **2014**, *23*, 3596–3606. [CrossRef] [PubMed]
158. Garcia, L.; Welchen, E.; Gonzalez, D.H. Mitochondria and copper homeostasis in plants. *Mitochondrion* **2014**, *19*, 269–274. [CrossRef] [PubMed]
159. Leary, S.C.; Cobine, P.A.; Nishimura, T.; Verdijk, R.M.; de Krijger, R.; de Coo, R.; Tarnopolsky, M.A.; Winge, D.R.; Shoubridge, E.A. COX19 mediates the transduction of a mitochondrial redox signal from SCO1 that regulates ATP7A-mediated cellular copper efflux. *Mol. Biol. Cell* **2013**, *24*, 683–691. [CrossRef] [PubMed]
160. Mayorga, J.; Camacho-Villasana, Y.; Shingú-Vázquez, M.; García-Villegas, R.; Zamudio-Ochoa, A.; García-Guerrero, A.; Hernández, G.; Pérez-Martínez, X. A novel function of Pet54 in regulation of Cox1 synthesis in saccharomyces cerevisiae mitochondria. *J. Biol. Chem.* **2016**, *291*, 9343–9355. [CrossRef] [PubMed]
161. Ito, J.; Taylor, N.L.; Castleden, I.; Weckwerth, W.; Millar, A.H.; Heazlewood, J.L. A survey of the *Arabidopsis thaliana* mitochondrial phosphoproteome. *Proteomics* **2009**, *9*, 4229–4240. [CrossRef] [PubMed]
162. Covian, R.; Balaban, R. Cardiac mitochondrial matrix and respiratory complex protein phosphorylation. *Am. J. Physiol. Heart Circ. Physiol.* **2012**, *303*, H940–H966. [CrossRef] [PubMed]
163. Helling, S.; Hüttemann, M.; Ramzan, R.; Kim, S.H.; Lee, I.; Müller, T.; Langenfeld, E.; Meyer, H.E.; Kadenbach, B.; Vogt, S.; et al. Multiple phosphorylations of cytochrome *c* oxidase and their functions. *Proteomics* **2012**, *12*, 950–959. [CrossRef] [PubMed]
164. Roitinger, E.; Hofer, M.; Köcher, T.; Pichler, P.; Novatchkova, M.; Yang, J.; Schlögelhofer, P.; Mechtler, K. Quantitative Phosphoproteomics of the Ataxia Telangiectasia-Mutated (ATM) and Ataxia Telangiectasia-Mutated and Rad3-related (ATR) Dependent DNA Damage Response in *Arabidopsis thaliana*. *Mol. Cell. Proteom.* **2015**, *14*, 556–557. [CrossRef] [PubMed]
165. Engelsberger, W.R.; Schulze, W.X. Nitrate and ammonium lead to distinct global dynamic phosphorylation patterns when resupplied to nitrogen-starved *Arabidopsis* seedlings. *Plant J.* **2012**, *69*, 978–995. [CrossRef] [PubMed]
166. Lurin, C.; Andres, C.; Aubourg, S.; Bellaoui, M.; Bitton, F.; Bruyère, C.; Caboche, M.; Debast, C.; Gualberto, J.; Hoffmann, B.; et al. Genome-wide analysis of Arabidopsis pentatricopeptide repeat proteins reveals their essential role in organelle biogenesis. *Plant Cell* **2004**, *16*, 2089–2103. [CrossRef] [PubMed]
167. Rüdinger, M.; Funk, H.T.; Rensing, S.A.; Maier, U.G.; Knoop, V. RNA editing: Only eleven sites are present in the *Physcomitrella patens* mitochondrial transcriptome and a universal nomenclature proposal. *Mol. Genet. Genom.* **2009**, *281*, 473–481. [CrossRef] [PubMed]
168. Yagi, Y.; Tachikawa, M.; Noguchi, H.; Satoh, S.; Obokata, J.; Nakamura, T. Pentatricopeptide repeat proteins involved in plant organellar RNA editing. *RNA Biol.* **2013**, *10*, 1419–1425. [CrossRef] [PubMed]
169. Barkan, A.; Small, I. Pentatricopeptide repeat proteins in plants. *Annu. Rev. Plant Biol.* **2014**, *65*, 415–442. [CrossRef] [PubMed]
170. Jonietz, C.; Forner, J.; Holzle, A.; Thuss, S.; Binder, S. RNA PROCESSING FACTOR2 Is Required for 59 End Processing of *nad9* and *cox3* mRNAs in Mitochondria of *Arabidopsis thaliana*. *Plant Cell* **2010**, *22*, 443–453. [CrossRef] [PubMed]
171. Takenaka, M.; Verbitskiy, D.; Zehrmann, A.; Brennicke, A. Reverse genetic screening identifies five E-class PPR proteins involved in RNA editing in mitochondria of *Arabidopsis thaliana*. *J. Biol. Chem.* **2010**, *285*, 27122–27129. [CrossRef] [PubMed]
172. Arenas-M, A.; Zehrmann, A.; Moreno, S.; Takenaka, M.; Jordana, X. The pentatricopeptide repeat protein MEF26 participates in RNA editing in mitochondrial *cox3* and *nad4* transcripts. *Mitochondrion* **2014**, *19*, 126–134. [CrossRef] [PubMed]
173. Wang, X.; Auwerx, J. Systems Phytohormone Responses to Mitochondrial Proteotoxic Stress. *Mol. Cell* **2017**, *68*, 540–551. [CrossRef] [PubMed]

174. Howell, K.A.; Narsai, R.; Carroll, A.; Ivanova, A.; Lohse, M.; Usadel, B.; Millar, A.H.; Whelan, J. Mapping metabolic and transcript temporal switches during germination in rice highlights specific transcription factors and the role of RNA instability in the germination process. *Plant Physiol.* **2009**, *149*, 961–980. [CrossRef] [PubMed]
175. Law, S.R.; Narsai, R.; Taylor, N.L.; Delannoy, E.; Carrie, C.; Giraud, E.; Millar, A.H.; Small, I.; Whelan, J. Nucleotide and RNA metabolism prime translational initiation in the earliest events of mitochondrial biogenesis during Arabidopsis germination. *Plant Physiol.* **2012**, *158*, 1610–1627. [CrossRef] [PubMed]
176. Logan, D.C.; Millar, A.H.; Sweetlove, L.J.; Hill, S.A.; Leaver, C.J. Mitochondrial biogenesis during germination in maize embryos. *Plant Physiol.* **2001**, *25*, 662–672. [CrossRef]
177. Howell, K.A.; Millar, A.H.; Whelan, J. Ordered assembly of mitochondria during rice germination begins with pro-mitochondrial structures rich in components of the protein import apparatus. *Plant Mol. Biol.* **2006**, *60*, 201–223. [CrossRef] [PubMed]
178. Paszkiewicz, G.; Gualberto, J.M.; Benamar, A.; Macherel, D.; Logan, D.C. *Arabidopsis* seed mitochondria are bioenergetically active inmediatly upon imbibition and specialize via biogenesis in preparation for autotrophic growth. *Plant Cell* **2017**, *29*, 109–128. [CrossRef] [PubMed]

© 2018 by the authors. Licensee MDPI, Basel, Switzerland. This article is an open access article distributed under the terms and conditions of the Creative Commons Attribution (CC BY) license (http://creativecommons.org/licenses/by/4.0/).

Article

Nitrogen Source Dependent Changes in Central Sugar Metabolism Maintain Cell Wall Assembly in Mitochondrial Complex I-Defective *frostbite1* and Secondarily Affect Programmed Cell Death

Anna Podgórska [1,*], Monika Ostaszewska-Bugajska [1], Agata Tarnowska [1], Maria Burian [1], Klaudia Borysiuk [1], Per Gardeström [2] and Bożena Szal [1]

[1] Institute of Experimental Plant Biology and Biotechnology, Faculty of Biology, University of Warsaw, I. Miecznikowa 1, 02-096 Warsaw, Poland; m.ostaszewska@biol.uw.edu.pl (M.O.-B.); atarnowska@biol.uw.edu.pl (A.T.); mburian@biol.uw.edu.pl (M.B.); k.borysiuk@biol.uw.edu.pl (K.B.); szal@biol.uw.edu.pl (B.S.)

[2] Umeå Plant Science Centre, Department of Plant Physiology, Umeå University, SE-90187 Umeå, Sweden; per.gardestrom@umu.se

* Correspondence: apodgorski@biol.uw.edu.pl; Tel.: +48-22-55-43-009

Received: 11 June 2018; Accepted: 24 July 2018; Published: 28 July 2018

Abstract: For optimal plant growth, carbon and nitrogen availability needs to be tightly coordinated. Mitochondrial perturbations related to a defect in complex I in the *Arabidopsis thaliana frostbite1* (*fro1*) mutant, carrying a point mutation in the 8-kD Fe-S subunit of NDUFS4 protein, alter aspects of fundamental carbon metabolism, which is manifested as stunted growth. During nitrate nutrition, *fro1* plants showed a dominant sugar flux toward nitrogen assimilation and energy production, whereas cellulose integration in the cell wall was restricted. However, when cultured on NH_4^+ as the sole nitrogen source, which typically induces developmental disorders in plants (i.e., the ammonium toxicity syndrome), *fro1* showed improved growth as compared to NO_3^- nourishing. Higher energy availability in *fro1* plants was correlated with restored cell wall assembly during NH_4^+ growth. To determine the relationship between mitochondrial complex I disassembly and cell wall-related processes, aspects of cell wall integrity and sugar and reactive oxygen species signaling were analyzed in *fro1* plants. The responses of *fro1* plants to NH_4^+ treatment were consistent with the inhibition of a form of programmed cell death. Resistance of *fro1* plants to NH_4^+ toxicity coincided with an absence of necrotic lesion in plant leaves.

Keywords: cell wall synthesis; complex I defect; *frostbite1*; mitochondrial mutant; NDUFS4; necrosis; sugar catabolism; sugar signaling; programmed cell death; reactive oxygen species

1. Introduction

Plants are autotrophic organisms that use assimilated nitrogen and carbon for the biosynthesis of proteins and other organic compounds in order to fulfil the developmental needs of their organs. It must be noted that the assimilation of nitrogen is one of the most energy-consuming cellular processes for plants. Indeed, the reduction of nitrate (NO_3^-) to ammonium (NH_4^+) and its incorporation into amino acids consumes the equivalent of 12 ATP molecules [1,2]. Like all living organisms, plants require energy (in the form of ATP) and reductants (mainly NADH and NADPH) for their maintenance. Plant mitochondria carry out the final step of respiration to integrate sugar catabolism with ATP production. Therefore, mitochondria drive metabolism throughout the cell, since they can regulate energy and redox balance [3–5]. Furthermore, the central metabolic position of mitochondria, and their key roles in bioenergetics, mean that they are ideally placed to act as sensors and integrators of the biochemical status of the cell. On the other hand, a close and active communication between mitochondria and other organelles and the nucleus (retrograde signaling) exists to adjust correct metabolic function of plants in response to different environmental conditions [6,7]. Additionally, a significant role in the response of mitochondria to stress conditions has been proposed for reactive oxygen species (ROS), which can be very effective signaling molecules [8,9]. Thus, mitochondrial retrograde signals can mediate diverse developmental processes, from growth regulation to programmed cell death (PCD).

The classical mitochondrial electron-transport chain (mtETC) is composed of four respiratory oxidoreductases (complexes I–IV), which couple redox energy recycling with ATP synthesis catalyzed by the ATP synthase complex (complex V) [10]. The main entry point for electrons to the mtETC is complex I, which functions as an NADH dehydrogenase. Complex I oxidizes matrix NADH, which can be supplied primarily by the tricarboxylic acid (TCA) cycle, by glycine decarboxylation or cytosolic NADH generated mainly during glycolysis and further shuttled into mitochondria by the oxaloacetate (OAA)-malate valve. Loss of complex I activity in mutant plants lowers the efficiency of oxidative phosphorylation by more than 30% [11]. Nevertheless, the activity of specific plant type II dehydrogenases allows electrons from NAD(P)H to enter the mtETC, which enables mutants to survive without a functional complex I [12,13]. Complex I, which is composed of many subunits, is the largest transmembrane, proton-pumping complex in the mtETC. Mutations in any of these subunits can severely hinder or even inhibit complex I assembly or activity. To date, only four mutants in mitochondrially encoded subunits have been characterised—cytoplasmic male sterility in tobacco (*Nicotiana sylvestris*), CMSII, [14–16], non-chromosomal stripe in maize (*Zea mays*), NCS2, [17,18], mosaic phenotype (MSC16) in cucumber (*Cucumis sativus*) [19,20]. In addition, nuclear-gene encoded mutants having defects in complex I have been identified in *Arabidopsis*, e.g., *ndufs4* and *ndufv1* [11,21]. Moreover, there are several *Arabidopsis thaliana* complex I mutants that are defective in other complex I-connected subunits, including *ca1ca2* [22] and *atcib22* [23], or are connected to splicing factors such as *otp43* [24], *css1* [25], *nMat2* [26], *rug3* [27], *mterf15* [28], and *bir6* [29], as well as an *N. sylvestris* mutant, NMS1 [30,31]. In the last decade, several of these complex I mutants have been characterized (reviewed by [32,33]).

All mutants with dysfunction or loss of complex I exhibit reorganized respiratory metabolism, which may affect their redox and energy status. MSC16 plants showed lower NAD(P)H availability [34] and lower respiratory rates, which resulted in lower ATP contents [20,35]. Similarly, the NMS1 and NCS2 mutants showed reduced respiratory capacity but no data about their adenylate or nucleotide status is available [17,31]. Even though *ndufs4* showed normal respiratory capacity, the mutant produced only limited amounts of ATP [11]. The exception is the CMSII mutant, which had a higher content of adenylates and NAD(P)H [34,36], concomitantly with unchanged respiratory fluxes [16,29]. Overall, research using complex I mutants indicates that complex I defects in plants are compensated by reorganization of respiration, although oxidative phosphorylation rates are not fully restored, and most mutant plants are energy deficient. Because of their altered metabolic status, most complex I mutants examined so far showed retarded growth and developmental disorders, in comparison to wild-type (WT) plants. Moreover, a defect in the mtETC often correlates with the occurrence of oxidative stress [11,36,37], and mitochondria were mainly highlighted in these mutants as a primary source of the observed higher rates of ROS generation [37]. Furthermore, a reduced complex I abundance was also found to affect mitochondrial biogenesis. Mutants plants were characterized by altered mitochondrial transcription, translation, and showed altered protein uptake capacities [27,28,38,39].

Interestingly, many complex I mutants apparently have high tolerance to stress conditions. In CMSII plants, higher tolerance to ozone and to the tobacco mosaic virus was detected [16,40–42]. The MSC16 mutant showed an increased resistance to chilling stress and high irradiance conditions [35,38]. In NCS2 plants, improved tolerance to oxidative stress was observed, which limited initiation of PCD [43,44]. In a study of several types of stress (drought, osmotic, chilling, freezing, paraquat, NaCl, H_2O_2), *ndufs4* mutant plants showed improved resistance to abiotic stress conditions in comparison to the WT [11,45]. Similarly, the *nMat22* mutant showed improved tolerance to ethanol treatment [23] and *bir6* was resistant to salt and osmotic stress [29]. Another complex I mutant was discovered by chance when looking for genes involved in stress signal transduction in an ethyl methanesulfonate-mutagenized population under different stress conditions and was named *frostbite1* (*fro1*), because of its susceptibility to chilling temperatures [46]. It was shown that *fro1* plants had a single point mutation in the nuclear-encoded 18-kDa Fe–S subunit of complex I, which concerned a G-to-A change at an intron–exon junction at the start codon resulting in missplicing and a premature stop codon [46]. Consequently, the lack of NDUFS4 led to the disassembly of complex I [47]. Moreover, the *fro1* mutation reduced the expression of stress-inducible genes during chilling conditions, which impaired cold acclimation, whereby mutants also became sensitive to other stress factors like NaCl and osmotic stress [46]. In contrast to these responses, in our recent study, *fro1* plants showed improved resistance to ammonium nutrition [47].

Cultivation using NH_4^+ as the sole nitrogen source for many plants, including *Arabidopsis*, leads to severe toxicity symptoms known as the "ammonium syndrome" [48,49]. Ammonium regulates many physiological processes, ranging from mitosis and cell elongation to senescence and death; hence, ammonium availability may act as a major determinant of plant morphogenesis [50,51]. During NH_4^+ nutrition, nitrate reduction reactions catalyzed by nitrate reductase (NR) and nitrite reductase (NiR) are bypassed, resulting in a surplus of reductants in the cytosol and chloroplasts. Therefore, in terms of energy economy, NH_4^+ would seem to be a better source of nitrogen for plants, as its assimilation requires less energy than that for NO_3^- [1,2]. However, plants cultured on NH_4^+ as a sole nitrogen source often exhibit serious growth disorders; still, despite many years of research concerning this phenomenon, the cause is still not well understood [52,53]. Plant mitochondria are a source of metabolites needed during NH_4^+ assimilation, particularly the TCA cycle, which is the origin of the necessary 2-oxoglutarate (2-OG) for amino acid synthesis [2,52,54]. Elevated activity of the TCA cycle during NH_4^+ nutrition increases mitochondrial matrix NADH production, which must be oxidized by the mtETC [55,56]. Therefore, ammonium nutrition may primarily affect plant mitochondria, since the increased load of redox equivalents to the mtETC and the consequent high respiratory capacity leads to elevated ROS levels in mitochondria [57].

Furthermore, use of the *fro1* mutant revealed that the combined effect of an impairment of complex I and NH_4^+ treatment not only affects mitochondrial functioning in plants, but also changes their extracellular metabolism [47]. It is known that higher cell wall stiffness in response to NH_4^+ nutrition, resulting from altered cell wall modifying enzyme activities, can restrict expansion growth of plant cells [58]. Thus, the aim of this study was to investigate the interplay between mitochondrial functioning and cell wall-related processes in response to NH_4^+ nutrition. To understand changes in the growth rate of *fro1* when cultured on NH_4^+, properties of cell walls and sugar metabolism were examined. Moreover, the role of plant mitochondria in retrograde signalling and PCD was analyzed.

2. Results

2.1. Characterization of fro1 Plants Cultured on Different Nitrogen Sources

The consequences of limited ability to oxidize cellular oxidants in mutants carrying a point mutation in NDUFS4 (AT5G67590), affecting complex I assembly—*frostbite1* [46,47], on plant growth was observed under NH_4^+ and NO_3^- (control) nutrition. During NO_3^- assimilation, *fro1* plants showed strong growth retardation, compared to WT plants (*Arabidopsis thaliana*, ecotype C24) (Figure 1). However, under NH_4^+ nutrition *fro1* plants grew overall bigger rosettes than the same plants cultured in NO_3^- conditions, while WT plants displayed growth inhibition in response to the NH_4^+ treatment (Figure 1) (similarly to a previous report by [47]).

Figure 1. Phenotype of *frostbite1* (*fro1*) or wild-type (WT) *Arabidopsis* ecotype C24 plants cultured hydroponically for 8 weeks on nutrient medium containing either 5 mM nitrate (NO_3^-) or 5 mM ammonium (NH_4^+) as the only nitrogen source.

2.2. Sugar Metabolism in fro1

We investigated whether the changes brought about by disabled function of mtETC in *fro1* are connected to changes in sugar metabolism. *Fro1* plants showed higher sucrose (Suc) and glucose (Glc) contents in leaf tissue when cultured on NO_3^--containing medium as compared to WT plants. In contrast, growth on ammonium led to an increase in soluble sugar content in WT but not in *fro1* plants (Figure 2A,B). Hexokinase (HXK) activity was almost 3 times higher in *fro1* than in WT plants grown under NO_3^- conditions. On the other hand, it remained unchanged in WT plants under NH_4^+ treatment, while it decreased in *fro1*, although it was still higher than in WT plants (Figure 2C). Protein level of UDP-Glc pyrophosphorylase (UGPase) was not statistically different between *fro1* and WT under control growth conditions but increased significantly more in *fro1* as compared to WT plants under NH_4^+ treatment (Figure 2D and Figure S1).

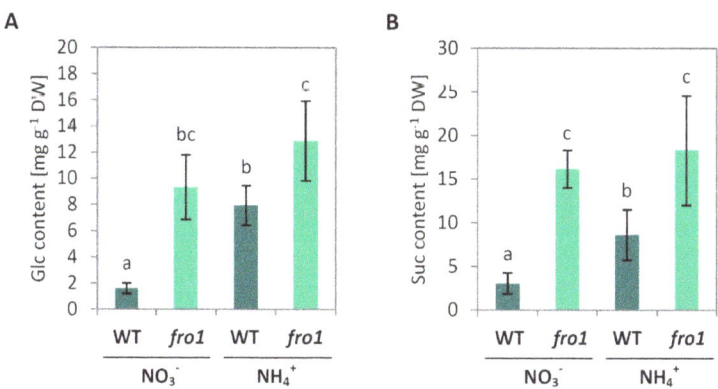

Figure 2. *Cont.*

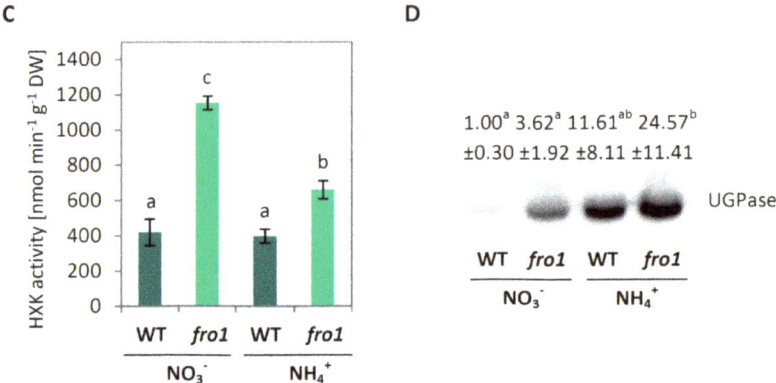

Figure 2. Polysaccharide metabolism in leaves of *frostbite1* (*fro1*) or wild-type (WT) *Arabidopsis* ecotype C24 plants cultured on NH_4^+ and NO_3^- as the only nitrogen source. (**A**) Content of glucose (Glc) and (**B**) Sucrose (Suc); (**C**) activity of hexokinase (HXK); (**D**) protein level of UDP-glucose pyrophosphorylase (UGPase) in leaf tissue extracts. Representative blot is shown. Values are the mean ± standard deviation (SD) of three biological and two technical replicates. Means with different letters are significantly different ($p < 0.05$) by ANOVA followed by Tukey's test.

2.3. Analysis of Cell Size and Cell Wall Design in fro1

Cell wall in palisade cells was visualized using Calcofluor White staining (Figure 3A). As observed for changes in rosette size, *fro1* showed a smaller cell size under NO_3^- nutrition than WT plants, as determined by the cross-sectional area of individual cells. Surprisingly, the cell area of *fro1* plants remained unchanged under NH_4^+ nutrition, while in WT plants it decreased by 35% (Figure 3B). To better understand the growth rate of *fro1* plants when cultured on NH_4^+, we analyzed cell wall properties. Cell wall thickness exhibited a similar trend as cell size (Figure 3C). However, the ratio of cell area to cell wall thickness was 40% higher in NO_3^--treated *fro1* plants than in WT (Figure 3D). Ammonium nutrition led to its decrease only in WT plants, while in *fro1*, it was maintained at a similar level as under NO_3^- nutrition (Figure 3D). Further, the correlation of cell wall assembly and cell wall sensing receptor-like kinases was determined [59]. Expression of *Feronia* (*FER*, AT3G51550), which is thought to be associated with cell wall-dependent regulation of cell elongation, was lower in *fro1* than in WT plants, while NH_4^+ treatment lowered the expression in both genotypes significantly (Figure 3E). The expression of *Thesseus1* (*THE1*, AT5G54380), a cell wall integrity sensor-kinase, was similar in *fro1*, as compared to WT plants under NO_3^- nutrition, but it decreased about 70% during the NH_4^+ growth-regime in both genotypes (Figure 3F).

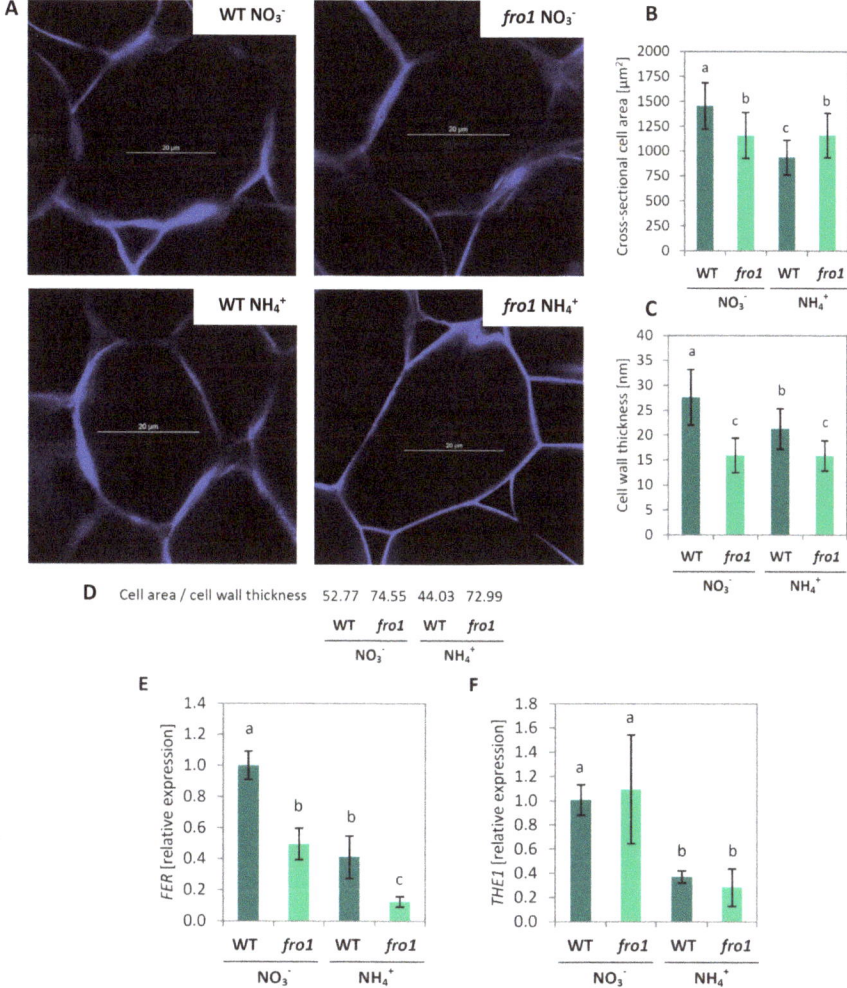

Figure 3. Cell wall characterization of leaf cells of *frostbite1* (*fro1*) or wild-type (WT) *Arabidopsis* ecotype C24 plants cultured on NH_4^+ and NO_3^- as the only nitrogen source. (**A**) Cell wall visualization of individual palisade cells by Calcofluor White staining in CLSM (representative images are shown; scale bar = 20 μm); (**B**) cross-sectional cell area measured from 8 independent biological replicates; (**C**) cell wall thickness calculated from 10 independent images; (**D**) ratio of cell area to cell wall thickness; (**E**) transcript levels for *Feronia* (*FER*) and (**F**) *Thesseus1* (*THE1*) determined from three biological and two technical replicates. Values are the mean ± standard deviation (SD). Means with different letters are significantly different ($p < 0.05$) by ANOVA followed by Tukey's test.

2.4. Cell Wall Composition in fro1

The analysis of cell wall building components showed decreased cellulose incorporation in cell walls of *fro1*, compared to WT plants during NO_3^- feeding (Figure 4A). However, NH_4^+ nutrition did not affect the cellulose content in *fro1* in contrast to WT plants, which exhibited a decrease in its level (Figure 4A). To explain those differences between genotypes, the expression of selected cellulose synthase (*CesA*) genes was examined. The genes *CesA1* (AT4G32410), *CesA3* (AT5G05170), and *CesA6* (AT5G64740) are specifically involved in cellulose synthesis during primary wall formation, while *CesA4* (AT5G44030), *CesA7* (AT5G17420), and *CesA8* (AT4G18780) are involved in cellulose synthesis during secondary cell wall assembly [60]. The expression of all analyzed genes associated with cellulose synthesis in primary cell walls exhibited a similar pattern. *CesA1*, *CesA3*, and *CesA6* relative transcript level was about 50% lower in *fro1* than in WT plants under NO_3^- nutrition; further, the treatment of plants with NH_4^+ led to decreases (by approx. 50%) in expression of these genes in both genotypes (Figure 4B–D). Transcript level of *CesA4* and *CesA7* was similar in *fro1* plants, while *CesA8* was about 60% lower than in WT plants during NO_3^- nutrition (Figure 4E–G). All genes were down-regulated in WT plants in response to NH_4^+ nutrition, in contrast to NH_4^+-treated *fro1*, in which their expression was up-regulated, except for the gene *CesA4* whose expression remained similar to that under NO_3^- growth conditions (Figure 4B–D).

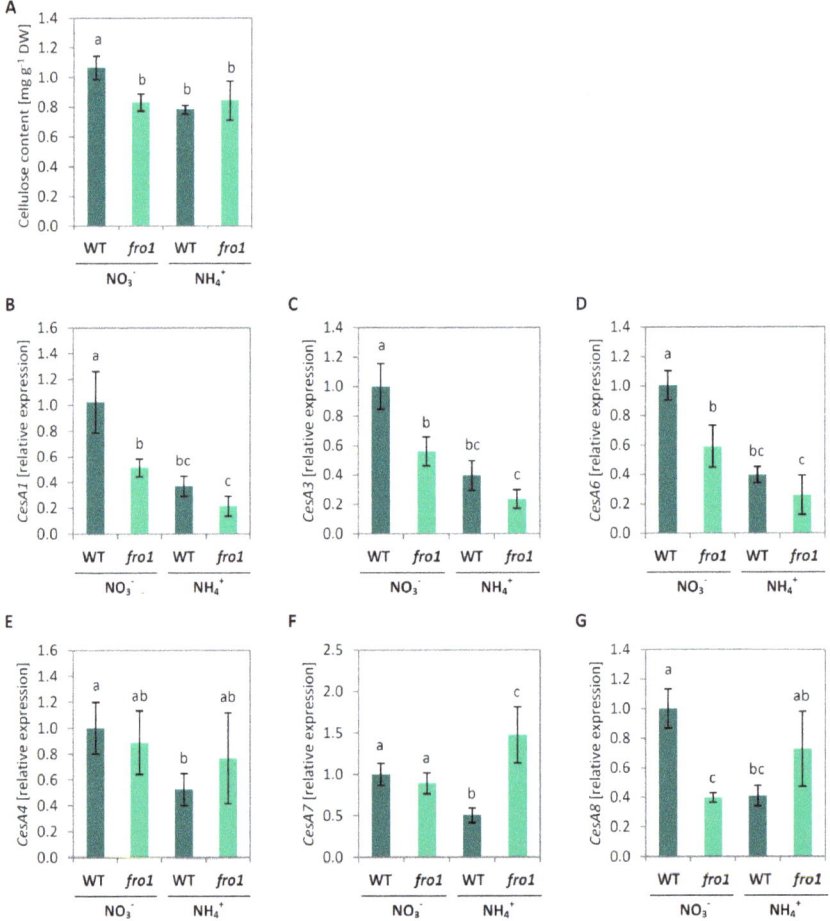

Figure 4. Cellulose metabolism in leaves of *frostbite1* (*fro1*) or wild-type (WT) *Arabidopsis* ecotype C24 plants cultured on NH_4^+ and NO_3^- as the only nitrogen source. (**A**) Cellulose levels measured in dried cell walls. Transcript level for (**B**) cellulose synthase (Ces) *A1*, (**C**) *CesA3*, (**D**) *CesA6*, (**E**) *CesA4*, (**F**) *CesA7*, (**G**) *CesA8*. Values are the mean ± standard deviation (SD) of three biological and two technical replicates. Means with different letters are significantly different ($p < 0.05$) by ANOVA followed by Tukey's test.

Lignin, a highly cross-linked polymer, is formed to support the structure of the secondary cell wall in plants. The content of lignin in cell walls was similar in *fro1* and WT plants. Ammonium nutrition lowered lignification of cell walls in *fro1* plants but it did not influence lignin level in WT plants (Figure 5A). As lignin is composed of phenolic polymers, we analyzed total content of phenolics in the cell wall, and found it was higher in *fro1*, compared to WT plants under NO_3^- nutrition. Ammonium nutrition caused further increase in phenolic levels in plants (Figure 5B). Primary genes involved in lignin biosynthesis are cinnamyl alcohol dehydrogenase genes (CAD) *CAD1* (AT1G72680), *CAD4* (AT3G19450), and *CAD5* (AT4G34230) [61]. The *fro1* mutant did not show significant changes in the expression of any of the analyzed *CAD* genes as compared to WT under the NO_3^- growth regime (Figure 6C–E). In response to NH_4^+ nutrition, *CAD4* expression was inhibited by more than 50% in both genotypes, while *CAD1* and *CAD5* was up-regulated only in *fro1* plants (Figure 5C–E). Further, we analyzed the expression patterns of cell wall peroxidases (POX) related to cell wall lignification. The transcript level of *POX64* (AT5G42180) and *POX72* (AT5G66390) was unchanged between *fro1* and WT plants during NO_3^- nutrition, but it showed an increase in *fro1* in response to NH_4^+ treatment (Figure 5F,G).

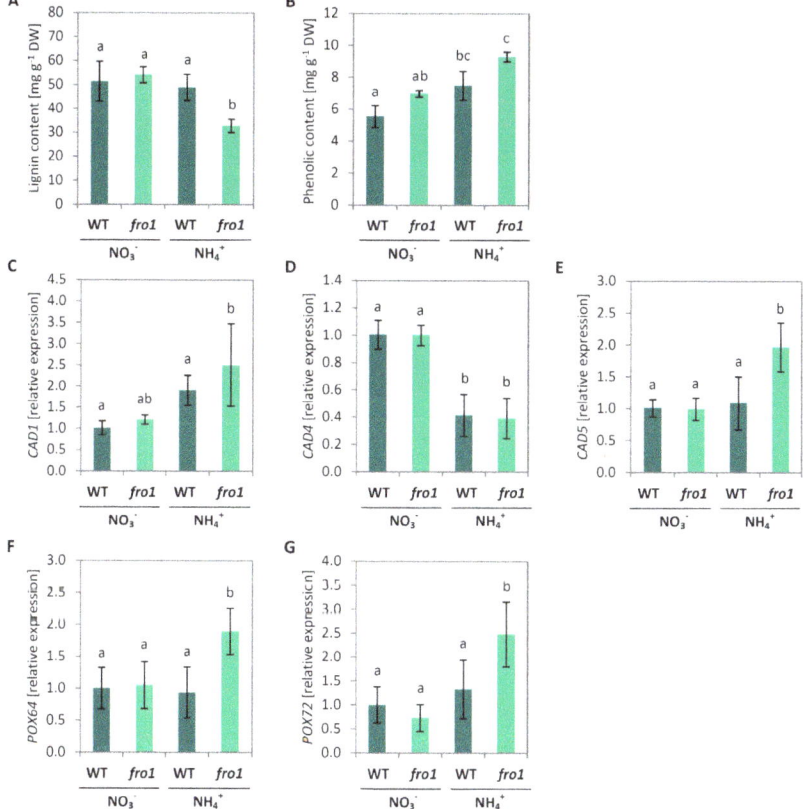

Figure 5. Lignin metabolism in leaves of *frostbite1* (*fro1*) or wild-type (WT) *Arabidopsis* ecotype C24 plants cultured on NH_4^+ and NO_3^- as the only nitrogen source. (**A**) Lignin and (**B**) phenolic content in cell walls. Transcript level for (**C**) cinnamyl alcohol dehydrogenase (CAD) 1, (**D**) CAD4, (**E**) CAD5, and peroxidases (POX) related to cell wall lignification: (**F**) POX64 and (**G**) POX72. Values are the mean ± standard deviation (SD) of three biological and two technical replicates. Means with different letters are significantly different ($p < 0.05$) by ANOVA followed by Tukey's test.

2.5. Analysis of Programmed Cell Death Markers in fro1

Visual examination of leaf blades of *fro1* or WT plants revealed the N source-dependent presence of lesions. In WT plants, ammonium caused emergence of few lesions. Conversely, in the case of *fro1* plants, lesions appeared in NO_3^--supplied mutants and were absent in NH_4^+-treated plants (Figure 6A and Figure S3). Furthermore, trypan blue staining of leaves was performed to specifically indicate necrotic areas. Characteristic blue spot appearance on leaves revealed strong development of necrotic areas on leaves of *fro1* when grown on NO_3^- (Figure 6B and Figure S4), while NH_4^+ nutrition induced the occurrence of some necrotic areas in WT plants, which could mostly not be identified in NH_4^+-grown *fro1*.

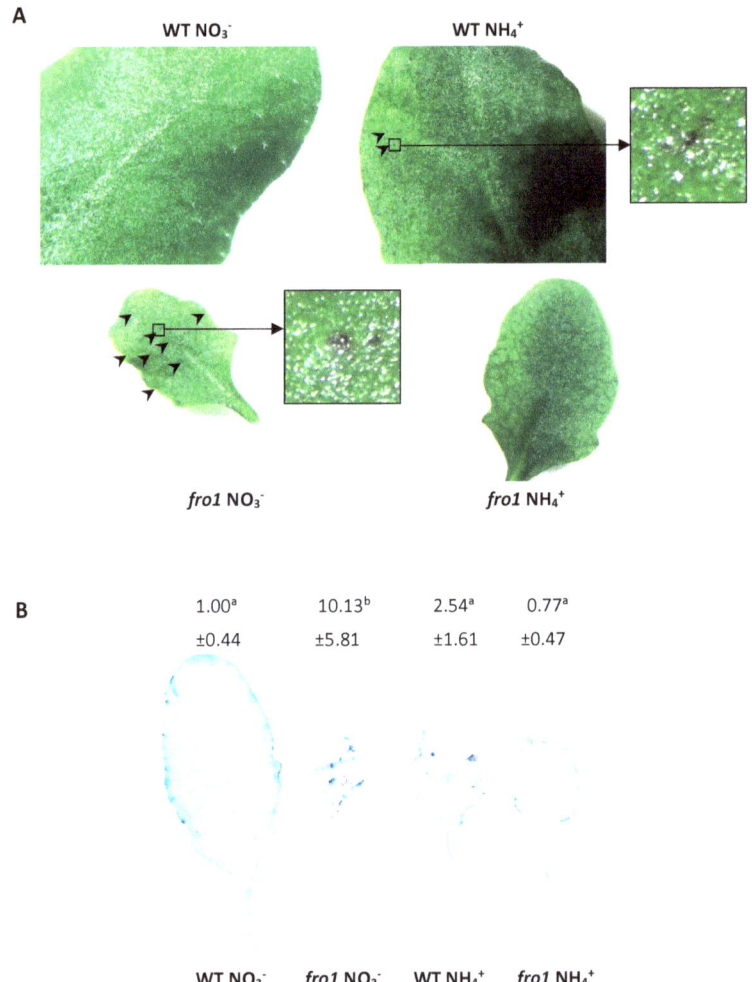

Figure 6. Visualization of lesions on leaves of *frostbite1* (*fro1*) or wild-type (WT) *Arabidopsis* ecotype C24 plants cultured on NH_4^+ and NO_3^- as the only nitrogen source. (**A**) Appearance of necrosis on leaf blades indicated by arrows and (**B**) necrosis stained with trypan blue. Representative leaf blades from three independent plant cultures are shown. Means with different letters are significantly different ($p < 0.05$) by ANOVA followed by Tukey's test.

The marker gene for PCD called *Kiss-of-death* (*KOD*) [62] showed a lower expression in *fro1* plants under NO_3^- nutrition and was up-regulated by NH_4^+ treatment in both genotypes (Figure 7A). At the same time, down-regulation of expression of *Bax inhibitor 1* (*BI-1*) (AT5G47120) under NH_4^+ nutrition was observed in both genotypes. The expression of *BI-1* was not changed in response to mtETC dysfunction (Figure 7B). Expression of the autophagy-related gene *ATG5* was induced in *fro1* during NO_3^- nutrition as compared to WT. While the expression was unchanged in response to the NH_4^+ treatment in WT plants, in *fro1* the transcript level was approximately 60% lower (Figure 7D). Additionally, *fro1* plants showed 50% higher cytochrome *c* (cyt *c*) levels than WT under NO_3^- conditions. Similar to previous observations in WT *Arabidopsis thaliana* [63], NH_4^+ nutrition led to approximately 70% higher cyt *c* level. The NH_4^+ treatment had no influence on cyt *c* level in *fro1* plants in our experiments (Figure 7C and Figure S2).

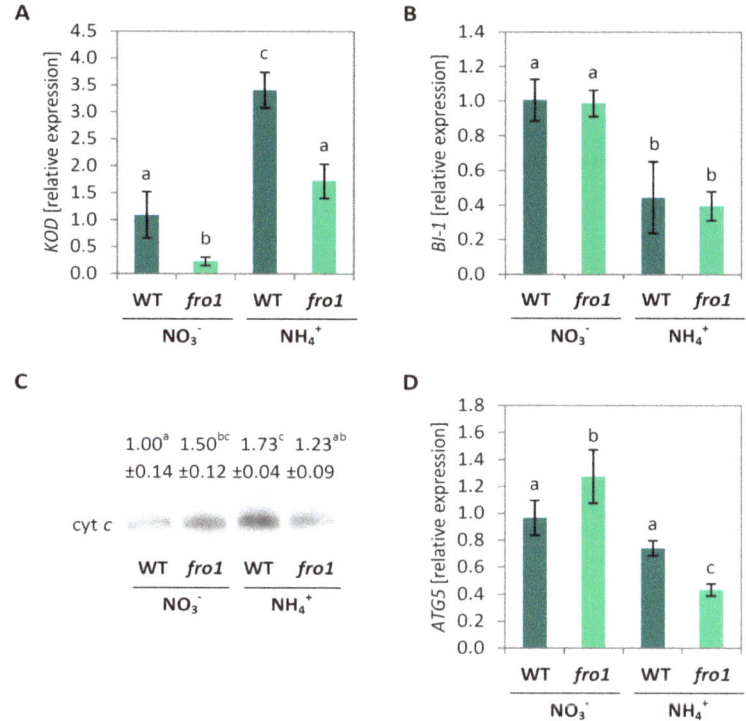

Figure 7. Programmed cell death markers in *frostbite1* (*fro1*) or wild-type (WT) *Arabidopsis* ecotype C24 plants cultured on NH_4^+ and NO_3^- as the only nitrogen source. Transcript levels for (**A**) Kiss-of-death (*KOD*); (**B**) Bax inhibitor 1 (*BI-1*) genes; (**C**) Cytochrome *c* (cyt *c*) protein level in isolated mitochondria and (**D**) transcript level for autophagy 5 (*ATG5*) gene. Values are the mean ± standard deviation (SD) of three biological and two technical replicates. Means with different letters are significantly different ($p < 0.05$) by ANOVA followed by Tukey's test. Representative blot is shown.

2.6. Reactive Oxygen Species Localization in fro1

Hydrogen peroxide (H_2O_2) levels in plant tissues were visualized in situ via 3,3′-diaminobenzidine (DAB) staining. Brownish yellow color development in leaves of *fro1* was slightly more intense than in WT when grown on NO_3^- as nitrogen source (Figure 8A). WT plants showed stronger coloration under NH_4^+ treatment, but *fro1* developed the most intense staining among all the analyzed leaves. The acute staining intensity implies that these plants had the highest H_2O_2 level in leaf tissues. Analysis of the presence of H_2O_2 in leaf tissues by DAB staining was performed simultaneously with a respiratory burst oxidase homolog (RBOH) and POX inhibitor diphenylene iodonium chloride (DPI) to eliminate apoplastic-generated ROS contents during the staining procedure. Color development in *fro1* leaves incubated with DPI was less intense than in WT plants under NO_3^- nutrition (Figure 8A). On the other hand, under NH_4^+ nutrition, DAB staining in DPI-treated WT plants showed slightly reduced coloring, while the least coloration by DPI treatment was found in *fro1*. The difference in DAB staining with and without DPI enables the estimation of the amount of ROS accumulated in the apoplastic space. Results indicated that significant H_2O_2 accumulation in the apoplast was associated with NH_4^+ treatment, especially in *fro1* leaf cells (Figure 8A).

Next, we examined the expression of genes related to extracellular ROS metabolism. The expression of peroxidase 33 (*POX33*, AT3G49110), one of the POXs responsible for apoplastic ROS production, was slightly down-regulated in *fro1* plants, as compared to NO_3^--treated WT plants. Ammonium nutrition induced *POX33* expression only in *fro1* plants (Figure 8B). Further, transcript level of oxidation-related zinc finger 1 (*OZF1*, AT2G19810), a plasma membrane protein involved in oxidative stress [64], was lower in *fro1* plants compared to WT under NO_3^- treatment, but was stimulated in both genotypes under NH_4^+ supply (Figure 8C).

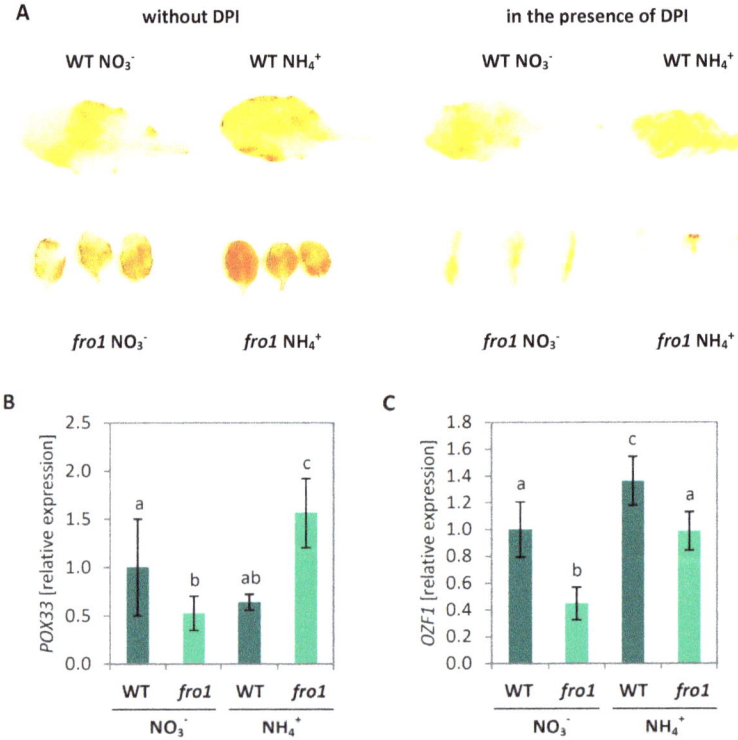

Figure 8. Extracellular ROS metabolism in *frostbite1* (*fro1*) or wild-type (WT) *Arabidopsis* ecotype C24 plants cultured on NH_4^+ and NO_3^- as the only nitrogen source. (**A**) Visualization of leaf H_2O_2 content by 3,3-diaminobenzidine (DAB) staining in the presence or without diphenylene iodonium chloride (DPI). Representative results are shown. Transcript level for (**B**) peroxidase 33 (*POX33*) and (**C**) oxidation-related zinc finger 1 (*OZF1*). Values are the mean ± standard deviation (SD) of three biological and two technical replicates. Means with different letters are significantly different ($p < 0.05$) by ANOVA followed by Tukey's test.

2.7. Changes in Mitochondria-Related Signaling in fro1

We determined transcript levels for marker genes of sugar signaling and retrograde signaling. First, we analyzed the expression of hexokinase 1 (*HXK1*), which is associated with the mitochondria, acts as a sugar sensor, and may regulate Glc-dependent gene expression [65]. Expression of *HXK1* in *fro1* mutants was similar to that in NO_3^--supplied WT plants, regardless of the nitrogen source on which the mutants were grown (Figure 9A). Transcript level of *HXK1* decreased in WT plants when grown on NH_4^+. The expression of sucrose non-fermenting 1–related kinase 1 (*SnRK1.1*, AT3G01090), involved in sugar signaling pathways that responds to the availability of carbohydrates [66] was lower in *fro1* plants compared to WT under NO_3^- conditions. Ammonium nutrition led to a decrease in *SnRK1.1* transcript level in both genotypes (Figure 9B).

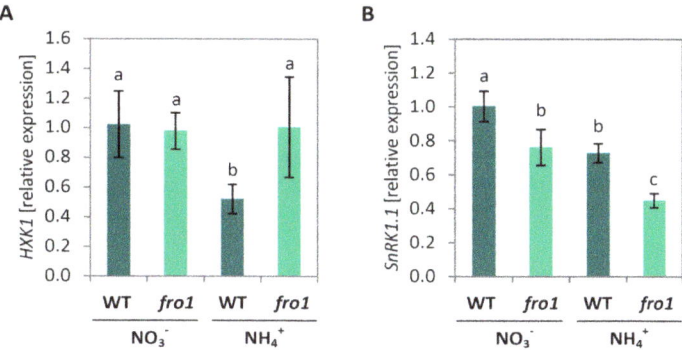

Figure 9. Marker genes for sugar signaling in *frostbite1* (*fro1*) or wild-type (WT) *Arabidopsis* ecotype C24 plants cultured on NH_4^+ and NO_3^- as the only nitrogen source. (**A**) Transcript levels for hexokinase 1 (*HXK1*) and (**B**) sucrose non-fermenting 1–related kinase 1 (*SnRK1.1*). Values are the mean ± standard deviation (SD) of three biological and two technical replicates. Means with different letters are significantly different ($p < 0.05$) by ANOVA followed by Tukey's test.

3. Discussion

A major challenge for complex I mutant plants is to retain high energy levels required for maintenance and biosynthetic reactions. Accordingly, the complex I defect in *fro1* is associated with decreased biomass production in plants (Figure 1). In order to prevent the stunted growth phenotype, *fro1* plants strive to maintain constantly high ATP levels. In this regard, altered sugar catabolism might, to some extent, counteract the energy deficiency. The higher Suc and Glc (Figure 2A,B) contents in *fro1* plants may be used to produce energy in substrate-level phosphorylation, which confirms, for example, increased HXK activity (Figure 2C). In addition, NAD(P)H produced in the glycolytic pathway is channeled toward up-regulated type II dehydrogenases [47] to generate ATP in oxidative phosphorylation. Nevertheless, the lower ratio of ATP to ADP in *fro1* plants indicates that these plants cannot fully restore the energy-deficient status of cells [47].

3.1. Sugar Availability under Ammonium Nutrition May Limit Cell Wall Synthesis in WT but not in fro1 Plants

Sugars are not only the ultimate source of energy and carbon skeletons for intracellular biomolecules but also provide the material used by plants to produce cell walls. While the plant cell is growing, an extensible primary cell wall is formed, the layers of which consist of cellulose microfibrils embedded in a matrix of cross-linked carbohydrates (hemicelluloses and pectin). Among the wall polysaccharides, cellulose, a polymer derived from β-1,4-linked Glc units, is the main load-bearing wall component [67,68]. The high input of sugars related to the energy-conserving phase in *fro1* plants might limit sugar availability for cell wall synthesis (Figure 10). In the present study, we detected lower cellulose synthase gene expression and decreased cellulose content in *fro1* plants, in particular, the expression of *CesA1, CesA3,* and *CesA6* (Figure 4B–D), which have been proposed to be connected with primary cell wall biosynthesis [60,69]. Similarly, as observed by Lee [46], the low incorporation of cellulose results in the generally thinner cell walls of these mutants (Figure 3C). Disturbed cell wall assembly might be a universal response in mitochondrial complex I mutants. In a proteomics study—an analysis of the functional context of altered proteins in the *ca1ca2* mutant line with impaired complex I—a major cell wall response was observed, although carbohydrate metabolism was affected to a lesser extent [22]. Alterations in sugar content have also been detected in another complex I mutant, *css1*, which was further characterized because of its lower cellulose synthesis [25]. Dysfunctional mitochondria of the *css1* mutant were proposed to compete with cell wall synthesis reactions for carbon, highlighting the branched pathways at the level of sucrose synthase (SuSy). UGPase and SuSy are involved in the synthesis of UDP-Glc in source tissues for cellulose production, and an unchanged UGPase protein level in *fro1* plants (Figure 2D) prevented restoration of the low cellulose synthesis in these mutants. It should be noted that UGPases have a dual function and might also promote the accumulation of free cytosolic UDP-Glc [70], and thus sugar breakdown instead of cell wall synthesis might be favored in *fro1* plants. In general, a high cell area to cell wall thickness ratio (Figure 3D) indicates that the cell walls in *fro1* plants might be weakened due to higher sugar flux towards catabolism. In *fro1* plants, NH_4^+ nutrition has the opposite effect on cell wall plasticity compared with that observed under control conditions, that is, WT plants have a lower cell area to cell wall thickness ratio in response to NH_4^+ supply (Figure 3D). This is because NH_4^+-grown plants are characterized by smaller cells (Figure 3A,B), and therefore the thin cell walls are relatively stronger compared to those of the small cells. In *Arabidopsis thaliana*, NH_4^+ nutrition has been found to increase cell wall firmness [58]. Despite a lower total cellulose content and *CesA* expression in WT plants during NH_4^+ nutrition (Figure 4), the cell wall thickness is not appreciably decreased as in *fro1* plants by the inactivation of complex I (Figure 3C). Substrate availability in the form of Suc and Glc, together with higher UGPase engagements (Figure 2A,B,D), might maintain cell wall synthesis at a level sufficient for small cells to grow. It can be assumed that NH_4^+-based changes in carbohydrate metabolism might compensate for the weak cell walls in *fro1* plants, thereby promoting better growth of these plants in the presence of NH_4^+. Therefore, in NH_4^+-grown *fro1* plants, a large proportion of soluble sugar (Glc and Suc, Figure 2A,B) might not be channeled to energy-producing processes (since HXK activity is decreased, Figure 2C), but rather toward cellulose synthesis due to higher UGPase engagement (Figure 2D). Consequently, in contrast to WT plants, the thickness of cell walls in *fro1* plants is not decreased in response to NH_4^+ nutrition (Figure 3C).

On completion of expansion, the structure of plant cells need to be strengthened, which is facilitated by the generation of a secondary cell wall that is mainly composed of cellulose, hemicelluloses, and lignin. The expression of *CesA4, CesA7,* and *CesA8*, which are genes associated with cellulose synthesis for secondary cell wall formation [71,72], were induced in *fro1* plants when grown on NH_4^+ and might be associated with a mechanism that compensates for the low cellulose deposition in these plants (Figure 4E–G). Plant growth is generally not directly related to cellulose availability but might be limited to some degree by cell wall rigidification. In the process of lignification, phenolic polymers are cross-linked to provide mechanical strength as a defense against different environmental stress conditions. In this regard, POXs have been found to catalyze the polymerization of a wide variety of small phenolic compounds [73]. For example, the cell wall-localized *POX64* and *POX72* isoforms have been shown to participate in this process [74,75], and we found that the expression of these two genes was increased in NH_4^+-grown *fro1* plants (Figure 5). Consistent with the previously observed higher expression of major CAD isoforms related to phenolic synthesis [76], we found that the expression of *CAD1* and *CAD5* was correlated with higher phenolic resources in cell walls (Figure 5C,E). Interestingly, despite higher substrate availability for lignification in NH_4^+-grown *fro1* plants, these plants had lower contents of lignin (Figure 5A) and showed lower POX activity [47].

Therefore, we did not expect the stiffening of the cell walls in *fro1* plants when treated with NH_4^+, which may therefore favor cell expansion. It should also be noted that POX activity might be involved not only in cell wall stiffening but also in contrasting processes such as cell wall loosening. In the hydroxylic cycle, POX can produce HO^- from the superoxide anion and hydrogen peroxide [77,78]. In response to NH_4^+ treatment, the selected *POX33* isoform showed an increased expression in *fro1* plants (Figure 8B), indicating that POX may play a role in non-enzymatic cell wall loosening, thereby enabling growth. Cvetkowska et al. [8] have proposed a relationship between defective mitochondrial functioning and processes occurring in the extracellular space associated with ROS-triggered signaling. Consistent with this supposition, we detected higher H_2O_2 levels in *fro1* plants, primarily within the apoplast (Figure 8A). The apoplastic ROS pool was even increased in *fro1* plants when grown on NH_4^+ (Figure 8A). A ROS-related response associated with the plasmalemma was also indicated by the induced expression of the marker gene *OZF1* in response to NH_4^+ in both *fro1* and WT plants (Figure 8C). Thus, we speculate that the ROS burst in the apoplastic space in response to NH_4^+ might activate signaling events [79,80].

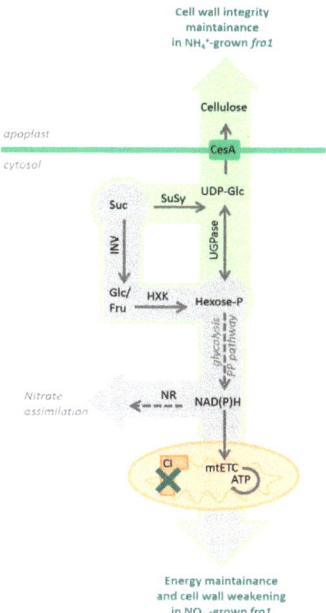

Figure 10. Carbohydrate metabolism in *frostbite1* (*fro1*) mutants lacking complex I (CI), when cultured on NH_4^+ or NO_3^- as the sole nitrogen source. Sucrose, the major fixed carbon in plants, is channeled toward sugar catabolism via hexokinase (HXK) activity to generate hexose-phosphates (Hexose-P). Further, the glycolytic or pentose phosphate (PP) pathways provide reductants that can be oxidized in the mitochondrial electron transport chain (mtETC) to produce ATP. Alternatively to dissipation of reductants in the mtETC, high NAD(P)H expenditure is necessary for NO_3^- assimilation catalyzed by nitrate reductase (NR). As indicated by grey arrows both energy fluxes are required to maintain the growth of *fro1* plants on NO_3^-. In contrast, when *fro1* is grown on NH_4^+, the reaction catalyzed by NR is omitted, resulting in a surplus of reductants. Therefore, the lower energy flux towards energy synthesis in *fro1* during NH_4^+ nutrition, allows sugars to be available for cell wall synthesis indicated as green arrow. Cytosolic sugars can provide a substrate for sucrose synthase (SuSy) or UDP-glucose phosphorylase (UGPase) to produce UDP-Glc, which is a precursor for cellulose synthesis. The cellulose synthetizing complex at the plasma membrane (containing cellulose synthase subunits, CesA) is responsible for the incorporation of carbohydrates into the cell wall.

The plant cell wall is an active structure that can respond to environmental cues, integrate signaling pathways, and regulate cell physiology and growth [81,82]. Perception of the integrity of cell wall cellulose can be ensured by dedicated cell wall sensor receptors such as kinases [83–85]. Cell wall remodeling in response to either NH_4^+ nutrition or a defect in complex I is reflected in the decreased expression of *FER* and *THE1* (Figure 3E,F). However, the effect of both sources of stress is to trigger a strong down-regulation of *FER* in *fro1* plants when grown on NH_4^+. Although *FER* is essential for expansion growth of cells, the biomass production of *fro1* plants is increased when cultivated on NH_4^+ (Figure 1). Mitochondrial dysfunction can in some cases induce tolerance against cellulose deficiency. In this regard, it has previously been shown that suppressed mitochondrial PPR-like protein induces retrograde signaling, resulting in a resistance to cellulose synthesis inhibition [86]. Accordingly, mutants can reconstruct weak cell walls and overcome growth suppression.

3.2. Fro1 Does not Show Significant Differences in the Pattern of Sugar Signaling

Sugars are probably the most important metabolites in the energy economy of living cells, and therefore cells need to have a precise system for monitoring sugar levels. Signalling pathways for sucrose, glucose, trehalose-6-phosphate, and fructose have previously been described [87–89]. HXK1 plays a dual role in cell metabolism, in addition to its enzymatic function of promoting hexose phosphorylation in glycolytic pathway, it can also act as a sugar sensor. Although *Arabidopsis* HXK1 is mostly associated with mitochondrial membranes, it is also expressed in the nucleus [87], where it forms a complex with specific subunits of other proteins and modulates the transcription of target genes. The expression level of *HXK1* has been demonstrated to be positively correlated with sensitivity to Glc [90–92]. In the present study, we found that expression of the *HXK1* gene in WT plants was decreased in response to NH_4^+ (Figure 9A). A lower expression of *HXK1* may be a mechanism whereby WT plants grown on NH_4^+ show a reduced sensitivity to increased levels of Glc. Additionally, it should be noted that HXK1-dependent Glc sensing is modulated by nitrogen availability [91] and in ammonium-stressed plants, nitrogen content is substantially increased (results not published). Although HXK activity in leaves of WT plants was not altered under NH_4^+ nutrition, it was relatively high in *fro1* plants under both growth conditions (Figure 2C). The energy metabolism of *fro1* plants depends largely on substrate-level phosphorylation, and this necessitates an up-regulated glycolytic flux in these plants. Indeed, HXK activity (Figure 2C) and soluble sugar content were increased (Figure 2A,B) in *fro1* plants. Furthermore, *HXK1* transcript levels remained unchanged and were at similar levels to those in NO_3^--grown WT plants (Figure 9A), therefore the regulatory role of HXK1 in plants with dysfunction of the mtETC remains elusive.

The second well-described protein involved in sugar sensing in plant cells is the SnRK1 complex. This complex has kinase activity and is assumed to be regulated by sugar availability. However, recently, SnRK1.1 has been recognized as playing a role in sugar-signaling, hub-regulating metabolism in response to changes in cellular energy status [93,94]. Among SnRK1-activated (and sugar-repressed) genes are those associated with catabolic pathways (cell wall, starch, Suc, amino acids, and protein degradation), which provide substrates for generating energy [95–97]. According to Baena et al. [95], SnRK1 senses stress-associated energy deprivation and reprograms metabolism to restore homeostasis and promote plant stress tolerance. *SnRK1.1* (also referred to as KIN10) is one of the catalytic subunits of the heterotrimeric SnRK1 complex in plants [98]. Surprisingly, we found that *SnRK1.1* transcript levels appear to decrease in response to NH_4^+ stress and mitochondrial complex I dysfunction (Figure 9B) when there is a cellular energy deficit [47]. Furthermore, SnRK1 activity was recently shown to be redox state-dependent [99]. Since the redox state of *Arabidopsis* leaf cells is increased in response to both mtETC dysfunction and NH_4^+ nutrition [47,57], we cannot exclude the possibility that this may induce SnRK1 activity despite lower transcript/protein levels.

3.3. Ammonium Nutrition Mitigates PCD Occurrence in fro1 Plants

Abiotic stress signaling or the energy status of cells can activate processes leading to PCD in plant cells. KOD [62] induces depolarization of the mitochondrial membrane, and constitutes an early step in plant PCD. Simultaneously, *KOD*-promoted PCD can be suppressed by the highly conserved survival factor *BI-1* which can delay the onset of PCD upon stress signaling [100,101]. Besides, a direct link between HXK1 activity and PCD has been proposed [91]. HXK1 inhibits PCD via binding to the voltage-dependent anion channel (VDAC) in plant mitochondrial membranes and inhibiting cyt *c* translocation from mitochondria in response to cellular stress [65]. In *fro1* plants, cyt *c* level was higher than in WT plants, (Figure 7C) however, a lower *KOD* transcript level and no changes in *HXK1* and *BI-1* expression (Figure 7A,B and Figure 9A) indicate that cell death in *fro1* plants under nitrate conditions (Figure 6) appears to be induced by other stimuli than the analysed genes. On the other hand, NH_4^+ nutrition has the opposite trend on marker gene expression, which correlates with unchanged cyt *c* abundance and the lack of lesion development (Figures 6 and 7). Distinct differences in the molecular responses at the transcript and protein levels of *fro1* plants indicate that multiple pathways may be involved in mediating the progression or inhibition of PCD due to functional changes in mtETC or varying nitrogen supply.

Recently, Van Doorn [102] postulated the occurrence of two morphological classes of PCD: necrosis and vacuolar cell death. Necrosis is typically found under conditions of abiotic stress. Although necrosis is no longer considered to be an un-programmed process, it remains poorly characterized at the biochemical and genetic levels, and yet no associated molecular markers have been identified. Mitochondrial changes related to necrotic cell death include respiratory decline, the production of ROS, a decrease in ATP levels, and mitochondrial membrane permeabilization, most of which have been observed in *fro1* mitochondria [47]. Autophagy is an intracellular process involved in the vacuolar degradation of cytoplasmic components, and although it has yet to be determined whether autophagic pathways are required for the progression of vacuolar cell death, *ATG5*, one of the *ATG* genes that are essential for autophagosome formation, has recently been found to be involved in developmental vacuolar cell death of *Arabidopsis* [103]. In the present study, the induced expression of *ATG5* in *fro1* plants (Figure 7D) may thus indicate that lesions emerging on the leaf blades of *fro1* plants under NO_3^- nutrition (Figure 6A) have the vacuolar cell death origin. Moreover, the decreased expression of *ATG5* which correlates with lack of lesions and fewer areas of dead cells in the leaf blades of NH_4^+-grown *fro1* plants, compared with those in plants grown under NO_3^- nutrition (Figure 6), indicates that two different mechanisms underlie the responses of *fro1* plants to the nitrogen status. However, since the plants used in our experiments were long-term grown and PCD is a rapidly developing process, it is not possible to distinguish the exact morphological symptoms characteristic of both types of PCD.

4. Materials and Methods

4.1. Plant Material and Growth Conditions

Experiments were performed on *Arabidopsis thaliana* plants of ecotype C24 (WT) and *frostbite1* mutants, which were derived through chemical mutagenesis as described by Lee et al. [46]. Plants were grown hydroponically using an Araponics system (Liège, Belgium) as described in Podgórska et al. [17]. The nutrient medium (according to [104]), containing 5 mM NO_3^- or 5 mM NH_4^+ as nitrogen source was renewed twice a week. NO_3^--treated WT plants were used as controls. Plants were grown for 8 weeks until they reached growth stage 5.10 according to [105]. The culture was conducted under an 8 h photoperiod at 150 µmol m^{-2} s^{-1} photosynthetically active radiation (PAR, daylight and warm white 1:1, LF-40W, Piła, Poland), day/night temperature of 21 °C/18 °C, and approximately 70% relative humidity.

4.2. Phenotype Analysis

Representative rosettes were photographed. Plant leaves were stained with 0.5 mg/mL Calcofluor White (Sigma Aldrich, Darmstadt, Germany) as previously described in Podgórska et al. [58]. The cross-section area of cells were determined using the Nikon A1R MP confocal laser scanning microscope (Nikon, Tokyo, Japan). Eight to 10 plants analyzed from each variant were randomly selected from 3 independent plant cultures. Cell size was calculated on micrographs using the Nis-Elements 3.22 imaging software (Nikon). The thickness of the cell walls was measured on micrographs obtained by transmission electron microscopy (TEM) (as previously described by [47]) according to Podgórska et al. [58]. The thickness of a double layer of cell walls was measured using the Image Processing and Analysis in Java software (ImageJ, v.1.51f, https://imagej.nih.gov/ij/).

4.3. Lesions Identification

Selected leaves were photographed using a binocular to show lesion spots. The occurrence of spots on leaves was counted. The precise location of necrosis within leaf blades was analyzed using trypan blue staining [106]. The trypan blue solution was composed of 10% phenol, 10% glycerol, 10% lactic acid in 60% ethanol and 0.02% trypan blue [107]. Whole leaves were immersed in the trypan blue solution for 5 min at 35 °C; next, leaves were cleared with a distaining solution (40% methanol, 10% acetic acid, 10% glycerol) at 60 °C, and photographed. The staining intensity of trypan blue on leaves was quantified using ImageJ software.

4.4. Cell Wall Preparations, Cellulose, Lignin, and Phenol Content Assay

Cell walls were prepared from around 2 g of frozen leaf tissue as described by Solecka et al. [108]. The resulting precipitate containing the cell wall was air dried and used for cellulose and lignin determination. Cellulose content was measured via the colorimetric Anthrone protocol according to Updegraff [109]. Lignin content was determined by the acetyl bromide method [110] as described in Hatfield et al. [111]. The amount of phenolics bound to cell walls was measured using a method described in Forrest and Bendall [112], as described earlier in Solecka et al. [113]. Phenolics were released from the cell wall preparations by alkaline hydrolysis and their content was determined spectrophotometrically using the Folin reagent.

4.5. Determination of Sugars and Protein Level

Soluble sugars were extracted as described in Szal et al. [114]. Glucose content was determined by the glucose oxidase-peroxidase reaction [115]. Sucrose concentration was determined after degradation to Glc and fructose. Protein level was measured as described by Bradford [116] using BSA as a standard.

4.6. Enzyme Activity Measurement and Protein Level Determination

Hexokinase activity was assayed according to the method described in Huber and Akazawa [117]. Protein extracts for enzyme activity determination and Western-blotting were prepared from 100 mg of leaf tissue which was homogenized with 2.5 volumes of extraction buffers.

Cytochrome *c* level determination in mitochondrial samples was done as described in Borysiuk et al. [63] and resulting bands were normalized on the basis of the mitochondrial marker protein voltage-dependent anion-selective channel protein 1 (VDAC1, Agrisera, Vännäs, Sweden, Figure S2). For other protein level analyses protein extracts (5 µL of protein extracts) (corresponding to 20 µg of protein) were separated in 10% sodium dodecyl sulfate-polyacrylamide gel electrophoresis (SDS-PAGE). Anti-UGPase [70], were used as primary antibodies (diluted 1:1000), and anti-rabbit antibodies (Bio-Rad, Hercules, CA, USA) were used as secondary antibodies. Immuno-blotting was performed according to standard protocols. Visualization was performed using a chemiluminescence kit (Clarity™ Western ECL, Bio-Rad, Hercules, CA, USA), and signals were detected using a Chemi-Doc imaging system (Bio-Ra). Bands (located at approximately 12 kDa for cyt *c* and 51 kDa for UGPase) were determined based on a pre-stained protein marker (Bio-Rad) as reference. Relative protein levels were quantified by densitometry analysis using Image-Lab 5.2. software (Bio-Rad).

4.7. Quantitative RT-PCR Analyses

Total RNA was extracted using a Syngen Plant RNA Mini kit (Syngen Biotech, Wrocław, Poland). DNAse digestion was performed using a RNase-free DNAse Set (Qiagen, Hilden, Germany). cDNA was synthesized using a RevertAid H minus first-strand cDNA synthesis kit (Thermo Fisher Scientific, Inc., Waltham, MA, USA) and RNAse H digestion was performed according to the procedure described in Escobar et al. [118]. The transcript levels were determined using iTaq Universal SYBR Green Supermix (Bio-Rad). Quantitative RT-PCR reactions were performed using a thermo cycler (CFX Content™, Bio-Rad) at 60 °C for annealing temperature. Reference protein phosphatase 2A (*PP2A*, AT1G13320, [119]) gene was used to normalize results. Transcript levels and qRT-PCR efficiency of genes were quantified as described in Pfaffl [120]. Results are expressed in relation to those in control plants. PCR primer pairs have been previously described for *FER* (AT3G51550), *THE1* (AT5G54380) [58], and *KOD* (AT4G22970) [62]. New primers were designed for: *OZF1* (AT2G19810), *CesA1* (AT4G32410), *CesA3* (AT5G05170), *CesA4* (AT5G44030), *CesA6* (AT5G64740), *CesA7* (AT5G17420), *CesA8* (AT4G18780), *SnRK1.1* (AT3G01090), *HXK1* (AT4G29130), *CAD1* (AT1G72680), *CAD4* (AT3G19450), *CAD5* (AT4G34230), *POX33* (AT3G49110), *POX34* (AT3G49120), *POX64* (AT5G42180), *POX72* (AT5G66390), *BI-1* (AT5G47120), and *ATG5* (AT5G17290) (Supplementary Materials Table S1), in which one sequence spanned always an exon–exon border if the gene had at least one intron.

4.8. Statistical Analysis

Results were expressed as means and standard deviations (SD) from 3 to 10 measurements taken from at least three independent plant cultures. One-way analysis of variance (ANOVA) and Tukey's post-hoc test at p-values ≤ 0.05 were performed to analyze statistical significance of observed differences, using the Statistica 13.1 software (StatSoft, Inc., Tulsa, OK, USA).

Supplementary Materials: Supplementary materials can be found at http://www.mdpi.com/1422-0067/19/8/2206/s1.

Author Contributions: A.P. and B.S. conceived and designed the experiments; A.P., A.T., K.B., and M.O.-B. carried out qRT-PCR analysis; A.T., B.S., and M.B. measured metabolites and enzyme activities; A.P. performed CLSM microscopy and cytochemical staining; B.S. and M.O.-B. performed immunoblotting; A.P., A.T., B.S., K.B., and M.O.-B. wrote the paper; P.G. revised the manuscript.

Funding: This work was partially supported by grant 2014/13/B/NZ3/00847 and 2014/14/E/NZ3/00155 from the National Science Centre (NCN, Kraków, Poland) given to Bożena Szal.

Acknowledgments: Seeds of WT and *fro1* plants were kindly donated by Jian-Kang Zhu (Purdue University). Anti-UGPase antibodies were kindly provided by Leszek Kleczkowski (Umeå Plant Science Centre). The authors thank Bohdan Paterczyk from the Faculty of Biology (University of Warsaw) for support during CLSM analysis. We are grateful to Anna Książek from the Faculty of Biology (University of Warsaw) for assistance with qRT-PCR studies. Anna Podgórska and Monika Ostaszewska-Bugajska were the beneficiaries of a scholarship from the Polish Minister of Science and Higher Education.

Conflicts of Interest: The authors declare no conflict of interest.

Abbreviations

ATP	adenosine triphosphate
cyt *c*	cytochrome *c*
fro1	*frostbite1*
Glc	glucose
HXK	hexokinase
mtETC	mitochondrial electron transport chain
NADH	nicotinamide adenine dinucleotide, reduced

NADPH	nicotinamide adenine dinucleotide phosphate, reduced
PCD	programmed cell death
POX	peroxidase
ROS	reactive oxygen species
Suc	sucrose
TCA	tricarboxylic acid
UGPase	UDP-glucose pyrophosphorylase
WT	wild-type

References

1. Bloom, A.J. Photorespiration and nitrate assimilation: A major intersection between plant carbon and nitrogen. *Photosynth. Res.* **2015**, *123*, 117–128. [CrossRef] [PubMed]
2. Noctor, G.; Foyer, C.H. A re-evaluation of the ATP:NADPH budget during C3 photosynthesis: A contribution from nitrate assimilation and its associated respiratory activity? *J. Exp. Bot.* **1998**, *49*, 1895–1908. [CrossRef]
3. Fernie, A.R.; Carrari, F.; Sweetlove, L.J. Respiratory metabolism: Glycolysis, the TCA cycle and mitochondrial electron transport. *Curr. Opin. Plant Biol.* **2004**, *7*, 254–261. [CrossRef] [PubMed]
4. Sweetlove, L.J.; Fait, A.; Nunes-Nesi, A.; Williams, T.; Fernie, A.R. The mitochondrion: An integration point of cellular metabolism and signalling. *CRC Crit. Rev. Plant Sci.* **2007**, *26*, 17–43. [CrossRef]
5. Gardeström, P.; Igamberdiev, A.U. The origin of cytosolic ATP in photosynthetic cells. *Physiol. Plant.* **2016**, *157*, 367–379. [CrossRef] [PubMed]
6. Ng, S.; De Clercq, I.; Van Aken, O.; Law, S.R.; Ivanova, A.; Willems, P.; Giraud, E.; Van Breusegem, F.; Whelan, J. Anterograde and retrograde regulation of nuclear genes encoding mitochondrial proteins during growth, development, and stress. *Mol. Plant* **2014**, *7*, 1075–1093. [CrossRef] [PubMed]
7. Crawford, T.; Lehotai, N.; Strand, Å. The role of retrograde signals during plant stress responses. *J. Exp. Bot.* **2017**, *69*, 2783–2795. [CrossRef] [PubMed]
8. Cvetkovska, M.; Vanlerberghe, G.C. Alternative oxidase impacts the plant response to biotic stress by influencing the mitochondrial generation of reactive oxygen species. *Plant Cell Environ.* **2013**, *36*, 721–732. [CrossRef] [PubMed]
9. Huang, S.; Van Aken, O.; Schwarzländer, M.; Belt, K.; Millar, A.H. The roles of mitochondrial reactive oxygen species in cellular signaling and stress response in plants. *Plant Physiol.* **2016**, *171*, 1551–1559. [CrossRef] [PubMed]
10. Schertl, P.; Braun, H.P. Respiratory electron transfer pathways in plant mitochondria. *Front. Plant Sci.* **2014**, *5*, 163. [CrossRef] [PubMed]
11. Meyer, E.H.; Tomaz, T.; Carroll, A.J.; Estavillo, G.; Delannoy, E.; Tanz, S.K.; Small, I.D.; Pogson, B.J.; Millar, A.H. Remodeled respiration in ndufs4 with low phosphorylation efficiency suppresses *Arabidopsis* germination and growth and alters control of metabolism at night. *Plant Physiol.* **2009**, *151*, 603–619. [CrossRef] [PubMed]
12. Rasmusson, A.G.; Geisler, D.A.; Møller, I.M. The multiplicity of dehydrogenases in the electron transport chain of plant mitochondria. *Mitochondrion* **2008**, *8*, 47–60. [CrossRef] [PubMed]
13. Klodmann, J.; Braun, H.P. Proteomic approach to characterize mitochondrial complex I from plants. *Phytochemistry* **2011**, *72*, 1071–1080. [CrossRef] [PubMed]
14. Pla, M.; Mathieu, C.; De Paepe, R.; Chétrit, P.; Vedel, F. Deletion of the last two exons of the mitochondrial *nad7* gene results in lack of the NAD7 polypeptide in a *Nicotiana sylvestris* CMS mutant. *Mol. Gen. Genet.* **1995**, *248*, 79–88. [CrossRef] [PubMed]
15. Gutierres, S.; Sabar, M.; Lelandais, C.; Chetrit, P.; Diolez, P.; Degand, H.; Boutry, M.; Vedel, F.; de Kouchkovsky, Y.; De Paepe, R. Lack of mitochondrial and nuclear-encoded subunits of complex I and alteration of the respiratory chain in *Nicotiana sylvestris* mitochondrial deletion mutants. *Proc. Natl. Acad. Sci. USA* **1997**, *94*, 3436–3441. [CrossRef] [PubMed]
16. Vidal, G.; Ribas-Carbó, M.; Garmier, M.; Dubertret, G.; Rasmusson, A.G.; Mathieu, C.; Foyer, C.H.; De Paepe, R. Lack of respiratory chain complex I impairs alternative oxidase engagement and modulates redox signaling during elicitor-induced cell death in *Tobacco*. *Plant Cell* **2007**, *19*, 640–655. [CrossRef] [PubMed]

17. Marienfeld, J.R.; Newton, K.J. The maize NCS2 abnormal growth mutant has a chimeric nad4-nad7 mitochondrial gene and is associated with reduced complex I function. *Genetics* **1994**, *138*, 855–863. [CrossRef] [PubMed]
18. Yamato, K.T.; Newton, K.J. Heteroplasmy and homoplasmy for maize mitochondrial mutants: A rare homoplasmic *nad4* deletion mutant plant. *J. Hered.* **1999**, *90*, 369–373. [CrossRef]
19. Lilly, J.W.; Bartoszewski, G.; Malepszy, S.; Havey, M.J. A major deletion in the cucumber mitochondrial genome sorts with the MSC phenotype. *Curr. Genet.* **2001**, *40*, 144–151. [CrossRef] [PubMed]
20. Juszczuk, I.M.; Flexas, J.; Szal, B.; Dąbrowska, Z.; Ribas-Carbó, M.; Rychter, A.M. Effect of mitochondrial genome rearrangement on respiratory activity, photosynthesis, photorespiration and energy status of MSC16 cucumber (*Cucumis sativus*) mutant. *Physiol. Plant.* **2007**, *131*, 527–541. [CrossRef] [PubMed]
21. Kühn, K.; Obata, T.; Feher, K.; Bock, R.; Fernie, A.R.; Meyer, E.H. Complete mitochondrial complex I deficiency induces an up-regulation of respiratory fluxes that is abolished by traces of functional complex I. *Plant Physiol.* **2015**, *168*, 1537–1549. [CrossRef] [PubMed]
22. Fromm, S.; Senkler, J.; Eubel, H.; Peterhänsel, C.; Braun, H.P. Life without complex I: Proteome analyses of an *Arabidopsis* mutant lacking the mitochondrial NADH dehydrogenase complex. *J. Exp. Bot.* **2016**, *67*, 3079–3093. [CrossRef] [PubMed]
23. Han, L.; Qin, G.; Kang, D.; Chen, Z.; Gu, H.; Qu, L.J. A nuclear-encoded mitochondrial gene AtCIB22 is essential for plant development in *Arabidopsis*. *J. Genet. Genom.* **2010**, *37*, 667–683. [CrossRef]
24. de Longevialle, A.F.; Meyer, E.H.; Andrés, C.; Taylor, N.L.; Lurin, C.; Millar, A.H.; Small, I.D. The pentatricopeptide repeat gene OTP43 is required for trans-splicing of the mitochondrial nad1 intron 1 in *Arabidopsis thaliana*. *Plant Cell* **2007**, *19*, 3256–3265. [CrossRef] [PubMed]
25. Nakagawa, N.; Sakurai, N. A mutation in At-nMat1a, which encodes a nuclear gene having high similarity to group II intron maturase, causes impaired splicing of mitochondrial NAD4 transcript and altered carbon metabolism in *Arabidopsis thaliana*. *Plant Cell Physiol.* **2006**, *47*, 772–783. [CrossRef] [PubMed]
26. Keren, I.; Bezawork-Geleta, A.; Kolton, M.; Maayan, I.; Belausov, E.; Levy, M.; Mett, A.; Gidoni, D.; Shaya, F.; Ostersetzer-Biran, O. AtnMat2, a nuclear-encoded maturase required for splicing of group-II introns in *Arabidopsis* mitochondria. *RNA* **2009**, *15*, 2299–2311. [CrossRef] [PubMed]
27. Kühn, K.; Carrie, C.; Giraud, E.; Wang, Y.; Meyer, E.H.; Narsai, R.; des Francs-Small, C.C.; Zhang, B.; Murcha, M.W.; Whelan, J. The RCC1 family protein RUG3 is required for splicing of nad2 and complex I biogenesis in mitochondria of *Arabidopsis thaliana*. *Plant J.* **2011**, *67*, 1067–1080. [CrossRef] [PubMed]
28. Hsu, Y.W.; Wang, H.J.; Hsieh, M.H.; Hsieh, H.L.; Jauh, G.Y. Arabidopsis mTERF15 is required for mitochondrial nad2 intron 3 splicing and functional complex I activity. *PLoS ONE*. **2014**, *9*, e112360. [CrossRef] [PubMed]
29. Koprivova, A.; des Francs-Small, C.C.; Calder, G.; Mugford, S.T.; Tanz, S.; Lee, B.-R.; Zechmann, B.; Small, I.; Kopriva, S. Identification of a pentatricopeptide repeat protein implicated in splicing of intron 1 of mitochondrial nad7 transcripts. *J. Biol. Chem.* **2010**, *285*, 32192–32199. [CrossRef] [PubMed]
30. Brangeon, J.; Sabar, M.; Gutierres, S.; Combettes, B.; Bove, J.; Gendy, C.; Chetrit, P.; des Francs-Small, C.C.; Pla, M.; Vedel, F.; et al. Defective splicing of the first nad4 intron is associated with lack of several complex I subunits in the *Nicotiana sylvestris* NMS1 nuclear mutant. *Plant J.* **2000**, *21*, 269–280. [CrossRef] [PubMed]
31. Sabar, M.; De Paepe, R.; de Kouchkovsky, Y. Complex I impairment, respiratory compensations, and photosynthetic decrease in nuclear and mitochondrial male sterile mutants of *Nicotiana sylvestris*. *Plant Physiol.* **2000**, *124*, 1239–1250. [CrossRef] [PubMed]
32. Juszczuk, I.M.; Szal, B.; Rychter, A.M. Oxidation–reduction and reactive oxygen species homeostasis in mutant plants with respiratory chain complex I dysfunction. *Plant Cell Environ.* **2012**, *35*, 296–307. [CrossRef] [PubMed]
33. Subrahmanian, N.; Remacle, C.; Hamel, P.P. Plant mitochondrial complex I composition and assembly: A review. *Biochim. Biophys. Acta* **2016**, *1857*, 1001–1014. [CrossRef] [PubMed]
34. Szal, B.; Dąbrowska, Z.; Malmberg, G.; Gardeström, P.; Rychter, A.M. Changes in energy status of leaf cells as the consequence of mitochondrial genome rearrangement. *Planta* **2008**, *227*, 697–706. [CrossRef] [PubMed]
35. Florez-Sarasa, I.; Ostaszewska, M.; Galle, A.; Flexas, J.; Rychter, A.M.; Ribas-Carbó, M. Changes of alternative oxidase activity, capacity and protein content in leaves of *Cucumis sativus* wild-type and MSC16 mutant grown under different light intensities. *Physiol. Plant.* **2009**, *137*, 419–426. [CrossRef] [PubMed]

36. Dutilleul, C.; Lelarge, C.; Prioul, J.L.; De Paepe, R.; Foyer, C.H.; Noctor, G. Mitochondria-driven changes in leaf NAD status exert a crucial influence on the control of nitrate assimilation and the integration of carbon and nitrogen metabolism. *Plant Physiol.* **2005**, *139*, 64–78. [CrossRef] [PubMed]
37. Szal, B.; Łukawska, K.; Zdolińska, I.; Rychter, A.M. Chilling stress and mitochondrial genome rearrangement in the MSC16 cucumber mutant affect the alternative oxidase and antioxidant defense system to a similar extent. *Physiol. Plant.* **2009**, *137*, 435–445. [CrossRef] [PubMed]
38. Wang, Y.; Carrie, C.; Giraud, E.; Elhafez, D.; Narsai, R.; Duncan, O.; Whelan, J.; Murcha, M.W. Dual location of the mitochondrial preprotein transporters B14.7 and Tim23-2 in complex I and the TIM17:23 complex in Arabidopsis links mitochondrial activity and biogenesis. *Plant Cell* **2012**, *24*, 2675–2695. [CrossRef] [PubMed]
39. Wang, Y.; Lyu, W.; Berkowitz, O.; Radomiljac, J.D.; Law, S.R.; Murcha, M.W.; Carrie, C.; Teixeira, P.F.; Kmiec, B.; Duncan, O.; et al. Inactivation of mitochondrial complex I induces the expression of a twin cysteine protein that targets and affects cytosolic, chloroplastidic and mitochondrial function. *Mol. Plant* **2016**, *9*, 696–710. [CrossRef] [PubMed]
40. Garmier, M.; Dutilleul, C.; Mathieu, C.; Chétrit, P.; Boccara, M.; De Paepe, R. Changes in antioxidant expression and harpin-induced hypersensitive response in a *Nicotiana sylvestris* mitochondrial mutant. *Plant Physiol. Biochem.* **2002**, *40*, 561–566. [CrossRef]
41. Dutilleul, C.; Garmier, M.; Noctor, G.; Mathieu, C.; Chétrit, P.; Foyer, C.H.; de Paepe, R. Leaf mitochondria modulate whole cell redox homeostasis, set antioxidant capacity, and determine stress resistance through altered signaling and diurnal regulation. *Plant Cell* **2003**, *15*, 1212–1226. [CrossRef] [PubMed]
42. Liu, Y.J.; Norberg, F.E.B.; Szilágyi, A.; De Paepe, R.; Åkerlund, H.E.; Rasmusson, A.G. The mitochondrial external NADPH dehydrogenase modulates the leaf NADPH/NADP$^+$ ratio in transgenic *Nicotiana sylvestris*. *Plant Cell Physiol.* **2008**, *49*, 251–263. [CrossRef] [PubMed]
43. Karpova, O.V.; Kuzmin, E.V.; Elthon, T.E.; Newton, K.J. Differential expression of alternative oxidase genes in maize mitochondrial mutants. *Plant Cell* **2002**, *14*, 3271–3284. [CrossRef] [PubMed]
44. Kuzmin, E.V.; Karpova, O.V.; Elthon, T.E.; Newton, K.J. Mitochondrial respiratory deficiencies signal up-regulation of genes for heat shock proteins. *J. Biol. Chem.* **2004**, *279*, 20672–20677. [CrossRef] [PubMed]
45. Tarasenko, V.I.; Garnik, E.Y.; Konstantinov, Y.M. Characterization of *Arabidopsis* mutant with inactivated gene coding for Fe-S subunit of mitochondrial respiratory chain complex I. *Russ. J. Plant Physiol.* **2010**, *57*, 392–400. [CrossRef]
46. Lee, B.H.; Lee, H.; Xiong, L.; Zhu, J.K. A mitochondrial complex I defect impairs cold-regulated nuclear gene expression. *Plant Cell* **2002**, *14*, 1235–1251. [CrossRef] [PubMed]
47. Podgórska, A.; Ostaszewska, M.; Gardeström, P.; Rasmusson, A.G.; Szal, B. In comparison with nitrate nutrition, ammonium nutrition increases growth of the *frostbite1* Arabidopsis mutant. *Plant Cell Environ.* **2015**, *38*, 224–237. [CrossRef] [PubMed]
48. Britto, D.T.; Kronzucker, H.J. NH$_4^+$ toxicity in higher plants: A critical review. *J. Plant Physiol.* **2002**, *159*, 567–584. [CrossRef]
49. Bittsánszky, A.; Pilinszky, K.; Gyulai, G.; Komives, T. Overcoming ammonium toxicity. *Plant Sci.* **2015**, *231*, 184–190. [CrossRef] [PubMed]
50. Walch-Liu, P.; Neumann, G.; Bangerth, F.; Engels, C. Rapid effects of nitrogen form on leaf morphogenesis in tobacco. *J. Exp. Bot.* **2000**, *51*, 227–237. [CrossRef] [PubMed]
51. Liu, Y.; von Wirén, N. Ammonium as a signal for physiological and morphological responses in plants. *J. Exp. Bot.* **2017**, *68*, 2581–2592. [CrossRef] [PubMed]
52. Gerendás, J.; Zhu, Z.; Bendixen, R.; Ratcliffe, R.G.; Sattelmacher, B. Physiological and biochemical processes related to ammonium toxicity in higher plants. *J. Plant Nutr. Soil Sci.* **1997**, *160*, 239–251. [CrossRef]
53. Britto, D.T.; Kronzucker, H.J. Ecological significance and complexity of N-source preference in plants. *Ann. Bot.* **2013**, *112*, 957–963. [CrossRef] [PubMed]
54. Szal, B.; Podgórska, A. The role of mitochondria in leaf nitrogen metabolism. *Plant Cell Environ.* **2012**, *35*, 1756–1768. [CrossRef] [PubMed]
55. Guo, S.; Schinner, K.; Sattelmacher, B.; Hansen, U.P. Different apparent CO_2 compensation points in nitrate-and ammonium-grown *Phaseolus vulgaris* and the relationship to non-photorespiratory CO_2 evolution. *Physiol. Plant.* **2005**, *123*, 288–301. [CrossRef]

56. Escobar, M.A.; Geisler, D.A.; Rasmusson, A.G. Reorganization of the alternative pathways of the *Arabidopsis* respiratory chain by nitrogen supply: Opposing effects of ammonium and nitrate. *Plant J.* **2006**, *45*, 775–788. [CrossRef] [PubMed]
57. Podgórska, A.; Gieczewska, K.; Łukawska-Kuźma, K.; Rasmusson, A.G.; Gardeström, P.; Szal, B. Long-term ammonium nutrition of *Arabidopsis* increases the extrachloroplastic NAD(P)H/NAD(P)$^+$ ratio and mitochondrial reactive oxygen species level in leaves but does not impair photosynthetic capacity. *Plant Cell Environ.* **2013**, *36*, 2034–2045. [CrossRef] [PubMed]
58. Podgórska, A.; Burian, M.; Gieczewska, K.; Ostaszewska-Bugajska, M.; Zebrowski, J.; Solecka, D.; Szal, B. Altered cell wall plasticity can restrict plant growth under ammonium nutrition. *Front. Plant Sci.* **2017**, *8*, 1344. [CrossRef] [PubMed]
59. Cheung, A.Y.; Wu, H.M. THESEUS 1, FERONIA and relatives: A family of cell wall-sensing receptor kinases? *Curr. Opin. Plant Biol.* **2011**, *14*, 632–641. [CrossRef] [PubMed]
60. Desprez, T.; Juraniec, M.; Crowell, E.F.; Jouy, H.; Pochylova, Z.; Parcy, F.; Hofte, H.; Gonneau, M.; Vernhettes, S. Organization of cellulose synthase complexes involved in primary cell wall synthesis in *Arabidopsis thaliana*. *Proc. Natl. Acad. Sci. USA* **2007**, *104*, 15572–15577. [CrossRef] [PubMed]
61. Ma, Q.H. Functional analysis of a cinnamyl alcohol dehydrogenase involved in lignin biosynthesis in wheat. *J. Exp. Bot.* **2010**, *61*, 2735–2744. [CrossRef] [PubMed]
62. Blanvillain, R.; Young, B.; Cai, Y.; Hecht, V.; Varoquaux, F.; Delorme, V.; Lancelin, J.-M.; Delseny, M.; Gallois, P. The *Arabidopsis* peptide kiss of death is an inducer of programmed cell death. *EMBO J.* **2011**, *30*, 1173–1183. [CrossRef] [PubMed]
63. Borysiuk, K.; Ostaszewska-Bugajska, M.; Vaultier, M.N.; Hasenfratz-Sauder, M.P.; Szal, B. Enhanced formation of methylglyoxal-derived advanced glycation end products in *Arabidopsis* under ammonium nutrition. *Front. Plant Sci.* **2018**, *9*, 667. [CrossRef] [PubMed]
64. Huang, P.; Chung, M.S.; Ju, H.W.; Na, H.S.; Lee, D.J.; Cheong, H.S.; Kim, C.S. Physiological characterization of the *Arabidopsis thaliana* oxidation-related zinc finger 1, a plasma membrane protein involved in oxidative stress. *J. Plant Res.* **2011**, *124*, 699–705. [CrossRef] [PubMed]
65. Kim, M.; Lim, J.H.; Ahn, C.S.; Park, K.; Kim, G.T.; Kim, W.T.; Pai, H.S. Mitochondria-associated hexokinases play a role in the control of programmed cell death in *Nicotiana benthamiana*. *Plant Cell* **2006**, *18*, 2341–2355. [CrossRef] [PubMed]
66. Jossier, M.; Bouly, J.-P.; Meimoun, P.; Arjmand, A.; Lessard, P.; Hawley, S.; Grahame Hardie, D.; Thomas, M. SnRK1 (SNF1-related kinase 1) has a central role in sugar and ABA signalling in *Arabidopsis thaliana*. *Plant J.* **2009**, *59*, 316–328. [CrossRef] [PubMed]
67. Endler, A.; Persson, S. Cellulose synthases and synthesis in *Arabidopsis*. *Mol. Plant* **2011**, *4*, 199–211. [CrossRef] [PubMed]
68. Verbančič, J.; Lunn, J.E.; Stitt, M.; Persson, S. Carbon supply and the regulation of cell wall synthesis. *Mol. Plant* **2018**, *11*, 75–94. [CrossRef] [PubMed]
69. Persson, S.; Paredez, A.; Carroll, A.; Palsdottir, H.; Doblin, M.; Poindexter, P.; Khitrov, N.; Auer, M.; Somerville, C.R. Genetic evidence for three unique components in primary cell-wall cellulose synthase complexes in *Arabidopsis*. *Proc. Natl. Acad. Sci. USA* **2007**, *104*, 15566–15571. [CrossRef] [PubMed]
70. Kleczkowski, L.A.; Geisler, M.; Ciereszko, I.; Johansson, H. UDP-glucose pyrophosphorylase. An old protein with new tricks. *Plant Physiol.* **2004**, *134*, 912–918. [CrossRef] [PubMed]
71. Taylor, N.G. Cellulose biosynthesis and deposition in higher plants. *New Phytol.* **2008**, *178*, 239–252. [CrossRef] [PubMed]
72. Timmers, J.; Vernhettes, S.; Desprez, T.; Vincken, J.P.; Visser, R.G.; Trindade, L.M. Interactions between membrane-bound cellulose synthases involved in the synthesis of the secondary cell wall. *FEBS Lett.* **2009**, *583*, 978–982. [CrossRef] [PubMed]
73. Marjamaa, K.; Kukkola, E.M.; Fagerstedt, K.V. The role of xylem class III peroxidases in lignification. *J. Exp. Bot.* **2009**, *60*, 367–376. [CrossRef] [PubMed]
74. Tokunaga, N.; Kaneta, T.; Sato, S.; Sato, Y. Analysis of expression profiles of three peroxidase genes associated with lignification in *Arabidopsis thaliana*. *Physiol. Plant.* **2009**, *136*, 237–249. [CrossRef] [PubMed]
75. Fernández-Pérez, F.; Pomar, F.; Pedreño, M.A.; Novo-Uzal, E. Suppression of *Arabidopsis* peroxidase 72 alters cell wall and phenylpropanoid metabolism. *Plant Sci.* **2015**, *239*, 192–199. [CrossRef] [PubMed]

76. Kim, S.J.; Kim, K.W.; Cho, M.H.; Franceschi, V.R.; Davin, L.B.; Lewis, N.G. Expression of cinnamyl alcohol dehydrogenases and their putative homologues during *Arabidopsis thaliana* growth and development: Lessons for database annotations? *Phytochemistry* **2007**, *68*, 1957–1974. [CrossRef] [PubMed]
77. Liszkay, A.; Kenk, B.; Schopfer, P. Evidence for the involvement of cell wall peroxidase in the generation of hydroxyl radicals mediating extension growth. *Planta* **2003**, *217*, 658–667. [CrossRef] [PubMed]
78. Passardi, F.; Penel, C.; Dunand, C. Performing the paradoxical: How plant peroxidases modify the cell wall. *Trends Plant Sci.* **2004**, *9*, 534–540. [CrossRef] [PubMed]
79. Baxter, A.; Mittler, R.; Suzuki, N. ROS as key players in plant stress signalling. *J. Exp. Bot.* **2014**, *65*, 1229–1240. [CrossRef] [PubMed]
80. Podgórska, A.; Burian, M.; Szal, B. Extra-cellular but extra-ordinarily important for cells: Apoplastic reactive oxygen species metabolism. *Front. Plant Sci.* **2017b**, *8*, 1353. [CrossRef] [PubMed]
81. Seifert, G.J.; Blaukopf, C. Irritable walls: The plant extracellular matrix and signaling. *Plant Physiol.* **2010**, *153*, 467–478. [CrossRef] [PubMed]
82. Tenhaken, R. Cell wall remodeling under abiotic stress. *Front. Plant Sci.* **2014**, *5*, 771. [CrossRef] [PubMed]
83. Wolf, S.; Höfte, H. Growth control: A saga of cell walls, ROS, and peptide receptors. *Plant Cell* **2014**, *26*, 1848–1856. [CrossRef] [PubMed]
84. Li, C.; Wu, H.-M.; Cheung, A.Y. FERONIA and her pals: Functions and mechanisms. *Plant Physiol.* **2016**, *171*, 2379–2392. [CrossRef] [PubMed]
85. Nissen, K.S.; Willats, W.G.T.; Malinovsky, F.G. Understanding CrRLK1L function: Cell walls and growth control. *Trends Plant Sci.* **2016**, *21*, 516–527. [CrossRef] [PubMed]
86. Hu, Z.; Vanderhaeghen, R.; Cools, T.; Wang, Y.; De Clercq, I.; Leroux, O.; Nguyen, L.; Belt, K.; Millar, A.H.; Audenaert, D.; et al. Mitochondrial defects confer tolerance against cellulose deficiency. *Plant Cell* **2016**, *28*, 2276–2290. [CrossRef] [PubMed]
87. Harrington, G.N.; Bush, D.R. The bifunctional role of hexokinase in metabolism and glucose signaling. *Plant Cell* **2003**, *15*, 2493–2496. [CrossRef] [PubMed]
88. Sheen, J. Master regulators in plant glucose signaling networks. *J. Plant Biol.* **2014**, *57*, 67–79. [CrossRef] [PubMed]
89. Cho, Y.H.; Yoo, S.D.; Sheen, J. Regulatory functions of nuclear hexokinase1 complex in glucose signaling. *Cell* **2006**, *127*, 579–589. [CrossRef] [PubMed]
90. Jang, J.C.; León, P.; Zhou, L.; Sheen, J. Hexokinase as a sugar sensor in higher plants. *Plant Cell* **1997**, *9*, 5–19. [CrossRef] [PubMed]
91. Moore, B.; Zhou, L.; Rolland, F.; Hall, Q.; Cheng, W.H.; Liu, Y.X.; Hwang, I.; Jones, T.; Sheen, J. Role of the *Arabidopsis* glucose sensor HXK1 in nutrient, light, and hormonal signaling. *Science* **2003**, *300*, 332–336. [CrossRef] [PubMed]
92. Kim, Y.; Heinzel, N.; Giese, J.; Koeber, J.; Melzer, M.; Rutten, T.; Von Wirén, N.; Sonnewald, U.; Hajirezaei, M. Dual function of hexokinases in tobacco plants. *Plant Cell Environ.* **2013**, *36*, 1311–1327. [CrossRef] [PubMed]
93. Simon, N.M.L.; Kusakina, J.; Fernández-López, Á.; Chembath, A.; Belbin, F.E.; Dodd, A.N. The energy-signaling hub SnRK1 is important for sucrose-induced hypocotyl elongation. *Plant Physiol.* **2018**, *176*, 1299–1310. [CrossRef] [PubMed]
94. Wurzinger, B.; Nukarinen, E.; Nägele, T.; Weckwerth, W.; Teige, M. The SnRK1 kinase as central mediator of energy signaling between different organelles. *Plant Physiol.* **2018**, *176*, 1085–1094. [CrossRef] [PubMed]
95. Baena-González, E.; Rolland, F.; Thevelein, J.M.; Sheen, J. A central integrator of transcription networks in plant stress and energy signalling. *Nature* **2007**, *448*, 938–942. [CrossRef] [PubMed]
96. Chen, L.; Su, Z.-Z.; Huang, L.; Xia, F.-N.; Qi, H.; Xie, L.-J.; Xiao, S.; Chen, Q.-F. The AMP-activated protein kinase KIN10 is involved in the regulation of autophagy in *Arabidopsis*. *Front. Plant Sci.* **2017**, *8*, 1201. [CrossRef] [PubMed]
97. Pedrotti, L.; Weiste, C.; Nägele, T.; Wolf, E.; Lorenzin, F.; Dietrich, K.; Mair, A.; Weckwerth, W.; Teige, M.; Baena-González, E.; et al. Snf1-RELATED KINASE1-controlled C/S$_1$-bZIP signaling activates alternative mitochondrial metabolic pathways to ensure plant survival in extended darkness. *Plant Cell* **2018**, *30*, 495–509. [CrossRef] [PubMed]
98. Ghillebert, R.; Swinnen, E.; Wen, J.; Vandesteene, L.; Ramon, M.; Norga, K.; Rolland, F.; Winderickx, J. The AMPK/SNF1/SnRK1 fuel gauge and energy regulator: Structure, function and regulation. *FEBS J.* **2011**, *278*, 3978–3990. [CrossRef] [PubMed]

99. Wurzinger, B.; Mair, A.; Fischer-Schrader, K.; Nukarinen, E.; Roustan, V.; Weckwerth, W.; Teige, M. Redox state-dependent modulation of plant SnRK1 kinase activity differs from AMPK regulation in animals. *FEBS Lett.* **2017**, *591*, 3625–3636. [CrossRef] [PubMed]
100. Watanabe, N.; Lam, E. *Arabidopsis* Bax inhibitor-1 functions as an attenuator of biotic and abiotic types of cell death. *Plant J.* **2006**, *45*, 884–894. [CrossRef] [PubMed]
101. Watanabe, N.; Lam, E. BAX inhibitor-1 modulates endoplasmic reticulum stress-mediated programmed cell death in *Arabidopsis*. *J. Biol. Chem.* **2008**, *283*, 3200–3210. [CrossRef] [PubMed]
102. Van Doorn, W.G.; Beers, E.P.; Dangl, J.L.; Franklin-Tong, V.E.; Gallois, P.; Hara-Nishimura, I.; Jones, A.M.; Kawai-Yamada, M.; Lam, E.; Mundy, J.; et al. Morphological classification of plant cell deaths. *Cell Death Differ.* **2011**, *18*, 1241. [CrossRef] [PubMed]
103. Kwon, S.I.; Cho, H.J.; Jung, J.H.; Yoshimoto, K.; Park, O.K. The RabGTPase RabG3b functions in autophagy and contributes to tracheary element differentiation in *Arabidopsis*. *Plant J.* **2010**, *64*, 151–164. [CrossRef] [PubMed]
104. Lasa, B.; Frechilla, S.; Apricio-Tejo, P.M.; Lamsfus, C. Alternative pathway respiration is associated with ammonium ion sensitivity in spinach and pea plants. *Plant Growth Regul.* **2002**, *37*, 49–55. [CrossRef]
105. Boyes, D.C.; Zayed, A.M.; Ascenzi, R.; McCaskill, A.J.; Hoffman, N.E.; Davis, K.R.; Görlach, J. Growth stage-based phenotypic analysis of Arabidopsis: A model for high throughput functional genomics in plants. *Plant Cell* **2002**, *13*, 1499–1510. [CrossRef]
106. Pogány, M.; von Rad, U.; Grün, S.; Dongó, A.; Pintye, A.; Simoneau, P.; Bahnweg, G.; Kiss, L.; Barna, B.; Durner, J. Dual roles of reactive oxygen species and NADPH oxidase RBOHD in an Arabidopsis-Alternaria pathosystem. *Plant Physiol.* **2009**, *151*, 1459–1475. [CrossRef] [PubMed]
107. Weremczuk, A.; Ruszczyńska, A.; Bulska, E.; Antosiewicz, D.M. NO-Dependent programmed cell death is involved in the formation of Zn-related lesions in tobacco leaves. *Metallomics* **2017**, *19*, 924–935. [CrossRef] [PubMed]
108. Solecka, D.; Zebrowski, J.; Kacperska, A. Are pectins involved in cold acclimation and de-acclimation of winter oil-seed rape plants? *Ann. Bot.* **2008**, *101*, 521–530. [CrossRef] [PubMed]
109. Updegraff, D. Semimicro determination of cellulose in biological materials. *Anal. Biochem.* **1969**, *32*, 420–424. [CrossRef]
110. Johnson, D.B.; Moore, W.E.; Zank, L.C. The spectrophotometric determination of lignin in small wood samples. *Tappi* **1961**, *44*, 793–798. [CrossRef]
111. Hatfield, R.D.; Grabber, J.; Ralph, J.; Brei, K. Using the acetylbromide assay to determine lignin concentrations in herbaceous plants: Some cautionarynotes. *J. Agric. Food Chem.* **1999**, *47*, 628–632. [CrossRef] [PubMed]
112. Forrest, G.I.; Bendall, D.S. The distribution of polyphenols in the tea plant (*Camellia sinensis* L.). *Biochem. J.* **1969**, *113*, 741–755. [CrossRef] [PubMed]
113. Solecka, D.; Boudet, A.M.; Kacperska, A. Phenylpropanoid and anthocyanin changes in low-temperature treated winter ilseed rape leaves. *Plant Physiol. Biochem.* **1999**, *37*, 491–496. [CrossRef]
114. Szal, B.; Jastrzębska, A.; Kulka, M.; Leśniak, K.; Podgórska, A.; Pärnik, T.; Ivanova, H.; Keerberg, O.; Gardeström, P.; Rychter, A.M. Influence of mitochondrial genome rearrangement on cucumber leaf carbon and nitrogen metabolism. *Planta* **2010**, *232*, 1371–1382. [CrossRef] [PubMed]
115. Kunst, A.; Draeger, B.; Ziegenhorn, J. Colorimertic methods with glucose oxidase and peroxidase. In *Methods in Enzymatic Analysis, Metabolites: Carbohydrates*; Bergmeyer, H.U., Bergmeyer, J., Grassl, M., Eds.; Verlag Chemie: Weinheim, Germany, 1985; pp. 178–185.
116. Bradford, M.M. A rapid and sensitive method for quantification of microgram quantities of protein utilizing the principle of protein-dye binding. *Anal. Biochem.* **1976**, *72*, 248–254. [CrossRef]
117. Huber, S.C.; Akazawa, T. A novel sucrose synthase pathway for sucrose degradation in cultured cells. **1986**, *81*, 1008–1013. [CrossRef] [PubMed]
118. Escobar, M.A.; Franklin, K.A.; Svensson, A.S.; Salter, M.G.; Whitelam, G.C.; Rasmusson, A.G. Light regulation of the *Arabidopsis* respiratory chain. Multiple discrete photoreceptor responses contribute to induction of type II NAD(P)H dehydrogenase genes. *Plant Physiol.* **2004**, *136*, 2710–2721. [CrossRef] [PubMed]

119. Czechowski, T.; Stitt, M.; Altmann, T.; Udvardi, MK.; Scheible, W.R. Genome-wide identification and testing of superior reference genes for transcript normalization in *Arabidopsis*. *Plant Physiol.* **2005**, *139*, 5–17. [CrossRef] [PubMed]
120. Pfaffl, M.W. A new mathematical model for relative quantification in real-time RT-PCR. *Nucleic Acids Res.* **2001**, *29*, e45. [CrossRef] [PubMed]

© 2018 by the authors. Licensee MDPI, Basel, Switzerland. This article is an open access article distributed under the terms and conditions of the Creative Commons Attribution (CC BY) license (http://creativecommons.org/licenses/by/4.0/).

Article

AOX1-Subfamily Gene Members in *Olea europaea* cv. "Galega Vulgar"—Gene Characterization and Expression of Transcripts during IBA-Induced In Vitro Adventitious Rooting

Isabel Velada [1], Dariusz Grzebelus [2], Diana Lousa [3], Cláudio M. Soares [3], Elisete Santos Macedo [4], Augusto Peixe [1], Birgit Arnholdt-Schmitt [4,5,6,*] and Hélia G. Cardoso [1,*]

1. Departamento de Fitotecnia, ICAAM—Instituto de Ciências Agrárias e Ambientais Mediterrânicas, Universidade de Évora, Pólo da Mitra, Ap. 94, 7006-554 Évora, Portugal; ivelada@uevora.pt (I.V.); apeixe@uevora.pt (A.P.)
2. Institute of Plant Biology and Biotechnology, Faculty of Biotechnology and Horticulture, University of Agriculture in Kraków, 31-120 Kraków, Poland; d.grzebelus@ogr.ur.krakow.pl
3. ITQB NOVA, Instituto de Tecnologia Química e Biológica António Xavier, Universidade Nova de Lisboa, Av. da República, 2780-157 Oeiras, Portugal; dlousa@itqb.unl.pt (D.L.); claudio@itqb.unl.pt (C.M.S.)
4. Functional Cell Reprogramming and Organism Plasticity (FunCrop), EU Marie Curie Chair, ICAAM, Universidade de Évora, 7006-554 Évora, Portugal; elisetemcd@gmail.com
5. Functional Genomics and Bioinformatics, Department of Biochemistry and Molecular Biology, Federal University of Ceará, 60020-181Fortaleza, Brazil
6. Science and Technology Park Alentejo (PACT), 7005-841 Évora, Portugal
* Correspondence: eu_chair@uevora.pt (B.A.-S.); hcardoso@uevora.pt (H.G.C.); Tel.: +351-266760800 (B.A.-S. & H.G.C.)

Received: 21 November 2017; Accepted: 8 February 2018; Published: 17 February 2018

Abstract: Propagation of some *Olea europaea* L. cultivars is strongly limited due to recalcitrant behavior in adventitious root formation by semi-hardwood cuttings. One example is the cultivar "Galega vulgar". The formation of adventitious roots is considered a morphological response to stress. Alternative oxidase (AOX) is the terminal oxidase of the alternative pathway of the plant mitochondrial electron transport chain. This enzyme is well known to be induced in response to several biotic and abiotic stress situations. This work aimed to characterize the alternative oxidase 1 (AOX1)-subfamily in olive and to analyze the expression of transcripts during the indole-3-butyric acid (IBA)-induced in vitro adventitious rooting (AR) process. *OeAOX1a* (acc. no. MF410318) and *OeAOX1d* (acc. no. MF410319) were identified, as well as different transcript variants for both genes which resulted from alternative polyadenylation events. A correlation between transcript accumulation of both *OeAOX1a* and *OeAOX1d* transcripts and the three distinct phases (induction, initiation, and expression) of the AR process in olive was observed. Olive *AOX1* genes seem to be associated with the induction and development of adventitious roots in IBA-treated explants. A better understanding of the molecular mechanisms underlying the stimulus needed for the induction of adventitious roots may help to develop more targeted and effective rooting induction protocols in order to improve the rooting ability of difficult-to-root cultivars.

Keywords: vegetative propagation; olive; adventitious rooting; auxins; IBA; plant mitochondria; alternative oxidase; alternative polyadenylation; transposable elements; gene expression

1. Introduction

Olive (*Olea europaea* L.) is one of the oldest agricultural fruit crops worldwide and is mostly cultivated for olive oil production. Olive orchards are predominantly concentrated in the

Mediterranean basin [1], although they have recently expanded to new regions due to the importance of olive oil in the human diet. Portugal has a production area of 430,000 ha of olive orchards, which represents about 5% of the world olive oil production. Portuguese olive oils are known worldwide for their exceptional organoleptic characteristics. Nowadays, olive plants are mostly propagated by semi-hardwood cuttings, a process in which adventitious root formation is a key factor. However, some of the agronomically interesting Portuguese olive cultivars used for oil production have been revealed to be recalcitrant to adventitious rooting (AR), which leads to a reduced availability of those varieties in the nurseries that are to be used in new orchard plantations. For example, "Galega vulgar" usually presents average rooting rates of 5–20% when semi-hardwood cuttings are used, being considered a difficult-to-root cultivar [2]. Similar recalcitrant behaviour has been described for autochthone cultivars with high agronomical interest in different countries (for review see [3]). In this frame, the study of AR in *O. europaea*, in view of the optimization of the process in stem cuttings of recalcitrant olive cultivars, has become an important research topic, which requires fundamental and applied research at different levels.

The process of AR at the base of stem cuttings is considered a morphological response to stress [4] that can be influenced by a large number of interacting internal and external factors. It involves hormone-transmitted metabolic changes, molecular transduction pathway activation, protein degradation, and protein de novo synthesis, as well as adaptive global genome regulation (for review see [3]). The cutting's removal from the mother tree and subsequent treatment with auxin are both stress factors that are highly involved in cell response towards AR. AR, as a directed growth response, can be supposed as a plant strategy to diminish stress exposure [5].

Mitochondria, as a physical platform for networks, signal perception, and signal canalization, play a central role in plant cell response to fluctuating cellular conditions, such as is often seen in environmental stresses, and in the further reacquisition of metabolic homeostasis [6–8]. The alternative respiratory pathway, localized in mitochondria, has been a relevant research topic regarding plant stress acclimation and adaptation in many reports focused on the involvement of the alternative oxidase (AOX; EC 1.10.3.11 ubiquinol:O_2 oxidoreductase id IPR002680) gene family members [6–11]. Clifton et al. [12,13] pointed to the importance of this alternative respiratory pathway as an early-sensing system for cell programming. The involvement of AOX in AR has been also taken as an important research topic, not only in view of understanding the role of the genes during the AR process, but also to further develop functional markers in order to be able to select genotypes that show efficient cell reprogramming [4,14,15].

AOX is a terminal quinol oxidase located on the matrix side of the inner mitochondrial membrane, and it works as the key enzyme in the alternative respiratory pathway. It has been described in a wide variety of species from different kingdoms, like plants, protists, fungi, and also in some animals [16]. However, AOX has been best studied in the plant kingdom, particularly in angiosperm plant species [17–19], where it is often encoded by a small multigene family composed of one to six genes distributed in two discrete gene subfamilies termed *AOX1* and *AOX2* [7,17,19–21]. The number of *AOX* genes and their distribution within the two subfamilies is species-specific [19]. Due to the high diversity in terms of gene duplication pattern [17], some *AOX* classification schemes for angiosperm plant species have been developed [17,22,23]. In dicot plant species, genes belonging to both subfamilies have been described. Only recently, the *AOX2*-subfamily was identified in species within the *Araceae* family [18], which is due to the availability of increasing information regarding monocot whole genome sequencing data.

AOX can play a number of roles in the optimization of the respiratory metabolism and in the integration of the respiratory metabolism with other metabolic pathways that impact the supply of or demand for carbon skeletons, reducing power and ATP [6,10,11]. This enzyme also modulates the levels of signalling molecules, thus supporting the crosstalk between the metabolic status of mitochondria and the nucleus that regulates gene expression [8]. For a long time, genes belonging to the *AOX1*-subfamily have been implicated in plant responses to a diversity of abiotic and biotic stresses (for reviews see [7,8,24]), while *AOX2*-subfamily members were described as housekeeping genes or more involved in plant development [13,25,26]. Nevertheless, the paradigm that *AOX1*-subfamily members are the only ones related to the stress response has been challenged [12,25–27].

In the context of AR in olive, the *AOX2*-subfamily gene member has been the main focus of AOX research. In addition to understanding the involvement of *AOX* gene members in the AR process, and in view of the development of further functional markers that are able to discriminate between genotypes with different potential to develop adventitious roots, gene sequence variability has been investigated [4,14,15]. *AOX* sequence variability, located in the protein coding and non-coding regions, has been reported in different plant species [4,28–32]. However, despite the studies carried out by direct mutagenesis to investigate the effect that a specific single nucleotide polymorphism (SNP) has on the protein functionality (see overview in [33]), there are few reports in natural systems showing the link between sequence polymorphisms and changes in the phenotype. Abe et al. [34] were the first research group to indicate the relevance of AOX polymorphisms in abiotic stress tolerance by identifying in the *Oryza sativa AOX1a* a SNP that mapped to a region of a QTL for low temperature tolerance in anthers at the booting stage. More recently, Hedayati et al. [15] reported the existence of two SNPs located at intron 3 of *OeAOX2* that correlate with differences in rooting ability.

In addition to the SNPs present within *AOX* gene sequences, other forms of sequence variability have been reported, whose differential processing can be influenced by physiological conditions, such as cell growth, differentiation, development, or pathological events [35]. Variability at the 3′-UTR sequence that encompasses sequence and length variability due to alternative polyadenylation (APA) events has been reported in *AOX* gene members from different plant species [4,36]. APA allows a single gene to encode multiple mRNA transcripts. Depending on the location of the alternative poly(a) signal (PAS), APA events may affect gene expression qualitatively by the production of different protein isoforms, or quantitatively if miRNAs binding sites and/or other regulatory elements are concerned that can act as negative regulators.

Considering that stress stimulus is a key factor for the AR process to be successful, it becomes relevant to investigate the involvement of the *AOX1*-subfamily members in the process, as well as to explore whether mechanisms related to the gene expression regulation might be involved in AR. Here, we characterize the gene members of the *AOX1* sub-family at the cDNA and gDNA levels, and we analyze the expression of its transcripts during AR in olive.

2. Results

2.1. Characterization of AOX1-Subfamily Members

In a first attempt to clarify the information about the composition of the *AOX1*-subfamily in olive, a blast search was carried out at the web page of the olive whole genome sequencing project that uses the *O. europaea* L. cv. "Farga" as target genome (Oe6 browser at http://denovo.cnag.cat/genomes/olive/). For that search, *AOX1a* sequence from *Arabidopsis thaliana* L. deposited at the NCBI GenBank (acc. no. AT3G22370) was used. Three sequences with high similarity were identified. A blast search made at the wgs NCBI database also identified those sequences, corresponding to the scaffolds Oe6_s00216c09 (acc. no. FKYM01004812.1), Oe6_s05781c34 (acc. no. FKYM01030627.1), and Oe6_s00133c14 (acc. no. FKYM01003481.1). Additionally, a blast search using the same *AtAOX1a* sequence was made at the NCBI databases nr/nt and transcriptome shotgun assembly (TSA), which allowed for the identification of complete *OeAOX* sequences from cv. "Leccino" (acc. no. GCJV01040584, KM514918, and KM514919), cv. "Picual" (acc. no. GBKW01105538), and cv. "Dolce Agogia" (acc. no. KM514920 and KM514921). Extracted sequences were used to construct a dendrogram using the in silico translated sequences together with sequences retrieved from 56 plant species at Phytozome and Plaza databases, which include monocot and eudicot plant species. To determine the relationship between the OeAOX from *O. europaea* and those retrieved sequences, a NJ tree was constructed using the translated sequences. There are clearly two different clusters composed by OeAOX1-subfamily members and a single cluster corresponding to the OeAOX2-subfamily members from all *O. europaea* cultivars (in yellow the three clusters that include OeAOX sequences) (Figure 1). Members from cv. "Galega vulgar" belonging to the AOX1 clusters were named as OeAOX1a (acc. no. MF410314) and OeAOX1d (acc. no. MF410315 and JX912721), the later one considering the high similarity with the AOX1d members from different plant species, and the AOX2 as OeAOX2 (acc. no. JX912722). A clear separation of AOX2-subfamily members can be seen (cluster in green). AOX1 members from monocot plant species form a separated group within the AOX1-subfamily (cluster in light blue). *AOX1*-subfamily members identified from other olive cultivars, not annotated as *AOX* gene members, appeared deposited as TSA in both cases transcribed upon cold stress.

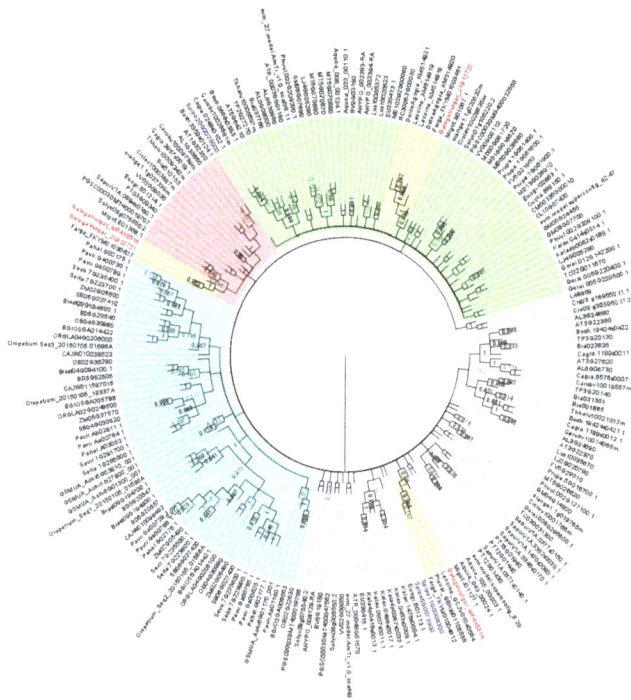

Figure 1. Neighbor-Joining (NJ) tree showing the relationships among deduced AOX sequences from 56 plant species, including monocot and eudicot plant species. Putative peptide sequences corresponding to the isolated *AOX1*-subfamily members of *Olea europaea* L. were included (shown in red). 206 AOX sequences from higher plants were included (correspondence of accession numbers and the plant species is included in supplementary Tables S3 and S4). The NJ tree was obtained using the complete peptide sequences. The alignments were bootstrapped with 1000 replicates by the NJ method using the MEGA 7 software. AOX sequence from *Neurospora crassa* and two sequences of *Chlamydomonas reinhardtii* were used as outgroups. The scale bar indicates the relative amount of change along branches. In green: the branch corresponding to the AOX2-subfamily members. AOX1d members are in red and AOX1 members from monocot plant species are in the branch colored in light blue. Clusters grouping olive AOX members are in yellow and accessions corresponding to AOX from cv. "Galega vulgar" are in red (*OeAOX1a*, acc. no. MF410314; *OeAOX1d*, acc. no. MF410315 and JX912721; OeAOX2, acc. no. JX912722).

Based on the available information [14], it was possible to successfully produce the 5′ and 3′ ends of both *AOX1*-subfamily members in cv. "Galega vulgar". RACE work developed for the isolation of 3′ ends of both *OeAOX1* gene members allowed the identification of high sequence variability due to different APA events. *OeAOX1a* transcripts presented the coding region unchanged but the 3′-UTR length variable, which ranged between 144 and 229 bp (Figure 2). Figures S1 and S2 show the complete cDNA sequences of *OeAOX1a*_transcript variant X1 with 1462 bp (deposited at the NCBI as acc. no. MF410314) and *OeAOX1a*_transcript variant X2 with 1249 bp (acc. no. MG208095). *OeAOX1a* sequences present an open reading frame (ORF) of 1086 bp, which encodes a putative polypeptide of 362 amino acid residues, that corresponds to a putative peptide with 40.6 kDa and a pI of 8.19.

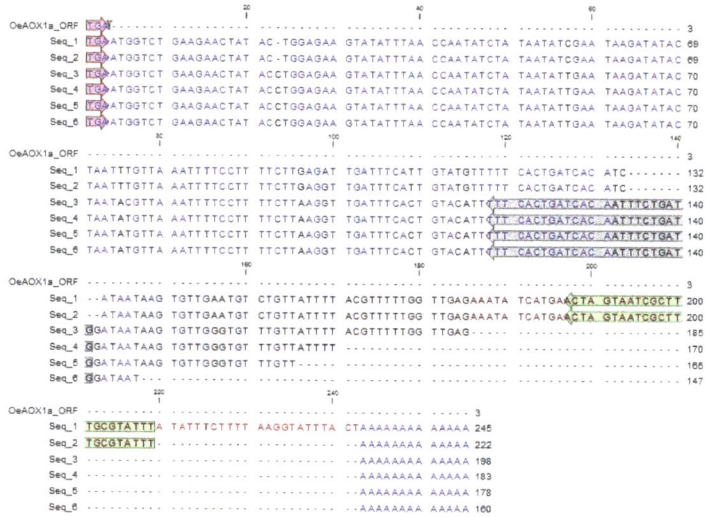

Figure 2. Alignment of the six different isolated sequences corresponding to the 3′-UTR of *OeAOX1a* gene. The sequences are presented starting at the stop codon TGA shown in red. The reverse primers used in RT-qPCR analysis for each transcript variant are shown in different colors (green: variant X1, acc. no. MF410314; grey: variant X2, acc. no. MG208095) (for primers sequence see Table S2).

Variability at the 3′ end of *OeAOX1d* was not due to 3′-UTR length size variability but to differences on the protein coding sequence, also due to an APA event. This event gave rise to two different transcripts named as variant X1 and variant X2 (deposited at the NCBI with the acc. no. MF410315 and JX912721, respectively) with 1277 bp (Figure S3) and 1597 bp (Figure S4), respectively. While *OeAOX1d*_transcript variant X1 results from the transcription of the four exons, typical of the general structure of plant *AOX* composed by four exons interrupted by three introns, transcript variant X2 includes the transcription of the N-terminal region of intron 3 with transcript cleavage located 125 bp downstream the 5′ conserved GT dinucleotide at splice site (described below). This sequence variability can lead to different putative peptide sequences. Transcript variant X1 is characterized by an ORF with 996 bp, which encodes a polypeptide with 332 aa and the variant X2 with an ORF with 1062 bp that encodes a polypeptide with 354 aa, which will give a putative peptide with 38 (pI of 6.7) and 40.7 kDa (pI 7.79), respectively.

Figures S1–S4 indicate the cDNA sequences for *AOX1* genes including the putative translated peptide and the conserved sites for intron positions. The difference in the overall length for the complete ORF sequences between *OeAOX1a* and transcript variant X1 of *OeAOX1d* (considering for this last one the transcript with the conserved gene structure) is due to the size variability at the N-terminal region of exon 1. The first exon has a size of 411 bp for *OeAOX1a* and 318 bp for *OeAOX1d*_transcript variant X1. Variation on both *OeAOX1d* complete ORF sequences is due to the exon 4, which shows 57 bp in transcript variant X1, leading to a peptide sequence homologous to the most common sequence across higher plants (see polypeptide alignment in Figure 3).

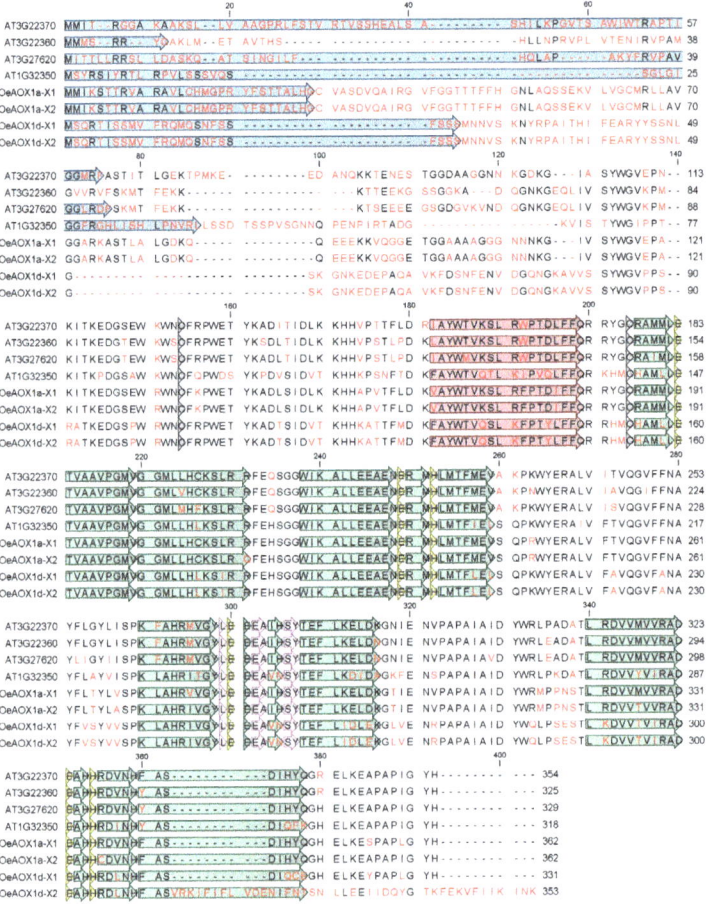

Figure 3. Multiple alignment of putative amino acid translated sequences of previously reported AOX proteins from *A. thaliana* (*AtAOX1a*_AT3G22370, *AtAOX1b*_AT3G22360, *AtAOX1c*_AT3G27620, *AtAOX1d*_AT1G32350) and AOX from *O. europaea* L. cv. "Galega vulgar" (*OeAOX1a*_transcript variant X1_MF410314 and *OeAOX1a*_transcript variant X2_MG208095, *OeAOX1d*_transcript variant X1_MF410315 and *OeAOX1d*_ transcript variant X2_JX91272). The alignment was performed using CLC Main Workbench 6.7.1 software. The data were retrieved from public web-based database Plaza v2.5, freely available at http://bioinformatics.psb.ugent.be/plaza/versions/plaza_v2_5/. Amino acid residues differing are shown in red, deletions are shown by minus signs. The putative mitochondrial transit peptides (mTP) are shown in blue boxes. The sites of two conserved cysteines (CysI and CysII) involved in dimerization of the AOX protein by S–S bond formation [37] are indicated in dark grey boxes. Helices α1 and α4, which form the hydrophobic region on the AOX molecular surface and are involved in membrane binding, are shown in red; helices α2, α3, α5, and α6, which form the four-helix bundle accommodating the diiron center, are shown in green [38]. Amino acids residues that coordinate the diiron center (E, glutamate and H, histidine) and those that interact with the inhibitor are in yellow and light pink boxes, respectively.

Forward and reverse gene specific primers located at the 5′ and 3′ gene ends, respectively, were used at the genomic level and allowed the isolation of both *OeAOX1* gene members: *OeAOX1a* with 2215 bp and *OeAOX1d* with 2054 bp length (from start to stop codon). To identify gene structure, genomic and transcript sequences were used at the Splign software. A four exons structure showing size conservation at the three last exons (exon 2: 129, exon 3: 489 and exon 4: 57 bp), interrupted by three size variable introns, was identified at the *OeAOX1a* and *OeAOX1d* (Figures S5 and S6). In both *OeAOX1* genes, introns were flanked by a GT sequence at the 5′ end and an AG at the 3′ end known as donor and acceptor splicing sites, respectively.

Comparing the isolated sequences from cv. "Galega vulgar" with the sequences available at the whole genome databases from cv. "Farga", high conservation was found at the protein encoding sequence. However, high variability was found at intonic regions (see Figure S7). In silico analysis also revealed that *OeAOX1d*_transcript variant X2 cannot be transcribed on cv."Farga" due to the existence of a stop codon previous to the polyadenylation site (also known as the poly(A) site—PAS) [39].

A search for the identification of putative sequences coding for miRNAs located at the intronic regions revealed their absence in both *OeAOX1a* and *OeAOX1d*. Also, no other regulatory elements related to transposable elements and repetitive sequences were identified within gene sequences of cv. "Galega vulgar". In *OeAOX1d* genomic sequence of cv. "Farga", a putative transposable element (data not shown) was identified. The availability of upstream and downstream sequences of both *OeAOX1* genes in the cv. "Farga" allowed us to search for transposable elements located in the vicinity of both genes and to perform the analysis of promoter region to scan for *cis*-elements regulated by auxins. The in silico analysis allowed the identification of several copies of copia and gypsy LTR retrotransposons. Details of the identified full length LTR insertions are presented in Table 1. Directly upstream to *OeAOX1a*, there were two gypsy elements; a copy of 84856_A was nested within 95401_A (Figure S8). The latter element was inserted ca. 1 Kb upstream from the start codon of *OeAOX1a*. Interestingly, another copy of 95401_A was found directly upstream from the *OeAOX1d* gene (Figure S9), albeit at a larger distance (ca. 3.5 Kb) to the start codon and in the opposite orientation as related to the copy associated with *OeAOX1a*. Repetitive sequences were present directly downstream *OeAOX1d*; however, owing to the lack of contiguous assembly of that region, a detailed analysis was not performed.

Table 1. Information regarding the full length elements identified in the upstream and downstream region of *OeAOX1* genes in cv. "Farga". For visualization of elements position within genomic sequence see Figures S7 and S8.

Element	Length (bp)	LTR Length	TSD	Position Relative to Start Codon
OeAOX1a				
isolate 84856_A retrotransposon gypsy-type [KM577525]	13,288	1791 (left) 1752 (right)	GAAAG	−18,801/−5513
isolate 95401_B retrotransposon gypsy-type [KM577546]	12,998	782 (left) 773 (right)	GTCAT	−27,411/−1125
OeAOX1d				
isolate 95401_B retrotransposon gypsy-type [KM577546]	12,948	767 (left) 773 (right)	CAATT	−16,375/−3427
isolate 70744_E retrotransposon copia-type [KM577454]	6023	750 (left) 750 (right)	TTATC	undetermined (downstream)
unknown, copia-type, similar to Copia-63_VV-I	4959	275 (left) 283 (right)	[A/G]TAGC	undetermined (downstream)

OeAOX1a and OeAOX1d promoter sequences up to 1.5 kbp upstream from the translation start site were scanned using PlantCARE and New Place software's for the identification of auxin *cis*-acting regulatory elements (CAREs). Four CAREs were identified in the OeAOX1a's promoter region and 3 CAREs were identified in the OeAOX1d (see details in Table S5). From the four CAREs identified in OeAOX1a, three were located at a region prior to −500 bp upstream of the translation start site (New PLACE IDs: NTBBF1ARROLB, ARFAT and SURECOREATSULTR11). In the case of OeAOX1d, from the three motifs identified, only one is located closer to the start codon, prior to −500 bp upstream of the translation start site (New PLACE ID: ASF1MOTIFCAM, with sequence TGACG, at position −164). Comparing CAREs of the promoter regions for both OeAOX1 genes, only one is common, the New PLACE ID: NTBBF1ARROLB, with sequence ACTTTA. However, in terms of its location, the sequence ACTTTA is not conserved between the two genes. It is located at −266 upstream from the translation start site for OeAOX1a, while for OeAOX1d, it is located at −668 and repeated at +1109. In addition, it is interesting to note that the sequence of auxin response factor binding site found in promoters of primary/early auxin responsive genes (New PLACE ID: ARFAT, with sequence TGTCTC) was only identified in OeAOX1a (at positions +305 and −306).

A multiple sequence alignment including AOX1-translated peptides from *A. thaliana* and *O. europaea* cv. "Galega vulgar" was used to highlight similarities and differences in the putative protein sequences (Figure 3). OeAOX1a and OeAOX1d encoded by both transcript variants (X1 and X2) revealed structural features usually found in most of the higher plants' AOX with the identification of two conserved cysteines (CystI and CystII) and di-iron-binding sites. Sequence diversity at the C-terminal region is here restricted to putative peptide of OeAOX1d_transcript variant X2. This change is implicated in the sequence of helice $\alpha 6$, one of the four-helix bundles accommodating the diiron center.

In order to gain insight into the possible protein structural effects of the sequence change present in OeAOX1d_transcript variant X2, we turned to the structure of the AOX from *Trypanosoma brucei* (PDB ID: 3VV9) [38], which is the only homologous protein whose structure is available. Unfortunately, the N- and C-terminal regions of OeAOX1 do not align well with the sequence of AOX from *T. brucei* and, therefore, we could only model the region between helices $\alpha 2$ and $\alpha 6$ (Figure 4A). However, based on the alignment of OeAOX1d and AOX from *T. brucei*, we can infer that the sequence change of OeAOX1d_transcript variant X2 is located between the end of helix $\alpha 6$ (one of the four-helix bundle accommodating the diiron center) and the C-terminus (Figure 4B). The region that is affected by this change is close to the diiron center and is also implicated in inter-subunit interactions in the dimeric form of AOX.

High sequence diversity was detected at the N-terminal region, which consequently lead to high diversity on the mitochondrial transit peptide (see Figure 3). Both putative OeAOX1 translated peptides were predicted to be localized in mitochondria (mTP score of 0.672 and 0.604 regarding the OeAOX1a and OeAOX1d, respectively). The predicted length of the cleavage site of the mitochondrial targeting sequence for OeAOX1a and OeAOX1d is 28 and 45 amino acid residues, respectively. The predicted mitochondrial transit peptide is shown in the alignment of Figure 3, in which no conservation across protein sequences is visible, not within the AOX1-subfamily members across species and not even within the AOX1-subfamily genes from a single plant species.

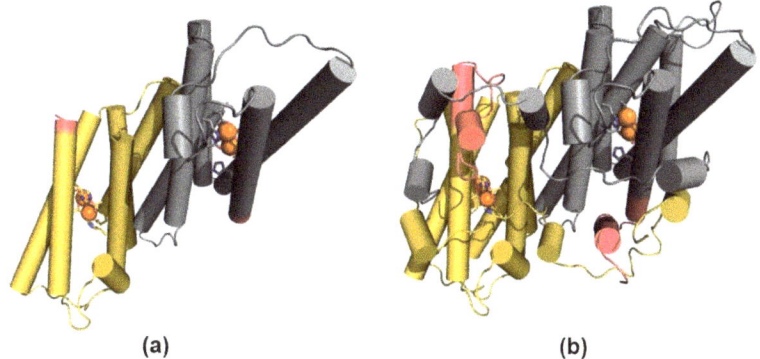

Figure 4. Structural mapping of the sequence diversity of OeAOX1d. The homology-based model of (**a**) OeAOX1d_transcript variant X2 and the structure of (**b**) AOX from *T. brucei* are displayed using a cartoon representation, with the helices shown as cylinders. The two identical functional subunits are colored in yellow (subunit A) and grey (subunit B), and the iron atoms that form the diiron center are represented by orange spheres, with the coordinating residues displayed using sticks. The region that corresponds to the sequence change in OeAOX1d_transcript variant X2 is highlighted in pink.

2.2. Analysis of Transcript Expression

2.2.1. *OeAOX1* Genes are Differentially Expressed during IBA-induced AR

In order to verify whether the expression levels of *OeAOX1a* and *OeAOX1d* genes were changed during IBA-induced rooting, quantitative real time PCR was performed. Both *OeAOX1a* and *OeAOX1d* genes showed a similar expression pattern throughout the rooting assay (Figure 5A,B). The maximum peak of up-regulation for both genes occurred at 8 h after IBA treatment. *OeAOX1a* showed, however, higher levels of expression at this time point (36.0-fold change, $p \leq 0.01$) (Figure 5A) than *OeAOX1d* (13.2-fold change, $p \leq 0.001$) (Figure 5B) when compared to the levels observed at the corresponding controls (0 h, without IBA treatment). Looking at days 1 and 2, which can be seen as the recovery time point after the maximum peak of expression, it can be observed that both genes decreased drastically reaching even, at day 2, expression values close to the ones observed at 0 h. A second increment of expression for both genes, although lower than the first increment, was observed at day 4. Here, and again, the expression levels at this time point, and compared with the ones observed at 0 h, were higher for *OeAOX1a* (3.2-fold change, $p \leq 0.001$) than for *OeAOX1d* (1.8-fold change, not statistically significant). Day 4 corresponds to the end of the induction phase and beginning of the initiation phase in AR process in olive [40]. From this time point forward, the expression levels decreased again reaching the minimum peak at day 8, for both genes. A third increment occurs for both genes around days 10–14. Around days 12–14 corresponds to the time when calli formation becomes apparent before root emergence. From these time points on and until the end of the rooting trial (30 days), and in opposition to what was previously observed, the expression levels for *OAOX1a* were lower than the ones observed for *OeAX1d*. For example, at day 22, which corresponds to the end of the initiation phase and the beginning of the expression phase of the AR process [40], *OeAOX1a* had a slight 1.8-fold ($p \leq 0.01$) increase, whereas *OeAOX1d* had a higher increment of 3.7-fold ($p \leq 0.001$) when compared to the expression levels of the corresponding controls.

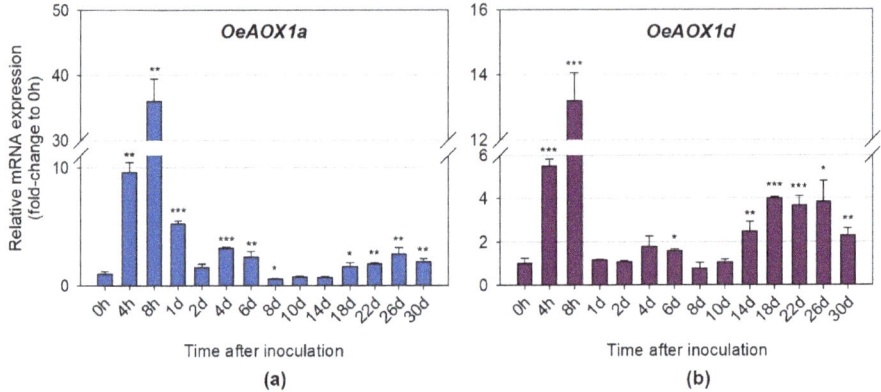

Figure 5. Relative mRNA expression of (**a**) *OeAOX1a* and (**b**) *OeAOX1d* in stem basal segments of *O. europaea* L. microcuttings during IBA-induced adventitious rooting. *OeACT* and *OeEF1a* were used as reference genes in data normalization. The relative expression values are depicted as the mean ± standard deviation of four biological replicates for each time point. The bars represent the fold-change related to the time point 0 hours after microcuttings treatment and inoculation, which was set to 1. Statistical significances (* $p \leq 0.05$, ** $p \leq 0.01$, and *** $p \leq 0.001$) between the two means were determined by the *t*-test using IBM® SPSS® Statistics version 22.0 (SPSS Inc., Armonk, NY, USA), h: hours, d: days.

2.2.2. Distinct Transcripts Variants Show Different Expression Levels

Quantitative real time PCR was also performed to further investigate whether the distinct transcript variants (primers were designed for a specific region of each transcript variant) for each gene (*OeAOX1a* and *OeAOX1d*), produced due to APA events, were differentially expressed during IBA-induced AR. *OeAOX1a* transcripts with longer 3′-UTRs (*OeAOX1a*_transcript variant X1) (Figure 6A) showed higher expression levels at the time points corresponding to the first and second increments than the transcripts with shorter 3′-UTRs (*OeAOX1a*_transcript variant X2) (Figure 6B). For example, the expression levels at 8 h in relation to the control (time point 0 h) for *OeAOX1a*_transcript variant X1 were about 48.7–fold higher ($p \leq 0.05$), while for *OeAOX1a*_transcript variant X2 was around 33.5-fold higher ($p \leq 0.01$). At day 4, variant X1 showed a 4.1-fold change ($p \leq 0.05$) and variant X2 a 2.9-fold change ($p \leq 0.05$) compared with the time point 0 h. A very similar expression pattern throughout the rooting assay was observed for both transcript variants. On the contrary, the shorter *OeAOX1d* transcripts composed by the four exons (*OeAOX1d*_transcript variant X1) (Figure 6C) showed higher expression levels at these time points (8 h: 16.1-fold change, $p \leq 0.001$; 4 days: 1.8-fold change, $p \leq 0.001$) than the longer transcripts with an alternative PAS located at the intron 3 (*OeAOX1d*_transcript variant X2) (Figure 6D) and lacking the exon 4 sequence (8 h: 6.9-fold change, $p \leq 0.001$; 4 days: 1.3-fold change, not statistically significant). As for *OeAOX1a*, a similar expression profile over all the time points tested was observed for both variants. Additionally, the expression profile of transcript variants was very similar to the expression profile exhibited by the *OeAOX1a* and *OeAOX1d* genes (including all sets of transcripts, since primers were designed in a common region). From day 14 onwards, which corresponds to the time point when roots start to emerge, both *OeAOX1a* transcript variants (with shorter (variant X2) and longer (variant X1) 3′-UTRs) showed a similar level of expression between them when compared to the levels of the corresponding controls. On the other hand, *OeAOX1d* transcripts corresponding to the variant X1 were more expressed than the variant X2. While the expression levels of *OeAOX1a* gene correlate to the ones shown by each transcript variant, in the case of *OeAOX1d* gene this does not happen. The expression levels of *OeAOX1d* gene were higher (almost double) than the expression levels of the most expressed transcript variant (variant X1). This result suggests that *OeAOX1d* gene may have other transcript variants that were not analysed here separately, despite the fact that they have been detected by the primers designed to a common region among the transcript variants.

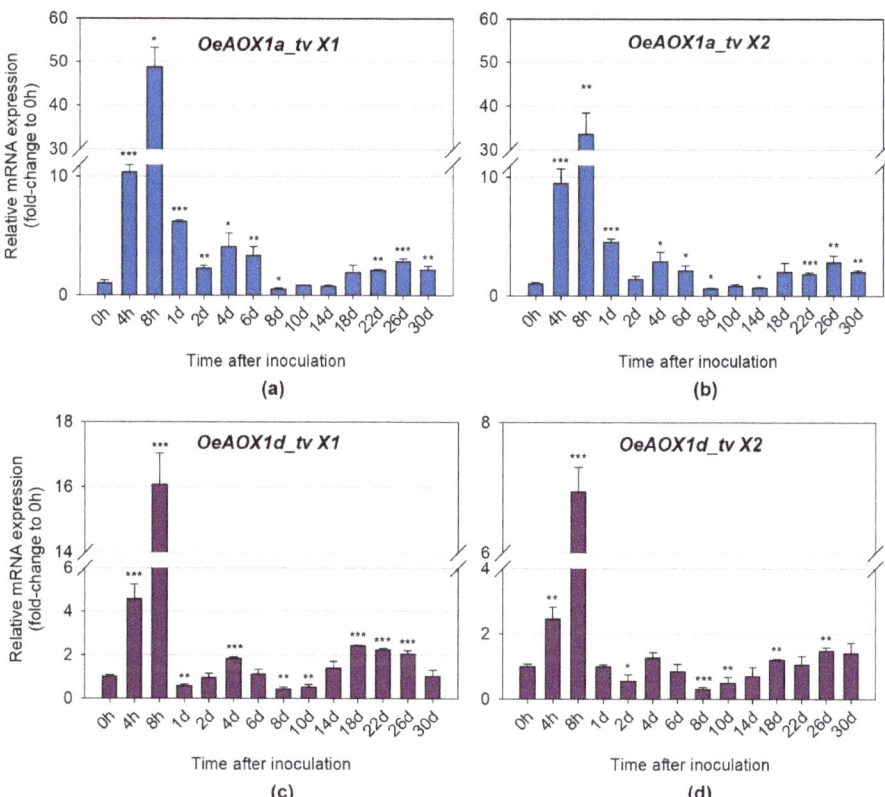

Figure 6. Relative mRNA expression of (**a**) *OeAOX1a*_transcript variant X1, (**b**) *OeAOX1a*_transcript variant X2, (**c**) *OeAOX1d*_transcript variant X1, and (**d**) *OeAOX1d*_transcript variant X2 (D) in stem basal segments of *O. europaea* L. microcuttings during IBA-induced adventitious rooting. *OeACT* and *OeEF1a* were used as reference genes in data normalization. The relative expression values are depicted as the mean + standard deviation of four biological replicates for each time point. The bars represent the fold-change related to the time point 0 hours after stem microcuttings treatment and inoculation, which was set to 1. Statistical significances (* $p \leq 0.05$, ** $p \leq 0.01$ and *** $p \leq 0.001$) between the two means were determined by the *t*-test using IBM® SPSS® Statistics version 22.0 (SPSS Inc., Armonk, NY, USA), h: hours, d: days, tv: transcript variant.

3. Discussion

Correct classification of homologous *AOX* genes across plant species is challenging [32]. However, it gains high importance when the physiological role of those genes should be comparable across species. Thus, classification is a dynamic process that needs regular updates to develop knowledge [23]. Based on the most recent classification system available for *AOX*, basal angiosperms and eudicots contain both *AOX*-subfamilies (*AOX1* and *AOX2*) subdivided into specific types (*AOX1a-c/e*, *AOX1d*, *AOX2a-c*, and *AOX2d*) [17]. Here, 206 sequences from public data bases were analysed. The distribution of the putative translated peptide encoded by the isolated *OeAOX1* sequences in two different clusters within the main cluster of AOX1-subfamily revealed the existence of two *AOX1*-subfamily members in olive. One of those sequences clustered together with the *AOX1d* member of *A. thaliana* (AT1G32350) and the sequence of *Solanum lycopersium* available at the NCBI as *AOX1b* (NP_001234120.1) but renamed by Costa et al. [17] as *AOX1d*. Considering this homology, the olive member was named as *OeAOX1d* and submitted to the NCBI databases at cDNA and gDNA levels (acc. no. MF410315, JX912721, and MF410319). Putative AOX2 translated peptides from different *O. europaea* cultivars clustered together within the AOX2-subfamily, confirming a single *AOX2* member in this species.

Across kingdoms, there is a lack of a general pattern with respect to intron/exon structure in *AOX* genes [16]. However, within plants, the most common gene structure described for *AOX* comprises four exons interrupted by three introns [41,42]. Exceptions to this gene structure have been reported in some *AOX* gene members and different plant species, due to events of intron loss and gain [19]. From the known examples, an intron loss leads to a structure of three exons, and an intron gain to a structure of five-exons. Despite the typical structure of four exons, *AOX* gene members are well known by exons size conservation achieved at the three last exons (129, 489 and 57 bp, respectively). The combination between gene homology, gene structure, and exons size allows us to be more confident about the classification of a gene as a member of the *AOX* family. Cases of miss-annotation regarding *AOX* gene members and genes from another membrane-bound di-iron carboxylate protein, the plastid terminal oxidase (PTOX; EC 1.10.3.11 ubiquinol:O_2 oxidoreductase id IPR002680), are still common, since *AOX* and *PTOX* share high level of homology [32]. In general, the exon size conservation in the 4-exons structure of *AOX* gene members is one of the factors that contributes to the low variability in protein size. Normally, exon 1 is the one that is variable not only in length but in sequence composition as well. Protein size typically ranges between 32–41 kDa depending on the species [22,43,44]. Putative translated OeAOX1 peptides from cv. "Galega vulgar" showed, by in silico analysis, a size that is in range from 38–40.7 kDa. Both OeAOX1a and OeAOX1d were predicted, with high score, to be located in the mitochondria. However, high variability in the N-terminal region was observed within AOX1 members. Sequence variability located in that gene region was previously reported across AOX members within and between species [31,41,42]. Nevertheless, how this variability can affect the regulation of gene expression and/or the protein transport and activity is still not known. The N-terminal region determines interaction with the protein transport system that regulates integration into the organelle. In many cases, amino acids comprising the signal peptide are cleaved off the protein once they reach their final destination. A comparison between *A. thaliana* and *O. sativa* using a high set of proteins showed high variability at that region, going from 19 to 109 amino acids in *A. thaliana*, and from 18 to 117 amino acids in *O. sativa* [45]. Specifically on AOX, Campos et al. [41] described mTP sequence length variability across plant species and between protein isoforms within the same plant species. More recently, a study that aimed the identification of allelic variation within the *AOX1* gene member considering 39 carrot genotypes described high variability at that region, going from a mTP sequence with 20 to 41 amino acid residues [31].

Contrarily to the conservation at AOX protein coding sequences, high variability can be seen in protein non-coding regions, which include introns and untranslated regions (5'-UTR and 3'-UTR) [4,15,28–32,42]. Size and sequence variability located at these regions can have an important physiological role. Gene architecture, which considers not only the number but also the length of exons and introns, is nowadays considered as one important regulatory player. Several studies indicate that gene architecture toward short genes with few introns allows for efficient expression during short cell cycles. In contrast, genes composed by long introns can be expected to exhibit delayed expression [46]. Long introns are described as a timing mechanism that works for biological signal feedback regulatory networks [47]. Despite this role in regulation of gene expression, which is associated with gene expression delays, it is also known that the presence of introns could enhance gene transcription [48]. Some introns harbour non-coding RNAs (e.g., miRNAs and snoRNAs) for which the processing from introns can speed up or slow down the rate of expression of the host gene [49]. Despite the fact that no miRNAs were identified within the *OeAOX1* sequences, an in silico analysis, performed to search for TEs, revealed the existence of several putative retrotransposons located in the adjacent regions upstream and downstream of the AOX gene position. TEs located in intergenic regions (up- or downstream target genes) or within a gene sequence (promoter or intron sequences) may provide regulatory elements affecting gene expression through a variety of mechanisms (for review see [50]). In plants, TEs can constitute from ca. 10% of *A. thaliana* genome (Arabidopsis Genome Initiative 2000) to 85% of the B73 *Zea mays* genome [51]. Several reports point out the existence of TEs within *AOX* gene sequences [52–54]. These results lead us to hypothesize that expression of *AOX* genes might be influenced by adjacent TEs.

Simultaneously, introns offer the potential for regulatory functions such as alternative splicing and APA events. It is nowadays evident that APA acts as a major mechanism of gene regulation being widespread across all eukaryotic species [39]. In plants, it was reported that 70% of *A. thaliana* genes and around 50% of *O. sativa* genes have at least one alternative poly(A) site [55]. In some specific cases, variability in transcripts is associated with regulation of flowering time, growth, and developmental processes [56,57]. As a result of APA events, a single pre-mRNA can produce more than one mRNA. If alternative PAS is located in internal introns/exons, APA events may lead to the production of different protein isoforms with differences in subcellular localization, stability, or function by changing or completely removing functional domains (for review see [39]). It may also result in unstable mRNA isoforms with a negative feedback on gene expression by generation of truncated transcripts that are recognized and degraded by a specific mechanism named nonsense-mediated decay (NMD) [58]. From *OeAOX1d*, two putative protein isoforms could be produced due to an alternative PAS located at intron 3: (a) *OeAOX1d*_transcript variant X1 with four exons that encodes the structural feature typical of AOX in plants, and (b) *OeAOX1d*_transcript variant X2 with three exons and a partial sequence of intron 2 that replaced exon 4. This latter transcript encodes a putative truncated protein that lacks 19 amino acids at the C-terminal end. When this transcript variant was analyzed during IBA-induced rooting assay, it demonstrated lower differential expression levels (compared to the control) than variant X1. The sequence alteration present in the *OeAOX1d*_transcript variant X2 will affect the structure of helix α6, which can have an impact on the coordination of the diiron center. This can be predicted by referring to the structure of AOX from *T. brucei* [38]. In AOX from *T. brucei*, the C-terminal region is involved in important inter-subunit interactions. Therefore, it is possible that the sequence alteration present in the *OeAOX1d*_transcript variant X2 will also affect interaction between the two polypeptide chains. This can have implications on the stability of the dimer. However, further studies will be required to investigate whether both transcripts would be translated to two different protein isoforms, and if so, whether both are functional. Additionally, if PAS are located in the 3'-UTRs, APA events will lead to the synthesis of transcripts conserving unchanged the protein coding sequence but presenting different 3'-UTR lengths. 3'-UTR length can affect the transcript stability, localization, transport, and translational properties [35]. 3'-UTRs often harbor miRNAs binding sites and/or other regulatory elements [59] that can act as negative regulators, mostly of larger transcripts. Many mRNAs use 3'-UTR alternative PAS to achieve tissue-specific expression and function [60,61]. We found in our study alternative PAS located at the 3'-UTR of the *OeAOX1a*, which generates short and long 3'-UTRs. The identification of *OeAOX1a* sequences carrying 3'-UTR regions with different sizes may suggest the possibility of differential post-transcriptional regulation. It should be noted that in our system (IBA-induced AR in olive explants), although both *OeAOX1a* transcript variants X1 and X2 (carrying 3'-UTR with different sizes) showed a similar expression pattern throughout the process, indicating co-regulation, *OeAOX1a*_transcript variant X1 showed more pronounced differential expression levels (compared to the control) than variant X2, up to day 8. Differential 3'-UTR sizes of a gene can have positive, negative, or even neutral effects on mRNA stability and on the resulting protein levels depending on whether the availability of RNA-binding sites, such as miRNA-binding sites, is influenced [62–65]. Many different RNA-binding proteins (RBPs) and a variety of signals located at that transcript region can regulate mRNA localization, decay, and translation [65].

The expression of both genes, *OeAOX1a* and *OeAOX1d*, was dramatically increased in the early stages of rooting with the maximum peak of transcript accumulation at 8 h after IBA treatment. Three previous studies addressed the involvement of *AOX* genes in the process of IBA-induced AR [4,14,15]. They were based on the earlier raised hypothesis, which proposed *AOX* as a functional marker candidate for efficient adventitious rooting of *O. europaea* L. [66,67]. Santos Macedo et al. [4,14] showed first in semi-hardwood shoot cuttings of an easy-to-root olive cultivar (cv. "Cobrançosa") that root induction was significantly reduced by treatment with an inhibitor of AOX activity (salicylhydroxamic acid—SHAM). This observation could be confirmed in an in vitro system for olive propagation (cv. "Galega vulgar"), thus pointing to the general importance of *AOX* genes during the process of induced rooting [14]. In the latter work, it could be shown that SHAM-inhibition was in fact specific to rooting and did not interfere with preceding callus formation. This observation was later confirmed by Porfirio et al. [68] using the same experimental system. In a first approach, *OeAOX2* was identified as a promising gene candidate for functional marker development towards improving rooting efficiency in olive [4]. First evidence of an association between rooting ability and *OeAOX2* gene expression in olive cuttings was then provided by Hedayati et al. [15]. This group also confirmed the presence of polymorphisms in *OeAOX2* with a possible correlation to distinct rooting behavior. The present work adds new information to the rooting system of olive by showing the expression of the *OeAOX1a* and *OeAOX1d* genes by quantitative real time PCR during IBA-induced AR in microcuttings.

Adventitious rooting is considered a developmental process organized in a sequence of interdependent stages [69–74]. It includes three phases: (1) induction, corresponding to the period preceding any visible histological occurrence, with molecular and biochemical events; (2) initiation, starting with the first histological events, like root primordia organization; here, small cells with large nuclei and dense cytoplasm start to be apparent; and (3) expression, involving the development of the typical dome shape structures, intra-stem growth, and emergence of root primordial [75–78]. In olive, induction phase corresponds to the first 4 days after microcuttings treatment and inoculation, when cells regain meristematic features. From 4 until 14 days, the first meristemoids and morphogenetic root zones were observed, events corresponding to the initiation phase. These events are followed by high mitotic activity that eventually leads to the expression phase, which starts at 22 days after the root-inducing treatment [40]. From our results, it can be seen that both *OeAOX1a* and *OeAOX1d* genes exhibited three increments in their expressions throughout the rooting assay. The first one, as mentioned above, was the most pronounced, and occurred at 8 h after microcuttings treatment and inoculation, and corresponded to the beginning of the induction phase. The second increment was observed at 4 days, which corresponds to the end of the induction phase and beginning of the initiation phase. The third increment was observed from the end (14 days for *OeAOX1d* and 18 days for *OeAOX1a*) of the initiation phase onwards. It is likely that the first observed increment may be related to the stress associated with the cut injury and auxin treatment. In fact, as suggested by Santos Macedo et al. [4], the initial cut of olive microcuttings and its subsequent treatment with auxins may constitute a stress to the involved cells, and therefore olive AR can be seen as stress-induced reprogramming of shoot cells [4,66,67]. The second and third increments may be related to the role that AOX might have on cell differentiation and growth/development. The link between AOX and cell differentiation and plant growth has been indicated by several reports [7,67,79,80].

Interestingly, Porfirio et al. [68], using the same experimental system, observed elevated free IAA (indole-3-acetic acid) levels in the first hours after treatment, peaking also at 8 h. It would be worth investigating further this correlation between the levels of free IAA observed by Porfirio et al. [68] and the accumulation of *AOX1* transcripts found in the present study during the induction phase of olive rooting. It is likely that a link may exist between altered auxin homeostasis and induced *OeAOX* gene expression, as suggested by others [81,82], during AR in olive. In this context, it would be desirable to investigate also genes involved in the auxin signaling and transport.

It is tempting to speculate that, in our study, *AOX* genes were highly induced, probably by increased levels of ROS as a consequence of a stressful situation (cut injury plus auxin treatment) and by IBA application to promote rooting. Indeed, it has been reported that different abiotic stresses are likely to cause the formation of different ROS signatures in plant cells (for review see [83]). Moreover, the production of ROS by mitochondria was suggested to be a critical factor for the induction of AOX [13,84], which in turn regulates the amounts of ROS, and therefore AOX has a large impact on redox regulation at a cellular level on environmental stresses [7]. Auxins can induce the production of ROS [85] and regulate ROS homeostasis [86], hinting at the relationship between auxin signaling and oxidative stress [85]. Thus, the marked accumulation of *OeAOX* transcripts at 8 h observed in our study may also be the result of the increased levels of free IAA detected by Porfírio et al. [68], which possibly contributed to elevate the levels of ROS. Moreover, in silico analysis at the *OeAOX1* promoter region in search of cis-acting regulation elements identified different auxin responsive elements (AuxREs) that could be involved in regulation of *AOX* gene expression. The identification of different AuxREs in the promoter region of both *OeAOX1* genes allows us to speculate that regulation might be done in a differential way. Presence of the TGTCTC-motif in *OeAOX1a*, which belongs to the family of Auxin Response Factors (ARFs) (see review in [87]), could explain the higher increase in gene expression in comparison to *OeAOX1d* that lacks this motif.

In summary, two genes were identified as members of the *AOX1*-subfamily in the cv."Galega vulgar", with both showing the most common structure of *AOX* gene members with four exons interrupted by three introns. Alternative polyadenylation (APA) events were responsible for the production of transcript variants in both genes. *OeAOX1a* transcript variants show different 3'-UTR lengths with no changes at the protein coding sequence. *OeAOX1d* shows an alternative PAS located at the intron 3 that leads to the synthesis of one transcript variant showing a truncated protein coding sequence that lacks the exon 4 sequence. The sequence alteration found in the *OeAOX1d*_transcript variant X2 will prevent the structure of helix α6 having an impact on the coordination of the diiron center and also it can affect the interaction between the two polypeptide chains having implications in the stability of the dimer. Our findings showed a strong correlation between *OeAOX1a* and *OeAOX1d* transcripts accumulation and the three distinct phases (induction, initiation, and expression) of the AR process in olive, with the expression of these genes more pronounced at the induction phase. The elevated IBA-induced expressions at this phase may be related to the stressful conditions associated with AR process and the application of IBA for rooting induction. A possible link between OeAOX1 induction and altered auxin homeostasis in olive AR may exist, since *OeAOX1* transcripts were increased at the same time point for which earlier studies showed elevated levels of free auxins. Additionally, different transcript variants for each gene studied here, although showing a similar expression pattern, demonstrated different levels of expression during AR, which would be worth exploring further. Further studies will be required to clarify whether the diverse transcripts encountered may give rise to distinct functional protein isoforms and also to understand the physiological role of such variability. Taken together, these results contribute to a better understanding of the molecular mechanisms underlying the stress stimulus needed for the induction of adventitious roots. Thus, the results may allow us to develop more targeted and effective rooting induction protocols, which in turn can help to increase the rooting ability of difficult-to-root cultivars.

4. Materials and Methods

4.1. Characterization of the AOX Genes at the cDNA and Genomic Levels

4.1.1. Plant Material

Olea europaea L. explants of cv. "Galega vulgar" (clone 1441), established under in vitro conditions since 2005, were maintained until today following the procedure described by Peixe et al. [88]. The derived in vitro grown plantlets were used for gene isolation. Leaves were collected from a single plantlet and used for total RNA and genomic DNA (gDNA) extractions.

4.1.2. Isolation of Complete *OeAOX1*-Subfamily Gene Members

The isolation of complete gene sequences was performed in several steps. The first one was based on the protocol described by Saisho et al. [21] for isolation of the *AOX* gene members in *A. thaliana* and further referred to by different authors for gene isolation in different plant species [41,42,79]. The isolation of mainly two different sequences belonging to the *AOX1*-subfamily was previously reported by Santos Macedo et al. [14]. Based on that information, gene specific primers were designed in order to isolate gene ends of the identified *OeAOX* gene fragments by 5′ and 3′ RACE-PCRs. Total RNA used for cDNA synthesis was previously extracted using the RNeasy Plant Mini Kit (Qiagen, Hilden, Germany) with on-column digestion of DNA applying the RNase-Free DNase Set (Qiagen, Hilden, Germany), according to the manufacturer's protocol. The concentration of total RNA and its purity was determined with the NanoDrop-2000C spectrophotometer (Thermo Scientific, Wilmington, DE, USA). For both ends, 1 µg of total RNA was used to synthesize the first-strand cDNA using the SMARTerTM RACE cDNA Amplification kit (Clontech Laboratories, Inc., Mountain View, CA, USA) according to the manufacturer's instructions. RACE-PCRs were carried out separately using 1 µL of the corresponding single strand cDNA as template and 0.2 µM of the forward/reverse gene specific primers (depending if 3′ or 5′ end isolation) (Table S1) combined with 0.2 µM of the Universal Primer Mix (provided with the kit) following the instructions recommended by the manufacturer. PCRs were all carried out in a 2770 thermocycler (Applied Biosystems, Foster City, CA, USA).

For complete gene isolation, gDNA was isolated using the DNeasy Plant Mini Kit (Qiagen, Hilden, Germany) according to the manufacturer's protocol. The amount of gDNA and its purity was determined with the NanoDrop-2000C spectrophotometer (Thermo Scientific, Wilmington, DE, USA). One gene-specific primer set was designed for each *OeAOX* gene (Table S1) based on the previously isolated 5′ and 3′-UTR sequences. Ten ng of gDNA were used as template with 0.2 µM of each specific primers. PCRs were performed using the Phusion™ High-Fidelity DNA Polymerase (Finnzymes, Espoo, Finland) according to the manufacturer's protocol. PCR was carried out in a 2770 thermocycler (Applied Biosystems, Foster City, CA, USA) running for 35 cycles each one consisting in 10 s at 98 °C, 30 s at 55 °C, and 2 min at 72 °C. An initial step at 98 °C for 30 s and a final step at 72 °C for 10 min were used.

4.1.3. Cloning and in Silico Sequence Analysis

PCR fragments were separately cloned into a pGem®-T Easy vector (Promega, Madison, WI, USA) and used to transform *E. coli* JM109 (Promega, Madison, WI, USA) competent cells. Plasmid DNA was further extracted from putative recombinant clones by using the GeneJET Plasmid Miniprep kit (Thermo Scientific, Vilnius, Lithuania) and further sequenced in sense and antisense strands (Macrogen company, Seoul, South Korea: www.macrogen.com).

CLC Main Workbench 7.5.1 software (ClCbio, Aarhus N, Denmark) was used to edit sequences. Intron location was made using the software Splign (https://www.ncbi.nlm.nih.gov/sutils/splign/splign.cgi?textpage=online&level=form).

In order to clarify the question related to the number of genes that compose the *AOX1*-subfamily, a blast search using the *AOX1* from *A. thaliana* L. (acc. no.AT3G22370) deposited at the NCBI—National Center for Biotechnology Information (GenBank) was made at the web page of olive genome databases (http://denovo.cnag.cat/genomes/olive/) using the Oe6 browser. To get confirmation, the retrieved sequences were then blasted at the NCBI data bases using the BLAST algorithm [89] (http://www.ncbi.nlm.nih.gov/) (BLASTn) at the whole-genome shot gun contigs (wgs). To identify *AOX* sequences from other olive cultivars, a BLASTn search using the same sequence was made at different NCBI databases (nucleotide collection, nr/nt; transcriptome shotgun assembly, TSA; expressed sequence tags, est).

To perform a comparison between AOX proteins from higher plants, sequences were retrieved from the whole genomes available at the Plaza (http://bioinformatics.psb.ugent.be/plaza/) and the

Phytozome (https://phytozome.jgi.doe.gov/pz/portal.html) databases using a BLAST search analysis based on the exon 3 as the most conservative region across *AOX* genes and plant species.

Sequences retrieved were aligned in MUSCLE (http://www.ebi.ac.uk/Tools/msa/muscle/) following the standard parameters. Phylogenetic reconstruction was performed in MEGA 7 software [90] by Neighbor-Joining (NJ) and the inferred tree was tested by bootstrap analysis using 1000 replicates, "number of differences" as the substitution model, and "pairwise deletion" for gaps/missing data treatment. Graphical view was edited in the Fig Tree v14.0 software (http://tree.bio.ed.ac.uk/software/figtree/).

The freely available TargetP software [91] was used to predict the protein subcellular localization and the position of the cleavage sites of mitochondrial targeting signals (http://www.cbs.dtu.dk/services/TargetP/) using the translated peptide corresponding to exon 1. The prediction of putative isoelectric point (pI) and the molecular weight was obtained using the freely available tool PeptideMass at the Expasy software (http://web.expasy.org/peptide_mass/).

4.1.4. Homology-Based Model

The model of AOX1d was generated using the structure of AOX from *T. brucei* (PDB ID: 3VV9) [38] as a template. Only the region between residues 159 and 313 was modelled, since the N- and C-terminal regions did not align well with the template. The model was built using the software modeler [92], version 9.6, and setting the refinement degree to slow. The final model corresponds to the one with the lowest value of the objective function, out of 20 generated structures.

4.1.5. In Silico Identification of Regulatory Elements Located at the *AOX* Gene Boundaries

For the identification of putative miRNA precursor sequences located at the introns and UTRs of the isolated *OeAOX* genes, the publicly available software miR-abela (http://www.mirz.unibas.ch/cgi/pred_miRNA_genes.cgi) was used. Further steps related with pre-miRNAs validation, prediction of the secondary structure of predicted pre-miRNA, screening of potential miRNAs, and identification of target genes candidates were developed according to the procedure described by Velada et al. [42].

For identification of transposable elements (TE) in the vicinity of the *OeAOX* genes (using olive genome sequencing information), contigs encompassing these genes were self-aligned using Blast2Seq at the NCBI (blast.ncbi.nlm.nih.gov). The contigs were also used as queries to search GeneBank nucleotide database restricted for the *Olea* genus (taxid:4145). Hits matching known *O. europaea* TEs were retained. Regions flanked by putative LTRs not showing similarity to known *Olea* retrotransposons were used as queries for Censor (www.girinst.org) to search for the most similar elements in the RepBase. Target site duplications (TSDs) were identified manually. Additional manipulations and alignments were performed in BioEdit 7.2.5 [93]. All names of olive LTR retrotransposons are reported after Barghini et al. [94].

For identification of transposable elements (TE) in the vicinity of the *OeAOX* genes (using olive genome sequencing information), contigs encompassing these genes were self-aligned using Blast2Seq at the NCBI (blast.ncbi.nlm.nih.gov). The contigs were also used as queries to search GeneBank nucleotide database restricted for the *Olea* genus (taxid:4145). Hits matching known *O. europaea* TEs were retained. Regions flanked by putative LTRs not showing similarity to known *Olea* retrotransposons were used as queries for Censor (www.girinst.org) to search for the most similar elements in the RepBase. Target site duplications (TSDs) were identified manually. Additional manipulations and alignments were performed in BioEdit 7.2.5 [93]. All names of olive LTR retrotransposons are reported after Barghini et al. [94].

To screen for the presence of cis-regulatory elements located at the promoter region that could be related with regulation of gene expression by auxins, a region comprising 1.5 Kb upstream the translation start site of both AOX1 gene sequences was considered for analysis. Promoter region of *OeAOX1a* (Oe6_s00216) and *OeAOX1d* (Oe6_s05781) was retrieved from the whole genome sequencing project (http://denovo.cnag.cat/genomes/olive/). The freely available New PLACE tool—A Database of Plant Cis-acting Regulatory DNA elements (https://sogo.dna.affrc.go.jp/cgi-bin/sogo.cgi?lang=en&pj=640&action=page&page=newplace, [95]) and PlantCARE (http://bioinformatics.psb.ugent.be/webtools/plantcare/html/, [96]) were used.

4.2. Transcript Expression of the AOX1-subfamily Members during AR on in Vitro Cultured Stem Segments

4.2.1. Plant Material and in Vitro Rooting Experiments

The in vitro grown plantlets of cv. "Galega vulgar" (clone 1441), described above, were used as initial explant source for the AR experiment. Rooting treatments and culture conditions were adapted from Macedo et al. [40]. In brief, stem segments (microcuttings) with four-to-five nodes were prepared from the upper part of in vitro-cultured plantlets and all leaves were removed with the exception of the upper four. The base (approx. 1.0 cm) of each microcutting was immersed in a sterile solution of 14.7 mM IBA (indole-3-butyric acid) [14] for 10 s. Subsequently, the microcuttings were aseptically inoculated in 500 mL glass flasks containing 75 mL semi-solid olive culture medium (OM), without plant growth regulators and supplemented with 7 g/L commercial agar-agar, 30 g/L d-mannitol, and 2 g/L activated charcoal [97]. Medium pH was adjusted to 5.8 prior autoclaving (20 min at 121 °C, 1 kg/cm^2). Twenty microcuttings were inoculated per flask, and four flasks were used per time point (4 and 8 h and 1, 2, 4, 6, 8, 10, 14, 18, 22, 26, and 30 days). All cultures were kept in a plant growth chamber at 24 °C/21 °C (±1 °C) day/night temperatures, with a 15 h photoperiod, under cool-white fluorescent light at a photosynthetically active radiation (PAR) level of 36 µmol/m^2 s^{-2} at culture height. During the in vitro rooting experiment, the segments from the basal portion (approx. 1 cm from the base) of the microcuttings were collected from each flask at the time points mentioned above. Basal segments from microcuttings prepared as described above, but not immersed in IBA and not inoculated in OM medium, were also collected and these corresponded to the time point 0 h (control group). All samples were flash frozen in liquid nitrogen and stored at −80 °C for subsequent analyses.

4.2.2. RNA Isolation and First-Strand cDNA Synthesis

Total RNA was isolated with the Maxwell 16 LEV simplyRNA purification kit (Promega, Madison, WI, USA) on the Maxwell 16 Instrument (Promega, Madison, WI, USA) according to the supplier's instructions, and eluted in 50 µL volume of RNase-free water. The concentration of total RNA was determined with the NanoDrop-2000C spectrophotometer (Thermo Scientific, Wilmington, DE, USA), and the total RNA integrity was analyzed by agarose gel electrophoresis through visualization of the two ribosomal subunits in a Gene Flash Bio Imaging system (Syngene, Cambridge, UK). GoScript Reverse Transcription System (Promega, Madison, WI, USA) was used to synthesize complementary DNA (cDNA) from RNA samples (using 1 µg of total RNA), according to the manufacturer's instructions.

4.2.3. Quantitative Real-Time PCR

Real-time PCR was performed in the Applied Biosystems 7500 Real-Time PCR System (Applied Biosystems, Foster City, CA, USA). Real-time PCR reactions were carried out using 1X Maxima SYBR Green qPCR Master Mix, 10 ng of cDNA, and the primers for each target gene and the corresponding transcript variants, as well as for the reference genes used here, as shown in Table S2, in a total volume of 18 µL. Primers for the *OeAOX1a* and *OeAOX1d* genes were designed to a common region in the exon 4 in order to detect and amplify all set of distinct transcripts variants. Primers to detect and amplify transcript variants for each gene were designed for a specific region for each variant (see primers localization in Figure 2 (for *OeAOX1a*) and Figure S6 (for *OeAOX1d*)). Primers were designed based on the *OeAOX1a* and *OeAOX1d* sequences deposited in the NCBI with the Primer Express v3.0 (Applied Biosystems, Foster City, CA, USA), using the default properties given by the software. Regarding the reference genes, actin (*OeACT*) and the elongation factor 1a (*OeEF1a*) were selected in a previous analysis as the most stable genes for this experimental system, following an analysis described by

Velada et al. [27]. All primer pairs were checked for their probability to form dimers and secondary structures using the primer test tool of the software. The reactions were performed using the following thermal profile: 10 min at 95 °C, and 40 cycles of 15 s at 95 °C and 60 s at 60 °C. No-template controls (NTCs) were used to assess contaminations and primer dimers formation. A standard curve was performed using an undiluted pool containing all cDNA samples and three four-fold serial dilutions, with a total of four points. All samples were run in duplicate. Melting curve analysis was done to ensure amplification of the specific amplicon. Quantification cycle (Cq) values were acquired for each sample with the Applied Biosystems 7500 software (Applied Biosystems, Foster City, CA, USA).

4.2.4. Expression Analysis of Transcripts

For expression levels normalization of the transcripts under study, Cq values were converted into relative quantities (RQ) by the delta-Ct method described by Vandesompele and co-authors [98]. The normalization factor was determined by the GeNorm algorithm [98] and corresponds to the geometric mean between the RQ of the selected reference genes for each sample. For each gene of interest, calculating the ratio between the RQ for each sample and the corresponding normalization factor, a normalized gene expression value was obtained. The graphs show the mean ± standard deviation of four biological replicates, and the bars represent the fold-change related to the control group (0 h), which was set to 1. Statistical significances ($p \leq 0.05$, $p \leq 0.01$, and $p \leq 0.001$) between means were determined by the t-test method using the IBM® SPSS® Statistics version 22.0 (SPSS Inc., Armonk, NY, USA).

Supplementary Materials: Supplementary materials can be found at www.mdpi.com/1422-0067/19/2/597/s1.

Acknowledgments: This work was financially supported by the Project OLEAVALOR (ALT20-03-0145-FEDER-000014) funded by FEDER funds through the Program Alentejo 2020; by national funds through FCT (Foundation for Science and Technology) under the Project UID/AGR/00115/2013; by the Project LISBOA-01-0145-FEDER-007660 (Microbiologia Molecular, Estrutural e Celular) funded by FEDER funds through COMPETE2020—Programa Operacional Competitividade e Internacionalização (POCI); and by FCT. Hélia G. Cardoso and Diana Lousa were supported by the FCT post-doc fellowship SFRH/BPD/109849/2015 and SFRH/BPD/92537/2013, respectively. Isabel Velada was supported by the post-doc1_oleavalor fellowship within the Project OLEAVALOR (ALT20-03-0145-FEDER-000014). Dariusz Grzebelus was supported by the Ministry of Science and Higher Education of the Republic of Poland. The authors are grateful to Virginia Sobral for helpful technical assistance.

Author Contributions: Augusto Peixe, Birgit Arnholdt-Schmitt, and Hélia G. Cardoso conceived and designed the experiments; Isabel Velada, Elisete Santos Macedo, and Hélia G. Cardoso performed the experiments; Isabel Velada, Dariusz Grzebelus, Diana Lousa, Cláudio M. Soares, and Hélia G. Cardoso analyzed the data; Augusto Peixe, Dariusz Grzebelus, Diana Lousa, Cláudio M. Soares, and Hélia G. Cardoso contributed reagents/materials/analysis tools; Isabel Velada, Dariusz Grzebelus, Diana Lousa, and Hélia G. Cardoso wrote the paper. All co-authors read the text, made comments, and approved the final version.

Conflicts of Interest: The authors declare no conflict of interest.

Abbreviations

AOX	Alternative oxidase
AR	Adventitious rooting
IBA	indole-3-butyric acid
AR	adventitious rooting
SNP	single nucleotide polymorphism
QTL	Quantitative Trait Locus
UTR	Untranslated region
APA	alternative polyadenylation
mRNA	Messenger Ribonucleic acid
PAS	poly(a) site
gDNA	Genomic Deoxyribonucleic acid
NCBI	National Center for Biotechnology Information

TSA	transcriptome shotgun assembly
NJ	Neighbor-Joining
RACE	Rapid amplification of cDNA ends
RT-qPCR	Reverse transcription-quantitative real time polymerase chain reaction
ORF	Open reading frame
cDNA	complementary DNA
LTR	long terminal repeats
TSD	Target site duplications
PCR	polymerase chain reaction
mTP	mitochondrial transit peptides
snoRNAs	Small nucleolar RNAs
TEs	transposable elements
NMD	nonsense-mediated decay
RBPs	RNA-binding proteins
SHAM	salicylhydroxamic acid
ROS	Reactive oxygen species
ACT	actin
EF1a	elongation factor 1a
Cq	Quantification cycle
RQ	relative quantities

References

1. Baldoni, L.; Belaj, A. Olive. In *Oil Crops*; Handbook of Plant Breeding, Vollmann, J., Rajean, I., Eds.; Springe Science Business Media: New York, NY, USA, 2009; pp. 397–421.
2. Peixe, A.; Santos Macedo, E.; Vieira, C.M.; Arnholdt-Schmitt, B. A histological evaluation of adventitious root formation in olive (*Olea europaea* L. cv. *Galega vulgar*) microshoots cultured *in vitro*. In Proceedings of the 28th International Horticultural Congress, Lisbon, Portugal, 22–27 August 2010; pp. S08–S214.
3. Porfírio, S.; Gomes da Silva, M.D.R.; Cabrita, M.J.; Azadi, P.; Peixe, A. Reviewing current knowledge on olive (*Olea europaea* L.) adventitious root formation. *Sci. Hortic. (Amsterdam)* **2016**, *198*, 207–226. [CrossRef]
4. Santos MacEdo, E.; Cardoso, H.G.; Hernández, A.; Peixe, A.A.; Polidoros, A.; Ferreira, A.; Cordeiro, A.; Arnholdt-Schmitt, B. Physiologic responses and gene diversity indicate olive alternative oxidase as a potential source for markers involved in efficient adventitious root induction. *Physiol. Plant.* **2009**, *137*, 532–552. [CrossRef] [PubMed]
5. Potters, G.; Pasternak, T.P.; Guisez, Y.; Palme, K.J.; Jansen, M.A.K. Stress-induced morphogenic responses: Growing out of trouble? *Trends Plant Sci.* **2007**, *12*, 98–105. [CrossRef] [PubMed]
6. Vanlerberghe, G.C.; Cvetkovska, M.; Wang, J. Is the maintenance of homeostatic mitochondrial signaling during stress a physiological role for alternative oxidase? *Physiol. Plant.* **2009**, *137*, 392–406. [CrossRef] [PubMed]
7. Vanlerberghe, G.C. Alternative oxidase: A mitochondrial respiratory pathway to maintain metabolic and signaling homeostasis during abiotic and biotic stress in plants. *Int. J. Mol. Sci.* **2013**, *14*, 6805–6847. [CrossRef] [PubMed]
8. Saha, B.; Borovskii, G.; Panda, S.K. Alternative oxidase and plant stress tolerance. *Plant Signal. Behav.* **2016**, *11*, e1256530. [CrossRef] [PubMed]
9. Simons, B.H.; Lambers, H. The alternative oxidase: Is it a respiratory pathway allowing a plant to cope with stress? In *Plant Responses to Environmental Stress: From Phytohormones to Gene Reorganization*; Lerner, H.R., Ed.; Marcel Dekker Inc.: New York, NY, USA, 1999; pp. 265–286.
10. Finnegan, P.M.; Soole, P.M.; Umbach, A.L. Alternative electron transport proteins. In *Plant Mitochondria: From Gene to Function. Advances in Photosynthesis and Respiration*; Day, D., Millar, H., Whelan, J., Eds.; Kluwer Academic Publishers: Dordrecht, The Netherlands, 2004; p. 182.
11. Finnegan, P.M.; Soole, K.L.; Umbach, A.L. Alternative mitochondrial electron transport proteins in higher plants. In *Plant Mitochondria Advances in Photosynthesis and Respiration*; Day, D.A., Millar, A.H., Whelan, J., Eds.; Kluwer Academic Publishers: Dordrecht, The Netherlands, 2004; pp. 163–230.

12. Clifton, R.; Lister, R.; Parker, K.L.; Sappl, P.G.; Elhafez, D.; Millar, A.H.; Day, D.A.; Whelan, J. Stress-induced co-expression of alternative respiratory chain components in Arabidopsis thaliana. *Plant Mol. Biol.* **2005**, *58*, 193–212. [CrossRef] [PubMed]
13. Clifton, R.; Millar, A.H.; Whelan, J. Alternative oxidases in Arabidopsis: A comparative analysis of differential expression in the gene family provides new insights into function of non-phosphorylating bypasses. *Biochim. Biophys. Acta-Bioenerg.* **2006**, *1757*, 730–741. [CrossRef] [PubMed]
14. Santos Macedo, E.; Sircar, D.; Cardoso, H.G.; Peixe, A.; Arnholdt-Schmitt, B. Involvement of alternative oxidase (AOX) in adventitious rooting of *Olea europaea* L. microshoots is linked to adaptive phenylpropanoid and lignin metabolism. *Plant Cell Rep.* **2012**, *31*, 1581–1590. [CrossRef] [PubMed]
15. Hedayati, V.; Mousavi, A.; Razavi, K.; Cultrera, N.; Alagna, F.; Mariotti, R.; Hosseini-Mazinani, M.; Baldoni, L. Polymorphisms in the AOX2 gene are associated with the rooting ability of olive cuttings. *Plant Cell Rep.* **2015**, *34*, 1151–1164. [CrossRef] [PubMed]
16. McDonald, A.E.; Costa, J.H.; Nobre, T.; De Melo, D.F.; Arnholdt-Schmitt, B. Evolution of AOX genes across kingdoms and the challenge of classification. In *Alternative Respiratory Pathways in Higher Plants*; John Wiley & Sons, Inc.: Hoboken, NJ, USA, 2015; pp. 267–272. ISBN 9781118789971.
17. Costa, J.H.; McDonald, A.E.; Arnholdt-Schmitt, B.; Fernandes de Melo, D. A classification scheme for alternative oxidases reveals the taxonomic distribution and evolutionary history of the enzyme in angiosperms. *Mitochondrion* **2014**, *19*, 172–183. [CrossRef] [PubMed]
18. Costa, J.H.; dos Santos, C.P.; de Sousa e Lima, B.; Moreira Netto, A.N.; Saraiva, K.D.d.C.; Arnholdt-Schmitt, B. *In silico* identification of alternative oxidase 2 (AOX2) in monocots: A new evolutionary scenario. *J. Plant Physiol.* **2017**, *210*, 58–63. [CrossRef] [PubMed]
19. Cardoso, H.G.; Nogales, A.; Frederico, A.M.; Svensson, J.T.; Macedo, E.S.; Valadas, V.; Arnholdt-Schmitt, B. Exploring AOX gene diversity. In *Alternative Respiratory Pathways in Higher Plants*; John Wiley & Sons, Inc.: Hoboken, NJ, USA, 2015; pp. 239–254. ISBN 9781118789971.
20. Whelan, J.; Millar, A.H.; Day, D.A. The alternative oxidase is encoded in a multigene family in soybean. *Planta* **1996**, *198*, 197–201. [CrossRef] [PubMed]
21. Saisho, D.; Nambara, E.; Naito, S.; Tsutsumi, N.; Hirai, A.; Nakazono, M. Characterization of the gene family for alternativa oxidase from Arabidopsis thaliana. *Plant Mol. Biol.* **1997**, *35*, 585–596. [CrossRef] [PubMed]
22. Considine, M.J.; Holtzapffel, R.C.; Day, D.A.; Whelan, J.; Millar, A.H. Molecular distinction between alternative oxidase from monocots and dicots. *Plant Physiol.* **2002**, *129*, 949–953. [CrossRef] [PubMed]
23. Costa, J.H.; dos Santos, C.P.; da Cruz Saraiva, K.D.; Arnholdt-Schmitt, B. A step-by-step protocol for classifying AOX proteins in flowering plants. *Methods Mol. Biol.* **2017**, *1670*, 225–234. [PubMed]
24. Feng, H.; Guan, D.; Sun, K.; Wang, Y.; Zhang, T.; Wang, R. Expression and signal regulation of the alternative oxidase genes under abiotic stresses. *Acta Biochim. Biophys. Sin. (Shanghai)* **2013**, *45*, 985–994. [CrossRef] [PubMed]
25. Costa, J.H.; Mota, E.F.; Cambursano, M.V.; Lauxmann, M.A.; De Oliveira, L.M.N.; Silva Lima, M.D.G.; Orellano, E.G.; Fernandes De Melo, D. Stress-induced co-expression of two alternative oxidase (VuAox1 and 2b) genes in Vigna unguiculata. *J. Plant Physiol.* **2010**, *167*, 561–570. [CrossRef] [PubMed]
26. Cavalcanti, J.H.F.; Oliveira, G.M.; Saraiva, K.D.d.C.; Torquato, J.P.P.; Maia, I.G.; Fernandes de Melo, D.; Costa, J.H. Identification of duplicated and stress-inducible Aox2b gene co-expressed with Aox1 in species of the Medicago genus reveals a regulation linked to gene rearrangement in leguminous genomes. *J. Plant Physiol.* **2013**, *170*, 1609–1619. [CrossRef] [PubMed]
27. Velada, I.; Ragonezi, C.; Arnholdt-Schmitt, B.; Cardoso, H. Reference Genes Selection and Normalization of Oxidative Stress Responsive Genes upon Different Temperature Stress Conditions in *Hypericum perforatum* L. *PLoS ONE* **2014**, *9*, e115206. [CrossRef] [PubMed]
28. Guerra Cardoso, H.; Doroteia Campos, M.; Rita Costa, A.; Catarina Campos, M.; Nothnagel, T.; Arnholdt-Schmitt, B. Carrot alternative oxidase gene AOX2a demonstrates allelic and genotypic polymorphisms in intron 3. *Physiol. Plant.* **2009**, *137*, 592–608. [CrossRef] [PubMed]
29. Cardoso, H.; Campos, M.D.; Nothnagel, T.; Arnholdt-Schmitt, B. Polymorphisms in intron 1 of carrot AOX2b–a useful tool to develop a functional marker? *Plant Genet. Resour.* **2011**, *9*, 177–180. [CrossRef]
30. Ferreira, A.O.; Cardoso, H.G.; Macedo, E.S.; Breviario, D.; Arnholdt-Schmitt, B. Intron polymorphism pattern in AOX1b of wild St John's wort (*Hypericum perforatum*) allows discrimination between individual plants. *Physiol. Plant.* **2009**, *137*, 520–531. [CrossRef] [PubMed]

31. Nogales, A.; Nobre, T.; Cardoso, H.G.; Muñoz-Sanhueza, L.; Valadas, V.; Campos, M.D.; Arnholdt-Schmitt, B. Allelic variation on DcAOX1 gene in carrot (*Daucus carota* L.): An interesting simple sequence repeat in a highly variable intron. *Plant Gene* **2016**, *5*, 49–55. [CrossRef]
32. Nobre, T.; Campos, M.D.; Lucic-Mercy, E.; Arnholdt-Schmitt, B. Misannotation Awareness: A Tale of Two Gene-Groups. *Front. Plant Sci.* **2016**, *7*, 868. [CrossRef] [PubMed]
33. Albury, M.S.; Elliott, C.; Moore, A.L. Towards a structural elucidation of the alternative oxidase in plants. *Physiol. Plant.* **2009**, *137*, 316–327. [CrossRef] [PubMed]
34. Abe, F.; Saito, K.; Miura, K.; Toriyama, K. A single nucleotide polymorphism in the alternative oxidase gene among rice varieties differing in low temperature tolerance. *FEBS Lett.* **2002**, *527*, 181–185. [CrossRef]
35. Di Giammartino, D.C.; Nishida, K.; Manley, J.L. Mechanisms and Consequences of Alternative Polyadenylation. *Mol. Cell* **2011**, *43*, 853–866. [CrossRef] [PubMed]
36. Polidoros, A.N.; Mylona, P.V.; Arnholdt-Schmitt, B. Aox gene structure, transcript variation and expression in plants. *Physiol. Plant.* **2009**, *137*, 342–353. [CrossRef] [PubMed]
37. Umbach, A.L.; Siedow, J.N. Covalent and Noncovalent Dimers of the Cyanide-Resistant Alternative Oxidase Protein in Higher Plant Mitochondria and Their Relationship to Enzyme Activity. *Plant Physiol.* **1993**, *103*, 845–854. [CrossRef] [PubMed]
38. Moore, A.L.; Shiba, T.; Young, L.; Harada, S.; Kita, K.; Ito, K. Unraveling the Heater: New Insights into the Structure of the Alternative Oxidase. *Annu. Rev. Plant Biol.* **2013**, *64*, 637–663. [CrossRef] [PubMed]
39. Tian, B.; Manley, J.L. Alternative polyadenylation of mRNA precursors. *Nat. Rev. Mol. Cell Biol.* **2017**, *18*, 18–30. [CrossRef] [PubMed]
40. Macedo, E.; Vieira, C.; Carrizo, D.; Porfírio, S.; Hegewald, H.; Arnholdt-Schmitt, B.; Calado, M.L.; Peixe, A. Adventitious root formation in olive (*Olea europaea* L.) microshoots: Anatomical evaluation and associated biochemical changes in peroxidase and polyphenoloxidase activities. *J. Hortic. Sci. Biotechnol.* **2013**, *88*, 53–59. [CrossRef]
41. Campos, M.D.; Cardoso, H.G.; Linke, B.; Costa, J.H.; De Melo, D.F.; Justo, L.; Frederico, A.M.; Arnholdt-Schmitt, B. Differential expression and co-regulation of carrot AOX genes (*Daucus carota*). *Physiol. Plant.* **2009**, *137*, 578–591. [CrossRef] [PubMed]
42. Velada, I.; Cardoso, H.G.; Ragonezi, C.; Nogales, A.; Ferreira, A.; Valadas, V.; Arnholdt-Schmitt, B. Alternative Oxidase Gene Family in *Hypericum perforatum* L.: Characterization and Expression at the Post-germinative Phase. *Front. Plant Sci.* **2016**, *7*, 1–16. [CrossRef] [PubMed]
43. Liang, W.-S. Respiratory pathways in bulky tissues and storage organs. In *Alternative Respiratory Pathways in Higher Plants*; Gupta, K.J., Mur, L.A.J., Neelwarne, B., Eds.; John Wiley & Sons, Ltd.: Chichester, UK, 2015; pp. 221–232. ISBN 9781118789971.
44. Vishwakarma, A.; Dalal, A.; Tetali, S.D.; Kirti, P.B.; Padmasree, K. Genetic engineering of AtAOX1a in Saccharomyces cerevisiae prevents oxidative damage and maintains redox homeostasis. *FEBS Open Bio* **2016**, *6*, 135–146. [CrossRef] [PubMed]
45. Huang, S.; Taylor, N.L.; Whelan, J.; Millar, A.H. Refining the Definition of Plant Mitochondrial Presequences through Analysis of Sorting Signals, N-Terminal Modifications, and Cleavage Motifs. *Plant Physiol.* **2009**, *150*, 1272–1285. [CrossRef] [PubMed]
46. Heyn, P.; Kalinka, A.T.; Tomancak, P.; Neugebauer, K.M. Introns and gene expression: Cellular constraints, transcriptional regulation, and evolutionary consequences. *BioEssays* **2015**, *37*, 148–154. [CrossRef] [PubMed]
47. Oswald, A.; Oates, A.C. Control of endogenous gene expression timing by introns. *Genome Biol.* **2011**, *12*, 107. [CrossRef] [PubMed]
48. Brinster, R.L.; Allen, J.M.; Behringer, R.R.; Gelinas, R.E.; Palmiter, R.D. Introns increase transcriptional efficiency in transgenic mice. *Proc. Natl. Acad. Sci. USA* **1988**, *85*, 836–840. [CrossRef] [PubMed]
49. Pawlicki, J.M.; Steitz, J.A. Primary microRNA transcript retention at sites of transcription leads to enhanced microRNA production. *J. Cell Biol.* **2008**, *182*, 61–76. [CrossRef] [PubMed]
50. Hirsch, C.D.; Springer, N.M. Transposable element influences on gene expression in plants. *Biochim. Biophys. Acta-Gene Regul. Mech.* **2017**, *1860*, 157–165. [CrossRef] [PubMed]
51. Schnable, P.S.; Ware, D.; Fulton, R.S.; Stein, J.C.; Wei, F.; Pasternak, S.; Liang, C.; Zhang, J.; Fulton, L.; Graves, T.A.; et al. The B73 Maize Genome: Complexity, Diversity, and Dynamics. *Science* **2009**, *326*, 1112–1115. [CrossRef] [PubMed]

52. Ohtsu, K.; Hirano, H.Y.; Tsutsumi, N.; Hirai, A.; Nakazono, M. Anaconda, a new class of transposon belonging to the Mu superfamily, has diversified by acquiring host genes during rice evolution. *Mol. Genet. Genom.* **2005**, *274*, 606–615. [CrossRef] [PubMed]
53. Costa, J.H.; De Melo, D.F.; Gouveia, Z.; Cardoso, H.G.; Peixe, A.; Arnholdt-Schmitt, B. The alternative oxidase family of Vitis vinifera reveals an attractive model to study the importance of genomic design. *Physiol. Plant.* **2009**, *137*, 553–565. [CrossRef] [PubMed]
54. Macko-Podgorni, A.; Nowicka, A.; Grzebelus, E.; Simon, P.W.; Grzebelus, D. DcSto: Carrot Stowaway-like elements are abundant, diverse, and polymorphic. *Genetica* **2013**, *141*, 255–267. [CrossRef] [PubMed]
55. Li, Q.Q.; Liu, Z.; Lu, W.; Liu, M. Interplay between Alternative Splicing and Alternative Polyadenylation Defines the Expression Outcome of the Plant Unique OXIDATIVE TOLERANT-6 Gene. *Sci. Rep.* **2017**, *7*, 2052. [CrossRef] [PubMed]
56. Tantikanjana, T. An Alternative Transcript of the S Locus Glycoprotein Gene in a Class II Pollen-Recessive Self-Incompatibility Haplotype of *Brassica oleracea* Encodes a Membrane-Anchored Protein. *Plant Cell Online* **1993**, *5*, 657–666. [CrossRef] [PubMed]
57. Simpson, G.G.; Dijkwel, P.P.; Quesada, V.; Henderson, I.; Dean, C. FY is an RNA 3′ end-processing factor that interacts with FCA to control the Arabidopsis floral transition. *Cell* **2003**, *113*, 777–787. [CrossRef]
58. Lykke-Andersen, S.; Jensen, T.H. Nonsense-mediated mRNA decay: An intricate machinery that shapes transcriptomes. *Nat. Rev. Mol. Cell Biol.* **2015**, *16*, 665–677. [CrossRef] [PubMed]
59. Fabian, M.R.; Sonenberg, N.; Filipowicz, W. Regulation of mRNA Translation and Stability by microRNAs. *Annu. Rev. Biochem.* **2010**, *79*, 351–379. [CrossRef] [PubMed]
60. Lianoglou, S.; Garg, V.; Yang, J.L.; Leslie, C.S.; Mayr, C. Ubiquitously transcribed genes use alternative polyadenylation to achieve tissue-specific expression. *Genes Dev.* **2013**, *27*, 2380–2396. [CrossRef] [PubMed]
61. Smibert, P.; Miura, P.; Westholm, J.O.; Shenker, S.; May, G.; Duff, M.O.; Zhang, D.; Eads, B.D.; Carlson, J.; Brown, J.B.; et al. Global Patterns of Tissue-Specific Alternative Polyadenylation in Drosophila. *Cell Rep.* **2012**, *1*, 277–289. [CrossRef] [PubMed]
62. Sandberg, R.; Neilson, J.R.; Sarma, A.; Sharp, P.A.; Burge, C.B. Proliferating cells express mRNAs with shortened 3′ UTRs and fewer microRNA target sites. *Science* **2008**, *320*, 1643–1647. [CrossRef] [PubMed]
63. Mayr, C.; Bartel, D.P. Widespread Shortening of 3′UTRs by Alternative Cleavage and Polyadenylation Activates Oncogenes in Cancer Cells. *Cell* **2009**, *138*, 673–684. [CrossRef] [PubMed]
64. Gupta, I.; Clauder-Münster, S.; Klaus, B.; Järvelin, A.I.; Aiyar, R.S.; Benes, V.; Wilkening, S.; Huber, W.; Pelechano, V.; Steinmetz, L.M. Alternative polyadenylation diversifies post-transcriptional regulation by selective RNA-protein interactions. *Mol. Syst. Biol.* **2014**, *10*. [CrossRef] [PubMed]
65. Gruber, A.R.; Martin, G.; Müller, P.; Schmidt, A.; Gruber, A.J.; Gumienny, R.; Mittal, N.; Jayachandran, R.; Pieters, J.; Keller, W.; et al. Global 3′UTR shortening has a limited effect on protein abundance in proliferating T cells. *Nat. Commun.* **2014**, *5*, 5465. [CrossRef] [PubMed]
66. Arnholdt-Schmitt, B.; Santos Macedo, E.; Peixe, A.; Cardoso, H.; Cordeiro, A. AOX—A potential functional marker for efficient rooting in olive shoot cuttings. In Proceedings of the Second International Seminar Oliviobioteq, Mazara del Vallo, Italy, 5–10 November 2006; pp. 249–254
67. Arnholdt-Schmitt, B.; Costa, J.H.; de Melo, D.F. AOX-a functional marker for efficient cell reprogramming under stress? *Trends Plant Sci.* **2006**, *11*, 281–287. [CrossRef] [PubMed]
68. Porfirio, S.; Calado, M.L.; Noceda, C.; Cabrita, M.J.; da Silva, M.G.; Azadi, P.; Peixe, A. Tracking biochemical changes during adventitious root formation in olive (*Olea europaea* L.). *Sci. Hortic. (Amsterdam)* **2016**, *204*, 41–53. [CrossRef]
69. De Klerk, G. Markers of adventitious root formation. *Agronomie* **1996**, *16*, 609–616. [CrossRef]
70. Gaspar, T.; Kevers, C.; Hausman, J.; Berthon, J.; Ripetti, V. Practical uses of peroxidase activity as a predictive marker of rooting performance of micropropagated shoots. *Agronomie* **1992**, *12*, 757–765. [CrossRef]
71. Gaspar, T.; Kevers, C.; Hausman, J.-F. *Indissociable Chief Factors in the Inductive Phase of Adventitious Rooting BT-Biology of Root Formation and Development*; Altman, A., Waisel, Y., Eds.; Springer: Boston, MA, USA, 1997; pp. 55–63. ISBN 978-1-4615-5403-5.
72. Jarvis, B.C. Endogenous Control of Adventitious Rooting in Non-Woody Cuttings. In *BT-New Root Formation in Plants and Cuttings*; Jackson, M.B., Ed.; Springer: Dordrecht, The Netherlands, 1986; pp. 191–222. ISBN 978-94-009-4358-2.

73. Kevers, C.; Hausman, J.F.; Faivre-Rampant, O.; Evers, D.; Gaspar, T. Hormonal control of adventitious rooting: Progress and questions. *J. Appl. Bot.* **1997**, *71*, 71–79.
74. Rout, G.R.; Samantaray, S.; Das, P. In vitro rooting of Psoralea corylifolia Linn: Peroxidase activity as a marker. *Plant Growth Regul.* **2000**, *30*, 215–219. [CrossRef]
75. Berthon, J.Y.; Ben Tahar, S.; Gaspar, T.; Boyer, N. Rooting phases of shoots of Sequoiadendron giganteum *in vitro* and their requirements. *Plant Physiol. Biochem.* **1990**, *28*, 631–638.
76. Heloir, M.C.; Kevers, C.; Hausman, J.F.; Gaspar, T. Changes in the concentrations of auxins and polyamines during rooting of in-vitro-propagated walnut shoots. *Tree Physiol.* **1996**, *16*, 515–519. [CrossRef] [PubMed]
77. Li, S.W.; Xue, L.; Xu, S.; Feng, H.; An, L. Mediators, genes and signaling in adventitious rooting. *Bot. Rev.* **2009**, *75*, 230–247. [CrossRef]
78. Pacurar, D.I.; Perrone, I.; Bellini, C. Auxin is a central player in the hormone cross-talks that control adventitious rooting. *Physiol. Plant.* **2014**, *151*, 83–96. [CrossRef] [PubMed]
79. Campos, D.M.; Nogales, A.; Cardoso, H.G.; Kumar, S.R.; Nobre, T.; Sathishkumar, R.; Arnholdt-Schmitt, B. Stress-induced accumulation of DcAoX1 and DcAoX2a transcripts coincides with critical time point for structural biomass prediction in carrot primary cultures (*Daucus carota* L.). *Front. Genet.* **2016**, *7*. [CrossRef] [PubMed]
80. Frederico, A.M.; Campos, M.D.; Cardoso, H.G.; Imani, J.; Arnholdt-Schmitt, B. Alternative oxidase involvement in *Daucus carota* somatic embryogenesis. *Physiol. Plant.* **2009**, *137*, 498–508. [CrossRef] [PubMed]
81. Ivanova, A.; Law, S.R.; Narsai, R.; Duncan, O.; Lee, J.-H.; Zhang, B.; Van Aken, O.; Radomiljac, J.D.; van der Merwe, M.; Yi, K.; et al. A Functional Antagonistic Relationship between Auxin and Mitochondrial Retrograde Signaling Regulates Alternative Oxidase1a Expression in Arabidopsis. *Plant Physiol.* **2014**, *165*, 1233–1254. [CrossRef] [PubMed]
82. Kerchev, P.I.; De Clercq, I.; Denecker, J.; Mühlenbock, P.; Kumpf, R.; Nguyen, L.; Audenaert, D.; Dejonghe, W.; Van Breusegem, F. Mitochondrial perturbation negatively affects auxin signaling. *Mol. Plant* **2014**, *7*, 1138–1150. [CrossRef] [PubMed]
83. Choudhury, F.K.; Rivero, R.M.; Blumwald, E.; Mittler, R. Reactive oxygen species, abiotic stress and stress combination. *Plant J.* **2017**, *90*, 856–867. [CrossRef] [PubMed]
84. Rhoads, D.M.; Umbach, A.L.; Subbaiah, C.C.; Siedow, J.N. Mitochondrial reactive oxygen species. Contribution to oxidative stress and interorganellar signaling. *Plant Physiol.* **2006**, *141*, 357–366. [CrossRef] [PubMed]
85. Tognetti, V.B.; Mühlenbock, P.; van Breusegem, F. Stress homeostasis-the redox and auxin perspective. *Plant Cell Environ.* **2012**, *35*, 321–333. [CrossRef] [PubMed]
86. Pasternak, T.; Potters, G.; Caubergs, R.; Jansen, M.A.K. Complementary interactions between oxidative stress and auxins control plant growth responses at plant, organ, and cellular level. *J. Exp. Bot.* **2005**, *56*, 1991–2001. [CrossRef] [PubMed]
87. Guilfoyle, T.J.; Hagen, G. Auxin response factors. *Curr. Opin. Plant Biol.* **2007**, *10*, 453–460. [CrossRef] [PubMed]
88. Peixe, A.; Raposo, A.; Lourenço, R.; Cardoso, H.; Macedo, E. Coconut water and BAP successfully replaced zeatin in olive (*Olea europaea* L.) micropropagation. *Sci. Hortic. (Amsterdam)* **2007**, *113*, 1–7. [CrossRef]
89. Karlin, S.; Altschul, S.F. Applications and statistics for multiple high-scoring segments in molecular sequences. *Proc. Natl. Acad. Sci. USA* **1993**, *90*, 5873–5877. [CrossRef] [PubMed]
90. Kumar, S.; Stecher, G.; Tamura, K. MEGA7: Molecular Evolutionary Genetics Analysis Version 7.0 for Bigger Datasets. *Mol. Biol. Evol.* **2016**, *33*, 1870–1874. [CrossRef] [PubMed]
91. Emanuelsson, O.; Nielsen, H.; Brunak, S.; von Heijne, G. Predicting Subcellular Localization of Proteins Based on their N-terminal Amino Acid Sequence. *J. Mol. Biol.* **2000**, *300*, 1005–1016. [CrossRef] [PubMed]
92. Šali, A.; Blundell, T.L. Comparative Protein Modelling by Satisfaction of Spatial Restraints. *J. Mol. Biol.* **1993**, *234*, 779–815. [CrossRef] [PubMed]
93. Hall, T.A. BioEdit: A user-friendly biological sequence alignment editor and analysis program for Windows 95/98/NT. *Nucleic Acids Symp. Ser.* **1999**, *41*, 95–98.
94. Barghini, E.; Natali, L.; Giordani, T.; Cossu, R.M.; Scalabrin, S.; Cattonaro, F.; Šimková, H.; Vrána, J.; Doležel, J.; Morgante, M.; et al. LTR retrotransposon dynamics in the evolution of the olive (*Olea europaea*) genome. *DNA Res.* **2015**, *22*, 91–100. [CrossRef] [PubMed]

95. Higo, K.; Ugawa, Y.; Iwamoto, M.; Korenaga, T. Plant cis-acting regulatory DNA elements (PLACE) database: 1999. *Nucleic Acids Res.* **1999**, *27*, 297–300. [CrossRef] [PubMed]
96. Lescot, M.; Déhais, P.; Thijs, G.; Marchal, K.; Moreau, Y.; Van de Peer, Y.; Rouzé, P.; Rombauts, S. PlantCARE, a database of plant cis-acting regulatory elements and a portal to tools for *in silico* analysis of promoter sequences. *Nucleic Acids Res.* **2002**, *30*, 325–327. [CrossRef] [PubMed]
97. Rugini, E. *In vitro* propagation of some olive (*Olea europaea sativa* L.) cultivars with different root-ability, and medium development using analytical data from developing shoots and embryos. *Sci. Hortic. (Amsterdam)* **1984**, *24*, 123–134. [CrossRef]
98. Vandesompele, J.; De Preter, K.; Pattyn, F.; Poppe, B.; Van Roy, N.; De Paepe, A.; Speleman, F. Accurate normalization of real-time quantitative RT-PCR data by geometric averaging of multiple internal control genes. *Genome Biol.* **2002**, *3*. [CrossRef]

© 2018 by the authors. Licensee MDPI, Basel, Switzerland. This article is an open access article distributed under the terms and conditions of the Creative Commons Attribution (CC BY) license (http://creativecommons.org/licenses/by/4.0/).

Article

Alternative Respiratory Pathway Component Genes (*AOX* and *ND*) in Rice and Barley and Their Response to Stress

Vajira R. Wanniarachchi [†], Lettee Dametto [†], Crystal Sweetman, Yuri Shavrukov, David A. Day, Colin L. D. Jenkins and Kathleen L. Soole *

College of Science and Engineering, Flinders University of South Australia, GPO Box 5100, Adelaide, SA 5001, Australia; Vajira.Wanniarachchi@anu.edu.au (V.R.W.); dame0006@flinders.edu.au (L.D.); Crystal.Sweetman@flinders.edu.au (C.S.); Yuri.Shavrukov@flinders.edu.au (Y.S.); David.Day@flinders.edu.au (D.A.D.); Colin.Jenkins@flinders.edu.au (C.L.D.J.)
* Correspondence: Kathleen.Soole@flinders.edu.au; Tel.: +61-(08)-8201-2030
† These authors contributed equally to this work.

Received: 14 February 2018; Accepted: 16 March 2018; Published: 20 March 2018

Abstract: Plants have a non-energy conserving bypass of the classical mitochondrial cytochrome c pathway, known as the alternative respiratory pathway (AP). This involves type II NAD(P)H dehydrogenases (NDs) on both sides of the mitochondrial inner membrane, ubiquinone, and the alternative oxidase (AOX). The AP components have been widely characterised from Arabidopsis, but little is known for monocot species. We have identified all the genes encoding components of the AP in rice and barley and found the key genes which respond to oxidative stress conditions. In both species, AOX is encoded by four genes; in rice *OsAOX1a, 1c, 1d* and *1e* representing four clades, and in barley, *HvAOX1a, 1c, 1d1* and *1d2*, but no *1e*. All three subfamilies of plant *ND* genes, *NDA*, *NDB* and *NDC* are present in both rice and barley, but there are fewer *NDB* genes compared to Arabidopsis. Cyanide treatment of both species, along with salt treatment of rice and drought treatment of barley led to enhanced expression of various AP components; there was a high level of co-expression of *AOX1a* and *AOX1d*, along with *NDB3* during the stress treatments, reminiscent of the co-expression that has been well characterised in Arabidopsis for *AtAOX1a* and *AtNDB2*.

Keywords: alternative oxidase; NADH dehydrogenase; rice; barley; oxidative stress

1. Introduction

The alternative respiratory pathway of plant mitochondria is an important bypass of the energy-conserving classical electron transport (ETC) pathway [1]. In the classical ETC, oxidation of mitochondrial NADH occurs via rotenone-sensitive Complex I and cytochrome *c* (cyt *c*) oxidase, concomitantly generating a proton motive force that drives ATP synthesis via ATP synthase. Plants have a non-phosphorylating (non-energy conserving) bypass of the cyt *c* pathway, evading adenylate control of cellular redox poise [2–4]. This bypass, known as the alternative respiratory pathway (AP), involves rotenone-insensitive type II NAD(P)H dehydrogenases (NDs) on both the outer and inner surface of the mitochondrial inner membrane, ubiquinone, and the alternative oxidase (AOX). Activity of the AP, however, does not generate a proton gradient and excess energy is released as heat. There is strong evidence that operation of the AP prevents over-reduction of the ETC [5–8], thereby minimizing the generation of reactive oxygen species (ROS) and allowing a degree of metabolic flexibility in the cell. A large number of publications have reported that this pathway responds to temperature, nutrients, heavy metals, high light, drought and oxidative stress [4].

In Arabidopsis, members of the AP include seven type II NDs (NDB1–4, NDA1–2, and NDC1) and five AOXs (AOX1a–d and AOX2) [9]. NDB1-4 have been localized to the external face of the mitochondrial inner membrane, while NDA1 and 2, and NDC1 were determined to be internal (facing the matrix), based on import assays [10,11]. Transcript, protein and activity levels of NDs and AOX appear to be regulated in a tissue, developmental, and stress-inducible manner [11–15]. This suggests specialization of function and possible formation of specific ND/AOX pathways in response to particular stimuli and there is evidence that an alternative electron transport chain of NDA2, NDB2 and AOX1a forms in Arabidopsis during many stresses [14].

The majority of studies to date on the response of the AP to stresses have focussed on Arabidopsis and other dicot plants, while there have been few studies on monocots, including crops such as rice and barley. In rice, *Oryza sativa*, Ito, et al. [16] first identified *OsAOX1a* and *OsAOX1b* isoforms, which were up-regulated in whole seedlings during cold stress. Considine, et al. [17] and Saika, et al. [18] identified a third isoform *OsAOX1c*, which showed developmentally regulated expression patterns and was not stress-responsive. Considine, et al. [17] also identified an additional gene *OsAOX1d* that was thought to be not expressed, but more recently Costa, et al. [19] reported a number of corresponding ESTs found in germinating tissue. *OsAOX* genes have been shown to respond to different stresses such as cold, drought, dehydration, and salt [16,20–23], but not all studies have assessed all three isoforms and very few have assessed the corresponding protein and activity levels. In contrast, while rice orthologues for the ND family have been described [3,24] there have been no studies on this gene family's response to stress.

Even less is known about the AP components in barley, *Hordeum vulgare*. The *HvAOX* gene family has been described by Costa et al. [19], identifying four isoforms, *HvAOX1a*, *HvAOX1c*, *HvAOX1d1* and *HvAOX1d2*, but the orthologues/homologues of the *ND* family are as yet unidentified. There are two reports showing that *AOX* responds to stress in barley. An *AOX*-like gene in *H. spontaneum* (wild barley) was up-regulated in leaves during cold stress, and *HvAOX1a* in *H. vulgare*, cv. Golden Promise during short-term heat stress in spikes at anthesis, and AOX protein levels were up-regulated under nutrient stress and UVB stress in highland barley [25,26]. Clearly, the stress-responsiveness of AP components needs further examination in this important crop.

It has been suggested that *AOX* sequence variability has the potential to be used as a marker for selective breeding [27], but it is likely that other components such as the *NDs*, which are required for better oxidative stress responses, also need to be considered. *AOX* has been well studied in many species, especially in Arabidopsis, but the *NDs* have received less attention. In this study, we expand these studies into the cereal crops, rice and barley, identifying the alternative respiratory pathway components in *O. sativa* and *H. vulgare*, and investigate the expression of *AOX* and *ND* genes in different tissue types at both the transcript and protein level under mitochondrial ETC-inhibiting and ROS-generating conditions. The results provide a platform for investigations of these genes as potential selective markers for rice and barley breeding.

2. Results

2.1. Identification of the Genes Encoding Alternative Respiratory Pathway (AP) Components in Rice and Barley

2.1.1. Alternative Oxidase

Previous studies have identified the gene family encoding *AOX* in rice [16–18] and recently Costa et al. [19] have reclassified these genes as *OsAOX1a*, *OsAOX1c*, *OsAOX1d* (previously *OsAOX1b*) and *OsAOX1e*. This nomenclature was based on the fact that *OsAOX1b* was more like an orthologue of the eudicot *AOX1d* gene. Sequence searching (BLAST) of the Rice Genome Annotation Project database (http://rice.plantbiology.msu.edu) identified four gene sequences homologous to Arabidopsis: *OsAOX1a* (LOC_Os04g51150), *OsAOX1d* (LOC_Os04g51160), *OsAOX1c*, (LOC_Os02g47200), and *OsAOX1e* (LOC_Os02g21300) in agreement with Costa et al. [19]. Homologous sequences for all *OsAOX* genes identified in japonica rice were also present in indica rice (Table S1). Pairwise alignment of CDS sequences revealed that all *OsAOX* isoforms from japonica rice were very similar or identical to those of indica rice (Table S1) consistent with them being closely related.

The isoforms of *HvAOX* in barley, *H. vulgare*, were recently reported in the *AOX* phylogeny study of Costa et al. [19], where four *AOX* partial genomic sequences were identified from a few ESTs. Prior to this, only one report of an *AtAOX1a*-like gene had been identified during a microarray analysis of high temperature stress in barley floral tissue [28]. To confirm these findings, the three cDNA sequences of *OsAOX* (Table S1) were used as queries to conduct BLASTN searches within *H. vulgare* mRNA sequence databases, using NCBI (https://www.ncbi.nlm.nih.gov) and IPK (http://webblast.ipk-gatersleben.de/barley_ibsc); four putative cDNAs were found in *H. vulgare*, cv. Haruna Nijo (Table 1). These were of similar size except for *HvAOX1c*. The putative *HvAOX* cDNA sequences were used to conduct BLASTN search against the assembly WGS-Morex database on IPK to confirm correct identification. Genomic contig sequences from cv. Morex demonstrate high sequence identity with sequences identified by Costa et al. [19] and provide an estimation of chromosomal location (Table 1). Two of the rice genes are located on the same chromosome, where a common tandem position for *OsAOX1a* and *OsAOX1d* genes was identified in rice chromosome 4 (Figure 1A, [19]). A strong colinearity with rice chromosome 4 was found in barley chromosome 2H with an additional likely duplication leading to tandem repeat *HvAOX1d1* and *HvAOX1d2* in close proximity on the chromosome (Figure 1A). In contrast, two other *AOX* genes in rice, *OsAOX1c* and *OsAOX1e*, were localised on different arms of rice chromosome 2 (Table S1), while only a single *HvAOX1c* was identified on barley chromosome 6H, in a strongly colinear genetic region to those for rice *OsAOX1c* (Figure 1B, Table 1). No barley gene was identified homologous to rice *OsAOX1e*.

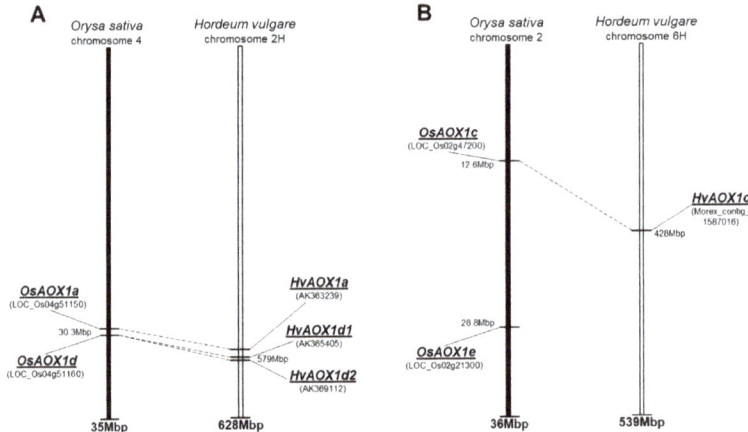

Figure 1. Rice and barley comparative maps with positions of identified *AOX* genes. (**A**) *AOX1a* and *AOX1d*; (**B**) *AOX1c* and *AOX1e*. Corresponding genes are indicated by dashed lines. Information about rice and barley genes are their locations was extracted from web-sites, respectively: http://rice.plantbiology.msu.edu and http://pgsb.helmholtz-muenchen.de/plant/barley.

Table 1. Alternative respiratory pathway components identified in barley. *HvAOX1a*, *HvAOX1d1* and *HvAOX1d2* cDNA sequences as well as a *HvAOX1c* EST sequence were identified in Costa et al., 2014. BLAST servers from NCBI (National Centre for Biotechnology Information) and IBSC (International Barley Sequencing Consortium) were used to find barley sequences with high homology with the rice alternative dehydrogenase sequences identified in [3,24]. Chromosome location as well as genomic sequences (contigs) were found using IPBC. * are partial sequences.

Gene	Chromosome Location chr (Chromosome Number) cM (centi-Morgan)	Contig ID (Genomic Sequence/s)	Relevant cDNA Accession Number/s	Relevant Protein Accession Number/s (Predicted from cDNA Sequence)	Predicted Protein Size (kDa)
HvAOX1A	chr = 2H cM = 105.37 chr2H:695364063-695501552	morex_contig_38523	AK363239	AK363239 (on IBSC) HORVU2Hr1G101980.1	36.36
HvAOX1C	chr = 6H cM = 53.67 chr = 6H cM = 53.46 chr6H:472176203-472177532	morex_contig_1576377 morex_contig_1587016	HORVU6Hr1G068150.2 * GH226429.1 * AK251266.1 *	HORVU6Hr1G068150.2 *	36.13
HvAOX1D1	chr = 2H cM = 106.86	morex_contig_99492	AK365405.1	BAJ96608.1 HORVU0Hr1G005420.4	36.38
HvAOX1D2	chr = 2H cM = 126.86 chr2H:695390581-695392039	morex_contig_242686	AK369112.1	BAK00314.1 HORVU2Hr1G101990.1	37.08
HvNDA1	chr = 2H cM = 39.5 chr7H:445843858-445845235	morex_contig_2565566 morex_contig_363823	AK376348.1	BAK07543.1 HORVU7Hr1G076330.1	60.06
HvNDA2	chr = 3H cM = 93.58	morex_contig_7977 morex_contig_348032 morex_contig_189545 morex_contig_2081103	AK354319.1	BAJ85538.1 HORVU3Hr1G087850.1 HORVU4Hr1G058410.1 HORVU5Hr1G020920.2 HORVU7Hr1G040660.3 HORVU0Hr1G000930.1 HORVU2Hr1G101830.2 HORVU2Hr1G040130.1	54.21
HvNDB1	chr = 7H	morex_contig_1569455	HORVU7Hr1G101500.3	HORVU7Hr1G101500.3	65.08
HvNDB2	chr3H:570908294-570913236	morex_contig_36877	AK367948.1	BAJ99151.1 HORVU3Hr1G076920.2	64.23
HvNDB3	chr = 7H cM = 7C.61	morex_contig_47987	AK354220.1 AK362739.1 AK371807.1 AK252091.1 AK371114.1	BAJ85439.1 BAJ93943.1 BAK03005.1 BAK02312.1	64.59
HvNDC1	chr = 7H cM = 61.76	morex_contig_49824 morex_contig_60206	AK364677.1	AK364677 (on IBSC) HORVU7Hr1G035810.4	59.33

2.1.2. Type-II NADH Dehydrogenase Family

The gene family of Type-II NAD(P)H dehydrogenases in plants is divided into three subfamilies: *NDA*, *NDB* and *NDC* [10,29], with NDA and NDC proteins located on the inside of the inner mitochondrial membrane, facing the matrix space, and NDB proteins on the outside of the inner membrane, facing the intermembrane space [11]. Orthologues of these genes in rice have been identified previously [3,24] and their designations correspond to the gene sequences used in this study (Table S1). Sequence alignment studies showed that, similarly to the *AOX* genes, the *ND* genes from japonica rice are nearly identical to those from indica rice except for NDB3, which in japonica appears to have a shorter sequence compared to that in indica rice. An analysis of the gene structure for rice *ND* genes shows that generally they contain a similar number of exons, however the intronic regions are often longer than members of the *ND* gene family in Arabidopsis [29]. Importantly, the discrepancy between japonica and indica can be explained by an additional intron present in exon 5 in the japonica *OsNDB3* gene, which has been documented incorrectly in current databases (http://rice.plantbiology.msu.edu/cgi-bin/ORF_infopage.cgi) as a coding region. The reading of this intronic region as coding sequence results in a premature termination codon, giving a 357 aa protein rather than the 580 aa protein predicted from the indica *OsNDB3* gene (Figure 2 and Figure S1). Evidence for this insertion region being an intron comes from its absence in rice ESTs found in NCBI for this region of the genome.

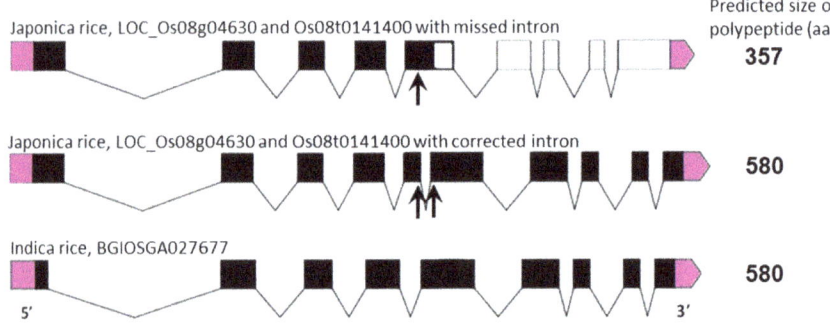

Figure 2. Identification of an additional intron in the sequence of *OsNDB3*, japonica rice, which encodes a predicted OsNDB3 polypeptide of 580 aa, correcting the annotated size of 357 aa. A schematic presentation of the *OsNDB3* gene structures in both japonica and indica rice with the missing or corrected intron. Exons are shown as black boxes or as clear boxes after the premature stop-codon of the missed intron. Start-codons on 5′-end and stop-codons on 3′-end are shown in purple. The predicted sizes of the polypeptides are indicated.

The *O. sativa ND* cDNA sequences (Table S1) were used to conduct BLASTN searches on NCBI revealing 6 barley *ND* sequences (Table 1). A BLASTP search of the predicted *O. sativa* protein sequences against NCBI and IPK databases revealed the corresponding protein sequences. All *ND* genes are predicted to encode proteins in the range of 54 to 65 kDa, similarly to Arabidopsis. Chromosome location was also identified for these genes (Table 1).

An analysis of the sequence identity between Arabidopsis and rice *ND* genes revealed that *OsNDB2* and *OsNDB3* have highest identity with *AtNDB3* (Table S2); however, a previous study [24] used *OsNDB2* notation for the gene (LOC_05g26660), so we have retained this designation here. A relationship tree of sequences from Arabidopsis, rice and barley grouped the corresponding proteins into three clades, NDA, NDB and NDC, but it also confirmed the close sequence relationship between NDB2 and NDB3 in both rice and barley (Figure 3). No orthologue for AtNDB4 was apparent in these monocots.

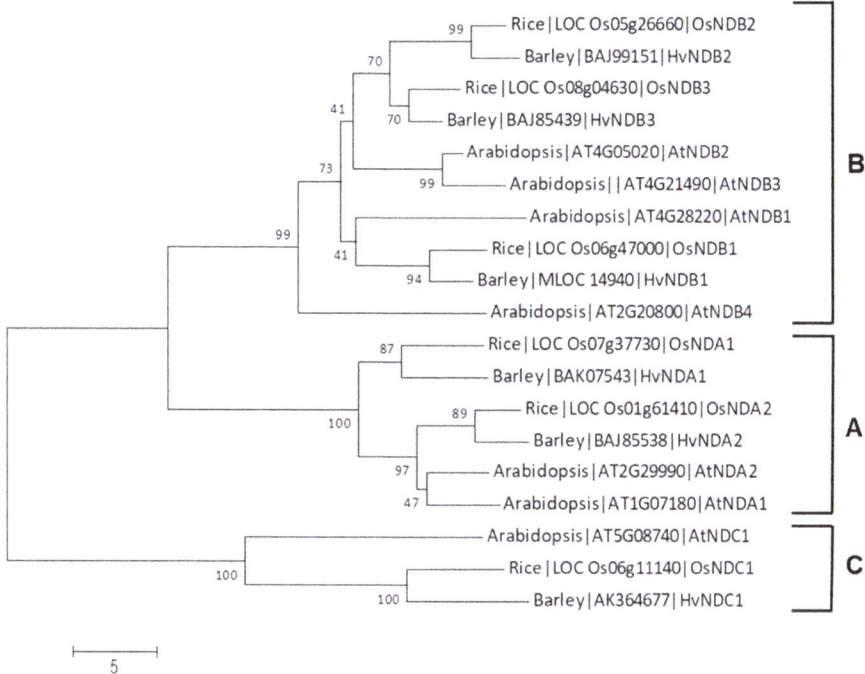

Figure 3. Relationship tree that shows the sequence homology of alternative dehydrogenases from rice, barley and Arabidopsis. Clades A, B and C correspond to NDA, NDB and NDC, respectively. The tree was constructed using the Neighbour-Joining method in MEGA 7. Numbers at each node are the percentage bootstrap values of 1000 replicates. The scale bar indicates the number of amino acid substitutions at each site.

2.2. Subcellular Localisation and Structural Similarities

Using a bioinformatics approach, the subcellular location of rice and barley AOX was confirmed to be mitochondrial. The situation with the ND proteins is not so clear. Arabidopsis NDs contain signals in their sequences that can target these proteins to subcellular compartments other than the mitochondria; for example, AtNDC1 can be targeted to both the chloroplast and the mitochondrion [30]. More recently, Xu et al. [24] indicated that OsNDB1 and OsNDB2 contain both mitochondrial and peroxisomal (PTS1) targeting signals, and showed that when the C terminal sequence of these proteins was fused to GFP they were targeted to the peroxisome in vivo. OsNDB3 was not included in that study and so its location remains uncertain. We used a number of bioinformatic targeting programs to reassess the putative subcellular location of all the rice and barley ND proteins (Table S3); all contained mitochondrial targeting information and HvNDB1 and HvNDB2, like their rice counterparts, were predicted also to contain a PTS1 sequence, a small amino acid sequence found at the C-terminal end of peroxisomal proteins [11,30]. Since the rice proteins have been demonstrated to be targeted to the peroxisome [24], it is likely that this is also true for HvNDB1 and HvNDB2. Interestingly, OsNDB3 and HvNDB3 were not predicted to target to the peroxisome, reminiscent of AtNDB2 and AtNDB4 that are also confined to the mitochondria [30].

AOX proteins are characterised by conserved cysteine residues (Cys_I and Cys_{II}), four helical regions, and conserved glutamine and histidine residues that act as iron ligands [31]. The cysteines have been linked to post-translational regulation of AOX, being involved in the formation of an inactive dimer via a disulphide bond, which when reduced allows the enzyme to be active ([31] and references therein). Cys$_I$ and Cys$_{II}$ [31–34] and more recently Cys$_{III}$ [35], have been shown also to be important in the response of these AOX isoforms to metabolites that activate AOX, such as pyruvate and glyoxylate [36]. As previously reported, OsAOX1d (formerly OsAOX1b) has a serine substitution at Cys$_I$ and Cys$_{II}$ positions [16], indicating that it cannot form a covalent disulphide-linked dimer and therefore may not be inactivated in vivo. However, it may have a rather low activity. When AtAOX1a was mutated to replace Cys$_I$ with a serine, it no longer responded to pyruvate but was instead activated by succinate [34,35]. This is also true of a naturally occurring tomato AOX [37]. However, when serine was substituted for Cys$_I$ in AtAOX1c or d, succinate did not stimulate [35], presumably because of other sequence differences in these proteins. Whether OsAOX1d is activated by succinate is not known and has to be experimentally determined before we can judge whether it is regulated in vivo.

Interestingly, while no AOX isoform in barley has a serine substitution at the Cys$_I$ position, HvAOX1d1 and HvAOX1d2 have serine at the Cys$_{II}$ position. This is not expected to affect the enzyme's regulation by dimerization via oxidation-reduction, but it may affect its stimulation by glyoxylate [31]. All AOX isoforms in rice and barley have leucine at the Cys$_{III}$ position, which is similar to AtAOX1d, not AtAOX1a (Figure 4).

HvAOX1d1 and 2 are particularly interesting because they have a C, S, L (Cys$_I$, Cys$_{II}$, Cys$_{III}$) configuration (Figure 4). This configuration has not been reported for any other AOX protein to date, but when AtAOX1d was mutated to a C, S, L form, its basal activity (i.e., without an organic acid activator) increased quite dramatically [35]. It is possible, therefore, that the HvAOX1d isoforms are also super active, but again, this needs to be experimentally proven in a heterologous assay system.

Figure 4. Alignment of a section of AOX isoforms in rice and barley. The amino acid sequences were aligned using the Clustal program. The region showing all the cysteine residues of these AOX proteins is shown. OsAOX1d has a serine at Cys$_I$; HvAOX1d1, HvAOX1d2 and OsAOX1d have a serine at Cys$_{II}$; all rice and barley AOX isoforms have a leucine at Cys$_{II}$ instead of a cysteine or phenylalanine.

An alignment of the rice and barley ND protein sequences showed that they all contain both of the conserved dinucleotide folds (DNF) that are involved in Flavin and NAD(P)H binding. It has been shown that the DNF sequence at the C-terminal end of these proteins is involved in NADH and NADPH binding and an analysis of this DNF structure, which has a βαβ motif, has shown that a residue at the end of the second β-pleated sheet can influence substrate specificity and the key residues involved have been denoted in Figure 5 [38]. A glutamate (E) in this position enables NADH binding and repels the phosphate on NADPH. Isoforms that show specificity for NADPH have either an uncharged glutamine (Q) or asparagine (N) at this position. These residues have been experimentally confirmed to be involved in NADPH binding via recombinant expression of the Arabidopsis NDs [39]. On this basis, a structural analysis of the rice and barley ND sequences suggests that NDB1 and NDC1 are NADPH specific, while all the others are likely to be NADH specific (Figure 5). Recently, it has been shown that *AtNDC1* encodes an enzyme involved in vitamin K synthesis in the chloroplast [40]. Its role in mitochondria is unknown, but it is targeted to the matrix side of the IM and may be the protein responsible for matrix NADPH oxidation, detected in some plant mitochondria [41].

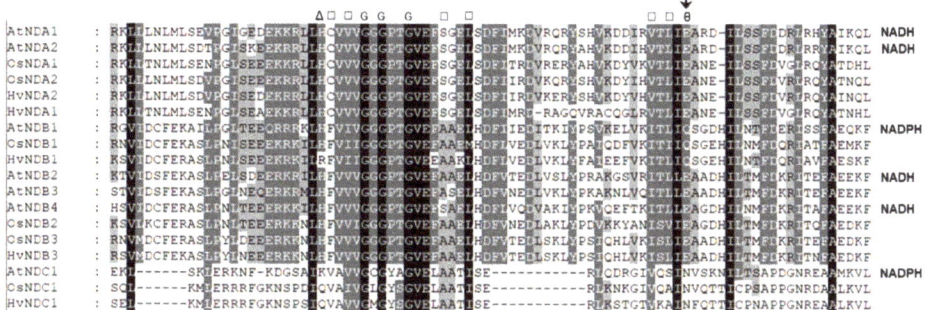

Figure 5. An alignment of the NAD(P)H binding domain in rice and barley ND families with Arabidopsis homologues. The sequences were aligned using the Clustal program. Substrate specificity is given on the right for each enzyme that has been experimentally determined [10,39,42]. Residues are Δ, basic or hydrophilic; □, hydrophobic; θ, acidic; and G, Glycine. The acidic residue alters the specificity for NADH- and NADPH-specific proteins and is indicated as an arrow.

The NDB proteins, when aligned with NDAs and NDC, contain additional amino acid sequence in which are located EF-hand motifs that have been linked to calcium binding. The conserved N-terminal EF-hand motif has been shown to be conserved in Arabidopsis NDB where it has been experimentally demonstrated to bind calcium [39,43]. All rice and barley NDBs contained this EF hand, which suggests that these dehydrogenases may be regulated by calcium. In Arabidopsis, AtNDB4 lacks critical key residues in its EF-hand motif and has been shown to not bind calcium [39], but rice and barley do not appear to have this calcium-independent NDB protein (Table 1, Figure 2).

2.3. Expression of AP Components under Chemical-Induced Stress

A focus of this work was to determine the stress responsive components of the AP in the monocots rice and barley. Initially, plants were exposed to KCN and transcript abundance of AP component genes determined and compared to untreated controls. KCN inhibits cytochrome oxidase leading to an over-reduction of the electron transport chain and accumulation of ROS [44]. In this study, *OsAOX1e* expression was not assessed. While Costa et al. [19] found a small number of ESTs with partial sequence corresponding to this gene, they were isolated from germinating seeds and are more likely to be linked to development. In untreated tissues, *OsAOX1c* was the most highly expressed *AOX* gene in shoots but not roots (Table S4). In contrast, in untreated barley, *HvAOX1c*, was the lowest expressed gene in both shoots (not detectable) and root tissue (Table S4). After 6 h of KCN treatment, all *AOX* isoforms in both rice and barley increased in expression in both shoots and roots, except for *AOX1c* (Tables 2 and 3). In KCN-treated rice, *OsAOX1d* was the most responsive of the AP genes, in both roots and shoots. *OsAOX1a* transcript returned to lower levels 24 h after treatment, whereas the increase in expression was sustained for *OsAOX1d* in shoots (Table 2). *HvAOX1a*, *HvAOX1d1* and *HvAOX1d2* were expressed at similar levels in control roots and shoots (Table S4), and were very responsive to KCN treatment, but this was not sustained at 24 h in any tissue (Table 3 and Figure 6). An analysis of AOX protein levels in barley tissue total protein extracts showed that AOX protein was detectable in control root samples, but not in shoot tissue. However, AOX protein increased in both shoot and root tissue after 24 h exposure to KCN (Figure 6).

Table 2. Expression of rice AP genes under chemical stress. Exposure to potassium cyanide (5 mM) up-regulates *AOX* and *ND* gene expression in rice (*O. sativa* ssp. *japonica*, cv. Nipponbare). Gene expression analyses were carried out using qRT-PCR and transcript levels were determined in both shoot and root tissues exposed to treatments for 6 h and 24 h or grown under control conditions. Data are shown as mean fold change ± SE of three biological replicates relative to the control at each point set at 1.00. * indicates significant transcript level changes relative to the control ($p \leq 0.05$).

Tissue	OsAOX1a	OsAOX1c	OsAOX1d	OsNDA1	OsNDA2	OsNDB1	OsNDB2	OsNDB3	OsNDC1
6 h shoots	58.2 ± 15.51 *	0.3 ± 0.08	248.6 ± 63.97 *	17.4 ± 3.13 *	0.2 ± 0.04	0.2 ± 0.11	5.2 ± 0.91 *	18.3 ± 0.75 *	0.6 ± 0.07
24 h shoots	1.0 ± 0.29	0.1 ± 0.02	348.6 ± 78.37 *	0.7 ± 0.05	1.1 ± 0.20	0.9 ± 0.66	1.9 ± 0.42	0.7 ± 0.05	0.2 ± 0.11
6 h roots	44.0 ± 4.52 *	1.7 ± 0.25	355.4 ± 46.55 *	56.9 ± 13.04 *	3.0 ± 0.81	1.6 ± 1.13	2.0 ± 0.37	29.7 ± 5.64 *	1.0 ± 0.19
24 h roots	5.1 ± 1.01	0.1 ± 0.02	11.3 ± 1.36	2.0 ± 0.15	6.0 ± 1.33 *	0.9 ± 0.23	7.1 ± 1.63 *	1.0 ± 0.38	0.4 ± 0.14

Table 3. Expression of barley AP genes under chemical stress. Fourteen-day-old barley seedlings (cv. Golden Promise) were exposed to 5 mM potassium cyanide for 6 and 24 h. Transcript amount was determined using qRT-PCR. Data were normalized with Housekeeping genes, *Actin*, *Ubiquitin* and *Pdf* (Protein phosphatase) and shown as mean fold change ± SE of four biological replicates relative to the control at each point set at 1.00. ND refers to being at the limit of detection for the assay used. After KCN treatment, *HvNDA1* did increase but to a very low level, and the absence of expression in the control tissue made this difficult to express as a fold change. * denotes significantly different ($p < 0.01$) from the control.

Tissue	HvAOX1a	HvAOX1c	HvAOX1d1	HvAOX1d2	HvNDA1	HvNDA2	HvNDB1	HvNDB2	HvNDB3	HvNDC1
6 h shoots	16.5 ± 7.2	0.6 ± 0.3	23 ± 5 *	80 ± 14 *	N.D.	26 ± 19	1.4 ± 0.25	2.5 ± 1.0	17 ± 1.8 *	1.2 ± 0.6
24 h shoots	10.1 ± 4.7	0.3 ± 0.1 *	3.5 ± 1.7 *	3.5 ± 1.5	N.D.	1.2 ± 0.1	3.3 ± 1.0	3.2 ± 1.2	2.2 ± 0.5 *	0.1 ± 0.02
6 h roots	27 ± 1.6 *	N.D.	17 ± 3.4 *	14 ± 3 *	N.D.	4.8 ± 1.0	0.8 ± 0.2	1.6 ± 0.3 *	4.5 ± 1.3 *	3.4 ± 1.8
24 h roots	6.7 ± 1.8	N.D.	4.1 ± 1.7	2.2 ± 1.2	N.D.	1.8 ± 0.2 *	0.7 ± 0.1	1.5 ± 0.3	1.7 ± 0.3	0.2 ± 0.14

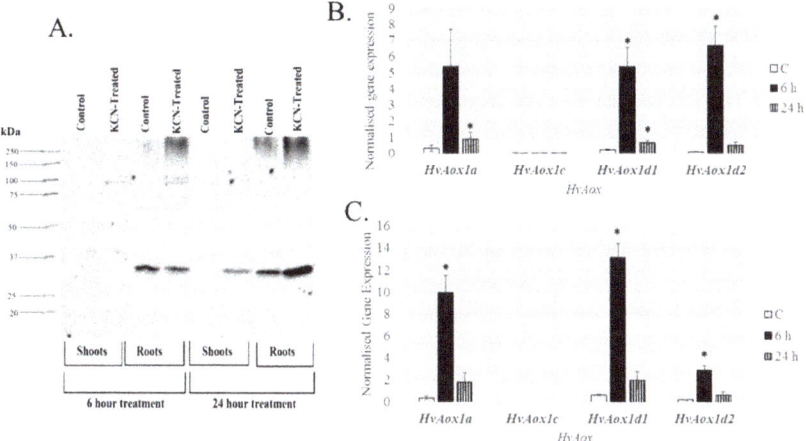

Figure 6. Chemical treatment increases AOX protein levels in barley shoot and root tissue. Barley seedlings (14 days old) were exposed to 5 mM potassium cyanide (KCN) for up to 24 h, then root and shoot tissue were used for crude protein extractions. (**A**) Immunoblot of crude protein extracts under reduction conditions. The expected size for HvAOX is 35–37 kDa. Expression levels of *AOX* isoforms were determined using qRT-PCR in shoots (**B**) and roots (**C**). Expression data were normalized with reference genes *Actin*, *Ubiquitin* and *Pdf* (Protein phosphatase) and shown as mean of 4 biological replicates. Gene expression data were derived from the same experiment as presented in Table 3, but shown as normalised expression instead of fold-change, to demonstrate the relative level of expression between genes. * denotes significantly different ($p < 0.01$) from the control (C).

Prior to KCN treatment, *OsNDA1*, *OsNDB2*, *OsNDB3* were the most highly expressed of the *ND* genes in both roots and shoots (Table S4). In contrast, in untreated barley tissue the most highly expressed *ND* genes were *HvNDA2*, *HvNDB3* and *HvNDB1*, while *HvNDA1* was not detectable (Table S4). In rice, *OsNDA1* and *OsNDB3* expression was upregulated within 6 h of KCN treatment in shoots and roots, but was not sustained at 24 h. *OsNDB2* was upregulated in shoots after 6 h and in root tissue after 24 h. *OsNDA2* expression also increased in roots (Table 2). After KCN treatment, *HvNDA2* and *HvNDB3* were upregulated in shoots and roots and *HvNDB2* to a lesser extent in shoots after 6 h (Table 3). Again, the *HvNDB2* response was sustained in shoots at 24 h. *HvNDA1* was not detectable in control tissue, but showed a little expression, at the limit of detection after KCN treatment, therefore a fold-change could not be calculated although this gene may also be stress-responsive (Table 3).

2.4. Expression of Rice and Barley AP Components under Abiotic Stress

Data from Genevestigator show that the same stress-responsive genes identified in the KCN treatment in rice are up-regulated under many abiotic and biotic stresses, including cold, drought, nutrient stress, and arsenate treatment (Table S5). Interestingly, in the salt studies available, *OsAOX1a*, *OsAOX1d*, *OsNDB3* and *OsNDB2* showed the highest up-regulation in shoot tissue, but were down-regulated in root tissue. We explored this further in rice exposed to 120 mM NaCl over a 12 days period to examine the specific AP response. In shoots, *OsAOX1a*, was up-regulated early in the stress while *OsAOX1d*, showed a later response and *OsAOX1c* was only up-regulated at 12 days (Figure 7A). In the roots, there was a decline in the abundance of all AOX isoform transcripts after the stress (Figure 7B). To elaborate this response further, mitochondria were isolated from shoot and root tissue at 9 days after salt application and respiratory activities determined. In both shoot and root mitochondria there was an increase in AOX capacity. In shoots this was reflected in an increase in AOX protein, especially as the monomeric active form (Figure 7C). The two bands present may reflect different gene products, but only N-terminal sequencing would confirm this. However, in soybean, different apparent Mr of AOX proteins in immunoblots was not indicative of different size proteins [45] In root mitochondria, overall AOX protein abundance did not change but there was a shift from the oxidised dimer to the monomeric form, consistent with the increase in AOX capacity. All the data for roots are consistent with the view that in response to salt stress, existing AOX protein is activated post-translationally, while in shoots there may be an increase in overall AOX protein as well as a post-translational activation.

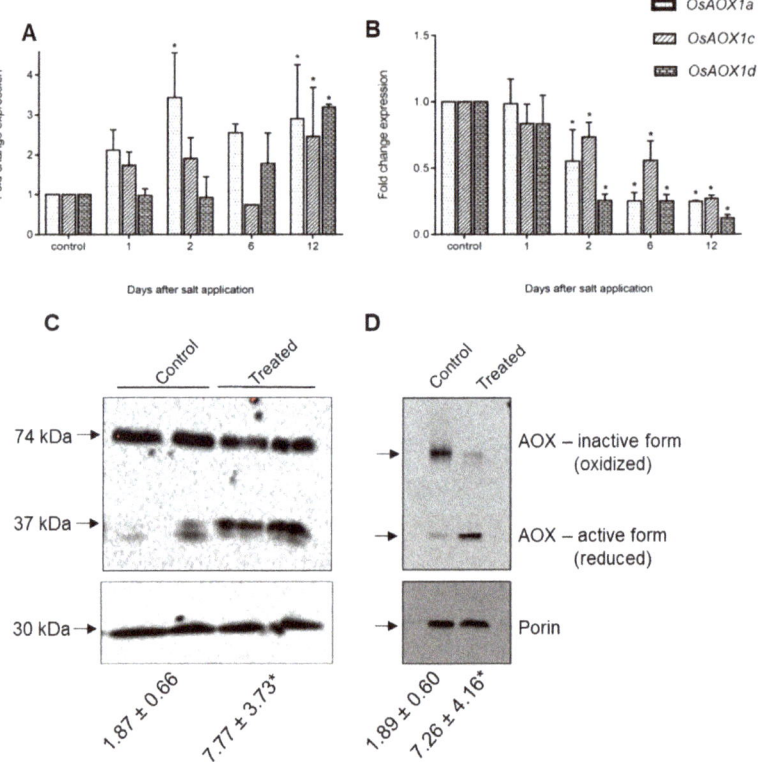

Figure 7. Response of alternative oxidases (AOXs) in rice under salt stress. Fold change expression of *AOX* isoforms, *AOX1a*, *AOX1c* and *AOX1d* in seedling shoots (**A**) and roots (**B**) were analysed in response to 120 mM NaCl using qRT-PCR over a period of 12 days. Data are shown as the mean ± SE of three biological replicates relative to the control at each time point set as 1.0. Significant differences are indicated by asterisks (*) at $p < 0.01$ (*t*-test). Immunoblots of the AOX protein present in purified mitochondria from salt-treated and non-treated (control) shoots (**C**) and roots (**D**) harvested 9 days after the start of salt application. The numbers below each lane of the blot, represent the AOX capacity determined for the purified mitochondria for that treatment.

In this study, in shoots from salt stressed plants, *OsNDB3* was the first *ND* gene to be up-regulated (day 6) while by day 12, *OsNDA1*, *OsNDB1*, *OsNDB2* and *OsNDC1* were all up-regulated, in agreement with Genevestigator data (Figure 8; Table S5). In roots, *OsNDB3* was up-regulated at the earliest time-point, followed by *OsNDB1* at day 2 and *OsNDB2* at day 6 and *OsNDC1* at day 12 (Figure 8). Further, there was a significant decrease in *OsNDA1* and *OsNDA2* levels in roots, reminiscent of the *OsAOX* gene response in root tissue. It is unknown whether this change in *OsNDA* transcript levels reflects a change in corresponding protein levels or activities. Antibodies for each ND in rice or barley are currently not available, but an antibody raised to a peptide-specific for the *OsNDB2* isoform was developed and used to detect NDB2 protein in mitochondria isolated from shoot and root tissue. OsNDB2 protein was detectable in control shoot but not in root mitochondria (Figure 9C). There was an increase in OsNDB2 protein in mitochondria from day 9 salt treated shoot tissue but not root mitochondria, consistent with transcript data collected for this tissue (Figure 9A,B).

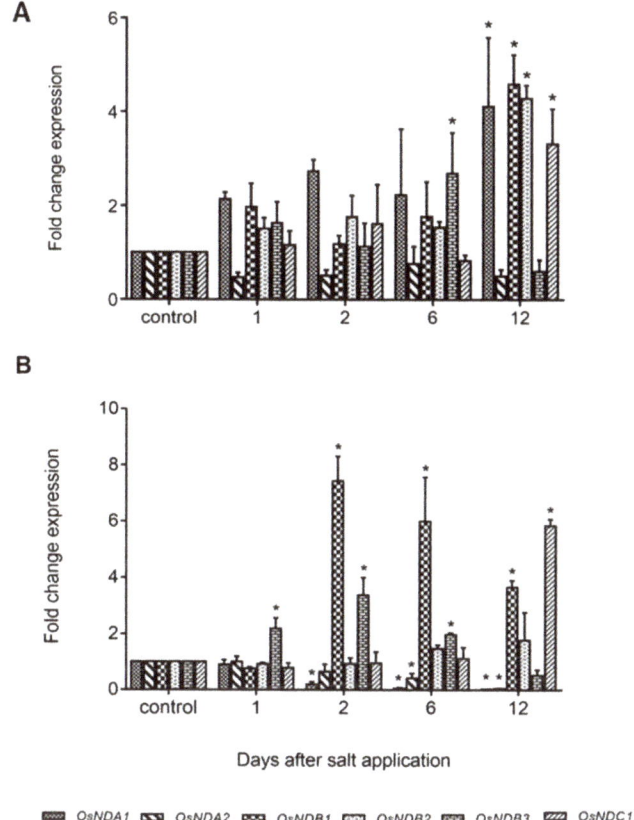

Figure 8. Expression of alternative dehydrogenases (NDs) in rice under salt stress. Fold change expression of *ND* isoforms, *OsNDA1*, *OsNDA2*, *OsNDB1*, *OsNDB2*, *OsNDB3* and *OsNDC1*, in seedling shoots (**A**) and roots (**B**) were analysed in response to 120 mM NaCl using qRT-PCR over a period of 12 days. Data are shown as the mean ± SE of three biological replicates relative to the control at each time point set as 1.00. Significant differences are indicated by asterisks (*) at $p < 0.01$ (*t*-test).

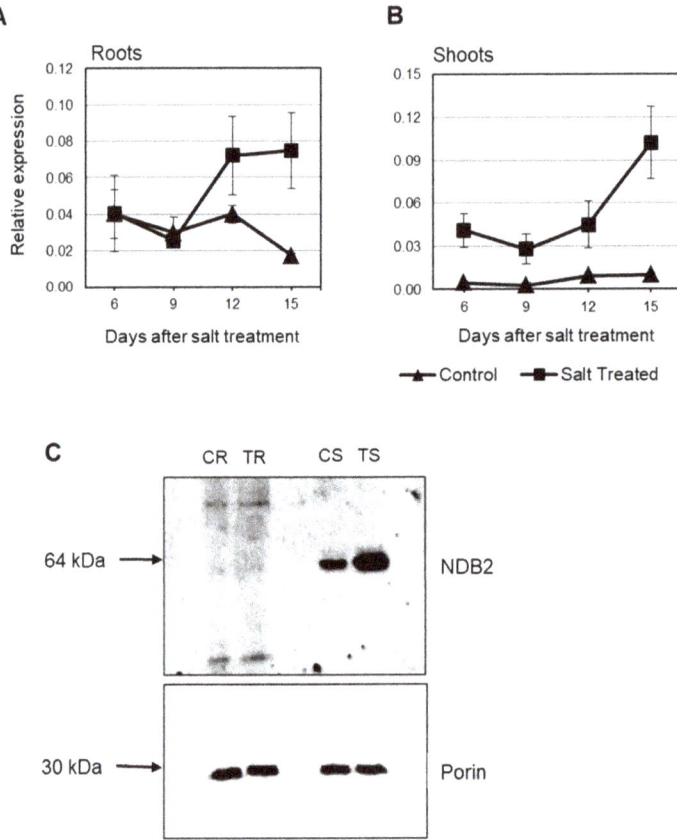

Figure 9. Expression of *OsNDB2* in rice under salt stress. Expression was analysed in seedling roots (**A**) and shoots (**B**) exposed to 120 mM NaCl over the period of 15 days using qRT-PCR. Data are shown as the mean relative gene expression ± SE of three biological replicates; (**C**). An immunoblot of the OsNDB2 protein present in mitochondria isolated from salt-treated and control shoots harvested 9 days after the start of salt application. CR-control roots; TR-treated roots; CS-control shoots; TS-treated shoots.

Genevestigator data showed that the same stress-responsive genes identified in the KCN treatment of barley were also up-regulated by many abiotic and biotic stresses (Table 3 and Table S6). *HvAOX1a*, *HvAOX1d1* and *HvAOX1d2* were the stress-responsive *AOX* isoforms, along with *HvNDB3*. No information for *HvNDB2* is available in Genevestigator. Interestingly, *HvNDA1* was reported to increase in some perturbations, but as the *HvNDA1* and *HvNDA2* show high sequence identity, it is possible that there may be a mis-identification of these genes in Genevestigator. Increases in abundance of AP gene transcripts occur with biotic stress in barley and are also seen in abiotic stresses such as cold and drought. To elaborate on the response to drought, total protein extracts were made from shoot tissue of barley grown under normal or moderate drought conditions and probed with the AOX antibody. Shoots from drought treated plants had higher AOX protein levels than control samples, indicating transcriptional and/or translational control (Figure 10). Since these protein extracts were obtained under reducing conditions, it is not clear if there is also a post-translational level of activation via disulphide reduction under these conditions, and this is worth further investigation.

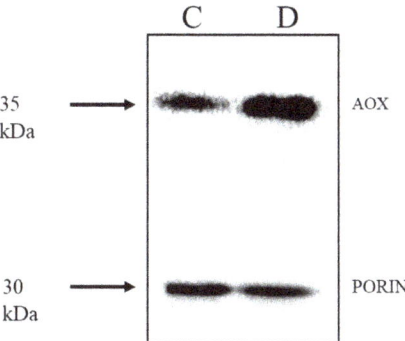

Figure 10. Detection of AOX protein in barley exposed to moderate drought stress. Soil grown 12 days old barley seedlings (*Hordeum vulgare*, cv. Golden Promise) were exposed to drought stress by withholding water for 14 days under greenhouse conditions. AOX protein was detected in total protein extracts of shoot tissue in plants grown in well-watered controls (C) and drought, withhold conditions (D) resolved under reducing conditions. The relative water contents for the plant were 83% and 69% in control and drought conditions, respectively.

It is well documented that *AtAOX1a* and *AtNDB2* show a high level of co-expression under various stress conditions [14]. Transcriptomic datasets for perturbations in Genevestigator were scanned to determine the genes showing the highest correlation of expression with *AtAOX1a* and *AtNDB2* as well as all rice and barley AP genes. Table S7 shows the level of correlation between AP genes. For example, *AtNDB2* is the sixth most correlated gene expressed with *AtAOX1a*, confirming the close co-expression relationship between these genes during stress. Interestingly, *AtNDA2* is also highly expressed with *AtNDB2* and *AtAOX1d* lies within the top 100 genes for co-expression with *AtAOX1a*.

A similar analysis of rice and barley AP genes suggests that co-expression of *AOX* and *ND* genes also occurs in monocots. In both rice and barley, the genes with strongest correlated expression are *AOX1a* and *AOX1d* (*HvAOX1d1* and *HvAOX1d2*, in barley). In rice, *OsNDB3* and *OsNDA1* are the isoforms most highly co-expressed with *AOX* isoforms. For barley, a close co-expression relationship exists between *HvNDB3*, *HvAOX1d1* and *HvAOX1d2* (Table S7).

3. General Discussion

Plant mitochondria contain a complex electron transport chain, characterised by ND and AOX proteins that constitute an alternative pathway, AP, which is not linked to energy conservation. Evidence suggests that the AP plays a role in plant abiotic stress tolerance through minimising ROS [44,46], influencing stress related amino acid synthesis [47], protecting cytochrome oxidase from ROS under photoinhibitory conditions, such as drought [48], and influencing stomatal function [49].

Many studies that have examined the role of the AP in plant growth and stress responses have focused on AOX, with the NDs often overlooked, and have rarely used monocot species. In this study, we have identified all the genes associated with the components of the AP in rice and barley, examined their expression under stress and identified the key genes involved in stress response.

As described previously, AOX is encoded by four genes in rice and barley, all clustering within the *AOX1-like* gene clade and none in the *AOX2-like* gene clade [19]. Recently, *AOX2-like* genes have been found in ancient monocot ancestors [50], but they are yet to be found in more recent monocots, such as rice. *OsAOX1a* and *OsAOX1d* were found to be located in a tandem arrangement on the same chromosome, and we now show that this section of the rice genome maps to a similar region in barley where a gene duplication has resulted in formation of *HvAOX1d1* and *HvAOX1d2*, which are closely related in sequence. These closely located genes are highly co-expressed in both rice (*OsAOX1a* and *OsAOX1d*) and barley (*HvAOX1a*, *HvAOX1d1* and *HvAOX1d2*) in the stress experiments in this study and in previous studies reported in Genevestigator data. While this co-expression was evident in previous studies where rice plants were exposed to stresses such as chilling, drought and salinity [20,22,51], this is the first report of this response for barley (Tables S6 and S7). *AOX1c* was not stress responsive in either rice or barley in our experiments. In other species there are some *AOX1* genes that do not respond to stress (e.g., *AtAOX1b* and *1c* in Arabidopsis, based on observations from the Genevestigator Perturbations data set), so this is not unique. However, the level of expression of *AOX1c* differed greatly between the two species. *OsAOX1c* was constitutively expressed in both shoot and root tissue of rice. In contrast, *HvAOX1c* was expressed at very low levels in shoots and was not detectable in roots of 12 days old barley plants, suggesting a species difference in the regulatory elements present in the promoter regions of this gene and tissue-specific roles not yet uncovered.

We also report for the first time a preliminary characterisation of the barley *ND* genes. All three subfamilies of plant *ND* genes, *NDA*, *NDB* and *NDC* are present in both rice and barley, but both species have fewer *NDB* genes than the well characterised Arabidopsis. The amino acid sequence can be a guide to substrate specificity of ND proteins. On this basis, apart from NDB1 and NDC1, all the rice and barley NDs apparently use NADH as a substrate, as described for Arabidopsis [39]. All the NDBs we have identified contain a conserved EF hand motif and so have the potential to be regulated by calcium binding, but rice and barley lack an equivalent to AtNDB4, an NADH-specific ND that does not bind calcium. The role of this protein in Arabidopsis has not yet been clarified, but its absence in rice and barley suggests a metabolic difference between these species. Further analysis revealed that the barley and rice NDB3 proteins lack the peroxisomal targeting sequences previously identified for OsNDB1 and OsNDB2 [24], suggesting that NDB3 in rice and barley is orthologous to AtNDB2.

Experiments with chemical and abiotic stress treatments show that a number of *ND* genes in both rice and barley are responsive to oxidative stress along with AOXs. The most stress-responsive of these genes was *NDB3* in both monocots, but *NDA1* or *2* and *NDB2* also responded. However, *NDB2* showed a different temporal response during KCN and salinity treatment of rice with lower, but sustained expression, possibly reflecting a response to different signals during the stress. Experiments with an antibody specific for OsNDB2 showed that the protein also increased in roots of stressed plants. Data in Genevestigator confirms the transcriptional response of these *ND* genes to a large number of stresses, both biotic and abiotic. Taken together, these observations indicate that *NDB3* is the main stress-responsive *ND* gene in rice and barley, further suggesting that it serves a similar functional role as *AtNDB2*.

A more detailed analysis of the response of rice to saline conditions also suggested post-translational regulation of AOX by stress. Transcript abundance of various *AOX* genes, but especially *OsAOX1a* and *OsAOX1d*, gradually increased in shoots but decreased in roots. The results with shoots agree with those of Ohtsu et al. [20] and Feng et al. [21]. In contrast, Feng et al. [51] reported an increase of transcript abundance of *AOX1a* and *AOX1d* when roots were exposed to 200 mM and 300 mM salt stress for 12 h, but their experimental conditions may have caused a "salt shock", which was avoided in our experiments. Ohtsu et al. [20] also reported an increase in transcript abundance of *OsAOX1a* and *OsAOX1d* under a similar 'salt shock' experiment, which then decreased gradually. In our experiments, salt concentration was gradually increased over time with several increments, perhaps resulting in different signals. Microarray data from Genevestigator clearly demonstrate the tissue-specific nature of *OsAOX1* expression in rice in response to salinity stress and the patterns of *OsAOX1a* and *OsAOX1d* expression in various tissues from different cultivars are consistent with our results. In particular, there was a reduction of *OsAOX1a* and *OsAOX1d* transcript abundance in roots from the cultivars Pokkali, IR63731, IR29 and FL478, eight days after salt application (Table S5). Despite this decrease in *AOX* gene transcript over the long term, AOX capacity increased in mitochondria isolated from salt stressed rice root tissue. This may reflect post-translational regulation by reduction of AOX from its inactive oxidised dimer to an active reduced form. Such post-translational regulation of AOX capacity has also been observed during high light stress in the shade plant *Allocasia* [52] and in soybean leaves [53]. Since OsAOX1d has a serine at the Cys$_I$ site involved in redox control of AOX activity, which will prevent its conversion to an inactive covalently linked dimer, it is likely that the protein we detected immunologically in roots was another AOX isoform.

The difference in results between the studies of Feng et al. [51], Ohtsu et al. [20] and the present study, suggests that different signals were involved in the roots under the different conditions imposed, one leading to a transcriptional upregulation of *AOX* genes, the other to post-translational regulation of existing protein. This may reflect different roles of AOX in response to a direct oxidative stress (KCN) versus salt stress where it might play a role in metabolite synthesis, but this remains to be proven. However, it does highlight the complexity of regulation of AOX and AP in plants.

Less is known currently about the response of barley AP components to stress. Up-regulation of an *AOX* gene in response to cold has been reported for wild barley [54] and heat stress (*HvAOX1a*) for the commercial cultivar Golden Promise [28]. More recently, Zhao et al. [25] and Wang et al. [26] reported an increase in AOX protein linked to varieties of highland barley showing improved tolerance to low N, UVB and salt. We have undertaken a more detailed analysis of the *AOX* and *ND* gene family to oxidative, abiotic and biotic stresses, building on the previous work and confirming that these transcriptional responses are translated into higher AOX protein levels under drought stress. This suggests that AOX may play a similar role during drought to that shown in tobacco [48]. Whether ND protein increases as well awaits the development of antibodies to detect these proteins.

In conclusion, we have shown that there is a high level of co-expression of *AOX1a* and *AOX1d*, together with *NDB3*, in response to chemical and environmental stresses in rice and barley. This is reminiscent of the co-expression that has been well characterised in Arabidopsis for *AtAOX1a* and *AtNDB2*. Our results pave the way for an assessment of the role the AP path in these important cereal crops plants and the possible use of AP components as molecular markers in breeding programs.

4. Materials and Methods

4.1. Plant Material

Oryza sativa ssp. *japonica*, cv. Nipponbare or *Hordeum vulgare*, cv. Golden Promise were used for transcript level, protein level and enzyme activity analyses as described individually for each experiment.

4.2. Gene Identification

Nucleotide and amino acid sequences for all Arabidopsis *AOXs* (*AtAOX1a*, At3g22370; *AtAOX1b*, At3g22360; *AtAOX1c*, At3g27620; *AtAOX1d*, At1g32350; *AtAOX2*, At5g64210) and NDs (*AtNDA1*, At1g07180; *AtNDA2*, At2g29990; *AtNDB1*, At4g28220; *AtNDB2*, At4g05020; *AtNDB3*, At4g21490; *AtNDB4*, At2g20800; *AtNDC1*, At5g08740) were retrieved from the The Arabidopsis Information Resource [55] and used for BLASTp/BLASTn searches against protein/genomic reference sequences in the MSU RGAP Release 7 on Rice Genome Annotation Project database [56] to identify alternative pathway homologs in japonica rice. Putative japonica rice AP genes were then used in BLASTp/BLASTn searches against *O. sativa* ssp. *indica* group sequences on the Gramene database [57] to identify AOX and ND homologs in indica rice. The rice AP sequences (Table S1) were also used for BLASTp/BLASTn searches against reference sequences of genes from NCBI [58] and the International Barley Sequencing Consortium database [59] to identify AP homologs in barley. Protein sizes were determined using the predicted amino acid sequence and the ExPASy tool [60].

4.3. Protein Relationship Analyses

A protein relationship tree of NDs from rice, barley and Arabidopsis, was constructed using the neighbour-joining method [61] in the MEGA7 computer program [62]. Protein sequences identified above (Table 1 and Table S1) were aligned in Clustal Omega [63] following standard parameters. Phylogenetic tree construction was carried out using the Neighbour-joining method and Bootstrap of 1000 replications; "number of differences" as the substitution model and "complete deletion" for gaps/missing data treatment.

4.4. Subcellular Localization

Computational determination of the subcellular localization of identified AP genes from rice and barley was carried out using publicly available subcellular localization prediction programs; TargetP [64], Predotar [65], ChloroP 1.1 [66] and Plant-mPLOC [67]. Proteins targeting to the peroxisomes were predicted using PredPlantPTS1 [68].

4.5. Genevestigator Expression Analyses

Microarray probe-set IDs with high homology were retrieved for all Arabidopsis, rice and barley AP genes, except barley HvNDB2, which was not represented on the Affymetrix Barley Genome Array chip. At the time of analysis, the transcriptomic databases for rice and barley contained data for 2656 and 1822 samples, respectively, and featured many different stress experiments. The 'Perturbation' dataset was screened using the Condition Search Tool within Genevestigator [69] to assess the stress-responsiveness of each gene in all microarray datasets available on Genevestigator. Fold changes for each AP gene in response to selected biotic and abiotic stress treatments are given in Tables S5 (rice) and S6 (barley). The Similarity Search Tool for Co-expression was also applied to the "Perturbation" dataset to identify genes that may be co-expressed with particular AP genes from Arabidopsis, rice and barley during stress (Table S7).

4.6. Screening for Responsive AP Genes under Oxidative Stress

To generate oxidative stress, rice seedlings (*O. sativa* ssp. *japonica*, cv. Nipponbare) were grown on Petri dishes for three weeks, then exposed to the respiratory inhibitor, potassium cyanide (KCN, 5 mM) and barley seedlings (*H. vulgare*, cv. Golden Promise) were grown for two weeks on damp paper towel in petri dishes, then exposed to KCN (5 mM). RNA for gene expression analysis was extracted from shoot and root tissue collected at 6 h and 24 h for both KCN treated and control (no KCN) tissue.

O. sativa (cv. Nipponbare) seedlings grown in a hydroponic system, were exposed to salinity stress following transplantation of the seedlings (7–10 days old) to 12 L hydroponic tanks filled with constantly aerated nutrient solution [70], at 26 °C/22 °C (day/night), 12 h photoperiod and 75% relative humidity, 400–500 µE m^{-2}·s^{-1}. The solution was changed after two weeks and 120 mM salt stress was applied in two increments, 50 mM followed by a further 70 mM NaCl two days later. Supplementary CaCl$_2$, 0.75 and 1 mM, was added respectively to keep constant activity of calcium. Salt was applied gradually for several increments, to avoid a salt "shock" and calcium was present to remove any complications from an induced calcium deficiency that can occur in the presence of sodium ions [71]. Root and shoot tissues were harvested from salt-treated and control tanks at pre-determined time points, and either snap-frozen in liquid nitrogen for gene expression analyses or used fresh for mitochondrial isolation and enzyme activity assays.

Drought tolerance assays were carried out using barley seedlings grown in soil. Seedlings (7 days old) of *H. vulgare* (cv. Golden Promise) were transferred from Petri dishes into Osmocote soil in pots. Each pot contained the same dry weight of soil and the same volume of water. The pants were grown in a glasshouse for 7 days at 23 °C/15 °C (day/night), 13 h photoperiod (1500 µmol m^{-2}·s^{-1}), with equal watering (to weight) before the drought treatment was initiated by completely withholding water for 14 days. Control plants continued to be watered as necessary. Shoot tissues were harvested from drought treated and control plants, immediately snap-frozen in liquid nitrogen and subsequently used for protein extraction and immunoblotting analyses.

4.7. Gene Expression Analysis Using qRT-PCR

Samples were snap frozen in liquid nitrogen then stored at −80 °C until use. In experiments with rice, the leaves (two youngest) and root were sampled separately. In experiments with barley the entire shoot and root were harvested separately. Total RNA was extracted using the TRIZOL-like extraction method [72] following recent modifications [71]. Genomic DNA contamination was eliminated by DNase treatment using DNase I (Invitrogen-ThermoFisher, Waltham, MA, USA). Single stranded cDNA synthesis was performed on 1 μg of DNase-treated RNA using iScript cDNA Synthesis Kit (Bio-Rad, Hercules, CA, USA). The qRT-PCR analysis was performed with a CFX96 Real Time PCR Detection System (Bio-Rad). For rice, reactions (10 μL) included SsoFast EvaGreen Supermix Master Mix (Bio-Rad) and 0.6 μM gene-specific primers (Table S8). For barley, reactions (10 μL) included KAPA SYBR-Fast qPCR Universal ReadyMix (Geneworks, Thebarton, Australia) and 0.2 μM gene-specific primers (Table S8).

The gene expression data obtained were normalized using the geometric average expression of three stable reference genes. For rice these were *OsElF1* [70], *OsPplase* [73] and *OsActin* [73]. For barley, these were *HvActin*, *HvUbiquitin* and *HvPdf* (Protein phosphatase) based on their stable expression levels according to data from Genevestigator.

4.8. Protein and Enzyme Activity Assays in Isolated Rice Tissue Mitochondria

Mitochondrial isolation was carried out following the method described by Kristensen, et al. [74] with slight modifications. Purification of isolated mitochondria was carried out using a discontinuous Percoll gradient, which consisted of four layers: 6 mL of 60% (v/v) Percoll, 8 mL of 45% (v/v) Percoll, 10 mL of 28% (v/v) Percoll and 10 mL of 5% (v/v) Percoll.

4.9. Oxygen Electrode Assays

The oxygen uptake rate of mitochondria isolated from rice shoots or roots was measured in a standard reaction medium (0.3 M mannitol, 10 mM TES, 10 mM KH_2PO_4, 2 mM $MgCl_2$, pH 7.2 at 25 °C using a Clark-type Oxygraph Plus oxygen electrode system (Hansatech Instruments Ltd., Pentney, UK). Total oxygen consumption rate of AOX was measured in the presence of the following components; 10 mM succinate, 5 mM pyruvate, 1 mM NADH, 1mM ATP, 1 mM ADP, 5 mM DTT. The cyanide resistant and the residual oxygen consumption rates were determined in the presence of 1 mM KCN and 250 nM OG (Octyl Gallate), respectively.

4.10. Immunoblot Analysis

Detection of AP protein levels in salt-treated and control rice shoot and root tissues was carried out on isolated mitochondria resolved on a 10% (v/v) Tris-glycine SDS-PAGE followed by immunoblotting. For barley tissue samples, protein was extracted from whole shoot or root tissue following the procedure of Umbach, et al. [75]. For both rice mitochondria and barley total protein extracts, samples were prepared either in the presence or absence of 200 mM DTT as indicated. Blots were probed with the monoclonal AOA antibody, with the peptide recognition sequence of RADEAHHRDVNH [76], or anti-porin antibody to show equal loading of samples on gel (supplied by Professor Harvey Millar, University of Western Australia).

The abundance of NDB2 protein was estimated by probing immunoblots with 1:8000 diluted polyclonal antibody OsNDB2, that was prepared by Biomatik (http://www.biomatik.com) with the peptide recognition sequence of CQDNQVLQINDGTGKKR.

4.11. Statistical Analyses

Data are generally expressed as mean ± SEM (Standard Error of the Mean). Analysis of variance or *t*-test was performed to determine whether there were significant differences between means of control and treated samples at 95% confidence level using GraphPad Prism (La Jolla, CA, USA).

Supplementary Materials: Supplementary materials can be found at http://www.mdpi.com/1422-0067/19/3/915/s1.

Author Contributions: Vajira R. Wanniarachchi and Lettee Dametto performed all the experimental work and conducted the bioinformatic and statistical analyses. Crystal Sweetman, Colin Jenkins and Yuri Shavrukov carried out bioinformatics analyses. Colin L. D. Jenkins, Kathleen L. Soole, David A. Day and Yuri Shavrukov conceived and designed the study.

Conflicts of Interest: The authors declare no conflict of interest.

References

1. Millar, A.H.; Whelan, J.; Soole, K.L.; Day, D.A. Organization and Regulation of Mitochondrial Respiration in Plants. In *Annual Review of Plant Biology*; Merchant, S.S., Briggs, W.R., Ort, D., Eds.; Annual Reviews: Palo Alto, CA, USA, 2011; Volume 62, pp. 79–104.
2. Finnegan, P.M.; Soole, K.L.; Umbach, A.L. Alternative Mitochondrial Electron Transport Proteins in Higher Plants. In *Plant Mitochondria: From Genome to Function*; Day, D.A., Millar, H., Whelan, J., Eds.; Kluwer Academic Publishers: Dordrecht, The Netherlands, 2004; pp. 163–230.
3. Rasmusson, A.G.; Soole, K.L.; Elthon, T.E. Alternative NAD(P)H dehydrogenases of plant mitochondria. *Annu. Rev. Plant Biol.* **2004**, *55*, 23–39. [CrossRef] [PubMed]
4. Vanlerberghe, G.C. Alternative oxidase: A mitochondrial respiratory pathway to maintain metabolic and signaling homeostasis during abiotic and biotic stress in plants. *Int. J. Mol. Sci.* **2013**, *14*, 6805–6847. [CrossRef] [PubMed]
5. Millar, A.H.; Wiskich, J.T.; Whelan, J.; Day, D.A. Organic-acid activation of the alternative oxidase of plant-mitochondria. *FEBS Lett.* **1993**, *329*, 259–262. [CrossRef]
6. Hoefnagel, M.H.N.; Millar, A.H.; Wiskich, J.T.; Day, D.A. Cytochrome and alternative respiratory pathways compete for electrons in the presence of pyruvate in soybean mitochondria. *Arch. Biochem. Biophys.* **1995**, *318*, 394–400. [CrossRef] [PubMed]
7. Purvis, A.C. Role of the alternative oxidase in limiting superoxide production by plant mitochondria. *Physiol. Plant.* **1997**, *100*, 165–170. [CrossRef]
8. Popov, V.N.; Purvis, A.C.; Skulachev, V.P.; Wagner, A.M. Stress-induced changes in ubiquinone concentration and alternative oxidase in plant mitochondria. *Biosci. Rep.* **2001**, *21*, 369–379. [CrossRef] [PubMed]
9. Bailey, C.D.; Carr, T.G.; Harris, S.A.; Hughes, C.E. Characterization of angiosperm nrDNA polymorphism, paralogy, and pseudogenes. *Mol. Phylgenet. Evol.* **2003**, *29*, 435–455. [CrossRef]
10. Moore, C.S.; Cook-Johnson, R.J.; Rudhe, C.; Whelan, J.; Day, D.A.; Wiskich, J.T.; Soole, K.L. Identification of AtNDI1, an internal non-phosphorylating NAD(P)H dehydrogenase in Arabidopsis mitochondria. *Plant Physiol.* **2003**, *133*, 1968–7198. [CrossRef] [PubMed]
11. Elhafez, D.; Murcha, M.W.; Clifton, R.; Soole, K.L.; Day, D.A.; Whelan, J. Characterization of mitochondrial alternative NAD(P)H dehydrogenases in arabidopsis: Intraorganelle location and expression. *Plant Cell Physiol.* **2006**, *47*, 43–54. [CrossRef] [PubMed]
12. Saish, D.; Nakazono, M.; Lee, K.H.; Tsutsumi, N.; Akita, S.; Hirai, A. The gene for alternative oxidase-2 (AOX2) from Arabidopsis thaliana consists of five exons unlike other AOX genes and is transcribed at an early stage during germination. *Genes Genet. Syst.* **2001**, *76*, 89–97. [CrossRef] [PubMed]
13. Borecky, J.; Nogueira, F.T.; de Oliveira, K.A.; Maia, I.G.; Vercesi, A.E.; Arruda, P. The plant energy-dissipating mitochondrial systems: Depicting the genomic structure and the expression profiles of the gene families of uncoupling protein and alternative oxidase in monocots and dicots. *J. Exp. Bot.* **2006**, *57*, 849–864. [CrossRef] [PubMed]
14. Clifton, R.; Lister, R.; Parker, K.L.; Sappl, P.G.; Elhafez, D.; Millar, A.H.; Day, D.A.; Whelan, J. Stress-induced co-expression of alternative respiratory chain components in *Arabidopsis thaliana*. *Plant Mol. Biol.* **2005**, *58*, 193–212. [CrossRef] [PubMed]
15. Wang, J.; Rajakulendran, N.; Amirsadeghi, S.; Vanlerberghe, G.C. Impact of mitochondrial alternative oxidase expression on the response of *Nicotiana tabacum* to cold temperature. *Physiol. Plant.* **2011**, *142*, 339–351. [CrossRef] [PubMed]
16. Ito, Y.; Saisho, D.; Nakazono, M.; Tsutsumi, N.; Hirai, A. Transcript levels of tandem-arranged alternative oxidase genes in rice are increased by low temperature. *Gene* **1997**, *203*, 121–129. [CrossRef]

17. Considine, M.J.; Holtzapffel, R.C.; Day, D.A.; Whelan, J.; Millar, A.H. Molecular distinction between alternative oxidase from monocots and dicots. *Plant Physiol.* **2002**, *129*, 949–953. [CrossRef] [PubMed]
18. Saika, H.; Ohtsu, K.; Hamanaka, S.; Nakazono, M.; Tsutsumi, N.; Hirai, A. *AOX1c*, a novel rice gene for alternative oxidase; comparison with rice *AOX1a* and *AOX1b*. *Genes Genet. Syst.* **2002**, *77*, 31–38. [CrossRef] [PubMed]
19. Costa, J.H.; McDonald, A.E.; Arnholdt-Schmitt, B.; de Melo, D.F. A classification scheme for alternative oxidases reveals the taxonomic distribution and evolutionary history of the enzyme in angiosperms. *Mitochondrion* **2014**, *19*, 172–183. [CrossRef] [PubMed]
20. Ohtsu, K.; Ito, Y.; Saika, H.; Nakazono, M.; Tsutsumi, N.; Hirai, A. ABA-independent expression of rice alternative oxidase genes under envinronmental stresses. *Plant Biotechnol.* **2002**, *19*, 187–190. [CrossRef]
21. Feng, H.Q.; Wang, Y.F.; Li, H.Y.; Wang, R.F.; Sun, K.; Jia, L.Y. Salt stress-induced expression of rice AOX1a is mediated through an accumulation of hydrogen peroxide. *Biologia* **2010**, *65*, 868–873. [CrossRef]
22. Li, C.R.; Liang, D.D.; Xu, R.F.; Li, H.; Zhang, Y.P.; Qin, R.Y.; Li, L.; Wei, P.C.; Yang, J.B. Overexpression of an alternative oxidase gene, OsAOX1a, improves cold tolerance in *Oryza sativa* L. *Genet. Mol. Res.* **2013**, *12*, 5424–5432. [CrossRef] [PubMed]
23. Oono, Y.; Yazawa, T.; Kawahara, Y.; Kanamori, H.; Kobayashi, F.; Sasaki, H.; Mori, S.; Wu, J.Z.; Handa, H.; Itoh, T.; et al. Genome-wide transcriptome analysis reveals that cadmium stress signaling controls the expression of genes in drought stress signal pathways in rice. *PLoS ONE* **2014**, *9*, e96946. [CrossRef] [PubMed]
24. Xu, L.; Law, S.R.; Murcha, M.W.; Whelan, J.; Carrie, C. The dual targeting ability of type II NAD(P)H dehydrogenases arose early in land plant evolution. *BMC Plant Biol.* **2013**, *13*, 100. [CrossRef] [PubMed]
25. Zhao, C.Z.; Wang, X.M.; Wang, X.Y.; Wu, K.L.; Li, P.; Chang, N.; Wang, J.F.; Wang, F.; Li, J.L.; Bi, Y.R. Glucose-6-phosphate dehydrogenase and alternative oxidase are involved in the cross tolerance of highland barley to salt stress and UV-B radiation. *J. Plant Physiol.* **2015**, *181*, 83–95. [CrossRef] [PubMed]
26. Wang, F.; Wang, X.M.; Zhao, C.Z.; Wang, J.F.; Li, P.; Dou, Y.Q.; Bi, Y.R. Alternative pathway is involved in the tolerance of highland barley to the low-nitrogen stress by maintaining the cellular redox homeostasis. *Plant Cell Rep.* **2016**, *35*, 317–328. [CrossRef] [PubMed]
27. Arnholdt-Schmitt, B.; Costa, J.H.; de Melo, D.F. AOX—A functional marker for efficient cell reprogramming under stress? *Trends Plant Sci.* **2006**, *11*, 281–287. [CrossRef] [PubMed]
28. Mangelsen, E.; Kilian, J.; Harter, K.; Jansson, C.; Wanke, D.; Sundberg, E. Transcriptome analysis of high-temperature stress in developing barley caryopses: Early stress responses and effects on storage compound biosynthesis. *Mol. Plant* **2011**, *4*, 97–115. [CrossRef] [PubMed]
29. Michalecka, A.M.; Svensson, A.S.; Johansson, F.I.; Agius, S.C.; Johanson, U.; Brennicke, A.; Binder, S.; Rasmusson, A.G. Arabidopsis genes encoding mitochondrial type II NAD(P)H dehydrogenases have different evolutionary orgin and show distinct responses to light. *Plant Physiol.* **2003**, *133*, 642–652. [CrossRef] [PubMed]
30. Carrie, C.; Murcha, M.W.; Kuehn, K.; Duncan, O.; Barthet, M.; Smith, P.M.; Eubel, H.; Meyer, E.; Day, D.A.; Millar, A.H.; et al. Type II NAD(P)H dehydrogenases are targeted to mitochondria and chloroplasts or peroxisomes in *Arabidopsis thaliana*. *FEBS Lett.* **2008**, *582*, 3073–3079. [CrossRef] [PubMed]
31. Umbach, A.L.; Ng, V.S.; Siedow, J.N. Regulation of plant alternative oxidase activity: A tale of two cysteines. *Biochimica Et Biophysica Acta-Bioenergetics* **2006**, *1757*, 135–142. [CrossRef] [PubMed]
32. Rhoads, D.M.; Umbach, A.L.; Sweet, C.R.; Lennon, A.M.; Rauch, G.S.; Siedow, J.N. Regulation of the cyanide-resistant alternative oxidase of plant mitochondria—Identification of the cysteine residue involved in α-keto acid stimulation and intersubunit disulfide bond formation. *J. Biol. Chem.* **1998**, *273*, 30750–30756. [CrossRef] [PubMed]
33. Vanlerberghe, G.C.; Yio, J.Y.H.; Parsons, H.L. In organello and in vivo evidence of the importance of the regulatory sulfhydryl/disulfide system and pyruvate for alternative oxidase activity in tobacco. *Plant Physiol.* **1999**, *121*, 793–803. [CrossRef] [PubMed]
34. Djajanegara, I.; Holtzapffel, R.; Finnegan, P.M.; Hoefnagel, M.H.N.; Berthold, D.A.; Wiskich, J.T.; Day, D.A. A single amino acid change in the plant alternative oxidase alters the specificity of organic acid activation. *FEBS Lett.* **1999**, *454*, 220–224. [PubMed]
35. Selinski, J.; Hartmann, A.; Kordes, A.; Deckers-Hebestreit, G.; Whelan, J.; Scheibe, R. Analysis of posttranslational activation of alternative oxidase isoforms. *Plant Physiol.* **2017**, *174*, 2113–2127. [CrossRef] [PubMed]

36. Millar, A.H.; Hoefnagel, M.H.N.; Day, D.A.; Wiskich, J.T. Specificity of the organic acid activation of alternative oxidase in plant mitochondria. *Plant Physiol.* **1996**, *111*, 613–618. [CrossRef] [PubMed]
37. Holtzapffel, R.C.; Castelli, J.; Finnegan, P.M.; Millar, A.H.; Whelan, J.; Day, D.A. A tomato alternative oxidase protein with altered regulatory properties. *Biochim. Biophys. Acta* **2003**, *1606*, 153–162. [CrossRef]
38. Michalecka, A.M.; Agius, S.C.; Møller, I.M.; Rasmusson, A.G. Identification of a mitochondrial external NADPH dehydrogenase by overexpression in transgenic Nicotiana sylvestris. *Plant J.* **2004**, *37*, 415–425. [CrossRef] [PubMed]
39. Geisler, D.A.; Broselid, C.; Hederstedt, L.; Rasmusson, A.G. Ca^{2+}-binding and Ca^{2+}-independent respiratory NADH and NADPH dehydrogenases of *Arabidopsis thaliana*. *J. Biol. Chem.* **2007**, *282*, 28455–28464. [CrossRef] [PubMed]
40. Fatihi, A.; Latimer, S.; Schmollinger, S.; Block, A.; Dussault, P.H.; Vermaas, W.F.J.; Merchant, S.S.; Basset, G.J. A dedicated type II NADPH dehydrogenase performs the penultimate step in the biosynthesis of Vitamin K-1 in Synechocystis and Arabidopsis. *Plant Cell* **2015**, *27*, 1730–1741. [CrossRef] [PubMed]
41. Melo, A.M.P.; Roberts, T.H.; Moller, I.M. Evidence for the presence of two rotenone-insensitive NAD(P)H dehydrogenases on the inner surface of the inner membrane of potato tuber mitochondria. *Biochim. Biophys. Acta* **1996**, *1276*, 133–139. [CrossRef]
42. Sweetman, C.; Waterman, C.D.; Rainbird, B.M.; Smith, P.M.C.; Jenkins, C.L.D.; Day, D.A.; Soole, K.L. AtNDB2 is the main external NADH dehydrogenase in Arabidopsis mitochondria and is important for tolerance to environmental stress. Unpublished work, 2018.
43. Rasmusson, A.G.; Svensson, A.S.; Knoop, V.; Grohmann, L.; Brennicke, A. Homologues of yeast and bacterial rotenone-insensitive NADH dehydrogenases in higher eukaryotes: Two enzymes are present in potato mitochondria. *Plant J.* **1999**, *20*, 79–87. [CrossRef] [PubMed]
44. Maxwell, D.P.; Wang, Y.; McIntosh, L. The alternative oxidase lowers mitochondrial reactive oxygen production in plant cells. *Proc. Natl. Acad. Sci. USA* **1999**, *96*, 8271–8276. [CrossRef] [PubMed]
45. Tanudji, M.; Djajanegara, I.N.; Daley, D.O.; McCabe, T.C.; Finnegan, P.M.; Day, D.A.; Whelan, J. The multiple alternative oxidase proteins of soybean. *Aust. J. Plant Physiol.* **1999**, *26*, 337–344. [CrossRef]
46. Smith, C.A.; Melino, V.J.; Sweetman, C.; Soole, K.L. Manipulation of alternative oxidase can influence salt tolerance in *Arabidopsis thaliana*. *Physiol. Plant.* **2009**, *137*, 459–472. [CrossRef] [PubMed]
47. Florez-Sarasa, I.; Ribas-Carbo, M.; Del-Saz, N.F.; Schwahn, K.; Nikoloski, Z.; Fernie, A.R.; Flexas, J. Unravelling the in vivo regulation and metabolic role of the alternative oxidase pathway in C-3 species under photoinhibitory conditions. *New Phytol.* **2016**, *212*, 66–79. [CrossRef] [PubMed]
48. Dahal, K.; Vanlerberghe, G.C. Alternative oxidase respiration maintains both mitochondrial and chloroplast function during drought. *New Phytol.* **2017**, *213*, 560–571. [CrossRef] [PubMed]
49. Cvetkovska, M.; Dahal, K.; Alber, N.A.; Jin, C.; Cheung, M.; Vanlerberghe, G.C. Knockdown of mitochondrial alternative oxidase induces the 'stress state' of signaling molecule pools in *Nicotiana tabacum*, with implications for stomatal function. *New Phytol.* **2014**, *203*, 449–461. [CrossRef] [PubMed]
50. Costa, J.H.; Santos, C.P.; de Sousa E Lima, B.; Moreira Netto, A.N.; Saraiva, K.D.; Arnholdt-Schmitt, B. In silico identification of alternative oxidase 2 (AOX2) in monocots: A new evolutionary scenario. *J. Plant Physiol.* **2017**, *210*, 58–63. [CrossRef] [PubMed]
51. Feng, H.Q.; Hou, X.L.; Li, X.; Sun, K.; Wang, R.F.; Zhang, T.G.; Ding, Y.P. Cell death of rice roots under salt stress may be mediated by cyanide-resistant respiration. *Z. Naturforsch. C* **2013**, *68*, 39–46. [CrossRef] [PubMed]
52. Noguchi, K.; Taylor, N.L.; Millar, A.H.; Lambers, H.; Day, D.A. Response of mitochondria to light intensity in the leaves of sun and shade species. *Plant Cell Environ.* **2005**, *28*, 760–771. [CrossRef]
53. Ribas-Carbo, M.; Taylor, N.L.; Giles, L.; Busquets, S.; Finnegan, P.M.; Day, D.A.; Lambers, H.; Medrano, H.; Berry, J.A.; Flexas, J. Effects of water stress on respiration in soybean leaves. *Plant Physiol.* **2005**, *139*, 466–473. [CrossRef] [PubMed]
54. Abu-Romman, S.; Shatnawi, M.; Hasan, M.; Qrunfleh, I.; Omar, S.; Salem, N. cDNA cloning and expression analysis of a putative alternative oxidase HsAOX1 from wild barley (*Hordeum spontaneum*). *Genes Genom.* **2012**, *34*, 59–66. [CrossRef]
55. The Arabidopsis Information Resource. Available online: https://www.arabidopsis.org (accessed on 19 March 2018).

56. Kawahara, Y.; de la Bastide, M.; Hamilton, J.P.; Kanamori, H.; McCombie, W.R.; Ouyang, S.; Schwartz, D.C.; Tanaka, T.; Wu, J.; Zhou, S.; et al. Improvement of the Oryza sativa Nipponbare reference genome using next generation sequence and optical map data. *Rice.* **2013**, *6*, 4. [CrossRef] [PubMed]
57. Gramene. Available online: www.gramene (accessed on 19 March 2018).
58. National Center for Biotechnology Information. Available online: https://www.ncbi.nlm.nih.gov (accessed on 19 March 2018).
59. IPK Barley Blast Server (International Barley Sequencing Consortium). Available online: http://webblast.ipk-gatersleben.de/barley_ibsc (accessed on 19 March 2018).
60. ExPASy Bioinformatics Resource Portal. Available online: https://web.expasy.org (accessed on 19 March 2018).
61. Saitou, N.; Nei, M. The neighbor-joining method—A new method for reconstructing phylogenetic trees. *Mol. Biol. Evol.* **1987**, *4*, 406–425. [PubMed]
62. Kumar, S.; Stecher, G.; Tamura, K. Mega7: Molecular evolutionary genetics analysis version 7.0 for bigger datasets. *Mol. Biol. Evol.* **2016**, *33*, 1870–1874. [CrossRef] [PubMed]
63. Clustal Omega Tool. Available online: https://www.ebi.ac.uk/Tools/msa/clustalo (accessed on 19 March 2018).
64. Emanuelsson, O.; Nielsen, H.; Brunak, S.; von Heijne, G. Predicting subcellular localization of proteins based on their N-terminal amino acid sequence. *J. Mol. Biol.* **2000**, *300*, 1005–1016. [CrossRef] [PubMed]
65. Predotar Tool, Plant and Fungi Data Integration Web Site. Available online: https://urgi.versailles.inra.fr/Tools/Predotar (accessed on 19 March 2018).
66. Emanuelsson, O.; Nielsen, H.; von Heijne, G. ChloroP, a neural network-based method for predicting chloroplast transit peptides and their cleavage sites. *Protein Sci.* **1999**, *8*, 978–984. [CrossRef] [PubMed]
67. Chou, K.; Shen, H. Plant-mPLoc: A top-down strategy to augment the power for predicting plant protein subcellular localization. *PLoS ONE* **2010**, *5*, e11335. [CrossRef] [PubMed]
68. Reumann, S.; Buchwald, D.; Lingner, T. PredPlantPTS1: A web server for the prediction of plant peroxisomal proteins. *Front. Plant. Sci.* **2012**, *3*, 194. [CrossRef] [PubMed]
69. Hruz, T.; Laule, O.; Szabo, G.; Wessendorp, F.; Bleuler, S.; Oertle, L.; Widmayer, P.; Gruissem, W.; Zimmermann, P. Genevestigator V3: A reference expression database for the meta-analysis of transcriptomes. *Adv. Bioinform.* **2008**, *420747*. [CrossRef] [PubMed]
70. Cotsaftis, O.; Plett, D.; Johnson, A.A.T.; Walia, H.; Wilson, C.; Ismail, A.M.; Close, T.J.; Tester, M.; Baumann, U. Root-Specific Transcript Profiling of Contrasting Rice Genotypes in Response to Salinity Stress. *Mol. Plant* **2011**, *4*, 25–41. [CrossRef] [PubMed]
71. Shavrukov, Y.; Bovill, J.; Afzal, I.; Hayes, J.E.; Roy, S.J.; Tester, M.; Collins, N.C. HVP10 encoding V-PPase is a prime candidate for the barley *HvNax3* sodium exclusion gene: Evidence from fine mapping and expression analysis. *Planta* **2013**, *237*, 1111–1122. [CrossRef] [PubMed]
72. Chomczynski, P.; Sacchi, N. Single-step method of RNA isolation by acid guanidinium thiocyanate phenol chloroform extraction. *Anal. Biochem.* **1987**, *162*, 156–159. [CrossRef]
73. Kim, B.-R.; Nam, H.-Y.; Kim, S.-U.; Kim, S.-I.; Chang, Y.-J. Normalization of reverse transcription quantitative-PCR with housekeeping genes in rice. *Biotechnol. Lett.* **2003**, *25*, 1869–1872. [CrossRef] [PubMed]
74. Kristensen, B.K.; Askerlund, P.; Bykova, N.V.; Egsgaard, H.; Moller, I.M. Identification of oxidised proteins in the matrix of rice leaf mitochondria by immunoprecipitation and two-dimensional liquid chromatography-tandem mass spectrometry. *Phytochemistry* **2004**, *65*, 1839–1851. [CrossRef] [PubMed]
75. Umbach, A.L.; Fiorani, F.; Siedow, J.N. Characterization of transformed Arabidopsis with altered alternative oxidase levels and analysis of effects on reactive oxygen species in tissue. *Plant Physiol.* **2005**, *139*, 1806–1820. [CrossRef] [PubMed]
76. Finnegan, P.M.; Wooding, A.R.; Day, D.A. An alternative oxidase monoclonal antibody recognises a highly conserved sequence among alternative oxidase subunits. *FEBS Lett.* **1999**, *447*, 21–24. [CrossRef]

© 2018 by the authors. Licensee MDPI, Basel, Switzerland. This article is an open access article distributed under the terms and conditions of the Creative Commons Attribution (CC BY) license (http://creativecommons.org/licenses/by/4.0/).

Article

Suppression of External NADPH Dehydrogenase—NDB1 in *Arabidopsis thaliana* Confers Improved Tolerance to Ammonium Toxicity via Efficient Glutathione/Redox Metabolism

Anna Podgórska [1,*], Monika Ostaszewska-Bugajska [1], Klaudia Borysiuk [1], Agata Tarnowska [1], Monika Jakubiak [1], Maria Burian [1], Allan G. Rasmusson [2] and Bożena Szal [1,*]

[1] Institute of Experimental Plant Biology and Biotechnology, Faculty of Biology, University of Warsaw, I. Miecznikowa 1, 02-096 Warsaw, Poland; m.ostaszewska@biol.uw.edu.pl (M.O.-B.); k.borysiuk@biol.uw.edu.pl (K.B.); atarnowska@biol.uw.edu.pl (A.T.); monika.jakubiak@student.uw.edu.pl (M.J.); mburian@biol.uw.edu.pl (M.B.)

[2] Department of Biology, Lund University, Sölvegatan 35B, SE-223 62 Lund, Sweden; allan.rasmusson@biol.lu.se

* Correspondence: apodgorski@biol.uw.edu.pl (A.P.); szal@biol.uw.edu.pl (B.S.); Tel.: +48-22-55-43-009 (A.P.); +48-22-55-43-005 (B.S.)

Received: 17 April 2018; Accepted: 2 May 2018; Published: 9 May 2018

Abstract: Environmental stresses, including ammonium (NH_4^+) nourishment, can damage key mitochondrial components through the production of surplus reactive oxygen species (ROS) in the mitochondrial electron transport chain. However, alternative electron pathways are significant for efficient reductant dissipation in mitochondria during ammonium nutrition. The aim of this study was to define the role of external NADPH-dehydrogenase (NDB1) during oxidative metabolism of NH_4^+-fed plants. Most plant species grown with NH_4^+ as the sole nitrogen source experience a condition known as "ammonium toxicity syndrome". Surprisingly, transgenic *Arabidopsis thaliana* plants suppressing *NDB1* were more resistant to NH_4^+ treatment. The *NDB1* knock-down line was characterized by milder oxidative stress symptoms in plant tissues when supplied with NH_4^+. Mitochondrial ROS accumulation, in particular, was attenuated in the *NDB1* knock-down plants during NH_4^+ treatment. Enhanced antioxidant defense, primarily concerning the glutathione pool, may prevent ROS accumulation in NH_4^+-grown *NDB1*-suppressing plants. We found that induction of glutathione peroxidase-like enzymes and peroxiredoxins in the *NDB1*-surpressing line contributed to lower ammonium-toxicity stress. The major conclusion of this study was that NDB1 suppression in plants confers tolerance to changes in redox homeostasis that occur in response to prolonged ammonium nutrition, causing cross tolerance among plants.

Keywords: ammonium toxicity; external type II NADPH dehydrogenase; glutathione metabolism; reactive oxygen species; redox homeostasis

---

1. Introduction

Cellular oxidation-reduction status functions as an integrator of subcellular metabolism and responds to signals from the external environment [1–3]. NADP(H) and NAD(H) mediate the flow of reductive power between cellular processes [4], and NAD(P)H can be oxidized by the mitochondrial electron transport chain (mtETC), located in the inner mitochondrial membrane. In plants, mtETC is composed of four multi-subunit complexes: complex I (NADH dehydrogenase), complex II (succinate dehydrogenase), complex III (cytochrome *c* reductase), and complex IV (cytochrome *c* oxidase). Plant mitochondria also possess unique electron routes that can bypass the pathway from complexes

I or II to complex IV. These additional electron pathways include external and internal type II dehydrogenases (NDex and NDin, respectively) and the alternative oxidase (AOX) [5,6]. NAD(P)H dehydrogenases are encoded by three gene families (NDA, NDB and NDC). NDex enzymes encoded by *NDB1* utilize NADPH, whereas the enzymes encoded by *NDB2*, *NDB3*, and *NDB4* utilize NADH; NDB1 and NDB2 isoforms are also Ca^{2+}-dependent or Ca^{2+}-stimulated [6–8]. The NDin isoforms NDA1, NDA2, and NDC1 utilize NADH, while the NDC1 isoform may also utilize NADPH. AOX genes belong to two subfamilies, which in *Arabidopsis* are composed of *AOX1a-d* and *AOX2* [9].

The path of electrons from complex I through ubiquinone to complexes III and IV (cytochrome pathway) couples the oxidation of reductants to the reduction of O_2 to H_2O. The accompanying translocation of protons builds an electrochemical proton gradient that drives oxidative phosphorylation by ATP synthase (ATPase) to produce energy-rich ATP molecules. In contrast with the cytochrome pathway, whose activity is under adenylate control, the additional "alternative" routes are not linked to ATP synthesis or controlled by adenylates. This is because the AOX, NDin, and NDex pathways dissipate excess reductants without proton motive force generation. This prevents over-reduction of the ubiquinone pool or hyperpolarization of the mitochondrial membrane potential, which could inhibit respiration. Plant mitochondria play a central role as reductant sinks, regulating cellular redox homeostasis and preventing redox poise in cells [4,10]. Although the additional pathways do not contribute to ATP production, their activity is crucial to minimize the production of reactive oxygen species (ROS). Alternative pathway activity is induced when superfluous reductants require oxidation [11,12]. Plant mitochondria can oxidize NAD(P)H from the matrix via complex I or via NDin, and can also oxidize cytosolic NAD(P)H present in the intermembrane space via NDex activity. NDex and NDin have low substrate affinity for NAD(P)H [13] and may be important when excessive reducing power must be oxidized [14]. Whereas, the AOX pathway is important for preventing over-reduction in chloroplasts and to balance redox partitioning during photosynthesis, photorespiration, and respiration [12,15]. Research has indicated the importance of NDin and NDex enzymes in regulating mitochondrial and cytosolic reductant pools. In particular, two *Arabidopsis* RNA interference (RNAi) lines, suppressing the *NDB1* gene for external Ca^{2+}-dependent NADPH oxidation by 90%, have highly similar effects on cellular NADP(H) homeostasis and consequentially on growth rate, respiratory metabolism and defense signaling, as shown by metabolome, transcriptome and flux analysis [16].

A disruption in cellular redox homeostasis, which might occur under stress conditions, is deleterious to plant cells because the production of ROS can be dramatically enhanced [17,18]. In plant tissues, ROS are formed by the inevitable leakage of electrons to O_2 from the electron transport chains of chloroplasts, mtETC, or as a by-product of metabolic pathways, which may be localized in cellular compartments such as peroxisomes. ROS can damage membranes, proteins, and nucleic acids; inhibit enzymes; and even induce programmed cell death in plants [19]. When ROS production is not effectively balanced by scavenging mechanisms, oxidative stress may occur.

To buffer the toxic effects of ROS, plants have developed several antioxidative systems [20]. The major plant low-mass antioxidants include glutathione and ascorbate, which are present in all cellular compartments. The production of ascorbate (the reduced form AsA) in plants is associated with the mtETC, i.e., the last step of synthesis is catalyzed by L-galactono-γ-lactone dehydrogenase, an enzyme attached to complex I [21]. The biosynthesis of glutathione, a tripeptide, requires two enzyme-mediated steps, as characterized in *Arabidopsis*. The enzymes are glutamate-cysteine ligase (GSH1), which is localized in plastids, and glutathione synthetase (GSH2), which is found in plastids and in the cytosol [22,23].

Glutathione is the most abundant non-protein thiol in cells and its nucleophilic activity is exploited in several stress response pathways to detoxify ROS [23,24]. Reduced glutathione (GSH) can directly reduce all kinds of ROS ($^1O_2$, $O_2^{·-}$, HO·, H_2O_2), while itself being oxidized and forming a disulfide form (GSSG). Moreover, GSH can form disulfides with Cys-residues in proteins (S-glutathionylation), protecting them from irreversible oxidation and regulating their activity [25]. Glutathione also has a critical function as an electron donor in other enzymatic detoxification systems. In the ascorbate-glutathione cycle (composed of the enzymes ascorbate peroxidase (APX), dehydroascorbate (DHA) reductase (DHAR), monodehydroascorbate reductase (MDHAR), and glutathione reductase (GR)), APX is ultimately responsible for the decomposition of H_2O_2 [26]. The ascorbate-glutathione cycle depends on three redox couples: AsA/DHA, GSH/GSSG, and NAD(P)H/NAD(P), the latter providing the reductant for the former two. In the process of ROS elimination, the network of glutathione peroxidase-like enzymes (GPX), peroxiredoxins (Prx), thioredoxins (TRX) and glutaredoxins (Grx), play an important role [27]. Briefly, GPX and Prx, in their catalytic activity, reduce H_2O_2 but also show a strong preference for organic hydroperoxides and peroxinitrite. As a result, GPX and Prx can protect the plant cell not only from ROS but also directly from other stress symptoms such as lipid peroxidation. Further, these enzymes are reduced by TRX in a NADPH- or reduced ferredoxin-dependent manner [28]. Grx, however, can reduce disulfide bonds in protein, a process which is dependent on GSH and that regulates the oxidation state of proteins. Overall, glutathione metabolism requires sulfide group reduction, which is essentially achieved by GR activity using electrons from NADPH. Therefore, GR has been proposed to recycle GSH and to regulate the glutathione redox state in cells [23]. Accordingly, GSH and GSSG form a redox couple that mediate the transfer of reducing equivalents from NADPH to ROS. It is well established that glutathione and ascorbate have a redox potential and are involved in the redox regulation of cells [29,30]. In recent years, the contribution of these low-mass antioxidants to redox-connected retrograde signaling has been identified.

Changes in redox state can be evaluated in plant tissues using different approaches. The role of mitochondria and chloroplasts in maintaining cellular redox balance has been demonstrated in studies in which the activity of the mtETC was altered using inhibitors [31,32]. Mutants are useful tools for investigating the influence of specific proteins in vivo on redox metabolism in plant cells. Another strategy is to investigate stress conditions affecting the intracellular redox balance. The responses of plants to diverse abiotic and biotic stress factors have been analyzed extensively in recent years. In their natural environment, plants must adapt to fluctuations in the availability of nutrients, including nitrogen; this implies the potential existence of a method of sensing mineral availability and executing a rapid metabolic response [2,33]. Plants can utilize two different inorganic nitrogen sources, mostly in the forms of ammonium (NH_4^+) or nitrate (NO_3^-) [34]. It has been proposed that the application of NH_4^+ as the sole source of nitrogen affects redox homeostasis in plant cells [35,36]. When comparing NH_4^+ with NO_3^- nutrition, a surplus of reductants may be expected in plant tissues; this is because NO_3^- must first be reduced to NH_4^+ in a two-step reduction reaction before its incorporation into amino acids. In the first step, NO_3^- must be reduced to nitrite (using NADH) in a reaction catalyzed by cytosolic nitrate reductase (NR), and then to NH_4^+ (using NADPH or reduced ferredoxin) via nitrite reductase (NiR) in the plastids [37,38]. In contrast, when NH_4^+ is supplied as the exclusive nitrogen source, thus NO_3^- reduction is omitted, the consumption of reducing equivalents is lower in the cytosol and chloroplasts. In terms of energy economy, NH_4^+ seems to represent a better source of nitrogen nutrition for plants, as its assimilation requires less energy than NO_3^- [37,39]. Ammonium nutrition can, however, potentially lead to an accumulation of reductants in these compartments, and consequentially NDex, NDin and AOX enzymes have been observed to be acutely induced by short-term NH_4^+ supply [38]. Nevertheless, plants cultured on NH_4^+ as the sole nitrogen source exhibit symptoms of toxicity (e.g., growth inhibition), commonly referred to as "ammonium syndrome" [40,41]. We have previously shown that an intracellular redox imbalance during NH_4^+ assimilation may lead to oxidative stress in *Arabidopsis* plant tissues [42]. The increased mtETC activity and possible over-reduction of the mtETC in response to NH_4^+ nutrition can also induce excessive mitochondrial ROS generation. Plant mitochondria have subsequently been identified as the primary source of ROS during extended NH_4^+ nutrition [42,43].

Extensive research has been devoted to the identification of the mechanisms behind the causes of NH_4^+ toxicity in plants, as it would be advantageous to minimize the effects of ammonium syndrome as a way to increase crop biomass production [44–46]. Mitochondria-associated metabolic reactions such as respiration may potentially increase nitrogen use efficiency [47–49]. It can therefore be expected that, for the proper functioning of plants grown on NH_4^+ as the sole source of nitrogen, it is important to dissipate excess reducing power arising in the cytosol or chloroplast. Since NDex are the major entry point of cytosolic NAD(P)H not used in metabolic reactions, their function may be primarily important during NH_4^+ nutrition. This was also indicated, in our previous study, where the main isoform of additional dehydrogenases elevated under growth on NH_4^+ was NDB1 [42].

The aim of the present study was to investigate the role of the Ca^{2+}-dependent external NADPH dehydrogenase, NDB1, during NH_4^+ nutrition. Transgenic *Arabidopsis* RNA interference lines suppressing NDB1 [16] were utilized in these studies. In the absence of NDex, an over-reduction of the redox-state might represent a burden for plants during NH_4^+ nutrition. However, NDB1 knock-downs were surprisingly less sensitive to prolonged NH_4^+ nutrition. In this study, we investigated redox/ROS homeostasis in plants grown on different nitrogen sources. It was shown that the improved tolerance to NH_4^+ in *NDB1* suppressing plants may be connected to elevated glutathione metabolism, including S-glutathionylation of proteins or altered responses of the redox regulatory network.

2. Results

NDB1 knock-down plants grown for 8 weeks in hydroponic culture under a short-day photoperiod had a similar rosette size and fresh weight as wild-type (WT) plants grown on the same nitrogen source (Figure 1A,B). Ammonium nutrition caused approximately 90% growth inhibition in both *NDB1* and WT plants (Figure 1B).

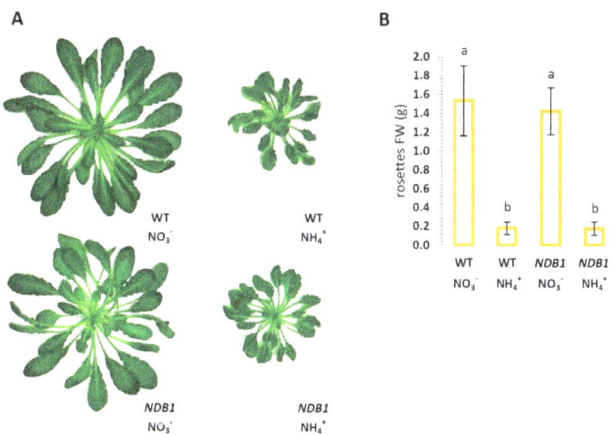

Figure 1. Phenotype of *Arabidopsis thaliana NDB1*-suppressing line 1.5 (*NDB1*) and wild type ecotype Col-0 (WT) after 8 weeks' growth in hydroponic cultures on 5 mM NO_3^- or 5 mM NH_4^+ as the sole nitrogen source. (**A**) Visual appearance of representative plants. All images are in the same scale; (**B**) Fresh weight (FW) of rosettes of WT and *NDB1* knock-down plants on respective nitrogen sources. Bars with different letters are statistically different ($p < 0.05$) by ANOVA.

2.1. Changes in Respiration and Pyridine Nucleotide Status in NDB1 Suppression Plants Cultured on NO_3^- or NH_4^+

In NH_4^+-grown plants, excess reductants not used in cytosolic reactions can be oxidized by type II dehydrogenases to prevent over-reduction of the mtETC. In general, ammonium nutrition induced higher total respiration (V_t), cytochrome pathway capacity (V_{cyt}), and alternative pathway capacity (V_{alt}) in both *NDB1* and WT plants (Table 1). Growth on NH_4^+ led to a slightly lower V_t in

NDB1 knock-down than in WT plants (Table 1). However, V_{cyt} and V_{alt} were similar between NDB1 knock-down and WT plants for each nitrogen treatment (Table 1).

Table 1. Respiratory parameters in *NDB1*-suppressing and WT *Arabidopsis* plants grown on NH_4^+ and NO_3^- (control) as the only source of nitrogen. Values are shown for total respiration (V_t), cytochrome pathway capacity (V_{cyt}), and alternative pathway capacity (V_{alt}) in dark-adapted leaves. Results with different letters are statistically different ($p < 0.05$) by ANOVA.

	WT NO_3^-	WT NH_4^+	NDB1 NO_3^-	NDB1 NH_4^+
	Oxygen consumption (nmol O_2 min^{-1} g^{-1} FW)			
Total respiration (V_t)	40.54 ^{a} ± 9.93	93.16 ^{c} ± 3.67	41.38 ^{a} ± 5.99	72.92 ^{b} ± 2.96
+SHAM (V_{cyt})	37.31 ^{a} ± 7.65	66.09 ^{b} ± 6.39	36.46 ^{a} ± 10.84	55.45 ^{ab} ± 4.62
+KCN (V_{alt})	5.94 ^{a} ± 3.11	38.98 ^{b} ± 10.58	5.82 ^{a} ± 1.38	46.47 ^{b} ± 2.74

We next examined how silencing of *NDB1* affects the redox state of pyridine nucleotides. Under nitrate nutrition in *NDB1* knock-down plants, we observed a slightly increased NADPH concentration, similar to that shown in [16]. Overall, however, *NDB1* silencing did not significantly perturb the redox state of NADP(H) in leaves in both growth conditions (Figure 2). The pyridine nucleotide pool was only affected by the applied nitrogen source; ammonium nutrition increased the NADPH content up to four-fold in both genotypes, which resulted in a higher NADPH/NADP ratio (Figure 2).

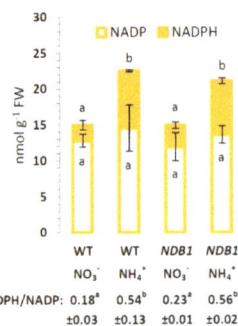

Figure 2. Phosphorylated pyridine nucleotide concentrations in *NDB1*-suppressing and WT plants growing on NH_4^+ and NO_3^- (control) as the only source of nitrogen. Bars with different letters are statistically different ($p < 0.05$) by ANOVA.

2.2. Influence of NDB1 Suppression on ROS Content and Oxidative Damage in Tissues of Plants Fed NO_3^- or NH_4^+

Altered redox homeostasis in plants may consequently affect ROS production in tissues. In WT plants, the content of H_2O_2 was elevated more than 20% in response to NH_4^+ nutrition. However, in *NDB1* knock-downs, which had a similar H_2O_2 level as WT plants, NH_4^+ nutrition did not increase tissue H_2O_2 content (Figure 3A). In order to identify the source of ROS in leaf cells, the distribution of H_2O_2 in mesophyll tissues was visualized by 2′,7′-dichlorodihydrofluorescein diacetate (DCF-DA) fluorescence. The co-localization of DCF-dependent fluorescence with mitochondria (green/red channels) or with chloroplasts (green/far red channels) (Figure 3C) showed the amount of ROS produced in the respective organelles (Figure 3B). Chloroplasts revealed an obvious fluorescent overlap that co-localized in the DCF-dependent fluorescence, showing a Pearson's coefficient higher than 0.7 in all analyzed variants (Figure 3B). The detection rate of co-localization between DCF- and mitochondria-dependent fluorescence was the highest in NH_4^+-exposed WT cells (Figure 3B), which suggests higher mitochondrial ROS production in these organelles. The defect in NDB1 in the mtETC of knock-down plants did not change the degree of DCF-fluorescence as compared with that of mitochondria of WT plants grown on NO_3^-. However, during the NH_4^+ growth regime, DCF-staining connected to mitochondria was only slightly increased in the *NDB1* knock-down (Figure 3B).

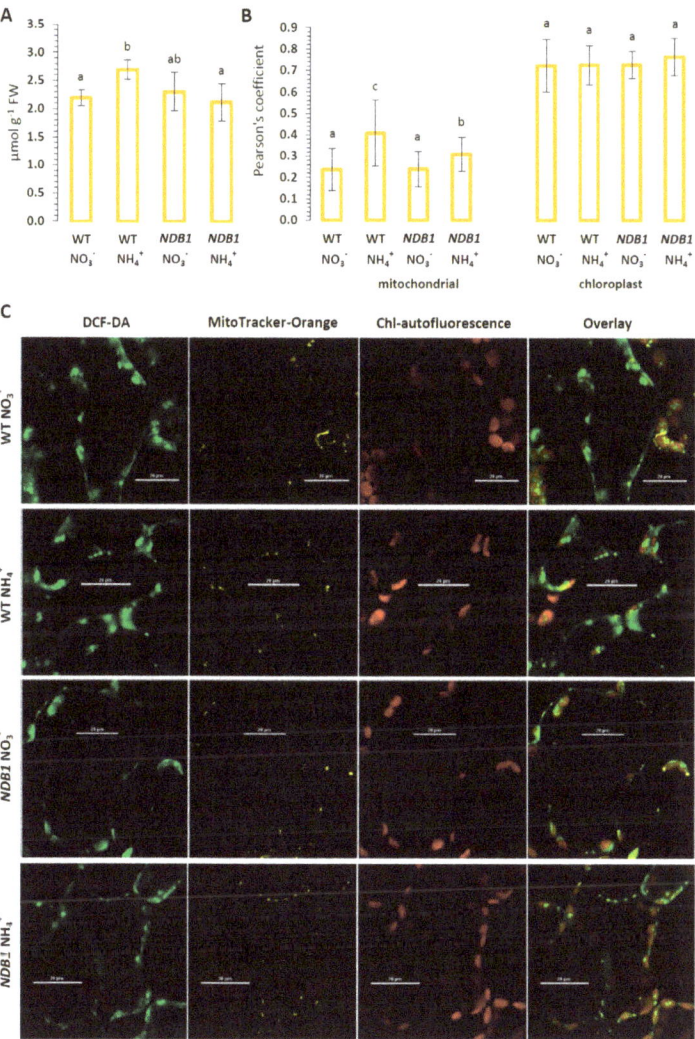

Figure 3. Reactive oxygen species (ROS) in *NDB1*-suppressed and WT plants growing on NH_4^+ and NO_3^- (control) as the only source of nitrogen. (**A**) Hydrogen peroxide content; (**B**) sub-cellular ROS distribution in leaf segments. Pearson's co-localization coefficient shows the proportion of DCF-labelling with mitochondria-dependent fluorescence or chlorophyll autofluorescence. Bars with different letters are statistically different ($p < 0.05$) by ANOVA; (**C**) The mesophyll cells of WT and *NDB1* knock-downs were double-labelled with 2',7'-dichlorodihydrofluorescein diacetate (DCF-DA, Green) and MitoTracker Orange (reduced CM-H_2 TMRos, yellow), and chloroplast autofluorescence (far red) was observed; representative images from four independent biological replicates; scale bar = 20 μm.

Since high ROS levels may lead to oxidative modifications of cell components, we estimated malondialdehyde (MDA) content as a marker of peroxidation of membrane lipids, expression of the specific stress response protein up-regulated by oxidative stress (UPOX, [50]), and protein carbonylation level. In WT plants, NH_4^+ nutrition, as compared with NO_3^- nutrition, led to a 20% increase in lipid peroxidation (Figure 4A). Lipid peroxidation in *NDB1* knock-down plants grown on NO_3^- was similar to that in WT plants (Figure 4A), but the treatment of *NDB1* knock-down plants with NH_4^+ resulted in an 15% lower MDA content in tissues (Figure 4A).

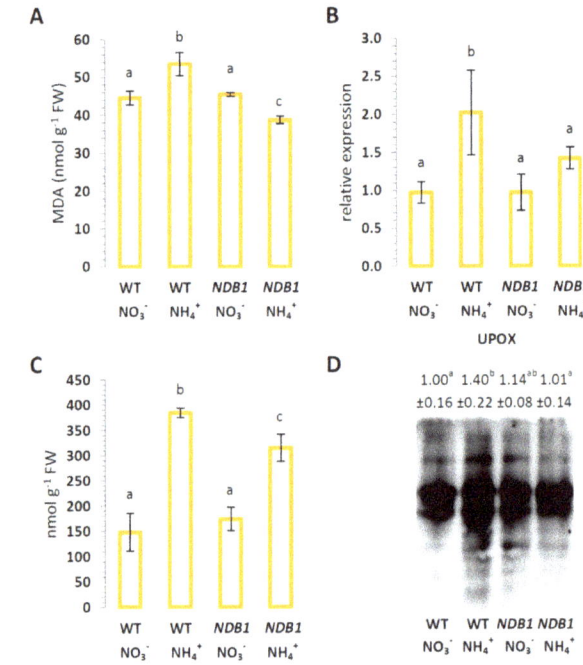

Figure 4. Oxidative stress markers in *NDB1*-suppressed and WT plants grown on NH_4^+ and NO_3^- (control) as the only source of nitrogen. (**A**) Lipid peroxidation estimated in leaves as MDA content; (**B**) transcript level for mitochondria-localized protein *UPOX* up-regulated by oxidative stress; (**C**) concentration of carbonylated proteins; and (**D**) immunoblot analysis of protein-bound carbonyls. Bars or bands with different letters are statistically different ($p < 0.05$) by ANOVA.

Ammonium nutrition induced a higher *UPOX* transcript level in WT plants grown on NO_3^-, but in the *NDB1* knock-down line, its expression was similar to that of the control, independent of the nitrogen source (Figure 4B). The level of protein carbonylation was more than 2.5-times higher in WT plants in response to NH_4^+ nutrition (Figure 4C). In *NDB1* knock-down plants, which had a similar extent of lipid oxidation to control plants, when grown on NO_3^-, the content of carbonylated proteins was elevated less than two-fold during NH_4^+ nutrition. The profile of proteins with carbonyl groups was also traced by immunoblotting. Ammonium nutrition induced a strong increase in protein carbonyls in WT plants, being 40% elevated within the whole protein spectrum (Figure 4D). However, the *NDB1* knock-down plants had an unchanged pattern of carbonylated proteins compared with control plants, and this pattern was not further increased by the NH_4^+ treatment.

Overall, our data indicate that unlike WT, *NDB1* knock-down plants do not develop major oxidative stress symptoms in response to ammonium nutrition, an effect that may be attributable to efficient antioxidant defense systems.

2.3. Ascorbate Level and Ascorbate-Related Antioxidant Defense in NDB1 Knock-Down Plants in Response to NO_3^- or NH_4^+ Nutrition

Ascorbate is a major low-mass antioxidant. The content of AsA was stable in WT and *NDB1* knock-down plants during growth on different nitrogen sources (Figure 5A). However, the DHA content decreased in the *NDB1* knock-down line during NO_3^- growth compared with WT, which resulted in an elevated AsA/DHA redox state (Figure 5A). In contrast, during growth on NH_4^+, an increase in DHA content was detected in the *NDB1* knock-downs compared with WT, but the AsA/DHA ratio remained similar to that of WT plants grown on NH_4^+ (Figure 5A). All APX isoforms had similar protein levels in *NDB1* knock-downs compared with WT plants and showed high accumulation under ammonium nutrition in both genotypes (Figure 5B). The same trend was observed for MDHAR and DHAR activities, which were similar between *NDB1* knock-downs and WT plants and were highly induced under ammonium nutrition (Figure 5C).

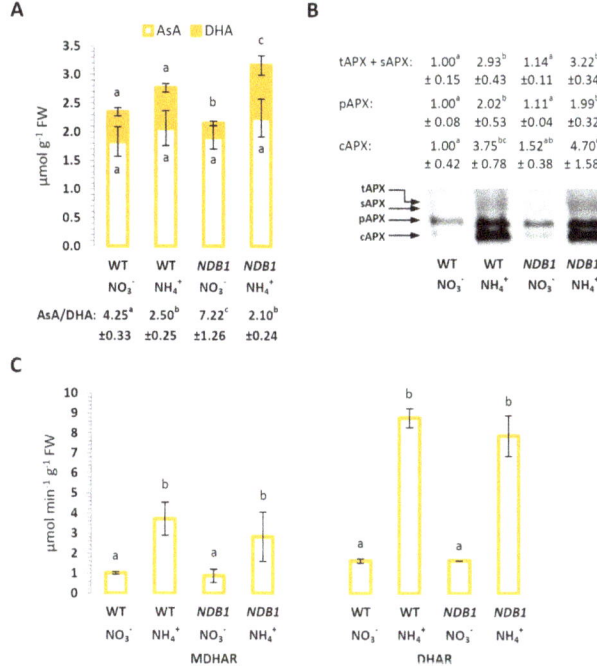

Figure 5. Content of ascorbate and ascorbate-utilizing enzymes in *NDB1*-suppressing and WT plants grown on NH_4^+ and NO_3^- (control) as the only source of nitrogen. (**A**) Concentration of reduced (AsA) and oxidized (DHA) ascorbate and derived AsA/DHA ratio; (**B**) ascorbate peroxidase (APX) protein levels. Thylakoid (tAPX, 38 kDa), stromal (sAPX, 33 kDa); peroxisomal (pAPX, 31 kDa), and cytoplasmic (cAPX, 25 kDa) forms of ascorbate peroxidases; (**C**) monodehydroascorbate reductase (MDHAR) and dehydroascorbate reductase (DHAR) activities. Bars or bands with different letters are statistically different ($p < 0.05$) by ANOVA.

2.4. Glutathione Metabolism in NDB1 Knock-Down Plants during NO_3^- or NH_4^+ Growth

Ammonium nutrition in WT plants resulted in higher levels of GSH and GSSG and a more oxidized GSH/GSSG ratio compared with NO_3^--grown plants. During NO_3^- nutrition, the *NDB1* knock-down line also showed increased GSH and GSSG content compared with WT plants and the GSH/GSSG ratio was similar in both genotypes (Figure 6A). When grown on NH_4^+, the *NDB1* knock-downs had even higher glutathione levels, showing an increase of 90% in the content of GSH and 50% in GSSG (Figure 6A). However, the GSH/GSSG ratio was unchanged compared with the NO_3^- nutrition of *NDB1* knock-down plants. As glutathione is considered the major cellular redox buffer, we can conclude from the redox state of the glutathione pool that *NDB1* knock-down plants are not adversely affected by modifications in their cellular redox state.

The transcript level of both genes from the glutathione biosynthetic pathway, *GSH1* and *GSH2*, was similar between different nitrogen treatments in WT plants, but in *NDB1* knock-down plants, ammonium nutrition stimulated *GSH1* and *GSH2* expression (Figure 6B), leading to a significantly higher glutathione pool (Figure 6A).

A high rate of glutathione reduction in *NDB1* knock-down plants could be explained by the up-regulation of glutathione reductase. We therefore analyzed GR protein levels. Densitometric analysis revealed that GR levels were similar between *NDB1* knock-downs and WT plants grown under control conditions (Figure 6C). Ammonium nutrition resulted in a two-fold increase in GR protein in both WT and *NDB1* knock-down plants (Figure 6C). To further elucidate the involvement of GR in GSH reduction in NH_4^+-grown *NDB1* knock-down lines, we analyzed the expression of two GR-encoding genes: *GR1*, which encodes the cytosolic isozyme, and *GR2*, which encodes an isoform dual-targeted to chloroplasts and mitochondria. In *NDB1* knock-down plants grown on NO_3^-, the transcript level for both genes was similar to that in WT plants (Figure 6D). Interestingly, *GR1* and *GR2* showed opposite trends of expression between treatments: *GR1* was induced while *GR2* was suppressed by ammonium nutrition in both genotypes.

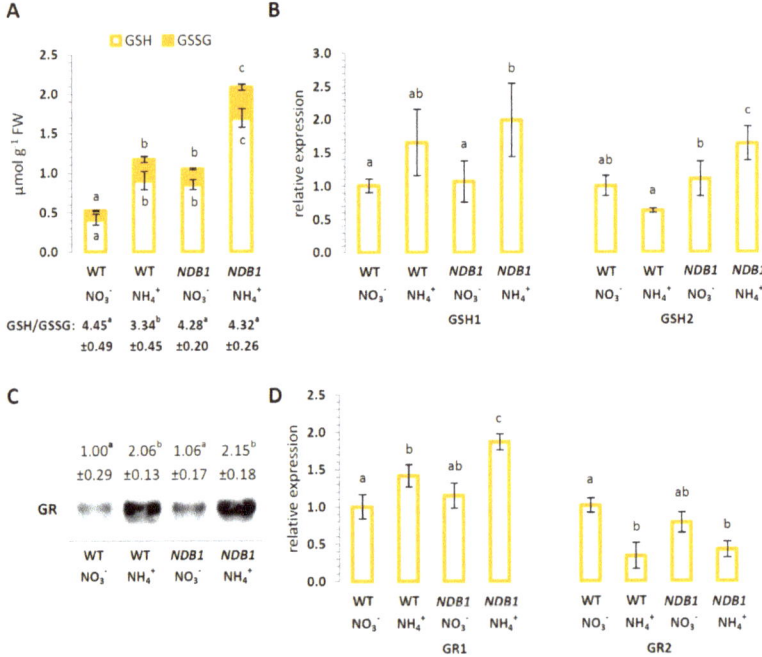

Figure 6. Glutathione contents and glutathione metabolism-related enzymes in *NDB1*-suppressing and WT plants growing on NH_4^+ and NO_3^- (control) as the only source of nitrogen. (**A**) Concentration of reduced (GSH) and oxidized (GSSG) glutathione and derived GSH/GSSG ratio; (**B**) transcript levels for *GSH1* and *GSH2*; (**C**) GR protein levels; (**D**) transcript levels for GR isoforms, *GR1* and *GR2*. Bars or bands with different letters are statistically different ($p < 0.05$) by ANOVA.

An induction of the glutathione biosynthetic pathway and increase in glutathione content may lead to higher S-glutathionylation of proteins. Therefore, we determined the level of this modification in leaf protein extracts by western blotting with anti-GSH antibodies. Densitometry analysis of the entire blot lane revealed that, in WT plants, ammonium nutrition increased the level of protein S-glutathionylation compared with NO_3^- nutrition (Figure 7). *NDB1* silencing alone had no effect on S-glutathionylation, while in knock-down plants under NH_4^+ nutrition, the increase in S-glutathionylation of proteins was even more pronounced than in WT plants (Figure 7).

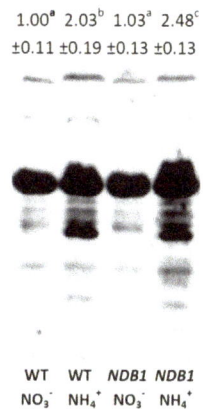

Figure 7. S-glutathionylated protein level in *NDB1*-suppressing and WT plants grown on NH_4^+ and NO_3^- (control) as the only source of nitrogen. Bands with different letters are statistically different ($p < 0.05$) by ANOVA.

2.5. The Effect of NDB1 Suppression on Redox-Related Enzymes in Plants Grown on NO_3^- or NH_4^+

We analyzed the expression profile of different redox related enzymes from the peroxiredoxin, glutathione peroxidase-like, thioredoxin, and glutaredoxin network. In WT plants, chloroplastic isoforms were generally either down-regulated (GPX1, GrxS14, NTRC, TRXx, TRXy2, and 2Cys PrxA) or unchanged (GPX7) in response to NH_4^+ nutrition (Figure 8A,D,F,G). The defect in *NDB1* did not alter the expression of most analyzed isoforms in knock-down plants. However, the treatment of *NDB1* knock-down plants with NH_4^+ led to an induction of GPX1 and GPX7, while GrxS14, NTRC, TRXx, TRXy2, and 2Cys PrxA remained at a stable level, as in *NDB1* knock-down plants grown on NO_3^- (Figure 8A,D,F,G). The protein level of the chloroplastic peroxidase PrxQ was unchanged between WT and *NDB1* knock-down plants but showed a major increase during NH_4^+ nutrition, especially in the *NDB1* knock-down line (Figure 8H). The expression of mitochondrial GPX6 was strongly induced in WT under NH_4^+ nutrition, and was also elevated in NH_4^+-grown *NDB1* knock-down plants (Figure 8C). Mitochondrial PrxIIF was down-regulated in response to NH_4^+ nutrition and in *NDB1* knock-down plants (Figure 8E). Ammonium supply to the *NDB1* knock-down plants increased the expression of mitochondrial PrxIIF (Figure 8E). Furthermore, the expression of cytosolic PrxIIc was induced in *NDB1* knock-down plants when grown on NH_4^+ (Figure 8G). The abundance of cytosolic GPX2 (Figure 8B) and GPX8 (Supplementary Materials Figure S2) transcripts was unchanged throughout.

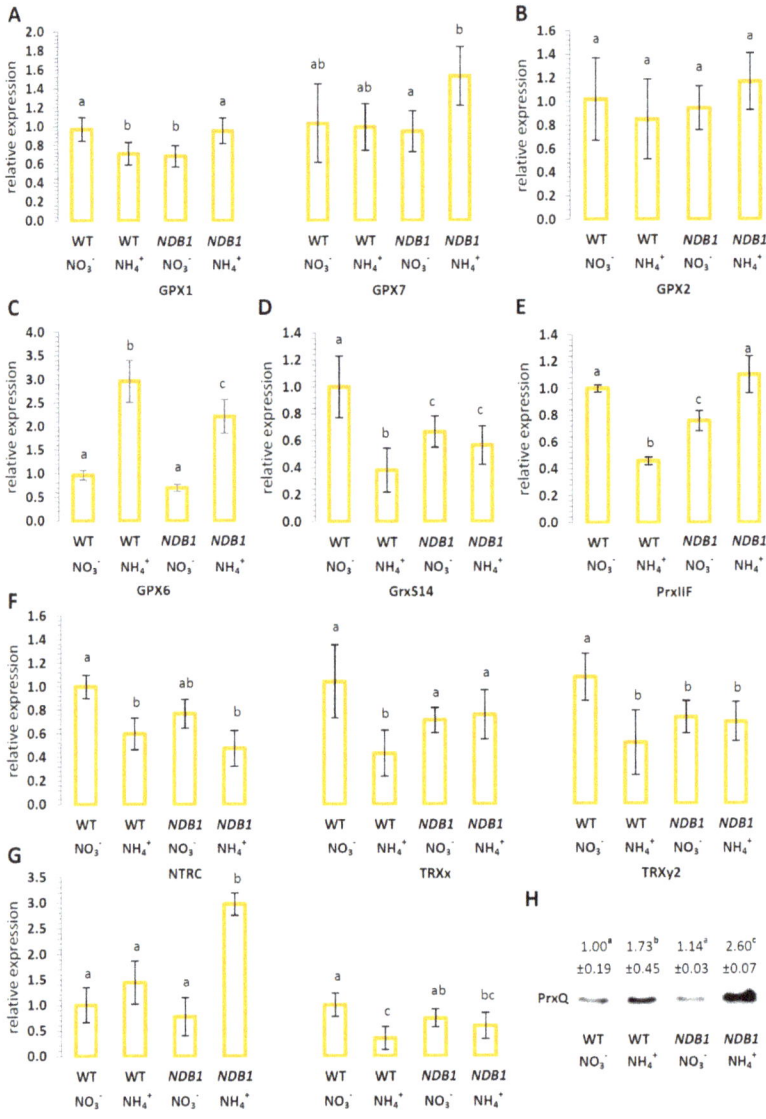

Figure 8. Expression of redox sensors and transmitters in *NDB1*-suppressing and WT plants grown on NH_4^+ and NO_3^- (control) as the only source of nitrogen. Transcript levels for (**A**) chloroplast glutathione peroxidase-like (GPX) *GPX1* and *GPX7*; (**B**) cytosolic *GPX2*; (**C**) mitochondrial *GPX6*; (**D**) chloroplast glutaredoxin (Grx) *GrxS14*; (**E**) mitochondrial *PrxIIF*; (**F**) chloroplast NADPH-dependent thioredoxin reductase C (NTRC) and thioredoxins (TRX) *TRXx* and *TRXy2*; (**G**) cytosolic peroxiredoxin (Prx) *PrxIIc* and chloroplast *2Cys PrxA*; and (**H**) PrxQ protein levels. Bars or bands with different letters are statistically different ($p < 0.05$) by ANOVA.

3. Discussion

3.1. NDB1-Suppressed Line Does Not Show a Growth Phenotype or Over-Reduction under Ammonium Nutrition

NDB1 knock-down plants do not show a growth phenotype, but have similar growth to WT plants in either NO_3^- or NH_4^+-supplied hydroponic growth conditions (Figure 1A,B). Previously, it has been shown that the *NDB1* suppressor line 1.5 (used in this study) grown on soil exhibited decreased biomass [16]. These inconsistent phenotypic responses can be attributed to different growth conditions, e.g., nutrient availability, daylight period (10 h versus 8 h in the present study), or light intensity (80 versus 150 µmol m^{-2} s^{-1}).

Under nitrate conditions, the NADP(H) pool showed a reduced trend in the *NDB1* suppressor line (Figure 2). Since ammonium nutrition greatly increases cell redox state [42], plants with impaired *NDB1* were expected to show cellular over-reduction. Surprisingly, under ammonium nutrition, *NDB1* knock-down plants exhibited a similar increase in redox state as WT plants (Figure 2). Maintenance of redox homeostasis is central to plant survival, especially under conditions where elevated redox input from the cytosol to the mtETC is expected. We have previously shown that growth of *A. thaliana* under long-term NH_4^+ supply results in up-regulation of *NDB1* expression, among all additional dehydrogenases [42]. Therefore, to compensate for *NDB1* suppression, induction of other type II dehydrogenases might be suspected. However, a lower total leaf respiratory rate in *NDB1* knock-down plants under ammonium nutrition compared with WT (Table 1) strongly suggests that *NDB1* knock-down plants do not fully compensate for the deficiency in NADPH-dependent NDex under these conditions. Noteworthy, the differences in total respiration between genotypes grown on NH_4^+ did not result from lower activity of terminal oxidases, since both V_{cyt} and V_{alt} were comparable between WT and *NDB1* knock-down plants (Table 1).

3.2. Ammonium Nutrition Causes Less Oxidative Injury in NDB1 Knock-Down Plants

Previously, we have shown that long-term ammonium nutrition results in reductive stress leading to increased ROS production, and consequently resulting in oxidative injury to biomolecules [42]. NDB1 activity seems to be especially important for plant growth under stress conditions, including ammonium nutrition [42], but the opposite was observed, in that the analyzed *NDB1* suppressor plants appeared to be more resistant to NH_4^+ treatment. This observation is supported by data on the second examined transgenic line (8.7), similarly, suppressed for *NDB1* [16], which we present in Supplementary Materials Figures S2–S7. All measured stress parameters, including lipid peroxidation, protein carbonylation, and the expression of the stress marker UPOX (Figure 4), indicated milder oxidative stress in tissues of *NDB1* knock-downs than in WT plants grown on NH_4^+. This may be the consequence of lower ROS generation, higher capacity of antioxidant systems, or both in the *NDB1* knock down plants during NH_4^+ nutrition as compared with WT plants.

In contrast to what was observed in WT plants [42] under ammonium nutrition, H_2O_2 content in the *A. thaliana NDB1* knock-down line was unchanged (Figure 3A). A defect in the mtETC often correlates with alterations in ROS metabolism, mainly concerning complex I mutants [51–54]. In contrast to complex I dysfunction, genetic modifications of the alternative pathways result in substantially milder phenotypic expression, being mainly affected during stress conditions, as seen in AOX suppressor plants [55–57]. Furthermore, in *Nicotiana sylvestris* NDB1-suppressor and *A. thaliana* NDB4 knock-down plants, no elevated ROS levels were observed [58,59]. Since long-term ammonium nutrition primarily induces mitochondrial ROS production [42,43], the lower ROS content in NH_4^+-treated *NDB1* knock-down plants could be attributable to lower ROS generation in this compartment. Indeed, lower ROS localization was detected in mitochondria during ammonium nutrition of *NDB1* knock-down plants compared with WT (Figure 3B). Under optimal growth conditions, the rate of mitochondrial ROS production is approximately 20 times lower than in chloroplasts [60], but may be more substantial under stress conditions that create a mitochondrial ROS burst and lead to oxidative damage in tissues [61]. An interesting observation is that dysfunction in one alternative dehydrogenase (mtETC component), whose activity is considered a mechanism that greatly reduces ROS production, does not lead to higher mtROS content under specific stress conditions.

3.3. Improved Resistance of the NDB1 Knock-Down Line to Ammonium Stress Is Not Related to Ascorbate-Dependent Antioxidant Systems but May Be Attributable to a Glutathione-Dependent System

To determine why the *NDB1* suppressor line shows less oxidative injury under ammonium nutrition, we analyzed changes in antioxidant system functioning between both genotypes under stress conditions. Foyer–Halliwell–Asada cycle function did not appear to be significantly affected by *NDB1* dysfunction, since MDHAR and DHAR activities (Figure 5C), and APX protein level (Figure 5B) were similar in WT and *NDB1* knock-down plants. Furthermore, we did not observe any significant differences in the activity of SOD isoenzymes (Supplementary Materials Figure S1).

We observed that, under nitrate nutrition, the ascorbate redox state in *NDB1* knock-down plants was even further reduced compared with WT plants (Figure 5A). Redox-related metabolic changes in transgenic *N. sylvestris NDB1* sense-suppression plants have previously been shown to correspond with altered ascorbate content [62]. In a metabolomics study, a negative correlation between NADPH level and DHA content was observed [62], which is in line with the decreased DHA levels observed in *NDB1 A. thaliana* knock-down plants in the present study (Figure 5A). Changes in ascorbate redox state may be the result of increased availability of substrate (NADPH) for MDHAR. It was proposed in [62] that changes in ascorbate level may be connected downstream to the NDB1 defect, because ascorbate synthesis takes place in the mtETC. However, it is possible that the changes in ascorbate content/reduction state may instead reflect the chloroplastic pool, since this is the main ascorbate reservoir induced under stress conditions [63]. Although no changes in Foyer–Halliwell–Asada cycle function were detected in tissue extracts between analyzed genotypes (Figure 5B,C), local changes in the activity of enzymes, affecting low-mass antioxidant reduction status in organelles, cannot be excluded.

Clearly, *NDB1* knock-down *A. thaliana* had a marked elevated total glutathione content (Figure 6A). Comparing the rate of oxidation of different ROS forms by ascorbate and glutathione, it appears that glutathione might represent the more potent antioxidant [30] and therefore plays a key role in plant stress tolerance. This effect may be of significant importance since, in *Arabidopsis* cells, the highest glutathione content was found in mitochondria [64], and thus GSH is presumably the primary mitochondrial ROS scavenger. In conclusion, the high glutathione levels in tissues of *NDB1* knock-down plants could primarily be responsible for efficient ROS detoxification in mitochondria (Figure 3B), which was shown to be the predominant ROS source during NH_4^+ nutrition.

Elevated total glutathione content may be achieved by the induction of glutathione-synthesizing enzymes located in chloroplasts [65]. Accordingly, NH_4^+ grown *NDB1* knock-down plants showed an up-regulation of *GSH1* and *GSH2* (Figure 6B). However, glutathione antioxidant potential depends not only on absolute glutathione concentration, but also on its redox state. The glutathione pool of *NDB1* knock-down plants showed a more reduced redox state than that in WT plants when nourished on NH_4^+ (Figure 6A). A high level of glutathione reduction is primarily triggered by the activity of GR, the main regulatory enzyme [66]. Despite a lack of change in GR protein level (Figure 6C), the NH_4^+-grown *NDB1* knock-down line showed an induction of cytosolic GR1 on the transcript level (Figure 6D). GR1 has been proposed as the major isoform of GR in plants responsive to stress factors [67] and therefore may be responsible for the reduction state of glutathione in NH_4^+-grown *NDB1* knock-down plants.

The glutathione pool is also implicated in the direct protection of proteins against irreversible oxidative injury due to the *S*-glutathionylation process. Cysteine (Cys) residues in proteins are exposed to modification by ROS in a three-step reaction to successively form sulphenic acid (Cys–SOH), sulphinic acid (Cys–SO_2H), and sulphonic acid (Cys–SO_3H), which may lead to protein inactivation. Advanced oxidation of Cys residues is irreversible and leads to inevitable protein degradation, although the Cys–SOH group may be protected against further oxidation by reversible S-glutathionylation [68]. Glutathionylation typically occurs under oxidative stress conditions and plays an important role in regulation and signaling [69]. Some studies indicate that, under oxidative stress conditions, the activity of enzymes associated with primarily carbon metabolism may be down-regulated due to glutathionylation [69], allowing more effective antioxidant protection of cells. Additionally, reversible S-glutathionylation is part of the catalytic cycle of some glutathione-dependent enzymes, including DHAR [23]. Ammonium nutrition results in increased protein glutathionylation in both genotypes, although a higher level of S-glutathionylated proteins was observed in *NDB1* knock-down plants (Figure 7). Unfortunately, we cannot determine whether increased S-glutathionylation is the result of the protection of Cys-SOH groups in oxidized proteins or whether it is aimed at the regulatory adjustment of metabolism; however, both processes may be responsible for improved tolerance of *NDB1* knock-down plants to NH_4^+. The regeneration of native Cys residues in proteins depends on GSH but requires GRXs activity. Interestingly, ammonium treatment results in the specific down-regulation of expression of several members of the GRX gene family [70]. In the present study, we measured only the transcript level of GrxS14, confirming the influence of NH_4^+ on Grx expression (Figure 8D). Moreover, we observed that in *NDB1* knock-down plants under nitrate conditions, the transcript level of GrxS14 was lower than in WT plants but was not regulated by NH_4^+ (Figure 8D). This observation may suggest that, under ammonium nutrition, Cys residues in *NDB1* knock-downs grown on NH_4^+ are more efficiently regenerated to their native forms, meaning that *NDB1*-suppressor lines have more efficient protection against oxidative damage of proteins. However, a reduced GrxS14 transcript level in *NDB1* knock-down plants may indicate that the down-regulation of GRX expression is not specifically in response to NH_4^+ nutrition, as was suggested previously [70], but rather to changes in the tissue redox state. The influence of *NDB1* suppression on the expression of other GRX genes and on GRX activity therefore requires further research.

3.4. NDB1 Deletion Leads to Changes in the Expression of Genes Involved in the Redox Regulatory Network

Following the observation of different effects of ammonium nutrition on the level of GRX transcripts in WT and *NDB1* knock-down plants (Figure 8D), we analyzed the expression of other genes engaged in redox signal integration. Redox-sensitive proteins were classified into two classes: redox sensors, including Prx and GPX, and redox transmitters, containing a large family of Grxs and TRXs [71]. In general, we can conclude that the expression of all redox-sensitive genes evaluated was regulated differently by NH_4^+ in WT and in *NDB1* knock-down plants (Figure 8A,B,D–F), with the exception of GPX2 and GPX8 isoforms, whose expression was unchanged irrespective of growth conditions and genotype (Figure 8B, Supplementary Materials Figure S7C).

Among the four Prx, which are targeted to the plastids, we estimated the expression of 2Cys PrxA and PrxQ (Figure 8G,H). 2Cys PrxA was strongly down-regulated by NH_4^+ in WT plants but not in *NDB1* knock-down plants. In contrast, in both genotypes, the protein level of PrxQ was increased in response to NH_4^+, but a greater increase was observed in the *NDB1*-suppressing line. Those observations suggest that both Prx can have markedly different roles in leaf cell metabolism, especially under stress conditions. In plants, the highly abundant 2Cys Prx is involved in protecting photosynthetic ETC functioning as reduced amounts of 2Cys Prx were shown to result in decreased levels of D1 protein and of the light-harvesting protein complex associated with photosystem II [72]. 2Cys Prx activity is related to the functioning of intermembrane protein complexes, which enforces that 2Cys Prx (at least in the majority of its conformational states) is associated with thylakoid membranes [73]. By contrast, PrxQ of *Arabidopsis* was shown to be a soluble protein in the lumen of the thylakoid membranes [74]. All Prx are thiol peroxidases and might function in oxidant detoxification [71]. However, according to available data, 2Cys Prx, in addition to a role in photosynthetic energy dissipation [73], and depending on conformational state, may also function as chaperones [75]. The precise function of the 2Cys Prx chaperone activity is not yet known, but it has been suggested that its binding to stromal fructose-1,6-bisphosphatase allows Calvin cycle activity to be maintained under conditions of excessive ROS production [76]. In addition to substrate specificity, both evaluated Prx may differ; 2Cys Prx catalyzes the detoxification of H_2O_2, alkyl hydroperoxides, and reactive nitrogen peroxides [77], while PrxQ has been demonstrated as an H_2O_2 peroxidase with very low activity toward lipid peroxides [78]. In the context of our results, high light conditions (resulting in higher production of NADPH) have been shown to lead to decreased 2Cys PrxA transcription but increased PrxQ [79]. Furthermore, PrxQ but not 2Cys PrxA was highly induced by the addition of oxidant to *Arabidopsis* tissues [79]. Finally, the expression of PrxQ was shown to be greatly increased in the hypersensitive response to pathogen treatment [80]. On the basis of these observations alongside our results, we may conclude that 2Cys PrxA is down-regulated in response to changes in chloroplastic NADPH and PrxQ is up-regulated in response to H_2O_2 and may be an effective organelle-localized antioxidant under stress conditions. However, both Prx are involved in redox sensing and signaling [71].

Interestingly, similarly to PrxQ, the expression of type II Prx was also increased in response to infection by *Melampsora larici-populina* [80] and both Prx were induced by (pro-) oxidants in the same manner [79]. Also, in our study, changes in cytosolic PrxIIC expression were similar to changes in PrxQ (Figure 8G,H). Therefore, we suggest that PrxIIC also protects *NDB1* knock-down plants against oxidative stress under ammonium nutrition.

The regeneration of oxidized Prx requires a supply of electrons from redox transmitters, but target selectivity between chloroplast redox-sensitive proteins exists in this process; NTRC is responsible mainly for 2Cys Prx reduction and PrxQ is regenerated mainly by the proteins of the TRX family [81]. The expression of NTRC and 2Cys Prx has been shown to be strictly co-regulated to maintain proper functioning of the photosynthetic apparatus and the entire chloroplast metabolism [82]. Our results confirm this observation, since the expression of NTRC and 2Cys PrxA is similarly regulated in both genotypes (Figure 8F,G). However, the changes in expression observed in TRXx and TRXy2 do not reflect the altered expression of PrxQ (Figure 8F,G). Furthermore, it should be noted that the Trx-PrxQ system does not show the same specificity as the NTRC-2Cys PrxA system does. 2Cys Prx may also be regenerated by TRX [71]; therefore, the decrease in TRX expression observed in WT plants under ammonium nutrition and in *NDB1* knock-down plants may reflect the down-regulation of a subclass of 2Cys Prx (Figure 8F–H).

PrxIIF is recognized as an important component of the mitochondrial defense system against peroxide stress, and is regenerated in a GSH-dependent manner. The specific substrate for PrxIIF is H_2O_2, and the reduction of lipid peroxides is catalyzed through this enzyme approximately 10 times less efficiently [83]. Studies on the response of PrxIIF to stress conditions have classified *PrxIIF* as non-responsive to high light and salt stress [79,84], but it's up-regulation has been shown as a result of cadmium treatment [83]. Ammonium stress differentially affects PrxIIF expression in WT and *NDB1* knock-down plants (Figure 8E). Lower PrxIIF abundance might have diminished the antioxidant protection of mitochondria of WT plants grown on NH_4^+, leading to higher H_2O_2 content in those organelles (Figure 3B). Indeed, it was shown previously that under optimal growth conditions, defects in PrxIIF may be compensated for by other compounds of the antioxidant defense system. However, such compensation is insufficient under stress conditions [83], which suggests an essential role for PrxIIF in organellar redox homeostasis.

Plant GPXs catalyze the reduction of H_2O_2 as well as different kinds of lipid peroxides by using TRX as an electron donor. The observed changes in GPX1 and GPX7 expression are relatively small (Figure 8A), with the mitochondrial isoform of GPX showing the only stress-responsive change (Figure 8C). Altogether, among redox sensors, which inactivate ROS, mostly GPX (cytosolic GPX1 and GPX7, and mitochondrial GPX6) and Prx (PrxIIF, PrxIIC, and PrxQ), are up-regulated in *NDB1* knock-down plants under ammonium nutrition. We hypothesize that this up-regulation increases the efficiency of ROS detoxification (including organic peroxide detoxification detected as lower levels of lipid peroxidation, Figure 4A) while enabling plants to better adjust their metabolism to stress conditions (Figure 9).

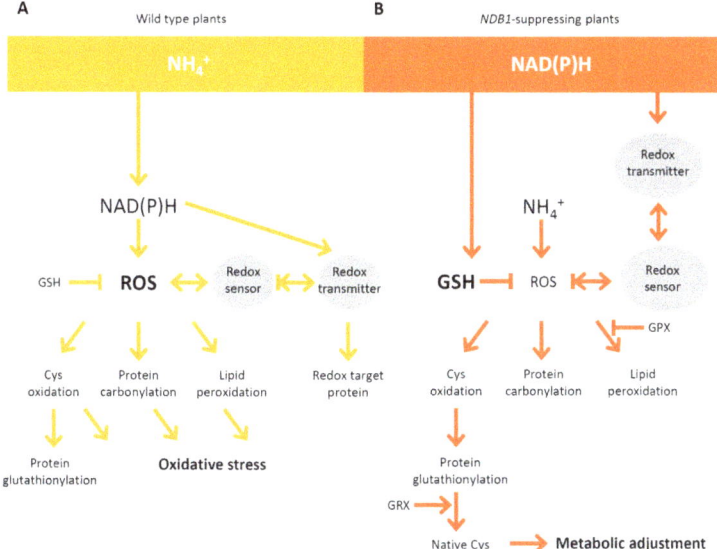

Figure 9. Schematic comparison between redox-related changes occurring in WT and *NDB1* suppressor plants grown on ammonium as the sole nitrogen source. (**A**) Changes in redox homeostasis in WT plants are a direct consequence of ammonium assimilation, and a secondary effect is the induction of redox transmitters/sensors and ROS accumulation leading to protein carbonylation, lipid peroxidation, and oxidation of Cys residues, which causes oxidative stress in NH_4^+-grown WT plants; (**B**) *NDB1* knock-down plants have an initial higher redox potential, and can rapidly induce changes in redox-related sensors/transmitters and glutathione content; in these plants, NH_4^+ nutrition is an additional factor that does not affect ROS content in tissues. Instead of exhibiting symptoms of oxidative stress symptoms, *NDB1* knock-down plants undergo metabolic adjustment to a high redox input during NH_4^+ nutrition. Sharp arrows indicate stimulatory influences, whereas blunt arrows represent inhibitory effects.

4. Material and Methods

4.1. Plant Material and Growth Conditions

Arabidopsis thaliana plants of ecotype Columbia-0 (WT) and their derivatives, *NDB1*-suppressed by RNA interference line 1.5 (named *NDB1*) and 8.7 (*NDB1* 8.7, Supplementary Materials Figure S2) [16], were grown hydroponically using an Araponics SA system (Liège, Belgium). Seeds were sown in half-strength [85] basal salt mixture with 1% agar, and after germination (1 week), deionized water in the hydroponic box was replaced with the nutrient solution described in [42], which was constantly aerated and replaced twice a week thereafter. The constant sole source of nitrogen for the plants was 5 mM NO_3^- or 5 mM NH_4^+. NO_3^--nourished plants were used as controls. Plants were grown for 8 weeks in a growth chamber under an 8-h photoperiod at 150 µmol m^{-2} s^{-1} photosynthetically active radiation (PAR, daylight and warm white, 1:1; LF-40W, Phillips, Pila, Poland), day/night temperature of 21 °C/18 °C, and approximately 70% relative humidity. All assays were carried out on leaf samples collected in the middle of the light period.

4.2. Phenotype Analysis

Ten to 25 plants were randomly selected from two plant cultures (grown on NO_3^- or NH_4^+, respectively) and weighed for fresh weight determination (FW). Representative rosettes were photographed.

4.3. Respiratory Measurements

Leaves were weighed and pre-incubated before respiratory measurement in an assay solution containing 30 mM 2-(N-morpholino)ethanesulfonic acid (MES), pH 6.2 supplemented with 0.2 mM $CaCl_2$. Oxygen consumption rates were measured in 3.0 mL assay solution using a Clark-type oxygen electrode (Rank Brothers Ltd., Cambridge, UK) in the dark at a constant temperature of 25 °C [86]. For inhibitor treatments, 10 mM KCN or 20 mM salicylhydroxamic acid (SHAM) in DMSO were used. To measure residual respiration, both inhibitors were added. Total respiration (V_t), alternative pathway capacity (V_{alt}), and cytochrome pathway capacity (V_{cyt}) values were determined after the subtraction of the value for residual respiration.

4.4. Quantitative RT–PCR Analyses

Total RNA was extracted from 100 mg of leaf tissue using a Syngen Plant RNA Mini kit (Syngen Biotech, Wrocław, Poland). DNase digestion was performed using an RNase-free DNase Set (Qiagen, Hilden, Germany). Complementary DNA was synthesized from 1 µg of total RNA as a template using reverse transcriptase and oligo-dT primers from a Revert Aid H minus first-strand cDNA synthesis kit, according to the manufacturer's protocol (Thermo Fisher Scientific Inc., Waltham, MA, USA). Thereafter, RNA digestion was performed as described in [87]. Relative transcript abundance was quantified by comparative quantitation analysis. Transcript content was analyzed using iTaq Universal SYBR Green Supermix (Bio-Rad, Hercules, CA, USA) according to the manufacturer's instructions.

Primer pairs for *GPX1* (At2g25080), *GPX6* (At4g11600), and *GR1* (At3g24170), were as described in [42], and for *UPOX* (At2g21640; [50]), as described previously in [88]. New primers were designed for *GrxS14* (At3g54900), *2Cys PrxA* (At3g11630), *PrxIIC* (At1g65970), *PrxIIF* (At3g06050), *TRXx* (At1g50320), *TRXy2* (At1g43560), *NTRC* (At2g41680), *GR2* (At3g54660), *GSH1* (At4g23100), *GSH2* (At5g27380), *GPX2* (At2g31570), *GPX7* (At4g31870) and *GPX8* (At1g63460) (Supplementary Materials Table S1), with one sequence covering an exon–exon junction if present in the gene sequence. Quantitative RT-PCR reactions were performed using a thermo cycler (CFX Connect™, Bio-Rad) with an annealing temperature of 60 °C. Transcript abundance was normalized to the expression of the reference *PP2A* (At1g13320) gene [89]. Transcript content and qRT-PCR efficiency of target genes were quantified as described in [90]. Results were expressed in relation to those in NO_3^--grown plants (value of 1).

4.5. Analysis of Metabolites

A previously reported method was used for the extraction and analysis of NAD(P)(H) [91]. Hydrogen peroxide content was determined according to [92] and quantified with reference to an internal standard (5 nmol H_2O_2). Ascorbate was extracted from leaf tissue using 0.1 M hydrochloric acid, and the levels of the reduced (AsA) and oxidized (DHA) forms of ascorbate were assayed according to the [93] method. Glutathione extraction was performed in 5% meta phosphoric acid and GSH was quantified using the enzymatic assay of [94]. Lipid peroxidation was estimated by measuring the secondary oxidation product MDA as described in [95]. Oxidized proteins were labeled with 2,4-dinitrophenylhydrazine (DNPH) and quantified by measuring carbonylated protein derivatives in tissue extracts according to [96]. Soluble protein content in the samples was estimated by the [97] method using bovine serum albumin (BSA) as the standard.

4.6. Confocal Fluorescent Imaging of Intracellular ROS

In vivo ROS localization was determined by detecting the fluorescence of DCF and colocalization with mitochondria and chloroplasts using confocal laser scanning microscopy according to the method described in [98], with minor modifications. The upper epidermis was removed from fresh plant leaves, and tissues were cut into small sections (approx. 5 mm). Samples were double-stained with cell-permeant 100 nM MitoTracker Orange (reduced CM-H_2 TMRos, Molecular Probes, Thermo

Fisher Scientific, Waltham, MA, USA) and 20 µM 2′,7′-dichlorodihydrofluorescein diacetate (DCF-DA, Molecular Probes, Thermo Fisher Scientific) under dark incubation for 15 min at room temperature (RT). The leaf pieces were washed in water three times for 5 min and then placed on a glass coverslip. MitoTracker Orange fluorescence was induced with a 561 nm helium-neon laser at emission of 570–620 nm. DCF-fluorescence was measured using the 488 nm line of an argon laser and was monitored at 500–550 nm. Chlorophyll fluorescence was induced with a 633 nm helium-neon laser and detected at 663–738 nm. Images were acquired using a NIKON A1R MP confocal laser scanning system (LSM-510, Carl Zeiss, Oberkochen, Germany) at 60× (numerical aperture 1.2) water immersion objective. Negative control images were obtained by omitting fluorescent probes to remove all signal from tissues except for chlorophyll autofluorescence. For individual cells, Pearson's colocalization coefficient was calculated between the green/red and green/far red channels using the Nis-Elements 3.22 imaging software (Nikon Co., Tokyo, Japan).

4.7. Western Blotting Analyses

To determine protein levels, tissue was homogenized with 2.5 volumes of extraction buffer and 5 µL of the resulting extract was used for GR, 15 µL for APX, 10 µL for PrxQ, and 3 µL for MnSOD. Extracts were subjected to sodium dodecyl sulfate-polyacrylamide gel electrophoresis (SDS-PAGE). The polypeptides were electroblotted onto nitrocellulose membranes using wet transfer (Bio-Rad, Hercules, CA, USA) and probed with anti-PrxQ (diluted 1:5000; Agrisera, Vännäs, Sweden), anti-APX (diluted 1:2000; Agrisera), anti-GR (diluted 1:4000; Agrisera), or anti-MnSOD (diluted 1:1000; Sigma-Aldrich, St. Louis, MO, USA) primary antibodies overnight at 4 °C, followed by anti-rabbit secondary antibodies conjugated to horseradish peroxidase (diluted 1:10,000 for APX and MnSOD, 1:40,000 for GR, and 1:20,000 for PrxQ determination; Bio-Rad). The quantification of protein S-glutathionylation was achieved after the detection of glutathione-protein complexes. Tissue extracts (20 µL) were mixed with Laemmli sample buffer containing N-ethylmaleimide (NEM, final concentration 5 mM); NEM (2.5 mM) was also present in the blocking buffer to prevent reduction of GSH adducts by thiol-containing proteins [99]. For immunoblotting, anti-GSH (diluted 1:1000; Abcam, Cambridge, UK) antibodies were used, with anti-mouse antibodies (diluted 1:20,000, Bio-Rad) as secondary antibodies. For the determination of carbonylated proteins in electrophoretically separated protein extracts (5 µL), an OxyBlot detection kit (Millipore, Billerica, MA, USA) was used according to the manufacturer's protocol. Visualization was performed using a chemiluminescence kit (Clarity™ Western ECL, Bio-Rad), and signals were detected using a Chemi-Doc imaging system (Bio-Rad). Specific bands were referred to pre-stained protein markers (Bio-Rad). The densitometry of bands for analyzed proteins or of the whole blot lane for carbonylated and glutathionylated proteins was quantified using Image-Lab 5.2 software (Bio-Rad) after background correction.

4.8. Enzyme Activity Assay

For the measurement of ascorbate-glutathione cycle enzymes activity, leaves were homogenized in 50 mM phosphate buffer, pH 7.0, 1 M NaCl, 1% (w/v) PVP, and 1 mM EDTA (as described in [100]). DHAR was determined according to the method of [101]. MDHAR and GR were assayed as in [102]. Superoxide dismutase (SOD) isoforms were visualized in gel by the method of [103] after electrophoretic separation of protein extracts (6 µL) on 12% polyacrylamide gels.

4.9. Statistical Analysis

Results are expressed as the mean value ± standard deviation (SD) of n measurements ($n = 3$–25) taken from at least two independent plant cultures. To analyze the statistical significance of observed differences, a one-way analysis of variance (ANOVA) with Tukey's post-hoc test at p-values ≤ 0.05 was performed using Statistica 13.1 software (StatSoft, Inc., Tulsa, OK, USA).

5. Conclusions

We have shown that the impairment of NDB1 in mitochondrial ETC does not trigger major oxidative stress in plants challenged with NH_4^+ treatment. Moreover, high light conditions were not a burden for plants with suppressed *NDB1* [16,62]. On the basis of a microarray study to identify a response overlap, the *NDB1 Arabidopsis thaliana* suppressor line exhibited a similar profile to OsHsfA21 plants, which have been characterized by improved tolerance to stress, including pathogen resistance [16]. It has also been shown that the suppression of another NADH-dependent NDex isoform, NDB4, improved salinity stress tolerance [59]. Altogether, these findings may lead to the assumption that the suppression of NDex and subsequent changes in redox metabolism induce cross tolerance among plants to withstand other stresses. This effect may be attributed to the observation that various stresses produce a similar effect to oxidative stress at the cellular level. An epigenetic response in plants may transmit the acquired stress resistance to their progeny [104,105]. This study shows that improved sensitivity to changes connected with altered redox/ROS status in *NDB1* knock-down plants may be beneficial during stress related to ammonium nutrition.

Supplementary Materials: Supplementary materials can be found at http://www.mdpi.com/1422-0067/19/5/1412/s1.

Author Contributions: A.P., B.S., and M.O.-B. conceived and designed the experiments; M.O.-B. performed respiratory measurements; A.P., K.B., M.J., and M.O.-B. carried out RT-PCR analysis; A.P., A.T., B.S., M.J., and M.O.-B. measured metabolites and enzyme activities; A.P. performed CLSM microscopy; A.T., K.B., M.B., M.J., and M.O.-B. performed immunoblotting; A.P., B.S., K.B., and M.O.-B. wrote the paper; and A.G.R revised the manuscript.

Acknowledgments: This work was partially supported by grant 2011/01/N/NZ3/02953 from the National Science Centre (NCN, Poland) given to A.P. A.P. and M.O.-B. were the beneficiaries of a scholarship from the Polish Minister of Science and Higher Education. The authors thank Bohdan Paterczyk from the Faculty of Biology (University of Warsaw) for support during CLSM analysis. We are grateful to Anna Książek from the Faculty of Biology (University of Warsaw) for assistance with RT-PCR studies.

Conflicts of Interest: The authors declare no conflict of interest. The funding sponsors had no role in the design of the study; in the collection, analyses, or interpretation of data; in the writing of the manuscript; and in the decision to publish the results.

Abbreviations

APX	ascorbate peroxidase
AsA	reduced form of ascorbate
DHA	oxidized form of ascorbate (dehydroascorbate)
DHAR	dehydroascorbate reductase
GPX	glutathione peroxidase-like
GR	glutathione reductase
GRX	glutaredoxin
GSH	reduced form of glutathione
GSSG	oxidized form of glutathione (glutathione disulfide)
MDHAR	monodehydroascorbate reductase
NDB1	external NADPH dehydrogenase
NTRC	NAPDH-dependent thioredoxin reductase C
Prx	peroxiredoxin
ROS	reactive oxygen species
SOD	superoxide dismutase
TRX	thioredoxin
UPOX	protein up-regulated by oxidative stress

References

1. Noctor, G. Metabolic signalling in defence and stress: The central roles of soluble redox couples. *Plant Cell Environ.* **2006**, *29*, 409–425. [CrossRef] [PubMed]
2. Suzuki, N.; Miller, G.; Morales, J.; Shulaev, V.; Torres, M.A.; Mittler, R. Respiratory burst oxidases: The engines of ROS signaling. *Curr. Opin. Plant Biol.* **2011**, *14*, 691–699. [CrossRef] [PubMed]
3. Gakière, B.; Fernie, A.R.; Pétriacq, P. More to NAD+ than meets the eye: A regulator of metabolic pools and gene expression in Arabidopsis. *Free Radic. Biol. Med.* **2018**. [CrossRef] [PubMed]
4. Noctor, G.; De Paepe, R.; Foyer, C.H. Mitochondrial redox biology and homeostasis in plants. *Trends Plant Sci.* **2007**, *12*, 125–134. [CrossRef] [PubMed]
5. Møller, I.M. Plant mitochondria and oxidative stress: Electron transport, NADPH turnover, and metabolism of reactive oxygen species. *Annu. Rev. Plant Biol.* **2001**, *52*, 561–591. [CrossRef] [PubMed]
6. Rasmusson, A.G.; Geisler, D.A.; Møller, I.M. The multiplicity of dehydrogenases in the electron transport chain of plant mitochondria. *Mitochondrion* **2008**, *8*, 47–60. [CrossRef] [PubMed]
7. Michalecka, A.M.; Agius, S.C.; Møller, I.M.; Rasmusson, A.G. Identification of a mitochondrial external NADPH dehydrogenase by overexpression in transgenic *Nicotiana sylvestris*. *Plant J.* **2004**, *37*, 415–425. [CrossRef] [PubMed]
8. Elhafez, D.; Murcha, M.W.; Clifton, R.; Soole, K.L.; Day, D.A.; Whelan, J. Characterization of mitochondrial alternative NAD(P)H dehydrogenases in *Arabidopsis*: Intraorganelle location and expression. *Plant Cell Physiol.* **2006**, *47*, 43–54. [CrossRef] [PubMed]
9. Considine, M.J.; Holtzapffel, R.C.; Day, D.A.; Whelan, J.; Millar, A.H. Molecular distinction between alternative oxidase from monocots and dicots. *Plant Physiol.* **2002**, *129*, 949–953. [CrossRef] [PubMed]
10. Rasmusson, A.G.; Wallström, S.V. Involvement of mitochondria in the control of plant cell NAD(P)H reduction levels. *Biochem. Soc. Trans.* **2010**, *38*, 661–666. [CrossRef] [PubMed]
11. Padmasree, K.; Padmavathi, L.; Raghavendra, A.S. Essentiality of mitochondrial oxidative metabolism for photosynthesis: Optimization of carbon assimilation and protection against photoinhibition. *Crit. Rev. Biochem. Mol. Biol.* **2002**, *37*, 71–119. [CrossRef] [PubMed]
12. Noguchi, K.; Yoshida, K. Interaction between photosynthesis and respiration in illuminated leaves. *Mitochondrion* **2007**, *8*, 87–99. [CrossRef] [PubMed]
13. Møller, I.M.; Rasmusson, A.G.; Fredlund, K.M. NAD(P)H-ubiquinone oxidoreductases in plant mitochondria. *J. Bioenerg. Biomemb.* **1993**, *25*, 377–384. [CrossRef]
14. Potters, G.; Horemans, N.; Jansen, M.A. The cellular redox state in plant stress biology—A charging concept. *Plant Physiol. Biochem.* **2010**, *48*, 292–300. [CrossRef] [PubMed]
15. Nunes-Nesi, A.; Araújo, W.L.; Fernie, A.R. Targeting mitochondrial metabolism and machinery as a means to enhance photosynthesis. *Plant Physiol.* **2011**, *155*, 101–107. [CrossRef] [PubMed]
16. Wallström, S.V.; Florez-Sarasa, I.; Araújo, W.L.; Aidemark, M.; Fernández-Fernández, M.; Fernie, A.R.; Ribas-Carbó, M.; Rasmusson, A.G. Suppression of the external mitochondrial NADPH dehydrogenase, NDB1, in *Arabidopsis thaliana* affects central metabolism and vegetative growth. *Mol. Plant* **2014**, *7*, 356–368. [CrossRef] [PubMed]
17. Suzuki, N.; Koussevitzky, S.; Mittler, R.; Miller, G. ROS and redox signalling in the response of plants to abiotic stress. *Plant Cell Environ.* **2012**, *35*, 259–270. [CrossRef] [PubMed]
18. Noctor, G.; Lelarge-Trouverie, C.; Mhamdi, A. The metabolomics of oxidative stress. *Phytochemistry* **2015**, *112*, 33–53. [CrossRef] [PubMed]
19. Gill, S.S.; Tuteja, N. Reactive oxygen species and antioxidant machinery in abiotic stress tolerance in crop plants. *Plant Physiol. Biochem.* **2010**, *48*, 909–930. [CrossRef] [PubMed]
20. Espinosa-Diez, C.; Miguel, V.; Mennerich, D.; Kietzmann, T.; Sánchez-Pérez, P.; Cadenas, S.; Lamas, S. Antioxidant responses and cellular adjustments to oxidative stress. *Redox Biol.* **2015**, *6*, 183–197. [CrossRef] [PubMed]
21. Schertl, P.; Braun, H.P. Respiratory electron transfer pathways in plant mitochondria. *Front. Plant Sci.* **2014**, *5*, 163. [CrossRef] [PubMed]
22. Meyer, A.J. The integration of glutathione homeostasis and redox signaling. *J. Plant Physiol.* **2008**, *165*, 1390–1403. [CrossRef] [PubMed]

23. Noctor, G.; Mhamdi, A.; Chaouch, S.; Han, Y.I.; Neukermans, J.; Marquez-Garcia, B.; Foyer, C.H. Glutathione in plants: An integrated overview. *Plant Cell Environ.* **2012**, *35*, 454–484. [CrossRef] [PubMed]
24. Deponte, M. Glutathione catalysis and the reaction mechanisms of glutathione-dependent enzymes. *Biochim. Biophys. Acta* **2013**, *1830*, 3217–3266. [CrossRef] [PubMed]
25. Dixon, D.P.; Skipsey, M.; Grundy, N.M.; Edwards, R. Stress-induced protein S-glutathionylation in *Arabidopsis*. *Plant Physiol.* **2005**, *138*, 2233–2244. [CrossRef] [PubMed]
26. Foyer, C.H.; Halliwell, B. The presence of glutathione and glutathione reductase in chloroplasts: A proposed role in ascorbic acid metabolism. *Planta* **1976**, *133*, 21–25. [CrossRef] [PubMed]
27. Sevilla, F.; Jiménez, A.; Lázaro, J.J. What do the plant mitochondrial antioxidant and redox systems have to say under salinity, drought, and extreme temperature? In *Reactive Oxygen Species and Oxidative Damage in Plants under Stress*, 1st ed.; Gupta, D.K., Palma, J.M., Corpas, F.J., Eds.; Springer International Publishing: Cham, Switzerland, 2015; pp. 23–55.
28. Geigenberger, P.; Thormählen, I.; Daloso, D.M.; Fernie, A.R. The unprecedented versatility of the plant thioredoxin system. *Trends Plant Sci.* **2017**, *22*, 249–262. [CrossRef] [PubMed]
29. Foyer, C.H.; Noctor, G. Ascorbate and glutathione: The heart of the redox hub. *Plant Physiol.* **2011**, *155*, 2–18. [CrossRef] [PubMed]
30. Rahantaniaina, M.S.; Tuzet, A.; Mhamdi, A.; Noctor, G. Missing links in understanding redox signaling via thiol/disulfide modulation: How is glutathione oxidized in plants? *Front. Plant Sci.* **2013**, *4*, 477. [CrossRef] [PubMed]
31. Zhang, J.; Kirkham, M.B. Antioxidant responses to drought in sunflower and sorghum seedlings. *New Phytol.* **1996**, *132*, 361–373. [CrossRef] [PubMed]
32. Garmier, M.; Carroll, A.J.; Delannoy, E.; Vallet, C.; Day, D.A.; Small, I.D.; Millar, A.H. Complex I dysfunction redirects cellular and mitochondrial metabolism in Arabidopsis. *Plant Physiol.* **2008**, *148*, 1324–1341. [CrossRef] [PubMed]
33. Vidal, E.A.; Gutierrez, R.A. A systems view of nitrogen nutrient and metabolite responses in *Arabidopsis*. *Curr. Opin. Plant Biol.* **2008**, *11*, 521–529. [CrossRef] [PubMed]
34. Bloom, A.J. Photorespiration and nitrate assimilation: A major intersection between plant carbon and nitrogen. *Photosynth. Res.* **2015**, *123*, 117–128. [CrossRef] [PubMed]
35. Zhu, Z.; Gerendás, J.; Bendixen, R.; Schinner, K.; Tabrizi, H.; Sattelmacher, B.; Hansen, U.P. Different Tolerance to Light Stress in NO_3^--and NH_4^+-Grown *Phaseolus vulgaris* L. *Plant Biol.* **2000**, *2*, 558–570. [CrossRef]
36. Guo, S.; Schinner, K.; Sattelmacher, B.; Hansen, U.P. Different apparent CO_2 compensation points in nitrate-and ammonium-grown *Phaseolus vulgaris* and the relationship to non-photorespiratory CO_2 evolution. *Physiol. Plant.* **2005**, *123*, 288–301. [CrossRef]
37. Noctor, G.; Foyer, C.H. A re-evaluation of the ATP: NADPH budget during C3 photosynthesis: A contribution from nitrate assimilation and its associated respiratory activity? *J. Exp. Bot.* **1998**, *49*, 1895–1908. [CrossRef]
38. Escobar, M.A.; Geisler, D.A.; Rasmusson, A.G. Reorganization of the alternative pathways of the *Arabidopsis* respiratory chain by nitrogen supply: Opposing effects of ammonium and nitrate. *Plant J.* **2006**, *45*, 775–788. [CrossRef] [PubMed]
39. Bloom, A.J.; Sukrapanna, S.S.; Warner, R.L. Root respiration associated with ammonium and nitrate absorption and assimilation by barley. *Plant Physiol.* **1992**, *99*, 1294–1301. [CrossRef] [PubMed]
40. Gerendás, J.; Zhu, Z.; Bendixen, R.; Ratcliffe, R.G.; Sattelmacher, B. Physiological and biochemical processes related to ammonium toxicity in higher plants. *J. Plant Nutr. Soil Sci.* **1997**, *160*, 239–251. [CrossRef]
41. Britto, D.T.; Kronzucker, H.J. Ecological significance and complexity of N-source preference in plants. *Ann. Bot.* **2013**, *112*, 957–963. [CrossRef] [PubMed]
42. Podgórska, A.; Gieczewska, K.; Łukawska-Kuźma, K.; Rasmusson, A.G.; Gardeström, P.; Szal, B. Long-term ammonium nutrition of *Arabidopsis* increases the extrachloroplastic NAD(P)H/NAD(P)+ ratio and mitochondrial reactive oxygen species level in leaves but does not impair photosynthetic capacity. *Plant Cell Environ.* **2013**, *36*, 2034–2045. [CrossRef] [PubMed]
43. Podgórska, A.; Szal, B. The role of reactive oxygen species under ammonium nutrition. In *Reactive Oxygen and Nitrogen Species Signaling and Communication in Plants*; Gupta, K., Igamberdiev, A., Eds.; Springer International Publishing: Cham, Switzerland, 2015; Volume 23, pp. 133–153.
44. Britto, D.T.; Kronzucker, H.J. NH_4^+ toxicity in higher plants: A critical review. *J. Plant Physiol.* **2002**, *159*, 567–584. [CrossRef]

45. Bittsánszky, A.; Pilinszky, K.; Gyulai, G.; Komives, T. Overcoming ammonium toxicity. *Plant Sci.* **2015**, *231*, 184–190. [CrossRef] [PubMed]
46. Esteban, R.; Ariz, I.; Cruz, C.; Moran, J.F. Review: Mechanisms of ammonium toxicity and the quest for tolerance. *Plant Sci.* **2016**, *248*, 92–101. [CrossRef] [PubMed]
47. Taylor, N.L.; Day, D.A.; Millar, A.H. Targets of stress-induced oxidative damage in plant mitochondria and their impact on cell carbon/nitrogen metabolism. *J. Exp. Bot.* **2004**, *55*, 1–10. [CrossRef] [PubMed]
48. Szal, B.; Podgórska, A. The role of mitochondria in leaf nitrogen metabolism. *Plant Cell Environ.* **2012**, *35*, 1756–1768. [CrossRef] [PubMed]
49. Foyer, C.H.; Noctor, G.; Hodges, M. Respiration and nitrogen assimilation: Targeting mitochondria-associated metabolism as a means to enhance nitrogen use efficiency. *J. Exp. Bot.* **2011**, *62*, 1467–1482. [CrossRef] [PubMed]
50. Van Aken, O.; Zhang, B.; Carrie, C.; Uggalla, V.; Paynter, E.; Giraud, E.; Whelan, J. Defining the mitochondrial stress response in *Arabidopsis thaliana*. *Mol. Plant* **2009**, *2*, 1310–1324. [CrossRef] [PubMed]
51. Lee, B.H.; Lee, H.; Xiong, L.; Zhu, J.K. A mitochondrial complex I defect impairs cold-regulated nuclear gene expression. *Plant Cell* **2002**, *14*, 1235–1251. [CrossRef] [PubMed]
52. Dutilleul, C.; Garmier, M.; Noctor, G.; Mathieu, C.; Chétrit, P.; Foyer, C.H.; De Paepe, R. Leaf mitochondria modulate whole cell redox homeostasis, set antioxidant capacity, and determine stress resistance though altered signaling and diurnal regulation. *Plant Cell* **2003**, *15*, 1212–1226. [CrossRef] [PubMed]
53. Meyer, E.H.; Tomaz, T.; Carroll, A.J.; Estavillo, G.; Delannoy, E.; Tanz, S.K.; Small, I.D.; Pogson, B.J.; Millar, A.H. Remodeled respiration in ndufs4 with low phosphorylation efficiency suppresses *Arabidopsis* germination and growth and alters control of metabolism at night. *Plant Physiol.* **2009**, *151*, 603–619. [CrossRef] [PubMed]
54. Szal, B.; Łukawska, K.; Zdolińska, I.; Rychter, A.M. Chilling stress and mitochondrial genome rearrangement in the MSC16 cucumber mutant affect the alternative oxidase and antioxidant defense system to a similar extent. *Physiol. Plant.* **2009**, *137*, 435–445. [CrossRef] [PubMed]
55. Maxwell, D.P.; Wang, Y.; McIntosh, L. The alternative oxidase lowers mitochondrial reactive oxygen production in plant cells. *Proc. Natl. Acad. Sci. USA* **1999**, *96*, 8271–8276. [CrossRef] [PubMed]
56. Umbach, A.L.; Fiorani, F.; Siedow, J.N. Characterization of transformed *Arabidopsis* with altered oxidase levels and analysis of effect on reactive oxygen species in tissue. *Plant Physiol.* **2005**, *139*, 1806–1820. [CrossRef] [PubMed]
57. Giraud, E.; Ho, L.H.; Clifton, R.; Carroll, A.; Estavillo, G.; Tan, Y.F.; Howell, K.A.; Ivanova, A.; Pogson, B.J.; Millar, A.H.; et al. The absence of alternative oxidase 1a in *Arabidopsis* results in acute sensitivity to combined light and drought stress. *Plant Physiol.* **2008**, *147*, 595–610. [CrossRef] [PubMed]
58. Liu, Y.J.; Norberg, F.E.B.; Szilágyi, A.; De Paepe, R.; Åkerlund, H.E.; Rasmusson, A.G. The mitochondrial external NADPH dehydrogenase modulates the leaf NADPH/NADP$^+$ ratio in transgenic *Nicotiana sylvestris*. *Plant Cell Physiol.* **2008**, *49*, 251–263. [CrossRef] [PubMed]
59. Smith, C.; Barthet, M.; Melino, V.; Smith, P.; Day, D.; Soole, K. Alterations in the mitochondrial alternative NAD(P)H dehydrogenase NDB4 lead to changes in mitochondrial electron transport chain composition, plant growth and response to oxidative stress. *Plant Cell Physiol.* **2011**, *52*, 1222–1237. [CrossRef] [PubMed]
60. Foyer, C.H.; Noctor, G. Redox sensing and signalling associated with reactive oxygen in chloroplasts, peroxisomes and mitochondria. *Physiol. Plant.* **2003**, *119*, 355–364. [CrossRef]
61. Huang, S.; Van Aken, O.; Schwarzländer, M.; Belt, K.; Millar, A.H. The roles of mitochondrial reactive oxygen species in cellular signaling and stress response in plants. *Plant Physiol.* **2016**, *171*, 1551–1559. [CrossRef] [PubMed]
62. Liu, Y.J.; Nunes-Nesi, A.; Wallström, S.V.; Lager, I.; Michalecka, A.M.; Norberg, F.E.; Widell, S.; Fredlund, K.M.; Fernie, A.R.; Rasmusson, A.G. A redox-mediated modulation of stem bolting in transgenic *Nicotiana sylvestris* differentially expressing the external mitochondrial NADPH dehydrogenase. *Plant Physiol.* **2009**, *150*, 1248–1259. [CrossRef] [PubMed]
63. Zechmann, B.; Stumpe, M.; Mauch, F. Immunocytochemical determination of the subcellular distribution of ascorbate in plants. *Planta* **2011**, *233*, 1–12. [CrossRef] [PubMed]
64. Zechmann, B. Compartment-specific importance of glutathione during abiotic and biotic stress. *Front. Plant Sci.* **2014**, *5*, 566. [CrossRef] [PubMed]

65. Queval, G.; Thominet, D.; Vanacker, H.; Miginiac-Maslow, M.; Gakière, B.; Noctor, G. H_2O_2-activated up-regulation of glutathione in *Arabidopsis* involves induction of genes encoding enzymes involved in cysteine synthesis in the chloroplast. *Mol. Plant* **2009**, *2*, 344–356. [CrossRef] [PubMed]
66. Gill, S.S.; Anjum, N.A.; Hasanuzzaman, M.; Gill, R.; Trivedi, D.K.; Ahmad, I.; Pereira, E.; Tuteja, N. Glutathione and glutathione reductase: A boon in disguise for plant abiotic stress defense operations. *Plant Physiol. Biochem.* **2013**, *70*, 204–212. [CrossRef] [PubMed]
67. Marty, L.; Siala, W.; Schwarzländer, M.; Fricker, M.D.; Wirtz, M.; Sweetlove, L.J.; Meyer, Y.; Meyer, A.J.; Reichheld, J.P.; Hell, R. The NADPH-dependent thioredoxin system constitutes a functional backup for cytosolic glutathione reductase in *Arabidopsis*. *Proc. Natl. Acad. Sci. USA* **2009**, *106*, 9109–9114. [CrossRef] [PubMed]
68. Zagorchev, L.; Seal, C.E.; Kranner, I.; Odjakova, M. A central role for thiols in plant tolerance to abiotic stress. *Int. J. Mol. Sci.* **2013**, *14*, 7405–7432. [CrossRef] [PubMed]
69. Zaffagnini, M.; Bedhomme, M.; Marchand, C.H.; Couturier, J.; Gao, X.H.; Rouhier, N.; Trost, P.; Lemaire, S.D. Glutaredoxin S12: Unique properties for redox signaling. *Antioxid. Redox Sign.* **2012**, *16*, 17–32. [CrossRef] [PubMed]
70. Patterson, K.; Walters, L.A.; Cooper, A.M.; Olvera, J.G.; Rosas, M.A.; Rasmusson, A.G.; Escobar, M.A. Nitrate-regulated glutaredoxins control Arabidopsis primary root growth. *Plant Physiol.* **2016**, *170*, 989–999. [CrossRef] [PubMed]
71. Dietz, K.-J. Redox signal integration: From stimulus to networks and genes. *Physiol. Plant.* **2008**, *133*, 459–468. [CrossRef] [PubMed]
72. Baier, M.; Dietz, K.-J. Protective function of chloroplast 2-cysteine peroxiredoxin in photosynthesis. Evidence from transgenic *Arabidopsis*. *Plant Physiol.* **1999**, *119*, 1407–1414. [CrossRef] [PubMed]
73. König, J.; Baier, M.; Horling, F.; Kahmann, U.; Harris, G.; Schürmann, P.; Dietz, K.-J. The plant-specific function of 2-Cys peroxiredoxin-mediated detoxification of peroxides in the redox-hierarchy of photosynthetic electron flux. *Proc. Natl. Acad. Sci. USA* **2002**, *99*, 5738–5743. [CrossRef] [PubMed]
74. Petersson, U.A.; Kieselbach, T.; García-Cerdán, J.G.; Schröder, W.P. The Prx Q protein of *Arabidopsis thaliana* is a member of the luminal chloroplast proteome. *FEBS Lett.* **2006**, *580*, 1873–3468. [CrossRef] [PubMed]
75. König, J.; Galliardt, H.; Jütte, P.; Schäper, S.; Dittmann, L.; Dietz, K.-J. The conformational bases for the two functionalities of 2-cysteine peroxiredoxins as peroxidase and chaperone. *J. Exp. Bot.* **2013**, *64*, 3483–3497. [CrossRef] [PubMed]
76. Muthuramalingam, M.; Seidel, T.; Laxa, M.; Nunes de Miranda, S.M.; Gartner, F.; Stroher, E.; Kandlbinder, A.; Dietz, K.-J. Multiple redox and non-redox interactions define 2-Cys peroxiredoxin as a regulatory hub in the chloroplast. *Mol. Plant* **2009**, *2*, 1273–1288. [CrossRef] [PubMed]
77. Dietz, K.-J.; Horling, F.; König, J.; Baier, M. The function of the chloroplast 2-cysteine peroxiredoxin in peroxide detoxification and its regulation. *J. Exp. Bot.* **2002**, *53*, 1321–1329. [CrossRef]
78. Lamkemeyer, P.; Laxa, M.; Collin, V.; Li, W.; Finkemeier, I.; Schöttler, M.A.; Holtkamp, V.; Tognetti, V.B.; Issakidis-Bourguet, E.; Kandlbinder, A.; et al. Peroxiredoxin Q of *Arabidopsis thaliana* is attached to the thylakoids and functions in context of photosynthesis. *Plant J.* **2006**, *45*, 968–981. [CrossRef] [PubMed]
79. Horling, F.; Lamkemeyer, P.; König, J.; Finkemeier, I.; Kandlbinder, A.; Baier, M.; Dietz, K.-J. Divergent light-, ascorbate-, and oxidative stress-dependent regulation of expression of the peroxiredoxin gene family in Arabidopsis. *Plant Physiol.* **2003**, *131*, 317–325. [CrossRef] [PubMed]
80. Rouhier, N.; Gelhaye, E.; Gualberto, J.M.; Jordy, M.-N.; De Fay, E.; Hirasawa, M.; Duplessis, S.; Lemaire, S.D.; Frey, P.; Martin, F.; et al. Poplar Peroxiredoxin Q. A thioredoxin-linked chloroplast antioxidant functional in pathogen defense. *Plant Physiol.* **2004**, *134*, 1027–1038. [CrossRef] [PubMed]
81. Yoshida, K.; Hisabori, T. Two distinct redox cascades cooperatively regulate chloroplast functions and sustain plant viability. *Proc. Natl. Acad. Sci. USA* **2016**, *113*, 3967–3976. [CrossRef] [PubMed]
82. Pérez-Ruiz, J.M.; Naranjo, B.; Ojeda, V.; Guinea, M.; Cejudo, F.J. NTRC-dependent redox balance of 2-Cys peroxiredoxins is needed for optimal function of the photosynthetic apparatus. *Proc. Natl. Acad. Sci. USA* **2017**, *114*, 12069–12074. [CrossRef] [PubMed]
83. Finkemeier, I.; Goodman, M.; Lamkemeyer, P.; Kandlbinder, A.; Sweetlove, L.J.; Dietz, K.J. The mitochondrial type II peroxiredoxin F is essential for redox homeostasis and root growth of *Arabidopsis thaliana* under stress. *J. Biol. Chem.* **2005**, *280*, 12168–12180. [CrossRef] [PubMed]

84. Horling, F.; König, J.; Dietz, K.-J. Type II peroxiredoxin C, a member of the peroxiredoxin family of *Arabidopsis thaliana*: Its expression and activity in comparison with other peroxiredoxins. *Plant Physiol. Biochem.* **2002**, *40*, 491–499. [CrossRef]
85. Murashige, T.; Skoog, F. A revised medium for rapid growth and bioassay with tobacco tissue cultures. *Physiol. Plant.* **1962**, *15*, 473–497. [CrossRef]
86. Florez-Sarasa, I.; Ostaszewska, M.; Galle, A.; Flexas, J.; Rychter, A.M.; Ribas-Carbó, M. Changes of alternative oxidase activity, capacity and protein content in leaves of *Cucumis sativus* wild-type and MSC16 mutant grown under different light intensities. *Physiol. Plant.* **2009**, *137*, 419–426. [CrossRef] [PubMed]
87. Escobar, M.A.; Franklin, K.A.; Svensson, A.S.; Salter, M.G.; Whitelam, G.C.; Rasmusson, A.G. Light regulation of the *Arabidopsis* respiratory chain. Multiple discrete photoreceptor responses contribute to induction of type II NAD(P)H dehydrogenase genes. *Plant Physiol.* **2004**, *136*, 2710–2721. [CrossRef] [PubMed]
88. Podgórska, A.; Ostaszewska, M.; Gardeström, P.; Rasmusson, A.G.; Szal, B. In comparison with nitrate nutrition, ammonium nutrition increases growth of the *frostbite1 Arabidopsis* mutant. *Plant Cell Environ.* **2015**, *38*, 224–237. [CrossRef] [PubMed]
89. Czechowski, T.; Stitt, M.; Altmann, T.; Udvardi, M.K.; Scheible, W.R. Genome-wide identification and testing of superior reference genes for transcript normalization in *Arabidopsis*. *Plant Physiol.* **2005**, *139*, 5–17. [CrossRef] [PubMed]
90. Pfaffl, M.W. A new mathematical model for relative quantification in real-time RT-PCR. *Nucleic Acids Res.* **2001**, *29*, e45. [CrossRef] [PubMed]
91. Szal, B.; Dąbrowska, Z.; Malmberg, G.; Gardeström, P.; Rychter, A.M. Changes in energy status of leaf cells as the consequence of mitochondrial genome rearrangement. *Planta* **2008**, *227*, 697–706. [CrossRef] [PubMed]
92. Veljović-Jovanović, S.; Noctor, G.; Foyer, C.H. Are leaf hydrogen peroxide concentrations commonly overestimated? The potential influence of artefactual interference by tissue phenolics and ascorbate. *Plant Physiol. Biochem.* **2002**, *40*, 501–507. [CrossRef]
93. Okamura, M. An improved method for determination of L-ascorbic acid and L-dehydroascorbic acid in blood plasma. *Clin. Chim. Acta* **1980**, *103*, 259–268. [CrossRef] [PubMed]
94. Brehe, J.E.; Burch, H.B. Enzymatic assay for glutathione. *Anal. Biochem.* **1976**, *74*, 189–197. [CrossRef]
95. Hodges, D.; DeLong, J.; Forney, C.F.; Prange, R.K. Improving the thiobarbituric acid-reactive-substances assay for estimating lipid peroxidation in plant tissues containing anthocyanin and other interfering compounds. *Planta* **1999**, *207*, 604–611. [CrossRef]
96. Ostaszewska-Bugajska, M.; Rychter, A.M.; Juszczuk, I.M. Antioxidative and proteolytic systems protect mitochondria from oxidative damage in S-deficient *Arabidopsis thaliana*. *J. Plant Physiol.* **2015**, *186–187*, 25–38. [CrossRef] [PubMed]
97. Bradford, M.M. A rapid and sensitive method for quantification of microgram quantities of protein utilizing the principle of protein-dye binding. *Anal. Biochem.* **1976**, *72*, 248–254. [CrossRef]
98. Kuznetsov, A.V.; Kehrer, I.; Kozlov, A.V.; Haller, M.; Redl, H.; Hermann, M.; Grimm, M.; Troppmair, J. Mitochondrial ROS production under cellular stress: Comparison of different detection methods. *Anal. Bioanal. Chem.* **2011**, *400*, 2383–2390. [CrossRef] [PubMed]
99. Hill, B.G.; Ramana, K.V.; Cai, J.; Bhatnagar, A.; Srivastava, S.K. Measurement and identification of S-glutathiolated proteins. In *Methods in Enzymology*; Cadenas, E., Packer, L., Eds.; Academic Press: Cambridge, MA, USA, 2010; Volume 473, pp. 179–197.
100. Podgórska, A.; Burian, M.; Rychter, A.M.; Rasmusson, A.G.; Szal, B. Short-term ammonium supply induces cellular defence to prevent oxidative stress in *Arabidopsis* leaves. *Physiol. Plant.* **2017**, *160*, 65–83. [CrossRef] [PubMed]
101. Hossain, M.A.; Asada, K. Inactivation of ascorbate peroxidase in spinach chloroplasts on dark addition of hydrogen peroxide: Its protection by ascorbate. *Plant Cell Physiol.* **1984**, *25*, 85–92. [CrossRef]
102. Kuźniak, E.; Skłodowska, M. The effect of *Botrytis cinerea* infection on ascorbate-glutathione cycle in tomato leaves. *Plant Sci.* **1999**, *148*, 69–76. [CrossRef]
103. Sehmer, L.; Fontaine, V.; Antoni, F.; Dizengremel, P. Effects of ozone and elevated atmospheric carbon dioxide on carbohydrate metabolism of spruce needles. Catabolic and detoxification pathways. *Physiol. Plant.* **1998**, *102*, 605–611. [CrossRef]

104. Gutzat, R.; Mittelsten Scheid, O. Epigenetic responses to stress: Triple defense? *Curr. Opin. Plant. Biol.* **2012**, *15*, 568–573. [CrossRef] [PubMed]
105. Foyer, C.H.; Rasool, B.; Davey, J.W.; Hancock, R.D. Cross-tolerance to biotic and abiotic stresses in plants: A focus on resistance to aphid infestation. *J. Exp. Bot.* **2016**, *67*, 2025–2037. [CrossRef] [PubMed]

© 2018 by the authors. Licensee MDPI, Basel, Switzerland. This article is an open access article distributed under the terms and conditions of the Creative Commons Attribution (CC BY) license (http://creativecommons.org/licenses/by/4.0/).

Article

Decoding the Divergent Subcellular Location of Two Highly Similar Paralogous LEA Proteins

Marie-Hélène Avelange-Macherel, Adrien Candat, Martine Neveu, Dimitri Tolleter and David Macherel *

IRHS, Agrocampus-Ouest, INRA, Université d'Angers, SFR 4207 Quasav, 42 rue George Morel, 49071 Beaucouzé, France; marie-helene.macherel@agrocampus-ouest.fr (M.-H.A.-M.); adrienpel@gmail.com (A.C.); martine.neveu@inra.fr (M.N.); dimitri.tolleter@univ-angers.fr (D.T.)
* Correspondence: david.macherel@univ-angers.fr; Tel.: +33-241-225-531

Received: 12 May 2018; Accepted: 28 May 2018; Published: 31 May 2018

Abstract: Many mitochondrial proteins are synthesized as precursors in the cytosol with an N-terminal mitochondrial targeting sequence (MTS) which is cleaved off upon import. Although much is known about import mechanisms and MTS structural features, the variability of MTS still hampers robust sub-cellular software predictions. Here, we took advantage of two paralogous late embryogenesis abundant proteins (LEA) from Arabidopsis with different subcellular locations to investigate structural determinants of mitochondrial import and gain insight into the evolution of the *LEA* genes. LEA38 and LEA2 are short proteins of the LEA_3 family, which are very similar along their whole sequence, but LEA38 is targeted to mitochondria while LEA2 is cytosolic. Differences in the N-terminal protein sequences were used to generate a series of mutated LEA2 which were expressed as GFP-fusion proteins in leaf protoplasts. By combining three types of mutation (substitution, charge inversion, and segment replacement), we were able to redirect the mutated LEA2 to mitochondria. Analysis of the effect of the mutations and determination of the LEA38 MTS cleavage site highlighted important structural features within and beyond the MTS. Overall, these results provide an explanation for the likely loss of mitochondrial location after duplication of the ancestral gene.

Keywords: late embryogenesis abundant protein; mitochondrion; mitochondrial import; gene duplication; paralog

---

1. Introduction

Mitochondria are key organelles of eukaryotic cells involved in energy production, metabolism, and signaling. Since only a few dozen proteins are encoded in the mitochondrial genome, several thousand proteins need to be imported into the organelle [1,2]. Mitochondrial targeted proteins are first synthesized in the cytosol as precursors and often display a mitochondrial targeting sequence (MTS) in their N-terminus. However, precursors for inner membrane proteins do not exhibit an MTS but internal targeting signals [3,4]. Precursor proteins are translocated as unfolded polypeptides across mitochondrial membranes via the transporter of the outer membrane complex (TOM) and the transporter of the inner membrane complex (TIM) [3,4]. According to the binding chain hypothesis [5], import proceeds through successive MTS binding to higher affinity sites, until the protein is trapped in the mitochondrial matrix by mitochondrial chaperones. While crossing the outer membrane does not require energy, an electrochemical proton gradient and ATP are needed for inner membrane pre-protein translocation [6]. MTS are generally cleaved off during the passage through mitochondrial membranes, or in the matrix, by the mitochondrial processing peptidase (MPP) or other peptidases, and further degraded by proteases [7]. Although it was generally considered that MTS were systematically cleaved upon import, which happens for the majority of matrix proteins, there is now increasing evidence that possibly half of mitochondrial proteins could be imported without cleavage of the MTS [2]. The overall mechanisms of mitochondrial protein import are similar in animals, yeast, and plants, although plant Tom20 does not appear to be orthologous to the animal and yeast Tom20 proteins [8]. Besides, the plant MPP peptidase is not localized in the matrix but is part of the cytochrome bc1 complex of the respiratory chain [9]. Protein import in plastids proceeds in a similar way as in mitochondria, but with different translocators (for review, [10]), and it is therefore not surprising that mitochondrial and plastidial pre-sequences share some common features and are not easily distinguished, although mitochondrial MTS generally contain more arginine [11]. Moreover, there is an increasing number of proteins which appear dually targeted to both organelles [12]. MTS share common properties because they carry essential information for proper targeting of several hundred mitochondrial proteins. However, they are still insufficiently characterized, partly because of their structural diversity, but also because of their similarity to plastid transit peptides. MTS length was found to be highly variable: 6–94 amino acids in yeast, 19–109 amino acids in Arabidopsis, and 1–122 in *Oryza sativa* [13]. Regarding amino acid composition, MTS contain a high proportion of positively charged, hydroxylated, and hydrophobic residues, but few acidic amino acids [14], and plant pre-sequences are noticeably enriched in serine [15].

MTS have the propensity to fold into an amphiphilic alpha-helix (with opposing positively charged and hydrophobic faces), a structure that would favor the interactions with TOM components and was considered to be necessary and sufficient for import [14,16,17]. Interestingly, in spite of the functional similarity between mitochondrial and plastid import systems, chloroplast transit peptides are usually unstructured although they could also form a helix upon contact with membranes [18]. No clear consensus sequence has been found for MTS, but loosely conserved motifs with Arg at the -2, -3, or -10 position from the processing site have been identified: -2R motif {R-X↓X}, -3R motif {R-X-(F/Y/L)↓(A/S)-X}, and -10R motif {R-X-(F/L/I)-X_2-(T/S/G)-X_4↓X} [7,14,19]. The Arg residue was experimentally shown to be important for cleavage processing [20]. Indeed, MPP recognises basic amino-acids and cleaves the sequence motif {(R/K)-Xn-R-X↓Φ-Ψ-Ψ} (with Φ and Ψ hydrophobic and hydrophilic residues, respectively) more efficiently [21]. However, pre-sequences without any conserved arginine close to the cleavage site (no-R motif) were also reported [22]. Studies in plants revealed a conserved motif {(F/Y)↓(S/A)} for the no-R group of proteins [13]. The -2R, -3R, and no-R motifs were found in pre-sequences of all organisms, whereas the -10R motif was not found in plants [13,22]. MTS features around the cleavage site have been thoroughly investigated in Arabidopsis and rice [13]. In these species, the -3R motif occurs more frequently in Arabidopsis MTS with Phe in the -1 position with respect to cleavage, or Phe, Tyr, and Leu in the case of rice MTS. Less flexibility was observed for plant MTS compared to non-plant MTS. Indeed, the presence of plastids may require a higher specificity of plant MTS in order to prevent mistargeting [23].

A number of publicly available online computer programs have been developed to predict mitochondrial targeted proteins, but they display poor consensus when comparing their predictions for a large number of protein sequences [24]. The programs are trained by using only a small number of proteins and their often ambiguous conclusions are due, in part, to the high diversity of MTS. A deeper understanding of the functional features of MTS, as well as larger bodies of experimental MTS data, are therefore needed to identify the key determinants for mitochondrial targeting and improve predictors.

In *Arabidopsis thaliana* Col-0, the subcellular locations of 51 proteins belonging to the Late Embryogenesis Abundant (LEA) protein family were determined experimentally using translational fusions with fluorescent proteins [25]. LEA proteins are characterized by a low sequence complexity, repeated motifs, and high hydrophilicity, and are often intrinsically disordered [26]. Among the 51 LEA proteins, five were found to be targeted to mitochondria. Two of them (LEA42, At4g15910; LEA48, At5g44310) belong to the PFAM family LEA_4, whose members were found to be distributed in many cellular compartments, and are expected to protect various cellular membranes during desiccation. LEA42 and LEA48 were dually targeted to mitochondria and chloroplasts [25]. They are expected to play a similar role to their orthologous pea protein LEAM in protecting the inner mitochondrial membrane during desiccation [27–29]. Three other LEA proteins (LEA37, At3g53770; LEA38, At4g02380; LEA41, At4g15910) were found to be exclusively mitochondrial. They belong to the PFAM LEA_3 family, which encompasses four members in Arabidopsis. Interestingly, this fourth member (LEA2, At1g02820), which is a paralogous protein to LEA38, was localized to the cytosol [25]. In contrast with proteins from the LEA_4 family which appears to be involved in seed desiccation tolerance, little is known about the function of proteins from the LEA_3 family [26]. *LEA38* was first identified as a senescence associated gene (*SAG21*) [30], and then as a gene (*AtLEA5*) able to complement an oxidant-stress sensitive yeast mutant [31]. Further work using overexpressor and antisense lines led to the proposal that LEA38 is involved in root development and abiotic stress tolerance [32]. However, to our knowledge, there is still no clue about the molecular function of any protein of the LEA_3 family. Excitingly, a genome-wide association study (GWAS) aiming to identify *Arabidopsis thaliana* loci involved in local geographic adaptation highlighted *LEA38* among the four best hits, suggesting a key role for the LEA38 protein in environmental adaptation [33].

Whatever their function, LEA38 and LEA2 are two paralogous proteins of a small size (around 10 kDa), with highly similar sequences but different subcellular locations [25]. These proteins therefore represent an original model to uncover MTS specific motifs, and to explore how these two paralogous genes evolved to encode proteins with different subcellular targeting. Here, bioinformatics, genetic engineering, and subcellular localization of synthetic proteins were used to identify the key determinants for mitochondrial localization of LEA38, which were likely lost in LEA2 after gene duplication.

2. Results

2.1. Proteins of the LEA_3 Family Are Expected to Be Mitochondrial

In Arabidopsis, the LEA_3 family comprises four proteins, with three of them (including LEA38) being exclusively mitochondrial, while the fourth (LEA2) is cytosolic [25]. This suggests that LEA2, which is paralogous to LEA38, could have lost its mitochondrial localization during evolution. To support this hypothesis, we first examined whether mitochondrial localization could possibly be the rule for LEA_3 proteins. The LEA38 sequence was used as a query in a BLASTP search of the NCBI Reference Sequence protein database. With an E-value cut-off of 10, the analysis yielded 390 hits (370 hits with *e*-value < 0.01). All the sequences were from higher plants, and a search for PFAM matches confirmed that all these sequences belonged to the PFAM family LEA_3 (PF03242). This strongly suggests that this protein family is specific to higher plants. The 390 protein sequences were then subjected to subcellular localization prediction using PProwler, which proved to be the more accurate prediction program for LEA proteins [25].

The program computes probabilities for localization in the secretory pathway, mitochondrion, chloroplast, peroxisome, and others (nucleus, cytoplasmic, or otherwise). The box plot shown in Figure 1 clearly reveals that the mitochondrion is by far the most probable subcellular destination of LEA_3 proteins. Therefore, it is likely that within the LEA38/LEA2 couple of paralogs, LEA38 has retained its mitochondrial localization, while LEA2 has lost it.

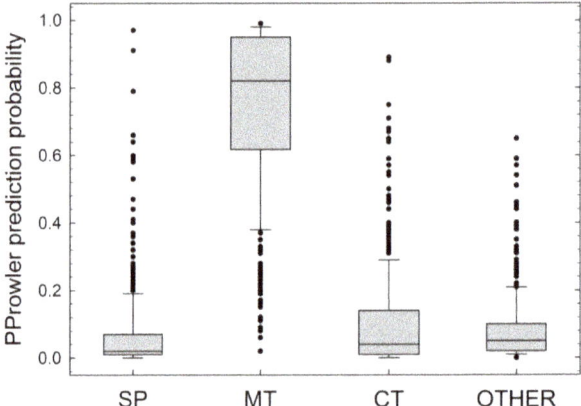

Figure 1. Subcellular prediction for proteins of the LEA_3 family. The graph shows the probability of subcellular targeting for 390 proteins of the LEA_3 family to different compartments (SP, secretory pathway; MT, mitochondria targeting; CT, chloroplast targeting) according to the PProwler software. Peroxisomal targeting probability is not indicated (null for all proteins). The 390 proteins selected are the best matches retrieved using the LEA38 protein sequence as the query in a BLASTP search against the NCBI Reference Sequence protein database, and all belong to the LEA_3 family. The boundary of the box closest to zero indicates the 25th percentile, a line within the box marks the median, and the boundary of the box farthest from zero indicates the 75th percentile. Whiskers (error bars) above and below the box indicate the 90th and 10th percentile, respectively.

2.2. Determinants of LEA38 Mitochondrial Targeting

As shown in Figure 2, LEA2 and LEA38, which share 60.8% identity and 80.4% similarity, as calculated by LALIGN, are very similar proteins, even in their N-terminal part, which is a critical region for organelle targeting. However, four differences are deemed of interest and are highlighted in Figure 2: a change in polarity at position 11 (Gln in LEA2, Val in LEA38), a charge inversion at position 17 and 18 (E17-K18 in LEA2, R17-E18 in LEA38), a stretch of 11 amino acids (^{33}AQGSVSSGGRS) specific to LEA38, and a segment of five amino acids (^{33}KTALD) specific to LEA2 in the same region.

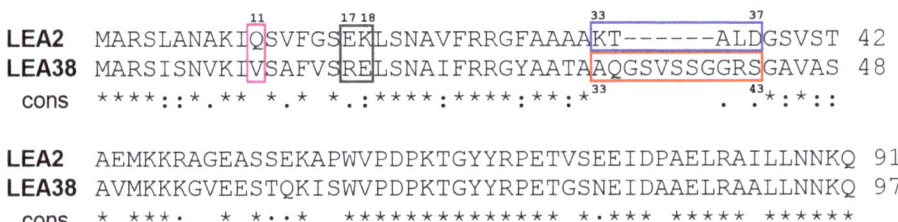

Figure 2. Comparison of LEA2 and LEA38 protein sequences. Amino acid alignment was performed with LALIGN. The degree of similarity between amino acids occupying the same position is indicated by the following symbols: "*" identity; ":" strong similarity; "." weak similarity. Sequence differences selected to generate mutated LEA2 proteins are indicated by colored boxes.

To assess the importance of these modifications with respect to mitochondrial targeting of LEA38, we generated a series of synthetic genes encoding mutated LEA2 and LEA38 proteins (Table 1).

Table 1. Description of the different mutated protein constructs.

Mutation	Description
LEA2del	LEA2 without residues 33 to 37 (KTALD)
LEA38del	LEA38 without residues 33 to 43 (AQGSVSSGGRS)
LEA2.1	LEA38 (residues 1 to 43) + LEA2 (residues 38 to 91)
LEA2.2	LEA2 with (KTALD) replaced by (AQGSVSSGGRS)
LEA2.3	Same as LEA2.2 with Q11V mutation
LEA2.4	Same as LEA2.2 with inversion at positions 17 and 18 (EK into KE)
LEA2.5	Same as LEA2.2 with Q11V mutation and inversion at position 17 and 18

The mutated genes were cloned into an expression vector in order to express the proteins of interest as translational fusions with GFP at their C-terminus. The constructs were used to transiently transform protoplasts from an Arabidopsis line expressing a mitochondrial mCherry marker [25]. LEA2-GFP and LEA38-GFP, used as controls, displayed their expected cytosolic and mitochondrial localizations, respectively (Figure 3).

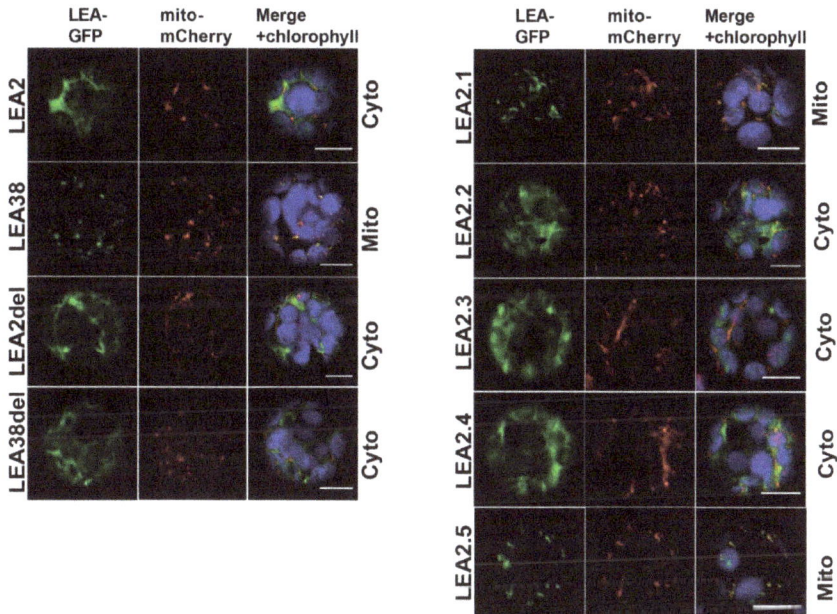

Figure 3. Impact of the mutations on the subcellular localization of LEA2-GFP and LEA38-GFP. The different proteins were expressed in fusion with GFP in Arabidopsis leaf protoplasts, using a line constitutively expressing a mitochondrial mCherry fluorescent protein. Observations were performed with a confocal microscope to visualize fluorescence from GFP, mCherry, and chlorophyll. First column: GFP, green fluorescence; second column: mCherry, red fluorescence; third column: merging of GFP and mCherry signals with chlorophyll autofluorescence in blue. Cyto, cytosolic; Mito, mitochondrial. Bars = 10 µm.

LEA2del-GFP, in which the ^{33}KTALD segment has been deleted, remained cytosolic, indicating that this sequence specific of LEA2 was not preventing mitochondrial targeting, having apparently no detectable effect on subcellular localization (Figure 3). LEA2.1 was constructed by replacing the first 37 amino acids of LEA2 with the first 43 amino acids of LEA38, including its specific ^{33}AQGSVSSGGRS segment (Figure 2; Table 1). LEA2.1-GFP systematically accumulated in mitochondria (Figure 3), suggesting that critical information for mitochondrial targeting is indeed enclosed in the first 43 amino acids of LEA38. The LEA38del-GFP construct, in which the specific ^{33}AQGSVSSGGRS segment was deleted, was no longer targeted to mitochondria, and remained cytosolic (Figure 3). This motif is therefore essential for mitochondrial targeting of LEA38. To determine if this motif could confer mitochondrial targeting to LEA2, LEA2 was modified to replace its ^{33}KTALD segment with the ^{33}AQGSVSSGGRS motif of LEA38, yielding the LEA2.2 construct (Figure 2; Table 1). However, the modified LEA2.2-GFP remained localized in the cytosol. Together with the LEA38del-GFP construct, these results indicate that the ^{33}AQGSVSSGGRS sequence is essential but not sufficient to confer mitochondrial targeting to LEA38 or LEA2.

To take into account the possible role of other sequence differences between LEA2 and LEA38, three additional constructs were generated from LEA2.2 (Table 1). First, the hydrophilic Gln residue at position 11 of LEA2 was replaced with the hydrophobic Val residue found in LEA38 to yield the LEA2.3 construct (Table 1). Secondly, the two residues E17-K18 in the LEA2 sequence were inversed into K17-E18 to mimic the corresponding charge arrangement found in LEA38 (R17-E18), yielding the construct LEA2.4 (Table 1). Finally, these two mutations were combined, yielding the LEA2.5 construct. When expressed in protoplasts, LEA2.3-GFP and LEA2.4-GFP systematically remained cytosolic (Figure 3). However, in the case of LEA2.5-GFP, which combines all candidate mutations, mitochondrial localization could finally be demonstrated (Figure 3). With other constructs (LEA2del; LEA38del; LEA2.1; LEA2.2; LEA2.3; LEA2.4), all transformed protoplasts showed a reporter GFP with cytosolic localization. However, in the case of LEA2.5-GFP, mitochondrial localization was not observed in all transformed protoplasts. Instead, approximately half of the transformed protoplasts (85/177 protoplasts from seven experiments) showed mitochondrial localization, while the others showed a diffuse cytosolic localization or large aggregates of GFP. We never observed a dual localization of LEA2.5 in mitochondria and cytosol (or aggregates). Although it is difficult to provide a rationale for this heterogeneity, it could be due to the fact that more mutations in the N-terminal sequence could be required to fully restore the efficiency of import, especially in the protoplast assay in which transgene expression is very strong. Nevertheless, the results indicate that while the specific KTALD segment of LEA2 is not involved in subcellular localization of the protein, the addition of the LEA38 specific motif AQGSVSSGGRS, the replacement of Gln11 with Val, and the inversion of E17-K18 are sufficient altogether to target LEA2 to the mitochondrial compartment. It should be noted that none of the synthetic proteins were found to be targeted to the chloroplast, although common features exist for protein import in chloroplasts and mitochondria [18].

Since MTS are expected to adopt an amphiphilic helix conformation, the secondary structures of LEA2, LEA38, and mutated proteins were investigated. As shown in Figure 4, all native and mutant proteins were predicted to harbor alpha helix domains in their N-terminal region, irrespective of their subcellular localization. Both LEA2 and LEA38 sequences display two alpha-helices; the first from residues 5–23 and 7–24, respectively, and a second from residues 28–35 and 33–41, respectively, which is preceded by a short extended strand sequence of four residues in LEA38. The five and eleven residue long segments specific to LEA2 or LEA38 are located in this second alpha-helix region. In LEA2del, deletion of the five residues generated a longer helix, and the protein remained cytosolic. In LEA38del, for which mitochondrial localization was abolished, the deletion of the eleven residues removed the small extended strand found in LEA38, yielding a longer helix (Figure 4). This suggests that the extended strand-helical conformation in this region could be important for mitochondrial localization, but the hypothesis can be dismissed because this secondary structure is detected in LEA2.4 (cytosolic) but not in LEA2.5 (mitochondrial). Since the other mutations introduced to redirect LEA2 to mitochondria were localized in the first alpha-helix, helical projections with the Heliquest program were performed to address the structural impact of mutations (Figure 4). The projection of residues 7–24 of LEA38 revealed an amphiphilic helix, with two positively charged residues in the hydrophilic face (VAVVL) and a negatively charged residue (Glu) at the border between the hydrophobic and hydrophilic faces. In LEA2, the projection of residues 5–23 did not reveal a clear amphiphilic face, in spite of the grouping of four residues (AGAL). As found for the corresponding helix in LEA38, three charged residues are present but with a different disposition (Figure 4). The negatively charged residue (Glu) was located at the opposite end of the AGAL region, and the positively charged residues were facing each other, at 90° of the Glu. The Q11V mutation in LEA2.3 restored an amphiphilic alpha-helix with a hydrophobic face (VAGAL), but with a disposition of charged residues similar to LEA2 (Figure 4). The charge inversion (EK/KE) in LEA2.4 restored a disposition of the three charged residues very similar to that in LEA38, but like in LEA2, a hydrophobic face could not be identified (Figure 4). In LEA2.5, in which these two latter mutations were combined, the alpha-helix shows a clear amphiphilic profile with a charged residue distribution almost identical to that of LEA38 (Figure 4). Since the LEA2.5-GFP construct was targeted to mitochondria, this indicates that an alpha-helix displaying a hydrophobic face of five residues, with a negatively charged residue at the interface, and two positively charged residues (separated by three residues in the projection) at the opposite end of the hydrophobic face, are essential conditions for mitochondrial targeting.

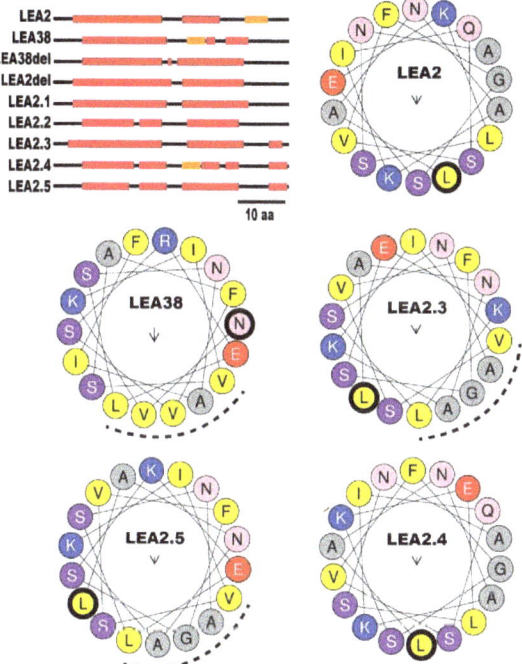

Figure 4. Secondary structure predictions for LEA2, LEA38, and mutated proteins. Secondary structure predictions for the first 50 amino acids (top left part of the figure) were performed using JPRED4 software. Red box: alpha-helix: orange box: extended strand; black line: random coil. Helical projections of α-helices were obtained using 18 amino-acids at the beginning of the first helix in the indicated proteins. The helical projections of LEA38, LEA38del, and LEA2.1, which share the same sequence in the analyzed region, are identical, as well as those of LEA2, LEA2del, and LEA2.2. The first residue is circled in black. Positively charged residues (K, R) are color coded in blue, while negatively charged residues (D, E) are in red. Non-polar residues are shown in yellow or grey and others in purple, light blue, or light pink colors. The arrow of variable size in the center of the helix shows the weight and orientation of the hydrophobic moment, and the dotted line identifies the hydrophobic face.

2.3. MTS Cleavage Site Determination for LEA38

Since mitochondrial transit peptides (30–50 residues) are often cleaved during the import process, in the case of LEA38, which has 97 amino acids, this could remove up to half of the precursor proteins, yielding a rather short mature protein. Another possibility is the absence of cleavage of the pre-sequence, which is often the case for mitochondrial membrane proteins [4]. The sub-mitochondrial localization of LEA38 has not yet been established, but its small size and the lack of predicted transmembrane helices (Table S1) do not favor a membrane localization. The existence of a cleavage site was questionable for LEA38, and we therefore attempted to determine the N-terminus of the protein using an Arabidopsis transgenic line constitutively expressing LEA38-GFP in mitochondria [25]. The LEA38-GFP was immunopurified from leaves and analysed by western blot using a specific anti-LEA38 antibody. The molecular mass of the protein isolated from leaves was compared with that of a recombinant LEA38-GFP synthetized in vitro (Figure 5). The recombinant LEA38-GFP precursor, produced by an *in-vitro* translation assay, displayed an apparent molecular mass of 39 kDa, consistent with its theoretical mass. LEA38-GFP which was purified from leaves exhibited a significantly lower apparent molecular mass (35 kDa), indicating that the protein was cleaved after mitochondrial import. The LEA2-GFP precursor synthesized in vitro was only slightly detected by the anti-LEA38 antibody, confirming its specificity (Figure 5).

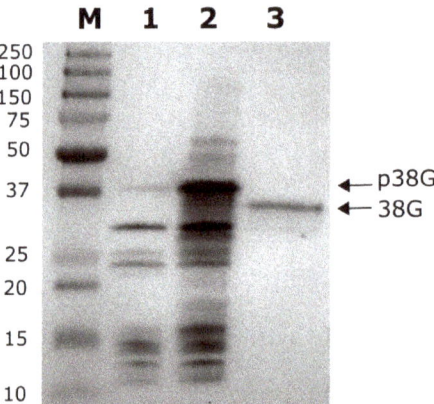

Figure 5. Comparison of the apparent molecular mass of the LEA38-GFP precursor and the mature LEA38-GFP. Precursors for LEA2-GFP and LEA38-GFP were synthesized in vitro. The mature LEA38-GFP (38G) was immunopurified from the leaves of an Arabidopsis overexpressor. Proteins were analyzed by Western blot using the anti-LEA38 antibody. This antibody was very specific to LEA38 and did not cross-react significantly with LEA2. 1, in vitro synthesized LEA2-GFP precursor (above 37 kDa); 2, in vitro synthesized LEA38-GFP precursor; 3, immunopurified mature LEA38-GFP. Arrows indicate the LEA38-GFP precursor (p38G) and mature LEA38-GFP (38G); M: molecular mass markers, with mass indicated in kDa.

Higher amounts of mature LEA38-GFP protein were immunopurified from leaf mitochondria of the overexpressing line, and the N-terminus was determined by Edman microsequencing. The results revealed that cleavage of the presequence occurred at two adjacent sites, after Y28 and A29, and thus upstream of the 11 amino acid segment (^{33}AQGSVSSGGRS), which is specific to LEA38 (see Figure 1). The theoretical molecular mass of the mature LEA38GFP (35.7 kDa) is thus consistent with the western blot results (Figure 5).

2.4. Comparison of Experimental Data and In Silico Predictions

Six online subcellular location prediction programs (iPSORT, MITOPROTII, MitoFates, PProwler, Predotar, and TargetP1.1) were used to predict the location of LEA2 and LEA38 and related GFP constructs (Table 2). Most programs suggested mitochondrial localization (32 out of 42 combinations in Table 2) for the seven proteins, although four of these are empirically cytosolic. More of the programs predicted a mitochondrial localization for LEA2, which is cytosolic, than for the mitochondrial LEA38 (5 and 4, respectively). The predictions for the native LEA proteins and their translational GFP fusions are logically very similar, with only some slight differences in probabilities, and a single difference in the decision for TargetP1.1 (LEA2 chloroplastic/LEA2-GFP mitochondrial), but both had a rather low score (0.36). This program was the least accurate with this set of proteins since LEA2, LEA38, LEA38-GFP, and LEA2.5-GFP were predicted as chloroplastic proteins and LEA38del-GFP as a secreted protein. The overall comparison of predictions with experimental localizations, including the series of mutant proteins, confirms that none of the programs are sufficiently accurate to predict the subcellular localisation of polypeptides with confidence (Table 2). Three of the programs include the predicted length of pre-sequences in their analysis. As shown in Table 2, predicted MTS displayed various lengths (28 to 43 amino-acids), with cleavage sites upstream or downstream from the 11 amino acid sequence that is required for correct mitochondrial targeting of LEA38-GFP. The MitoFates program provided the best results with the prediction of the genuine cleavage site of LEA38 (after Y28).

Table 2. Subcellular localisation and pre-sequence length predicted for LEA2, LEA38, and related mutated proteins. Six programs were used to predict the protein subcellular localisation of the indicated proteins (M, mitochondria; Ct, chloroplast; SP, secretory pathway). When available, probability is indicated in parentheses. The length of the predicted targeting sequence is indicated below each prediction for the three programs providing this analysis.

	IPsort	MitoFates	MITOPROT II v1.101	PProwler 1.2	Predotar	TargetP 1.1
LEA2	M -	M (0.839) 28	M (0.958) 35	M (0.84) -	M (0.43) -	Ct (0.364) 48
LEA38	Ct -	M (0.544) 28	M (0.986) 43	M (0.90) -	M (0.31) -	Ct (0.493) 46
LEA2-GFP	M -	M (0.889) 28	M (0.949) 35	M (0.85) -	M (0.43) -	M (0.357) 28
LEA38-GFP	Ct -	M (0.672) 28	M (0.995) 43	M (0.85) -	M (0.31) -	Ct (0.429) 46
LEA38del-GFP	Ct -	M (0.760) 28	M (0.984) 28	M (0.92) -	M (0.32) -	SP (0.499) 13
LEA2.1-GFP	Ct -	M (0.493) 28	M (0.984) 43	M (0.88) -	M (0.31) -	Ct (0.539) 54
LEA2.5-GFP	M -	M (0.500) 28	M (0.924) 43	M (0.73) -	M (0.35) -	Ct (0.657) 54

3. Discussion

LEA2 and LEA38 are two paralogous proteins of the LEA_3 family with very similar primary sequences but different subcellular locations. The LEA_3 family is specific to higher plants, and both experimental data in Arabidopsis [25] and targeting predictions for 390 proteins of the family strongly suggest that mitochondrial localization is a characterizing feature of this family. Thus, it is reasonable to assume that LEA2 has lost its original mitochondrial localization. The primary objective of this study was to uncover the structural changes that occurred during evolution by identifying the key structural features of LEA38 that, when incorporated into LEA2 by mutation, are able to redirect the mutated LEA2 to mitochondria. We also expected this experimental model to provide original information about the import of mitochondrial proteins.

The main difference between LEA38 and LEA2 sequences is an eleven residue-long stretch of aminoacids (^{33}AQGSVSSGGRS) which is specific to LEA38, and which was therefore a good candidate for a role in mitochondrial targeting. Indeed, its deletion abolished the mitochondrial localization of LEA38. However, its insertion in LEA2 had no effect on the subcellular targeting of the protein, which remained cytosolic; two additional mutations of LEA2 (Q11V and a charge inversion E17K/K18E) were required to redirect the protein to mitochondria. This confirmed that the ^{33}AQGSVSSGGRS sequence was essential but not sufficient for mitochondrial targeting. Interestingly, we could determine that this stretch of amino acids was not included in the presequence, but located a few amino acids after the MTS cleavage site (occurring after residues 28 and 29). This sequence could therefore be involved in the recognition of the pre-sequence cleavage site by proteases. Its enrichment in Ser residues supports this hypothesis, since Ser residues were frequently found downstream of presequence cleavage sites which have been established for 62 Arabidopsis and 52 rice mitochondrial proteins [13].

In agreement with previous studies [17,34–36], we found that the distribution of charged and apolar residues in the N-terminus was critical for mitochondrial import, and the results further support the requirement of an amphiphilic alpha helix structure with positive charges in the MTS. The analysis of helical projections in the MTS of the native and mutant proteins allowed us to identify critical structural features for the mitochondrial import of LEA38. The hydrophobic face must comprise five residues (four is not enough) and the hydrophilic face must harbor two positively charged residues, separated by three residues. Thus, the positive charges are positioned at the opposite end of the hydrophobic-hydrophilic interfaces of the helix. Finally, the single negatively charged residue (Glu) must be adjacent to the hydrophobic face of the helix. In this context, the effect of the charge inversion (^{17}EK to ^{17}KE) which was required for mitochondrial targeting of the LEA2 mutant is well justified. It has a major effect on the helical projection, restoring the original charge distribution in LEA38 MTS. All these features are necessarily important for mitochondrial import. The amphiphilic helix with its positive charges constitutes a target for interaction with cytosolic and mitochondrial HSP70 chaperones, with the latter being involved in the translocation of precursors [4,11]. The hydrophobic side and the positively charged side of the amphiphilic alpha-helix might be recognized sequentially by distinct TOM receptors [17]. Eventually, positively charged residues in the MTS could contribute to the subsequent translocation across the inner membrane, which is driven by membrane potential [4].

Two adjacent cleavage sites of the MTS were determined by Edman sequencing in LEA38, at positions ^{28}Y↓^{29}A and ^{29}A↓^{30}A (arrow indicates cleavage). For the first cleavage site, the surrounding sequence {^{24}F-R-R-G-Y↓A↓A-T-A} conforms well with the consensus motif {R-X-(F/Y/L)↓(S/A)-(S/A/T)-X} of the "-3R group" plant MTS [13]. In the processing of -3R group proteins, after cleavage by the MPP, the additional protease ICP55 (Intermediate Cleaving Peptidase of 55 kDa) cleaves the N-terminal amino acid from the intermediate first generated by MPP [37]. In the case of LEA38, we postulate that the presence of two successive Ala residues in the N-terminal sequence ^{28}YAATA of the protein after cleavage by MPP lowers the peptidase specificity of ICP55, releasing two mature proteins, each with an Ala N-terminus. Our results also highlight the importance of residues beyond the MTS cleavage site since the LEA38del mutant, in which 11 residues were deleted from position 33, lost its mitochondrial location. Further mutation studies would be required to predict how many residues downstream of the cleavage site are essential.

Since the vast majority of LEA_3 proteins were predicted to be mitochondrial, we built sequence logos using the first 40 amino acids of the 100 LEA_3 proteins with the best BLASTP score among the 390 sequences used for subcellular prediction analysis (Figure 6). The sequence logo illustrates the high sequence conservation in the N-terminal part of the proteins, and allows the establishment of a more precise consensus around the cleavage site {^{25}R-R-G-(Y/F)↓A↓A-(A/T)-(A/S)} for the LEA_3 family. Interestingly, the corresponding sequence of LEA2 {^{25}R-R-G-F-A-A-A-A-K-T} perfectly matches the consensus, and therefore the loss of mitochondrial location for the paralog of LEA38 is essentially due to a loss of targeting information, which does not affect the cleavage site motif. This appears logical since processing occurs within mitochondria during or upon import [7]. However, since LEA2-GFP expressed in protoplasts is apparently not cleaved [25], it also indicates that there is no protease activity in the cytosol with specificity equivalent to those of the mitochondrial processing peptidases. Limiting the processing peptidases to organellar locations should be crucial for the efficiency of import, and the high stability of LEA2-GFP in protoplasts supports the total lack of mitochondrial-like processing peptidase activity in the cytosol.

Figure 6. Sequence logo analysis of the N-terminal part of orthologs of LEA38. The first 40 amino-acid of 100 LEA_3 proteins sequences with the best scores in a BLASTP search were used to build the sequence logo. The mitochondrial targeting cleavage sites of LEA38 are indicated with the red dotted lines.

The coding sequences of LEA2 and LEA38 are very similar (71% identity), and both the length (96 vs. 100 nucleotides) and position (72 vs. 74) of the single intron are well conserved (Figure S1). The intron sequences are more divergent than the coding sequences, which reflects the lower evolutionary constraint on introns [38]. The high sequence similarity of sequences strongly supports the idea that LEA38 and LEA2 genes are paralogs resulting from one of the several whole genome duplication events in Arabidopsis [39]. The fact that all members of the LEA_3 family were predicted with high confidence to be mitochondrial strongly suggests that the ancestor of *LEA2* and *LEA38* genes encoded a mitochondrial protein, and that the protein encoded by the *LEA2* gene lost its mitochondrial localization after duplication. Gene duplication has been intensively reported as a very important evolution process leading to neofunctionalization (new function and/or expression pattern of one of the two duplicates) or subfunctionalization (division of ancestral functions and/or expression pattern between the two paralogs) [39]. Protein subcellular relocalization of duplicated genes has been observed in yeast [40], mammals [41], and plants [42]. Neolocalization (new localization for the copy product) or sublocalization (division of ancestral subcellular localizations between the paralogs) may contribute to the maintenance and functional divergence of genes pairs, and changes of expression patterns associated with protein relocalization were observed in Arabidopsis [42]. This is also the case for *LEA2* and *LEA38* genes that differ in their expression at the transcript level (data available at http://bar.utoronto.ca/efp2/Arabidopsis/Arabidopsis_eFPBrowser2.html). *LEA2* appears poorly expressed compared to *LEA38* in all developmental stages but its expression increases upon cold stress. *LEA38* is highly expressed within mature pollen, and in other tissues, its expression can be induced in response to osmotic stress and biotic stress. We can conclude that evolution of the paralogs led to a cytosolic sublocalization of LEA2, likely by the loss of mitochondrial targeting resulting from a few mutations, according to our experiments. Whether the neolocalization of LEA2 paralogs was associated with a new function for LEA2 in the cytosol is a difficult question to answer, because there is little information about the role of LEA_3 proteins. LEA38 (referred as AtLEA5) was shown earlier to complement an oxidant-sensitive yeast mutant, and its overexpression of *AtLEA5* in Arabidopsis increased oxidative stress tolerance [31]. In another study using over-expressor and anti-sense lines, LEA38 (referred as SAG21) was shown to interfere with leaf senescence, root development, and pathogen defense, which led to the proposal that LEA38 was involved in ROS signaling [32].

Interestingly, *SAG21* emerged from a genome-wide associated study as a major candidate gene for the local adaptation of Arabidopsis ecotypes [33]. We found that the discriminating single nucleotide polymorphism (SNP) of *SAG21* in this study was associated with an F/I substitution in position 24, in the MTS and just a few amino acids before the cleavage site (see Figure 6). It would be of interest in this context to determine if this substitution could have an effect on the import of LEA38. Still, the molecular function of LEA38 remains enigmatic, and to our knowledge, nothing is known about LEA2. If the latter performed a similar function as LEA38 in the cytosol, it would have to do so with an additional N-terminal extension (the part corresponding to LEA38 MTS) representing more than 30% of the length of mature LEA38 polypeptide. It must be recalled that, besides LEA38, two other very similar LEA_3 proteins (LEA37, LEA41) are also targeted to mitochondria [25], and therefore the loss of mitochondrial targeting for the *LEA2* gene product could possibly be compensated for by the other proteins. In conclusion, our results provide a rationale for the divergent subcellular localization of two LEA_3 protein paralogs with highly similar sequences, and suggest a scenario in which a few mutations resulted in the loss of mitochondrial targeting for one of the gene products, while the other one remained mitochondrial. We also provide arguments that support mitochondrial localization as a signature for LEA_3 proteins, and the first evidence, to our knowledge, that their MTS are cleaved upon import, releasing short mature proteins of around seven kDa. More research will be required to elucidate the molecular function of these small and intriguing plant mitochondrial proteins.

4. Materials and Methods

4.1. Plant Culture, Protoplast Isolation, and Transformation

Arabidopsis thaliana (Columbia-0 ecotype) wild type and transgenic lines were grown in potting compost Klassman 15 (Klasmann-Deilmann France SARL, Bourgoin-Jallieu, France) in a growth chamber (23 °C, 75% RH, 16 h light with an intensity of 100 µmol·m^{-2}·s^{-1}). Arabidopsis mesophyll protoplasts were isolated from a transgenic line expressing a mitochondrial mCherry protein and transformed according to the procedure described in [25].

4.2. Expression of Mutated Proteins

Synthetic genes encoding coding sequences of LEA2 (At1g02820) or LEA38 (At4g02380) with defined mutations indicated in Table 1 were obtained from GeneCust (Ellange, Luxembourg). LEA2, LEA38, and the mutated proteins coding sequence were cloned in the p2GWF7,0 vector from Plant System Biology (Ghent University, Ghent, Belgium) following the procedure described in [25]. These vectors were then introduced in Arabidopsis mesophyll protoplasts using polyethylene glycol mediated transformation [25].

4.3. Microscopy

The subcellular localition of fluorescent protein fusions in Arabidopsis mesophyll protoplasts were observed with a Nikon A1 laser scanning confocal microscope (Nikon France S.A, Champigny sur Marne, France). GFP, mCherry, and chlorophyll were excited with 488, 561, and 638 nm laser lines, respectively, with an emission band of 500 to 550 nm for GFP, 570 to 620 nm for mCherry, and 662 to 737 nm for chlorophyll autofluorescence.

4.4. In Vitro Production of Recombinant Proteins

LEA2-GFP and LEA38-GFP full-length proteins were obtained by coupled transcription-translation in vitro using the PURExpress kit (NEBS, Ipswich, MA, USA) using the recommended protocol.

4.5. Crude Mitochondria Isolation, and N-Terminus Sequencing

Mitochondria were isolated from rosette leaves of four week-old transgenic Arabidopsis plants expressing LEA38-GFP in mitochondria [25]. Leaves were grinded with a Waring blender in 10 mL isolation buffer (30 mM MOPS pH 7.8, 330 mM sorbitol, 2 mM EDTA, 1.5% w/v BSA) and then filtrated through eight layers of Miracloth. Chloroplasts and nuclei were removed by centrifugation at 1200× g for 45 s. The supernatant was further centrifuged at 5000× g for 5 min to obtain a crude mitochondrial fraction. GFP-tagged proteins were purified from crude mitochondria with the µMACS Epitope Tag Protein Isolation Kit (Miltenyi Biotec, Bergisch Gladbach, Germany) using the standard elution protocol. After separation by SDS-PAGE and transfer to the PVDF membrane, protein bands visualized by Coomassie Blue staining were excised for Edman microsequencing [43], which was performed by Proteome Factory AG (Berlin, Germany).

4.6. Protein Analysis by Western Blot

Proteins were separated by SDS PAGE and protein blotting was performed on PVDF membranes (Immobilon PSQ 0.2 µm, Merck KGaA, Darmstadt, Germany) using a transblot apparatus (Bio-Rad, Marnes-La-Coquette, France) and 10 mM CAPS pH 11, 10% (v/v) methanol as a transfer buffer [43]. The anti-LEA38 antibody (Genscript Biotech, Piscataway, NY, USA) was raised again the peptide (KKKGVEESTQKI) and used as a primary antibody at a dilution of 1:1000. As a secondary antibody, we used an anti-rabbit IgG coupled to horseradish peroxidase (Merck KGaA, Darmstadt, Germany) at a dilution of 1:50,000. The immunodetection was performed by incubating membranes in the Clarity™ Western ECL reagent (Bio-Rad, Marnes-La-Coquette, France) and the emitted chemiluminescence was recorded by a Chemidoc Imager (Bio-Rad). Molecular masses were estimated using Precision Plus Protein Dual Color Standards (Bio-Rad).

4.7. Bioinformatics

Sequence alignments were performed with the LALIGN software (http://www.ebi.ac.uk/Tools/psa/lalign/) [44], Clustal Omega software (http://www.ebi.ac.uk/Tools/msa/clustalo/) [45], and EMBOSS Needle (https://www.ebi.ac.uk/Tools/psa/emboss_needle/nucleotide.html) [46]. Search for orthologs of LEA38 within the NCBI Reference Sequence protein database was performed with BLASTP 2.8.0 (https://blast.ncbi.nlm.nih.gov/Blast.cgi) using BLOSUM62 matrix and default parameters [47]. The HMMER software (https://www.ebi.ac.uk/Tools/hmmer/) was used to search for Pfam matches [48]. Subcellular localisations and pre-sequence cleavage site predictions were performed using IPSORT (http://psort.hgc.jp/form.html) [49], MITOPROT II (http://ihg.gsf.de/ihg/mitoprot.html) [50], TargetP 1.1 (http://www.cbs.dtu.dk/services/TargetP/) [51], Predotar (https://urgi.versailles.inra.fr/predotar/) [52], PProwler (http://bioinf.scmb.uq.edu.au:8080/pprowler_webapp_1-2/) [53], and MitoFates (http://mitf.cbrc.jp/MitoFates/cgi-bin/top.cgi) [54] software with default settings. Molecular weight and theoretical pI were calculated using ProMoST (http://proteomics.mcw.edu/promost.html) [55]. Secondary structure predictions were performed with Jpred4 (http://www.compbio.dundee.ac.uk/jpred/) [56]. Alpha-helix projections were obtained with the HeliQuest Web server (http://heliquest.ipmc.cnrs.fr/) [57], using the Analysis module with default parameters (helix type set to alpha) and a window size set to FULL.

Supplementary Materials: Supplementary materials can be found at http://www.mdpi.com/1422-0067/19/6/1620/s1.

Author Contributions: A.C., M.-H.A.-M., and D.M. conceived and designed the experiments; A.C., M.-H.A.-M., D.T., and M.N. performed the experiments; A.C., M.-H.A.-M., D.T., and D.M. analyzed the data; M.-H.A.-M. and D.M. wrote the paper.

Funding: This research was funded by Agence Nationale de la Recherche (ANR, France), grant number MITOZEN ANR-12-BSV8-002.

Acknowledgments: We thank Aurélia Rolland and Fabienne Simonneau for their skillful assistance at the IMAC/SFR QUASAV facility. We are thankful to David C. Logan for invaluable suggestions and proofreading of the manuscript.

Conflicts of Interest: The authors declare no conflict of interest. The founding sponsors had no role in the design of the study; in the collection, analyses, or interpretation of data; in the writing of the manuscript, and in the decision to publish the results.

Abbreviations

MTS	Mitochondrial targeting sequence
GFP	Green fluorescent protein
LEA	Late Embryogenesis Abundant

References

1. Rao, R.S.P.; Salvato, F.; Thal, B.; Eubel, H.; Thelen, J.J.; Møller, I.M. The proteome of higher plant mitochondria. *Mitochondrion* **2017**, *33*, 22–37. [CrossRef] [PubMed]
2. Murcha, M.W.; Kmiec, B.; Kubiszewski-Jakubiak, S.; Teixeira, P.F.; Glaser, E.; Whelan, J. Protein import into plant mitochondria: Signals, machinery, processing, and regulation. *J. Exp. Bot.* **2014**, *65*, 6301–6335. [CrossRef] [PubMed]
3. Bolender, N.; Sickmann, A.; Wagner, R.; Meisinger, C.; Pfanner, N. Multiple pathways for sorting mitochondrial precursor proteins. *EMBO Rep.* **2008**, *9*, 42–49. [CrossRef] [PubMed]
4. Chacinska, A.; Koehler, C.M.; Milenkovic, D.; Lithgow, T.; Pfanner, N. Importing mitochondrial proteins: Machineries and mechanisms. *Cell* **2009**, *138*, 628–644. [CrossRef] [PubMed]
5. Pfanner, N.; Geissler, A. Versatility of the mitochondrial protein import machinery. *Nat. Rev. Mol. Cell Biol.* **2001**, *2*, 339–349. [CrossRef] [PubMed]
6. Mokranjac, D.; Neupert, W. Energetics of protein translocation into mitochondria. *Biochim. Biophys. Acta Bioenerg.* **2008**, *1777*, 758–762. [CrossRef] [PubMed]
7. Teixeira, P.F.; Glaser, E. Processing peptidases in mitochondria and chloroplasts. *Biochim. Biophys. Acta Mol. Cell Res.* **2013**, *1833*, 360–370. [CrossRef] [PubMed]
8. Perry, A.J.; Hulett, J.M.; Likić, V.A.; Lithgow, T.; Gooley, P.R. Convergent evolution of receptors for protein import into mitochondria. *Curr. Biol.* **2006**, *16*, 221–229. [CrossRef] [PubMed]
9. Duncan, O.; Murcha, M.W.; Whelan, J. Unique components of the plant mitochondrial protein import apparatus. *Biochim. Biophys. Acta Mol. Cell Res.* **2013**, *1833*, 304–313. [CrossRef] [PubMed]
10. Jarvis, P. Targeting of nucleus-encoded proteins to chloroplasts in plants. *New Phytol.* **2008**, *179*, 257–285. [CrossRef] [PubMed]
11. Zhang, X.P.; Glaser, E. Interaction of plant mitochondrial and chloroplast signal peptides with the Hsp70 molecular chaperone. *Trends Plant Sci.* **2002**, *7*, 14–21. [CrossRef]
12. Xu, L.; Carrie, C.; Law, S.R.; Murcha, M.W.; Whelan, J. Acquisition, conservation, and loss of dual-targeted proteins in land plants. *Plant Physiol.* **2013**, *161*, 644–662. [CrossRef] [PubMed]
13. Huang, S.; Taylor, N.L.; Whelan, J.; Millar, A.H. Refining the definition of plant mitochondrial presequences through analysis of sorting signals, N-terminal modifications, and cleavage motifs. *Plant Physiol.* **2009**, *150*, 1272–1285. [CrossRef] [PubMed]
14. Von Heijne, G.; Steppuhn, J.; Herrmann, R.G. Domain structure of mitochondrial and chloroplastic targeting peptides. *Eur. J. Biochem.* **1989**, *180*, 535–545. [CrossRef] [PubMed]
15. Peeters, N.; Small, I. Dual targeting to mitochondria and chloroplasts. *Biochim. Biophys. Acta Mol. Cell Res.* **2001**, *1541*, 54–63. [CrossRef]
16. Roise, D.; Horvath, S.J.; Tomich, J.M.; Richards, J.H.; Schatz, G. A chemically synthesized pre-sequence of an imported mitochondrial protein can form an amphiphilic helix and perturb natural and artificial phospholipid bilayers. *EMBO J.* **1986**, *5*, 1327–1334. [PubMed]
17. Saitoh, T.; Igura, M.; Obita, T.; Ose, T.; Kojima, R.; Maenaka, K.; Endo, T.; Kohda, D. Tom20 recognizes mitochondrial presequences through dynamic equilibrium among multiple bound states. *EMBO J.* **2007**, *26*, 4777–4787. [CrossRef] [PubMed]

18. Bhushan, S.; Kuhn, C.; Berglund, A.K.; Roth, C.; Glaser, E. The role of the N-terminal domain of chloroplast targeting peptides in organellar protein import and miss-sorting. *FEBS Lett.* **2006**, *580*, 3966–3972. [CrossRef] [PubMed]
19. Schneider, G.; Sjöling, S.; Wallin, E.; Wrede, P.; Glaser, E.; Von Heijne, G. Feature-extraction from endopeptidase cleavage sites in mitochondrial targeting peptides. *Proteins Struct. Funct. Bioinform.* **1998**, *30*, 49–60. [CrossRef]
20. Tanudji, M.; Sjöling, S.; Glaser, E.; Whelan, J. Signals required for the import and processing of the alternative oxidase into mitochondria. *J. Biol. Chem.* **1999**, *274*, 1286–1293. [CrossRef] [PubMed]
21. Kitada, S.; Yamasaki, E.; Kojima, K.; Ito, A. Determination of the cleavage site of the presequence by mitochondrial processing peptidase on the substrate binding scaffold and the multiple subsites inside a molecular cavity. *J. Biol. Chem.* **2003**, *278*, 1879–1885. [CrossRef] [PubMed]
22. Zhang, X.P.; Sjöling, S.; Tanudji, M.; Somogyi, L.; Andreu, D.; Göran Eriksson, L.E.; Gräslund, A.; Whelan, J.; Glaser, E. Mutagenesis and computer modelling approach to study determinants for recognition of signal peptides by the mitochondrial processing peptidase. *Plant J.* **2001**, *27*, 427–438. [CrossRef] [PubMed]
23. Staiger, C.; Hinneburg, A.; Klösgen, R.B. Diversity in degrees of freedom of mitochondrial transit peptides. *Mol. Biol. Evol.* **2009**, *26*, 1773–1780. [CrossRef] [PubMed]
24. Millar, A.H.; Heazlewood, J.L.; Kristensen, B.K.; Braun, H.P.; Møller, I.M. The plant mitochondrial proteome. *Trends Plant Sci.* **2005**, *10*, 36–43. [CrossRef] [PubMed]
25. Candat, A.; Paszkiewicz, G.; Neveu, M.; Gautier, R.; Logan, D.C.; Avelange-Macherel, M.H.; Macherel, D. The ubiquitous distribution of late embryogenesis abundant proteins across cell compartments in Arabidopsis offers tailored protection against abiotic stress. *Plant Cell* **2014**, *26*, 3148–3166. [CrossRef] [PubMed]
26. Tunnacliffe, A.; Hincha, D.K.; Leprince, O.; Macherel, D. LEA proteins: Versatility of form and function. In *Sleeping Beauties—Dormancy and Resistance in Harsh Environments*; Lubzens, E., Cerda, J., Clark, M., Eds.; Springer: Berlin/Heidelberg, Germany, 2010.
27. Tolleter, D.; Jaquinod, M.; Mangavel, C.; Passirani, C.; Saulnier, P.; Manon, S.; Teyssier, E.; Payet, N.; Avelange-Macherel, M.H.; Macherel, D. Structure and function of a mitochondrial late embryogenesis abundant protein are revealed by desiccation. *Plant Cell* **2007**, *19*, 1580–1589. [CrossRef] [PubMed]
28. Tolleter, D.; Hincha, D.K.; Macherel, D. A mitochondrial late embryogenesis abundant protein stabilizes model membranes in the dry state. *Biochim. Biophys. Acta* **2010**, *1798*, 1926–1933. [CrossRef] [PubMed]
29. Avelange-Macherel, M.-H.; Payet, N.; Lalanne, D.; Neveu, M.; Tolleter, D.; Burstin, J.; Macherel, D. Variability within a pea core collection of LEAM and HSP22, two mitochondrial seed proteins involved in stress tolerance. *Plant. Cell Environ.* **2015**, *38*, 1299–1311. [CrossRef] [PubMed]
30. Weaver, L.M.; Gan, S.; Quirino, B.; Amasino, R.M. A comparison of the expression patterns of several senescence-associated genes in response to stress and hormone treatment. *Plant Mol. Biol.* **1998**, *37*, 455–469. [CrossRef] [PubMed]
31. Mowla, S.B.; Cuypers, A.; Driscoll, S.P.; Kiddle, G.; Thomson, J.; Foyer, C.H.; Theodoulou, F.L. Yeast complementation reveals a role for an *Arabidopsis thaliana* late embryogenesis abundant (LEA) like protein in oxidative stress tolerance. *Plant J.* **2006**, *48*, 743–756. [CrossRef] [PubMed]
32. Salleh, F.M.; Evans, K.; Goodall, B.; Machin, H.; Mowla, S.B.; Mur, L.A.J.; Runions, J.; Theodoulou, F.L.; Foyer, C.H.; Rogers, H.J. A novel function for a redox-related LEA protein (SAG21/AtLEA5) in root development and biotic stress responses. *Plant Cell Environ.* **2012**, *35*, 418–429. [CrossRef] [PubMed]
33. Fournier-Level, A.; Korte, A.; Cooper, M.D.; Nordborg, M.; Schmitt, J.; Wilczek, A.M. A map of local adaptation in *Arabidopsis thaliana*. *Science* **2011**, *334*, 86–89. [CrossRef] [PubMed]
34. Abe, Y.; Shodai, T.; Muto, T.; Mihara, K.; Torii, H.; Nishikawa, S.; Endo, T.; Kohda, D. Structural basis of presequence recognition by the mitochondrial protein import receptor Tom20. *Cell* **2000**, *100*, 551–560. [CrossRef]
35. Duby, G.; Oufattole, M.; Boutry, M. Hydrophobic residues within the predicted N-terminal amphiphilic α-helix of a plant mitochondrial targeting presequence play a major role in in vivo import. *Plant J.* **2001**, *27*, 539–549. [CrossRef] [PubMed]
36. Liu, H.Y.; Liao, P.C.; Chuang, K.T.; Kao, M.C. Mitochondrial targeting of human NADH dehydrogenase (ubiquinone) flavoprotein 2 (NDUFV2) and its association with early-onset hypertrophic cardiomyopathy and encephalopathy. *J. Biomed. Sci.* **2011**, *18*, 29. [CrossRef] [PubMed]

37. Huang, S.; Nelson, C.J.; Li, L.; Taylor, N.L.; Ströher, E.; Petereit, J.; Millar, A.H. INTERMEDIATE CLEAVAGE PEPTIDASE55 modifies enzyme amino termini and alters protein stability in Arabidopsis mitochondria. *Plant Physiol.* **2015**, *168*, 415–427. [CrossRef] [PubMed]
38. Rodríguez-Trelles, F.; Tarrío, R.; Ayala, F.J. Origins and evolution of spliceosomal introns. *Annu. Rev. Genet.* **2006**, *40*, 47–76. [CrossRef] [PubMed]
39. Panchy, N.; Lehti-Shiu, M.D.; Shiu, S.-H. Evolution of gene duplication in plants. *Plant Physiol.* **2016**, *171*, 2294–2316. [CrossRef] [PubMed]
40. Qian, W.; Zhang, J. Protein subcellular relocalization in the evolution of yeast singleton and duplicate genes. *Genome Biol. Evol.* **2009**, *1*, 198–204. [CrossRef] [PubMed]
41. Wang, X.; Huang, Y.; Lavrov, D.V.; Gu, X. Comparative study of human mitochondrial proteome reveals extensive protein subcellular relocalization after gene duplications. *BMC Evol. Biol.* **2009**, *9*, 275. [CrossRef] [PubMed]
42. Liu, S.-L.; Pan, A.Q.; Adams, K.L. Protein subcellular relocalization of duplicated genes in Arabidopsis. *Genome Biol. Evol.* **2014**, *6*, 2501–2515. [CrossRef] [PubMed]
43. Candat, A.; Poupart, P.; Andrieu, J.-P.; Chevrollier, A.; Reynier, P.; Rogniaux, H.; Avelange-Macherel, M.-H.; Macherel, D. Experimental determination of organelle targeting-peptide cleavage sites using transient expression of green fluorescent protein translational fusions. *Anal. Biochem.* **2013**, *434*, 44–51. [CrossRef] [PubMed]
44. Huang, X.; Miller, W. A time-efficient, linear-space similarity algorithm. *Adv. Appl. Math.* **1991**, *12*, 337–357. [CrossRef]
45. Sievers, F.; Wilm, A.; Dineen, D.; Gibson, T.J.; Karplus, K.; Li, W.; Lopez, R.; McWilliam, H.; Remmert, M.; Söding, J.; et al. Fast, scalable generation of high-quality protein multiple sequence alignments using Clustal Omega. *Mol. Syst. Biol.* **2011**, *7*. [CrossRef] [PubMed]
46. Li, W.; Cowley, A.; Uludag, M.; Gur, T.; McWilliam, H.; Squizzato, S.; Park, Y.M.; Buso, N.; Lopez, R. The EMBL-EBI bioinformatics web and programmatic tools framework. *Nucleic Acids Res.* **2015**, *43*, W580–W584. [CrossRef] [PubMed]
47. Altschul, S.F.; Madden, T.L.; Schäffer, A.A.; Zhang, J.; Zhang, Z.; Miller, W.; Lipman, D.J. Gapped BLAST and PSI-BLAST: A new generation of protein database search programs. *Nucleic Acids Res.* **1997**, *25*, 3389–3402. [CrossRef] [PubMed]
48. Finn, R.D.; Clements, J.; Arndt, W.; Miller, B.L.; Wheeler, T.J.; Schreiber, F.; Bateman, A.; Eddy, S.R. HMMER web server: 2015 Update. *Nucleic Acids Res.* **2015**, *43*, W30–W38. [CrossRef] [PubMed]
49. Nakai, K.; Horton, P. PSORT: A program for detecting sorting signals in proteins and predicting their subcellular localization. *Trends Biochem. Sci.* **1999**, *24*, 34–36. [CrossRef]
50. Claros, M.G.; Vincens, P. Computational method to predict mitochondrially imported proteins and their targeting sequences. *Eur. J. Biochem.* **1996**, *241*, 779–786. [CrossRef] [PubMed]
51. Emanuelsson, O.; Nielsen, H.; Brunak, S.; von Heijne, G. Predicting subcellular localization of proteins based on their N-terminal amino acid sequence. *J. Mol. Biol.* **2000**, *300*, 1005–1016. [CrossRef] [PubMed]
52. Small, I.; Peeters, N.; Legeai, F.; Lurin, C. Predotar: A tool for rapidly screening proteomes for N-terminal targeting sequences. *Proteomics* **2004**, *4*, 1581–1590. [CrossRef] [PubMed]
53. Bodén, M.; Hawkins, J. Prediction of subcellular localization using sequence-biased recurrent networks. *Bioinformatics* **2005**, *21*, 2279–2286. [CrossRef] [PubMed]
54. Fukasawa, Y.; Tsuji, J.; Fu, S.C.; Tomii, K.; Horton, P.; Imai, K. MitoFates: Improved prediction of mitochondrial targeting sequences and their cleavage sites. *Mol. Cell. Proteom.* **2015**, *14*, 1113–1126. [CrossRef] [PubMed]
55. Halligan, B.D.; Ruotti, V.; Jin, W.; Laffoon, S.; Twigger, S.N.; Dratz, E.A. ProMoST (Protein Modification Screening Tool): A web-based tool for mapping protein modifications on two-dimensional gels. *Nucleic Acids Res.* **2004**, *32*, 638–644. [CrossRef] [PubMed]
56. Drozdetskiy, A.; Cole, C.; Procter, J.; Barton, G.J. JPred4: A protein secondary structure prediction server. *Nucleic Acids Res.* **2015**, *43*, W389–W394. [CrossRef] [PubMed]
57. Gautier, R.; Douguet, D.; Antonny, B.; Drin, G. HELIQUEST: A web server to screen sequences with specific alpha-helical properties. *Bioinformatics* **2008**, *24*, 2101–2102. [CrossRef] [PubMed]

© 2018 by the authors. Licensee MDPI, Basel, Switzerland. This article is an open access article distributed under the terms and conditions of the Creative Commons Attribution (CC BY) license (http://creativecommons.org/licenses/by/4.0/).

Communication

Identification of Physiological Substrates and Binding Partners of the Plant Mitochondrial Protease FTSH4 by the Trapping Approach

Magdalena Opalińska [1,*], Katarzyna Parys [1,2] and Hanna Jańska [1,*]

1. Faculty of Biotechnology, University of Wroclaw, Fryderyka Joliot-Curie 14A, 50-383 Wroclaw, Poland
2. Present address: Gregor Mendel Institute, Austrian Academy of Sciences, Vienna Biocenter, A-1030 Vienna, Austria; katarzyna.parys@gmi.oeaw.ac.at
* Correspondence: magdalena.opalinska@uwr.edu.pl (M.O.); hanna.janska@uwr.edu.pl (H.J.)

Received: 25 October 2017; Accepted: 16 November 2017; Published: 18 November 2017

Abstract: Maintenance of functional mitochondria is vital for optimal cell performance and survival. This is accomplished by distinct mechanisms, of which preservation of mitochondrial protein homeostasis fulfills a pivotal role. In plants, inner membrane-embedded *i*-AAA protease, FTSH4, contributes to the mitochondrial proteome surveillance. Owing to the limited knowledge of FTSH4's in vivo substrates, very little is known about the pathways and mechanisms directly controlled by this protease. Here, we applied substrate trapping coupled with mass spectrometry-based peptide identification in order to extend the list of FTSH4's physiological substrates and interaction partners. Our analyses revealed, among several putative targets of FTSH4, novel (mitochondrial pyruvate carrier 4 (MPC4) and Pam18-2) and known (Tim17-2) substrates of this protease. Furthermore, we demonstrate that FTSH4 degrades oxidatively damaged proteins in mitochondria. Our report provides new insights into the function of FTSH4 in the maintenance of plant mitochondrial proteome.

Keywords: AAA protease; ATP-dependent proteolysis; mitochondria; inner mitochondrial membrane proteostasis; carbonylated proteins

1. Introduction

Mitochondria are life-essential multifunctional organelles. In addition to their vital role in energy conversion, mitochondria are involved in diverse metabolic pathways including iron sulfur cluster biosynthesis, in cellular signaling and in the regulation of programmed cell death. Owing to the central role of these organelles, mitochondrial dysfunction is implicated in the onset and pathology of a myriad diseases and aging [1–3]. Distinct quality control pathways are engaged in mitochondrial surveillance, maintaining functional mitochondria and facilitating adaptation to stress conditions [4–6]. Mitochondrial proteases play a central role in these mechanisms, not only by the removal of damaged proteins or excess subunits, but also as regulatory components. To fully understand the spectrum of processes that rely on the action of mitochondrial proteases, detailed knowledge of their physiological substrates and interaction partners is required. In this regard plant mitochondrial proteases, including *i*-AAA protease—FTSH4, still remain poorly characterized.

i-AAA protease forms a homo-oligomeric ATP-dependent proteolytic complex that is embedded in the inner mitochondrial membrane (IM) with the catalytic sites exposed to the intermembrane space (IMS) [7]. In humans, a homozygous mutation in the gene encoding *i*-AAA protease (YME1L) causes mitochondriopathy with optic nerve atrophy [8]. Identification of proteolytic substrates and binding partners of mammalian *i*-AAA protease revealed that YME1L by proteolytic processing of fusion factors, selective removal of misfolded and unassembled subunits and turnover of specific regulatory proteins, influences a myriad of processes inside the mitochondria [9–11].

The molecular function of the plant YME1L counterpart—FTSH4 is less understood. Thus far, only one physiological substrate of FTSH4 has been described. FTSH4 is required for the turnover of the essential subunit of the TIM23 translocase, Tim17-2, indicating FTSH4-dependent proteolytic regulation of pre-protein influx into the plant mitochondria [12]. However, under stress conditions, loss of FTSH4 is linked to broad pleiotropic phenotypes including oxidative stress, imbalance in phospholipid metabolism, perturbation in oxidative phosphorylation system activity and aberrant mitochondrial morphology, suggesting that plant i-AAA protease, like its mammalian homologue, controls distinct processes inside the mitochondria [13,14]. Overall, to fully understand the role of FTSH4 in the maintenance of functional mitochondria, the in-depth knowledge of its in vivo targets is necessary.

Here, we applied an unbiased substrate trapping approach in order to identify potential physiological substrates and biding partners of FTSH4. Our strategy is based on the identification of proteins co-purifying with a proteolytically inactive mutant of the protease. The trapping method was successfully used to search for substrates and interactors of the bacterial homolog of FTSH4, FtsH [15]. However, to date this method has not been employed for the analysis of plant mitochondrial proteases. In the present study, we found 17 potential candidates, among which we identified known FTSH4's substrate, Tim17-2, and novel ones which are the subunit of the inner membrane translocase Pam18-2 and mitochondrial pyruvate transporter 4 (MPC4). Our analysis indicated that FTSH4 is predominantly linked to the mitochondrial protein import machinery, inner membrane organizing proteins and specific metabolic pathways. Furthermore, we provide evidence that oxidative protein damage might trigger FTSH4-dependent proteolysis.

2. Results and Discussion

2.1. Substrate Trapping Assay Reveals a List of Potential FTSH4 Targets and Its Interacting Partners inside Mitochondria

The substrate trapping assay is a broadly used method enabling identification of enzyme substrates in an unbiased manner [15–18]. This experimental strategy has been applied to successfully identify physiological targets of diverse proteases, including membrane-bound bacterial FtsH [15]. Here, we employed this strategy in order to expand the spectrum of in vivo substrates and interaction partners of FTSH4. Briefly, we utilized an *Arabidopsis thaliana* line that produces, instead of functional wild type FTSH4, a proteolytically inactivated tagged version of this protease (*ftsh4-1* FTSH4$^{TRAP.FLAG}$). Proteolytic activity of FTSH4 was abolished by a single point mutation in the catalytic center of the proteolytic chamber [12,15]. Since AAA domain of the FTSH4$^{TRAP.FLAG}$ mutant is active, the substrates are translocated into the proteolytic chamber, where they are not degraded and thus remain trapped [15]. To characterize proteins caught inside the protease, affinity purification of FTSH4$^{TRAP.FLAG}$ from mitochondria followed by mass spectrometry-based peptide identification was applied (Figure 1). Resulting proteins co-purifying with FTSH4$^{TRAP.FLAG}$ in two independent biological replicates are listed in Table 1 and Table S1. Obtained data were validated by immunoblotting analyses with available antibodies (Figure 2).

The vast majority of identified proteins were localized to the inner mitochondrial membrane and were assigned to the four main functional classes according to Uniprot; transmembrane transport, metabolism, mitochondrial protein import and membrane organization (Table 1). This list includes potential interaction partners (e.g., adaptor proteins) and substrates of the FTSH4 protease inside mitochondria. Since yeast or mammalian i-AAA proteases degrade damaged or excessive subunits, transport intermediates (preproteins) and regulatory proteins [9–11,19], it is conceivable to find all these diverse types of substrates amongst proteins co-purifying with FTSH4$^{TRAP.FLAG}$.

Several proteins identified in our assay reinforce the existence of a link between FTSH4 and mitochondrial ATP production by oxidative phosphorylation [14,20]. These include members of the mitochondrial substrate carrier family such as orthologues of mitochondrial NAD transporter (NDT2) [21] and mitochondrial pyruvate carrier (MPC4) [22]. In addition, amongst proteins related to mitochondria energy conversion that were found to co-purify with FTSH4$^{FLAG.TAG}$, we also identified components of succinate dehydrogenase flavoprotein complex; subunit 1 (SDH1-1) and subunit 5 (SDH5). These are peripheral membrane proteins localized to the matrix side of the inner membrane [23]. Based on the finding concerning the human counterpart of FTSH4 protease—YME1L [19], matrix localized proteins like SDH1-1, SDH5 and isocitrate dehydrogenase NAD regulatory subunit 1 (IDH1) could be FTSH4 substrates during import of their precursors from the cytosol into the mitochondrial matrix. Interestingly, SDH1-1 specifically accumulates in plants devoid of functional FTSH4 (*ftsh4* mutants) [14]. Alternatively, observed interactions with these proteins could occur through the N-terminal domain of FTSH4 that protrudes into the matrix. In addition, we found an accessory subunit of the respiratory complex dihydroorotate dehydrogenase [24] and intermembrane space localized l-galactono-1,4-lactone dehydrogenase (GLDH) [25] as potential FTSH4 targets. In *Arabidopsis thaliana*, GLDH not only catalyzes the last step of ascorbate biosynthesis pathway, but it also has a second non-enzymatic function in the assembly of respiratory complex I [25].

Furthermore, our data indicate a connection between FTSH4 and mitochondrial protein biogenesis pathways. We found an essential subunit of TIM17:23 translocase, Tim17-2, among proteins co-purifying with FTSH4$^{TRAP.FLAG}$. This protein was recently identified as a physiological substrate of the plant *i*-AAA protease [12]. FTSH4 protease regulates pre-protein influx into the mitochondria through the turnover of the Tim17-2 subunit. This function of *i*-AAA protease was also described in mammals, where YME1L controls mitochondrial protein import through the degradation of the Tim17A subunit, a homologue of Tim17-2 [10]. In addition, in the substrate trapping assay we found homologues of Pam18, an essential subunit of the pre-sequence translocase-associated (PAM) motor that is recruited by TIM23 translocase [26], and many other components involved in mitochondrial pre-protein uptake, implying the association of FTSH4 with these mitochondrial protein biogenesis machineries.

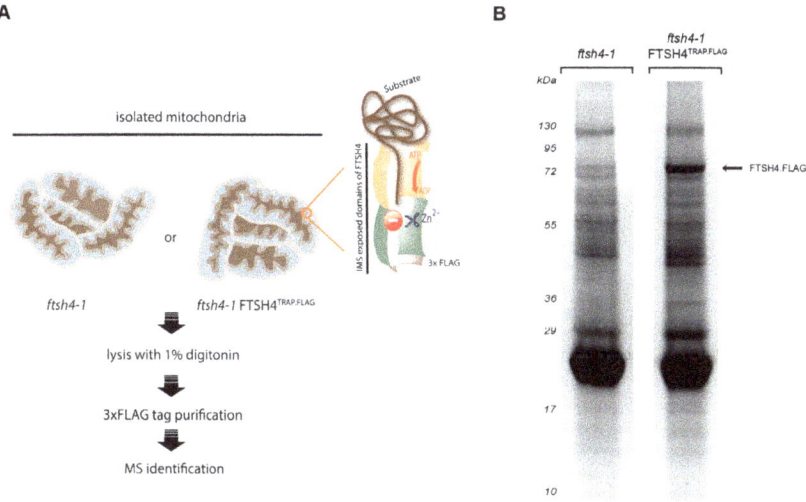

Figure 1. FTSH4 substrate trapping assay (**A**) Overview of FTSH4 substrate-trapping assay. Cartoon illustrating the experimental workflow; (**B**) Eluted fractions resolved on SDS-PAGE and stained with CBB. IMS—intermembrane space.

Our list of potential FTSH4 targets also includes *A. thaliana* homologue of highly conserved inner membrane protein Mitofilin (also called Mic60 or Fcj1 (in yeast)). Mic60 is a central component of the MICOS complex (mitochondrial contact site complex) that is essential for cristae junction formation [27].

Mammalian *i*-AAA protease, YME1L regulates Mic60/Mitofilin homeostasis, which is required for the maintenance of mitochondrial morphology and organization of mtDNA nucleoids. Impaired assembly of the MICOS complex is associated with the formation of giant mitochondria due to the dysregulated fusion and fission events [28]. Interestingly, also in *Arabidopsis* disturbance in Mic60 levels results in the formation of giant mitochondria [29]. Strikingly, similar mitochondrial morphology phenotype is observed in the case of *ftsh4* mutants [14].

Table 1. Proteins co-purifying with FTSH4$^{TRAP.FLAG}$. Listed are all proteins that were specifically co-purifying with proteolytically inactive FTSH4 (proteins enriched at least two times in two independent biological replicates over the background samples (Table S1)). Functional categories and mitochondrial sub-localization were assigned based on the Uniprot database and the literature. AGI—*Arabidopsis* Gene ID.

Functional Category	AGI	Protein	Localization
Transmembrane transport	AT1G25380.1	Nicotinamide adenine dinucleotide transporter 2	inner membrane
	AT4G22310.1	Mitochondrial pyruvate carrier 4 (MPC4)	inner membrane
Metabolism	AT4G35260.1	Isocitrate dehydrogenase NAD regulatory subunit 1	matrix
	AT1G47420.1	Succinate dehydrogenase flavoprotein 5	inner membrane
	AT5G66760.1	Succinate dehydrogenase flavoprotein 1	inner membrane
	AT3G47930.1	L-galactono-1,4-lactone dehydrogenase	intermembrane space
	AT5G23300.1	Dihydroorotate dehydrogenase	inner membrane
Mitochondrial protein import	AT1G53530.1	Mitochondrial inner membrane protease subunit 1	inner membrane
	AT2G46470.1	OXA1-like protein	inner membrane
	AT3G09700.1	Pam18-2	inner membrane
	AT5G05520.1	Sam50-2	outer membrane
	AT2G35795.1	Pam18-1	inner membrane
	AT2G37410.1	Tim17-2	inner membrane
Membrane organization	AT5G54100.1	Stomatin-like protein 2 (Slp2)	inner membrane
	AT4G27585.1	Stomatin-like protein 1 (Slp1)	inner membrane
	AT4G39690.1	Mic60	inner membrane
Unknown	AT5G62270.1	Gamete cell defective (GCD)	unknown

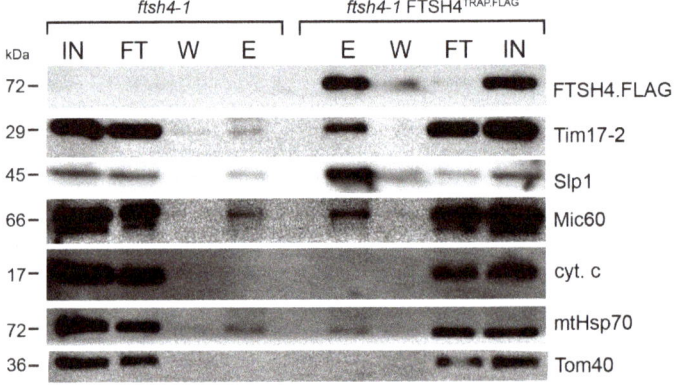

Figure 2. Immunoblot analysis of FTSH4 substrate trapping assay samples. Mitochondria from control (*ftsh4-1*) and *ftsh4-1* FTSH4$^{TRAP.FLAG}$ were solubilized with digitonin and subjected to immunoprecipitation with anti-FLAG affinity matrix. The precipitated proteins were immunoblotted with antibodies against the indicated proteins. IN—input (5%), FT—flow-through (5%), W—Wash, E—eluate (50%).

Among proteins co-purifying with FTSH4$^{TRAP.FLAG}$ also two stomatin-like proteins (Slp1 and Slp2) from mitochondrial SPFH family were identified. These proteins were implied in the organization of respiratory chain super-complexes in plant mitochondria [30]. Slp1 and Slp2 are homologues of mammalian SLP2 that is involved in mitochondrial fusion and formation of protein complexes in the inner membrane [31,32]. Recent data indicate that human SLP2 plays a role as a membrane scaffold required for the spatial organization of inner membrane proteases such as the FTSH4 counterpart, YME1L [32]. Our data provide an interesting link between FTSH4 and Slp1/Slp2, however it remains to be elucidated whether plant stomatin-like proteins, like their mammalian homolog SLP2, anchor a complex containing *i*-AAA protease in the inner mitochondrial membrane.

Overall, the identification of mitochondrial FTSH4's potential in vivo interaction partners and substrates gives more a detailed view of molecular functions of this protease in the mitochondria. Conversely, the further, independent validation of selected identified candidates is required.

2.2. Identification of Novel Physiological Substrates of FTSH4

To assess whether, among proteins co-purifying with FTSH4$^{TRAP.FLAG}$ there are novel proteolytic substrates of this *i*-AAA protease, we analyzed the "in organello" stability of selected proteins. We monitored the kinetics of degradation of candidate proteins inside mitochondria isolated from *ftsh4* mutant and wild type plants. It is expected that the stability of proteolytic substrates will be enhanced in the mitochondria of mutants devoid of functional protease required for their turnover [12,33]. This assay allowed us to identify two novel FTSH4 proteolytic substrates, such as the component of mitochondrial protein import machinery Pam18-2 and the processed form of mitochondrial pyruvate carrier (MPC4). Due to the inaccessibility of specific antibodies, both proteins were synthesized in a cell-free system in the presence of [^{35}S] methionine and imported post-translationally into mitochondria isolated from wild type and the *ftsh4* mutant (Figure 3). We found that newly imported Pam18-2 and the processed form of MPC4 were rapidly degraded in wild type mitochondria, but accumulated in *ftsh4* mitochondria, which is consistent with FTSH4-dependent proteolysis. Among tested candidates, there were also proteins (including Slp1, GCD, Mic60) characterized by slow turnover rates already in wild type mitochondria that precluded verification of these proteins in the context of this assay.

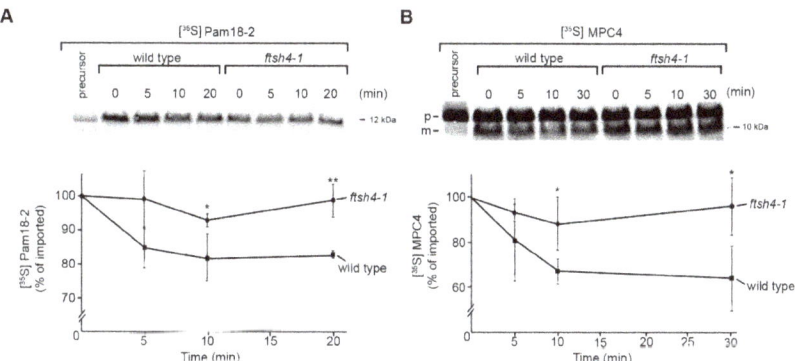

Figure 3. Novel proteolytic substrates of FTSH4 protease. (**A**) Degradation of Pam18-2 by FTSH4 following its in vitro import into mitochondria. Radiolabeled Pam18-2 was imported into mitochondria derived from either wild type or *ftsh4-1* plants. The stability of newly imported precursor upon further incubation at 26 °C was analyzed by SDS-PAGE and autoradiography. Quantification of [^{35}S] Pam18-2 in mitochondria is represented in the lower panel. Newly imported Pam18-2 was set to 100%. Data represent mean ± SD of three independent experiments. * $p < 0.02$ and ** $p < 0.003$ (*t*-student test); (**B**) Degradation of MPC4 processed form by FTSH4 following its in vitro import into mitochondria. Radiolabeled MPC4 was imported into mitochondria derived from either wild type or *ftsh4-1* plants. The stability of newly imported MPC4 upon further incubation at 26 °C was analyzed by SDS-PAGE and autoradiography. Quantification of [^{35}S] MPC4 processed form in mitochondria is represented in the lower panel. Newly imported MPC4 was set to 100%. Data represent mean ± SD of three independent experiments. * $p < 0.05$ (*t*-student test). m—MPC4 processed form, p—MPC4 precursor.

2.3. FTSH4 Degrades Mitochondrial Carbonylated Proteins

Across species, the lack of i-AAA protease is associated with the accumulation of oxidatively damaged proteins in the mitochondria [9,14]. Although it is very probable that these ATP-fueled proteolytic machineries can directly participate in the removal of carbonylated proteins from the mitochondria, so far no evidence has been provided. We found that amongst proteins co-purifying with FTSH4$^{\text{TRAP.FLAG}}$ in the substrate trapping assay were oxidatively damaged proteins, suggesting that modification of proteins by reactive oxygen species could trigger their recognition and degradation by the protease (Figure 4A). To test this hypothesis, we performed in vitro analysis employing mature FTSH4 protease produced in the insect-cell lysate [34]. We subjected mitochondrial fractions enriched in carbonylated proteins (derived from *ftsh4* mutant) to FTSH4 or control lysate and analyzed stability of oxidatively modified proteins using Oxiblot detection system. We found that upon FTSH4 addition, the stability of carbonylated proteins is significantly compromised (Figure 4B,C). These data indicate that FTSH4 degrades oxidatively damaged mitochondrial proteins.

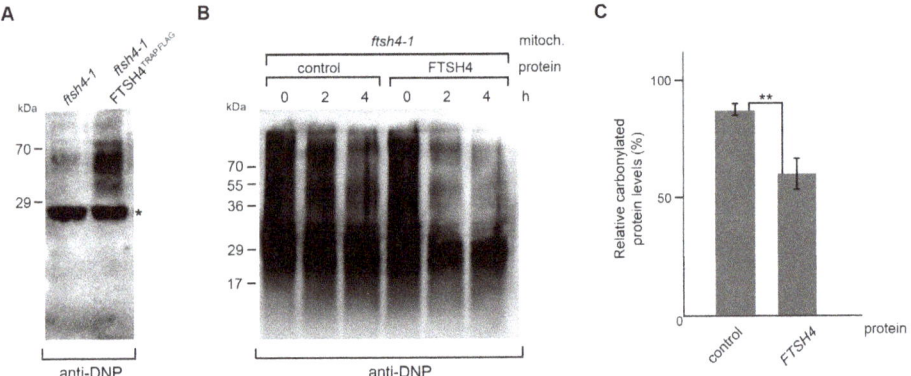

Figure 4. FTSH4 degrades oxidatively damaged mitochondrial proteins. (**A**) Anti-DNP (dinitrophenyl hydrazone) immunoblot detection of carbonylated proteins co-precipitating with FTSH4$^{\text{TRAP.FLAG}}$. The cross-reaction of IgG light chain is marked with asterisk. (**B**) The degradation of mitochondrial carbonylated proteins by FTSH4 protease was assessed by immunoblot analysis of the levels of carbonylated proteins in mitochondrial extract from *ftsh4-1* mutant incubated at 35 °C with or without FTSH4 protein. (**C**) Quantification of the remaining carbonylated proteins after 2 h incubation with or without FTSH4 protein, shown in (**B**). For each sample, the amount of carbonylated proteins was set to 100% at time point 0 h. Data represent mean ± SEM of three independent experiments. ** $p < 0.03$ (*t*-student test).

3. Materials and Methods

3.1. Plant Material and Growth Conditions

Arabidopsis thaliana lines: ecotype Col-0 wild type, T-DNA insertion lines *ftsh4-1* (SALK_035107/TAIR) [35] and *ftsh4-1* FTSH4$^{\text{TRAP.FLAG}}$ (*ftsh4-1* FTSH4(H486Y)) mutant line [12], were grown on 0.5 Murashige and Skoog (MS) medium supplemented with 3% (*w/v*) sucrose in chambers in a 16 h light/8 h dark (long day, LD) photoperiod at 22 °C for 2 weeks with a light intensity of 150 µmol m^{-2} s^{-1} [12].

3.2. Isolation of Mitochondria from Arabidopsis thaliana

Isolation of mitochondria from 14-day-old seedlings was performed as described in [36].

3.3. Substrate-Trapping Assay

Mitochondria of FTSH4$^{TRAP.FLAG}$ or control line were resuspended in digitonin solubilization buffer (1% digitonin, 20 mM Tris-HCl, 0.1 mM EDTA, 100 mM NaCl, 10% glycerol, pH 7.7) at 1 mg/mL. PMSF and EDTA-free protease inhibitor cocktail were added and samples were incubated for 30 min at 4 °C with mixing. After clarifying centrifugation, solubilized material was loaded on anti-FLAG affinity matrix and incubated under constant rotation for 1 h 30 min at 4 °C. After excessive washing steps proteins were eluted with Laemmli sample buffer and subjected to SDS–PAGE. Subsequently, protein bands were excised from the gel and analyzed by liquid chromatography coupled to the mass spectrometer in the Laboratory of Mass Spectrometry, Institute of Biochemistry and Biophysics, Polish Academy of Sciences (Warsaw, Poland). Samples were concentrated and desalted on a RP-C18 pre-column (Waters, Milford, MA, USA), and further peptide separation was achieved on a nano-Ultra Performance Liquid Chromatography (UPLC) RP-C18 column (Waters, BEH130 C18 column, 75 µm i.d., 250 mm long) of a nanoACQUITY UPLC system, using a 100 min linear acetonitrile gradient. Column outlet was directly coupled to the Electrospray ionization (ESI) ion source of the Orbitrap Velos type mass spectrometer (Thermo, San Jose, CA, USA), working in the regime of data dependent MS to MS/MS switch. An electrospray voltage of 1.5 kV was used. Raw data files were pre-processed with Mascot Distiller software (version 2.4.2.0, MatrixScience, London, UK). The obtained peptide masses and fragmentation spectra were matched to The Arabidopsis Information Resource (TAIR) database (35,386 sequences/14,482,855 residues) using the Mascot search engine (Mascot Daemon v. 2.4.0, Mascot Server v. 2.4.1, MatrixScience, London, UK). The following search parameters were applied: enzyme specificity was set to trypsin, peptide mass tolerance to ±20 ppm and fragment mass tolerance to ±0.1 Da. The protein mass was left as unrestricted, and mass values as monoisotopic with one missed cleavage being allowed. Alkylation of cysteine by carbamidomethylation as fixed, and oxidation of methionine was set as a variable modification. Protein identification was performed using the Mascot search engine (MatrixScience), with the probability based algorithm. The statistical significance of peptide identifications was estimated using a joined target/decoy database search approach, FDR (False Discovery Rate) was set below 1%. Proteins enriched at least two times in two independent biological replicates over the background samples were considered as co-purifying with FTSH4$^{TRAP.FLAG}$. Protein levels were estimated based on exponentially modified protein abundance index (emPAI) [37,38].

3.4. Immunoblot Analysis

Protein samples were resolved on SDS-PAGE and transferred on PVDF membrane. Immunodecoration was made with antibodies against AtMic60 [29], Slp1 [30], Tim17-2 [39], Tom40 [40], FLAG-tag (Sigma-Aldrich, Steinheim, Germany, F1804), mtHsp70 (Agrisera, Vännäs, Sweden, AS08 347), cytochrome c (Agrisera, AS08 343A), FTSH4 (Agrisera, AS07 205). Detection of carbonylated proteins was performed using OxiSelect Protein Carbonyl Immunoblot Kit (Cell Biolabs, San Diego, CA, USA) according to the manufacturer's instructions. Proteins were visualized with enhanced chemiluminescence [41] using a G-BOX ChemiXR5 (Syngene, Cambridge, UK).

3.5. Synthesis of Radiolabeled Precursor Proteins

Radiolabeling of selected mitochondrial proteins (mitochondrial pyruvate carrier 4 (MPC4; At4g22310.1) and Pam18-2 (At3g09700.1)) was performed using well-established procedures [42]. Briefly, for in vitro transcription, PCR products containing an SP6 promoter upstream the open reading frame were generated using cDNA as a template. Subsequently, in vitro transcription was carried out according to the manufacturer's recommendation (mMESSAGE mMACHINE SP6 kit; Ambion, Foster City, CA, USA), RNA was purified (MEGAclear kit; Ambion, Foster City, CA, USA) and used for in vitro translation in reticulocyte lysate (TNT kit; Promega, Mannheim, Germany) that was supplemented with [^{35}S] methionine.

3.6. [^{35}S]-Labelled Protein Uptake into the Mitochondria and in Organello Degradation Assay

Mitochondrial pre-proteins radiolabeled with [^{35}S] methionine were imported into mitochondria of wild type or *ftsh4* mutant according to [42,43]. Briefly, radiolabeled protein uptake into mitochondria was performed at 26 °C in the presence of 2 mM NADH, 5 mM sodium succinate, 2 mM ATP, 2.5 mM methionine, 5 mM creatine phosphate, and 100 µg/mL creatine kinase in import buffer (3% (*w/v*) BSA, 250 mM sucrose, 80 mM KCl, 5 mM MgCl$_2$, 10 mM MOPS/KOH, pH 7.2, and 2 mM KH$_2$PO$_4$). To analyze stability of imported protein import reaction was stopped after 20 min by placing samples on ice and addition of 1 µM valinomycine and 8 µM antimycine. After trypsin digestion (100 µg/mL, 20 min at 4 °C) of non-imported proteins, protease was inhibited by addition of 1.3 µg/µL trypsin inhibitor (5 min at 4 °C). Mitochondria were washed twice in import buffer containing trypsin inhibitor and 2 mM PMSF and resuspended in import buffer supplemented with 2 mM NADH, 5 mM sodium succinate, 2 mM ATP, 2.5 mM methionine, 5 mM creatine phosphate, and 100 µg/mL creatine kinase and 12.5 µM ZnSO$_4$. Samples were incubated at 26 °C for the indicated times to allow for proteolysis and subsequently resolved by SDS-PAGE. Radiolabeled proteins were visualized by digital autoradiography (PharosFX Plus Systems, Bio-Rad, Hercules, CA, USA) and analyzed by Quantity One software (Bio-Rad).

3.7. Cell-Free Synthesis of FTSH4

For in vitro transcription, PCR product was generated using RTS Linear Template Kit (Biotechrabbit, Hennigsdorf, Germany) and cDNA as a template. Expression of mature FTSH4 (54–717 aa) was performed with *Spodoptera frugiperda*-based cell-free expression system (RTS Insect Membrane Kit, Biotechrabbit) according to the provided manual.

3.8. In Vitro Degradation Assay of Carbonylated Proteins

To assess ability of FTSH4 to degrade carbonylated proteins, the in vitro degradation assay utilizing protease produced in cell-free expression system was performed [44]. Briefly, mitochondria isolated from *ftsh4-1* mutant grown under stress conditions (30 °C, long day (LD) photoperiod) were resuspended in assay buffer (25 mM HEPES/KOH pH 8.0, 1% digitonin, 100 mM KCl, 10 mM MgCl$_2$, 1 mM DTT, 10% glycerol, 25 µM ZnSO$_4$, 5 mM ATP, 5 mM creatine phosphate, and 100 µg/mL creatine kinase, 2 mM PMSF) with the addition of insect-cell lysate containing FTSH4 or control lysate. Samples were incubated at 35 °C for indicated time points; followed by SDS-PAGE and immunodetection of carbonylated proteins.

4. Conclusions

Here, we provide a list of potential substrates and binding partners of FTSH4 inside mitochondria. Among them, we found two novel proteolytic targets of this plant *i*-AAA protease. Noteworthy, part of the proteins that are designated for FTSH4-dependent removal are oxidatively damaged. Overall, our findings suggest that FTSH4 is linked to respiratory function, protein biogenesis pathways and inner membrane organization. Our data constitute a significant resource for studies addressing the role of plant *i*-AAA in these diverse aspects.

Supplementary Materials: Supplementary materials can be found at www.mdpi.com/1422-0067/18/11/2455/s1.

Acknowledgments: This work was supported by grant from the National Science Centre 2015/16/S/NZ3/00364 awarded to MO. This publication was supported by Wroclaw Centre of Biotechnology, the Leading National Research Centre (KNOW) for years 2014–2018. We thank Monika W. Murcha, Lee Sweetlove and Juliette Jouhet for providing antibodies.

Author Contributions: Magdalena Opalińska and Hanna Jańska designed the study. Magdalena Opalińska and Katarzyna Parys performed the experiments. All authors analyzed the data. Magdalena Opalińska wrote the manuscript with contribution from other authors.

Conflicts of Interest: The authors declare no competing financial interests.

References

1. McBride, H.M.; Neuspiel, M.; Wasiak, S. Mitochondria: More than just a powerhouse. *Curr. Biol.* **2006**, *25*, R551–R560. [CrossRef] [PubMed]
2. Nunnari, J.; Suomalainen, A. Mitochondria: In sickness and in health. *Cell* **2012**, *148*, 1145–1159. [CrossRef] [PubMed]
3. Raimundo, N. Mitochondrial pathology: Stress signals from the energy factory. *Trends Mol. Med.* **2014**, *20*, 282–292. [CrossRef] [PubMed]
4. Szklarczyk, R.; Nooteboom, M.; Osiewacz, H.D. Control of mitochondrial integrity in ageing and disease. *Philos. Trans. R. Soc. Lond. B Biol. Sci.* **2014**, *369*. [CrossRef] [PubMed]
5. Quirós, P.M.; Langer, T.; López-Otín, C. New roles for mitochondrial proteases in health, ageing and disease. *Nat. Rev. Mol. Cell Biol.* **2015**, *16*, 345–359. [CrossRef] [PubMed]
6. Hamon, M.P.; Bulteau, A.L.; Friguet, B. Mitochondrial proteases and protein quality control in ageing and longevity. *Ageing Res. Rev.* **2015**, *23*, 56–66. [CrossRef] [PubMed]
7. Glynn, S.E. Multifunctional Mitochondrial AAA Proteases. *Front. Mol. Biosci.* **2017**, *4*, 34. [CrossRef] [PubMed]
8. Hartmann, B.; Wai, T.; Hu, H.; MacVicar, T.; Musante, L.; Fischer-Zirnsak, B.; Stenze, W.; Gräf, R.; van den Heuvel, L.; Ropers, H.H.; et al. Homozygous YME1L1 mutation causes mitochondriopathy with optic atrophy and mitochondrial network fragmentation. *eLife* **2016**, *5*. [CrossRef] [PubMed]
9. Stiburek, L.; Cesnekova, J.; Kostkova, O.; Fornuskova, D.; Vinsova, K.; Wenchich, L.; Houstek, J.; Zeman, J. YME1L controls the accumulation of respiratory chain subunits and is required for apoptotic resistance, cristae morphogenesis, and cell proliferation. *Mol. Biol. Cell* **2012**, *23*, 1010–1023. [CrossRef] [PubMed]
10. Rainbolt, K.; Atanassova, N.; Genereux, J.C.; Wiseman, R.L. Stress-Regulated Translational Attenuation Adapts Mitochondrial Protein Import through Tim17A Degradation. *Cell Metab.* **2013**, *18*, 908–919. [CrossRef] [PubMed]
11. Anand, R.; Wai, T.; Baker, M.J.; Kladt, N.; Schauss, A.C.; Rugarli, E.; Langer, T. The *i*-AAA protease YME1L and OMA1 cleave OPA1 to balance mitochondrial fusion and fission. *J. Cell Biol.* **2014**, *204*, 919–929. [CrossRef] [PubMed]
12. Opalińska, M.; Parys, K.; Murcha, M.W.; Jańska, H. The plant *i*-AAA protease controls the turnover of an essential mitochondrial protein import component. *J. Cell Sci.* **2017**. [CrossRef] [PubMed]
13. Dolzblasz, A.; Smakowska, E.; Gola, E.M.; Sokołowska, K.; Kicia, M.; Janska, H. The mitochondrial protease AtFTSH4 safeguards Arabidopsis shoot apical meristem function. *Sci. Rep.* **2016**, *6*. [CrossRef] [PubMed]
14. Smakowska, E.; Skibior-Blaszczyk, R.; Czarna, M.; Kolodziejczak, M.; Kwasniak-Owczarek, M.; Parys, K.; Funk, C.; Janska, H. Lack of FTSH4 Protease Affects Protein Carbonylation, Mitochondrial Morphology, and Phospholipid Content in Mitochondria of Arabidopsis: New Insights into a Complex Interplay. *Plant Physiol.* **2016**, *171*, 2516–2535. [CrossRef] [PubMed]
15. Westphal, K.; Langklotz, S.; Thomanek, N.; Narberhaus, F. A trapping approach reveals novel substrates and physiological functions of the essential protease FtsH in *Escherichia coli*. *J. Biol. Chem.* **2012**, *287*, 42962–42971. [CrossRef] [PubMed]
16. Fischer, F.; Langer, J.D.; Osiewacz, H.D. Identification of potential mitochondrial CLPXP protease interactors and substrates suggests its central role in energy metabolism. *Sci. Rep.* **2015**, *5*. [CrossRef] [PubMed]
17. Ben-Lulu, S.; Ziv, T.; Admon, A.; Weisman-Shomer, P.; Benhar, M. A substrate trapping approach identifies proteins regulated by reversible S-nitrosylation. *Mol. Cell. Proteom.* **2014**, *13*, 2573–2583. [CrossRef] [PubMed]
18. Blanchetot, C.; Chagnon, M.; Dubé, N.; Hallé, M.; Tremblay, M.L. Substrate-trapping techniques in the identification of cellular PTP targets. *Methods* **2005**, *35*, 44–53. [CrossRef] [PubMed]
19. König, T.; Tröder, S.E.; Bakka, K.; Korwitz, A.; Richter-Dennerlein, R.; Lampe, P.A.; Patron, M.; Mühlmeister, M.; Guerrero-Castillo, S.; Brandt, U.; et al. The m-AAA Protease Associated with Neurodegeneration Limits MCU Activity in Mitochondria. *Mol. Cell* **2016**, *64*, 148–162. [CrossRef] [PubMed]
20. Kolodziejczak, M.; Gibala, M.; Urantówka, A.; Jańska, H. The significance of Arabidopsis AAA proteases for activity and assembly/stability of mitochondrial OXPHOS complexes. *Physiol. Plant.* **2006**, *129*, 135–142. [CrossRef]

21. Palmieri, F.; Rieder, B.; Ventrella, A.; Blanco, E.; Do, P.T.; Nunes-Nesi, A.; Trauth, A.U.; Fiermonte, G.; Tjaden, J.; Agrimi, G.; et al. Molecular identification and functional characterization of Arabidopsis thaliana mitochondrial and chloroplastic NAD$^+$ carrier proteins. *J. Biol. Chem.* **2009**, *284*, 31249–31259. [CrossRef] [PubMed]
22. Wang, M.; Ma, X.; Shen, J.; Li, C.; Zhang, W. The ongoing story: The mitochondria pyruvate carrier 1 in plant stress response in Arabidopsis. *Plant Signal. Behav.* **2014**, *9*, e973810. [CrossRef] [PubMed]
23. Huang, S.; Millar, A.H. Succinate dehydrogenase: The complex roles of a simple enzyme. *Curr. Opin. Plant Biol.* **2013**, *16*, 344–349. [CrossRef] [PubMed]
24. Schertl, P.; Braun, H.P. Respiratory electron transfer pathways in plant mitochondria. *Front. Plant Sci.* **2014**, *5*. [CrossRef] [PubMed]
25. Schimmeyer, J.; Bock, R.; Meyer, E.H. l-Galactono-1,4-lactone dehydrogenase is an assembly factor of the membrane arm of mitochondrial complex I in Arabidopsis. *Plant Mol. Biol.* **2016**, *90*, 117–126. [CrossRef] [PubMed]
26. Murcha, M.W.; Kmiec, B.; Kubiszewski-Jakubiak, S.; Teixeira, P.F.; Glaser, E.; Whelan, J. Protein import into plant mitochondria: Signals, machinery, processing, and regulation. *J. Exp. Bot.* **2014**, *65*, 6301–6335. [CrossRef] [PubMed]
27. Huynen, M.A.; Mühlmeister, M.; Gotthardt, K.; Guerrero-Castillo, S.; Brandt, U. Evolution and structural organization of the mitochondrial contact site (MICOS) complex and the mitochondrial intermembrane space bridging (MIB) complex. *Biochim. Biophys. Acta* **2016**, *1863*, 91–101. [CrossRef] [PubMed]
28. Li, H.; Ruan, Y.; Zhang, K.; Jian, F.; Hu, C.; Miao, L.; Gong, L.; Sun, L.; Zhang, X.; Chen, S.; et al. Mic60/Mitofilin determines MICOS assembly essential for mitochondrial dynamics and mtDNA nucleoid organization. *Cell Death Differ.* **2016**, *23*, 380–392. [CrossRef] [PubMed]
29. Michaud, M.; Gros, V.; Tardif, M.; Brugière, S.; Ferro, M.; Prinz, W.A.; Toulmay, A.; Mathur, J.; Wozny, M.; Falconet, D.; et al. AtMic60 Is Involved in Plant Mitochondria Lipid Trafficking and Is Part of a Large Complex. *Curr. Biol.* **2016**, *26*, 627–639. [CrossRef] [PubMed]
30. Gehl, B.; Lee, C.P.; Bota, P.; Blatt, M.R.; Sweetlove, L.J. An Arabidopsis stomatin-like protein affects mitochondrial respiratory supercomplex organization. *Plant Physiol.* **2014**, *164*, 1389–1400. [CrossRef] [PubMed]
31. Tondera, D.; Grandemange, S.; Jourdain, A.; Karbowski, M.; Mattenberger, Y.; Herzig, S.; Da Cruz, S.; Clerc, P.; Raschke, I.; Merkwirth, C.; et al. SLP-2 is required for stress-induced mitochondrial hyperfusion. *EMBO J.* **2009**, *28*, 1589–1600. [CrossRef] [PubMed]
32. Wai, T.; Saita, S.; Nolte, H.; Müller, S.; König, T.; Richter-Dennerlein, R.; Sprenger, H.G.; Madrenas, J.; Mühlmeister, M.; Brandt, U.; et al. The membrane scaffold SLP2 anchors a proteolytic hub in mitochondria containing PARL and the *i*-AAA protease YME1L. *EMBO Rep.* **2016**, *17*, 1844–1856. [CrossRef] [PubMed]
33. Potting, C.; Wilmes, C.; Engmann, T.; Osman, C.; Langer, T. Regulation of mitochondrial phospholipids by Ups1/PRELI-like proteins depends on proteolysis and Mdm35. *EMBO J.* **2010**, *29*, 2888–2898. [CrossRef] [PubMed]
34. Quast, R.B.; Kortt, O.; Henkel, J.; Dondapati, S.K.; Wüstenhagen, D.A.; Stech, M.; Kubick, S. Automated production of functional membrane proteins using eukaryotic cell-free translation systems. *J. Biotechnol.* **2015**, *203*, 45–53. [CrossRef] [PubMed]
35. Gibala, M.; Kicia, M.; Sakamoto, W.; Gola, E.M.; Kubrakiewicz, J.; Smakowska, E.; Janska, H. The lack of mitochondrial AtFtsH4 protease alters Arabidopsis leaf morphology at the late stage of rosette development under short-day photoperiod. *Plant J.* **2009**, *59*, 685–699. [CrossRef] [PubMed]
36. Murcha, M.W.; Whelan, J. Isolation of Intact Mitochondria from the Model Plant Species Arabidopsis thaliana and Oryza sativa. *Methods Mol. Biol.* **2015**, *1305*, 1–12. [CrossRef] [PubMed]
37. Ishihama, Y.; Oda, Y.; Tabata, T.; Sato, T.; Nagasu, T.; Rappsilber, J.; Mann, M. Exponentially modified protein abundance index (emPAI) for estimation of absolute protein amount in proteomics by the number of sequenced peptides per protein. *Mol. Cell. Proteom.* **2005**, *4*, 1265–1272. [CrossRef] [PubMed]
38. Yang, G.; Chu, W.; Zhang, H.; Sun, X.; Cai, T.; Dang, L.; Wang, Q.; Yu, H.; Zhong, Y.; Chen, Z.; et al. Isolation and identification of mannose-binding proteins and estimation of their abundance in sera from hepatocellular carcinoma patients. *Proteomics* **2013**, *13*, 878–892. [CrossRef] [PubMed]

39. Murcha, M.W.; Elhafez, D.; Millar, A.H.; Whelan, J. The C-terminal region of TIM17 links the outer and inner mitochondrial membranes in Arabidopsis and is essential for protein import. *J. Biol. Chem.* **2005**, *280*, 16476–16483. [CrossRef] [PubMed]
40. Carrie, C.; Kühn, K.; Murcha, M.W.; Duncan, O.; Small, I.D.; O'Toole, N.; Whelan, J. Approaches to defining dual-targeted proteins in Arabidopsis. *Plant J.* **2009**, *57*, 1128–1139. [CrossRef] [PubMed]
41. Mruk, D.D.; Cheng, C.Y. Enhanced chemiluminescence (ECL) for routine immunoblotting: An inexpensive alternative to commercially available kits. *Spermatogenesis* **2011**, *1*, 121–122. [CrossRef] [PubMed]
42. Becker, T.; Wenz, L.S.; Krüger, V.; Lehmann, W.; Müller, J.M.; Goroncy, L.; Zufall, N.; Lithgow, T.; Guiard, B.; Chacinska, A.; et al. The mitochondrial import protein Mim1 promotes biogenesis of multispanning outer membrane proteins. *J. Cell Biol.* **2011**, *194*, 387–395. [CrossRef] [PubMed]
43. Duncan, O.; Carrie, C.; Wang, Y.; Murcha, M.W. In vitro and in vivo protein uptake studies in plant mitochondria. *Methods Mol. Biol.* **2015**, *1305*, 61–81. [CrossRef] [PubMed]
44. Mossmann, D.; Vögtle, F.N.; Taskin, A.A.; Teixeira, P.F.; Ring, J.; Burkhart, J.M.; Burger, N.; Pinho, C.M.; Tadic, J.; Loreth, D.; et al. Amyloid-β peptide induces mitochondrial dysfunction by inhibition of preprotein maturation. *Cell Metab.* **2014**, *20*, 662–669. [CrossRef] [PubMed]

© 2017 by the authors. Licensee MDPI, Basel, Switzerland. This article is an open access article distributed under the terms and conditions of the Creative Commons Attribution (CC BY) license (http://creativecommons.org/licenses/by/4.0/).

Article

The Roles of Mitochondrion in Intergenomic Gene Transfer in Plants: A Source and a Pool

Nan Zhao [1], Yumei Wang [2] and Jinping Hua [1,*]

1. Laboratory of Cotton Genetics, Genomics and Breeding/Key Laboratory of Crop Heterosis and Utilization of Ministry of Education, College of Agronomy and Biotechnology, China Agricultural University, Beijing 100193, China; Nan_Zhao@cau.edu.cn
2. Institute of Cash Crops, Hubei Academy of Agricultural Sciences, Wuhan 430064, China; yumeiwang001@126.com
* Correspondence: jinping_hua@cau.edu.cn; Tel.: +86-10-6273-4748

Received: 22 December 2017; Accepted: 6 February 2018; Published: 11 February 2018

Abstract: Intergenomic gene transfer (IGT) is continuous in the evolutionary history of plants. In this field, most studies concentrate on a few related species. Here, we look at IGT from a broader evolutionary perspective, using 24 plants. We discover many IGT events by assessing the data from nuclear, mitochondrial and chloroplast genomes. Thus, we summarize the two roles of the mitochondrion: a source and a pool. That is, the mitochondrion gives massive sequences and integrates nuclear transposons and chloroplast tRNA genes. Though the directions are opposite, lots of likenesses emerge. First, mitochondrial gene transfer is pervasive in all 24 plants. Second, gene transfer is a single event of certain shared ancestors during evolutionary divergence. Third, sequence features of homologies vary for different purposes in the donor and recipient genomes. Finally, small repeats (or micro-homologies) contribute to gene transfer by mediating recombination in the recipient genome.

Keywords: mitochondrion; intergenomic gene transfer; nucleus; chloroplast; genome evolution

1. Introduction

A billion years ago, a host cell engulfed a dependent bacteria, α-proteobacteria, which turned into a semi-autonomic organelle, a mitochondrion [1]. It delivers energy to the eukaryotic host cell in the diversifying evolution [2]. Meanwhile, mitochondrial sequences transferred among intracellular genomes [3–7], which is intergenomic gene transfer (IGT). On the one hand, nuclear transposons and chloroplast tRNA genes transferred into the mitochondrial genomes in most seed plants [5,7–12]. Nuclear sequences contributed to mitogenome expansion, contributing almost half in melons [10]. Among the nuclear-like sequences in the mitochondrial genome, the long terminal repeat retrotransposons (LTR-retro) ranked first [4,7–12]. In addition, chloroplast-like genes promoted the translation in mitochondrial genome [13–15]. The sequence states of chloroplast genes changed in the donor (chloroplast) and receptor (mitochondrion) genomes [16,17], which concerned their later roles [18]. Besides, DNA sequence microhomology played an important role in chloroplast DNA inserting into the mitochondrion, which might be the microhomology-mediated break-induced replication (MMBIR) [19] or non-homologous end joining (NHEJ) [20].

On the other hand, a large-scale of mitochondrial genes moved into the nucleus and chloroplast [21–25]. The prokaryotic genes (mitochondrial genes) converted to eukaryotic genes (nuclear genes) [26] to engage in sexual recombination [27]. Besides, RNA could mediate mitochondrion-to-nucleus transfers [28,29]. The mitochondrial genes preferentially inserted in the open nuclear chromosome regions [30]. These nuclear integrants of mitochondrial genes (*numts*) would gradually decay or transform to nuclear sequences [31]. A few *numts* received nuclear promoters and transit peptides [2,32] that guided their products to the mitochondrion [33,34]. Few nuclear homologies of organellar DNA could transcribe successfully [35]. However, mitochondrion-to-chloroplast transfer only occurred in a few angiosperms [16–18,36–41]. Perhaps because plastids were conservative [17,36] and lacked efficient DNA uptake setups [42]. During the evolution, mitochondrial sequences moved to the chloroplast genomes of the shared ancestors of certain relative species [17,18,38,39]. They preferentially inserted into the intergenic spacer [16,17,41] or the large single copy (LSC) region of the chloroplast genomes [38]. The insertion accompanied DNA repair by homologous recombination [17]. However, most chloroplast homologies of mitochondrial genes had low transcriptional levels [17]. Environmental stresses could promote chloroplast [43] and nucleus [44] to absorb exogenous DNA. Meanwhile, the loss of mitochondrial membrane proteins could facilitate the export of the mitochondrial genes [2].

Recently, rapid development of genomic sequencing technologies has made it feasible to approach more IGT events in plants. It enables us to look into the details of intergenomic gene transfer. In this paper, we unveil the IGT events related to the mitochondrion based on 24 sets of nuclear, mitochondrial and chloroplast genomic sequences in plants (Table S1). We expect these results will lay the foundation for further exploration of genome evolution.

2. Results and Discussion

2.1. The Role of Mitochondrion as a Gene Source: Intergenomic Gene Transfer from Mitochondrion

2.1.1. Intergenomic Gene Transfer from Mitochondrion to Nucleus

There exist a number of conserved genes during the mitochondrial genome evolution [45,46]. In the present study, we use 67 essential genes to study the gene loss and transfer about the mitochondrial genome. As a result, genes encoding complex II and ribosomal subunits have been lost massively in most of the higher plants (Figure 1, yellow cells). Genes encoding complexes III and V display much greater conservation. These gene losses are parts of the mitochondrial genome variations in plants. Our next goal is to elucidate where the lost genes transferred. Two of the main detectable destinations are nuclear and chloroplast genomes.

Transferred genes exist in two forms: remnants left in the mitochondrial genome [47] and fragments inserted into the nuclear genome (*numts*) [48,49]. Few *numts'* products returned to the mitochondrion and played a role [33,34]. Researchers have achieved the mitochondrion-to-nucleus transfer by experiments, whose flow was as follows: (1) introduce a silent selectable marker gene with a nuclear promoter and transit peptide-encoding sequence into the mitochondrial genome; (2) transform this recombinant mitochondrion into a new cell; (3) detect the phenotype related to the marker gene. This approach has been successful in the unicellular green alga *Chlamydomonas reinhardtii* [27]. However, there is no experimental report on the real mitochondrion-to-nucleus IGT. Mitochondrial genes transferred with prokaryotic signals, which needed a long time or a favorable evolutionary event to turn into the eukaryotic ones.

In present research, to identify possible *numts*, we carry on non-experimental analyses by performing the genome alignment between conserved mitochondrial genes and nuclear genomes in above 21 land plants. First, we find extensive gene transfer and gene loss in these plants (Figure 1). Second, the gene transfer is more popular in eudicots and monocots than that in bryophytes. Specifically, the latter is merely 1/20 of the former (Figure 2). Since the bryophytes with few mitochondria could not survive after vast transfer [24]. Third, we identify a number of full-length mitochondrial-like protein-coding genes in the nuclear genome (Figure 1, red cells), which may be useful candidate genes. Fourth, there are also mitochondrion-like truncated genes, which we define as pseudogenes (Figure 1, green cells).

Genes integrated by nuclear genome have different endings. Nearly all lost their original roles and became a part of new nuclear sequences [31]. A few could re-gain function by receiving nuclear promoter and transit peptide [2,32]. Others would suffer from irreversible decay with accumulating an increasing number of unfavorable mutations. These events allow prokaryotic gene(s) to turn into eukaryotic gene(s) [26] to join in sexual recombination [27]. As for the transferred forms, early studies displayed RNA-mediated gene transfers from the mitochondrion to the nucleus [28,29], while DNA-mediated gene transfer was rare in plants. In addition, the lack of integral mitochondrial membrane proteins could hasten the gene export from the mitochondrion [2].

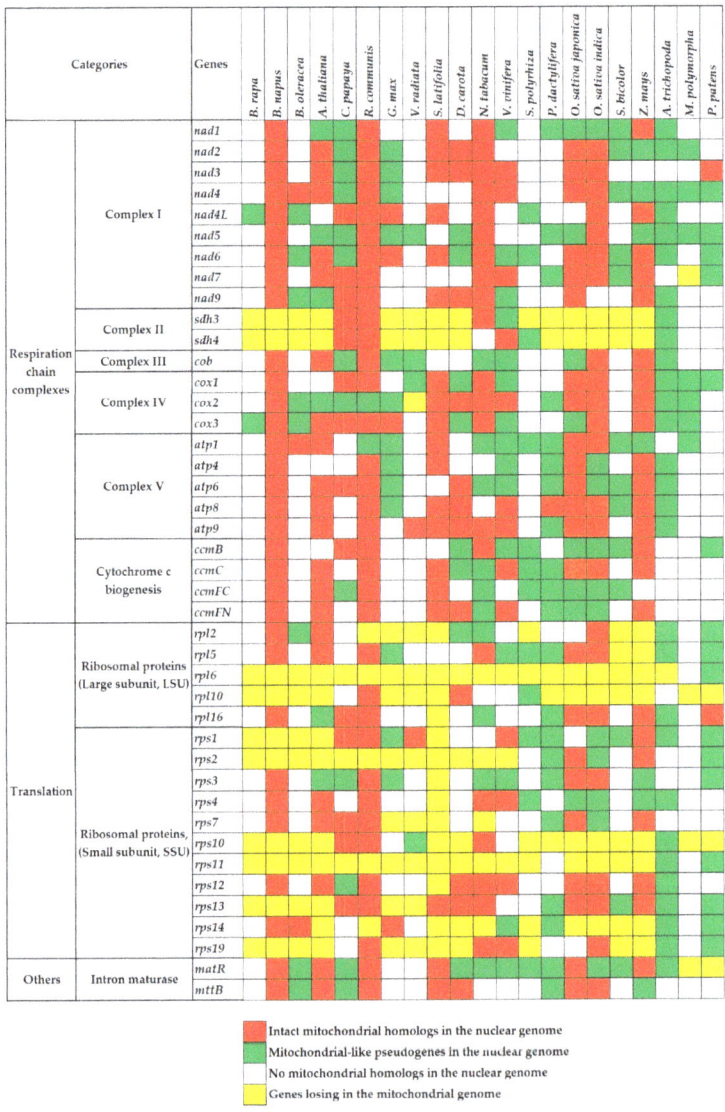

Figure 1. Genes identified to transfer in and out of the mitochondrial genome or genes lost from the mitochondrial genome of 21 land plants. The first two columns are mitochondrial protein-encoding genes (the second column) and their functional categories (the first column). The first line lists the names of plant species. The red and green cells represent mitochondrial full-length intact homologs and pseudogenes in nuclear genomes, respectively. The white and yellow cells represent no mitochondrial homologs in nuclear genomes and genes lost from mitochondrial genomes, respectively.

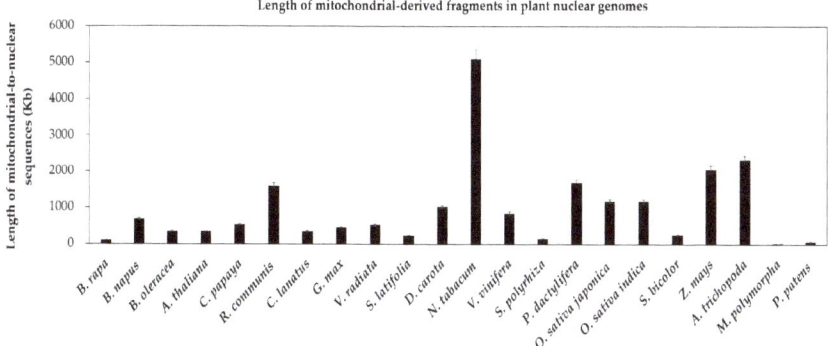

Figure 2. Length of mitochondrial-derived fragments in plant nuclear genomes. The plant species are arrayed on the horizontal axis. The total lengths of mitochondrial-to-nuclear sequences are along the vertical axis. The bars represent the lengths of sequences transferring from the mitochondrion to the nucleus in plant species. The error bars stand for the positive and negative deviations of 5.0%.

To dissect the mechanism of mitochondrion-to-nucleus gene transfers, we analyze the repeats in nuclear genomes of 22 land plants. The ratios of the repeat size to the genome size of the four species, including two bryophytes (*M. polymorpha* and *P. patens*) and two angiosperms (*A. thaliana* and *S. polyrhiza*), are less than 20% (Table 1). Meanwhile, these four species contain fewer *numts* than other species (Figure 1). And there is a positive correlation between *numts* and the repeats in the nuclear genome ($R^2 = 0.6321$) (Figure 3). The weak correlation may due to limited number of plant species used in present research. So, we consider that *numts* may become parts of the nuclear repeats to take part in repeat-mediated sexual recombination for a greater genetic diversity.

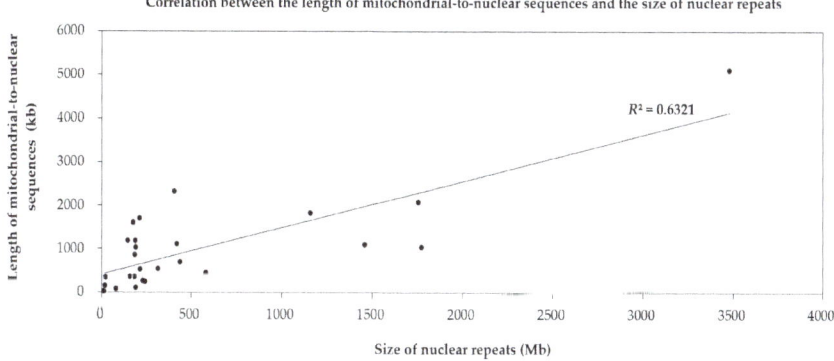

Figure 3. Correlation between the length of mitochondrial sequences transferring to the nucleus and repeat sizes of the nuclear genome in 22 land plants. Each dot represents a length value (X, Y). X refers to the size of the repeats in nuclear genomes of one species (based on the horizontal axis). Y means the length of mitochondrial-to-nuclear sequences in its corresponding species (based on the vertical axis). The slash represents the linear regression function of the distribution tendency of the dots. R^2 is the regression coefficient.

Table 1. Variation of repeats in nuclear genomes of 22 land plants.

	Species	Repeat Sizes (Mb)	Genome Sizes (Mb)	Repeat/Genome (%)	References
	Eudicots				
	B. rapa	191.63	284.13	67.44	[50]
	B. napus	441.77	930.51	47.48	[51]
	B. oleracea	185.43	539.91	34.34	[52]
	A. thaliana [1]	23.58	119.67	19.70	[53]
	C. papaya	316.53	369.78	85.60	[54]
	R. communis	176.00	350.62	50.20	[55]
	C. lanatus	159.80	321.05	49.77	[56]
	G. max	587.10	978.97	59.97	[57]
	V. radiata	216.17	548.08	39.44	[58]
	S. latifolia	244.82	665.28	36.80	[59]
Spermatophytes	D. carota	193.70	473.00	40.95	[60]
	N. tabacum	3479.49	4500.00	77.32	[61]
	V. vinifera	185.35	487.00	38.06	[62]
	Monocots				
	S. polyrhiza [1]	19.43	132.01	14.72	[63]
	P. dactylifera	214.34	558.02	38.41	[64]
	O. sativa japonica	188.00	374.42	50.21	[65]
	O. sativa indica	148.14	374.25	39.58	[66]
	S. bicolor	231.28	739.15	31.29	[67]
	Z. mays	1757.48	2067.62	85.00	[68]
	Basal Angiosperms				
	A. trichopoda	407.43	706.50	57.67	[69]
Bryophytes	M. polymorpha [1]	12.48	304.37	4.10	[70]
	P. patens [1]	79.37	477.95	16.61	[71]

[1] notes repeat content less than 20%.

2.1.2. Intergenomic Gene Transfer from Mitochondrion to Chloroplast

Given the prevailing mitochondrion-to-nucleus IGT, similar transfers into the chloroplast might be expected. However, mitochondrion-to-chloroplast IGT happened only in three angiosperms, *Apiaceae* [16,18,36–39], *Apocynaceae* [17] and *Poaceae* [40,41]. The first two families belong to the eudicots and the last to the monocots.

The existing forms of gene sequences in and out of both donor and receptor genomes altered after the transfer. *D. carota* Mitochondrial Plastid sequence (*Dc*MP)—presented three fragment sequences (*Dc*MP 1, −2 and −3 +4) in the plastid genome. The split probably arose from new DNA recombination that happened after one copy of *Dc*MP migrated into the mitochondrial genome [16]. Besides, mitochondrial-like *rpl2* only contained an exon in the plastid genome and two homologies in different regions of the mitochondrial genome in *A. syriaca* [17]. In addition, the traits of gene sequences in the plastid genome (recipient genomes) might affect their specialized roles. *Dc*MP inserted into two short direct repeats in the plastid genome, which suggested that it served as non-LTR retrotransposon [18]. For those mitochondrial-derived pseudogenes in the plastid, they contained nonsense mutations that would lead to a premature stop codon, which was consistent with the low transcriptional level of the plastid copy *rpl2* in *A. syriaca* [17].

From an evolutionary perspective, mitochondrion-to-chloroplast transfer occurred in the earlier common ancestor of certain relative species as a single event. For example, the homolog of mitochondrial gene, *Dc*MP, existed in the plastid genomes of *Daucus* and their close relative *Cuminum* [18]. Further studies showed that *Dc*MP moved to the shared ancestor of *Daucinae* Dumort and *Torilidinae* Dumort subtribes after they diverged from their ancestral tribe, *Scandiceae* Spreng [38,39]. Also, in *Apocynaceae*, mitochondrial *rpl2* transferred to the plastid genome of the common ancestor of the *Asclepiadeae* and *Eustegia* [17].

Mitochondrial sequences preferentially inserted into the intergenic spacer of plastid genomes. For instance, *Dc*MP inserted in the *rps12-trnV* intergenic spacer in the *D. carota* plastid genome [16]. There were also mitochondrial insertions in the *rps2-rpoC2* intergenic spacer of the plastid genome in *A. syriaca* [17] and in the *rpl23-ndhB* intergenic spacer of the plastid genome of *Parianinae* (*Eremitis* sp. and *Pariana radiciflora*) [41]. Besides, another mitochondrial-to-nuclear transfer appeared in the large single copy (LSC) region between the junction with inverted repeat A (IRA) and tRNA-His (GUG) (*trnH*-GUG) in limited *Apiaceae* species [38]. Additionally, insertion locations implied the roles of the transferred genes. *DcMP* was regarded as a non-LTR retrotransposon targeting tRNA-coding regions because it moved to the upstream of the *trnV* gene in the plastid genome. Otherwise, *DcMP* worked as three new promoters (P1–P3) that substituted two original promoters of the *trnV* gene (P4 and P5) [18]. More importantly, insertion typically came with DNA repair of a double-stranded break by homologous recombination. To create homologies, the plastid gene *rpoC2* preferentially inserted into the mitochondrial genome, just near the mitochondrial-native gene *rpl2*, then intact mitochondrial *rpl2* and part of *rpoC2* transferred together to the plastid of *A. syriaca* [17].

2.2. The Role of Mitochondrion as a Gene Pool: Intergenomic Gene Transfer into Mitochondrion

2.2.1. Intergenomic Gene Transfer from Nucleus to Mitochondrion

Compared with the conservative chloroplast genome, the mitochondrial genome diversified among plant species. The primary drivers of genome variations might be repetitive sequences and nuclear-derived DNA, which represented 42% and 47% of the total sequences in melon, respectively [10]. In present study, we analyze the nucleus-to-mitochondrion sequences of 23 plants. First, nuclear-derived sequences are widespread in all mitochondrial genomes of 23 plants (Figure 4). Second, among spermatophytes, total nuclear sequences in mitochondrial genomes range from a low of 7960 bp in *S. latifolia* to a high of 36,123 bp in *V. vinifera* (Table S2). Third, the nucleus-to-mitochondrion transferred sequences are less in bryophytes than in spermatophytes, 4249 bp and 4814 bp in *P. patens* and *M. polymorpha*, respectively (Figure 4).

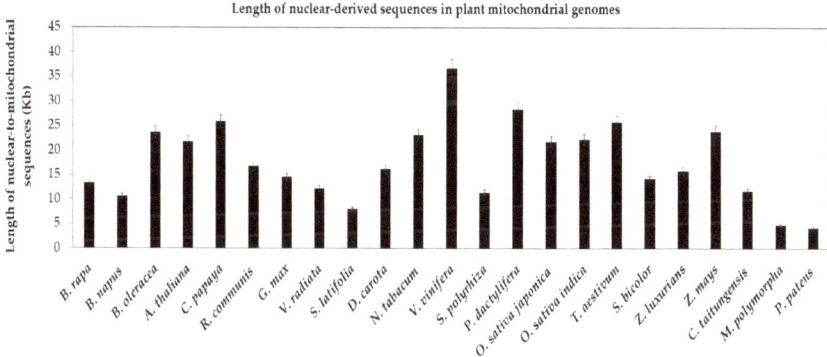

Figure 4. Length of nuclear-derived sequences in plant mitochondrial genomes. The plant species are arrayed on the horizontal axis. The total lengths of nuclear-to-mitochondrial sequences are along the vertical axis. The bars represent the lengths of sequences transferring from the nucleus to the mitochondrion in plant species. The error bars stand for the positive and negative deviations of 5.0%.

According to the different degrees of the matching and annotation, these nuclear-to-mitochondrial repetitive sequences fall into seven categories: copia, gypsy, low complexity, long terminal repeat retrotransposons (LTR-retro), simple repeat, transposable element (TE) and unspecified (Table S2). Copia and gypsy represent two main classes of LTR-retrotransposons that belong to Class 1 transposable elements [72]. Low-complexity DNA primarily include poly-purine/poly-pyrimidine stretches and regions of extremely high AT or GC content. First, the mean of each type in 21 spermatophytes is significantly larger than that in 2 bryophytes (Figure 5), which show most nucleus-to-mitochondrion transfers occurred after the differentiation of seed plants and bryophytes, at least, for the analyzed 2 bryophytes species. Second, the first three are LTR-retro, gypsy and copia in 23 plants (Figure 6, Table S2). This result conforms to the early discoveries in a number of plants, including the gymnosperm *Cycas taitungensis* [9], the monocot *Oryza sativa* [8] and the eudicots *Arabidopsis thaliana*, *Cucumis melo* and *Cucumis sativus* [4,7,10–12]. Third, the total length of transferred sequences correlates with the mitogenome size (Figure 7). This result supports the import of promiscuous DNA is a core mechanism for mitochondrial genome expansion in land plants [73].

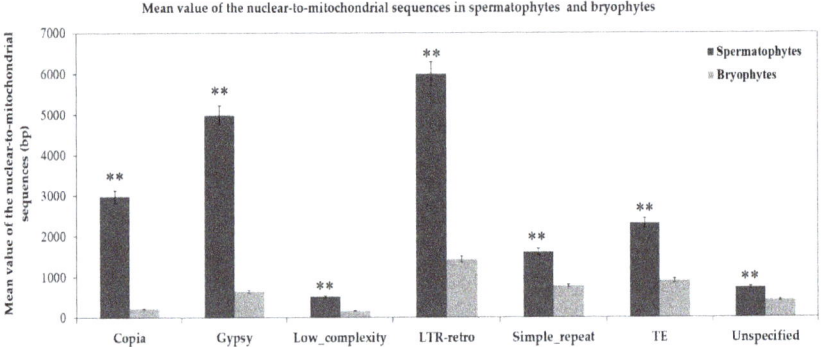

Figure 5. Mean value of the length of different nuclear sequences transferring to the mitochondrial genomes of spermatophytes and bryophytes. ** $p < 0.01$. The seven categories of repeats are arrayed on the horizontal axis. The total lengths of nuclear-to-mitochondrial repetitive sequences are along the vertical axis. The dark gray and light gray bars represent the mean values of repeats transferring from the nucleus to the mitochondrion in 21 spermatophytes and 2 bryophytes, respectively. The error bars stand for the positive and negative deviations of 5.0%.

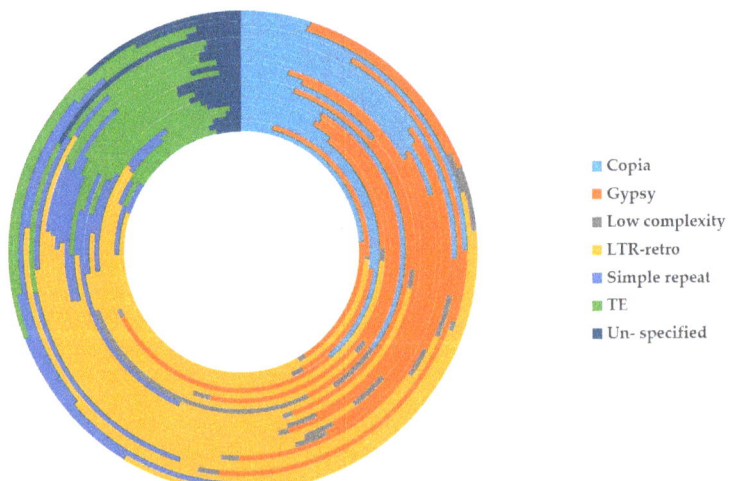

Figure 6. Percentages of each kind of repeats from all nuclear-to-mitochondrial repetitive sequences in 23 plants. The 23 circles represent the whole nuclear-to-mitochondrial repeats of 23 plants inside and out. The boxes in different colors on the right are the symbols of seven kinds of repetitive sequences. (From top to bottom) Light blue: copia; Orange: gypsy; Gray: low complex; Yellow: LTR-retro (long terminal repeat retrotransposons); Middle blue: simple repeat; Green: TE (transposable element); dark blue: un-specific.

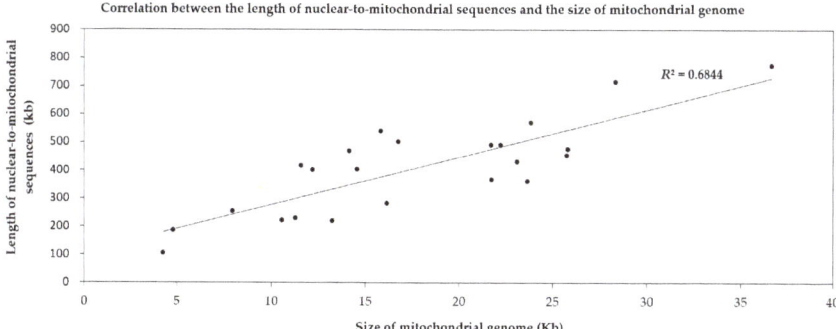

Figure 7. Correlation between the length of nuclear sequences transferring to the mitochondrion and the size of the mitochondrial genome in 23 land plants. Each dot represents a length value (X, Y). X refers to the length of the mitochondrial genome of one species (based on the horizontal axis). Y means the length of nuclear-to-mitochondrial sequences in this corresponding species (based on the vertical axis). The slash represents the linear regression function of the distribution tendency of the dots. R^2 is the regression coefficient.

2.2.2. Intergenomic Gene Transfer from Chloroplast to Mitochondrion

As with mitochondrial genomes, chloroplast genomes also contain a minimum set of largely conserved protein-encoding, rRNA and tRNA genes [21,74,75]. In contrast to the extensive gene loss of mitochondrial genomes, only few chloroplast-encoded genes have been lost in chloroplast genomes of specific plants (Figure S1, yellow cells). For example, three genes (*accD*, *ycf1* and *ycf2*) are lost in the grasses (*O. sativa japonica*, *O. sativa indica*, *S. bicolor*, *Z. mays*), another three genes (*ccsA*, *rpoA* and *rpl16*) are lost in the moss *P. patens* (Figure S1, yellow cells). Compared to a few gene loss, chloroplast genes transferring to nucleus and mitochondrion are richer (Figures S1 and S2). In our study, we unearth the enormous chloroplast-to-mitochondrion gene transfers in 24 land plants. Similar gene copies exist in two contemporary intracellular genomes simultaneously (Figure S1, the red and green cells). In two bryophytes, the total lengths of integrated sequences are close, 1.05 kb in *M. polymorpha* and 1.99 kb in *P. patens* (Figure 8). In addition, the variation range is greater in 22 seed plants, from 1.67 kb in *S. latifolia* to 130 kb in *A. trichopoda* (Figure 8). Besides, the chloroplast-to-mitochondrion fragments of most seed plants are more than that in bryophytes (Figure 8).

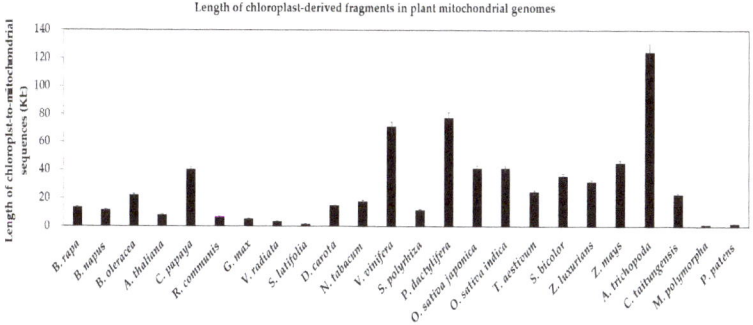

Figure 8. Length of chloroplast-derived fragments in plant mitochondrial genomes. The plant species are arrayed on the horizontal axis. The total lengths of chloroplast-to-mitochondrial sequences are along the vertical axis. The bars represent the lengths of sequences transferring from the chloroplast to the mitochondrion in species. The error bars stand for the positive and negative deviations of 5.0%.

Large parts of chloroplast tRNA genes immigrated into plant mitochondrial genomes [5,9]. These transfers were essential to the translation of the mitochondrial genes [13–15]. Here, we identify the chloroplast-like tRNA genes in the mitochondrial genome of 24 plants species using blast. And then we build a phylogenetic tree to elucidate the evolutionary implications. First, there is no chloroplast-derived tRNA gene in mitochondrial genomes of two bryophytes (Figure S2). Second, single or multiple chloroplast genes immigrated to the mitochondrial genomes of spermatophytes, at least, for the analyzed 21 angiosperms and 1 gymnosperms. For example, (1) chloroplast-like *trnM* gene appears in the mitochondrial genomes of all studied seed plants except Z. *mays*, which suggests that chloroplast *trnM* lost only in Z. *mays* during or after transferring to the mitochondrion and this transfer happened with spermatophytes and bryophytes diverging; (2) chloroplast *trnH* gene transferred to the mitochondrial genomes of most spermatophytes but lost in *P. dactylifera*, *T. aestivum* and *G. biloba*, which might be the random loss; (3) *trnN*, *trnP*, *trnS* and *trnW* transferred merely in angiosperms, despite parts of these four genes lost in a few species; (4) chloroplast *trnD* gene moved into the mitochondrion only in eudicots, which shows that *trnD* transferred when eudicots and monocots diverged; (5) chloroplast-like *trnC* gene and *trnF* gene transferred to the mitochondrion simply in *Gramineae* crops of monocots; (6) ten chloroplast-to-mitochondrion genes (*trnD*, *trnE*, *trnG*, *trnI*, *trnK*, *trnL*, *trnP*, *trnR*, *trnT* and *trnY*) transferred together in *V. vinifera* (Figure S2).

To infer the mechanism of chloroplast tRNA genes inserting into mitochondria, we analyze the flanking nucleotide sequences in insertion sites of mitochondrial genomes. *trnH* transferred in most spermatophytes (Figure 9). *trnD* moved specifically in eudicots (Figure S3). *trnC* and *trnF* migrated only in *Gramineae* crops (Figure S4). Taking together, we notice the micro-homologies (1 to 4 bp) among plant species in the breakpoint sequences of chloroplast-mitochondrial DNA fusion. The micro-homologies are the same adenine-thymine (AT) on the right of *trnH* in spermatophytes. But on the left are four short tandems Guanine (G) in eudicots, two repeated Guanine (G) in monocots and no microhomology in gymnosperms (Figure 9). Therefore, we confer that DNA sequence microhomology plays an important role in chloroplast DNA inserting into the mitochondrion, which may be the microhomology-mediated break-induced replication (MMBIR) [19] or non-homologous end joining (NHEJ) [20].

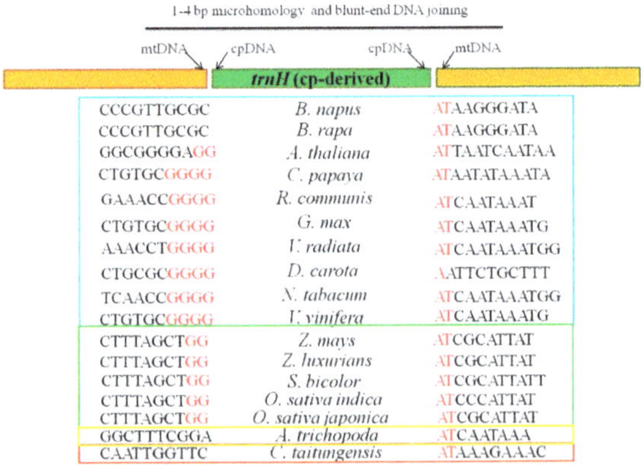

mtDNA / cpDNA	*trnH* (cp-derived)	cpDNA / mtDNA
CCCGTTGCGC	*B. napus*	ATAAGGGATA
CCCGTTGCGC	*B. rapa*	ATAAGGGATA
GGCGGGGAGG	*A. thaliana*	ATTAATCAATAA
CTGTGCGGGG	*C. papaya*	ATAATATAAATA
GAAACCGGGG	*R. communis*	ATCAATAAAT
CTGTGCGGGG	*G. max*	ATCAATAAATG
AAACCTGGGG	*V. radiata*	ATCAATAAATGG
CTGCGCGGGG	*D. carota*	AATTCTGCTTT
TCAACCGGGG	*N. tabacum*	ATCAATAAATGG
CTGTGCGGGG	*V. vinifera*	ATCAATAAATG
CTTTAGCTGG	*Z. mays*	ATCGCATTAT
CTTTAGCTGG	*Z. luxurians*	ATCGCATTAT
CTTTAGCTGG	*S. bicolor*	ATCGCATTATT
CTTTAGCTGG	*O. sativa indica*	ATCCCATTAT
CTTTAGCTGG	*O. sativa japonica*	ATCGCATTAT
GGCTTTCGGA	*A. trichopoda*	ATCAATAAA
CAATTGGTTC	*C. taitungensis*	ATAAAGAAAC

Figure 9. Nucleotide-resolution analysis on flanking sequences of the chloroplast-derived *trnH* gene in mitochondrial genomes of most spermatophytes. cpDNA and mtDNA are the abbreviations of chloroplast DNA and mitochondrial DNA. The yellow-green-yellow strip represents the fusion sequence of mtDNA-cpDNA-mtDNA. The sequences under the two yellow strips on the left and right are the flanking sequences of inserted chloroplast-like tRNA gene in the mitochondrial genomes. The red capital English letters close to cpDNA indicate the nucleotides of micro-homologies among the different species. The species in the blue, green, yellow and red boxes belong to eudicots, monocots, basal angiosperms and gymnosperms, respectively.

On top of it all, we infer the repeats in mitochondrial genomes have the potential to mediate DNA recombination, which contributes to gene transfer and reuse of the transferred genes in target genomes. Therefore, we analyze the repeats variation in recipient genomes (the mitochondrial genomes) of land plants (Table 2) to explain various rates of gene transfer to some extent. First, plants with smaller values of repeat size, repeat number (>1 kb) and repeat number (>100 bp) contain less gene transfer, among which the most obvious is a bryophyte *P. patens* (Table 2 and Figure 8). Second, small repeats (>100 bp) are more favorable to gene transfer than large repeats (>1 kb) (Table 2).

Table 2. Variation of repeats in mitochondrial genomes from 22 land plants.

	Species	Mitochondrial Genome		
		Repeat Size (Kb)	Repeat Number (>1 kb)	Repeat Number (>100 bp)
	Eudicots			
	B. rapa	3.80	1	9
	B. napus	4.62	1	17
	B. oleracea	152.00	2	24
	A. thaliana	15.63	2	25
	C. papaya	13.43	1	13
	R. communis	5.80	6	6
	G. max	60.67	13	68
	V. radiata	1.02	0	6
	S. latifolia	23.27	15	17
	D. carota	71.09	4	19
Spermatophytes	N. tabacum	42.07	3	22
	V. vinifera	5.77	0	26
	Monocots			
	S. polyrhiza	1.58	0	5
	P. dactylifera	3.03	1	12
	O. sativa japonica	141.19	12	39
	O. sativa indica	141.76	11	27
	S. bicolor	58.56	5	18
	Z. mays	51.94	4	19
	Basal Angiosperms			
	A. trichopoda	266.14	1	1811
	Gymnosperms			
	C. taitungensis	62.65	2	5070
Bryophytes	M. polymorpha	2.08	0	13
	P. patens	0	0	0

3. Materials and Methods

3.1. Availability of Chloroplast, Mitochondrial and Nuclear Genomes

We download all the chloroplast, mitochondrial and nuclear genome sequences and gene annotations from NCBI database. And then we list all the accession numbers in Table S1.

3.2. Detection of Total Intergenomic-Transfer DNA Sequences

For 24 land plants, we align the sequences of chloroplast and mitochondrial genomes to nuclear chromosomes to detect nuclear insertions of chloroplast DNA (*nupts*) and nuclear insertions of mitochondrial DNA (*numts*) using the BLAST program. We set e-value to $1e^{-5}$ [76]. The minimum length of an exact match (95%) is 100 bp. While identifying mitochondrial insertions of chloroplast DNAs (*mtpts*) by local BLASTN (version 2.2.23) [76], we set the minimum length of an exact match to be 50-bp.

3.3. Identification of Intergenomic-Transfer Homologies

Taking a set of essential chloroplast or mitochondrial genes as references (Table S3), we gain their copies in the donor and recipient genomes using the BLAST program with the same parameters above [76]. If there is no counterpart in the donor genomes (chloroplast or mitochondrial genomes), we would consider them as the lost genes (Figure 1 and Figure S1, the yellow cells). To those presented in the donor genomes but absent in the recipient genomes, we consider that they did not transfer between two genomes (Figure 1 and Figure S1, the white cells). For those appearing concurrently in both donor and recipient genomes, we consider that their copies moved into another genome after duplication in the original genome (Figure 1 and Figure S1, the red and green cells). Further, we define the full-length copies of the transferred genes in the recipient genomes as the intact homologies (Figure 1 and Figure S1, the red cells). Otherwise, we recognize the truncated copies as pseudogenes (Figure 1 and Figure S1, the green cells).

3.4. Detection of the Repeats in Mitochondrial Genomes

We detect nuclear-derived repetitive transposons using online software RepeatMasker (http://www.repeatmasker.org) in 24 land species and a custom repeats database. And then we use two-tailed *t*-tests to evaluate the significant difference of repeats between spermatophytes and bryophytes.

3.5. NHEJ Analysis

We perform the NHEJ analysis as previously described [77,78]. In short, *nupts*, *numts* or *mtpts* are inserted by NHEJ, like micro-homology or blunt end repair. If nucleotides close to the fusion point are similar in different land species, we would regard them as micro-homology. Otherwise, we would consider no micro-homology as blunt-end repair.

3.6. Phylogenetic Analysis

The phylogenetic analysis involves nucleotide sequences of 17 mitochondrial genes (*nad1–nad6*, *nad9*, *cob*, *cox1–cox3*, *atp1*, *atp4*, *atp6*, *atp8* and *atp9*). We use the maximum likelihood (ML) method with the model GTR + G + I in MEGA5.05 [79]. And then we perform phylogenetic analyses according to the same methods in previous studies [80,81].

4. Conclusions

With the rapid development of genomic sequencing technologies, nuclear and organellar genomes data became available for many plants. Here, based on 24 sets of genome data, we detect and analyze intergenomic gene transfers (IGT) related to the mitochondrion. Meanwhile, we review the research advances of intergenomic gene transfer. As a summary, we find mitochondrion mainly plays two essential roles in gene transfer: Source and pool. From the source perspective, massive mitochondrial genes transfer into nuclear and chloroplast genomes. For the role of the pool, the mitochondrion integrates enormous genes from the other two genomes. Except for the disparate orientation, a lot of likenesses emerge when bringing them together. First, gene transfer related to mitochondrial genomes is prevalent in plants, though few genes flow from the mitochondrion to the chloroplast. Second, specific IGT is a single event of certain shared ancestors, which is consistent with the divergence clade. Third, an intact gene usually changes existing forms after transferring in and out of both donor and recipient genomes, which agrees with their consequent roles, such as, functioning like before, reusing for new loci or decaying gradually. Fourth, most exogenous DNA preferentially inserts into the intergenic region. Besides, small repeats (or micro-homologies) may contribute to gene transfers by mediating recombination in the recipient genomes. In a word, mitochondrial gene transfers dedicate to the genome variation and evolutionary diversity.

Supplementary Materials: Supplementary materials can be found at http://www.mdpi.com/1422-0067/19/2/547/s1.

Acknowledgments: We thank Jonathan F. Wendel and Corrinne E. Grover (Iowa State University) for helpful suggestions. This work was supported by grants from the National Natural Science Foundation of China (31671741) to Jinping Hua.

Author Contributions: Nan Zhao substantively prepared the manuscript; Yumei Wang attended the bench work; Jinping Hua conceived and designed the experiments, provided research platform and revised the manuscript. All authors approved the final manuscript.

Conflicts of Interest: The authors declare no conflict of interest.

References

1. Gray, M.W. Mosaic nature of the mitochondrial proteome: Implications for the origin and evolution of mitochondria. *Proc. Natl. Acad. Sci. USA* **2015**, *112*, 10133–10138. [CrossRef] [PubMed]
2. Bock, R. Witnessing genome evolution: Experimental reconstruction of endosymbiotic and horizontal gene transfer. *Annu. Rev. Genet.* **2017**, *51*, 1–22. [CrossRef] [PubMed]
3. Stern, D.B.; Lonsdale, D.M. Mitochondrial and chloroplast genomes of maize have a 12-kilobase DNA-sequence in common. *Nature* **1982**, *299*, 698–702. [CrossRef] [PubMed]
4. Knoop, V.; Unseld, M.; Marienfeld, J.; Brandt, P.; Sunkel, S.; Ullrich, H.; Brennicke, A. Copia-, gypsy- and line-like retrotransposon fragments in the mitochondrial genome of *Arabidopsis thaliana*. *Genetics* **1996**, *142*, 579–585. [PubMed]
5. Wang, D.; Rousseau-Gueutin, M.; Timmis, J.N. Plastid sequences contribute to some plant mitochondrial genes. *Mol. Biol. Evol.* **2012**, *29*, 1707–1711. [CrossRef] [PubMed]
6. Liu, G.Z.; Cao, D.D.; Li, S.S.; Su, A.G.; Geng, J.N.; Grover, C.E.; Hu, S.N.; Hua, J.P. The complete mitochondrial genome of *Gossypium hirsutum* and evolutionary analysis of higher plant mitochondrial genomes. *PLoS ONE* **2013**, *8*, e69476. [CrossRef] [PubMed]
7. Tang, M.Y.; Chen, Z.W.; Grover, C.E.; Wang, Y.M.; Li, S.S.; Liu, G.Z.; Ma, Z.Y.; Wendel, J.F.; Hua, J.P. Rapid evolutionary divergence of *Gossypium barbadense* and *G. hirsutum* mitochondrial genomes. *BMC Genom.* **2015**, *16*, 770. [CrossRef] [PubMed]
8. Notsu, Y.; Masood, S.; Nishikawa, T.; Kubo, N.; Akiduki, G.; Nakazono, M.; Hirai, A.; Kadowaki, K. The complete sequence of the rice (*Oryza sativa* L.) mitochondrial genome: Frequent DNA sequence acquisition and loss during the evolution of flowering plants. *Mol. Genet. Genom.* **2002**, *268*, 434–445. [CrossRef] [PubMed]
9. Wang, D.; Wu, Y.W.; Shih, A.C.C.; Wu, C.S.; Wang, Y.N.; Chaw, S.M. Transfer of chloroplast genomic DNA to mitochondrial genome occurred at least 300 mya. *Mol. Biol. Evol.* **2007**, *24*, 2040–2048. [CrossRef] [PubMed]
10. Rodriguez-Moreno, L.; Gonzalez, V.M.; Benjak, A.; Marti, M.C.; Puigdomenech, P.; Aranda, M.A.; Garcia-Mas, J. Determination of the melon chloroplast and mitochondrial genome sequences reveals that the largest reported mitochondrial genome in plants contains a significant amount of DNA having a nuclear origin. *BMC Genom.* **2011**, *12*, 424. [CrossRef] [PubMed]
11. Alverson, A.J.; Rice, D.W.; Dickinson, S.; Barry, K.; Palmer, J.D. Origins and recombination of the bacterial-sized multichromosomal mitochondrial genome of cucumber. *Plant Cell* **2011**, *23*, 2499–2513. [CrossRef] [PubMed]
12. Chen, Z.W.; Nie, H.S.; Grover, C.E.; Wang, Y.M.; Li, P.; Wang, M.Y.; Pei, H.L.; Zhao, Y.P.; Li, S.S.; Wendel, J.F.; et al. Entire nucleotide sequences of *Gossypium raimondii* and *G. arboreum* mitochondrial genomes revealed a-genome species as cytoplasmic donor of the allotetraploid species. *Plant Biol.* **2017**, *19*, 484–493. [CrossRef] [PubMed]
13. Dietrich, A.; Small, I.; Cosset, A.; Weil, J.H.; Marechal-Drouard, L. Editing and import: Strategies for providing plant mitochondria with a complete set of functional transfer rnas. *Biochimie* **1996**, *78*, 518–529. [CrossRef]
14. Clifton, S.W.; Minx, P.; Fauron, C.M.R.; Gibson, M.; Allen, J.O.; Sun, H.; Thompson, M.; Barbazuk, W.B.; Kanuganti, S.; Tayloe, C.; et al. Sequence and comparative analysis of the maize NB mitochondrial genome. *Plant Physiol.* **2004**, *136*, 3486–3503. [CrossRef] [PubMed]
15. Sloan, D.B.; Alverson, A.J.; Storchova, H.; Palmer, J.D.; Taylor, D.R. Extensive loss of translational genes in the structurally dynamic mitochondrial genome of the angiosperm *Silene latifolia*. *BMC Evol. Biol.* **2010**, *10*, 1–15. [CrossRef] [PubMed]
16. Iorizzo, M.; Senalik, D.; Szklarczyk, M.; Grzebelus, D.; Spooner, D.; Simon, P. De novo assembly of the carrot mitochondrial genome using next generation sequencing of whole genomic DNA provides first evidence of DNA transfer into an angiosperm plastid genome. *BMC Plant Biol.* **2012**, *12*, 1–17. [CrossRef] [PubMed]

17. Straub, S.C.K.; Cronn, R.C.; Edwards, C.; Fishbein, M.; Liston, A. Horizontal transfer of DNA from the mitochondrial to the plastid genome and its subsequent evolution in milkweeds (*Apocynaceae*). *Genome Biol. Evol.* **2013**, *5*, 1872–1885. [CrossRef] [PubMed]
18. Iorizzo, M.; Grzebelus, D.; Senalik, D.; Szklarczyk, M.; Spooner, D.; Simon, P. Against the traffic: The first evidence for mitochondrial DNA transfer into the plastid genome. *Mob. Genet. Elem.* **2012**, *2*, 261–266. [CrossRef] [PubMed]
19. Liu, P.F.; Erez, A.; Nagamani, S.C.S.; Dhar, S.U.; Kolodziejska, K.E.; Dharmadhikari, A.V.; Cooper, M.L.; Wiszniewska, J.; Zhang, F.; Withers, M.A.; et al. Chromosome catastrophes involve replication mechanisms generating complex genomic rearrangements. *Cell* **2011**, *146*, 888–902. [CrossRef] [PubMed]
20. Hastings, P.J.; Lupski, J.R.; Rosenberg, S.M.; Ira, G. Mechanisms of change in gene copy number. *Nat. Rev. Genet.* **2009**, *10*, 551–564. [CrossRef] [PubMed]
21. Martin, W.; Stoebe, B.; Goremykin, V.; Hansmann, S.; Hasegawa, M.; Kowallik, K.V. Gene transfer to the nucleus and the evolution of chloroplasts. *Nature* **1998**, *393*, 162–165. [CrossRef] [PubMed]
22. Blanchard, J.L.; Schmidt, G.W. Pervasive migration of organellar DNA to the nucleus in plants. *J. Mol. Evol.* **1995**, *41*, 397–406. [CrossRef] [PubMed]
23. Bergthorsson, U.; Adams, K.L.; Thomason, B.; Palmer, J.D. Widespread horizontal transfer of mitochondrial genes in flowering plants. *Nature* **2003**, *424*, 197–201. [CrossRef] [PubMed]
24. Ku, C.; Nelson-Sathi, S.; Roettger, M.; Sousa, F.L.; Lockhart, P.J.; Bryant, D.; Hazkani-Covo, E.; McInerney, J.O.; Landan, G.; Martin, W.F. Endosymbiotic origin and differential loss of eukaryotic genes. *Nature* **2015**, *524*, 427–432. [CrossRef] [PubMed]
25. Chen, Z.W.; Nie, H.S.; Wang, Y.M.; Pei, H.L.; Li, S.S.; Zhang, L.D.; Hua, J.P. Rapid evolutionary divergence of diploid and allotetraploid gossypium mitochondrial genomes. *BMC Genom.* **2017**, *18*, 876. [CrossRef] [PubMed]
26. Bock, R.; Timmis, J.N. Reconstructing evolution: Gene transfer from plastids to the nucleus. *BioEssays* **2008**, *30*, 556–566. [CrossRef] [PubMed]
27. Bonnefoy, N.; Remacle, C.; Fox, T.D. Genetic transformation of *saccharomyces cerevisiae* and *chlamydomonas reinhardtii* mitochondria. *Methods Cell Biol.* **2007**, *80*, 525–548. [PubMed]
28. Covello, P.; Gray, M.W. Silent mitochondrial and active nuclear genes for subunit 2 of cytochrome c oxidase (cox2) in soybean: Evidence for rna-mediated gene transfer. *EMBO J.* **1992**, *11*, 3815–3820. [PubMed]
29. Nugent, J.M.; Palmer, J.D. RNA-mediated transfer of the gene coxII from the mitochondrion to the nucleus during flowering plant evolution. *Cell* **1991**, *66*, 473–481. [CrossRef]
30. Wang, D.; Timmis, J.N. Cytoplasmic organelle DNA preferentially inserts into open chromatin. *Genome Biol. Evol.* **2013**, *5*, 1060–1064. [CrossRef] [PubMed]
31. Kudla, J.; Albertazzi, F.; Blazević, D.; Hermann, M.; Bock, R. Loss of the mitochondrial cox2 intron 1 in a family of monocotyledonous plants and utilization of mitochondrial intron sequences for the construction of a nuclear intron. *Mol. Genet. Genom.* **2002**, *267*, 223–230.
32. Kadowaki, K.-I.; Kubo, N.; Ozawa, K.; Hirai, A. Targeting presequence acquisition after mitochondrial gene transfer to the nucleus occurs by duplication of existing target signals. *EMBO J.* **1997**, *15*, 6652–6661.
33. Adams, K.L.; Qiu, Y.L.; Stoutemyer, M.; Palmer, J.D. Punctuated evolution of mitochondrial gene content: High and variable rates of mitochondrial gene loss and transfer to the nucleus during angiosperm evolution. *Proc. Natl. Acad. Sci. USA* **2002**, *99*, 9905–9912. [CrossRef] [PubMed]
34. Adams, K.L.; Palmer, J.D. Evolution of mitochondrial gene content: Gene loss and transfer to the nucleus. *Mol. Phylogenet. Evol.* **2003**, *29*, 380–395. [CrossRef]
35. Wang, D.; Qu, Z.P.; Adelson, D.L.; Zhu, J.K.; Timmis, J.N. Transcription of nuclear organellar DNA in a model plant system. *Genome Biol. Evol.* **2014**, *6*, 1327–1334. [CrossRef] [PubMed]
36. Goremykin, V.V.; Salamini, F.; Velasco, R.; Viola, R. Mitochondrial DNA of *Vitis vinifera* and the issue of rampant horizontal gene transfer. *Mol. Biol. Evol.* **2009**, *26*, 99–110. [CrossRef] [PubMed]
37. Smith, D.R. Mitochondrion-to-plastid DNA transfer: It happens. *New Phytol.* **2014**, *202*, 736–738. [CrossRef] [PubMed]
38. Downie, S.; Jansen, R. A comparative analysis of whole plastid genomes from the *apiales*: Expansion and contraction of the inverted repeat, mitochondrial to plastid transfer of DNA and identification of highly divergent noncoding regions. *Syst. Bot.* **2015**, *40*, 336–351. [CrossRef]

39. Spooner, D.M.; Ruess, H.; Iorizzo, M.; Senalik, D.; Simon, P. Entire plastid phylogeny of the carrot genus (*Daucus*, *Apiaceae*): Concordance with nuclear data and mitochondrial and nuclear DNA insertions to the plastid. *Am. J. Bot.* **2017**, *104*, 296–312. [CrossRef] [PubMed]
40. Ma, P.F.; Zhang, Y.X.; Guo, Z.H.; Li, D.Z. Evidence for horizontal transfer of mitochondrial DNA to the plastid genome in a bamboo genus. *Sci. Rep.* **2015**, *5*, 1–9. [CrossRef] [PubMed]
41. Wysocki, W.P.; Clark, L.G.; Attigala, L.; Ruiz-Sanchez, E.; Duvall, M.R. Evolution of the bamboos (*Bambusoideae*; *Poaceae*): A full plastome phylogenomic analysis. *BMC Evol. Biol.* **2015**, *15*, 1–12. [CrossRef] [PubMed]
42. Smith, D.R. Extending the limited transfer window hypothesis to inter-organelle DNA migration. *Genome Biol. Evol.* **2011**, *3*, 743–748. [CrossRef] [PubMed]
43. Cerutti, H.; Jagendorf, A. Movement of DNA across the chloroplast envelope: Implications for the transfer of promiscuous DNA. *Photosynth. Res.* **1995**, *46*, 329–337. [CrossRef] [PubMed]
44. Wang, D.; Lloyd, A.H.; Timmis, J.N. Environmental stress increases the entry of cytoplasmic organellar DNA into the nucleus in plants. *Proc. Natl. Acad. Sci. USA* **2012**, *109*, 2444–2448. [CrossRef] [PubMed]
45. Kurland, C.G.; Andersson, S.G.E. Origin and evolution of the mitochondrial proteome. *Microbiol. Mol. Biol. Rev.* **2000**, *64*, 786–820. [CrossRef] [PubMed]
46. Kitazaki, K.; Kubo, T. Cost of having the largest mitochondrial genome: Evolutionary mechanism of plant mitochondrial genome. *J. Bot.* **2010**, *2010*, 620137. [CrossRef]
47. Ong, H.C.; Palmer, J.D. Pervasive survival of expressed mitochondrial *rps14* pseudogenes in grasses and their relatives for 80 million years following three functional transfers to the nucleus. *BMC Evol. Biol.* **2006**, *6*, 55. [CrossRef] [PubMed]
48. Pamilo, P.; Viljakainen, L.; Vihavainen, A. Exceptionally high density of numts in the honeybee genome. *Mol. Biol. Evol.* **2007**, *24*, 1340–1346. [CrossRef] [PubMed]
49. Timmis, J.N.; Ayliffe, M.A.; Huang, C.Y.; Martin, W. Endosymbiotic gene transfer: Organelle genomes forge eukaryotic chromosomes. *Nat. Rev. Genet.* **2004**, *5*, 123–135. [CrossRef] [PubMed]
50. Wang, X.; Wang, H.; Wang, J.; Sun, R.; Wu, J.; Liu, S.; Bai, Y.; Mun, J.H.; Bancroft, I.; Cheng, F.; et al. The genome of the mesopolyploid crop species *Brassica rapa*. *Nat. Genet.* **2011**, *43*, 1035–1039. [CrossRef] [PubMed]
51. Yang, J.; Liu, D.; Wang, X.; Ji, C.; Cheng, F. The genome sequence of allopolyploid *Brassica juncea* and analysis of differential homoeolog gene expression influencing selection. *Nat. Genet.* **2016**, *48*, 1225–1232. [CrossRef] [PubMed]
52. Liu, S.; Liu, Y.; Yang, X.; Tong, C.; Edwards, D.; Parkin, I.A.; Zhao, M.; Ma, J.; Yu, J.; Huang, S.; et al. The *Brassica oleracea* genome reveals the asymmetrical evolution of polyploid genomes. *Nat. Commun.* **2014**, *5*, 3930. [CrossRef] [PubMed]
53. Pucker, B.; Holtgrawe, D.; Rosleff Sorensen, T.; Stracke, R.; Viehover, P.; Weisshaar, B. A de novo genome sequence assembly of the *Arabidopsis thaliana* accession niederzenz-1 displays presence/absence variation and strong synteny. *PLoS ONE* **2016**, *11*, e0164321. [CrossRef] [PubMed]
54. Ming, R.; Hou, S.; Feng, Y.; Yu, Q.; Dionne-Laporte, A.; Saw, J.H.; Senin, P.; Wang, W.; Ly, B.V.; Lewis, K.L.; et al. The draft genome of the transgenic tropical fruit tree papaya (*Carica papaya* Linnaeus). *Nature* **2008**, *452*, 991–996. [CrossRef] [PubMed]
55. Chan, A.P.; Crabtree, J.; Zhao, Q.; Lorenzi, H.; Orvis, J.; Puiu, D.; Melake-Berhan, A.; Jones, K.M.; Redman, J.; Chen, G.; et al. Draft genome sequence of the oilseed species *Ricinus communis*. *Nat. Biotechnol.* **2010**, *28*, 951–956. [CrossRef] [PubMed]
56. Guo, S.; Zhang, J.; Sun, H.; Salse, J.; Lucas, W.J.; Zhang, H.; Zheng, Y.; Mao, L.; Ren, Y.; Wang, Z.; et al. The draft genome of watermelon (*Citrullus lanatus*) and resequencing of 20 diverse accessions. *Nat. Genet.* **2013**, *45*, 51–58. [CrossRef] [PubMed]
57. Schmutz, J.; Cannon, S.B.; Schlueter, J.; Ma, J.; Mitros, T.; Nelson, W.; Hyten, D.L.; Song, Q.; Thelen, J.J.; Cheng, J.; et al. Genome sequence of the palaeopolyploid soybean. *Nature* **2010**, *463*, 178–183. [CrossRef] [PubMed]
58. Kang, Y.J.; Kim, S.K.; Kim, M.Y.; Lestari, P.; Kim, K.H.; Ha, B.K.; Jun, T.H.; Hwang, W.J.; Lee, T.; Lee, J.; et al. Genome sequence of mungbean and insights into evolution within vigna species. *Nat. Commun.* **2014**, *5*, 5443. [CrossRef] [PubMed]

59. Cegan, R.; Vyskot, B.; Kejnovsky, E.; Kubat, Z.; Blavet, H.; Safar, J.; Dolezel, J.; Blavet, N.; Hobza, R. Genomic diversity in two related plant species with and without sex chromosomes—*Silene latifolia* and *S. vulgaris*. *PLoS ONE* **2012**, *7*, e31898. [CrossRef] [PubMed]
60. Iorizzo, M.; Ellison, S.; Senalik, D. A high-quality carrot genome assembly provides new insights into carotenoid accumulation and asterid genome evolution. *Nat. Genet.* **2016**, *48*, 657–666. [CrossRef] [PubMed]
61. Sierro, N.; Battey, J.N.; Ouadi, S.; Bakaher, N.; Bovet, L.; Willig, A.; Goepfert, S.; Peitsch, M.C.; Ivanov, N.V. The tobacco genome sequence and its comparison with those of tomato and potato. *Nat. Commun.* **2014**, *5*, 3833. [CrossRef] [PubMed]
62. Jaillon, O.; Aury, J.M.; Noel, B.; Policriti, A.; Clepet, C.; Casagrande, A.; Choisne, N.; Aubourg, S.; Vitulo, N.; Jubin, C.; et al. The grapevine genome sequence suggests ancestral hexaploidization in major angiosperm phyla. *Nature* **2007**, *449*, 463–467. [PubMed]
63. Wang, W.; Haberer, G.; Gundlach, H.; Glasser, C.; Nussbaumer, T.; Luo, M.C.; Lomsadze, A.; Borodovsky, M.; Kerstetter, R.A.; Shanklin, J.; et al. The *Spirodela polyrhiza* genome reveals insights into its neotenous reduction fast growth and aquatic lifestyle. *Nat. Commun.* **2014**, *5*, 3311. [CrossRef] [PubMed]
64. Al-Mssallem, I.S.; Hu, S.; Zhang, X.; Lin, Q.; Liu, W.; Tan, J.; Yu, X.; Liu, J.; Pan, L.; Zhang, T.; et al. Genome sequence of the date palm *Phoenix dactylifera* L. *Nat. Commun.* **2013**, *4*, 2274. [CrossRef] [PubMed]
65. Goff, S.A.; Ricke, D.; Lan, T.H.; Presting, G.; Wang, R.; Dunn, M.; Glazebrook, J.; Sessions, A.; Oeller, P.; Varma, H.; et al. A draft sequence of the rice genome (*Oryza sativa* L. ssp. *japonica*). *Science* **2002**, *296*, 92–100. [CrossRef] [PubMed]
66. Zhang, J.; Chen, L.L.; Xing, F.; Kudrna, D.A.; Yao, W.; Copetti, D.; Mu, T.; Li, W.; Song, J.M.; Xie, W. Extensive sequence divergence between the reference genomes of two elite indica rice varieties zhenshan 97 and minghui 63. *Proc. Natl. Acad. Sci. USA* **2016**, *113*, E5163–E5171. [CrossRef] [PubMed]
67. Paterson, A.H.; Bowers, J.E.; Bruggmann, R.; Dubchak, I.; Grimwood, J.; Gundlach, H.; Haberer, G.; Hellsten, U.; Mitros, T.; Poliakov, A.; et al. The *Sorghum bicolor* genome and the diversification of grasses. *Nature* **2009**, *457*, 551–556. [CrossRef] [PubMed]
68. Schnable, P.S.; Ware, D.; Fulton, R.S.; Stein, J.C.; Wei, F.; Pasternak, S.; Liang, C.; Zhang, J.; Fulton, L.; Graves, T.A.; et al. The B73 maize genome: Complexity, diversity and dynamics. *Science* **2009**, *326*, 1112–1115. [CrossRef] [PubMed]
69. Project, A.G. The *Amborella* genome and the evolution of flowering plants. *Science* **2013**, *342*, 1241089. [CrossRef] [PubMed]
70. Izuno, A.; Hatakeyama, M.; Nishiyama, T.; Tamaki, I.; Shimizu-Inatsugi, R.; Sasaki, R.; Shimizu, K.K.; Isagi, Y. Genome sequencing of *Metrosideros polymorpha* (Myrtaceae), a dominant species in various habitats in the hawaiian islands with remarkable phenotypic variations. *J. Plant Res.* **2016**, *129*, 727–736. [CrossRef] [PubMed]
71. Rensing, S.A.; Lang, D.; Zimmer, A.D.; Terry, A.; Salamov, A.; Shapiro, H.; Nishiyama, T.; Perroud, P.F.; Lindquist, E.A.; Kamisugi, Y.; et al. The *Physcomitrella* genome reveals evolutionary insights into the conquest of land by plants. *Science* **2008**, *319*, 64–69. [CrossRef] [PubMed]
72. Qiu, F.; Ungerer, M.C. Genomic abundance and transcriptional activity of diverse gypsy and copia long terminal repeat retrotransposons in three wild sunflower species. *BMC Plant Biol.* **2018**, *18*, 6. [CrossRef] [PubMed]
73. Goremykin, V.V.; Lockhart, P.J.; Viola, R.; Velasco, R. The mitochondrial genome of *Malus domestica* and the import-driven hypothesis of mitochondrial genome expansion in seed plants. *Plant J.* **2012**, *71*, 615–626. [CrossRef] [PubMed]
74. Zhang, J.; Ruhlman, T.A.; Sabir, J.; Blazier, J.C.; Jansen, R.K. Coordinated rates of evolution between interacting plastid and nuclear genes in Geraniaceae. *Plant Cell* **2015**, *27*, 563–573. [CrossRef] [PubMed]
75. Sugiura, C.; Kobayashi, Y.; Aoki, S.; Sugita, C.; Sugita, M. Complete chloroplast DNA sequence of the moss *Physcomitrella patens*: Evidence for the loss and relocation of *rpoa* from the chloroplast to the nucleus. *Nucleic Acids Res.* **2003**, *31*, 5324–5331. [CrossRef] [PubMed]
76. Altschul, S.F.; Gish, W.; Miller, W.; Myers, E.W.; Lipman, D.J. Basic local alignment search tool. *J. Mol. Biol.* **1990**, *215*, 403–410. [CrossRef]
77. Hazkani-Covo, E.; Covo, S. Numt-mediated double-strand break repair mitigates deletions during primate genome evolution. *PLoS Genet.* **2008**, *4*, e1000237. [CrossRef] [PubMed]

78. Ju, Y.S.; Tubio, J.M.C.; Mifsud, W.; Fu, B.Y.; Davies, H.R.; Ramakrishna, M.; Li, Y.L.; Yates, L.; Gundem, G.; Tarpey, P.S.; et al. Frequent somatic transfer of mitochondrial DNA into the nuclear genome of human cancer cells. *Genome Res.* **2015**, *25*, 814–824. [CrossRef] [PubMed]
79. Tamura, K.; Peterson, D.; Peterson, N.; Stecher, G.; Nei, M.; Kumar, S. Mega5: Molecular evolutionary genetics analysis using maximum likelihood, evolutionary distance and maximum parsimony methods. *Mol. Biol. Evol.* **2011**, *28*, 2731–2739. [CrossRef] [PubMed]
80. Chen, Z.W.; Feng, K.; Grover, C.E.; Li, P.B.; Liu, F.; Wang, Y.M.; Xu, Q.; Shang, M.Z.; Zhou, Z.L.; Cai, X.Y.; et al. Chloroplast DNA structural variation, phylogeny and age of divergence among diploid cotton species. *PLoS ONE* **2016**, *11*, e0157183. [CrossRef] [PubMed]
81. Chen, Z.W.; Grover, C.E.; Li, P.B.; Wang, Y.M.; Nie, H.S.; Zhao, Y.P.; Wang, M.Y.; Liu, F.; Zhou, Z.L.; Wang, X.X.; et al. Molecular evolution of the plastid genome during diversification of the cotton genus. *Mol. Phylogenet. Evol.* **2017**, *112*, 268–276. [CrossRef] [PubMed]

© 2018 by the authors. Licensee MDPI, Basel, Switzerland. This article is an open access article distributed under the terms and conditions of the Creative Commons Attribution (CC BY) license (http://creativecommons.org/licenses/by/4.0/).

Article

Impairment of Meristem Proliferation in Plants Lacking the Mitochondrial Protease AtFTSH4

Alicja Dolzblasz [1,*], Edyta M. Gola [1], Katarzyna Sokołowska [1], Elwira Smakowska-Luzan [2,†], Adriana Twardawska [1] and Hanna Janska [2]

1. Faculty of Biological Sciences, Institute of Experimental Biology, Kanonia 6/8, 50-328 Wroclaw, Poland; edyta.gola@uwr.edu.pl (E.M.G.); katarzyna.sokolowska@uwr.edu.pl (K.S.); adriana.twardawska@uwr.edu.pl (A.T.)
2. Faculty of Biotechnology, University of Wroclaw, F. Joliot-Curie 14A, 50-383 Wroclaw, Poland; elwira.smakowska@gmi.oeaw.ac.at (E.S.-L.); hanna.janska@uwr.edu.pl (H.J.)
* Correspondence: alicja.dolzblasz@uwr.edu.pl; Tel.: +48-71-375-4094
† Present Address: Gregor Mendel Institute (GMI), Austrian Academy of Sciences, Vienna Biocenter (VBC), Dr Bohr-Gasse 3, 1030 Vienna, Austria.

Received: 13 February 2018; Accepted: 9 March 2018; Published: 14 March 2018

Abstract: Shoot and root apical meristems (SAM and RAM, respectively) are crucial to provide cells for growth and organogenesis and therefore need to be maintained throughout the life of a plant. However, plants lacking the mitochondrial protease AtFTSH4 exhibit an intriguing phenotype of precocious cessation of growth at both the shoot and root apices when grown at elevated temperatures. This is due to the accumulation of internal oxidative stress and progressive mitochondria dysfunction. To explore the impacts of the internal oxidative stress on SAM and RAM functioning, we study the expression of selected meristem-specific (*STM*, *CLV3*, *WOX5*) and cell cycle-related (e.g., *CYCB1*, *CYCD3;1*) genes at the level of the promoter activity and/or transcript abundance in wild-type and loss-of-function *ftsh4-1* mutant plants grown at 30 °C. In addition, we monitor cell cycle progression directly in apical meristems and analyze the responsiveness of SAM and RAM to plant hormones. We show that growth arrest in the *ftsh4-1* mutant is caused by cell cycle dysregulation in addition to the loss of stem cell identity. Both the SAM and RAM gradually lose their proliferative activity, but with different timing relative to *CYCB1* transcriptional activity (a marker of G2-M transition), which cannot be compensated by exogenous hormones.

Keywords: Arabidopsis; cell divisions; mitochondria; oxidative stress; root apical meristem; shoot apical meristem

---

1. Introduction

Plant growth and development are enabled by the activity of the shoot and root apical meristems (SAM and RAM, respectively). The continuous maintenance of stem cells in the SAM and RAM is facilitated by signaling from the organizing center (OC) and quiescent center (QC), respectively [1,2]. The main roles of the SAM and RAM are analogous, but both meristems differ in terms of internal organization and localization of growth and organogenic activity [3]. Fundamental to SAM/RAM self-perpetuation are sustained cell divisions, as the meristem shape and size need to be maintained while continuously providing cells for development. Consequently, plant growth strongly depends on precisely coordinated cell proliferation and differentiation within various subdomains of the SAM and RAM [4–7]. Meristem regulators must be accurately interpreted by the cell cycle machinery, which in turn feedbacks on the production of meristem regulators (e.g., [2]). In addition, cell cycle perturbations impair the response of the meristem to extrinsic signals including hormones and metabolic sugars [8–11].

Cell cycle regulation in higher plants depends primarily on cyclins (CYCs) and their interacting partners including several cyclin-dependent kinases (CDKs) [12]. CDKA1 is a major cell-cycle controlling CDK in *Arabidopsis*, needed for both G1-S and G2-M transitions [13,14]. The most extensively studied cell cycle control proteins in reference to proper SAM functioning are the regulators of G1-S transition, CYCD3s, which extend the mitotic window of the cells [15,16], and the regulators of G2-M transition, CYCB1;1 and CDKB2 [8,16]. Surprisingly, the modification of cell division rates does not often result in pronounced architectural changes in the plant body [11,17], but mostly affects cell number and size within the SAM. Concomitantly, SAM size is usually altered, with internal organization either affected severely, as occurs after downregulation or overexpression of B-type CDKs, or not affected, as is observed after overexpression of CYCD3 and downregulation of CDKA1 or CYCD3 [8,11,15,18]. Studies also suggest that reduced division rates in the SAM exert a negative effect on proliferation outside of the meristem [18], and that root and shoot meristems may rely on different cell cycle regulators—for example, CDKB2 seems to act on SAM but not RAM [8].

The most notable function of mitochondria is the generation of ATP through oxidative phosphorylation (OXPHOS), and consequently mitochondria are strongly associated with the production of reactive oxygen species (ROS) [19]. Furthermore, cell division is highly energetically demanding and related to increased oxygen consumption and concomitant ROS production [20]. Within the SAM and RAM, undisturbed mitochondria functioning is therefore especially valuable as proliferation is crucial for SAM and RAM functionality. Different cell division ratios are preserved in separate meristematic zones, which seems to be regulated by the cells redox status [7]. Importantly, the integration of hormone homeostasis, the expression of meristematic genes, and the cell division rate all seem to be orchestrated by redox homeostasis [20]. Remarkably, SAM is also unique in terms of being characterized by the presence of one large mitochondrion surrounding the nucleus, with only a few smaller mitochondria also present in the cell [21]. Changes in the architecture of mitochondria relate to cell cycle-dependent mixing of mitochondrial DNA, which, together with the requirements for proper delivery of ATP during cell proliferation, further emphasizes the uniqueness and vulnerability of meristematic cells.

One of the factors enabling the undisturbed maintenance of the SAM in specific growth conditions is AtFTSH4 [22], a mitochondrial metalloprotease with proteolytic and chaperone-like domains facing the intermembrane space (i-AAA) [23]. While comparable to *Arabidopsis* WT plants in standard growth conditions (long day photoperiod (LD), 22 °C) under short day photoperiod (SD), or LD with elevated temperature (30 °C), *ftsh4* mutants display striking phenotypic features in both vegetative and generative development. Loss-of-function *ftsh4* mutants form aberrant vegetative rosettes and shorter precociously terminating inflorescences, with an irregular pattern of side branches and flowers [22,24]. AtFTSH4 was found to be particularly important for SAM function around flowering time, protecting the stem cells against internal oxidative stress and maintaining the functionality of mitochondria [22]. Studies at the molecular level indicate the accumulation of oxidatively damaged proteins and other markers of oxidative stress in the loss-of-function *ftsh4* plants [22,24–26]. This accumulation is an intrinsic response to the disturbed functionality of OXPHOS complexes, ineffective removal of oxidized proteins arising from reduced ATP production, and altered mitochondrial dynamics causing restricted mitophagy [24]. Recently, it was also documented that AtFTSH4 can degrade oxidatively damaged proteins in isolated mitochondria [27]. In addition, mitochondria lacking AtFTSH4 have the enhanced capacity of preprotein import through TIM17:23-dependent pathway [28]. Furthermore, the loss of AtFTSH4 also influences processes outside mitochondria–ROS generated in the mitochondria of *ftsh4* plants interact with the phytohormone auxin and affect plant architecture [29], and it seems that AtFTSH4 regulate the expression of WRKY transcription factors that control salicylic acid synthesis and signaling in autophagy and senescence [30].

Taken together, there is a growing body of evidence that AtFTSH4 displays pleiotropic functions during plant growth and development primarily associated with internal oxidative stress [22]. In the present study, we focus on the role of AtFTSH4 in the premature termination of shoot and root meristems. We test the following hypotheses, that abrupt shoot and root growth cessation is associated with: (i) the disrupted expression of key SAM and RAM related genes; (ii) disrupted cell cycle progression; and (iii) the depletion of hormones; and that (iv) the termination of both SAM and RAM results from a similar underlying mechanism.

2. Results

We have previously shown that in standard growth conditions (long day photoperiod (LD), 22 °C) wild-type (WT) and *ftsh4-1* mutant plants are alike [22,24–26]. Therefore, in this study, we focus only on plants grown at 30 °C, the conditions which terminate meristem in *ftsh4-1* mutant but not in WT plants.

2.1. SAM of ftsh4-1 Mutants Display Reduced Proliferative Activity at 30 °C Prior to Flowering

In this study, we grew plants under LD conditions at 30 °C, which induces internal oxidative stress accumulation in *ftsh4-1* mutants and therefore the phenotype of precocious shoot and root termination [22]. To analyze the cellular basis of the *ftsh4-1* mutant's main inflorescence stem shortening (Figure 1a), we analyzed the cell number and size in the first internode of flowering plants. In *ftsh4-1* mutants, the final cell number fell by over 50% when compared to WT plants, while cell size was negligibly affected (Figure 1b), indicating the impairment of proliferation in the SAM of mutants upon flowering. This result prompted us to monitor cell cycle progression directly in the SAM. For that purpose, we took the advantage of the Click-iT® EdU Imaging Kit, which facilitates the visualization of cells in S phase. No difference was observed between juvenile WT and mutant plants, but in the adult vegetative and bolting stages of growth, the number of cycling cells in *ftsh4-1* plants gradually decreased compared to WT plants (Figure 1c,d). The decrease was related to both the total number and percentage (mitotic index) of S-phase cells within the SAM (Figure 1c).

As cell cycle cessation was evident even prior to flowering, we examined the expression of several genes related to G1-S and/or G2-M transitions in the shoot apices of juvenile and adult plants. Transcript levels for the major cell-cycle controlling kinase *CDKA1* were similar between the mutant and WT plants, but transcript levels for the kinase *CDKB2* and of two cyclins *CYCB1* and *CYCD3;1* (all three of which are documented to have an impact on meristem maintenance) were significantly reduced in mutants (Figure 1e). Importantly, the activity of the *pCYCB1* and *pCYCD3;1* promoters, driving the expression of the *GUS* reporter gene (β-glucuronidase; WT/*ftsh4-1*;*pCYCB1:GUS* and WT/*ftsh4-1*;*pCYCD3;1:GUS*), was detected in the shoot apical region, including the SAM and the youngest leaves, of juvenile vegetative plants, and was reduced in *ftsh4-1* mutants (Figure 1g). In line with these results, the expression of *AtSTM*, analyzed with use of qRT-PCR and a transgenic lines WT/*ftsh4-1*;*pSTM:GUS*, was reduced in the shoot apices of the juvenile *ftsh4-1* mutants in comparison to WT plants (Figure 1f,h). On the other hand, expression of the *CLV3* promoter (a stem cell marker) driving the expression of *GUS* gene (WT/*ftsh4-1*;*pCLV3:GUS*) in the SAM, was comparable in juvenile WT and mutant plants (Figure 1i) but became reduced and more diffuse in adult vegetative mutant plants (Figure 1j), confirming the strong impairment of SAM identity.

In the shoot apices of the flowering WT plants (grown at 30 °C), promoter activity was detected for cell cycle genes (*CYCB1*) in the SAM and the youngest flowers and for stem cell marker genes (*STM*) in the SAM as indicated by GUS reporter intensity and localization (Figure 1k,l), which is consistent with the literature data on plants grown under LD conditions at 22 °C (e.g., [31,32]). In the mutant grown at 30 °C, on the other hand, which formed short and irregularly branched stems, the analyzed genes were variably expressed within an inflorescence, with stronger, weaker, or even absent activity of the GUS in shoot apices. In addition, variation was present between plants, with the strength of the GUS signal not necessarily weaker in comparison to that in the WT plants (Figure 1k,l).

These findings indicate that the short and precociously terminated inflorescences in the *ftsh4-1* mutant are mostly the result of reduced cell number. The expression of the cell cycle related genes (*CYCB1*, *CYCD3;1*, and *CDKB2*) and meristematic gene (*STM*) are reduced in the juvenile phase, preceding the reduced proliferation activity and more deteriorated SAM identity in the adult vegetative plants (shown by reduced *CLV3* promoter activity). In addition to the arrest of cellular proliferation, after transition to flowering the *ftsh4-1* mutant plants are characterized by strongly dysregulated expression of cell cycle and meristem genes.

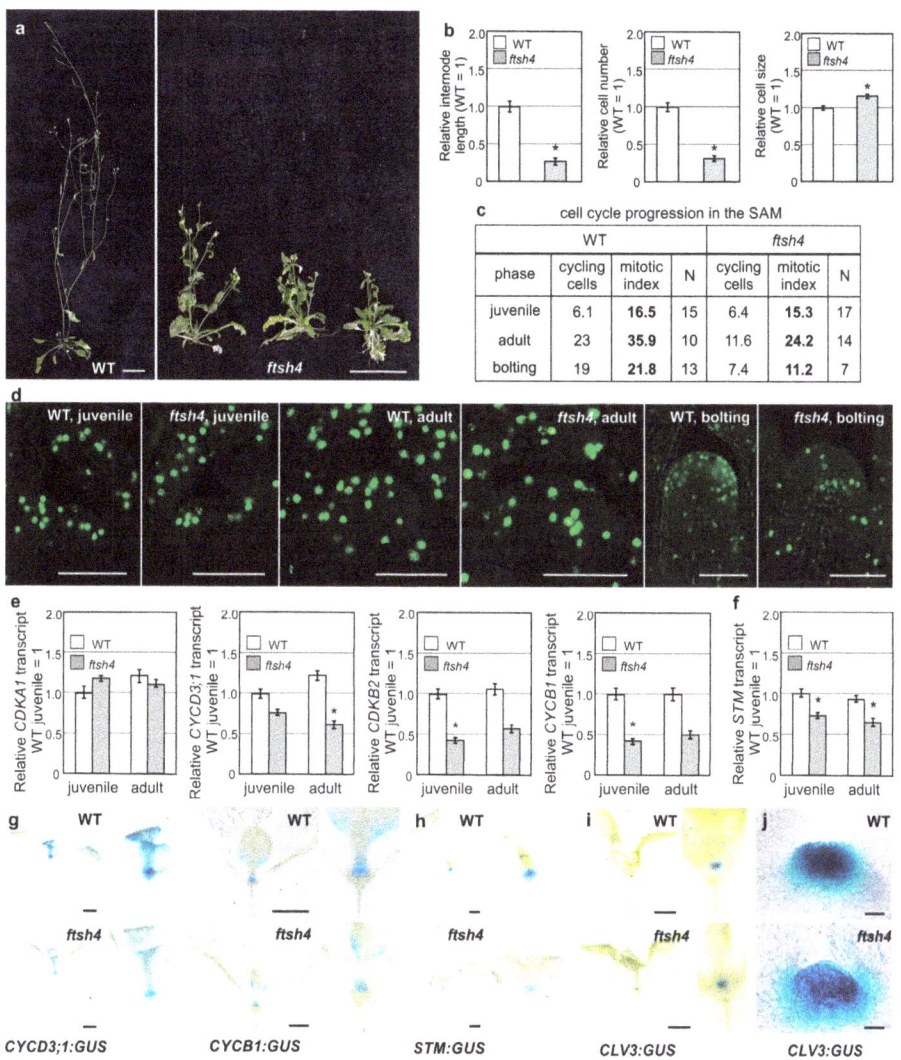

Figure 1. *Cont.*

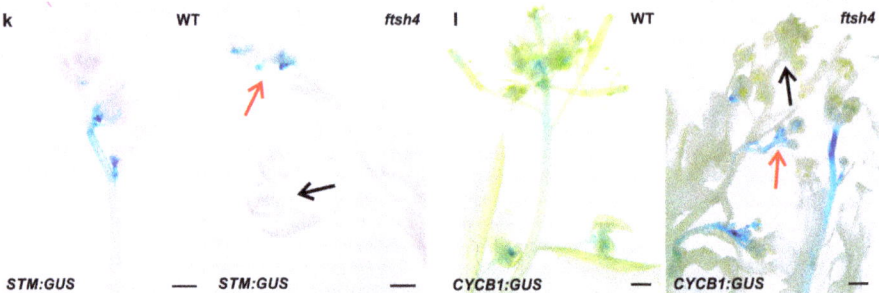

Figure 1. Impairment of proliferative activity and stem cell identity in shoots of *ftsh4-1* mutants grown at 30 °C. (**a**) Phenotypes of adult wild-type (WT) and *ftsh4-1* mutant plants grown under long day photoperiod (LD) conditions at 30 °C. The *ftsh4-1* mutants show a different degree of shortening of the main inflorescence stem and branching prior to growth cessation (note the drying rosette leaves). Scale bar: 20 mm; (**b**) lengths of the first internode and the number and size of cells in the first internode of WT and *ftsh4-1* mutant plants grown under LD conditions at 30 °C. The results are shown as relative to the WT samples (average values for WT plants are as follow: internode length 18.5 mm, cell number 54.3, elongation zone cells size 298.7 µm). Internode length and cell number strongly decreases, while cell size is less affected in the *ftsh4-1* mutants. Two biological replicates of the experiment were performed. The internode length and number of cells was measured in 10 plants of each genotype; average cell size was estimated from randomly measured 10 cells per internode. Mean values (±SE) are shown and significant differences between bars at $p < 0.05$ are denoted by asterisks; (**c,d**) comparison of S-phase progression (cell division) directly in the meristems of juvenile vegetative, adult vegetative and bolting wild-type and *ftsh4-1* mutants grown at 30 °C. The table shows the total number of cycling cells (i.e., in S-phase) and the mitotic index (percentage of cycling cells relative to the total number of SAM cells) in juvenile vegetative, adult vegetative and bolting wild-type and *ftsh4-1* mutant plants (**c**). In comparison to WT plants, a decrease in the number of S-phase cells (green signal) was detected only in adult and bolting *ftsh4-1* mutant plants. Scale bars: 50 µm (**d**); (**e**) comparison of At*CDKA1*, At*CYCD3;1*, At*CYCB1* and At*CDKB2;1* transcript levels analyzed in juvenile and adult vegetative shoot apices of WT and *ftsh4-1* mutant plants grown under LD conditions at 30 °C. Abundance of transcripts in each case is expressed relative to the juvenile WT tissue samples. Results from three experiments are shown; (**f**) the comparison of At*STM* transcript level analyzed in juvenile and adult vegetative shoot apices of WT and *ftsh4-1* mutant plants grown under LD conditions at 30 °C. Abundance of transcripts is expressed relative to the juvenile WT tissue samples. Results from three experiments are shown; (**g,h**) activity of the At*CYCD3;1* and At*CYCB1* (**g**) as well as At*STM* (**h**) promoters visualized by the activity of the GUS reporter protein (**blue**). Transgenic plants were grown under LD conditions at 30 °C and expression levels were analyzed during the juvenile vegetative stage of development of the WT (**upper panels**) and *ftsh4-1* mutants (**lower panels**). In each case, *ftsh4-1* mutants were characterized by weaker GUS activity. Scale bars: 3 mm; (**i,j**) activity of the At*CLV3* promoter visualized by the activity of the GUS reporter protein (**blue**). At 30 °C, in the juvenile SAM, GUS activity accumulates similarly in both genotypes (**i**), but in adult vegetative SAM (prepared with vibratome sections), *ftsh4-1* mutants are characterized by weaker and more diffusible GUS activity in comparison to the WT (**j**). Scale bars: 3 mm (**i**) and 20 µm (**j**); (**k,l**) activity of the At*STM* (**k**) and At*CYCB1* (**l**) promoters visualized by the activity of the GUS reporter protein (**blue**). Plants were grown under LD conditions at 30 °C and expression levels were analyzed during the generative stage of development. After transition to flowering, GUS activity in adult *ftsh4-1* mutant plants grown is variable (in terms of the strength of the signal) in the inflorescence apices (black arrows point exemplary shoot apices without GUS signal, red arrows—exemplary shoot apices with GUS signal). Scale bar: 15 mm.

2.2. Cessation of Root Growth in ftsh4-1 Mutants Is Related to Termination of the Cell Cycle

Proliferative activity was also analyzed in the other apical meristem, the RAM, as *ftsh4-1* mutant plants are also characterized by prematurely ceased root growth when grown at 30 °C [22]. In the root growth experiments, seedlings were germinated and grown for three days at the optimal temperature of 22 °C (S0 plants) and then transferred to 30 °C for additional three (S1 plants) and six days (S2 plants) to analyze the cumulative effect of elevated temperature. Root lengths, RAM sizes and RAM cell numbers were comparable between the WT and *ftsh4-1* mutant plants at the moment of the transfer to 30 °C (S0) and after three days of growth (S1). Interestingly, after a further three days at the elevated temperature (S2), the roots and meristems of the mutants were shorter and the number of cells in the RAM less than the WT plants (Figure 2a). In addition, at the S2 stage, cells within the elongation zone of mutant plants were shorter in comparison to WT plants (Figure 2a).

Next, we analyzed the proliferative activity of the RAM (as indicated by the number of cells in the S phase) with the fluorescent EdU kit. In WT plants at both analyzed time-points (S1 and S2), the proliferation was equally distributed throughout the RAM (Figure 2b) and maintained at a relatively high level as shown by the number and percentage (mitotic index) of cycling cells (77.7% in S1 and 88% in S2). On the contrary, the number and percentage of cycling cells were strongly reduced in the mutant plants at S1, with the proliferative activity concentrated closer to the root tip. Concomitantly, after subsequent three days at 30 °C (S2), cell division in the mutant plants had almost completely ceased (Figure 2b). Interestingly, the expression of pCYCB1:GUS was comparable between WT and *ftsh4-1* mutant plants in the S1 stage and only slightly weakened at S2 in some mutants (Figure 2c). On the other hand, the expression of the QC marker pWOX5:GUS was comparable between both genotypes at the S1 stage, but was noticeably weaker, more diffuse, or even almost completely gone in *ftsh4-1* mutant plants at S2 compared to the WT (Figure 2d).

In summary, after three days at 30 °C, root growth in the *ftsh4-1* mutant ceased, which is consistent with the observed limited cell cycle progression within the RAM, despite the maintenance of CYCB1:GUS expression. In addition, WOX5 expression weakened after six days, but not three days, of growth at 30 °C.

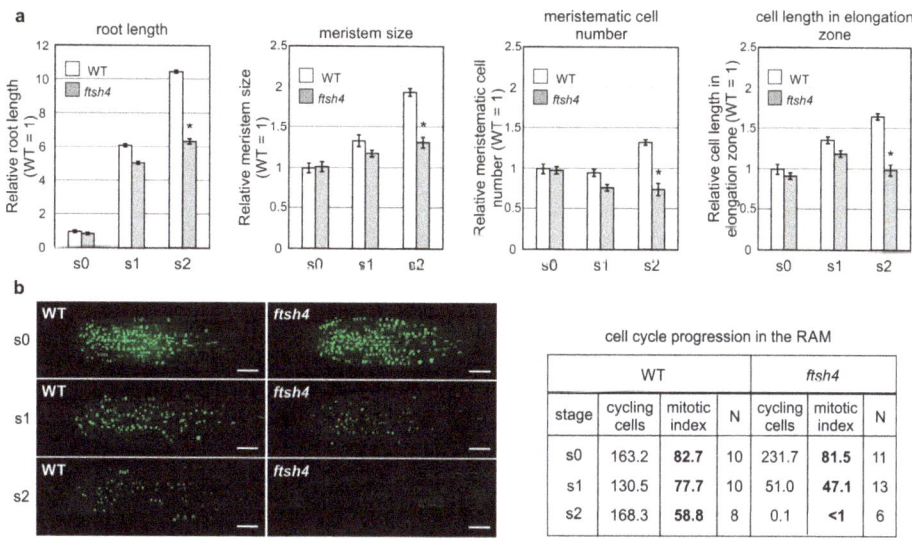

Figure 2. *Cont.*

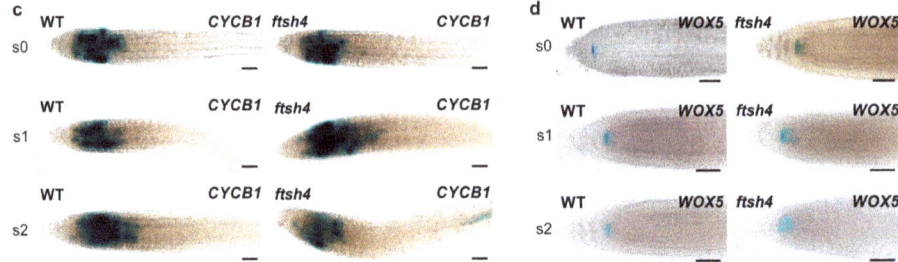

Figure 2. Impairment of proliferative activity and quiescence center (QC) cell identity in *ftsh4-1* mutant roots. (**a**) The length of the root and root apical meristem (RAM), the number of the RAM cells and the size of the cells in the elongation zone of wild-type (WT) and *ftsh4-1* mutant plants grown under long day photoperiod (LD) conditions at 30 °C. Seeds were germinated and grown for three days at 22 °C, and then transferred to 30 °C (S0) to continue growing for three days (S1) and six days (S2). Results are expressed relative to the WT S0 sample (average values for WT plants from S0 are as follow: root length 2.6 mm, meristem size 164 µm, meristematic cells number 14.4, cell length in elongation zone 127 µm). WT and mutant plants are comparable at S0 and S1 time-point, but then root and RAM size, RAM cell number and elongation cells size decreases in the *ftsh4-1* mutant S2 sample in comparison to the WT S2 sample. Two biological replicates of the experiment were performed. The root length, meristem length and meristematic cells number was measured in at least 10 plants of each genotypes; average size of cells in the elongation zone was estimated from randomly measured three cells per root. Mean values (±SE) are shown and significant differences between bars at $p < 0.05$ are denoted by asterisks; (**b**) comparison of S-phase progression (cell division) directly in the meristems of S0 (at the time of transfer), S1 and S2 RAM of wild-type and *ftsh4-1* mutants grown at 30 °C. A decreased in the number of S-phase cells (green signal) was detected in the S1 *ftsh4-1* mutant plants in comparison to WT plants. Scale bars: 40 µm. The table shows the total number of cycling cells (i.e., in S-phase) and the mitotic index (percentage of cycling cells relative to the total number of SAM cells) in S0, S1 and S2 wild-type and *ftsh4-1* mutant plants grown at 30 °C; (**c**,**d**) activity of the At*CYCB1* (**c**) and At*WOX5* (**d**) promoters visualized by the activity of the GUS reporter protein (blue). Transgenic plants were grown as described in (**a**) and expression levels were analyzed during the S0, S1 and S2 stages of development in the WT and *ftsh4-1* mutants. *ftsh4-1* mutants were characterized by comparable GUS activity for both promoters at S0 and S1 stages, only slightly weaker activity (not fully penetrant phenotype) in case of the At*CYCB1* promoter (**c**) and much weaker and diffusible in case of the At*WOX5* promoter (**d**) in the S2 stage. Scale bars: 40 µm.

2.3. Hormonal Insensitivity of the SAM and RAM of the ftsh4-1 Mutant

To test whether impaired SAM and RAM maintenance in plants lacking the At*FTSH4* gene is attributable to the deficiency of a particular hormone, we analyzed the impact of the exogenous application of hormones on shoot and root growth in plants grown at 30 °C.

When shoot growth was analyzed after treatment with auxin and cytokinin, no significant change in the WT plants or the mutant's short inflorescence was observed (Figures 3a and S1), while abscisic acid significantly reduced the inflorescence of both the mutant and WT plants (Figures 3a and S1). When gibberellic acid (GA$_3$) was applied, no change to WT plant's height was observed. Interestingly, the length of inflorescence in the *ftsh4-1* mutants increased, without significant increase in flower number (Figures 3a–c and S1). In addition, the application of GA$_3$ caused precocious flowering of both WT plants and *ftsh4-1* mutants, which occurred around six days earlier than in control plants (Figure S2).

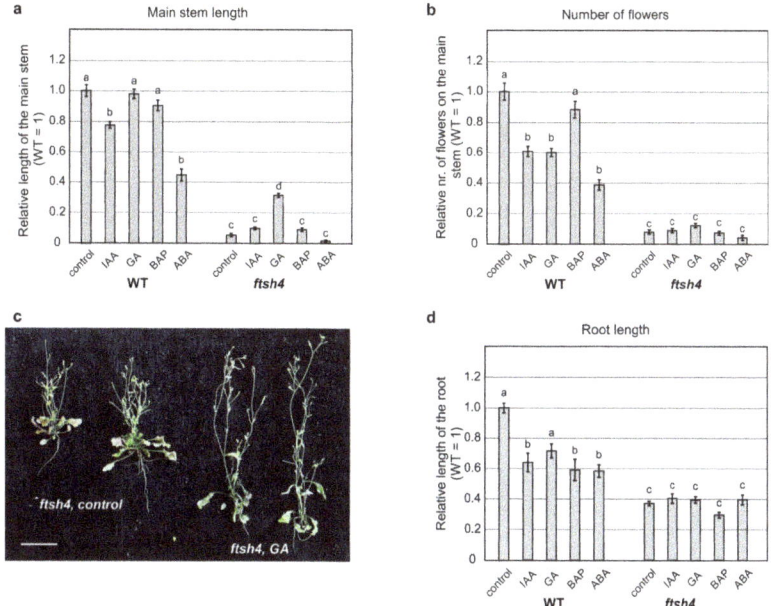

Figure 3. Lack of responsiveness to exogenous hormones in the shoots and roots of the *ftsh4-1* mutants grown at 30 °C. (**a,b**) Height of the main inflorescence (**a**) and the number of flowers on the main inflorescence (**b**) of wild-type (WT) and *ftsh4-1* mutant plants grown under long day photoperiod (LD) conditions at 30 °C. The results are expressed relative to the WT control sample (plants without exogenous hormone application). Average values for WT control plants are as follow: main stem length 35.6 mm, number of flowers 21. The main inflorescence height of the *ftsh4-1* mutants increases after gibberellic acid (GA$_3$) application (**a**), but the number of flowers is not significantly changed (**b**). Mean values (±SE) are shown; (**c**) the phenotype of adult (when rosette leaves are drying) *ftsh4-1* mutant plants grown under LD conditions at 30 °C. *ftsh4-1* mutants without exogenous hormone application are shown on the left, and those after gibberellic acid application are shown on the right. Scale bar: 20 mm; (**d**) the length of the main root of WT and *ftsh4-1* mutant plants, grown under LD conditions at 30 °C. Seeds were germinated and grown for three days at 22 °C, transferred to 30 °C to continue the growth for six days and then analyzed. The results are shown relative to the WT control sample (without exogenous hormone application). Mean values (±SE) are shown and significant differences in the WT and *ftsh4-1* mutant plants after various hormones application in comparison to their control plants (WT or *ftsh4-1*), at $p < 0.05$, are indicated with different letters: a and b between treated and not-treated WT plants; c and d between treated and not-treated *ftsh4-1* plants.

To test root growth, the same hormones mentioned above were applied after seed germination at 22 °C for three days and subsequent transfer to 30 °C. When growing at 30 °C, all hormones reduced the length of the main root in the WT plants (Figures 3d and S1). In contrast, *ftsh4-1* roots did not significantly respond at all to any of the hormones (Figures 3d and S1).

In summary, the size of the main inflorescence increased slightly in the *ftsh4-1* mutant after GA$_3$ application, with no increase in the organogenic activity of the SAM, while the length of the main root did not increase after application of the hormones.

3. Discussion

AtFTSH4 protease is essential for the maintenance of healthy mitochondria function and to counteract internal oxidative stress accumulation in plants growing under stress-inducing conditions.

It therefore plays an important role in the functioning of the whole plant. As previously reported, AtFTSH4 is particularly important for the maintenance of the meristematic identity [22]. This study highlights that the phenotype characterized by precocious cessation of shoot and root growth arises from accumulating defects in various processes occurring at the tissue and molecular levels, including reduced proliferative activity in the shoot and root apical meristems (SAM and RAM, respectively) and dysregulated expression in the SAM of meristematic genes and those related to cell cycle control (Figure 4).

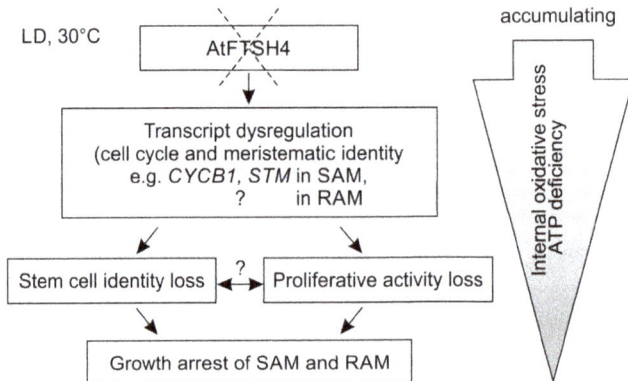

Figure 4. Developmental impairment of meristem proliferation and stem cell identity in plants lacking the mitochondrial protease AtFTSH4. In stress-inducing conditions of mildly elevated temperature of 30 °C, absence of AtFTSH4 in mitochondria (shown as dashed X) results in progressive accumulation of internal oxidative stress (ROS, carbonylated proteins, *AOX* transcripts) with time/plant age and ATP deficiency [22,24,25]. Stem cells within the meristems lose their meristematic characteristics and ability to proliferate. In the shoot apical meristem (SAM), this relates to the transcript dysregulation of cell cycle-related genes and those sustaining meristematic identity. Ultimately, growth of the SAM and RAM in the *ftsh4-1* mutant plants is precociously arrested. Question marks refer to probable, but not yet proven experimentally correlations.

Here, we showed that proliferation activity decreases with time/age in both meristems of mutant plants. In the SAM, the expression of *CYCB1*, *CYCD3;1* and *CDKB2* decreases in the *ftsh4-1* juvenile plants when compared to the wild-type (WT). At the same time point, cell cycle progression, visualized directly in the SAM, was not yet affected. Concomitantly, the proliferation rate gradually declines in the adult vegetative mutants, a phenomenon that continues after bolting. In addition, the expression of stem cell related genes, *STM* and *CLV3*, is reduced in the juvenile and adult stages, respectively. Similar to the SAM, growth cessation in the RAM was correlated to the gradually reduced proliferation. In this case, the termination of cell division was not preceded by the reduced level of *CYCB1* expression. The reduced expression of *CYCB1*, and also *WOX5* (which is a positive regulator of stem cell maintenance expressed in the quiescent center, QC), followed the loss of RAM proliferative activity and thus ongoing differentiation. We cannot, however, rule out the possibility that in the root a disturbance in the expression of genes other than *CYCB1* could drive the cessation of proliferation, as literature data suggest that shoots and roots rely on different cell cycle genes to integrate proliferation with meristem organization [8]. Moreover, most mitochondrial mutants affecting OXPHOS system are characterized by the phenotype of short roots and/or shoots, and that decreased growth rates can be linked to the reduced proliferation [33–35]. Thus, at least some aspects of SAM and RAM termination, as direct responses to accumulation of reactive oxygen species (ROS) and/or ATP deficiency, could be similar. It would be interesting in the future to compare in more detail the mechanisms behind SAM and RAM termination to decipher differences and similarities, also between the two meristems, across different mitochondria mutants affecting OXPHOS system.

These findings agree with our previous results showing that internal oxidative stress (ROS, carbonylated proteins, and *AOX* transcripts) and concomitant mitochondria dysfunction accumulate progressively [22]. The oxidative environment of the adult vegetative SAM thus causes a gradual loss of the SAM identity and proliferative ability. Nevertheless, upon transition to flowering, proliferative activity is preserved, though limited, and the stem cell character of the cells in the SAM is maintained sufficiently for generative development to progress. However, the inflorescence stems terminate precociously. In such plants, the expression of *CYCB1* and *STM* was not uniform, suggesting their significant dysregulation in mutant plants after or during flowering. The observed differential expression levels probably result from a varying degree of defects caused by accumulating internal oxidative stress, and reveals inherent variability in the fitness of individual meristems. From a whole plant perspective, such differences reflect a strong dysregulation of processes at multiple levels of organization.

Cell cycle and developmental programs need to be coordinated by hormonal signals, and hormonal outputs are coordinated by the cell cycle machinery [2,11]. Disrupted *CDKB2* function, for example, causes an elevation in auxin levels, a reduction in bioactive cytokinins, and an inability to properly interpret hormonal stimuli in a developmental output [8]; *CYCD3;1* is induced, among other stimuli, by sucrose and cytokinin [9,36,37]. Interestingly, two genes crucial for SAM maintenance, *STM* and *WUS*, are related to cytokinin [38,39], and maintenance of low mitotic activity in RAM QC cells was shown to depend on their highly oxidized status and proper auxin levels (which are dependent on redox homeostasis) [7]. The loss of apical dominance of the main inflorescence stem in *ftsh4* mutants was reported to be complemented with external auxin application [29]. However, the phenotype of the *ftsh4-1* mutant grown at 30 °C is rather pleiotropic and not purely the result of disturbed levels of or responses to cytokinin, auxin, or any other hormone. In our studies, only the application of gibberellic acid (GA_3) resulted in minor complementation of the shoot growth defects exhibited by *ftsh4-1* mutant plants grown under stress-inducing conditions. However, this might be simply the result of reduced accumulation of internal oxidative stress, and therefore less impact on SAM function, as these plants flowered earlier than the control *ftsh4-1* mutants. In addition, the roots of the *ftsh4-1* mutant were generally less responsive than WT to all exogenous hormones. These results support the assumption that the developmental defects observed in the *ftsh4-1* mutant induced by the accumulation of internal oxidative stress are multicomponent and cannot be explained by changes in any single process. At this stage, it is impossible to assess whether the driving force behind the impairment of proliferation is a disturbance in hormonal homeostasis interfering with cycling, or rather cell cycle related genes influencing hormonal homeostasis.

Our results show that growth arrest in the *ftsh4-1* mutant is caused not only by a loss of stem cell identity, but importantly, dysregulation of the cell cycle. SAM and RAM termination in plants lacking the mitochondrial protease AtFTSH4 is an outcome of progressively declining proliferation rates and stem cell maintenance due to progressive internal oxidative stress accumulation associated with ongoing mitochondria dysfunction. In a final attempt to survive, *ftsh4-1* mutants undergo transition to flowering, but form only defective inflorescences that fail to produce seeds and precociously cease development.

4. Materials and Methods

4.1. Plant Material and Growth Conditions

Arabidopsis thaliana (L.) Heynh. Columbia-0 (Col-0) was used as the wild type (WT) reference. The transgenic lines *ftsh4-1* (SALK_035107/TAIR) line was already previously characterized [25], and was originally obtained from the Salk Institute. Other transgenic lines used in this study were obtained: *pSTM:GUS* from Prof. W. Werr (University of Cologne, Cologne, Germany), *pCLV3:GUS* from Prof. T. Laux (University of Freiburg, Freiburg, Germany), *pCYCB1:GUS* from Prof. N. Dengler (University of Toronto, Toronto, ON, Canada), and *pCYCD3;1:GUS* from Prof. J. Murray (Cardiff School of Biosciences, Cardiff, UK). Plants were grown in a 16 h light/8 h dark (long day, LD) photoperiod at 22 °C and 30 °C. The transgenic lines used in that study were created by means of crossing the *ftsh4-1* mutant to the respective GUS reporter line. Concomitantly, double homozygous plants (homozygous for the GUS reporter and homozygous for wild-type or mutated allele of the At*FTSH4* gene) were selected. Hormones were exogenously applied at following concentrations: 100 µM indole-3-acetic acid (IAA), 100 µM gibberellic acid (GA_3), 1 µM 6-benzylaminopurine (BAP), and 10 µM abscisic acid (ABA) for the shoot activity; and 0.5 µM IAA, 50 µM GA, 0.5 µM BAP, and 10 µM ABA for root activity.

4.2. Histological Analyses

GUS gene activity, under the control of the At*STM*, At*CYCB1*, At*CYCD3;1*, and *pCLV3* promoters, was analyzed in *Arabidopsis* plants grown in LD at 30 °C at various developmental stages. Tissues were prefixed in ice-cold 90% (*v/v*) acetone, rinsed with 50 mM sodium phosphate buffer (pH 7.2) and then stained with 2 mM X-gluc (5-bromo-4-chloro-3-indolyl β-D-glucuronide cyclohexamine; Duchefa Biochemie, Haarlem, The Netherlands) for 3, 16, 16, and 7 h, respectively, in the dark, at 37 °C. Concomitantly, plants were treated with increasing ethanol concentrations, and fixed with the 50% ethanol-3.7% formaldehyde-5% acetic acid (FAA) solution or stored in ethanol. Longitudinal sections of the SAM (25 µm thick) were prepared on a vibratome (Leica VT 1200S; Leica Instruments GmbH, Wetzlar, Germany) using both juvenile and adult plants. Images were obtained using a stereomicroscope or light microscope (see Section 4.5). To observe cell number in the internodes, tissues from the *ftsh4-1* mutant and WT were dissected from fully adult flowering plants (around 20 plants for each genotype), fixed in FAA solution, hydrated in decreasing ethanol concentrations, and digested with 10% KOH for three days in 37 °C. The tissues were then rinsed extensively in water, dehydrated in increasing ethanol concentrations, and stained with nigrosine solution. The plant material was then documented using a stereomicroscope and an epi-fluorescence microscope (see Section 4.5).

4.3. Real-Time PCR

The shoot apices (SAM and youngest primordia) were hand dissected from at least 10 individual plants of each genotype, at the same time (9:00 a.m.), at the stage when the third or fifth leaf was visible (for juvenile and adult stages, respectively). Real-time PCR analyses were performed using a LightCycler480 (Roche, Penzberg, Germany) and Real-Time 2× PCR Master Mix SYBR version B (A&A Biotechnology, Gdynia, Poland) with a final primer concentration of 0.5 µM. Material from a pool of WT plants served as the calibrator, using the *PP2AA3* gene (At1g13320) as a reference. Amplification conditions comprised denaturation at 95 °C for 1 min followed by 45 cycles of amplification at 95 °C for 10 s, 55–65 °C (according to primer-specific annealing temperatures) for 10 s, and 72 °C for 20 s with single data acquisition, followed by cooling to 40 °C for 30 s. Melting curve analysis was performed as to test the specificity of the amplification products was. Real-time PCR analyses were performed on at least three independent biological replicates. The primers sequences are available upon request.

4.4. SAM in Vivo Fluorescent Analyses

Proliferation rates (S phase progression) were measured using a fluorescent Click-iT® EdU Imaging Kit (Thermo Fisher Scientific, Waltham, MA, USA) directly in the SAM and RAM. The nucleoside analog of thymidine, EdU (5-ethynyl-2′-deoxyuridine), was applied to the plant tissues for 1 h through the hypocotyl for the SAM and by submerging the roots in a solution in the case of RAM. EdU incorporated into DNA during DNA synthesis was labelled with Alexa Flour 488 according to the manufacturer's protocol. As a negative control, plants without the application of the Click-iT® reaction cocktail were analyzed. For SAM analysis, EdU was applied to juvenile vegetative, adult vegetative, and bolting (with the inflorescence stem showing early signs of growth) WT and *ftsh4-1* plants grown under LD at 30 °C. To analyze RAM, EdU was applied to the roots of six- and nine-day-old plants, which were first grown for three days at 22 °C and then transferred to 30 °C for three and six days to bypass defects related to germination. A confocal laser scanning microscope (CLSM, see Section 4.5) was used to detect the fluorescent signal of Alexa Flour 488 in at least 10 longitudinal median sections of individual SAM or whole mount roots. The mitotic index was calculated as the percentage of all cycling cells relative to the total number of the cells in the meristem.

4.5. Microscopy

The plant material and prepared slides were, after above-mentioned techniques, photographed by the following equipment: (a) a digital camera; (b) a stereomicroscope with a digital camera and DLTCam software (Delta Optical, Nowe Osiny, Poland); (c) an epi-fluorescence BX60 microscope with bright-field optics equipped with a digital camera DP73 and cellSens Entry software (Olympus Optical, Tokyo, Japan); and (d) an inverted CLSM (Fluo View100; Olympus Optical, Tokyo, Japan). Alexa Flour 488 was analyzed in CLSM using the excitation and emission wavelengths of 495 and 519 nm, respectively.

4.6. Statistical Analyses

The data did not have a normal distribution (Shapiro-Wilk test, $\alpha = 0.05$); thus, the significance of differences between independent groups was checked using Mann-Whitney U test. Statistical analyses were performed with Statistica 13 software (StatSoft; North Melbourne, Victoria, Australia).

Supplementary Materials: Supplementary materials can be found at http://www.mdpi.com/1422-0067/19/3/853/s1.

Acknowledgments: We thank the Editage service for English corrections. This work was supported by the National Science Centre (NCN) Grant No. 2012/07/D/NZ3/00501 to Alicja Dolzblasz and Grants No. 0401/0423/17 and No. 0401/0135/17. Publication cost was covered by Wrocław Center of Biotechnology program The Leading National Research Center (KNOW) for years 2014–2018.

Author Contributions: Alicja Dolzblasz and Hanna Janska formulated the research concept; Alicja Dolzblasz, Edyta M. Gola and Katarzyna Sokołowska performed all phenotyping and microscope-based analyses; Elwira Smakowska-Luzan performed the qRT-PCR; Alicja Dolzblasz and Adriana Twardawska performed hormonal analyses; Alicja Dolzblasz developed transgenic lines where *ftsh4-1* expressed GUS reporters; and Alicja Dolzblasz wrote the manuscript with help of Hanna Janska and Edyta M. Gola.

Conflicts of Interest: The authors declare no conflict of interest.

References

1. Aichinger, E.; Kornet, N.; Friedrich, T.; Laux, T. Plant Stem Cell Niches. *Annu. Rev. Plant Biol.* **2012**, *63*, 615–636. [CrossRef] [PubMed]
2. Gaillochet, C.; Lohmann, J.U. The Never-ending Story: From Pluripotency to Plant Developmental Plasticity. *Development* **2015**, *142*, 2237–2249. [CrossRef] [PubMed]
3. Evert, R.F.; Eichhorn, S.E. (Eds.) *Esau's Plant Anatomy: Meristems, Cells, and Tissues of the Plant Body: Their Structure, Function, and Development*, 3rd ed.; John Wiley & Sons: New York, NY, USA, 2007.
4. Den Boer, B.G.; Murray, J.A. Triggering the Cell Cycle in Plants. *Trends Cell Biol.* **2000**, *10*, 245–250. [CrossRef]
5. Donnelly, P.M.; Bonetta, D.; Tsukaya, H.; Dengler, R.E.; Dengler, N.G. Cell Cycling and Cell Enlargement in Developing Leaves of Arabidopsis. *Dev. Biol.* **1999**, *215*, 407–419. [CrossRef] [PubMed]
6. De Veylder, L.; Beeckman, T.; Beemster, G.T.S.; Krols, L.; Terras, F.; Landrieu, I.; Van Der Scheuren, E.; Maes, S.; Naudts, M.; Inzé, D. Functional Analysis of Cyclin-dependent Kinase Inhibitors of Arabidopsis. *Plant Cell* **2001**, *13*, 1653–1667. [CrossRef] [PubMed]
7. Jiang, K.; Ballinger, T.; Li, D.; Zhang, S.; Feldman, L. A Role for Mitochondria in the Establishment and Maintenance of the Maize Root Quiescent Center. *Plant Physiol.* **2006**, *140*, 1118–1125. [CrossRef] [PubMed]
8. Andersen, S.U.; Buechel, S.; Zhao, Z.; Ljung, K.; Novák, O.; Busch, W.; Schuster, C.; Lohmann, J.U. Requirement of B2-Type Cyclin-Dependent Kinases for Meristem Integrity in *Arabidopsis thaliana*. *Plant Cell* **2008**, *20*, 88–100. [CrossRef] [PubMed]

9. Riou-Khamlichi, C.; Menges, M.; Healy, J.M.S.; Murray, J.A.H. Sugar Control of the Plant Cell Cycle: Differential Regulation of Arabidopsis D-Type Cyclin Gene Expression. *Mol. Cell. Biol.* **2000**, *20*, 4513–4521. [CrossRef] [PubMed]
10. Skylar, A.; Wu, X. Regulation of Meristem Size by Cytokinin Signaling. *J. Integr. Plant Biol.* **2011**, *53*, 446–454. [CrossRef] [PubMed]
11. Dewitte, W.; Scofield, S.; Alcasabas, A.A.; Maughan, S.C.; Menges, M.; Braun, N.; Collins, C.; Nieuwland, J.; Prinsen, E.; Sundaresan, V.; et al. Arabidopsis CYCD3 D-type Cyclins Link Cell Proliferation and Endocycles and Are Rate-limiting for Cytokinin Responses. *Proc. Natl. Acad. Sci. USA* **2007**, *104*, 14537–14542. [CrossRef] [PubMed]
12. De Veylder, L.; Beeckman, T.; Inzé, D. The Ins and Outs of the Plant Cell Cycle. *Nat. Rev. Mol. Cell Biol.* **2007**, *8*, 655–665. [CrossRef] [PubMed]
13. Porceddu, A.; Stals, H.; Reichheld, J.-P.; Segers, G.; De Veylder, L.; De Barro, P.R.; Casteels, P.; Van Montagu, M. A Plant-specific Cyclin-dependent Kinase Is Involved in the Control of G_2/M Progression in Plants. *J. Biol. Chem.* **2001**, *276*, 36354–36360. [CrossRef] [PubMed]
14. Joubès, J.; De Schutter, K.; Verkest, A.; Inzé, D.; De Veylder, L. Conditional, Recombinase-mediated Expression of Genes in Plant Cell Cultures. *Plant J.* **2004**, *37*, 889–896. [CrossRef] [PubMed]
15. Boucheron, E.; Healy, J.H.S.; Bajon, C.; Sauvanet, A.; Rembur, J.; Noin, M.; Sekine, M.; Riou Khamlichi, C.; Murray, J.A.H.; Van Onckelen, H.; et al. Ectopic Expression of Arabidopsis CYCD2 and CYCD3 in Tobacco Has Distinct Effects on the Structural Organization of the Shoot Apical Meristem. *J. Exp. Bot.* **2005**, *56*, 123–134. [CrossRef] [PubMed]
16. Colon-Carmona, A.; You, R.; Haimovitch-Gal, T.; Doerner, P. Spatio-temporal Analysis of Mitotic Activity with a Labile Cyclin-GUS Fusion Protein. *Plant J.* **1999**, *20*, 503–508. [CrossRef] [PubMed]
17. Kaplan, D.R.; Hagemann, W. The Relationship of Cell and Organism in Vascular Plants. *Bioscience* **1991**, *41*, 693–703. [CrossRef]
18. Gaamouche, T.; Manes, C.-L.D.O.; Kwiatkowska, D.; Berckmans, B.; Koumproglou, R.; Maes, S.; Beeckman, T.; Vernoux, T.; Doonan, J.H.; Traas, J.; et al. Cyclin-dependent Kinase Activity Maintains the Shoot Apical Meristem Cells in an Undifferentiated State. *Plant J.* **2010**, *64*, 26–37. [CrossRef] [PubMed]
19. Dat, J.; Vandenabeele, S.; Vranová, E.; Van Montagu, M.; Inzé, D.; Van Breusegem, F. Dual Action of the Active Oxygen Species During Plant Stress Responses. *Cell. Mol. Life Sci.* **2000**, *57*, 779–795. [CrossRef] [PubMed]
20. Schippers, J.H.M.; Foyer, C.H.; van Dongen, J.T. Redox Regulation in Shoot Growth, SAM Maintenance and Flowering. *Curr. Opin. Plant Biol.* **2016**, *29*, 121–128. [CrossRef] [PubMed]
21. Seguí-Simarro, J.M.; Coronado, M.J.; Staehelin, L.A. The Mitochondrial Cycle of *Arabidopsis* Shoot Apical Meristem and Leaf Primordium Meristematic Cells Is Defined by a Perinuclear Tentaculate/Cage-like Mitochondrion. *Plant Physiol.* **2008**, *148*, 1380–1393. [CrossRef] [PubMed]
22. Dolzblasz, A.; Smakowska, E.; Gola, E.M.; Sokołowska, K.; Kicia, M.; Janska, H. The Mitochondrial Protease AtFTSH4 Safeguards Arabidopsis Shoot Apical Meristem Function. *Sci. Rep.* **2016**, *6*, 28315. [CrossRef] [PubMed]
23. Urantowka, A.; Knorpp, C.; Olczak, T.; Kolodziejczak, M.; Janska, H. Plant Mitochondria Contain at Least Two i-AAA-like Complexes. *Plant Mol. Biol.* **2005**, *59*, 239–252. [CrossRef] [PubMed]
24. Smakowska, E.; Skibior-Blaszczyk, R.; Czarna, M.; Kolodziejczak, M.; Kwasniak-Owczarek, M.; Parys, K.; Funk, C.; Janska, H. Lack of FTSH4 Protease Affects Protein Carbonylation, Mitochondrial Morphology, and Phospholipid Content in Mitochondria of Arabidopsis: New Insights into a Complex Interplay. *Plant Physiol.* **2016**, *171*, 2516–2535. [CrossRef] [PubMed]
25. Gibala, M.; Kicia, M.; Sakamoto, W.; Gola, E.M.; Kubrakiewicz, J.; Smakowska, E.; Janska, H. The Lack of Mitochondrial AtFtsH4 Protease Alters Arabidopsis Leaf Morphology at the Late Stage of Rosette Development under Short-Day Photoperiod. *Plant J.* **2009**, *59*, 685–699. [CrossRef] [PubMed]
26. Kicia, M.; Gola, E.M.; Janska, H. Mitochondrial Protease AtFtsH4 Protects Ageing Arabidopsis Rosettes against Oxidative Damage under Short-Day Photoperiod. *Plant Signal. Behav.* **2010**, *5*, 126–128. [CrossRef] [PubMed]
27. Opalińska, M.; Parys, K.; Janska, H. Identification of Physiological Substrates and Binding Partners of the Plant Mitochondrial Protease FTSH4 by the Trapping Approach. *Int. J. Mol. Sci.* **2017**, *18*, 2455. [CrossRef] [PubMed]

28. Opalińska, M.; Parys, K.; Murcha, M.W.; Janska, H. The Plant i-AAA Protease Controls the Turnover of an Essential Mitochondrial Protein Import Component. *J. Cell Sci.* **2018**, *131*, jcs200733. [CrossRef] [PubMed]
29. Zhang, S.; Wu, J.; Yuan, D.; Zhang, D.; Huang, Z.; Xiao, L.; Yang, C. Perturbation of Auxin Homeostasis Caused by Mitochondrial FtSH4 Gene-mediated Peroxidase Accumulation Regulates Arabidopsis Architecture. *Mol. Plant* **2014**, *7*, 856–873. [CrossRef] [PubMed]
30. Zhang, S.; Li, C.; Wang, R.; Chen, Y.; Shu, S.; Huang, R.; Zhang, D.; Xiao, S.; Yao, N.; Li, J.; et al. The Mitochondrial Protease FtSH4 Regulates Leaf Senescence via WRKY-dependent Salicylic Acid Signal. *Plant Physiol.* **2017**. [CrossRef] [PubMed]
31. Bulankova, P.; Akimcheva, S.; Fellner, N.; Riha, K. Identification of Arabidopsis Meiotic Cyclins Reveals Functional Diversification among Plant Cyclin Genes. *PLoS Genet.* **2013**, *9*, e1003508. [CrossRef] [PubMed]
32. Byrne, M.E.; Groover, A.T.; Fontana, J.R.; Martienssen, R.A. Phyllotactic Pattern and Stem Cell Fate Are Determined by the *Arabidopsis* Homeobox Gene *BELLRINGER*. *Development* **2003**, *130*, 3941–3950. [CrossRef] [PubMed]
33. Hsieh, W.-Y.; Liao, J.-C.; Hsieh, M.-H. Dysfunctional Mitochondria Regulate the Size of Root Apical Meristem and Leaf Development in Arabidopsis. *Plant Signal. Behav.* **2015**, *10*, e1071002. [CrossRef] [PubMed]
34. Kong, X.; Tian, H.; Yu, Q.; Zhang, F.; Wang, R.; Gao, S.; Xu, W.; Liu, J.; Shani, E.; Fu, C.; et al. PHB3 Maintains Root Stem Cell Niche Identity through ROS-Responsive AP2/ERF Transcription Factors in Arabidopsis. *Cell Rep.* **2018**, *22*, 1350–1363. [CrossRef] [PubMed]
35. Van Aken, O.; Pečenková, T.; Van De Cotte, B.; De Rycke, R.; Eeckhout, D.; Fromm, H.; De Jaeger, G.; Witters, E.; Beemster, G.T.S.; Inzé, D.; et al. Mitochondrial Type-I Prohibitins of *Arabidopsis thaliana* Are Required for Supporting Proficient Meristem Development. *Plant J.* **2007**, *52*, 850–864. [CrossRef] [PubMed]
36. Menges, M.; Murray, J.A.H. Synchronous *Arabidopsis* Suspension Cultures for Analysis of Cell-Cycle Gene Activity. *Plant J.* **2002**, *30*, 203–212. [CrossRef] [PubMed]
37. Oakenfull, E.A.; Riou-Khamlichi, C.; Murray, J.A.H. Plant D-type Cyclins and the Control of G1 Progression. *Philos. Trans. R. Soc. Lond. B Biol. Sci.* **2002**, *357*, 749–760. [CrossRef] [PubMed]
38. Leibfried, A.; To, J.P.; Busch, W.; Stehling, S.; Kehle, A.; Demar, M.; Kieber, J.J.; Lohmann, J.U. WUSCHEL Controls Meristem Function by Direct Regulation of Cytokinin-inducible Response Regulators. *Nature* **2005**, *438*, 1172–1175. [CrossRef] [PubMed]
39. Yanai, O.; Shani, E.; Dolezal, K.; Tarkows, P.; Sablowski, R.; Sandberg, G.; Samach, A.; Ori, N. Arabidopsis KNOXI Proteins Activate Cytokinin Biosynthesis. *Curr. Biol.* **2005**, *15*, 1566–1571. [CrossRef] [PubMed]

© 2018 by the authors. Licensee MDPI, Basel, Switzerland. This article is an open access article distributed under the terms and conditions of the Creative Commons Attribution (CC BY) license (http://creativecommons.org/licenses/by/4.0/).

Review

Plant Mitochondrial Inner Membrane Protein Insertion

Renuka Kolli [1], Jürgen Soll [1,2] and Chris Carrie [1,*]

1. Department of Biology I, Botany, Ludwig-Maximilians-Universität München, Großhaderner Strasse 2-4, D-82152 Planegg-Martinsried, Germany; renuka.kolli@biologie.uni-muenchen.de (R.K.); soll@lmu.de (J.S.)
2. Munich Center for Integrated Protein Science, CiPSM, Ludwig-Maximilians-Universität München, Feodor-Lynen-Strasse 25, D-81377 Munich, Germany
* Correspondence: christopher.carrie@lmu.de; Tel.: +49-89-2180-74-758

Received: 2 February 2018; Accepted: 22 February 2018; Published: 24 February 2018

Abstract: During the biogenesis of the mitochondrial inner membrane, most nuclear-encoded inner membrane proteins are laterally released into the membrane by the TIM23 and the TIM22 machinery during their import into mitochondria. A subset of nuclear-encoded mitochondrial inner membrane proteins and all the mitochondrial-encoded inner membrane proteins use the Oxa machinery—which is evolutionarily conserved from the endosymbiotic bacterial ancestor of mitochondria—for membrane insertion. Compared to the mitochondria from other eukaryotes, plant mitochondria have several unique features, such as a larger genome and a branched electron transport pathway, and are also involved in additional cellular functions such as photorespiration and stress perception. This review focuses on the unique aspects of plant mitochondrial inner membrane protein insertion machinery, which differs from that in yeast and humans, and includes a case study on the biogenesis of Cox2 in yeast, humans, two plant species, and an algal species to highlight lineage-specific similarities and differences. Interestingly, unlike mitochondria of other eukaryotes but similar to bacteria and chloroplasts, plant mitochondria appear to use the Tat machinery for membrane insertion of the Rieske Fe/S protein.

Keywords: plant mitochondria; membrane insertion; Oxa; twin-arginine translocation

1. Introduction

Mitochondria are essential organelles found in all eukaryotic cells. They are popularly known as the "powerhouse of the cell" for generating more than 90% of the cellular energy in the form of ATP. Additionally, they perform several vital cellular functions that include the synthesis of heme and iron-sulfur clusters, lipid metabolism, maintenance of calcium homeostasis, thermogenesis, innate immunity, and the activation of apoptosis [1]. Approximately 2 billion years ago, the ancestors of mitochondria were once free-living prokaryotes. The primordial eukaryote most likely arose from the symbiosis between a facultative anaerobic archaeon (host) and an α-proteobacterium (mitochondrial ancestor) [2]. Gradually over time, this single unicellular primitive eukaryote multiplied and evolved to form the wide range of complex multicellular life that we see today. Due to the monophyletic origin of mitochondria [3], the fundamental mitochondrial features and functions are conserved, despite the lineage-specific differences. Compared to mitochondria of other lineages, plant mitochondria have evolved several unique features: Plant mitochondrial genomes are larger and highly variable in size [4]. Their transcripts undergo extensive processing and editing [5,6]. They have a branched electron transport chain and the composition of the respiratory complexes is slightly different [7]. Moreover, plant mitochondria are involved in photorespiration and have evolved to coexist with chloroplasts, a second endosymbiotically-derived organelle.

The mitochondrial inner membrane (MIM) is extensively folded into cristae, which vastly increases the surface area [8]. Thus, the MIM is able to house more than 50% of the total mitochondrial proteins that are crucial for numerous aspects of mitochondrial functions, including oxidative phosphorylation, protein import, metabolite transport, maintenance of mitochondrial morphology, and nucleoid segregation [9,10]. Having descended from eubacterial ancestors, the MIM proteome is well-conserved across eukaryotic species; however, minor differences do occur [11]. A recent yeast mitochondrial proteomics approach assigned 245 integral and 236 peripheral MIM proteins that together represent 59% of the total mitochondrial proteome [12]. The protein composition of the plant MIM is likely to be roughly similar [13]. Respiratory complexes comprise 80% of the MIM proteins and are preferentially located in cristae membranes, whereas protein translocation complexes and carrier proteins are enriched in the inner boundary membrane [9]. The respiratory complexes I to V are involved in the main mitochondrial function of cellular energy generation. All of these complexes except for complex II are made up of a mosaic of both nuclear- and mitochondrial-encoded proteins. The respiratory complexes are not randomly distributed in the MIM but assemble into supramolecular structures [14–17]. While $I_1 + III_2$ is the most abundant supercomplex of plant mitochondria, other supercomplexes—$I_2 + III_4$, $III_2 + IV_{(1-2)}$, $I_1 + III_2 + IV_{(1-4)}$, and V_2—are found in lower abundance [18]. The MICOS (the mitochondrial contact site and cristae organizing system) complex, complex V dimer, and cardiolipin are crucial for cristae formation in the MIM [19–21].

Coordinated gene expression, insertion, and assembly of MIM proteins are essential for the survival of eukaryotic cells. This review provides an overview of presently known machineries and mechanisms for the insertion of proteins into plant MIM, with a particular focus on insertion from the matrix side. A case study comparing the biogenesis of Cox2 in yeast, humans, and three different plant species not only highlights some differences observed in plant mitochondria but also indicates that our understanding of biogenesis of plant MIM proteins is presently incomplete.

2. Mitochondrial Protein Import Machinery

During the course of eukaryotic evolution, most of the genes that were originally present in the genome of the α-proteobacterial endosymbiotic ancestor of mitochondria were gradually transferred to the host nucleus with only a very small number being retained in the genome of the mitochondria [22]. Consequently, the basic mitochondrial protein import machinery consisting of the core subunits had developed even before the divergence of eukaryotes into fungi, plants, and metazoans [23]. Later on, during the evolution of each species, a number of lineage-specific subunits were added while others were lost. Today, the vast majority of mitochondrial proteins are synthesized as precursors in the cytosol and are targeted to mitochondria via N-terminal mitochondrial-targeting sequences (Figure 1). At the mitochondrial outer membrane, the precursor proteins interact with receptor subunits of the translocase of the outer membrane (TOM) complex [24]. The TOM complex contains an aqueous pore through which all the nuclear-encoded mitochondrial proteins must pass in order to cross the outer membrane. After emerging from the TOM complex, the mitochondrial-targeting sequences interact with the TIM23 (translocase of the inner membrane) complex, which mediates the translocation of the precursor proteins across the inner membrane into the matrix in a membrane potential- and ATP-dependent manner [25–27] (Figure 1). In the matrix, the targeting sequences are proteolytically removed and the proteins are folded into their respective native structures [28]. For a more detailed description of the mitochondrial protein import machinery of yeast [29,30], humans [31,32], and plants [33], interested readers may refer to the recent reviews. While the majority of protein import components are conserved across lineages, in plants, the gene family members encoding the protein import components have expanded. For example, while yeast contain a single gene encoding each of Tim17, Tim23, and Tim22, all three of which belong to the PRAT (preprotein and amino acid transporters) family, the *Arabidopsis* genome contains 17 genes encoding different PRAT family members, of which ten are located in mitochondria, six in chloroplasts, and one is dual targeted [33].

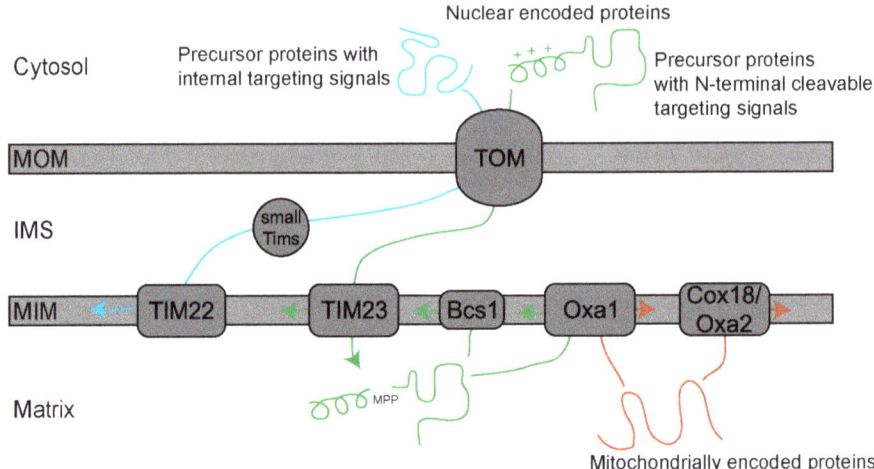

Figure 1. Mitochondrial protein transport and inner membrane insertion pathways. The MIM proteins can follow several routes for membrane insertion. There are the two translocase complexes in the inner membrane, TIM22 and TIM23, which laterally insert nuclear-encoded proteins into the MIM. Proteins with internal targeting signals are inserted into MIM by TIM22, while those with cleavable N-terminal targeting signals are inserted by TIM23. The conservatively sorted proteins are first targeted to the matrix by TIM23 and then inserted to the MIM in a manner reminiscent of bacterial inner membrane protein insertion. These conservatively sorted and mitochondrial-encoded proteins require Oxa1 or Cox18 (Oxa2) for membrane insertion. The newest member for conservative sorting is the Bcs1 protein for membrane insertion of the Rieske Fe/S protein. MOM = mitochondrial outer membrane, IMS = intermembrane space, MIM = mitochondrial inner membrane, MPP = mitochondrial processing peptidase, Oxa = oxidase assembly factor, TIM = translocase of the inner membrane.

3. Membrane Insertion of the MIM Proteins Synthesized in the Cytosol

Whereas all soluble matrix-localized proteins follow the route outlined above in the mitochondria of different organisms, MIM proteins have several different targeting and import pathways. First, the MIM proteins possessing N-terminal-targeting signals are arrested at the TIM23 complex, followed by lateral release into the membrane (Figure 1) [34]. So far, all proteins identified to be using this pathway contain only a single membrane-spanning transmembrane helix (TMH) [35]. The pore-forming subunit Tim23, its paralog Tim17, and the presequence receptor Tim50 form the essential catalytic core of the TIM23 complex [36]. A plant-specific C-terminal extension of *Arabidopsis* Tim17-2 is in contact with the outer membrane and is essential for protein import [37]. In yeast, the N-terminal extension of Tim23 links the inner and the outer membranes; however, all three isoforms of *Arabidopsis* Tim23 lack this domain [38]. Apart from being the subunits of TIM23, Tim23-2 and B14.7 are also present as part of the respiratory complex I and an inverse relationship between their abundance in the two complexes was shown to coordinate mitochondrial activity and biogenesis [39]. This appears to be a plant-specific mechanism. Interestingly, the presequence degrading peptidase of plants is dual targeted to both mitochondria and plastids to degrade mitochondrial-targeting sequences, as well as plastid-targeting signals [40,41].

The second group of MIM proteins can be inserted by the second TIM translocase, TIM22, also known as the carrier translocase as the majority of its substrates belong to the mitochondrial carrier protein family [42]. Carrier proteins are polytopic membrane proteins containing multiple hydrophobic internal signals and generally lacking N-terminal cleavable targeting signals [35]. They are transferred from TOM to the TIM22 via the heterohexameric chaperone complex that is composed of small

Tim proteins in the intermembrane space (IMS), which prevents aggregation or misfolding of the hydrophobic precursors [42] (Figure 1). In yeast, the TIM22 complex consists of the pore-forming Tim22 protein, Tim54 receptor, Sdh3, and Tim18 [43]. Since no plant homologs have been found for Tim54, Tim18, and Tim12 (a small Tim protein), which are essential for carrier import in yeast [44], it is highly likely that novel plant proteins perform similar functions. These additional components are yet to be identified. Upon activation of Tim22 by the membrane potential, substrate proteins are laterally released into the MIM (Figure 1). An exception to the general rule is that numerous plant carrier proteins possess a cleavable N-terminal-targeting sequence, despite displaying the typical tripartite structure and being homologous to other members of the carrier family [45]. These extensions appear to be non-essential for correctly targeting to the mitochondria. Nevertheless, they might be important for enhancing the import specificity and efficiency and might avoid mistargeting to chloroplasts [46]. After or during import, the extension is removed in a two-step process: the first processing by MPP (mitochondrial processing peptidase) and the second processing by a putative serine protease in the IMS [47].

The third group of MIM proteins displays a slightly more complicated sorting process. Firstly, they are translocated into the matrix and then integrated into the MIM from the matrix side of the membrane (Figure 1). This pathway is called the conservative sorting pathway because the direction of insertion is the same as that observed for bacterial membrane proteins [48]. While not just displaying a conserved direction, several aspects of this pathway also appear to be similar to that of bacteria, including the dependence on the membrane potential for insertion and adherence to the positive-inside rule for the charge distribution flanking the TMHs [49]. Since conservatively sorted proteins share similar pathways for membrane insertion as the mitochondrially encoded proteins, they will be discussed together in detail.

4. Membrane Insertion of MIM Proteins from the Matrix Side

4.1. Membrane Insertion of MIM Proteins Synthesized in the Matrix

The most central protein in the biogenesis of mitochondrial inner membrane proteins is Oxa1 (oxidase assembly factor 1) as it has been demonstrated to play critical roles in both co- and post-translational integration of MIM proteins [50]. Oxa1 was first identified in 1994 in a yeast mutant that failed to assemble cytochrome c oxidase (COX), resulting in a respiration-deficient phenotype [51,52]. Subsequent work has identified that Oxa1 is part of a larger family of proteins called the YidC/Oxa1/Alb3 family. The YidC/Oxa1/Alb3 family of membrane proteins plays a key role in the biogenesis of the respiratory chain complexes in bacteria (YidC) and mitochondria (Oxa1) and in the assembly of photosynthetic complexes in chloroplasts (Alb3). They insert proteins into membranes and help in protein folding and complex assembly [53,54]. Very recently, three members of this family—Get1 (guided entry of tail-anchored protein 1), EMC3 (Endoplasmic reticulum membrane complex 3), and TMCO1 (transmembrane and coiled coil domain-containing protein 1)—were identified to play roles in membrane insertion in the endoplasmic reticulum as well [55]. The conserved hydrophobic core domain consisting of five TMHs, which catalyzes membrane protein insertion, is present in each family member (Figure 2A). This domain can be functionally exchanged among different members of the protein family [56]. However, the soluble N- and C-terminal domains are variable and sometimes perform specialized functions (Figure 2A). Co-translational membrane insertion of substrate proteins is supported by the C-terminal ribosome-binding domain of the mitochondrial Oxa1, [57,58] whereas the cpSRP43-binding region at the C-terminus of the chloroplast Alb3 is crucial for targeting LHCPs (light-harvesting chlorophyll-binding proteins) to the thylakoid membrane [59] (Figure 2A).

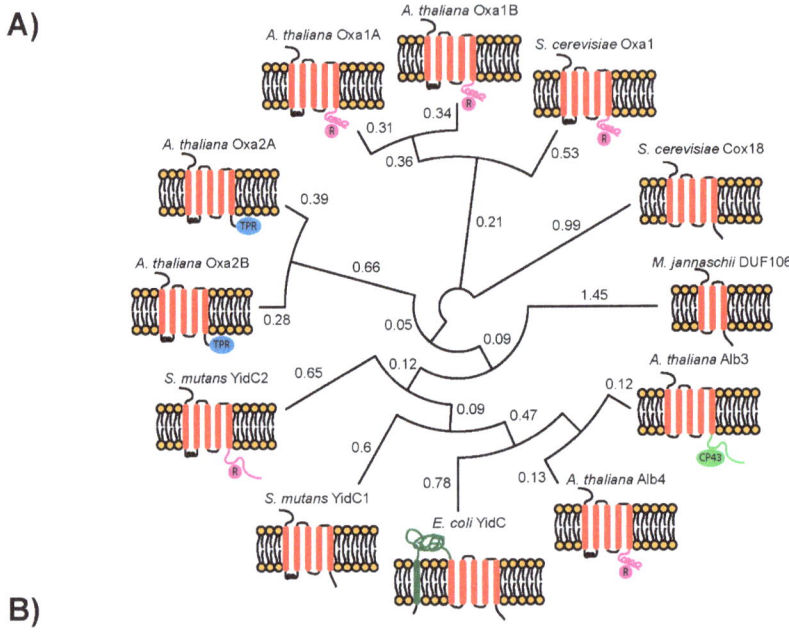

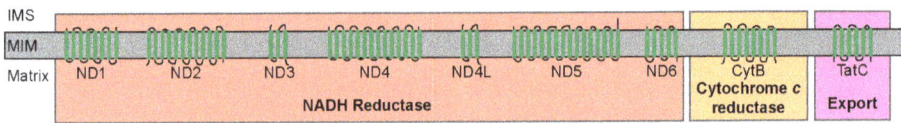

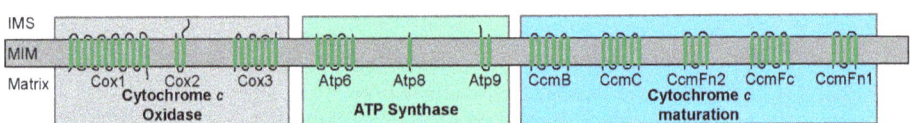

Figure 2. Phylogenetic and structural analysis of the YidC/Oxa1/Alb3 family and the mitochondrial-encoded membrane proteins of *Arabidopsis thaliana*. (**A**) A phylogenetic tree was generated by Clustal omega based on the full-length protein sequences for the indicated species of the YidC/Oxa1/Alb3 family [60]. The numbers next to each node represent the measure of support for the node. The conserved five TMH secondary structures are found in all members (except archaea) of the protein family. Differences at the C-terminal ends are indicated by different colors: pink = ribosome-binding, blue = TPR domain, green = CP43-interacting. (**B**) Transmembrane topologies of the 20 putative MIM proteins encoded in the mitochondrial genome of *Arabidopsis thaliana*. The proteins are displayed in an N- to C-terminus orientation going from left to right.

Usually, all organisms possess at least two homologs of the YidC/Oxa1/Alb3 family, performing non-redundant functions. Exceptionally, the Gram-negative bacteria have only a single YidC protein with an extra TMH0 near the N-terminus, an extended periplasmic loop between TMH0 and TMH1, and a unique cytoplasmic coiled-coil region between TMH1 and TMH2 [61] (*E. coli* YidC in Figure 2A).

YidC catalyzes the membrane insertion of several substrate proteins with one or two TMHs and lacking highly charged hydrophilic domains in the periplasmic side. YidC undergoes a conformational change upon co-translational substrate insertion [62]. The recent X-ray structures of YidC from *B. halodurans* and *E. coli* revealed a hydrophilic groove, which is open from the cytoplasm and the lipid bilayer but is closed from the periplasmic side. Moreover, the relatively shorter TMH3, 4, and 5 cause membrane thinning around YidC. Upon interaction with YidC in the groove region, the TMHs of the substrate slide into the lipid bilayer [63,64]. It is currently hypothesized that the mechanism of protein insertion by Oxa and Alb proteins is similar to YidC since they can be modeled on the known structure of YidC. YidC not only acts independently but also cooperates with SecYEG to facilitate the correct membrane integration and folding of membrane proteins [65]. Recently, it was shown that YidC contacts the interior of the SecY channel to form a ligand-activated and voltage-dependent complex [66]. Upon substrate binding, it facilitates the partitioning of nascent membrane proteins into the lipid environment by reducing the hydrophobicity of the SecY lateral gate [66]. Most Gram-positive bacteria have two YidC paralogs. In *Streptococcus mutans*, YidC2 has a ribosome-binding domain while YidC1 lacks it [67] (Figure 2A). The substrate profile of the bacterial YidC appears to be much larger than that of its organellar counterparts, Oxa1 and Alb3. These include the subunits a, b, and c of ATP synthase, TssL (a tail-anchored protein), and CyoA (cytochrome bo_3 oxidase subunit 2). With LacY (lactose permease) and MalF (maltose transporter), YidC also functions as a chaperone to assist in the folding and stability of the nascent polypeptide [68,69]. Similar to the bacterial YidC, the chloroplast Alb3 may also cooperate with cpSecY to perform co-translational membrane insertion of proteins into the thylakoid membrane [70], apart from its role in the post-translational insertion of LHCPs [71]. Its paralog Alb4 is involved in the assembly and stability of ATP synthase [72,73]. In *Chlamydomonas reinhardtii*, while the insertion of D1, a core subunit of photosystem II (PSII), is mediated by cpSecY, its assembly into PSII is strictly dependent on Alb3.1 [74]. Moreover, its paralog Alb3.2 is also exclusively involved in the assembly of PSI and II [75].

In mitochondria, Oxa1 is involved in the co-translational membrane insertion of polytopic membrane proteins, such as Cox2, into the MIM [57]. In cooperation with a peripheral membrane protein, Mba1 (multi-copy bypass of AFG3), the positively charged coiled-coil domain at the Oxa1 C-terminus interacts with the negatively charged 21S RNA of the large subunit of the ribosome, which is in proximity to the polypeptide exit tunnel [76,77] (Figure 2A). Thus, Oxa1 is able to directly contact the nascent polypeptide very early during the translation and inserts it efficiently into the membrane. The matrix-exposed loop between the TMHs 1 and 2 may also contribute to ribosome binding. The Oxa1–ribosome interaction not only promotes co-translational insertion but is also critical for the assembly of the COX subunits [78]. Similarly, during biogenesis of the F_o part of ATP synthase, although Oxa1 is not involved in the insertion of Atp9 into the MIM, it is required for the assembly of Atp9 with Atp6 [79]. Thus, Oxa1 can also function in a chaperone-like manner in addition to performing membrane insertion. While Oxa1 is conserved and Oxa1 proteins from different organisms are exchangeable, their specific functions can differ slightly: The yeast Oxa1 is involved in the biogenesis of complexes IV and V, whereas the human Oxa1 is required for the biogenesis of complexes I and V [50,80,81]. In contrast, Oxa1 depletion affects the levels of complexes I and IV in *Neurospora crassa* [82]. The paralog of Oxa1, Cox18 (Oxa2), lacks the ribosome-binding coiled-coil domain. Hence, with respect to the presence or absence of the ribosome-binding domain, the mitochondrial Oxa1 and Cox18 resemble the Gram-positive YidC2 and YidC1, respectively [67] (Figure 2A). Nevertheless, unlike other family members with several known substrate proteins, for Cox18, Cox2 is the only known substrate [83]. Cox18 performs a post-translational role in efficient topogenesis and stability of Cox2 during the assembly of COX.

In contrast to yeast and humans, due to independent gene duplications, there are four Oxa proteins in *Arabidopsis thaliana*: Oxa1a, Oxa1b, Oxa2a, and Oxa2b. Oxa1a and Oxa1b possess a coiled-coil domain similar to yeast Oxa1 (Figure 2A) [84]. Interestingly, Oxa2a and Oxa2b possess a tetratricopeptide repeat (TPR) domain consisting of four TPR motifs near the C-terminus, which is a plant-specific feature and is not found in any other known members of the protein family [85] (Figure 2A). TPR domains are involved in protein–protein interactions in a variety of cellular proteins. For example, the TPR domain of the TOM receptor, Tom70, interacts with the cytosolic chaperones Hsc70/Hsp70 and Hsp90, which guide precursor proteins [86]. This interaction is crucial for mitochondrial precursor targeting and translocation. Along these lines, it can be speculated that the TPR domains of Oxa2a and Oxa2b perhaps bind the matrix chaperones for facilitating the membrane protein insertion and/or help in complex assembly. Except for Oxa1b, the other three Oxa proteins are essential genes in *Arabidopsis*. This indicates that Oxa1a, Oxa2a, and Oxa2b play vital non-redundant roles in MIM protein biogenesis in all stages of plant development, while Oxa1b is probably not so important for normal plant growth and physiology [84].

In terms of plant Oxa proteins, two questions are immediately obvious: The first is why do plants have four Oxa proteins when most other organisms including humans and yeast have only two? Secondly, what is the role of the TPR domains found in the plant Oxa2 proteins? The answer to the first question may be just that plant mitochondria contain many different substrates compared to yeast and humans. This idea is supported by the fact that many plant mitochondrial genomes encode for more membrane proteins than those of yeast and humans [87]. For example, the *Arabidopsis* mitochondrial genome encodes for 20 putative membrane proteins displaying a wide range of membrane topologies (Figure 2B), whereas yeast and human mitochondrial genomes encode seven and thirteen membrane proteins, respectively. Although the major plant respiratory chain protein complexes are fairly well characterized and structurally resemble their counterparts in fungi and mammals, the electron transport chain of plant mitochondria differs significantly due to the presence of several plant-specific alternative oxidoreductases such as alternative oxidase (AOX) and rotenone-insensitive NAD(P)H dehydrogenases [88,89]. Furthermore, some of the respiratory complexes of plant mitochondria have numerous extra subunits that are not found in those of bacteria, fungi, and animals. To date, the functions of only some of them have been revealed. Complex I of plants is especially large, comprising of approximately 50 subunits [90]. In contrast to complex I of bacteria or other eukaryotic lineages, plant complex I has an additional spherical domain attached to the membrane arm on the matrix side, which includes five carbonic anhydrase-like proteins [91,92]. It was recently shown that these proteins are essential for complex I assembly and specifically influence central mitochondrial metabolism [93]. Nine more plant-specific subunits of complex I identified need further investigation. Complex II of flowering plants includes four additional subunits termed Sdh5, Sdh6, Sdh7, and Sdh8 [94]. While Sdh6 and Sdh7 represent the missing segments of plant Sdh3 and Sdh4, respectively, the functions of Sdh5 and Sdh8 are currently unknown [95]. Plant complex IV probably contains at least six additional putative plant-specific subunits that need further confirmation [94]. Similarly, plant complex V also includes several plant-specific subunits [96]. Any of the abovementioned plant-specific proteins of the respiratory chain could require either more diverse Oxa proteins or possibly require the plant-specific TPR domain for proper sorting and insertion.

In trying to answer the second question above, a specific role for the TPR domains found in Oxa2a and Oxa2b is hard to predict as these are the only known YidC/Oxa1/Alb3 family proteins to contain a TPR domain. As stated before, it is possible that they could interact with the chaperones of the matrix; however, since the mitochondrial Hsp70 lacks the classic EEVD motif required for TPR interactions, this hypothesis is weakened [97]. So far, in yeast and humans the only known substrate for Oxa2 (Cox18) is Cox2. More specifically, Oxa2 is required for the insertion of the second TMH of Cox2 and translocation of the C-terminus from the matrix to the IMS [83]. It is highly likely that either Oxa2a or Oxa2b or both of them play a similar role in plants. However, since both are essential, it is unlikely they have overlapping functions. The role of the TPR domain in Cox2 biogenesis could be similar to that of Mss2 (mitochondrial splicing system-related protein), which is a membrane-associated TPR-containing protein that cooperates with Cox18 during Cox2 biogenesis in yeast [98]. It is possible that plants have just formed a single protein of Cox18 and Mss2 via gene fusion. However, there is no significant similarity between the TPR domains of Oxa2 proteins and Mss2. Interestingly, no TPR protein has to date been found to be required for human Cox2 biogenesis. Considering that either Oxa2a or Oxa2b is possibly required for Cox2 biogenesis, then what is the other doing? Only future functional studies will be able to answer this question.

4.2. Membrane Insertion of MIM Proteins Imported into the Matrix

4.2.1. Oxa-Dependent Insertion

Oxa1 functions not only in the insertion of mitochondrial-encoded proteins but also in promoting the proper insertion of a variety of nuclear-encoded proteins containing multiple TMHs. While the role of Oxa1 in the insertion of mitochondrial-encoded proteins has been extensively studied, more recently there has been a focus on its role in the conservative sorting of nuclear-encoded MIM proteins. For the MIM insertion of nuclear-encoded proteins with a complex membrane topology, TIM23 and Oxa1 have to cooperate [99]. After translocation into the matrix via TIM23, some precursor proteins are inserted into the MIM with the help of Oxa1. Interestingly, many of the proteins sorted in this manner have bacterial homologs [48,100]. Oxa1 is required for its own biogenesis based on this mechanism [101]. In other cases, only some TMHs of the precursor proteins are laterally sorted by TIM23 while the remaining are first translocated into the matrix and then exported into the membrane by Oxa1. For instance, during the biogenesis of Mdl1, TIM23 laterally inserts the first pair of TMHs into the MIM while the subsequent pair is translocated into the matrix by TIM23 for MIM insertion by Oxa1 [102]. Generally, laterally inserted TMHs are more hydrophobic than those that are first translocated into the matrix. Another example is the biogenesis of Sdh4 in yeast. Sdh4 contains three TMHs, the first two are translocated through TIM23 into the matrix and exported by Oxa1, whereas the third is arrested in the TIM23 complex and laterally inserted into the MIM [103]. Interestingly, Sdh4 of flowering plants is truncated and only contains the third TMH found in yeast [104]. However, the additional subunit Sdh7 resembles the missing part and contains the two missing TMHs [95]. Thus, it can be speculated that plants simplified the complicated insertion process of Sdh4 by splitting it into two parts so that one part can follow the TIM23-dependent Oxa1 insertion while the other can be laterally inserted by TIM23. Since the two subunits of the carrier translocase Sdh3 and Tim18 also use the TIM23-Oxa1 pathway, depletion of Oxa1 leads to defects in the biogenesis of numerous carrier proteins, apart from a broad spectrum of other MIM proteins [105].

4.2.2. Oxa-Independent Insertion Using Bcs1 and Tat

The most recent sorting pathway to be identified in mitochondria is the Bcs1 pathway for the insertion of the Rieske Fe/S protein of complex III into the inner membrane [106,107]. The Rieske Fe/S protein undergoes a unique maturation and membrane insertion pathway and at this time is the only protein known to use this route. Similar to many other nuclear-encoded mitochondrial proteins, the Rieske Fe/S protein contains a cleavable N-terminal-targeting peptide and is translocated across the outer membrane through the TOM complex and through the inner membrane via the TIM23 complex [107]. Even though the Rieske Fe/S protein is a membrane protein, it is fully translocated into the matrix and during its maturation has a soluble intermediate [108]. Once inside the matrix, the N-terminal-targeting signal is cleaved off in two steps [108]. Then, the C-terminus is folded and the Fe/S cluster is inserted [106]. After C-terminal folding, the Rieske Fe/S protein is inserted into the MIM by Bcs1 (cytochrome bc_1 synthesis) so that it can reach its final topology of N-in and C-out [106]. Therefore, uniquely for mitochondria, the previously folded C-terminus of the Rieske Fe/S protein must be translocated back through the inner membrane into the IMS. This is in stark contrast to other conservative pathways where substrates remain in an unfolded state during membrane insertion.

Bcs1 is an AAA-type ATPase protein and has been demonstrated to be the only protein required to fully translocate the folded C-terminus of the Rieske Fe/S protein across the MIM. While this may be true in yeast and mammalian mitochondria, it has recently been demonstrated that the closest related protein in plants is located in the outer membrane—not the inner membrane—and plays no role in complex III biogenesis [109]. More recent work has sought to determine if plant mitochondria use a different pathway for Rieske Fe/S membrane insertion. Here, it could be demonstrated that plant mitochondria most likely use a twin-arginine translocation (Tat) pathway in place of the Bcs1 pathway [110]. This is interesting because in bacteria and chloroplasts a Tat pathway is also utilized for the insertion of Rieske Fe/S proteins into the cytochrome bc_1 complex and the cytochrome b_6f complex, respectively [111]. The evidence for plant mitochondria using a Tat pathway are that plant mitochondria contain two subunits of a potential Tat pathway (TatB and TatC) and that modification of the predicted Tat signal within the plant Rieske Fe/S protein blocks the assembly into complex III [110]. It was further shown that knockouts of the mitochondrial TatB were lethal in *Arabidopsis*. However, several questions still remain unanswered: First, where is the mitochondrial TatA? The majority of Tat pathways require TatA, TatB, and TatC proteins [111]. So far, only mitochondrial TatB and TatC proteins have been identified. Therefore, either the plant mitochondrial Tat pathway is highly unusual and functions without a TatA protein or TatA is yet to be identified. Second, why did plants retain a Tat pathway when it appears to have been lost in most other lineages (e.g., yeast, mammals, and some green algal species)? Are there more so-far unidentified substrates? Third, why do some species, such as *Chlamydomonas*, appear to lack both a Tat pathway and Bsc1? How is the Rieske Fe/S protein inserted in this situation? Are the Tat components from chloroplasts dual targeted to mitochondria or is there a completely different novel pathway for topogenesis of the Fe/S proteins in these organisms? The recent discovery that plant mitochondria contain a Tat pathway has opened up new and exciting areas of research and it will be fascinating to see how it develops in the future.

5. Case Study: A Comparison of Cox2 Biogenesis in Different Organisms

COX is the terminal enzyme in the respiratory electron transport chain. It catalyzes the transfer of electrons from soluble cytochrome *c* to molecular oxygen, coupled to protons getting pumped from the matrix to the IMS. The hydrophobic reaction center is made up of three mitochondrially-encoded subunits (Cox1, Cox2, and Cox3) and is surrounded by 11–13 nuclear-encoded subunits [112,113]. Post-assembly, COX forms a homodimer with cardiolipin as the connecting molecule, which is necessary for its activity and stability [114,115]. Electrons are transferred from the bivalent Cu_A site in Cox2 to heme *a* in Cox1 and then transferred intra-molecularly to the heme a_3-Cu_B site where oxygen is bound [116]. Cox3 possibly modulates oxygen access or coordinates proton pumping. The nuclear-encoded subunits surrounding the core are required for assembly, stability, and dimerization of the enzyme, protection from the ROS, and regulation of the catalytic activity. Additionally, approx. 30 more nuclear-encoded ancillary factors are required for translational regulation, membrane integration, heme and copper insertion, and assembly of the various COX subunits [113]. Studies with yeast and human cell lines have identified several of these factors, most of which are conserved, but relatively very little information regarding their plant counterparts is currently available.

Cox2 is an integral membrane protein containing two TMHs and both its N- and C-terminal regions are present in the IMS. The C-terminal region has a copper-binding Cx_3C motif, which upon maturation forms the Cu_A site. Generally, the mitochondrial-encoded membrane proteins have small hydrophilic regions and especially the IMS-exposed stretches are very short, except for Cox2 that has a large IMS-exposed hydrophilic domain [97]. This exceptional feature of Cox2 demands more than one insertase machinery and additional factors for its biogenesis. Only after successful export of the N-terminus does the membrane potential-dependent process of the C-terminus export occur [117]. Also, while the majority of organisms encode Cox2 in the mitochondrial genome, some plant species have moved the gene encoding Cox2 to the nucleus. Thus, a comparison of how different organisms perform Cox2 biogenesis makes for an interesting undertaking as it highlights how much we currently know about the MIM protein insertion in yeast and humans and how much information we are lacking in reference to plant mitochondria (Figure 3).

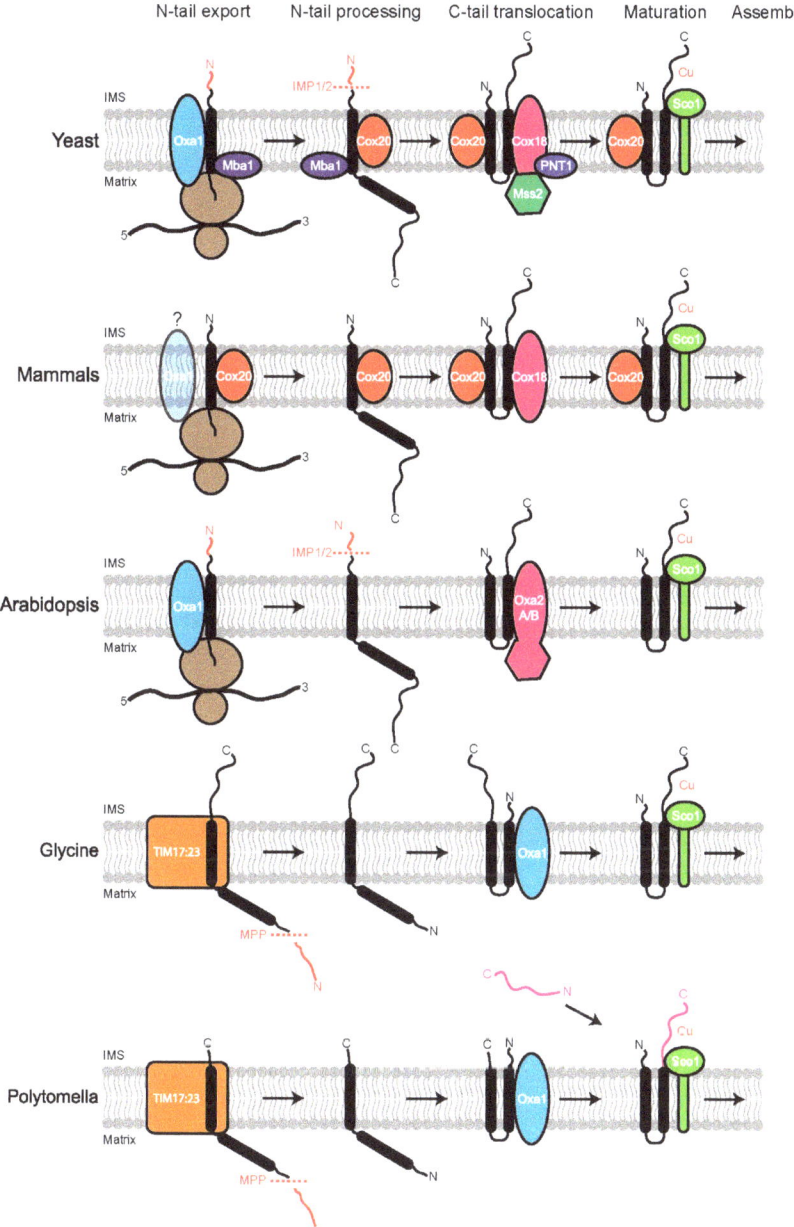

Figure 3. Cox2 biogenesis in yeast, mammals, *Arabidopsis*, *Glycine*, and *Polytomella*. Our current understanding of how Cox2 is inserted into the MIM of yeast (*Saccharomyces cerevisiae*) and mammals (*Homo sapiens*) is as depicted. The four steps of Cox2 assembly are N-tail export with TMH1 membrane insertion, N-tail processing, C-tail translocation with TMH2 membrane insertion and maturation by copper insertion. The last arrow represents the assembly step into complex IV. Based on this knowledge and some limited available studies with plant mitochondria, we have hypothesized the Cox2 insertion pathways for two plant species, *Arabidopsis thaliana*, *Glycine max*, and an algal species, *Polytomella* sp. In yeast, mammals, and *Arabidopsis*, Cox2 is synthesized by the mitochondrial ribosome whereas in *Glycine* and *Polytomella*, it is cytosolically synthesized and imported into the mitochondria. For full details, see the corresponding sections in the text.

5.1. Cox2 Biogenesis in Saccharomyces cerevisiae

Cox2 is the best studied membrane protein in yeast because the cleavage of its leader peptide serves as a simple indication for successful membrane insertion [76]. The *Cox2* gene has no introns and is transcribed by the general mitochondrial transcription machinery, followed by maturation with the help of 3′ processing machinery [113]. The translation requires a specific nuclear-encoded membrane-bound translational activator, Pet111 [118,119]. The coordination of Cox2 synthesis and the subsequent assembly is potentially regulated by the positively-charged N-terminal leader peptide and three additional downstream sequences [120,121]. Oxa1 interacts with the precursor of Cox2 (pCox2) and catalyzes the co-translational membrane insertion of the first TMH with the concomitant export of the N-terminal domain across the MIM into the IMS [117,122,123] (Figure 3). A specific chaperone called Cox20 interacts very early with pCox2 and makes it accessible to the membrane-bound IMP (inner membrane peptidase) complex for proteolytic processing. Cox20 also stabilizes and binds Cox2 until its assembly with the other subunits of COX [124] (Figure 3). Furthermore, the ribosome receptor Mba1 appears to stabilize the Cox20-ribosome complex and supports the transfer of Cox2 to the C-tail export module [125] (Figure 3). Then, Cox18, Mss2, and Pnt1 (pentamidine resistance factor) cooperatively facilitate the insertion of the second TMH with the simultaneous export of the C-tail [83] (Figure 3). While Pnt1 has no clear homologs in mammals and plants and appears to be a eukaryotic invention specific to fungi, Mss2 has a TPR domain resembling that of Tom70 [98,126], suggesting a prokaryotic origin. It has been suggested that Mss2 recognizes the Cox2 C-terminus in the matrix and promotes its translocation [98]. In the absence of Oxa1, Cox18, or Mss2, but not Pnt1, Cox2 proteins are rapidly degraded by proteases due to improper membrane insertion [127]. The last step in the biogenesis of Cox2 is its maturation by copper insertion at the $CxExCGx_2Hx_2M$ motif in the β-barrel structure of the Cox2 C-tail, to form a divalent $[Cu^{2+}/Cu^{1+}]$ complex called the Cu_A center [113]. Sco1 (suppressor of cytochrome oxidase deficiency 1) was identified as the metallochaperone and disulfide reductase which transfers copper to the Cu_A site in Cox2 [128–131] (Figure 3). Sco1 in turn receives copper from Cox17 [132,133]. It needs to be clarified whether the homolog of Sco1, Sco2, is also involved in the copper transfer. Although Sco2 deletion does not affect the biogenesis of COX, its overexpression partially rescues both Sco1 and Cox17 mutants [129]. After copper insertion, the mature Cox2 is ready to be incorporated into a COX assembly intermediate.

5.2. Cox2 Biogenesis in Homo sapiens

All thirteen mammalian mitochondrial-encoded genes lack introns and transcription is polycistronic with alternating structural genes and tRNAs. Processing of specific mRNAs is followed by polyadenylation and the mRNAs lack 5′ UTRs [113]. LRPPRC (leucine-rich PPR motif-containing protein) is the only known factor that stabilizes all mRNAs, including the COX core subunits [134]. The Cox2-specific translational activator in yeast, Pet111, is not conserved in human mitochondria [135]. It is still unclear how the mRNA translation in humans is activated. In contrast to the structure of yeast Cox2, the human Cox2 lacks a cleavable presequence. Oxa1L, the human homolog of Oxa1, is required for the biogenesis of complex I and V but not COX [81,136] (Figure 3). However, it can functionally complement the corresponding yeast mutant [137]. Hence, it is currently not clear whether Oxa1L actually performs the same role in Cox2 biogenesis as its yeast homolog. However, the role of Cox18 (Oxa2) appears to be conserved across species (Figure 3). During the insertion of its N-terminal TMH, Cox20 stabilizes Cox2 and subsequently Cox18 transiently interacts with Cox2 to promote the translocation of the Cox2 C-tail containing the apo-Cu_A site across the MIM (Figure 3). Then, the release of Cox18 from this complex coincides with the binding of the Sco1–Sco2–Coa6 (cytochrome oxidase assembly factor 6) copper metalation module to the Cox2–Cox20 intermediate [135,138] (Figure 3). Interestingly, although no homologs for Mss2 can be found in humans, the proteins corresponding to alternative splice variants of Cox18 [127] might perform a function similar to Mss2. Both Sco1 and Sco2 have a Cx_3C copper-binding motif and are essential for COX assembly [139]. Moreover, Sco2 has an additional role in the synthesis of Cox2 [140], implying that it probably coordinates the synthesis and maturation processes of Cox2. It was proposed that following Cox2 metalation, Sco2 re-oxidizes Sco1, which makes them both ready for the next round of copper transfer. Thus, contrary to yeast Sco proteins, human Sco proteins have independent cooperative functions in the maturation of Cox2 or in COX assembly [141]. Cu(I) can be easily transferred from Cox17 to the two Sco proteins since they have a higher copper affinity [142]. Apart from their role in the biogenesis of Cox2, Sco1 and Sco2 are also involved in cellular copper homeostasis [143]. Very recently, a new protein called TMEM177 (transmembrane protein 177), which lacks a clear homolog in yeast, was found to form a complex with Cox2, Cox20, and the copper chaperones. Although its depletion or increased level affected Cox20 abundance and Cox2 stability, the assembly, abundance, and function of COX remained unaffected [144].

5.3. Cox2 Biogenesis in Arabidopsis thaliana

Very little previous research has been done to investigate Cox2 biogenesis in plants. Hence, the following speculations are mostly based on the observations made in yeast and humans, mentioned above. Based on sequence alignment, the *Arabidopsis* Cox2 is predicted to be synthesized as a precursor protein with a presequence, similar to the yeast Cox2. Plant Oxa1a can functionally complement a yeast *oxa1* deletion mutant [145]. Moreover, Oxa1a is essential for embryogenesis whereas homozygous deletion of Oxa1b does not affect the plant phenotype [84]. Therefore, it is most likely that Oxa1a mediates the co-translational insertion of the first TMH with concomitant export of the N-tail into the IMS (Figure 3). The presequence appears to be cleaved off by an IMP-like protease. Then, either Oxa2a or Oxa2b are possibly involved in the membrane insertion of the second TMH, accompanied by the C-tail export into the IMS (Figure 3). No homolog of Cox20 nor Mss2 nor Pnt1 has been identified in plants so far. However, it has to be noted that Mss2 and Pnt1 appear to be fungi-specific and are also not found in humans. As mentioned previously, it is interesting to speculate that the plant Oxa2 TPR domains and yeast Mss2 possibly play the same role in Cox2 biogenesis since the C-termini of Oxa2 proteins harbor a predicted TPR domain and the yeast Mss2 also has a predicted TPR domain. However, since there is no evidence that the TPR domains of Oxa2 proteins and Mss2 are related, convergent evolution is more likely. Perhaps, the TPR domain of Oxa2a/Oxa2b stabilizes the Cox2 C-terminus in the matrix until its export into the IMS by the insertase domain of Oxa2a/Oxa2b. Alternatively, it is also possible that similar to the TPR domain of Tom70 [86], the Oxa2 TPR domains interact with mtHsp70 or other chaperones that help in Cox2 stability and/or the assembly process, thus eliminating the need for a dedicated chaperone such as Cox20. After successful translocation, Cox2 can proceed for maturation by copper insertion most likely with the help of a protein similar to Sco1, called HCC1 (homolog of the yeast copper chaperone 1) (Figure 3). Interestingly, another Sco1 homolog, HCC2 does not affect COX activity but appears to be important for UV-B stress response [146,147]. *Arabidopsis* also has two genes that encode putative Cox17 homologs that are able to complement a yeast *cox17* null mutant [148–150].

5.4. Cox2 Biogenesis in Glycine max

The *Cox2* gene typically found in the mitochondrial genome of nearly all plants was very recently transferred to the nucleus in legumes. The nuclear-encoded Cox2 acquired a mitochondrial-targeting sequence bordered by two introns [151] and largely decreased the hydrophobicity of the first TMH [152]. The unusually long targeting signal of 136 amino acids consists of three parts that play different roles: the first 20 amino acids for mitochondrial targeting, the central portion for efficient import, and the last 12 amino acids derived from the mitochondrial-encoded protein are required for correct maturation of the imported protein. During import, the presequence is cleaved in a three-step process, independent of assembly [151] (Figure 3). Membrane insertion of the TMHs most likely involves TIM23 and Oxa1 before proceeding to the maturation step (Figure 3). Interestingly, when Cox2 is nuclear encoded, its C-terminus can be directly imported into IMS and will not require an export step from the matrix.

Thus, the function of Oxa2 would not be required. However, the genome of *Glycine max* encodes for at least one Oxa2-like protein containing a similar TPR domain as the Oxa2 protein from *Arabidopsis*.

5.5. Cox2 Biogenesis in Chlorophycean algae

In contrast to the above case of legumes where the *Cox2* gene has been relocated to the nucleus as a single unit, in the chlorophycean algae *Chlamydomonas reinhardtii* and its colorless close relative *Polytomella* sp., the *Cox2* gene is split into complementary *Cox2a* and *Cox2b* genes before relocating independently to the nuclear genome and becoming lost from the mitochondrial genome. The *Cox2a* gene encodes the Cox2a protein that corresponds to the N-terminal half of the typical single polypeptide Cox2 and contains the two TMHs. The *Cox2b* gene encodes the Cox2b protein, which is equivalent to the C-terminal soluble IMS domain of the original protein. Both *Cox2a* and *Cox2b* are independently transcribed into mRNA and translated into separate polypeptides in the cytosol [153]. The mitochondria then import the two subunits and assemble them into the COX complex (Figure 3). Interestingly, in *Polytomella*, the two subunits are imported into mitochondria using different mechanisms. The Cox2a precursor exhibits a long cleavable mitochondrial-targeting sequence of 130 amino acids and appears to follow an energy-dependent import pathway involving mitochondrial-targeting sequence elimination by proteolytic processing and membrane integration of its two TMHs. The soluble Cox2b protein, lacking a cleavable targeting signal, is directly imported into the IMS via the TOM complex [154] (Figure 3). Researchers favor an import mechanism of Cox2a similar to the import route proposed for the soybean Cox2 precursor [151], described above. TIM23, after fully translocating the mitochondrial-targeting signal and the first TMH of Cox2a to the matrix, recognizes a stop transfer signal within the second TMH and thereby inserts it laterally into the MIM (Figure 3). Afterwards, the N-terminal part of the mature Cox2a would be translocated back in an export-like reaction by the Oxa1 machinery, in order to achieve the functional N-out C-out topology of Cox2a (Figure 3). It remains to be ascertained whether the mitochondrial disulfide relay through Mia40 and Erv1 is involved in the import of Cox2b. Probably after binding to the IMS-located Mia40 receptor, Cox2b interacts with the membrane-inserted Cox2a subunit to form the heterodimeric Cox2 subunit (Figure 3). Having derived from the process of split gene transfer, neither polypeptide has regions that are homologous to the known Cox2 proteins. While Cox2a has a 20 amino acid C-terminal extension, Cox2b has a 42 amino acid region exhibiting a high degree of charged amino acids at the N-terminus. According to the model proposed by Perez-Martinez et al., an interaction between the unique Cox2a C-terminal domain and the highly charged Cox2b N-terminal domain might stabilize the two Cox2 subunits in the COX complex [153]. However, it was recently shown that the C-terminal extension of Cox2b appears to be dispensable as it only weakly reinforces the Cox2a/Cox2b interaction. It was concluded that the hydrophilic domain of Cox2 is a highly stable structure that when split into Cox2a and Cox2b is able to maintain a strong interaction of the two fragments [155]. It is currently unknown whether the maturation step for the formation of the Cu_A center in Cox2b occurs before or after Cox2a/Cox2b interaction.

6. Outlook

While our understanding of the roles that plant mitochondria play in growth and development, stress responses and secondary metabolite production is ever increasing, our knowledge about how plant mitochondria insert and assemble their inner membrane proteins is based mainly on knowledge gained by research with yeast and mammalian mitochondria. Although many of the membrane protein insertion mechanisms identified in other organisms are also present in plant mitochondria, the fundamental differences with plant mitochondria—in containing double the normal number of Oxa protein homologs, having Oxa proteins with TPR domains, and also likely containing the most conserved of conservative sorting pathways, a Tat pathway—imply that the mechanisms of membrane insertion and assembly will also be different in plants. This necessitates specific studies on Oxa and Tat pathways in plant mitochondria to provide a better understanding of not only how plants insert and assemble the MIM proteins but also how these processes are regulated in different tissues and during various developmental growth stages. Plant mitochondria also appear to be missing several key proteins found in the mitochondria of other lineages, such as Cox20 or the components of the MITRAC (mitochondrial translation regulation assembly intermediate of COX) complex, which have been well studied in other organisms. This raises the possibility that plant mitochondria contain unique plant-specific proteins for performing these functions. Grouping all plants together is not a good idea because, considering the example of Cox2 biogenesis, there may be species-specific differences.

Acknowledgments: Financial support provided by the German Research Council (DFG, CA1775/1-1 to Chris Carrie) is acknowledged. Renuka Kolli is supported by the German Academic Exchange Service (DAAD) Scholarship.

Author Contributions: Renuka Kolli, Jürgen Soll, and Chris Carrie wrote the manuscript together. All authors have read and approved the final version.

Conflicts of Interest: The authors declare no conflict of interest.

References

1. Prasai, K. Regulation of mitochondrial structure and function by protein import: A current review. *Pathophysiology* **2017**, *24*, 107–122. [CrossRef] [PubMed]
2. Martin, W.F.; Garg, S.; Zimorski, V. Endosymbiotic theories for eukaryote origin. *Philos. Trans. R. Soc. Lond. B Biol. Sci.* **2015**, *370*. [CrossRef] [PubMed]
3. Gray, M.W. Mitochondrial evolution. *Cold Spring Harb. Perspect. Biol.* **2012**, *4*, a011403. [CrossRef] [PubMed]
4. Sloan, D.B.; Alverson, A.J.; Chuckalovcak, J.P.; Wu, M.; McCauley, D.E.; Palmer, J.D.; Taylor, D.R. Rapid evolution of enormous, multichromosomal genomes in flowering plant mitochondria with exceptionally high mutation rates. *PLoS Biol.* **2012**, *10*, e1001241. [CrossRef] [PubMed]
5. Binder, S.; Brennicke, A. Gene expression in plant mitochondria: Transcriptional and post-transcriptional control. *Philos. Trans. R. Soc. Lond. B Biol. Sci.* **2003**, *358*, 181–189. [CrossRef] [PubMed]
6. Gray, M.W. RNA editing in plant mitochondria: 20 years later. *IUBMB Life* **2009**, *61*, 1101–1104. [CrossRef] [PubMed]
7. Schertl, P.; Braun, H.P. Respiratory electron transfer pathways in plant mitochondria. *Front. Plant Sci.* **2014**, *5*, 163. [CrossRef] [PubMed]
8. Mannella, C.A.; Lederer, W.J.; Jafri, M.S. The connection between inner membrane topology and mitochondrial function. *J. Mol. Cell. Cardiol.* **2013**, *62*, 51–57. [CrossRef] [PubMed]
9. Taiz, L.; Zeiger, E.; Møller, I.M.; Murphy, A.S. *Plant Physiology and Development*, 6th ed.; Sinauer Associates, Inc.: Sunderland, MA, USA, 2015.
10. Gilkerson, R.; Bravo, L.; Garcia, I.; Gaytan, N.; Herrera, A.; Maldonado, A.; Quintanilla, B. The mitochondrial nucleoid: Integrating mitochondrial DNA into cellular homeostasis. *Cold Spring Harb. Perspect. Biol.* **2013**, *5*, a011080. [CrossRef] [PubMed]
11. Gray, M.W. Mosaic nature of the mitochondrial proteome: Implications for the origin and evolution of mitochondria. *Proc. Natl. Acad. Sci. USA* **2015**, *112*, 10133–10138. [CrossRef] [PubMed]
12. Vogtle, F.N.; Burkhart, J.M.; Gonczarowska-Jorge, H.; Kucukkose, C.; Taskin, A.A.; Kopczynski, D.; Ahrends, R.; Mossmann, D.; Sickmann, A.; Zahedi, R.P.; et al. Landscape of submitochondrial protein distribution. *Nat. Commun.* **2017**, *8*, 290. [CrossRef] [PubMed]
13. Rao, R.S.; Salvato, F.; Thal, B.; Eubel, H.; Thelen, J.J.; Moller, I.M. The proteome of higher plant mitochondria. *Mitochondrion* **2017**, *33*, 22–37. [CrossRef] [PubMed]
14. Schagger, H.; Pfeiffer, K. Supercomplexes in the respiratory chains of yeast and mammalian mitochondria. *EMBO J.* **2000**, *19*, 1777–1783. [CrossRef] [PubMed]
15. Dudkina, N.V.; Kouril, R.; Peters, K.; Braun, H.P.; Boekema, E.J. Structure and function of mitochondrial supercomplexes. *Biochim. Biophys. Acta* **2010**, *1797*, 664–670. [CrossRef] [PubMed]
16. Genova, M.L.; Lenaz, G. Functional role of mitochondrial respiratory supercomplexes. *Biochim. Biophys. Acta* **2014**, *1837*, 427–443. [CrossRef] [PubMed]
17. Letts, J.A.; Fiedorczuk, K.; Sazanov, L.A. The architecture of respiratory supercomplexes. *Nature* **2016**, *537*, 644–648. [CrossRef] [PubMed]

18. Eubel, H.; Heinemeyer, J.; Sunderhaus, S.; Braun, H.P. Respiratory chain supercomplexes in plant mitochondria. *Plant Physiol. Biochem.* **2004**, *42*, 937–942. [CrossRef] [PubMed]
19. Wollweber, F.; von der Malsburg, K.; van der Laan, M. Mitochondrial contact site and cristae organizing system: A central player in membrane shaping and crosstalk. *Biochim. Biophys. Acta* **2017**, *1864*, 1481–1489. [CrossRef] [PubMed]
20. Harner, M.E.; Unger, A.K.; Geerts, W.J.; Mari, M.; Izawa, T.; Stenger, M.; Geimer, S.; Reggiori, F.; Westermann, B.; Neupert, W. An evidence based hypothesis on the existence of two pathways of mitochondrial crista formation. *Elife* **2016**, *5*, e18853. [CrossRef] [PubMed]
21. Ikon, N.; Ryan, R.O. Cardiolipin and mitochondrial cristae organization. *Biochim. Biophys. Acta* **2017**, *1859*, 1156–1163. [CrossRef] [PubMed]
22. Timmis, J.N.; Ayliffe, M.A.; Huang, C.Y.; Martin, W. Endosymbiotic gene transfer: Organelle genomes forge eukaryotic chromosomes. *Nat. Rev. Genet.* **2004**, *5*, 123–135. [CrossRef] [PubMed]
23. Liu, Z.; Li, X.; Zhao, P.; Gui, J.; Zheng, W.; Zhang, Y. Tracing the evolution of the mitochondrial protein import machinery. *Comput. Biol. Chem.* **2011**, *35*, 336–340. [CrossRef] [PubMed]
24. Abe, Y.; Shodai, T.; Muto, T.; Mihara, K.; Torii, H.; Nishikawa, S.; Endo, T.; Kohda, D. Structural basis of presequence recognition by the mitochondrial protein import receptor Tom20. *Cell* **2000**, *100*, 551–560. [CrossRef]
25. Kang, P.J.; Ostermann, J.; Shilling, J.; Neupert, W.; Craig, E.A.; Pfanner, N. Requirement for HSP70 in the mitochondrial matrix for translocation and folding of precursor proteins. *Nature* **1990**, *348*, 137–143. [CrossRef] [PubMed]
26. Truscott, K.N.; Kovermann, P.; Geissler, A.; Merlin, A.; Meijer, M.; Driessen, A.J.; Rassow, J.; Pfanner, N.; Wagner, R. A presequence- and voltage-sensitive channel of the mitochondrial preprotein translocase formed by Tim23. *Nat. Struct. Biol.* **2001**, *8*, 1074–1082. [CrossRef] [PubMed]
27. Mokranjac, D.; Neupert, W. The many faces of the mitochondrial Tim23 complex. *Biochim. Biophys. Acta* **2010**, *1797*, 1045–1054. [CrossRef] [PubMed]
28. Hawlitschek, G.; Schneider, H.; Schmidt, B.; Tropschug, M.; Hartl, F.U.; Neupert, W. Mitochondrial protein import: Identification of processing peptidase and of pep, a processing enhancing protein. *Cell* **1988**, *53*, 795–806. [CrossRef]
29. Backes, S.; Herrmann, J.M. Protein translocation into the intermembrane space and matrix of mitochondria: Mechanisms and driving forces. *Front. Mol. Biosci.* **2017**, *4*, 83. [CrossRef] [PubMed]
30. Wiedemann, N.; Pfanner, N. Mitochondrial machineries for protein import and assembly. *Annu. Rev. Biochem.* **2017**, *86*, 685–714. [CrossRef] [PubMed]
31. Kang, Y.; Fielden, L.F.; Stojanovski, D. Mitochondrial protein transport in health and disease. In *Seminars in Cell & Developmental Biology*; Academic Press: Cambridge, MA, USA, 2017.
32. Sokol, A.M.; Sztolsztener, M.E.; Wasilewski, M.; Heinz, E.; Chacinska, A. Mitochondrial protein translocases for survival and wellbeing. *FEBS Lett.* **2014**, *588*, 2484–2495. [CrossRef] [PubMed]
33. Murcha, M.W.; Kmiec, B.; Kubiszewski-Jakubiak, S.; Teixeira, P.F.; Glaser, E.; Whelan, J. Protein import into plant mitochondria: Signals, machinery, processing, and regulation. *J. Exp. Bot.* **2014**, *65*, 6301–6335. [CrossRef] [PubMed]
34. Van der Laan, M.; Meinecke, M.; Dudek, J.; Hutu, D.P.; Lind, M.; Perschil, I.; Guiard, B.; Wagner, R.; Pfanner, N.; Rehling, P. Motor-free mitochondrial presequence translocase drives membrane integration of preproteins. *Nat. Cell Biol.* **2007**, *9*, 1152–1159. [CrossRef] [PubMed]
35. Herrmann, J.M.; Neupert, W. Protein insertion into the inner membrane of mitochondria. *IUBMB Life* **2003**, *55*, 219–225. [CrossRef] [PubMed]
36. Chacinska, A.; van der Laan, M.; Mehnert, C.S.; Guiard, B.; Mick, D.U.; Hutu, D.P.; Truscott, K.N.; Wiedemann, N.; Meisinger, C.; Pfanner, N.; et al. Distinct forms of mitochondrial TOM-TIM supercomplexes define signal-dependent states of preprotein sorting. *Mol. Cell. Biol.* **2010**, *30*, 307–318. [CrossRef] [PubMed]
37. Murcha, M.W.; Elhafez, D.; Millar, A.H.; Whelan, J. The c-terminal region of TIM17 links the outer and inner mitochondrial membranes in *Arabidopsis* and is essential for protein import. *J. Biol. Chem.* **2005**, *280*, 16476–16483. [CrossRef] [PubMed]
38. Murcha, M.W.; Lister, R.; Ho, A.Y.; Whelan, J. Identification, expression, and import of components 17 and 23 of the inner mitochondrial membrane translocase from *Arabidopsis*. *Plant Physiol.* **2003**, *131*, 1737–1747. [CrossRef] [PubMed]

39. Wang, Y.; Carrie, C.; Giraud, E.; Elhafez, D.; Narsai, R.; Duncan, O.; Whelan, J.; Murcha, M.W. Dual location of the mitochondrial preprotein transporters B14.7 and TIM23-2 in complex I and the TIM17:23 complex in *Arabidopsis* links mitochondrial activity and biogenesis. *Plant Cell* **2012**, *24*, 2675–2695. [CrossRef] [PubMed]
40. Stahl, A.; Moberg, P.; Ytterberg, J.; Panfilov, O.; Brockenhuus Von Lowenhielm, H.; Nilsson, F.; Glaser, E. Isolation and identification of a novel mitochondrial metalloprotease (prep) that degrades targeting presequences in plants. *J. Biol. Chem.* **2002**, *277*, 41931–41939. [CrossRef] [PubMed]
41. Bhushan, S.; Stahl, A.; Nilsson, S.; Lefebvre, B.; Seki, M.; Roth, C.; McWilliam, D.; Wright, S.J.; Liberles, D.A.; Shinozaki, K.; et al. Catalysis, subcellular localization, expression and evolution of the targeting peptides degrading protease, AtPrep2. *Plant Cell Physiol.* **2005**, *46*, 985–996. [CrossRef] [PubMed]
42. Sirrenberg, C.; Bauer, M.F.; Guiard, B.; Neupert, W.; Brunner, M. Import of carrier proteins into the mitochondrial inner membrane mediated by TIM22. *Nature* **1996**, *384*, 582–585. [CrossRef] [PubMed]
43. Gebert, N.; Gebert, M.; Oeljeklaus, S.; von der Malsburg, K.; Stroud, D.A.; Kulawiak, B.; Wirth, C.; Zahedi, R.P.; Dolezal, P.; Wiese, S.; et al. Dual function of SDH3 in the respiratory chain and TIM22 protein translocase of the mitochondrial inner membrane. *Mol. Cell* **2011**, *44*, 811–818. [CrossRef] [PubMed]
44. Carrie, C.; Murcha, M.W.; Whelan, J. An in silico analysis of the mitochondrial protein import apparatus of plants. *BMC Plant Biol.* **2010**, *10*, 249. [CrossRef] [PubMed]
45. Laloi, M. Plant mitochondrial carriers: An overview. *Cell. Mol. Life Sci.* **1999**, *56*, 918–944. [CrossRef] [PubMed]
46. Murcha, M.W.; Millar, A.H.; Whelan, J. The n-terminal cleavable extension of plant carrier proteins is responsible for efficient insertion into the inner mitochondrial membrane. *J. Mol. Biol.* **2005**, *351*, 16–25. [CrossRef] [PubMed]
47. Murcha, M.W.; Elhafez, D.; Millar, A.H.; Whelan, J. The N-terminal extension of plant mitochondrial carrier proteins is removed by two-step processing: The first cleavage is by the mitochondrial processing peptidase. *J. Mol. Biol.* **2004**, *344*, 443–454. [CrossRef] [PubMed]
48. Hartl, F.U.; Schmidt, B.; Wachter, E.; Weiss, H.; Neupert, W. Transport into mitochondria and intramitochondrial sorting of the Fe/S protein of ubiquinol-cytochrome c reductase. *Cell* **1986**, *47*, 939–951. [CrossRef]
49. Von Heijne, G. Control of topology and mode of assembly of a polytopic membrane protein by positively charged residues. *Nature* **1989**, *341*, 456–458. [CrossRef] [PubMed]
50. Hell, K.; Neupert, W.; Stuart, R.A. Oxa1p acts as a general membrane insertion machinery for proteins encoded by mitochondrial DNA. *EMBO J.* **2001**, *20*, 1281–1288. [CrossRef] [PubMed]
51. Bauer, M.; Behrens, M.; Esser, K.; Michaelis, G.; Pratje, E. Pet1402, a nuclear gene required for proteolytic processing of cytochrome oxidase subunit 2 in yeast. *Mol. Gen. Genet.* **1994**, *245*, 272–278. [CrossRef] [PubMed]
52. Bonnefoy, N.; Chalvet, F.; Hamel, P.; Slonimski, P.P.; Dujardin, G. Oxa1, a *Saccharomyces cerevisiae* nuclear gene whose sequence is conserved from prokaryotes to eukaryotes controls cytochrome oxidase biogenesis. *J. Mol. Biol.* **1994**, *239*, 201–212. [CrossRef] [PubMed]
53. Hennon, S.W.; Soman, R.; Zhu, L.; Dalbey, R.E. YidC/Alb3/Oxa1 family of insertases. *J. Biol. Chem.* **2015**, *290*, 14866–14874. [CrossRef] [PubMed]
54. Wang, P.; Dalbey, R.E. Inserting membrane proteins: The YidC/Oxa1/Alb3 machinery in bacteria, mitochondria, and chloroplasts. *Biochim. Biophys. Acta* **2011**, *1808*, 866–875. [CrossRef] [PubMed]
55. Anghel, S.A.; McGilvray, P.T.; Hegde, R.S.; Keenan, R.J. Identification of Oxa1 homologs operating in the eukaryotic endoplasmic reticulum. *Cell Rep.* **2017**, *21*, 3708–3716. [CrossRef] [PubMed]
56. Saller, M.J.; Wu, Z.C.; de Keyzer, J.; Driessen, A.J. The YidC/Oxa1/Alb3 protein family: Common principles and distinct features. *Biol. Chem.* **2012**, *393*, 1279–1290. [CrossRef] [PubMed]
57. Szyrach, G.; Ott, M.; Bonnefoy, N.; Neupert, W.; Herrmann, J.M. Ribosome binding to the Oxa1 complex facilitates co-translational protein insertion in mitochondria. *EMBO J.* **2003**, *22*, 6448–6457. [CrossRef] [PubMed]
58. Kohler, R.; Boehringer, D.; Greber, B.; Bingel-Erlenmeyer, R.; Collinson, I.; Schaffitzel, C.; Ban, N. YidC and Oxa1 form dimeric insertion pores on the translating ribosome. *Mol. Cell* **2009**, *34*, 344–353. [CrossRef] [PubMed]
59. Falk, S.; Ravaud, S.; Koch, J.; Sinning, I. The C terminus of the Alb3 membrane insertase recruits cpSRP43 to the thylakoid membrane. *J. Biol. Chem.* **2010**, *285*, 5954–5962. [CrossRef] [PubMed]

60. Sievers, F.; Wilm, A.; Dineen, D.; Gibson, T.J.; Karplus, K.; Li, W.; Lopez, R.; McWilliam, H.; Remmert, M.; Soding, J.; et al. Fast, scalable generation of high-quality protein multiple sequence alignments using clustal omega. *Mol. Syst. Biol.* **2011**, *7*, 539. [CrossRef] [PubMed]
61. Saaf, A.; Monne, M.; de Gier, J.W.; von Heijne, G. Membrane topology of the 60-kda Oxa1p homologue from *Escherichia coli*. *J. Biol. Chem.* **1998**, *273*, 30415–30418. [CrossRef] [PubMed]
62. Kedrov, A.; Wickles, S.; Crevenna, A.H.; van der Sluis, E.O.; Buschauer, R.; Berninghausen, O.; Lamb, D.C.; Beckmann, R. Structural dynamics of the YidC:Ribosome complex during membrane protein biogenesis. *Cell Rep.* **2016**, *17*, 2943–2954. [CrossRef] [PubMed]
63. Kumazaki, K.; Kishimoto, T.; Furukawa, A.; Mori, H.; Tanaka, Y.; Dohmae, N.; Ishitani, R.; Tsukazaki, T.; Nureki, O. Crystal structure of *Escherichia coli* YidC, a membrane protein chaperone and insertase. *Sci. Rep.* **2014**, *4*, 7299. [CrossRef] [PubMed]
64. Kumazaki, K.; Chiba, S.; Takemoto, M.; Furukawa, A.; Nishiyama, K.; Sugano, Y.; Mori, T.; Dohmae, N.; Hirata, K.; Nakada-Nakura, Y.; et al. Structural basis of sec-independent membrane protein insertion by YidC. *Nature* **2014**, *509*, 516–520. [CrossRef] [PubMed]
65. Sachelaru, I.; Petriman, N.A.; Kudva, R.; Kuhn, P.; Welte, T.; Knapp, B.; Drepper, F.; Warscheid, B.; Koch, H.G. Yidc occupies the lateral gate of the SecYEG translocon and is sequentially displaced by a nascent membrane protein. *J. Biol. Chem.* **2013**, *288*, 16295–16307. [CrossRef] [PubMed]
66. Sachelaru, I.; Winter, L.; Knyazev, D.G.; Zimmermann, M.; Vogt, A.; Kuttner, R.; Ollinger, N.; Siligan, C.; Pohl, P.; Koch, H.G. YidC and SecYEG form a heterotetrameric protein translocation channel. *Sci. Rep.* **2017**, *7*, 101. [CrossRef] [PubMed]
67. Wu, Z.C.; de Keyzer, J.; Berrelkamp-Lahpor, G.A.; Driessen, A.J. Interaction of streptococcus mutans YidC1 and YidC2 with translating and nontranslating ribosomes. *J. Bacteriol.* **2013**, *195*, 4545–4551. [CrossRef] [PubMed]
68. Nagamori, S.; Smirnova, I.N.; Kaback, H.R. Role of YidC in folding of polytopic membrane proteins. *J. Cell Biol.* **2004**, *165*, 53–62. [CrossRef] [PubMed]
69. Wagner, S.; Pop, O.I.; Haan, G.J.; Baars, L.; Koningstein, G.; Klepsch, M.M.; Genevaux, P.; Luirink, J.; de Gier, J.W. Biogenesis of MalF and the MalFGK(2) maltose transport complex in *Escherichia coli* requires YidC. *J. Biol. Chem.* **2008**, *283*, 17881–17890. [CrossRef] [PubMed]
70. Klostermann, E.; Droste Gen Helling, I.; Carde, J.P.; Schunemann, D. The thylakoid membrane protein ALB3 associates with the cpSecY-translocase in *Arabidopsis thaliana*. *Biochem. J.* **2002**, *368*, 777–781. [CrossRef] [PubMed]
71. Moore, M.; Harrison, M.S.; Peterson, E.C.; Henry, R. Chloroplast Oxa1p homolog albino3 is required for post-translational integration of the light harvesting chlorophyll-binding protein into thylakoid membranes. *J. Biol. Chem.* **2000**, *275*, 1529–1532. [CrossRef] [PubMed]
72. Benz, M.; Bals, T.; Gugel, I.L.; Piotrowski, M.; Kuhn, A.; Schunemann, D.; Soll, J.; Ankele, E. Alb4 of *Arabidopsis* promotes assembly and stabilization of a non chlorophyll-binding photosynthetic complex, the CF1CF0-ATP synthase. *Mol. Plant* **2009**, *2*, 1410–1424. [CrossRef] [PubMed]
73. Gerdes, L.; Bals, T.; Klostermann, E.; Karl, M.; Philippar, K.; Hunken, M.; Soll, J.; Schunemann, D. A second thylakoid membrane-localized Alb3/Oxai/YidC homologue is involved in proper chloroplast biogenesis in *Arabidopsis thaliana*. *J. Biol. Chem.* **2006**, *281*, 16632–16642. [CrossRef] [PubMed]
74. Ossenbuhl, F.; Gohre, V.; Meurer, J.; Krieger-Liszkay, A.; Rochaix, J.D.; Eichacker, L.A. Efficient assembly of photosystem II in *Chlamydomonas reinhardtii* requires Alb3.1p, a homolog of *Arabidopsis* ALBINO3. *Plant Cell* **2004**, *16*, 1790–1800. [CrossRef] [PubMed]
75. Gohre, V.; Ossenbuhl, F.; Crevecoeur, M.; Eichacker, L.A.; Rochaix, J.D. One of two Alb3 proteins is essential for the assembly of the photosystems and for cell survival in *Chlamydomonas*. *Plant Cell* **2006**, *18*, 1454–1466. [CrossRef] [PubMed]
76. Ott, M.; Herrmann, J.M. Co-translational membrane insertion of mitochondrially encoded proteins. *Biochim. Biophys. Acta* **2010**, *1803*, 767–775. [CrossRef] [PubMed]
77. Gruschke, S.; Grone, K.; Heublein, M.; Holz, S.; Israel, L.; Imhof, A.; Herrmann, J.M.; Ott, M. Proteins at the polypeptide tunnel exit of the yeast mitochondrial ribosome. *J. Biol. Chem.* **2010**, *285*, 19022–19028. [CrossRef] [PubMed]

78. Keil, M.; Bareth, B.; Woellhaf, M.W.; Peleh, V.; Prestele, M.; Rehling, P.; Herrmann, J.M. Oxa1-ribosome complexes coordinate the assembly of cytochrome c oxidase in mitochondria. *J. Biol. Chem.* **2012**, *287*, 34484–34493. [CrossRef] [PubMed]
79. Jia, L.; Dienhart, M.K.; Stuart, R.A. Oxa1 directly interacts with ATP9 and mediates its assembly into the mitochondrial F1FO-ATP synthase complex. *Mol. Biol. Cell* **2007**, *18*, 1897–1908. [CrossRef] [PubMed]
80. Stuart, R. Insertion of proteins into the inner membrane of mitochondria: The role of the Oxa1 complex. *Biochim. Biophys. Acta* **2002**, *1592*, 79–87. [CrossRef]
81. Stiburek, L.; Fornuskova, D.; Wenchich, L.; Pejznochova, M.; Hansikova, H.; Zeman, J. Knockdown of human Oxa1l impairs the biogenesis of F1FO-ATP synthase and NADH:Ubiquinone oxidoreductase. *J. Mol. Biol.* **2007**, *374*, 506–516. [CrossRef] [PubMed]
82. Nargang, F.E.; Preuss, M.; Neupert, W.; Herrmann, J.M. The Oxa1 protein forms a homooligomeric complex and is an essential part of the mitochondrial export translocase in *Neurospora crassa*. *J. Biol. Chem.* **2002**, *277*, 12846–12853. [CrossRef] [PubMed]
83. Saracco, S.A.; Fox, T.D. Cox18p is required for export of the mitochondrially encoded *Saccharomyces cerevisiae* Cox2p C-tail and interacts with Pnt1p and Mss2p in the inner membrane. *Mol. Biol. Cell* **2002**, *13*, 1122–1131. [CrossRef] [PubMed]
84. Benz, M.; Soll, J.; Ankele, E. *Arabidopsis thaliana* Oxa proteins locate to mitochondria and fulfill essential roles during embryo development. *Planta* **2013**, *237*, 573–588. [CrossRef] [PubMed]
85. Zhang, Y.J.; Tian, H.F.; Wen, J.F. The evolution of YidC/Oxa/Alb3 family in the three domains of life: A phylogenomic analysis. *BMC Evol. Biol.* **2009**, *9*, 137. [CrossRef] [PubMed]
86. Fan, A.C.; Young, J.C. Function of cytosolic chaperones in Tom70-mediated mitochondrial import. *Protein Pept. Lett.* **2011**, *18*, 122–131. [CrossRef] [PubMed]
87. Palmer, J.D.; Adams, K.L.; Cho, Y.; Parkinson, C.L.; Qiu, Y.L.; Song, K. Dynamic evolution of plant mitochondrial genomes: Mobile genes and introns and highly variable mutation rates. *Proc. Natl. Acad. Sci. USA* **2000**, *97*, 6960–6966. [CrossRef] [PubMed]
88. Liu, Y.J.; Norberg, F.E.; Szilagyi, A.; De Paepe, R.; Akerlund, H.E.; Rasmusson, A.G. The mitochondrial external NADPH dehydrogenase modulates the leaf NADPH/NADP$^+$ ratio in transgenic *Nicotiana sylvestris*. *Plant Cell Physiol.* **2008**, *49*, 251–263. [CrossRef] [PubMed]
89. Vanlerberghe, G.C.; Vanlerberghe, A.E.; McIntosh, L. Molecular genetic evidence of the ability of alternative oxidase to support respiratory carbon metabolism. *Plant Physiol.* **1997**, *113*, 657–661. [CrossRef] [PubMed]
90. Braun, H.P.; Binder, S.; Brennicke, A.; Eubel, H.; Fernie, A.R.; Finkemeier, I.; Klodmann, J.; Konig, A.C.; Kuhn, K.; Meyer, E.; et al. The life of plant mitochondrial complex I. *Mitochondrion* **2014**, *19* (Pt B), 295–313. [CrossRef] [PubMed]
91. Sunderhaus, S.; Dudkina, N.V.; Jansch, L.; Klodmann, J.; Heinemeyer, J.; Perales, M.; Zabaleta, E.; Boekema, E.J.; Braun, H.P. Carbonic anhydrase subunits form a matrix-exposed domain attached to the membrane arm of mitochondrial complex I in plants. *J. Biol. Chem.* **2006**, *281*, 6482–6488. [CrossRef] [PubMed]
92. Perales, M.; Parisi, G.; Fornasari, M.S.; Colaneri, A.; Villarreal, F.; Gonzalez-Schain, N.; Echave, J.; Gomez-Casati, D.; Braun, H.P.; Araya, A.; et al. Gamma carbonic anhydrase like complex interact with plant mitochondrial complex I. *Plant Mol. Biol.* **2004**, *56*, 947–957. [CrossRef] [PubMed]
93. Fromm, S.; Going, J.; Lorenz, C.; Peterhansel, C.; Braun, H.P. Depletion of the "gamma-type carbonic anhydrase-like" subunits of complex I affects central mitochondrial metabolism in *Arabidopsis thaliana*. *Biochim. Biophys. Acta* **2016**, *1857*, 60–71. [CrossRef] [PubMed]
94. Millar, A.H.; Eubel, H.; Jansch, L.; Kruft, V.; Heazlewood, J.L.; Braun, H.P. Mitochondrial cytochrome c oxidase and succinate dehydrogenase complexes contain plant specific subunits. *Plant Mol. Biol.* **2004**, *56*, 77–90. [CrossRef] [PubMed]
95. Schikowsky, C.; Senkler, J.; Braun, H.P. SDH6 and SDH7 contribute to anchoring succinate dehydrogenase to the inner mitochondrial membrane in *Arabidopsis thaliana*. *Plant Physiol.* **2017**, *173*, 1094–1108. [CrossRef] [PubMed]
96. Zancani, M.; Casolo, V.; Petrussa, E.; Peresson, C.; Patui, S.; Bertolini, A.; De Col, V.; Braidot, E.; Boscutti, F.; Vianello, A. The permeability transition in plant mitochondria: The missing link. *Front. Plant Sci.* **2015**, *6*, 1120. [CrossRef] [PubMed]

97. Radons, J. The human HSP70 family of chaperones: Where do we stand? *Cell Stress Chaperones* **2016**, *21*, 379–404. [CrossRef] [PubMed]
98. Broadley, S.A.; Demlow, C.M.; Fox, T.D. Peripheral mitochondrial inner membrane protein, Mss2p, required for export of the mitochondrially Coded Cox2p C Tail in *Saccharomyces cerevisiae*. *Mol. Cell. Biol.* **2001**, *21*, 7663–7672. [CrossRef] [PubMed]
99. Bohnert, M.; Rehling, P.; Guiard, B.; Herrmann, J.M.; Pfanner, N.; van der Laan, M. Cooperation of stop-transfer and conservative sorting mechanisms in mitochondrial protein transport. *Curr. Biol.* **2010**, *20*, 1227–1232. [CrossRef] [PubMed]
100. Rojo, E.E.; Stuart, R.A.; Neupert, W. Conservative sorting of F0-ATpase subunit 9: Export from matrix requires delta pH across inner membrane and matrix ATP. *EMBO J.* **1995**, *14*, 3445–3451. [PubMed]
101. Reif, S.; Randelj, O.; Domanska, G.; Dian, E.A.; Krimmer, T.; Motz, C.; Rassow, J. Conserved mechanism of Oxa1 insertion into the mitochondrial inner membrane. *J. Mol. Biol.* **2005**, *354*, 520–528. [CrossRef] [PubMed]
102. Webb, C.T.; Lithgow, T. Mitochondrial biogenesis: Sorting mechanisms cooperate in ABC transporter assembly. *Curr. Biol.* **2010**, *20*, R564–R567. [CrossRef] [PubMed]
103. Park, K.; Botelho, S.C.; Hong, J.; Osterberg, M.; Kim, H. Dissecting stop transfer versus conservative sorting pathways for mitochondrial inner membrane proteins in vivo. *J. Biol. Chem.* **2013**, *288*, 1521–1532. [CrossRef] [PubMed]
104. Giege, P.; Knoop, V.; Brennicke, A. Complex ii subunit 4 (SDH4) homologous sequences in plant mitochondrial genomes. *Curr. Genet.* **1998**, *34*, 313–317. [CrossRef] [PubMed]
105. Hildenbeutel, M.; Theis, M.; Geier, M.; Haferkamp, I.; Neuhaus, H.E.; Herrmann, J.M.; Ott, M. The membrane insertase Oxa1 is required for efficient import of carrier proteins into mitochondria. *J. Mol. Biol.* **2012**, *423*, 590–599. [CrossRef] [PubMed]
106. Wagener, N.; Ackermann, M.; Funes, S.; Neupert, W. A pathway of protein translocation in mitochondria mediated by the AAA-ATPase BCS1. *Mol. Cell* **2011**, *44*, 191–202. [CrossRef] [PubMed]
107. Wagener, N.; Neupert, W. BCS1, a AAA protein of the mitochondria with a role in the biogenesis of the respiratory chain. *J. Struct. Biol.* **2012**, *179*, 121–125. [CrossRef] [PubMed]
108. Fu, W.; Japa, S.; Beattie, D.S. Import of the iron-sulfur protein of the cytochrome b.c1 complex into yeast mitochondria. *J. Biol. Chem.* **1990**, *265*, 16541–16547. [PubMed]
109. Zhang, B.; Van Aken, O.; Thatcher, L.; De Clercq, I.; Duncan, O.; Law, S.R.; Murcha, M.W.; van der Merwe, M.; Seifi, H.S.; Carrie, C.; et al. The mitochondrial outer membrane AAA ATPase ATOM66 affects cell death and pathogen resistance in *Arabidopsis thaliana*. *Plant J.* **2014**, *80*, 709–727. [CrossRef] [PubMed]
110. Carrie, C.; Weissenberger, S.; Soll, J. Plant mitochondria contain the protein translocase subunits TatB and TatC. *J. Cell Sci.* **2016**, *129*, 3935–3947. [CrossRef] [PubMed]
111. Berks, B.C. The twin-arginine protein translocation pathway. *Annu. Rev. Biochem.* **2015**, *84*, 843–864. [CrossRef] [PubMed]
112. Tsukihara, T.; Aoyama, H.; Yamashita, E.; Tomizaki, T.; Yamaguchi, H.; Shinzawa-Itoh, K.; Nakashima, R.; Yaono, R.; Yoshikawa, S. The whole structure of the 13-subunit oxidized cytochrome c oxidase at 2.8 A. *Science* **1996**, *272*, 1136–1144. [CrossRef] [PubMed]
113. Soto, I.C.; Fontanesi, F.; Liu, J.; Barrientos, A. Biogenesis and assembly of eukaryotic cytochrome c oxidase catalytic core. *Biochim. Biophys. Acta* **2012**, *1817*, 883–897. [CrossRef] [PubMed]
114. Pfeiffer, K.; Gohil, V.; Stuart, R.A.; Hunte, C.; Brandt, U.; Greenberg, M.L.; Schagger, H. Cardiolipin stabilizes respiratory chain supercomplexes. *J. Biol. Chem.* **2003**, *278*, 52873–52880. [CrossRef] [PubMed]
115. Musatov, A.; Robinson, N.C. Cholate-induced dimerization of detergent- or phospholipid-solubilized bovine cytochrome c oxidase. *Biochemistry* **2002**, *41*, 4371–4376. [CrossRef] [PubMed]
116. Brunori, M.; Giuffre, A.; Sarti, P. Cytochrome c oxidase, ligands and electrons. *J. Inorg. Biochem.* **2005**, *99*, 324–336. [CrossRef] [PubMed]
117. He, S.; Fox, T.D. Membrane translocation of mitochondrially coded Cox2p: Distinct requirements for export of N and C termini and dependence on the conserved protein Oxa1p. *Mol. Biol. Cell* **1997**, *8*, 1449–1460. [CrossRef] [PubMed]
118. Mulero, J.J.; Fox, T.D. PET111 acts in the 5′-leader of the *Saccharomyces cerevisiae* mitochondrial COX2 mRNA to promote its translation. *Genetics* **1993**, *133*, 509–516. [PubMed]
119. Poutre, C.G.; Fox, T.D. PET111, a *Saccharomyces cerevisiae* nuclear gene required for translation of the mitochondrial mRNA encoding cytochrome c oxidase subunit II. *Genetics* **1987**, *115*, 637–647. [PubMed]

120. Bonnefoy, N.; Bsat, N.; Fox, T.D. Mitochondrial translation of *Saccharomyces cerevisiae* COX2 mRNA is controlled by the nucleotide sequence specifying the pre-Cox2p leader peptide. *Mol. Cell. Biol.* **2001**, *21*, 2359–2372. [CrossRef] [PubMed]
121. Williams, E.H.; Fox, T.D. Antagonistic signals within the COX2 mRNA coding sequence control its translation in *Saccharomyces cerevisiae* mitochondria. *RNA* **2003**, *9*, 419–431. [CrossRef] [PubMed]
122. Hell, K.; Herrmann, J.M.; Pratje, E.; Neupert, W.; Stuart, R.A. Oxa1p, an essential component of the N-tail protein export machinery in mitochondria. *Proc. Natl. Acad. Sci. USA* **1998**, *95*, 2250–2255. [CrossRef] [PubMed]
123. Hell, K.; Herrmann, J.; Pratje, E.; Neupert, W.; Stuart, R.A. Oxa1p mediates the export of the N- and C-termini of pCoxII from the mitochondrial matrix to the intermembrane space. *FEBS Lett.* **1997**, *418*, 367–370. [CrossRef]
124. Hell, K.; Tzagoloff, A.; Neupert, W.; Stuart, R.A. Identification of Cox20p, a novel protein involved in the maturation and assembly of cytochrome oxidase subunit 2. *J. Biol. Chem.* **2000**, *275*, 4571–4578. [CrossRef] [PubMed]
125. Lorenzi, I.; Oeljeklaus, S.; Ronsor, C.; Bareth, B.; Warscheid, B.; Rehling, P.; Dennerlein, S. The ribosome-associated Mba1 escorts Cox2 from insertion machinery to maturing assembly intermediates. *Mol. Cell. Biol.* **2016**, *36*, 2782–2793. [CrossRef] [PubMed]
126. Fiumera, H.L.; Broadley, S.A.; Fox, T.D. Translocation of mitochondrially synthesized Cox2 domains from the matrix to the intermembrane space. *Mol. Cell. Biol.* **2007**, *27*, 4664–4673. [CrossRef] [PubMed]
127. Gaisne, M.; Bonnefoy, N. The COX18 gene, involved in mitochondrial biogenesis, is functionally conserved and tightly regulated in humans and fission yeast. *FEMS Yeast Res.* **2006**, *6*, 869–882. [CrossRef] [PubMed]
128. Nittis, T.; George, G.N.; Winge, D.R. Yeast SCO1, a protein essential for cytochrome c oxidase function is a Cu(I)-binding protein. *J. Biol. Chem.* **2001**, *276*, 42520–42526. [CrossRef] [PubMed]
129. Glerum, D.M.; Shtanko, A.; Tzagoloff, A. SCO1 and SCO2 act as high copy suppressors of a mitochondrial copper recruitment defect in *Saccharomyces cerevisiae*. *J. Biol. Chem.* **1996**, *271*, 20531–20535. [CrossRef] [PubMed]
130. Lode, A.; Kuschel, M.; Paret, C.; Rodel, G. Mitochondrial copper metabolism in yeast: Interaction between Sco1p and Cox2p. *FEBS Lett.* **2000**, *485*, 19–24. [CrossRef]
131. Rigby, K.; Cobine, P.A.; Khalimonchuk, O.; Winge, D.R. Mapping the functional interaction of SCO1 and COX2 in cytochrome oxidase biogenesis. *J. Biol. Chem.* **2008**, *283*, 15015–15022. [CrossRef] [PubMed]
132. Rentzsch, A.; Krummeck-Weiss, G.; Hofer, A.; Bartuschka, A.; Ostermann, K.; Rodel, G. Mitochondrial copper metabolism in yeast: Mutational analysis of Sco1p involved in the biogenesis of cytochrome c oxidase. *Curr. Genet.* **1999**, *35*, 103–108. [CrossRef] [PubMed]
133. Abajian, C.; Rosenzweig, A.C. Crystal structure of yeast Sco1. *J. Biol. Inorg. Chem.* **2006**, *11*, 459–466. [CrossRef] [PubMed]
134. Sasarman, F.; Brunel-Guitton, C.; Antonicka, H.; Wai, T.; Shoubridge, E.A.; Consortium, L. Lrpprc and slirp interact in a ribonucleoprotein complex that regulates posttranscriptional gene expression in mitochondria. *Mol. Biol. Cell* **2010**, *21*, 1315–1323. [CrossRef] [PubMed]
135. Bourens, M.; Barrientos, A. Human mitochondrial cytochrome c oxidase assembly factor COX18 acts transiently as a membrane insertase within the subunit 2 maturation module. *J. Biol. Chem.* **2017**, *292*, 7774–7783. [CrossRef] [PubMed]
136. Sato, T.; Mihara, K. Topogenesis of mammalian Oxa1, a component of the mitochondrial inner membrane protein export machinery. *J. Biol. Chem.* **2009**, *284*, 14819–14827. [CrossRef] [PubMed]
137. Bonnefoy, N.; Kermorgant, M.; Groudinsky, O.; Minet, M.; Slonimski, P.P.; Dujardin, G. Cloning of a human gene involved in cytochrome oxidase assembly by functional complementation of an Oxa1-mutation in *Saccharomyces cerevisiae*. *Proc. Natl. Acad. Sci. USA* **1994**, *91*, 11978–11982. [CrossRef] [PubMed]
138. Bourens, M.; Boulet, A.; Leary, S.C.; Barrientos, A. Human COX20 cooperates with SCO1 and SCO2 to mature COX2 and promote the assembly of cytochrome c oxidase. *Hum. Mol. Genet.* **2014**, *23*, 2901–2913. [CrossRef] [PubMed]
139. Horng, Y.C.; Leary, S.C.; Cobine, P.A.; Young, F.B.; George, G.N.; Shoubridge, E.A.; Winge, D.R. Human SCO1 and SCO2 function as copper-binding proteins. *J. Biol. Chem.* **2005**, *280*, 34113–34122. [CrossRef] [PubMed]

140. Leary, S.C.; Sasarman, F.; Nishimura, T.; Shoubridge, E.A. Human SCO2 is required for the synthesis of CO II and as a thiol-disulphide oxidoreductase for SCO1. *Hum. Mol. Genet.* **2009**, *18*, 2230–2240. [CrossRef] [PubMed]
141. Leary, S.C.; Kaufman, B.A.; Pellecchia, G.; Guercin, G.H.; Mattman, A.; Jaksch, M.; Shoubridge, E.A. Human SCO1 and SCO2 have independent, cooperative functions in copper delivery to cytochrome c oxidase. *Hum. Mol. Genet.* **2004**, *13*, 1839–1848. [CrossRef] [PubMed]
142. Banci, L.; Bertini, I.; Ciofi-Baffoni, S.; Hadjiloi, T.; Martinelli, M.; Palumaa, P. Mitochondrial copper(I) transfer from COX17 to SCO1 is coupled to electron transfer. *Proc. Natl. Acad. Sci. USA* **2008**, *105*, 6803–6808. [CrossRef] [PubMed]
143. Leary, S.C.; Cobine, P.A.; Kaufman, B.A.; Guercin, G.H.; Mattman, A.; Palaty, J.; Lockitch, G.; Winge, D.R.; Rustin, P.; Horvath, R.; et al. The human cytochrome c oxidase assembly factors SCO1 and SCO2 have regulatory roles in the maintenance of cellular copper homeostasis. *Cell Metab.* **2007**, *5*, 9–20. [CrossRef] [PubMed]
144. Lorenzi, I.; Oeljeklaus, S.; Aich, A.; Ronsor, C.; Callegari, S.; Dudek, J.; Warscheid, B.; Dennerlein, S.; Rehling, P. The mitochondrial TMEM177 associates with COX20 during COX2 biogenesis. *Biochim. Biophys. Acta* **2018**, *1865*, 323–333. [CrossRef] [PubMed]
145. Hamel, P.; Sakamoto, W.; Wintz, H.; Dujardin, G. Functional complementation of an Oxa1-yeast mutation identifies an *Arabidopsis thaliana* cDNA involved in the assembly of respiratory complexes. *Plant J.* **1997**, *12*, 1319–1327. [CrossRef] [PubMed]
146. Attallah, C.V.; Welchen, E.; Martin, A.P.; Spinelli, S.V.; Bonnard, G.; Palatnik, J.F.; Gonzalez, D.H. Plants contain two SCO proteins that are differentially involved in cytochrome c oxidase function and copper and redox homeostasis. *J. Exp. Bot.* **2011**, *62*, 4281–4294. [CrossRef] [PubMed]
147. Steinebrunner, I.; Gey, U.; Andres, M.; Garcia, L.; Gonzalez, D.H. Divergent functions of the *Arabidopsis* mitochondrial SCO proteins: HCC1 is essential for cox activity while HCC2 is involved in the UV-B stress response. *Front. Plant Sci.* **2014**, *5*, 87. [CrossRef] [PubMed]
148. Wintz, H.; Vulpe, C. Plant copper chaperones. *Biochem. Soc. Trans.* **2002**, *30*, 732–735. [CrossRef] [PubMed]
149. Balandin, T.; Castresana, C. Atcox17, an *Arabidopsis* homolog of the yeast copper chaperone COX17. *Plant Physiol.* **2002**, *129*, 1852–1857. [CrossRef] [PubMed]
150. Garcia, L.; Welchen, E.; Gonzalez, D.H. Mitochondria and copper homeostasis in plants. *Mitochondrion* **2014**, *19* (Pt B), 269–274. [CrossRef] [PubMed]
151. Daley, D.O.; Adams, K.L.; Clifton, R.; Qualmann, S.; Millar, A.H.; Palmer, J.D.; Pratje, E.; Whelan, J. Gene transfer from mitochondrion to nucleus: Novel mechanisms for gene activation from COX2. *Plant J.* **2002**, *30*, 11–21. [CrossRef] [PubMed]
152. Daley, D.O.; Clifton, R.; Whelan, J. Intracellular gene transfer: Reduced hydrophobicity facilitates gene transfer for subunit 2 of cytochrome c oxidase. *Proc. Natl. Acad. Sci. USA* **2002**, *99*, 10510–10515. [CrossRef] [PubMed]
153. Perez-Martinez, X.; Antaramian, A.; Vazquez-Acevedo, M.; Funes, S.; Tolkunova, E.; d'Alayer, J.; Claros, M.G.; Davidson, E.; King, M.P.; Gonzalez-Halphen, D. Subunit II of cytochrome c oxidase in chlamydomonad algae is a heterodimer encoded by two independent nuclear genes. *J. Biol. Chem.* **2001**, *276*, 11302–11309. [CrossRef] [PubMed]
154. Jimenez-Suarez, A.; Vazquez-Acevedo, M.; Rojas-Hernandez, A.; Funes, S.; Uribe-Carvajal, S.; Gonzalez-Halphen, D. In *Polytomella* sp. Mitochondria, biogenesis of the heterodimeric COX2 subunit of cytochrome c oxidase requires two different import pathways. *Biochim. Biophys. Acta* **2012**, *1817*, 819–827. [CrossRef] [PubMed]
155. Jimenez-Suarez, A.; Vazquez-Acevedo, M.; Miranda-Astudillo, H.; Gonzalez-Halphen, D. Cox2A/Cox2B subunit interaction in *Polytomella* sp. cytochrome c oxidase: Role of the Cox2B subunit extension. *J. Bioenerg. Biomembr.* **2017**, *49*, 453–461. [CrossRef] [PubMed]

© 2018 by the authors. Licensee MDPI, Basel, Switzerland. This article is an open access article distributed under the terms and conditions of the Creative Commons Attribution (CC BY) license (http://creativecommons.org/licenses/by/4.0/).

Review

Emerging Roles of Mitochondrial Ribosomal Proteins in Plant Development

Pedro Robles and Víctor Quesada *

Instituto de Bioingeniería, Universidad Miguel Hernández, Campus de Elche, 03202 Elche, Spain; probles@umh.es
* Correspondence: vquesada@umh.es; Tel.: +34-96-665-88-12; Fax: +34-96-665-85-11

Received: 8 November 2017; Accepted: 1 December 2017; Published: 2 December 2017

Abstract: Mitochondria are the powerhouse of eukaryotic cells because they are responsible for energy production through the aerobic respiration required for growth and development. These organelles harbour their own genomes and translational apparatus: mitochondrial ribosomes or mitoribosomes. Deficient mitochondrial translation would impair the activity of this organelle, and is expected to severely perturb different biological processes of eukaryotic organisms. In plants, mitoribosomes consist of three rRNA molecules, encoded by the mitochondrial genome, and an undefined set of ribosomal proteins (mitoRPs), encoded by nuclear and organelle genomes. A detailed functional and structural characterisation of the mitochondrial translation apparatus in plants is currently lacking. In some plant species, presence of small gene families of mitoRPs whose members have functionally diverged has led to the proposal of the heterogeneity of the mitoribosomes. This hypothesis supports a dynamic composition of the mitoribosomes. Information on the effects of the impaired function of mitoRPs on plant development is extremely scarce. Nonetheless, several works have recently reported the phenotypic and molecular characterisation of plant mutants affected in mitoRPs that exhibit alterations in specific development aspects, such as embryogenesis, leaf morphogenesis or the formation of reproductive tissues. Some of these results would be in line with the ribosomal filter hypothesis, which proposes that ribosomes, besides being the machinery responsible for performing translation, are also able to regulate gene expression. This review describes the phenotypic effects on plant development displayed by the mutants characterised to date that are defective in genes which encode mitoRPs. The elucidation of plant mitoRPs functions will provide a better understanding of the mechanisms that control organelle gene expression and their contribution to plant growth and morphogenesis.

Keywords: mitoribosomes; mitochondrial ribosomal proteins (mitoRPs); arabidopsis; ribosomal filter hypothesis; plant development; mutants

1. Introduction

Ribosomes are the cellular machinery that performs protein synthesis from translating the information contained in mRNA molecules. They are ribonucleoprotein complexes that comprise two subunits, one large (LSU) and one small (SSU), and consist of rRNAs and proteins. In a eukaryotic cell, ribosomes are found in the cytoplasm, mitochondria and plant chloroplasts. In evolutionary terms, chloroplasts and mitochondria derive from the ancestors of current cyanobacteria and α-proteobacteria, respectively, that established a symbiotic relationship with an ancestral eukaryote. During evolution, the number of genes in the endosymbiotic genomes drastically dropped as most were transferred to the nuclear genome. Hence they contain only a few dozen genes in the present-day. Transferred genes also include those that encode mitochondrial and plastid ribosomal proteins, although both organelles have retained in their genomes some genes encoding the ribonucleoprotein complexes.

A mitochondrion is a double-membrane organelle essential for life, and is present virtually in all eukaryotic cells, except for several protozoa, some fungi and mature red blood cells in mammals [1]. Widely known for its role in ATP production through oxidative phosphorylation, the mitochondrion also plays a key role in a wide range of cellular functions, such as fatty acid oxidation, amino acid biosynthesis, apoptosis and transduction of cellular signals [2]. All these processes require accurate protein synthesis inside the organelle.

Mitochondrial ribosomes, or mitoribosomes, are essential for the synthesis of oxidative phosphorylation machinery. They have been subjected to major research efforts in yeast and humans, in the former for being a model system for eukaryotic cell biology, and in the latter for mitoribosomes being implicated in human health. Both the composition and structure of mitoribosomes in both systems have been solved by cryo-EM [3,4]. In contrast, the precise structure and protein composition of plant mitoribosomes are not yet known [5], although they are bigger (around 78S) than mammalian mitoribosomes (55S) [6]. Regarding rRNA composition, plant mitoribosomes are constituted of three different molecules (5S, 18S and 26S), all of which are encoded by the mitochondrial genome [7]. In contrast, the genes that encode plant mitochondrial ribosomal proteins (hereafter mitoRPs) lie in both the nuclear and mitochondrial genomes, and their numbers vary from one species to another. Accordingly, Bonen and Calixte [8] identified in *Arabidopsis thaliana* (hereafter Arabidopsis) and rice nuclear genomes 46 and 48 genes, respectively, that encode mitoRPs were 11 of these genes present in multiple copies (2–4). Furthermore, these authors also identified seven additional mitoRP genes in the Arabidopsis mitochondrial genome. Sormani et al. [9] described 71 genes in Arabidopsis that encode mitoRPs, with 63 and 8 of them located in the nuclear and the mitochondrial genomes, respectively. Similar numbers were also reported for potato and broad bean with 68 to 80 mitoRPs [10,11]. In contrast, a typical eubacteria such as *Escherichia coli* contains 54 ribosomal proteins, 33 and 21 in the LSU and SSU subunits, respectively [12].

In ribosomes, each ribosomal protein type is represented by a single polypeptide. However, as stated above, several ribosomal proteins are encoded by two genes or more of the same family (paralogous genes), which results from gene duplications. In Arabidopsis, 13 plastid ribosomal proteins and 16 mitoRPs are encoded by small-multigenic families [9], whereas the 81 ribosomal protein types that integrate cytoplasmic ribosomes are encoded by 251 genes [13]. The expression patterns of paralogous genes may differ, as shown for members of the families that encode the Arabidopsis cytoplasmic S18, L16 and S15 proteins [14–16]. This suggests that they may be involved in different developmental processes and/or may act at distinct times in tissues or cell types. Furthermore, translation in plants may be regulated by modifying the composition of the proteins that form part of the ribosome. Accordingly, the abundance and composition of polysomes (groups of ribosomes that translate the same mRNA) vary while bean leaves grow and develop [17]. In addition, transcript profiling in *Brassica napus* has revealed the existence of functional divergence and expression networks among the paralogous genes that encode ribosomal proteins, which strongly suggests their participation in development, differentiation and/or tissue-specific processes [18].

The presence in plants of small gene families of mitoRPs, whose members are functionally divergent, has also been reported. In line with this, four paralogues of mitochondrial L12 protein in potato have been differentially associated with mitochondrial ribosomes [19] and eight members of the Arabidopsis L18 family have highly divergent sequences and specificities during plant growth and development [20]. This supports the hypothesis of the heterogeneity of plant mitoribosomes, which would allow a highly dynamic mitochondrial translational machinery composition [21], and constitutes the basis of the so-called ribosomal filter hypothesis proposed by Mauro and Edelman [22]. This hypothesis argues that ribosomes are not simply machines that carry out translation, but they are also able to regulate gene expression. Consequently, the ribosome would act as a filter that would select specific mRNA molecules for translation in response to different physiological conditions during development. Hence distinct populations of ribosomes would have varying abilities to translate particular mRNA molecules [5].

This review principally focuses on analysing the perturbed plant developmental processes and the resulting phenotypes hitherto described, caused by mutations in genes that encode mitoRPs, or in other genes that impair the mitoRP function.

2. Developmental Defects Caused by Mutations in Genes that Encode mitoRPs

Plant growth, including cell expansion and division, is fundamental for plant development and morphogenesis, and requires a substantial supply of energy and metabolites. This is in consonance with the increased number of mitochondria in cells observed during leaf and reproductive development [21]. Therefore, perturbed mitochondrial translation is expected to severely impair mitochondrial activity and, consequently, plant developmental processes will require this organelle to perform well. To date, mutations in both nuclear and mitochondrial genes that encode mitoRPs have been reported to affect plant growth and development. The phenotypic alterations described to date due to these mutations have clearly shown the involvement of mitoRPs in several aspects of plant development and different plant processes. Accordingly, mutations in some mitoRPs result in an embryo-lethal phenotype while the analysis of other mutants has revealed a role for some mitoRPs in leaf morphogenesis and in reproductive tissue formation (Table 1).

Table 1. Plant mitochondrial ribosomal proteins characterized from the analysis of developmental mutants.

Defects in	Gene	mitoRP [a]	Species	Mutant Phenotype
Embryo development	HEART STOPPER (HES) [b] AT1G08845 [d]	L18	Arabidopsis thaliana	Reduced proliferation of endosperm cells and arrested embryo development in the late globular stage [20]
Reproductive development	HUELLENLOS (HLL) [b] AT1G17560 [d]	L14	Arabidopsis thaliana	Early cellular degeneration of the eggs, characterised by arrested ovule development before or just after the formation of the integuments (hll-1) or after the integuments have begun to spread around the nucela (hll-2). hll-1 and hll-2 also show alterations in the gynoecium [23]
	NUCLEAR FUSION DEFECTIVE1 (NFD1) [b] AT4G30925 [d]	L21	Arabidopsis thaliana	Defective in kariogamy during fertilization and development of the female and male gametophytes [24]
	NFD3 [b] AT1G31817 [d]	S11	Arabidopsis thaliana	Defective in kariogamy during fertilization and development of the female gametophyte [24]
Vegetative development	rps3 [c] and rpl16 [c]	S3 and L16	Zea mays	Sectors of poorly developed tissue on leaves and ears, which result from the segregation of somatic wild-type and mutant mitochondria [25]
	rps3 [c] and rpl16 [c] AtMg00090 [d] and AtMg00080 [d]	S3 and L16	Arabidopsis thaliana	Distorted leaf phenotype [26]
	Rps10 [b] AT3G22300 [d]	S10	Arabidopsis thaliana	Plants homozygous for S10 silencing, show severe morphological alterations; they exhibit small, undulating, and yellowish leaves and died prior bolting [27]
	Mrpl11 [b] AT4G35490 [d]	L11	Arabidopsis thaliana	Stunted plant size and a darker leaf coloring than the wild type [28]

[a] Mitochondrial ribosomal protein; [b] Nuclear gene; [c] Mitochondrial gene; [d] AGI code.

2.1. Embryo-Lethal Mutations in mitoRPs

In Arabidopsis, the *hes* (*heart stopper*) mutant, which is affected in mitochondrial ribosomal protein L18, displays a low proliferation of seed endosperm cells and arrested embryo development in the late globular stage (Table 1) [20]. *hes* embryos have been cultured in vitro, but their phenotypic rescue has not yet been achieved. Although some give rise to callus, they do not differentiate into plants despite adding hormones to the culture medium. This indicates that HES is required for cell growth, differentiation and the establishment of organ patterns. Zhang et al. [20] identified eight genes that encode L18 ribosomal proteins in the Arabidopsis nuclear genome, five and two of them potentially located in the mitochondria and chloroplasts, respectively. The subcellular localisation of the remaining one is ambiguous. Interestingly, these authors found that the members of this small gene family markedly differ in their amino acid sequences. Besides, the *hes* mutant phenotype cannot be complemented by other L18 members. *HES* expression is restricted to tissues that undergo active cell division and differentiation, including the embryo and root tip. The spatial expression pattern of *HES* corresponds well to the mutant phenotypes of the *hes* individuals during seed development. In *E. coli*, L18 is an essential protein that forms part of the central protuberance of the 50S subunit of the ribosome and binds to 5S and 23S rRNAs [29]. The 3D modelling of mutant and wild-type L18 proteins suggests that the amino acid substitution present in the *hes* mutant protein might affect its binding to 5S rRNA and hence, the stability of the 50S subunit. However, the *hes* mutation does not alter the mitochondria morphology in the embryo or the endosperm. This made Zhang et al. [20] to propose that the effects on development caused by impaired *HES* function might be due to alterations in the mitochondrial metabolic processes affected by reduced mitochondrial translation, which would require L18. Consistently with this, these authors identified several marker genes of mitochondrial dysfunction, such as *ALTERNATIVE OXIDASE 1a* (*AOX1a*) and *NAD(P)H DEHYDROGENASE* (*NDB4*), which are overexpressed in the *hes* mutant compared with the wild type. They concluded that the strong divergence between the genes that encode L18 proteins, the restricted expression pattern of *HES* and the inability of other L18 proteins to complement the *hes* mutant phenotype all support the existence of heterogeneous mitoribosomes, which would consist in different L18 proteins with distinct functions. Heterogeneous mitoribosomes would likely have different properties and could modulate gene expression, which would affect the translation efficiency of certain types of mRNAs in response to different physiological requirements during development.

Interestingly, the loss of function of one of the Arabidopsis L18 proteins, EMB3105 encoded by the AT1G48350 gene, and putatively localised in plastids, causes embryonic lethality in the same developmental stage as the *hes* mutations does (the globular stage; [30,31]).

2.2. Effects of the Mutations in mitoRPs on Reproductive Tissues

The characterisation of plant mutants has revealed a role for some mitoRPs in reproductive tissue formation. Along these lines, the Arabidopsis *huellenlos-1* (*hll-1*) and *hll-2* mutants are good representatives (Table 1) [23]. *hll-1* and *hll-2* individuals carry point mutations which lead to truncated L14 mitoribosomal proteins and cause arrested ovule development before or immediately after the formation of integuments of ovules (*hll-1*), or after integuments have begun to spread around the nucela (*hll-2*) [23]. *hll-1* and *hll-2* also present alterations in the gynoecium, which is smaller than in the wild type and has a few ovules. In the Arabidopsis genome, Skinner et al. [23] identified a paralogous gene functionally related with *HLL*, *HUELLENLOS PARALOG* (*HLP*). The ectopic expression of *HLP* complements the *hll* mutant phenotype [23]. This contrast with the lack of complementation of the *hes* mutant phenotype by other L18 proteins (see above). Notwithstanding, both genes differ in their expression levels in organs because transcripts of the *HLP* and *HLL* genes are detected mostly in pistils and leaves, respectively. In addition, the HLL and HLP proteins fused to the green fluorescent protein (GFP) are targeted to mitochondria, which supports a role for both proteins in this organelle. In *E. coli*, the L14 ribosomal protein is an essential protein that binds to rRNA and participates in forming a bridge between the two ribosomal subunits [32]. This falls in line with the phenotype of gametic lethality found in *hll*. Skinner et al. [23] proposed that the phenotypic effect of *hll* mutations on reproductive development might be explained by carpels and ovules' considerable energy requirements. In agreement with this, an increase in the number of mitochondria in reproductive tissues and the specific degeneration of ovaries in transgenic plants with reduced activity of the citrate synthase enzyme, commonly used as a quantitative marker of the presence of intact mitochondria, have been reported [33,34].

Karyogamy, this being the fusion of two cellular nuclei to produce a single nucleus, is fundamental for the sexual reproduction of animals and plants [35]. An analysis of an array of Arabidopsis mutants, affected in the fusion of the polar nuclei during female gametophyte development, allowed Portereiko et al. [24] to identify six mutants, namely *nuclear fusion defective 1* (*nfd1*) to 6. One of these mutants, *nfd1*, is also affected in karyogamy during fertilisation and male gametophyte development (Table 1). Defective karyogamy is due to the non-fusion of outer nuclear membranes [24]. The nuclear *NFD1* gene encodes the L21 mitoRP of Arabidopsis, and the orthologous protein in *E. coli* is a component of the 50S subunit of the mitoribosome, which binds to 23S rRNA [36,37]. The Arabidopsis genome contains a single gene for the mitochondrial L21 protein, which is expressed in all the studied organs. Portereiko et al. [24] proposed that the *nfd1* mutation might impair nuclear fusion by altering the composition of the phospholipids of the nuclear membrane. The importance of mitochondria in karyogamy is further supported by the identification of four additional *nfd* mutants (*nfd3* to 6) which also carry T-DNA insertions in nuclear genes predicted to encode mitochondrial proteins [24]. One of them, *NFD3*, encodes S11 mitoRP of the 30S subunit (Table 1). Other Arabidopsis mutants affected in genes that encode mitochondrial proteins such as *gametophytic factor2* [38], *embryo sac development arrest28* (*eda28*) and *eda35* [39] are also defective in cellular nuclei fusion.

Remarkably, mutations in the plastid ribosomal L21 protein, the only homolog of Arabidopsis NFD1, cause embryonic lethality in the globular stage [40,41]. The different L21 proteins hitherto characterised in several species through the analysis of loss of function mutant alleles, suggest a key role for these proteins in ribosomal function. Nonetheless, their biological effects cannot be directly inferred [40]. Despite being conserved, these proteins might play different complex roles in plant development, partly due to their different subcellular localisation (cytoplasm, mitochondria or chloroplasts).

2.3. Mutations in mitoRP Genes Affect Vegetative Development

2.3.1. Alterations in Leaf Morphology

Defects in leaf development due to mutations in some mitoRPs have been reported: the maize "non-chromosomal stripe" NCS3 mutant displays sectors of poorly developed tissue on leaves and ears, which results from the segregation of somatic wild-type and mutant mitochondria (Table 1) [25]. The molecular nature of this phenotype is a deletion produced by a mitochondrial DNA (mtDNA) rearrangement of a region that contains genes *rps3* and *rpl16*, which respectively code for mitochondrial ribosomal proteins S3 and L16 [25]. Remarkably, Sakamoto et al. [26] also described a mtDNA rearrangement that affects Arabidopsis mitochondrial genes *rps3*–*rpl16*, caused by the recessive nuclear mutation *chloroplast mutator*, which results in a distorted leaf phenotype (Table 1). Genes *S3* and *L16* have proven to be essential in *E. coli* [42,43] and their protein products appear to function as assembly factors of their corresponding ribosomal subunits [44,45]. More recently, the analysis of the maize *mppr6* mutant impaired in the nuclear gene that encodes mitochondrial pentatricopeptide repeat protein (PPR) MPPR6, which is required for the posttranscriptional regulation of the mitochondrial *rps3* gene, suggests a role of the latter gene also in embryo and endosperm development [46].

Other phenotypes characterised by severe irregularities in leaf morphology have also been reported for defective nuclear genes that encode mitoRPs. Accordingly, the down-regulation by RNAi silencing of the Arabidopsis gene for S10 mitoRP causes severe leaf anomalies (Table 1) [27]. In bacteria, the orthologous protein of S10 is NusE, a multifunctional protein that recruits the ribosome to RNA polymerase [47]. In order to study the effect of S10 mitoRP silencing on vegetative growth, Majewski et al. [27] cultivated transgenic plants under short day conditions (SD) to favour plant growth on reproductive development because SD delays the onset of flowering. Transgenic plants exhibited vastly varying morphologies in relation to the homozygous vs. hemizygous state of the transgene used for gene silencing, and from the timing of its onset [27]. Accordingly, plants homozygous for

S10 silencing, showed severe morphological alterations and some even exhibited small, undulating yellowish leaves that died prior to bolting [27].

Kwasniak et al. [48] focused on studying the effects of silencing the Arabidopsis *S10* gene on the expression of the mitochondrial and nuclear genes that encode mitoRPs or proteins of the mitochondrial respiratory chain (Table 1) [48]. They concluded that the perturbation of *S10* alters the levels of the above-mentioned mitochondrial components, especially those encoded by the mitochondrial genome. Thus, in the transgenic plants with the *S10* silenced gene, the transcript levels of the mitochondrial genome genes increased, especially those that code for mitoRPs, whereas those transcribed from the nuclear genes barely alter. At the translational level, mitoRPs and respiratory chain proteins accumulate in the *S10* silenced plants at higher and lower levels than in the wild type, respectively [48]. This suggests the existence of differential changes in mitochondrial translation efficacy when the mitoribosomal function is compromised. The authors proposed that mitoribosomes can self-regulate their own biogenesis by translational control, as previously reported in bacteria and chloroplasts [49,50]. The results of Kwasniak et al. [48] support the ribosomal filter hypothesis proposed by Mauro and Edelman [22], which states that ribosomes are not simple machines for mRNA translation, but can act as regulators of gene expression by acting as a filter that differentially affects the translation of different transcripts. In line with this, defective mitoribosomes, due to the silencing of the S10 protein, would differentially affect the translation of different mRNA species.

Consistent with this view, Schippers and Mueller-Roeber [21] have reported that the expression of the genes that encode mitoRPs and the relative translational activity of different ribosomal protein transcripts in several leaf tissues are highly variable during leaf development in Arabidopsis.

2.3.2. Mutations in mitoRPs and the OGE Retrograde Signalling Pathway

Other mutations in *mitoRP* genes have a subtle effect on leaf development. For instance, the Arabidopsis mutant defective for the nuclear gene that encodes L11 mitoRP shows reduced mitochondrial respiratory proteins abundance, which suggests an alteration in mitochondrial activity. As a likely consequence, *mrpl11* plants display stunted plant size and a darker leaf colouring than the wild type (Table 1). However, no clear alteration in leaf morphology has been reported [28]. In *E. coli*, L11 is a non-essential protein [42] and constitutes one of the main anatomical features of the 50S ribosomal subunit, the L11 arm, which includes the binding site for the 23S rRNA [51], and may be important for translation termination [52]. Pesaresi et al. [53] had previously reported that the *prpl11* mutant, which is affected in the nuclear gene that encodes the plastid L11 protein, shows reduced growth and pale pigmentation in cotyledons and leaves. Interestingly, double mutant plants *mrpl11 prpl11*, but none of the single mutant plants, display a drastically reduced expression of nuclear genes that encode photosynthetic proteins targeted to chloroplasts (Table 1) [28]. The repression of nuclear photosynthetic genes may result from perturbed plastid and/or mitochondrial gene expressions due to the activation of the retrograde signalling pathway named OGE (organelle gene expression). Therefore, the results reported by Pesaresi et al. [28] indicate cooperation for the signals emitted by chloroplasts and mitochondria to regulate the expression of nuclear photosynthetic genes when translation in both organelles is disturbed. This down-regulation of nuclear photosynthetic genes is similar to that reported for the Arabidopsis *prors1-1* and *1-2* mutants affected in the nuclear gene that encodes the prolyl-tRNA synthetase protein targeted to both chloroplasts and mitochondria [28]. Remarkably, null mutant alleles *prors1-3* and *1-4* are embryonic-lethal as they arrest embryonic sac formation and, hence, embryo development [28].

3. Defective Mitoribosomal Function by Mutations in Mitochondrial Proteins Other than mitoRPs

The plant mitoribosome function can be modulated by the activity of nuclear genes that encode mitochondrial-targeted proteins apart from mitoRP. One example of this is the PPR family of proteins, a large group of eukaryotic-specific modular RNA proteins encoded by the nucleus that have undergone expansion in terrestrial plants [54]. PPR proteins are important for the expression of organelle genomes and organelle biogenesis because they are involved in transcription, and also in RNA stability, processing, splicing, editing and translation [54,55]. In line with this, the Arabidopsis PPR336 protein has been associated with mitochondrial polysomes and is required for the stability of mitoribosomes [54]. Notwithstanding, no morphological alterations have been described for the mutants affected in the *PPR336* gene. Despite this, the mitochondrial polysomes in these mutant plants have a lighter molecular weight than those of wild-type plants, which might have an effect on mitochondria protein translation [56]. Interestingly, Del Valle-Echevarria et al. [57] found that the MCS16 mosaic mutant of cucumber, which displays distorted cotyledons, chlorotic leaves, stunted growth and reduced fertility, also shows lower levels of the transcripts of the *rps7* mitochondrial gene, which codes for S7 mitoRP. These authors proposed the *PPR336* gene of cucumber to be the likely candidate responsible for the phenotype of the MCS16 mutant as PPR336 is required for the accurate processing of *rps7* transcripts [58]. The S7 protein is essential in *E. coli* [42,43] and, together with the S11 protein, forms the 30S E (exit) site [59]. Besides, S7 binds to 16S rRNA and functions as an assembly initiator of the 30S subunit in bacteria [60].

In Arabidopsis, another PPR protein, the product of the *PNM1* (*PPR protein localized to the nucleus and mitochondria 1*) gene, has also been reported to be associated with mitochondrial polysomes in an RNA-dependent manner [59]. Remarkably, impaired *PNM1* function in the mitochondria is embryo-lethal, although it has not been possible to identify the precise RNA targets of the PNM1 protein [61]. The null mutations in the *EMP5* gene (*EMPTY PERICARP5*) of maize, which encodes a DYW subgroup of PPR proteins involved in editing several mitochondrial transcripts, result in kernels devoid of embryo or endosperm structures, which reveals a role for this gene in seed development [62]. Interestingly, these defects are due mainly to the incorrect editing of *rpl16* mitochondrial transcripts by changing a leucine for a proline residue at position 153. This change may be critical for the L16 protein function, and hence for mitorribosome activity, as it alters organelle function and compromises seed development. This would extend the *rpl16* function to not only leaf morphogenesis, as previously mentioned (see Section 2.3.1), but also to seed development. The *EMP5* function seems conserved in rice because its down-regulation results in defective seeds and slower seedling growth, which indicates other roles for this protein in plant development apart from embryonic ones. Remarkably, the Arabidopsis *mef35* (*mitochondrial editing factor 35*) mutant, which is affected in a nuclear gene encoding, as *EMP5*, a DYW PPR protein, also displays a defect in the editing of the mitochondrial *rpl16* transcript by changing a very conserved threonine of the L16 protein for isoleucine [63]. Yet unlike *emp5*, this change has no phenotypic effects on *mef35* plants and questions whether the edition of *L16* mediated by MEF35 has any functional consequences.

4. Conclusions and Future Perspectives

In plants, only a few mutants affected in mitoRPs have been hitherto described and characterised phenotypically and molecularly. Therefore, information on the contribution of plant mitoRPs and, by extension, mitoribosomes, to plant growth and to different development stages is still scarce. Nevertheless, the results obtained in recent years by characterising several plant mutants defective in mitoRPs reveals a prominent role for these proteins in plant morphogenesis (Figure 1). Some of the results obtained to date support the participation of specific mitoRPs in different developmental processes, which might be interpreted as a result of the functional specialisation of distinct mitoRPs [20,23–28]. Accordingly, the modification of the protein composition of mitoribosomes in various plant tissues, organs or developmental stages may be a mechanism to help regulate its activity and, finally, the expression of the genes whose products are located in mitochondria. Consequently, mitochondrial activity would adjust to the needs of the biological processes that take place at specific times of development. If this were the case, it would support the plant mitoribosomes heterogeneity hypothesis, which is the basis of the so-called ribosomal filter hypothesis [22]. In this review, we focused on several pieces of genetic evidence that support this hypothesis in plant mitochondria. To strengthen such evidence, we consider it necessary to look in-depth into the isolation and characterisation of new mutants affected in mitoRP genes in Arabidopsis and other plant species. Special attention should be paid to the mutants defective in different members of gene families to identify differential phenotypic effects. In a plant model such as Arabidopsis, it is possible to screen collections of insertional mutations, mainly induced by T-DNA, to cover almost every gene [64]. This allows systematic screening for those mutants defective in each predicted *mitoRP* gene. Nonetheless, some genes may not be tagged and, even if they are, the insertion might not affect the function of the corresponding protein or cause a desirable structural or functional change. New genome editing tools based on the CRISPR/Cas system could overcome these limitations [65] and be used to generate new alleles of either previously described nuclear *mitoRP* genes or novel ones. In line with this, a mitochondria-targeted Cas9 (mitoCas9) protein has been designed and used in cultured human cells to edit the mitochondrial genome [66]. Gene editing might also be applied to create a series of hypomorphic alleles of mitoRP genes. To date, only null alleles of the *HES*, *HLL* and *NFD* genes causing embryonic, ovule or gametophyte lethality respectively, have already been described [20,23,24]. Therefore, the identification and characterisation of hypomorphic alleles of these genes should be instrumental to ascertain if the functions of the corresponding mitoRPs are restricted exclusively to early development. To define the post-embryonic functions of lethal genes, other genetic and molecular strategies, such as clonal analysis in post-embryonic tissues [67], lethality rescue based on inducible promoters [68] or post-embryonic knock-down mediated by tissue-specific [69] or inducible promoters [70], may also be used.

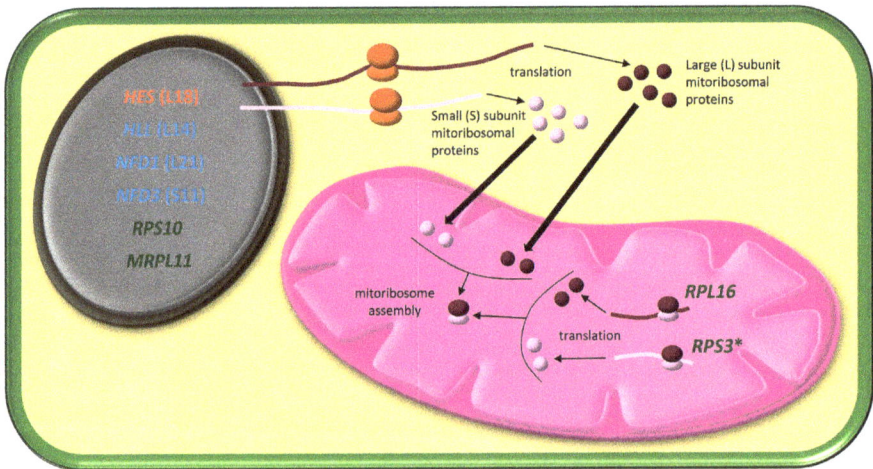

Figure 1. Genes that encode mitochondrial ribosomal proteins (mitoRPs) whose mutations cause developmental defects are shown in the diagrams for the nucleus (grey) and mitochondria (magenta). The mRNAs encoding proteins of the large (dark purple spheres) and small (light purple spheres) subunits are shown in dark purple and light purple, respectively. The genes characterised from the analysis of the mutants defective in embryonic, vegetative or reproductive development are respectively depicted in red, blue and green. When a gene was named according to a mutant phenotype, the encoded mitoRP is shown in parentheses. Cytosolic ribosomes are depicted in orange and mitorribosomes in purple. HES: HEART STOPPER; HLL: HUELLENLOS; NFD1 and 3: NUCLEAR FUSION DEFECTIVE 1 and 3. * The mutations that affect the genes in this figure were all characterised in *Arabidopsis thaliana*, except for *RPS3*, for which a mutant allele was also described in *Zea mays*.

Besides genetic evidence, it has been proposed that demonstration of the existence of specialized ribosomes will require resolving three main challenges: (a) the isolation of naturally-occurring specific homogenous ribosomes; (b) their structural, biochemical, molecular and cellular characterisation; (c) the identification and validation of the different substrates of the specialised ribosomes [71]. A plethora of new technical advances, such as single-particle cryo-electron microscopy [72] and serial femtosecond X-ray crystallography [73], among others, might contribute to characterise the different ribosomes found in a particular species, organ, tissue, cell or organelle, and to set up their unique structural and functional properties. This is particularly relevant in plants because to date, the cryo-EM structure of mitorribosomes is still lacking. A recent study into cytosolic ribosomes of mouse embryonic stem cells by quantitative mass spectrometry has revealed a functional link between ribosome heterogeneity, at the RPs composition level, and gene regulation [74]. Consequently, translating ribosomes lacking particular RPs associate with specific types of mRNAs. Similar studies of organelle ribosomes are expected to also reveal a functional relationship between its composition and the control of the gene expression in mitochondria and chloroplasts.

Acknowledgments: The research conducted in the laboratory of V. Q. has been supported by grants from the Conselleria d´ Educació of the Generalitat Valenciana (Spain) (GV/2009/058 and AICO/2015).

Conflicts of Interest: The authors declare no conflict of interest.

References

1. Selwood, S.P.; Chrzanowska-Lightowlers, Z.M.; Lightowlers, R.N. Does the mitochondrial transcription-termination complex play an essential role in controlling differential transcription of the mitochondrial DNA? *Biochem. Soc. Trans.* **2000**, *28*, 154–159. [CrossRef] [PubMed]
2. Bonawitz, N.D.; Clayton, D.A.; Shadel, G.S. Initiation and beyond: Multiple functions of the human mitochondrial transcription machinery. *Mol. Cell* **2006**, *24*, 813–825. [CrossRef] [PubMed]
3. Amunts, A.; Brown, A.; Toots, J.; Scheres, S.; Ramakrishnan, V. The structure of the human mitochondrial ribosome. *Science* **2015**, *348*, 95–98. [CrossRef] [PubMed]
4. Desai, N.; Brown, A.; Amunts, A.; Ramakrishnan, V. The structure of the yeast mitochondrial ribosome. *Science* **2017**, *355*, 528–531. [CrossRef] [PubMed]
5. Janska, H.; Kwasniak, M. Mitoribosomal regulation of OXPHOS biogenesis in plants. *Front. Plant Sci.* **2014**, *5*, 79. [CrossRef] [PubMed]
6. Breiman, A.; Fieulaine, S.; Meinnel, T.; Giglione, C. The intriguing realm of protein biogenesis: Facing the green co-translational protein maturation networks. *Biochim. Biophys. Acta* **2016**, *1864*, 531–550. [CrossRef] [PubMed]
7. Petersen, G.; Cuenca, A.; Moller, I.M.; Seberg, O. Massive gene loss in mistletoe (*Viscum*, Viscaceae) mitochondria. *Sci. Rep.* **2015**, *5*, 17588. [CrossRef] [PubMed]
8. Bonen, L.; Calixte, S. Comparative analysis of bacterial-origin genes for plant mitochondrial ribosomal proteins. *Mol. Biol. Evol.* **2006**, *23*, 701–712. [CrossRef] [PubMed]
9. Sormani, R.; Masclaux-Daubresse, C.; Daniel-Vedele, F.; Chardon, F. Transcriptional regulation of ribosome components are determined by stress according to cellular compartments in *Arabidopsis thaliana*. *PLoS ONE* **2011**, *6*, e28070. [CrossRef] [PubMed]
10. Pinel, C.; Douce, R.; Mache, R. A study of mitochondrial ribosomes from the higher plant *Solanum tuberosum* L. *Mol. Biol. Rep.* **1986**, *11*, 93–97. [CrossRef] [PubMed]
11. Maffey, L.; Degand, H.; Boutry, M. Partial purification of mitochondrial ribosomes from broad bean and identification of proteins encoded by the mitochondrial genome. *Mol. Gen. Genet.* **1997**, *254*, 365–371. [CrossRef] [PubMed]
12. Wittmann, H.G. Components of bacterial ribosomes. *Annu. Rev. Biochem.* **1982**, *51*, 155–183. [CrossRef] [PubMed]
13. Barakat, A.; Szick-Miranda, K.; Chang, I.F.; Guyot, R.; Blanc, G.; Cooke, R.; Delseny, M.; Bailey-Serres, J. The organization of cytoplasmic ribosomal protein genes in the Arabidopsis genome. *Plant Physiol.* **2001**, *127*, 398–415. [CrossRef] [PubMed]
14. Van Lijsebettens, M.; Vanderhaeghen, R.; De Block, M.; Bauw, G.; Villarroel, R.; Van Montagu, M. An S18 ribosomal protein gene copy at the Arabidopsis *PFL* locus affects plant development by its specific expression in meristems. *EMBO J.* **1994**, *13*, 3378–3388. [PubMed]

15. Williams, M.E.; Sussex, I.M. Developmental regulation of ribosomal protein L16 genes in *Arabidopsis thaliana*. *Plant J.* **1995**, *8*, 65–76. [CrossRef] [PubMed]
16. Hulm, J.L.; McIntosh, K.B.; Bonham-Smith, P.C. Variation in transcript abundance among the four members of the *Arabidopsis thaliana RIBOSOMAL PROTEIN S15a* gene family. *Plant Sci.* **2005**, *169*, 267–278. [CrossRef]
17. Makrides, S.C.; Goldthwaite, J. Biochemical changes during bean leaf growth, maturity and senescence. Content of DNA, polyribosomes, ribosomal RNA, protein and chlorophyll. *J. Exp. Bot.* **1981**, *32*, 725–735. [CrossRef]
18. Whittle, C.A.; Krochko, J.E. Transcript profiling provides evidence of functional divergence and expression networks among ribosomal protein gene paralogs in *Brassica napus*. *Plant Cell* **2009**, *21*, 2203–2219. [CrossRef] [PubMed]
19. Delage, L.; Giegé, P.; Sakamoto, M.; Maréchal-Drouard, L. Four paralogues of RPL12 are differentially associated to ribosome in plant mitochondria. *Biochimie* **2007**, *89*, 658–668. [CrossRef] [PubMed]
20. Zhang, H.; Luo, M.; Day, R.C.; Talbot, M.J.; Ivanova, A.; Ashton, A.R.; Chaudhury, A.M.; Macknight, R.C.; Hrmova, M.; Koltunow, A.M. Developmentally regulated *HEART STOPPER*, a mitochondrially targeted L18 ribosomal protein gene, is required for cell division, differentiation, and seed development in Arabidopsis. *J. Exp. Bot.* **2015**, *66*, 5867–5880. [CrossRef] [PubMed]
21. Schippers, J.; Mueller-Roeber, B. Ribosomal composition and control of leaf development. *Plant Sci.* **2010**, *179*, 307–315. [CrossRef]
22. Mauro, V.P.; Edelman, G.M. The ribosome filter redux. *Cell Cycle* **2007**, *6*, 2246–2251. [CrossRef] [PubMed]
23. Skinner, D.J.; Baker, S.C.; Meister, R.J.; Broadhvest, J.; Schneitz, K.; Gasser, C.S. The Arabidopsis *HUELLENLOS* gene, which is essential for normal ovule development, encodes a mitochondrial ribosomal protein. *Plant Cell* **2001**, *13*, 2719–2730. [CrossRef] [PubMed]
24. Portereiko, M.F.; Sandaklie-Nikolova, L.; Lloyd, A.; Dever, C.A.; Otsuga, D.; Drews, G.N. *NUCLEAR FUSION DEFECTIVE1* encodes the Arabidopsis RPL21M protein and is required for karyogamy during female gametophyte development and fertilization. *Plant Physiol.* **2006**, *141*, 957–965. [CrossRef] [PubMed]
25. Hunt, M.D.; Newton, K.J. The NCS3 mutation: Genetic evidence for the expression of ribosomal protein genes in *Zea mays* mitochondria. *EMBO J.* **1991**, *10*, 1045–1052. [PubMed]
26. Sakamoto, W.; Kondo, H.; Murata, M.; Motoyoshi, F. Altered mitochondrial gene expression in a maternal distorted leaf mutant of Arabidopsis induced by *Chloroplast mutator*. *Plant Cell* **1996**, *8*, 1377–1390. [CrossRef] [PubMed]
27. Majewski, P.; Wołoszyńska, M.; Jańska, H. Developmentally early and late onset of *Rps10* silencing in *Arabidopsis thaliana*: Genetic and environmental regulation. *J. Exp. Bot.* **2009**, *60*, 1163–1178. [CrossRef] [PubMed]
28. Pesaresi, P.; Masiero, S.; Eubel, H.; Braun, H.P.; Bhushan, S.; Glaser, E.; Salamini, F.; Leister, D. Nuclear photosynthetic gene expression is synergistically modulated by rates of protein synthesis in chloroplasts and mitochondria. *Plant Cell* **2006**, *18*, 970–991. [CrossRef] [PubMed]
29. Shajani, Z.; Sykes, M.T.; Williamson, J.R. Assembly of bacterial ribosomes. *Annu. Rev. Biochem.* **2011**, *80*, 501–526. [CrossRef] [PubMed]
30. Bryant, N.; Lloyd, J.; Sweeney, C.; Myouga, F.; Meinke, D. Identification of nuclear genes encoding chloroplast-localized proteins required for embryo development in Arabidopsis. *Plant Physiol.* **2011**, *155*, 1678–1689. [CrossRef] [PubMed]
31. Muralla, R.; Lloyd, J.; Meinke, D. Molecular foundations of reproductive lethality in *Arabidopsis thaliana*. *PLoS ONE* **2011**, *6*, e28398. [CrossRef] [PubMed]
32. Neidhardt, F.C.; Curtiss, R. *Escherichia coli and Salmonella: Cellular and Molecular Biology*, 2nd ed.; American Society of Microbiology Press: Washington, DC, USA, 1996; ISBN 155-5-81-084-5.
33. Wiegand, G.; Remington, S.J. Citrate synthase: Structure, control, and mechanism. *Annu. Rev. Biophys. Biophys. Chem.* **1986**, *15*, 97–117. [CrossRef] [PubMed]
34. Landschütze, V.; Willmitzer, L.; Müller-Röber, B. Inhibition of flower formation by antisense repression of mitochondrial citrate synthase in transgenic potato plants leads to a specific disintegration of the ovary tissues of flowers. *EMBO J.* **1995**, *14*, 660–666. [PubMed]
35. Van Went, J.L.; Willemse, M.T.M. Fertilization. In *Embryology of Angiosperms*; Johri, B., Ed.; Springer: Berlin, Germany, 1984; pp. 273–318.

36. Alexander, R.W.; Cooperman, B.S. Ribosomal proteins neighboring 23 S rRNA nucleotides 803–811 within the 50 S subunit. *Biochemistry* **1998**, *37*, 1714–1721. [CrossRef] [PubMed]
37. Vladimirov, S.N.; Druzina, Z.; Wang, R.; Cooperman, B.S. Identification of 50S components neighboring 23 SrRNA nucleotides A2448 and U2604 within the peptidyl transferase center of *Escherichia coli* ribosomes. *Biochemistry* **2000**, *39*, 183–193. [CrossRef] [PubMed]
38. Christensen, C.A.; Gorsich, S.W.; Brown, R.H.; Jones, L.G.; Brown, J.; Shaw, J.M.; Drews, G.N. Mitochondrial GFA2 is required for synergid cell death in *Arabidopsis*. *Plant Cell* **2002**, *14*, 2215–2232. [CrossRef] [PubMed]
39. Pagnussat, G.C.; Yu, H.J.; Ngo, Q.A.; Rajani, S.; Mayalagu, S.; Johnson, C.S.; Capron, A.; Xie, L.F.; Ye, D.; Sundaresan, V. Genetic and molecular identification of genes required for female gametophyte development and function in Arabidopsis. *Development* **2005**, *132*, 603–614. [CrossRef] [PubMed]
40. Yin, T.; Pan, G.; Liu, H.; Wu, J.; Li, Y.; Zhao, Z.; Fu, T.; Zhou, Y. The chloroplast ribosomal protein L21 gene is essential for plastid development and embryogenesis in Arabidopsis. *Planta* **2012**, *235*, 907–921. [CrossRef] [PubMed]
41. Savage, L.J.; Imre, K.M.; Hall, D.A.; Last, R.L. Analysis of essential Arabidopsis nuclear genes encoding plastid-targeted proteins. *PLoS ONE* **2013**, *8*, e73291. [CrossRef] [PubMed]
42. Baba, T.; Ara, T.; Hasegawa, M.; Takai, Y.; Okumura, Y.; Baba, M.; Datsenko, K.A.; Tomita, M.; Wanner, B.L.; Mori, H. Construction of *Escherichia coli* K-12 in-frame, single-gene knockout mutants: The Keio collection. *Mol. Syst. Biol.* **2006**, *2*, 2006.0008. [CrossRef] [PubMed]
43. Shoji, S.; Dambacher, C.M.; Shajani, Z.; Williamson, J.R.; Schultz, P.G. Systematic chromosomal deletion of bacterial ribosomal protein genes. *J. Mol. Biol.* **2011**, *413*, 751–761. [CrossRef] [PubMed]
44. Breitenreuter, G.; Lotti, M.; Stöffler-Meilicke, M.; Stöffler, G. Comparative electron microscopic study on the location of ribosomal proteins S3 and S7 on the surface of the *E. coli* 30S subunit using monoclonal and conventional antibody. *Mol. Gen. Genet.* **1984**, *197*, 189–195. [CrossRef] [PubMed]
45. Franceschi, F.J.; Nierhaus, K.H. Ribosomal proteins L15 and L16 are mere late assembly proteins of the large ribosomal subunit. Analysis of an *Escherichia coli* mutant lacking L15. *J. Biol. Chem.* **1990**, *265*, 16676–16682. [PubMed]
46. Manavski, N.; Guyon, V.; Meurer, J.; Wienand, U.; Brettschneider, R. An essential pentatricopeptide repeat protein facilitates 5′ maturation and translation initiation of *rps3* mRNA in maize mitochondria. *Plant Cell* **2012**, *24*, 3087–3105. [CrossRef] [PubMed]
47. Burmann, B.M.; Schweimer, K.; Luo, X.; Wahl, M.C.; Stitt, B.L.; Gottesman, M.E.; Rösch, P. A NusE:NusG complex links transcription and translation. *Science* **2010**, *328*, 501–504. [CrossRef] [PubMed]
48. Kwasniak, M.; Majewski, P.; Skibior, R.; Adamowicz, A.; Czarna, M.; Sliwinska, E.; Janska, H. Silencing of the nuclear *RPS10* gene encoding mitochondrial ribosomal protein alters translation in arabidopsis mitochondria. *Plant Cell* **2013**, *25*, 1855–1867. [CrossRef] [PubMed]
49. Nomura, M. Regulation of ribosome biosynthesis in *Escherichia coli* and *Saccharomyces cerevisiae*: Diversity and common principles. *J. Bacteriol.* **1999**, *181*, 6857–6864. [PubMed]
50. Fleischmann, T.T.; Scharff, L.B.; Alkatib, S.; Hasdorf, S.; Schöttler, M.A.; Bock, R. Nonessential plastid-encoded ribosomal proteins in tobacco: A developmental role for plastid translation and implications for reductive genome evolution. *Plant Cell* **2011**, *23*, 3137–3155. [CrossRef] [PubMed]
51. Schuwirth, B.S.; Borovinskaya, M.A.; Hau, C.W.; Zhang, W.; Vila-Sanjurjo, A.; Holton, J.M.; Cate, J.H. Structures of the bacterial ribosome at 3.5 A resolution. *Science* **2005**, *310*, 827–834. [CrossRef] [PubMed]
52. Van Dyke, N.; Xu, W.; Murgola, E.J. Limitation of ribosomal protein L11 availability in vivo affects translation termination. *J. Mol. Biol.* **2002**, *319*, 329–339. [CrossRef]
53. Pesaresi, P.; Varotto, C.; Meurer, J.; Jahns, P.; Salamini, F.; Leister, D. Knock-out of the plastid ribosomal protein L11 in Arabidopsis: Effects on mRNA translation and photosynthesis. *Plant J.* **2001**, *27*, 179–189. [CrossRef] [PubMed]
54. Schmitz-Linneweber, C.; Small, I. Pentatricopeptide repeat proteins: A socket set for organelle gene expression. *Trends Plant Sci.* **2008**, *13*, 663–670. [CrossRef] [PubMed]
55. Manna, S. An overview of pentatricopeptide repeat proteins and their applications. *Biochimie* **2015**, *113*, 93–99. [CrossRef] [PubMed]
56. Uyttewaal, M.; Mireau, H.; Rurek, M.; Hammani, K.; Arnal, N.; Quadrado, M.; Giegé, P. PPR336 is associated with polysomes in plant mitochondria. *J. Mol. Biol.* **2008**, *375*, 626–636. [CrossRef] [PubMed]

57. Del Valle-Echevarria, A.R.; Kiełkowska, A.; Bartoszewski, G.; Havey, M.J. The Mosaic Mutants of Cucumber: A Method to Produce Knock-Downs of Mitochondrial Transcripts. *G3 Genes Genomes Genet.* **2015**, *5*, 1211–1221. [CrossRef] [PubMed]
58. Del Valle-Echevarria, A.R.; Sanseverino, W.; Garcia-Mas, J.; Havey, M.J. Pentatricopeptide repeat 336 as the candidate gene for paternal sorting of mitochondria (*Psm*) in cucumber. *Theor. Appl. Genet.* **2016**, *129*, 1951–1959. [CrossRef] [PubMed]
59. Selmer, M.; Dunham, C.M.; Murphy, F.V.; Weixlbaumer, A.; Petry, S.; Kelley, A.C.; Weir, J.R.; Ramakrishnan, V. Structure of the 70S ribosome complexed with mRNA and tRNA. *Science* **2006**, *313*, 1935–1942. [CrossRef] [PubMed]
60. Nowotny, V.; Nierhaus, K.H. Assembly of the 30S subunit from *Escherichia coli* ribosomes occurs via two assembly domains which are initiated by S4 and S7. *Biochemistry* **1988**, *27*, 7051–7055. [CrossRef] [PubMed]
61. Hammani, K.; Gobert, A.; Hleibieh, K.; Choulier, L.; Small, I.; Giegé, P. An Arabidopsis dual-localized pentatricopeptide repeat protein interacts with nuclear proteins involved in gene expression regulation. *Plant Cell* **2011**, *23*, 730–740. [CrossRef] [PubMed]
62. Liu, Y.J.; Xiu, Z.H.; Meeley, R.; Tan, B.C. *Empty Pericarp5* encodes a pentatricopeptide repeat protein that is required for mitochondrial RNA editing and seed development in maize. *Plant Cell* **2013**, *25*, 868–883. [CrossRef] [PubMed]
63. Brehme, N.; Bayer-Császár, E.; Glass, F.; Takenaka, M. The DYW Subgroup PPR Protein MEF35 Targets RNA Editing Sites in the Mitochondrial *rpl16*, *nad4* and *cob* mRNAs in *Arabidopsis thaliana*. *PLoS ONE* **2015**, *10*, e0140680. [CrossRef] [PubMed]
64. Alonso, J.M.; Stepanova, A.N.; Leisse, T.J.; Kim, C.J.; Chen, H.; Shinn, P.; Stevenson, D.K.; Zimmerman, J.; Barajas, P.; Cheuk, R.; et al. Genome-wide insertional mutagenesis of *Arabidopsis thaliana*. *Science* **2003**, *301*, 653–657. [CrossRef] [PubMed]
65. Yin, K.; Gao, C.; Qiu, J.L. Progress and prospects in plant genome editing. *Nat. Plants* **2017**, *3*, 17107. [CrossRef] [PubMed]
66. Jo, A.; Ham, S.; Lee, G.H.; Lee, Y.I.; Kim, S.; Lee, Y.S.; Shin, J.H.; Lee, Y. Efficient Mitochondrial Genome Editing by CRISPR/Cas9. *Biomed. Res. Int.* **2015**, *2015*, 305716. [CrossRef] [PubMed]
67. Candela, H.; Pérez-Pérez, J.M.; Micol, J.L. Uncovering the post-embryonic functions of gametophytic- and embryonic-lethal genes. *Trends Plant Sci.* **2011**, *16*, 336–345. [CrossRef] [PubMed]
68. Chaiwongsar, S.; Strohm, A.K.; Su, S.H.; Krysan, P.J. Genetic analysis of the Arabidopsis protein kinases *MAP3Kε1* and *MAP3Kε2* indicates roles in cell expansion and embryo development. *Front. Plant Sci.* **2012**, *3*, 228. [CrossRef] [PubMed]
69. Burgos-Rivera, B.; Dawe, R.K. An *Arabidopsis* tissue-specific RNAi method for studying genes essential to mitosis. *PLoS ONE* **2012**, *7*, e51388. [CrossRef] [PubMed]
70. Fujii, S.; Kobayashi, K.; Nakamura, Y.; Wada, H. Inducible knockdown of *MONOGALACTOSYLDIACYLGLYCEROL SYNTHASE1* reveals roles of galactolipids in organelle differentiation in Arabidopsis cotyledons. *Plant Physiol.* **2014**, *166*, 1436–1449. [CrossRef] [PubMed]
71. Dinman, J.D. Pathways to Specialized Ribosomes: The Brussels Lecture. *J. Mol. Biol.* **2016**, *428*, 2186–2194. [CrossRef] [PubMed]
72. Liu, Z.; Gutierrez-Vargas, C.; Wei, J.; Grassucci, R.A.; Sun, M.; Espina, N.; Madison-Antenucci, S.; Tong, L.; Frank, J. Determination of the ribosome structure to a resolution of 2.5 Å by single-particle cryo-EM. *Protein Sci.* **2017**, *26*, 82–92. [CrossRef] [PubMed]
73. Sierra, R.G.; Gati, C.; Laksmono, H.; Dao, E.H.; Gul, S.; Fuller, F.; Kern, J.; Chatterjee, R.; Ibrahim, M.; Brewster, A.S.; et al. Concentric-flow electrokinetic injector enables serial crystallography of ribosome and photosystem II. *Nat. Methods* **2016**, *13*, 59–62. [CrossRef] [PubMed]
74. Shi, Z.; Fujii, K.; Kovary, K.M.; Genuth, N.R.; Röst, H.L.; Teruel, M.N.; Barna, M. Heterogeneous Ribosomes Preferentially Translate Distinct Subpools of mRNAs Genome-wide. *Mol. Cell* **2017**, *67*, 71–83. [CrossRef] [PubMed]

© 2017 by the authors. Licensee MDPI, Basel, Switzerland. This article is an open access article distributed under the terms and conditions of the Creative Commons Attribution (CC BY) license (http://creativecommons.org/licenses/by/4.0/).

Article

Analysis of the Roles of the Arabidopsis nMAT2 and PMH2 Proteins Provided with New Insights into the Regulation of Group II Intron Splicing in Land-Plant Mitochondria

Michal Zmudjak [1], Sofia Shevtsov [1], Laure D. Sultan [1], Ido Keren [1,2] and Oren Ostersetzer-Biran [1,*]

[1] Department of Plant and Environmental Sciences, The Alexander Silberman Institute of Life Sciences, The Hebrew University of Jerusalem, Givat-Ram, Jerusalem 91904, Israel; michalzm@gmail.com (M.Z.); sofia.shevtsov@gmail.com (S.S.); laure.sultan@gmail.com (L.D.S.); Ido.Keren@stonybrook.edu (I.K.)
[2] Department of Biochemistry and Cell Biology, State University of New York, Stony Brook, NY 11794, USA
* Correspondence: oren.ostersetzer@mail.huji.ac.il; Tel./Fax: +972-2-658-5191

Received: 28 September 2017; Accepted: 7 November 2017; Published: 17 November 2017

Abstract: Plant mitochondria are remarkable with respect to the presence of numerous group II introns which reside in many essential genes. The removal of the organellar introns from the coding genes they interrupt is essential for respiratory functions, and is facilitated by different enzymes that belong to a diverse set of protein families. These include maturases and RNA helicases related proteins that function in group II intron splicing in different organisms. Previous studies indicate a role for the nMAT2 maturase and the RNA helicase PMH2 in the maturation of different pre-RNAs in Arabidopsis mitochondria. However, the specific roles of these proteins in the splicing activity still need to be resolved. Using transcriptome analyses of Arabidopsis mitochondria, we show that nMAT2 and PMH2 function in the splicing of similar subsets of group II introns. Fractionation of native organellar extracts and pulldown experiments indicate that nMAT2 and PMH2 are associated together with their intron-RNA targets in large ribonucleoprotein particle in vivo. Moreover, the splicing efficiencies of the joint intron targets of nMAT2 and PMH2 are more strongly affected in a double *nmat2/pmh2* mutant-line. These results are significant as they may imply that these proteins serve as components of a proto-spliceosomal complex in plant mitochondria.

Keywords: group II introns; splicing; maturases; RNA helicases; mitochondria; Arabidopsis; angiosperms

1. Introduction

Plants are able to regulate and coordinate their energy demands during particular growth and developmental stages. These activities require complex cellular signaling between the nucleus and the mitochondrial genome (i.e., mitogenome (mtDNA)) (reviewed by e.g., [1,2]). Although mitochondria contain their own genetic material, encoding some proteins and structural RNAs, the vast majority of mitochondrial proteins are encoded by nuclear loci, and are imported from the cytosol post-translationally [3-6]. In fact, both the ribosomes and the respiratory machinery are composed of proteins encoded by both nuclear and organellar loci (see e.g., [7,8]). These necessitate complex mechanisms to allow the stoichiometric accumulation of subunits encoded by the two physically remote genetic compartments, through different biosynthetic pathways [9].

The expression of the mtDNA in plants is regulated mainly at the post-transcriptional level. High-throughput RNA-seq analyses provided with new insights into the complexity of RNA metabolism in plant mitochondria and indicated that the regulation of mtRNA processing plays a critical role in plant organellar gene-expression [1,10-15]. The importance of post-transcriptional regulation in plant mitochondria is further reflected by the extended half-lives of many of the organellar transcripts, and the fact that their translation seems uncoupled from their transcription [10,16–18]. One of the most notable features of plant mitochondria gene-expression involves the splicing of numerous intervening sequences (introns) that must be removed from the coding-regions post-transcriptionally [10–15]. The processing of the organellar pre-RNAs is essential for respiratory activities and relies on the activities of many different cofactors which belong to a diverse set of RNA-binding protein families.

The mitochondrial introns in angiosperms are classified as group II-type and are found mainly within protein-coding genes [14,19–21]. In Arabidopsis mitochondria, these include 23 group II-type introns found within the complex I *nad1*, *nad2*, *nad4*, *nad5* and *nad7* subunits, the cytochrome c biogenesis factor C (*ccmFc*), the *cox2* subunit of complex IV and the ribosomal proteins *rpl2* and *rps3* (Table 1 and [22]). Introns belonging to this class are large catalytic RNAs (and in some cases also mobile genetic elements) which are excised from the precursor RNAs by two sequential transesterification reactions, involving the release of the intron as an RNA lariat (reviewed by, e.g., [21]). Based on structural similarities and the catalytic activities, it is generally accepted that the nuclear spliceosomal introns have originated from bacterial group II-related introns, which were introduced to the eukaryotic genomes by the bacterial ancestor of the mitochondrion [23]. Although some group II introns are able to catalyze their own excision in vitro, the splicing of group II introns under native conditions (in vivo) is assisted by protein cofactors. In bacteria and yeast mitochondria, this involves proteins that are encoded within the introns themselves (i.e., Intron Encoded Proteins, IEPs; or maturases) [24–27]. Genetic and biochemical data indicate that group II intron-encoded maturases bind with high affinity and specificity towards the introns in which they are encoded from, and facilitate splicing by assisting the folding of the RNAs into their catalytically active forms [15,24–27].

The mitochondrial introns in plants are expected to have evolved from maturase-encoding group II intron RNAs. However, throughout the evolution of land-plants, these have diverged considerably from their related bacterial ancestors, such as they lack many sequence elements that are considered essential for the splicing activity, and also lost the vast majority of their related maturase ORFs [14,19,28,29]. Several of the mitochondrial introns in plants are transcribed as individual pieces that are assemble in *trans* through base-pairing interactions, to form a splicing-competent structure. Interestingly, this situation is reminiscent of the *trans*-interaction of spliceosomal RNAs with pre-mRNAs (reviewed in e.g., [14]). In Arabidopsis, the *trans*-spliced introns include the first and third introns in the NADH-dehydrogenase subunit 1 (i.e., *nad1* introns 1 and 3), *nad2* intron 2, and *nad5* introns 2 and 3 [22]. Due to their degenerate nature and the fact that the organellar introns have also lost their cognate maturase factors, both the cis- and trans-splicing reactions of plant mitochondrial group II introns rely upon the activities of different catalytic enzymes, most of which are encoded by nuclear loci (reviewed by [14,15,19]). However, the specific roles of these factors in the splicing activity is still under investigation.

Genetic screens have led to the identification of different splicing cofactors in plant mitochondria [12,14,15,30]. These include proteins which are closely related to maturases encoded within group II introns (i.e., the mitochondrial MatR encoded in *nad1* i4 and four nuclear-encoded maturase proteins, nMATs 1 to 4) [15,31–35], and at least two RNA helicases (i.e., PMH2 and ABO6) that are encoded in the nucleus and imported into the organelles [36–39]. Maturases and RNA helicases belong to ancient groups of RNA-binding proteins that facilitate intron splicing and were most likely inherited directly from the bacterial symbiont [14,15,21]. Other proteins that influence the splicing of group II introns in plant organelles include pentatricopeptide repeat (PPR) proteins, chloroplast RNA maturation (CRM)-related factors (i.e., the mCSF1 protein), members of the plant organellar RNA recognition (PORR) family, mitochondrial transcription termination factor (mTERF) related proteins and several other factors, which are seemingly unique to eukaryotes and are thus expected to evolve from the host genomes to function in organellar group II intron splicing in plants (reviewed in, e.g., [12,14,15,30]). Some of the proteins, such as nMAT2 [33], PMH2 [37] and mCSF1 [40], are required for the processing of a larger set of introns, while other factors such as PPR proteins (see, e.g., [41–49]) and mTERF15 [50] appear to be more specific, influencing the splicing of a single or only a few group II introns [14].

Table 1. List of group II intron and their splicing efficiencies in *nmat2*, *pmh2* and double *nmat2/pmh2* mutants.

Mitochondrial Group II Introns	Introns Configuration	nMAT2-Dependent	PMH2-Dependent	Splicing Affected in *nmat2/pmh2*
ccmFc i1	cis	No [a,c]	No [b], ambiguous [c]	No
cox2 i1	cis	Yes [a,c]	Yes [b,c]	Yes
nad1 i1	trans	No [a,c]	No [b,c]	No
nad1 i2	cis	Yes [a,c]	Yes [b,c]	Yes
nad1 i3	trans	ambiguous [a], Yes [c]	Yes [b,c]	Yes
nad1 i4	cis	No [a,c]	No [b,c]	ambiguous
nad2 i1	cis	ambiguous [a], Yes [c]	Yes [b,c]	Yes
nad2 i2	trans	No [a,c]	Yes [b,c]	Yes
nad2 i3	cis	No [a,c]	No [b,c]	No
nad2 i4	cis	ambiguous [a], Yes [c]	Yes [b,c]	Yes
nad4 i1	cis	No [a,c]	No [b,c]	No
nad4 i2	cis	ambiguous [a], Yes [c]	Yes [b,c]	Yes
nad4 i3	cis	No [a,c]	Yes [b,c]	Yes
nad5 i1	cis	ambiguous [a], Yes [c]	Yes [b,c]	Yes
nad5 i2	trans	ND [a], Yes [c]	Yes [b,c]	Yes
nad5 i3	trans	ND [a], Yes [c]	Yes [b,c]	Yes
nad5 i4	cis	ambiguous [a], No [c]	No [b], ambiguous [c]	No
nad7 i1	cis	No [a,c]	Yes [b,c]	Yes
nad7 i2	cis	Yes [a,c]	No [b], ambiguous [c]	Yes
nad7 i3	cis	ambiguous [a], No [c]	No [b,c]	ambiguous
nad7 i4	cis	ambiguous [a], No [c]	Yes [b,c]	Yes
rpl2 i1	cis	ambiguous [a,c]	Yes [b,c]	Yes
rps3 i1	cis	ambiguous [a], Yes [c]	Yes [b,c]	Yes

[a], splicing defects in *nmat2* mutants as indicated in Keren et al. [33]; [b], reduced splicing efficiencies in *pmh2* mutants as indicated by Köhler et al. [37]; [c], splicing defects supported by the transcriptome data and RT-qPCR analyses (this study); Grey shaded columns indicate to decreased splicing efficiencies in the double *nmat2/pmh2* mutants; ND, Not determined.

Here, we examined the functions of the nMAT2 maturase, encoded by the At5g46920 locus, and the RNA helicase PMH2 (also known as AtRH53), encoded by the At3g22330 gene, in mitochondrial RNA (mtRNA) metabolism in Arabidopsis plants. Published data have provided with evidence that the splicing of some group II introns (i.e., *cox2* i1 and *nad1* i2) in Arabidopsis mitochondria require both nMAT2 and PMH2 [33,37]. These results are interesting as they may imply that nMAT2 and PMH2 cooperate in the splicing of various organellar introns, in a similar manner to the roles of spliceosomal factors in the splicing of multiple introns in the nucleus. However, these studies which aimed to identify the intron targets of nMAT2 and PMH2 could not provide with definitive experimental data regarding to the roles of these proteins in the splicing of each of the 23 introns in Arabidopsis mitochondria [22]. Furthermore, although the main functions of nMAT2 and PMH2 are supporting RNA splicing, recent data indicate that some splicing cofactors may also affect other RNA processing events in plants mitochondria [41,51–53]. The effects of lowering the expression of nMAT2 and PMH2 on mitochondria functions, organelle gene-expression and the physiology of single *nmat2*, *pmh2* and double *nmat2/pmh2* mutant-lines in Arabidopsis plants are discussed.

2. Results

2.1. The Topology of nMAT2 and PMH2 Proteins

Proteins that interact with group II introns to facilitate their splicing are divided into two main categories, based on their topology and predicted evolutionary origins [14,15,54–56]: (*i*) proteins that are encoded within the introns themselves (i.e., Intron Encoded Proteins, IEPs; or maturases); and (*ii*) various '*trans*-acting' factors that function in the splicing of group II introns. These proteins typically contain motifs that are identified as nucleic-acid binding sites, and, in some cases, also seem to harbor regions likely to mediate protein–protein interactions (see, e.g., [14,15]). Here, we focus on two key plant members of the maturase and RNA helicase families, nMAT2 and PMH2, which have key roles in group II introns splicing in different biological systems [14,15]. While nMAT2 (At5g46920; Figures S1A and S2) is identified as a mitochondria-localized nuclear-encoded maturase protein [31,33], the PMH2 protein (At3g22330; Figures S1B and S2) is closely related to DEAD-box RNA helicases [36,37].

Maturases encoded within group II introns contain several functional domains that are required for both the splicing activity and intron mobility [15,28,57]. These include a retroviral-like reverse transcriptase (RT) domain and a carboxyl-termini region with a sequence similarity to DNA endonucleases (D/En). Domain search analysis, using SMART [58] and Conserved Domain Database (CDD) [59] servers, indicate that nMAT2 (735 amino acids) harbors the consensus fingers-palm (amino acids 104–461) and thumb (X) motifs (amino acids 486–625), typical to reverse-transcriptases found in group II intron maturases, but lacks the C-terminal D/En domain (Figures S1 and S2). Based on these data and previous reports [31,34,60], nMAT2 is categorized as a type-I maturase (i.e., maturases that retain the RT domain but lack the canonical D/En motif). Similar to other mitochondrial proteins that are encoded within the nucleus, nMAT2 harbors a short (i.e., 12 aa) N-terminal sequence which represents the mitochondrial targeting signal (Figures S1 and S2).

To gain structural insight, we used the Protein Homology/Analogy Recognition Engine (Phyre) server [61] to model the 3D-structure of nMAT2 protein (Figure S3A). Based on the "in silico" analysis, the nMAT2 protein shares a significant homology with the *L. lactis* LtrA maturase (c5g2xC; confidence 100%) [31,34,60], as well with the RT domains of P21 maturase (c5hhlA; confidence 100), a telomerase (c3du6A; confidence 99.7) and a retroviral-type RT region of HIV-1 (c1rthA; confidence 98.2).

Characterization of the deduced amino acid sequence of the Arabidopsis PMH2 protein (616 aa), using the PHYRE [61] and ROBETTA [62] servers, indicates that PMH2 shares high similarities with other RNA helicases, as the ATP-dependent RNA helicases PRP5 (c4ljyA; confidence 100) and PRP28 (c4w7sA; confidence 100), which are required for the assembly of the spliceosome and the splicing of spliceosomal introns within the nucleus [63]. Analysis of functional domains, implemented in the CDD [59] and SMART [58] servers, revealed that in addition to the predicted N-terminal mitochondrial targeting signal (25 aa long), PMH2 also harbors the two conserved motifs typical to DEAD-box RNA helicases: a DEXD-box (amino acids 124–326) domain, which also include the consensus ATP binding site (amino acids 148–154), and a C-terminal region with a great similarity to the "Helicase C-terminal domain" (amino acids 364–443) (Figures S1 and S2).

2.2. Mutations in the nMAT2 and PMH2 Gene-Loci Affect the Steady-State Levels of Many mRNAs in Arabidopsis Mitochondria

The homology of nMAT2 and PMH2 with known maturases and RNA-helicases, respectively [33,36,37], supports a role for these proteins in the processing of mtRNAs in plants. Accordingly, in our prior work, we showed that nMAT2 functions in the splicing of at least three mitochondrial introns, including *cox2* i1, *nad1* i2 and *nad7* i2 [33]. The RNase protection and Northern blot analyses also indicated to perturbations in the processing of several other pre-RNAs in *nmat2* mutants [33]. Likewise, using multiplex RT-PCR analyses, Köhler et al. [37] showed that mutations in the *PMH2* locus led to reduced splicing efficiencies of many introns in Arabidopsis mitochondria. However, the identity of the complete sets of intron targets and the specific roles of nMAT2 and PMH2 in the splicing of group II introns still needs to be resolved. It also remains possible that these factors will participate in aspects of RNA metabolism other than group II introns splicing [33,37].

To establish the roles of nMAT2 and PMH2 in mitochondrial RNA metabolism we analyzed several Arabidopsis plants affected in the *mMAT2* and *PMH2* gene-loci. These including a T-DNA knockout *nmat2* mutant (SALK-line 064659), a *pmh2* knockdown T-DNA line, which is strongly affected in the expression of PMH2 (i.e., SAIL-628C06, [37]), and a double mutant line, *nmat2/pmh2*, that is affected in both these loci (for some unknown reasons, we failed to establish a double mutant line using *nmat2* and a knockout *pmh2* mutant, the SALK-line 056387 [37]) (Figures S1 and S2). Analysis of the phenotypes associated with *nmat2* and *pmh2* mutants indicate that they are both hardly distinguishable from those of wild-type plants, grown under similar conditions (see "Materials and Methods") (Figure 1) [34,37]. To gain more insights into the roles of nMAT2 and PMH2 in mtRNA metabolism, and to address specific changes in the levels of the different mRNAs and pre-RNAs in Arabidopsis mitochondria, we used transcriptome analyses by quantitative reverse transcription PCRs (RT-qPCR) (see, e.g., [34,35,40,60,64]), of wild-type plants and *nmat2*, *pmh2* and *nmat2/pmh2* mutants.

Figure 1. Plant phenotypes associated with *nmat2*, *pmh2* and the double mutant line. The effects of loss of nMAT2 and PMH2 on seedling development (**A**), leaf (**B**) and flower (**C**) morphologies of Arabidopsis wild-type (Col-0) and knockout/knockdown mutant lines. The effects of higher light intensity (i.e., 300 µE m-2 sec-1) on three-week-old *nmat2/pmh2* plants is indicated in panel (**D**). Bars represent 5.0 mm in panel (**A**), 2.5 mm in panel (**B**) and 1.0 mm in panel (**C**).

Different from canonical maturases encoded within group II introns, the RNA profiles by RT-qPCR indicated that nMAT2 functions in the maturation of numerous different transcripts in Arabidopsis mitochondria. In *nmat2*, notable reductions in mRNA levels were seen in transcripts which correspond to different exons in complex I *nad1* (i.e., exons b–c and exons c–d), *nad2* (exons a–b and d–e), *nad4* (exons b–c), *nad5* (exons a–b, b–c and c–d) and *nad7* (exons b–c) genes, the *cox2* subunit of complex IV and the ribosomal *rps3* gene (Figure 2), that are all interrupted by group II intron sequences in Arabidopsis mitochondria [22]. Lower steady-state levels of *cox2*, *nad1* and *nad7* transcripts in *nmat2* coincide with the data shown in Keren et al. [34]. However, reduced transcript levels corresponding to *nad2*, *nad4*, *nad5* and *rps3* may indicate additional splicing defects in *nmat2*, as was previously suggested by the RNase protections and Northern blot analyses [34]. The expression of *ccmFc*, which is also interrupted by a group II intron sequence, was not (or only slightly) affected in *nmat2*. Similarly, the levels of mRNAs that correspond to mitochondrial genes which lack group II introns (i.e., "intron-less" transcripts) were not significantly affected by the mutation in the *nMAT2* gene-locus (Figure 3). These included *cox1* and *cox3* subunits of complex IV, different subunits of the ATP synthase enzyme (i.e., complex V), various cytochrome C biogenesis and maturation (*ccm*) factors, and many ribosomal genes (other than *rps3*), which their mRNAs levels in *nmat2* were comparable to those seen in wild-type plants (Figure 2).

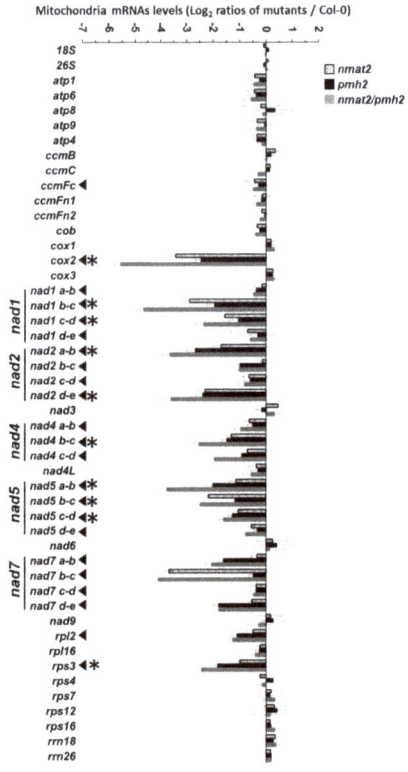

Figure 2. Transcript abundance of mitochondrial mRNAs in *nmat2*, *pmh2* and *nmat2/pmh2* mutants. Transcriptome analyses of mitochondria mRNAs levels in Arabidopsis plants by RT-qPCR was preformed essentially as described previously [35,40,60] (see also in "Materials and Methods", RNA extraction and analysis). RNA extracted from three-week-old seedlings of wild-type (Col-0) and mutant plants was reverse-transcribed, and the relative steady-state levels of cDNAs corresponding to the different organellar transcripts were evaluated by qPCR with primers which specifically amplified mRNAs (see "Materials and Methods"). The histogram shows the relative mRNAs levels (i.e., log2 ratios) in mutant lines versus those of wild-type plants. Arrows indicate to genes that are interrupted by group II intron sequences in Arabidopsis mitochondria [22], while asterisks indicate to transcripts where the mRNA levels were reduced in both *nmat2* and *pmh2* lines. The values are means of five biological replicates (error bars indicate one standard deviation).

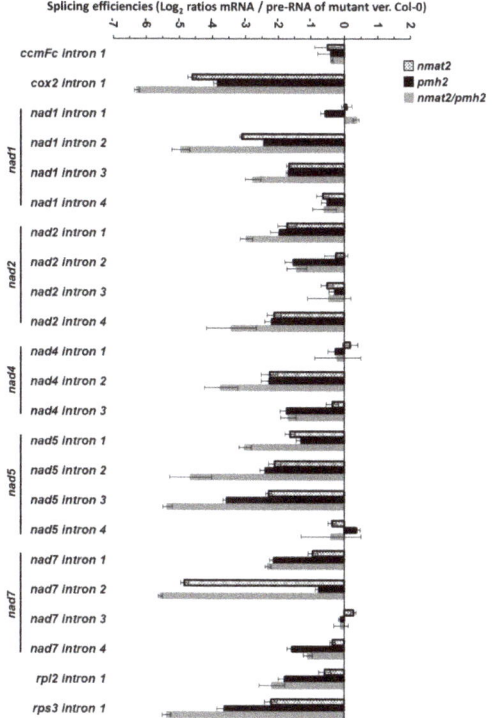

Figure 3. Splicing efficiencies and abundance of mitochondrial transcripts in *nmat2*, *pmh2* and *nmat2/pmh2* mutants. Transcriptome analyses of mitochondria gene-expression in Arabidopsis plants by RT-qPCR were preformed essentially as described previously [34,35,49] (see also in "Materials and Methods", RNA extraction and analysis). RNA extracted from three-week-old seedlings of wild-type (Col-0) and mutant plants was reverse-transcribed, and the relative steady-state levels of cDNAs corresponding to the different organellar transcripts were evaluated by qPCR with primers which specifically amplified pre-RNAs and mRNAs (see "Materials and Methods"). The histogram shows the splicing efficiencies as indicated by the log2 ratios of pre-RNA to mRNA transcript abundance in mutant lines compared with those of wild-type plants. The values are means of three biological replicates (error bars indicate one standard deviation).

The RNA profiles of *pmh2* mutants were similar to those previously reported in Köhler et al. [37]. These analyses indicated to reduced mRNA levels of multiple genes, including *cox2*, *nad1* exons b–c and c–d, *nad2* exons a–b, b–c and d–e, *nad4* exons a–b, b–c and c–d, *nad5* exons a–b, b–c, c–d, *nad7* exons a–b, as well as *rpl2* and *rps3* mRNAs (Figure 2). Similarly to *nmat2* mutant, the accumulation of "intron-less" transcripts in the mitochondria was not significantly affected in the *pmh2* mutant (Figure 2). Small changes in the abundances of various mtRNAs in *nmat2* and *pmh2* plants (Figure 2 and [37]) may be caused by compensatory effects during transcription or post-transcriptional processes. Together, these data strongly suggest that both nMAT2 and PMH2 function in the splicing of many group II introns in Arabidopsis mitochondria.

2.3. nMAT2 and PMH2 Function in the Splicing of Similar Subsets of Group II Introns in Arabidopsis Mitochondria

To better understand the roles of nMAT2 and PMH2 in the processing of group II introns in plants, we compared the splicing efficiencies (i.e., the ratios of pre-RNAs to mRNAs) of the 23 group II introns found in the mitochondria of *Arabidopsis thaliana* plants [22], between wild-type (Col-0) plants and *nmat2* and *pmh2* mutants. Splicing defects were determined to be present in cases where the accumulation of a specific pre-mRNA was correlated with a reduced level of its corresponding mRNA in each mutant line.

Although Northern blot and ribonuclease protection experiments indicated a role for nMAT2 in the splicing of *cox2* i1, *nad1* i2 and *nad7* i2, these analyses did not yield conclusive results with regard to the roles of nMAT2 in the processing of other transcripts in Arabidopsis mitochondria (Table 1 and [33]). To establish the roles of nMAT2 in the splicing of specific mitochondrial introns, we used transcriptome analyses by RT-qPCRs of total mtRNAs obtained from wild-type and mutant plants [34,35,40,60,64]. These analyses indicated to splicing defects in *cox2* i1, *nad1* i2, *nad1* i3, *nad2* introns 1 and 4, *nad4* i2, *nad5* introns 1,2 and 3, *nad7* i2, and the single intron within *rps3* (Figure 3). To some extent, reduced splicing efficiencies were also observed in the cases of *nad1* i4, *nad2* i3, *nad4* i3, *nad5* i4, *nad7* i4 and *rpl2* i1 (Figure 3). However, given the small degree to which their mRNAs levels were reduced in the mutant (Figures 2 and 3), it is difficult to draw firm conclusions about the roles of nMAT2 in the splicing of these introns (Table 1). The RNA profiles further indicate that the splicing of *ccmFc* i1, *nad1* i4, *nad2* introns 2 and 3, *nad4* introns 1 and 3, *nad5* i4 and *nad7* introns 1, 3 and 4 did not rely upon nMAT2 (Figures 2 and 3, and Table 1). The splicing of these introns is therefore expected to be facilitated by various other splicing cofactors.

Similar to the data shown in Köhler et al. [37], splicing defects in *pmh2* were supported in the cases of 15 mitochondrial introns (see Figure 3 and Table 1). These including *cox2* i1, *nad1* i2 and i3, *nad2* intron 1,2 and 4, *nad4* introns 2 and 3, *nad5* introns 1, 2 and 3, *nad7* introns 1 and 4, *rpl2* i1 and *rps3* i1, in which the accumulations of pre-RNAs were correlated with notable reductions in their corresponding mRNAs in *pmh2* mutants. However, we could not confidently draw any conclusions about the significance of PMH2 to the splicing of *ccmFc1* i1, *nad1* i4 and *nad7* i2, as although the steady-state levels of their pre-RNAs were higher in the mutant, the accumulation of their corresponding mRNAs (i.e., *ccmFc* mRNA, *nad1* exons d–e and *nad7* exons b–c) was not significantly affected in *pmh2* (Figure 2). Notably, the transcriptome profiles and splicing defects of *pmh2* mutants shared some similarities with those seen in *nmat2* mutants (Figures 2 and 3, and Table 1). Introns that their splicing was affected in both *nmat2* and *pmh2* include *cox2* i1, *nad1* i2, *nad1* i3, *nad2* i1, *nad2* i4, *nad4* i2, *nad5* i1, *nad5* i2, *nad5* i3 and *rps3* i1 (Figures 2 and 3, and Table 1).

Taken together, the RNA profiles (Figures 2 and 3), and the data shown in Keren et al. [34] and Köhler et al. [37] strongly indicate that nMAT2 and PMH2 function specifically in group II introns splicing. In light of the expected significance of splicing to the functionality of many of the organellar transcripts, and hence to respiratory functions and plant physiology, we speculate that the lack of gross phenotypes associated with mutations in *nMAT2* or *PMH2* gene-loci may relate to redundant functions with other organellar splicing cofactors (i.e., maturases, RNA helicases, PPRs, PORRs or CRM-related proteins) that exist in Arabidopsis mitochondria [14,15,35,40]. Accordingly, the splicing of individual group II introns in plant mitochondria seems to rely on the activities of different protein cofactors [14,15,30]. While the functions of RNA helicases, including PMH2, are expected to affect a broad set of introns, the roles of the maturase-type nMat2 factor, which evolved from a protein acting only on the intron carrying its own gene, in the splicing of many group II introns in plant mitochondria are quite remarkable. Similarly, the organellar-encoded maturases, MatR and MatK, also act on multiple intron targets in the mitochondria and chloroplasts (respectively) of land-plants [60,65].

2.4. nMAT2 Is Found in Large Ribonucleoprotein Particles that Also Contain PMH2

The experimental data indicate that nMAT2 and PMH2 are involved in the processing of similar subsets of group II introns in Arabidopsis mitochondria (Figures 2 and 3, and Table 1). We thus intended to investigate whether these factors may cooperate in regulating the splicing of many mitochondrial pre-RNAs. Native organellar complexes were fractionated by velocity centrifugation sedimentation throughout sucrose gradients, and each fraction was analyzed by immunoblot analyses with antibodies against nMAT2 [33] and PMH2 [36,37]. Immunoblots with antibodies against serine hydroxymethyltransferase 1 (SHMT1) were used as a control. In agreement with previous analyses [33,36,37], both nMAT2 and PMH2 proteins migrated through the sucrose gradients towards the bottom of the tubes, with apparent molecular masses of ≥1000 kDa, which are significantly higher than their monomeric sizes (74 and 61 kDa, respectively) (Figure 4A). Interestingly, the signals of nMAT2 and PMH2 were observed in the same fractions, following the sucrose gradient centrifugation (i.e., fractions 16, 17 and strongly enriched in fraction 18). Notably, the particle sizes of nMAT2 and PMH2 proteins were reduced following the ribonuclease A treatment, as observed by their appearance in fractions of lower molecular weights (i.e., lanes 12–18) after the sucrose gradient fractionation (Figure 4B). Thus, the RNase-sensitivity assays lend additional support to the association of nMAT2 and PMH2 with organellar transcript in Arabidopsis mitochondria.

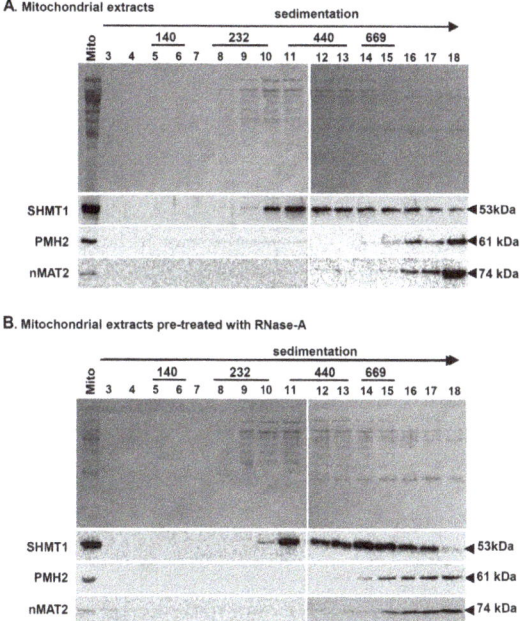

Figure 4. nMAT2 and PMH2 proteins are identified in large ribonucleoprotein particles in Arabidopsis mitochondria. Fractionation of wild-type and mutant mitochondria pre-treated with RNase inhibitor (**A**); or ribonuclease A (RNase-A) (**B**) by sucrose gradient centrifugation. Aliquots of crude mitochondria protein extract (Mito) were solubilized in n-dodecyl-β-maltoside (DDM) (1.5% (w/v)), and sucrose gradient fractionated mitochondria samples (Fractions 3–18) were subjected to immunoblot analysis with antibodies raised against different organellar proteins (see "Materials and Methods"), including serine hydroxymethyltransferase 1 (SHMT1) protein (53 kDa), PMH2 (61 kDa, [36,37]) and nMAT2 (74 kDa, [33]), as indicated in each blot. High molecular mass standards (GE Healthcare) sizes are given in kilodaltons.

The similarities in the RNA profiles of *nmat2* and *pmh2* mutants and the appearance of both of these proteins in ribonucleoprotein (RNP) particles of similar molecular masses may indicate that nMAT2 and PMH2 cooperate in the splicing of group II introns in Arabidopsis mitochondria. Analogously, the splicing of group II introns in yeast mitochondria was also found to be mediated by both maturases and RNA helicases [52,53,66], although the specific roles of these factors in the splicing of group II introns are still under investigation.

Using a transgenic approach, we showed that the expression of a recombinant nMAT2 protein, which contains a hemagglutinin (3HA) tag at its carboxyl terminus, restores the splicing defects of *nad1* i2, *nad7* i1 and *cox2* i1 in *nmat2* mutants (Figure S4) [33]. Here, we further applied this method to identify proteins and RNAs that are stably associated with nMAT2, in vivo. Affinity purifications with anti-HA antibodies following mass spectrometry analyses, were used to identify proteins that co-purified with nMAT2-HA that were absent in co-IPs of Arabidopsis plants transformed with an empty vector control (Figure S4). Antibodies against SHMT1 were used as a control for the integrity of the co-immunoprecipitations (co-IPs) assays. Intriguingly, in addition to nMAT2, the co-IPs followed by LC-MS/MS analyses also revealed the presence of various peptides which corresponded to PMH2 protein, as well as two ribosomal factors (At1g16870 and At1g31817) and four pentatricopeptide repeat (PPR) proteins (i.e., At1g26460, At1g10270, At1g55890 and At1g61870) (Table S1). Other proteins that co-purified with nMAT2 included several prohibitins (At4g28510, At2g20530 and At5g44140), which are highly abundant in Arabidopsis mitochondria and are found in high molecular weight complexes [67]. It remains possible, therefore, that these may have nonspecifically co-purified with the nMAT2 particles.

Published data, as well as our own studies, indicate that organellar splicing factors in plants are tightly associated with their intron targets [33,60,65,68–72]. Here, we used the co-IPs to identify mtRNAs that are stably associated with nMAT2-PMH2 particles, in vivo (see "Materials and Methods"). The association of group II intron-containing pre-RNAs with nMAT2-PMH2 particles was supported in the cases of *cox2* i1, *nad1* i2, *nad1* i3, *nad2* introns 1, 2 and 4, *nad4* intron 3, *nad5* introns 1, 2 and 3, *nad7* introns 1, 2, 3 and 4 and *rps3* i1 (Table S2). The sequencing data also indicated the presence of *nad2* i2, *nad4* i3, *nad7* introns 1 and 4 and *rpl2* i1 in the co-IPs of *nmat2/nMAT2-HA* plants (Table S2). As the splicing of these introns is not affected in *nmat2* (Figure 3), we thus assume that these intron are likely associated with PMH2 protein. No cDNAs that correspond to *ccmFc* i1, *nad1* i4 and *nad4* i1 transcripts were detected in the pelleted RNAs. As the splicing of these introns does not depend upon nMAT2 or PMH2 (Table 1), these results strongly support the specificity of the co-IP analyses.

2.5. Study of Double Mutants Affected in nMAT2 and PMH2 Gene-Loci

Mutations in nMAT2 and PMH2 affect the maturation of numerous pre-RNAs in Arabidopsis mitochondria, many of which are affected in both mutant lines (Figures 2 and 3, Table 1). Fractionation of native organellar complexes and pulldown assays indicate that nMAT2 and PMH2 are found in RNP complexes of similar molecular weights, in vivo (Figure 4 and, Tables S1 and S2). We therefore anticipated that mutants homozygous to both gene-loci would show strong defects in the maturation of pre-RNAs that their splicing is assisted by both nMAT2 and PMH2 proteins (i.e., *cox2*, *nad1*, *nad2*, *nad4*, *nad5* and *rps3* pre-RNAs). For this purpose, the *nmat2* and *pmh2* plants were crossed, and the resulting double mutant-line, *nmat2/pmh2* (Figure S1B), was analyzed for its associated growth phenotypes and organellar activities. Phenotypic examination of *nmat2/pmh2* seedlings suggested that under optimal growth conditions i.e., in either long (16:8-h) or short day conditions (8:16-h), 22 °C, 100 µE m^{-2} s^{-1}, primary root elongation, vegetative growth, flower morphology and fertility were only slightly affected in the double mutant, (Figure 1). However, we noticed that the leaves of *nmat2/pmh2* plants turn reddish when these are grown under higher light intensities (i.e., 300 µE m^{-2} s^{-1}), suggesting the accumulation of anthocyanin pigments in the leaves (Figure 1D). Leaf redness through anthocyanin is often considered as a stress response (Chalker-Scott 1999). Accordingly, accumulation of anthocyanin and ROS was observed in other mutants affected in mitochondrial RNA metabolism, as *nmat1* mutants [34].

2.6. Double nmat2/pmh2 Mutants Are Strongly Affected in the Maturation of cox2, nad1, nad5 and rps3 Pre-RNA Transcripts in Arabidopsis Mitochondria

To identify changes in transcript levels in the double mutant line, we examined the splicing efficiencies of each of the 23 mitochondrial group II introns in wild-type and *nmat2/pmh2* plants by

RT-qPCR analyses. Notable splicing defects were seen in *cox2* i1, *nad1* i2, *nad1* i3, *nad2* i1, *nad2* i4, *nad4* i2, *nad5* i1, *nad5* i2, *nad5* i3 and *rps3* i1. In many cases, the splicing in the double mutant line was more strongly affected than in the single *nmat2* or *pmh2* mutants (Figure 3), further supporting that the splicing of these introns depends upon the activities of both of these factors. Splicing defects were also apparent in the cases of *nad2* i2, *nad4* i3, *nad7* introns 1,2 and 4 and *rpl2* i1, which their splicing relies on either nMAT2 (*nad7* i2) or PMH2 (*nad2* i2, *nad4* i3, *nad7* i1, *nad7* i4 and *rpl2* i1). However, the splicing efficiencies of these introns in *nmat2/pmh2* were comparable with those seen in the single *nmat2* or *pmh2* mutant lines. To some extent, the levels of nad1 i4, nad2 i3 and nad7 i3 pre-RNAs were slightly higher in *nmat2/pmh2* plants compare to those of the wild-type plants (Figure 3). However, as their corresponding mRNAs levels were not significantly reduced in the mutant (Figure 2), it is difficult to draw firm conclusions to whether the splicing of these introns was affected in the double mutant. The splicing of introns that their splicing is not dependent upon nMAT2 or PMH2, including *ccmFc* i1, *nad1* i1, *nad4* i1 and *nad5* i4, was not significantly affected in *nmat2/pmh2* (Figure 3 and Table 1). The maturation of these RNAs is therefore facilitated by splicing cofactors other than nMAT2 or PMH2. Similarly to the single mutant lines, no significant difference in the relative accumulation of mRNAs corresponding to intron-less transcripts was observed between the wild-type plants and *nmat2/pmh2* mutant line (Figure 2).

2.7. Analysis of Mitochondrial Respiratory Activity and the Biogenesis of Organellar Respiratory Chain Complexes in nmat2, pmh2 and nmat2/pmh2 Mutants

The respiratory machinery is an aggregation of four major electron transport complexes (i.e., CI to CIV) and the ATP synthase enzyme (also denoted as CV), which function together in oxidative phosphorylation and drive the synthesis of ATP. Number of studies have shown that perturbation of splicing can affect mitochondria biogenesis and functions [14]. While the activity of complex IV is expected to be essential, complex I defects in plants result in a broad spectrum of phenotypes, ranging from mild to severe growth and developmental defects (reviewed by e.g., [73]). Interestingly, Arabidopsis mutants that are completely lacking complex I activity are strongly affected in their cellular physiology, but are viable when they are grown on sugar-containing MS media [74,75]. To analyze whether the respiratory activity was altered in *nmat2*, *pmh2* and *nmat2/pmh2* plants, the oxygen-uptake rates of wild-type and mutant-lines were monitored in the dark with a Clark-type electrode. When respiration was analyzed on three-week-old seedlings grown on MS-plates, the average O_2-uptake rates of *nmat2* and *pmh2* mitochondria (105.76 ± 11.25 and 99.11 ± 4.45 nmol O_2 min^{-1} gr FW^{-1}, respectively) were similar to those measured in wild-type plants (103.84 ± 8.79 nmol O_2 min^{-1} gr FW^{-1}) (Figure 5). However, inhibition of mitochondrial respiratory chain complex I by rotenone (+ROT) affected the respiration rates of wild-type and *pmh2* plants (i.e., 52.32 ± 7.25 and 61.68 ± 7.95 nmol O_2 min^{-1} gr FW^{-1}, respectively), whereas the inhibitor appeared to has a less effect on the respiratory activity of *nmat2* plants (i.e., $82.54 + 5.25$ nmol O_2 min^{-1} gr FW^{-1}) (Figure 5). Similarly, while the average O_2-uptake rates of *nmat2/pmh2* (104.29 ± 8.08 nmol O_2 min^{-1} gr FW^{-1}) were similar to those of wild-type plants, they were also less sensitive to inhibition by rotenone (91.06 ± 3.98 nmol O_2 min^{-1} gr FW^{-1}) (Figure 5). As the maturation of COX2 is affected in both the single and double mutant lines (Figures 2 and 3), we also measured the O_2-uptake rates in the presence of potassium cyanide (KCN), which inhibits electron transport through complex IV. The data shown in Figure 5 indicate that inhibition of mitochondrial respiration by KCN is more pronounced in wild-type plants than in each of the mutants. In summary, the respiration measurements suggest that complex IV is affected in *nmat2*, *pmh2* and the double *nmat2/pmh2* mutant-line, whereas complex I is more notably affected in the *nmat2* and *nmat2/pmh2* mutants. These results may correlate with the severity in the RNA metabolism defects observed in each of the mutants (Figures 2 and 3).

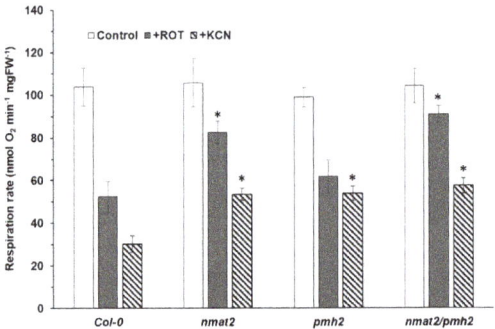

Figure 5. Respiration activities in wild-type and mutant lines. O_2-uptake rates of wild-type plants and *nmat2*, *pmh2* and *nmat2/pmh2* mutants were analyzed with a Clark-type electrode as we described previously [35]. For each assay, equal weight (i.e., 100 µg) three-week-old MS-grown Arabidopsis seedlings were submerged in 2.5–3.0 mL sterilized water and applied to the electrode in a sealed glass chamber in the dark. O_2-uptake rates were measured in the absence (Control) or in presence of rotenone (ROT, 50 µM) and potassium cyanide (KCN, 1 mM) which inhibit complexes I and IV activities, respectively. The values are means of four biological replicates (error bars indicate one standard deviation). The asterisk indicates a significant difference from wild-type plants (Student's *t*-test, p 0.05).

Protein accumulation depends on the balance between the rates of translation and protein degradation. Reduced mRNA transcript levels may also affect the organellar translation efficiencies. Accordingly, protein synthesis in the chloroplasts is correlated with transcript abundance [76]. The relative accumulation (i.e., steady-state levels) of different mitochondrial proteins in three-week-old *nmat2*, *pmh2* and *nmat2/pmh2* mutants was analyzed by immunoblot assays with antibodies raised against different organellar proteins (Table S3). These included the complex I subunits, 18-kDa (also termed as NDUFS4), NAD9, the Rieske iron-sulfur protein (RISP) of complex III, COX2 subunit of complex IV, the β subunit of the ATP synthase enzyme (AtpB, complex V), the plant mitochondrial voltage-dependent anion channel (VDAC, or Porin), the serine hydroxymethyltransferase 1 (SHMT1) protein, the mitochondrial alternative oxidase subunits 1 or 2 (AOX1/2) and the plastidial Rubisco enzyme (Figure 6). Relative protein levels were measured by densitometry of Western blots, and quantified using ImageJ software [77]. These analyses indicated that the abundances of the complex I 18-kDa subunit and Nad9 protein were similar in wild-type, *pmh2* and *nmat2* plants. The signals corresponding to the 18-kDa subunit and RISP protein seemed to be, at least to some extent, higher in the double mutant (1.40-fold and 2.11-fold, respectively; Figure 6A). However, the steady-state levels of COX2 subunit (Figure 6A), which its maturation was affected in *nmat2*, *pmh2* and the double mutant line (Figures 3 and 4, and Table 1), were found to be somewhat decreased in the mutants (i.e., between 31% and 56% lower) than in the wild-type plants. Several other proteins, including the complex V subunit AtpB (2-fold to 4-fold), Porin (1.23-fold to 2.40-fold), SHMT1 (1.65-fold to 2.27-fold), and more notably AOX1/2, accumulated to higher levels in the mutants (Figure 6A). Upregulation in AOX expression is tightly associated with mitochondrial dysfunction and stress in plants [3,5,78–82].

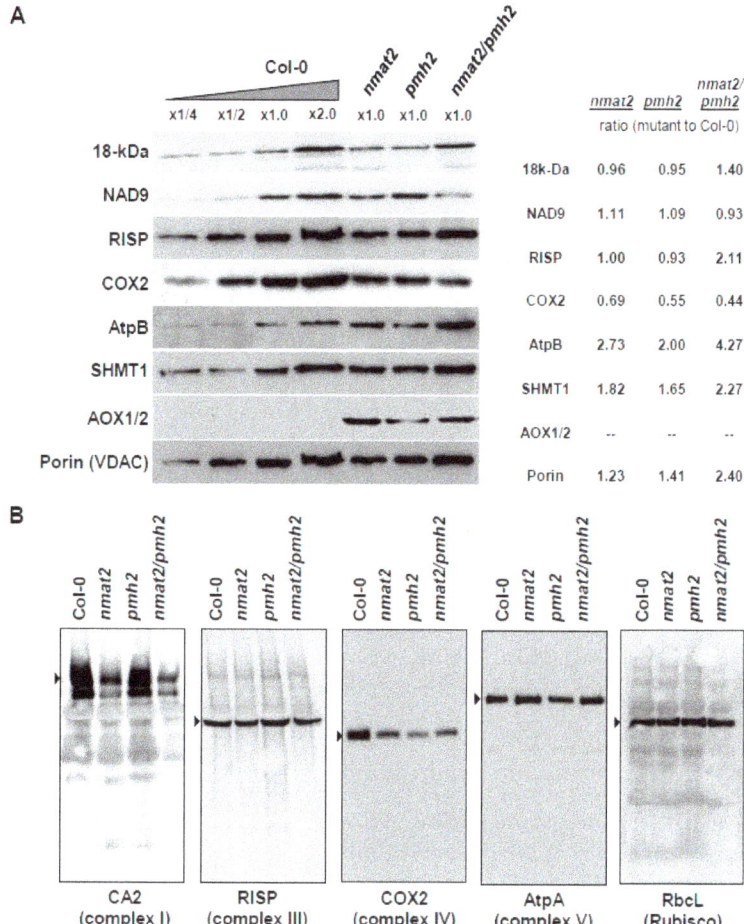

Figure 6. Relative accumulation of organellar proteins in wild-type plants and *nmat2*, *pmh2* and *nmat2/pmh2* mutants. (**A**) Immunoblots with total proteins (about 50 µg) extracted from three-week-old rosette leaves of wild-type plants, and homozygous *nmat2* and *pmh2* mutants. The blots were probed with polyclonal antibodies raised to γ-carbonic anhydrase-like subunit 2 (CA2), NADH-oxidoreductase subunit 9 (NAD9) and the 18-kDa subunits of complex I, Rieske iron-sulfur protein (RISP) of complex III, the cytochrome oxidase subunit 2 (COX2) of complex IV, mitochondrial ATP-synthase subunits and (AtpA and AtpB) of complex V, serine hydroxymethyltransferase 1 (SHMT1) protein and the mitochondrial voltage-dependent anion channel (VDAC, or Porin). Detection was carried out by chemiluminescence assays after incubation with HRP-conjugated secondary antibody; (**B**) BN-PAGE of crude mitochondria preparations was performed according to the method described previously [83]. Crude mitochondria preparations, obtained from three-weeks old Arabidopsis seedlings, were solubilized with DDM (1.5% (w/v)) and the organellar complexes were resolved by BN-PAGE. For immunodetections, the proteins were transferred from the native gels onto a PVDF membrane. The membranes were distained with ethanol before probing with specific antibodies (Table S3), as indicated below each blot. Arrows indicate to the native complexes I (~1000 kDa), III (dimer, ~500 kDa), IV (~220 kDa) and V (~600 kDa).

Blue Native Polyacrylamide Gel Electrophoresis (BN-PAGE) was used to determine the effects of the mutations in nMAT2 and PMH2 on the biogenesis of the respiratory machinery. Separation of native organellar complexes by BN-PAGE and Western blot analyses revealed that complex I (using antibodies against the and the γ-type carbonic anhydrase CA2) was affected in *nmat2* and more notably in the mitochondria of *nmat2/pmh2* plants, while the levels of complex I in *pmh2* mutant were similar to those observed in the wild-type plants (Figure 6B). As the splicing of various *nad* transcripts is affected in both *nmat2* and *pmh2*, we speculate the differences in complex I levels between the mutant lines may relate to the strong maturation defects in *nad7* pre-RNAs seen in the *nmat2* mutant. In accordance with the splicing defects (Figures 2 and 3) and reduced COX2 protein levels (Figure 6A), lower complex IV levels were evident in crude organellar preparations of *nmat2*, *pmh2* and the double mutant line. The respiratory chain complexes III and V were found in similar abundances in the mitochondria of wild-type and mutant plants (Figure 6B). Similarly, the accumulation of RuBisCO enzyme was also not significantly affected by the mutations in *nMAT2* or *PMH2* gene-loci. We, therefore, assume that the mtRNA maturation defects affect the biogenesis of the respiratory machinery in *nmat2*, *pmh2* and *nmat2/pmh2* plants. While the steady-state levels of COX2 and complex IV are reduced in the three mutant lines, the levels of the complex I 18-kDa and NAD9 subunits were not significantly affected in the mutant plants. Thus, reduced complex I levels, seen in *nmat2* and the double mutant line, may relate to altered translation rates or lower availability of different complex I subunits in these mutants.

3. Discussion

3.1. The Splicing of Group II Introns in Land-Plant Mitochondria Rely on the Activities of Different Nuclear-Encoded RNA-Binding Cofactors

The challenges of maintaining prokaryotic-type structures and functions within the cells are common to all eukaryotes. However, plants possess some of the most complex organelle compositions of all known eukaryotic cells. Plant mitochondrial genomes are unique in structural complexity, and gene-expression in plant mitochondria is highly complicated, involving multiple transcription initiation sites and extensive RNA processing steps [10–12,14,15]. These including the splicing of numerous group II-type introns that interrupt the coding regions of many essential genes. A major difference between bacterial group II introns and their counterparts in plant mitochondria resides in the nature of their target loci [21,57,84]. While in prokaryotes the group II introns often lie outside transcribed regions and are thus expected to have only a minor effect on bacterial fitness, the mitochondrial introns in plants reside within many genes required in both translation and respiratory-mediated functions [14,19,21]. These have diverged considerably from their bacterial ancestors, such as they lost many elements that are considered essential for splicing, and typically are also lacking their related maturase ORFs [19,20]. It is not surprising, therefore, that the splicing of the organellar introns in plants is accomplished largely by nuclear-encoded RNA-binding cofactors [14,15,60], which may also provide a means to link organellar functions with environmental and developmental signals. Here, we focus on the roles of the nuclear-encoded nMAT2 and PMH2 factors in the splicing of group II introns in Arabidopsis mitochondria.

3.2. nMAT2 and PMH2 Play Key Roles in the Splicing of Group II Introns in Plant Mitochondria

RNA-binding proteins play central roles in the post-transcriptional regulation of gene-expression in different biological systems. Accumulating data indicate that maturases and RNA helicases have pivotal roles in the splicing of group II introns (Reviewed by [14,15,21,57]). Model maturases encoded within group II introns function as intron-specific splicing cofactors, while RNA helicases belong to a large group of enzymes which carry multiple roles in gene-expression and RNA metabolism. To unravel the molecular basis of group II introns splicing processes in land-plant mitochondria, we utilized detailed expression analyses of wild-type, *nmat2*, *pmh2* and double *nmat2/pmh2* mutants.

Both nMAT2 and PMH2 function in group II introns splicing (Figures 2 and 3, and [33,37]). The mtRNA landscapes of *nmat2* and *pmh2* mutants indicated to perturbations in the maturation of many organellar pre-RNAs, all of which are containing group II intron sequences, while the steady-state levels of different intron-less transcripts were similar between wild-type plants and *nmat2* and *pmh2* mutants (Figures 2 and 3).

Significantly, the transcriptomes analyses revealed high similarities in the RNA profiles and splicing defects of *nmat2* and *pmh2* mutants. Introns that their splicing is notably affected in both *nmat2* and *pmh2* include *cox2* i1, *nad1* i2, *nad1* i3, *nad2* i1, *nad2* i4, *nad4* i2, *nad5* i1, *nad5* i2, *nad5* i3 and *rps3* i1 (Figures 2 and 3, and Table 1). When the activity of both these factors is affected (i.e., in the double *nmat2/pmh2* mutant line), the maturation defects of these pre-RNAs are more noticeable in *nmat2/pmh2* then in each individual line (Figures 2 and 3, and Table 1). Still, none of the analyzed processing events is completely abolished and mRNAs corresponding to the intron targets of nMAT2 and PMH2 are observed in the mutants (Figures 2 and 3) [33,37]. We thus speculate that these results indicate to redundant functions with other splicing cofactors that exist in plant mitochondria [14,15]. Accordingly, MatR [60], nMAT1 and nMAT4 [33–35], the CRM-related mCSF1 protein [40,85,86] and the RNA helicase ABO6 [39] also influence the splicing efficiencies of multiple pre-RNAs in Arabidopsis mitochondria. For example, the splicing efficiencies of at least 13 different introns is affected in the *mcsf1* mutants [40]. The splicing of several of these introns, including *cox2* i1, *nad1* i2 and i3, *nad2* i1 and i4 and *nad5* i1, i2 and i3, also depends upon the activities of both nMAT2 and PMH2. Such a general substrate recognition is typical for RNA chaperones, which in many cases were shown to facilitate the transitions of the RNA ligands from non-functional, or intermediate structures, into the functionally-active forms (see e.g., [87]).

3.3. nMAT2 and PMH2 Are Associated Together in Large RNP Complexes with Various Group II Introns in Arabidopsis Mitochondria

The RNA profiles indicate that nMAT2 is an atypical *trans*-acting maturase factor, which regulates the splicing of at least 11 out of the 23 group II introns in Arabidopsis mitochondria (Figures 2 and 3, and Table 1) [33]. In accordance with previous reports [37], PMH2 was found to function in the splicing of 15 mitochondrial group II introns (Figures 2 and 3, and Table 1). Interestingly, many of the introns that their splicing was affected in *nmat2* (beside *nad7* i2) are also identified as the intron targets of PMH2 (Table 1). The similarities in the RNA profiles of *nmat2* and *pmh2* mutants, and the fact that the splicing of these introns is more strongly affected in the double mutant-line (Figures 2 and 3, and Table 1), may indicate that nMAT2 and PMH2 function together in the splicing of organellar pre-RNAs in Arabidopsis. Accordingly, separation of native mitochondrial preparations by sucrose gradients revealed that nMAT2 and PMH2 are both parts of high molecular weight ribonucleoprotein particles (Figure 4) [33,37]. Pulldown experiments further showed that PMH2 and several other proteins co-purify with the recombinant nMAT2-HA protein, in vivo (Table S1). The co-IP's also indicated to the presence of various group II introns (or pre-RNAs) in the nMAT2/PMH2-associated particles (Table S2). However, it remains unclear whether nMAT2 and PMH2 operate independently, or whether their functions in the splicing of mitochondrial introns may be coordinated.

In an attempt to relate the experimental observations produced by the genetic and biochemical analyses with the overall tertiary structures of the mature forms of nMAT2 and PMH2 proteins (i.e., lacking their predicted N-termini targeting regions), we performed an atomic model of these proteins using the Phyre2 server [61]. The predicted structures of nMAT2 and PMH2 indicate the presence of positively charged surfaces that may serve as RNA-binding modules, while the uncharged or negatively charged regions may be required for protein–protein interactions (Figure S3). nMAT2 and PMH2 docking calculations were performed by Gramm-X [88], with no predetermined bias towards any specific residue interactions, and the predicted structures were visualized by PyMol [89]. The in silico analyses showed that PMH2 and nMAT2 may interact together throughout the association of amino acids found in a loop region between the two helicase domains of PMH2 (amino acids 327–363), and a short linker region (amino acids 461–485) between the fingers-palm and thumb domains of nMAT2 (Figure S5). However, this hypothetical model, which suggests an interaction between these two factors, needs to be supported experimentally. Nevertheless, the presence of nMAT2 and PMH2 together with group II introns in organellar RNPs (Figure 4 and, Tables S1 and S2) may relate to spliceosomal-like complexes in plant mitochondria [15]. Studies are under way in our laboratory to specify the binding characteristics of nMAT2 and PMH2 to group II intron RNAs and to examine whether these proteins associate with one another.

3.4. Mitochondrial Respiratory Activity Is Altered in nmat2 and pmh2 Mutants

Analysis of the protein profiles of wild-type and mutants plants indicated that complex I was affected in *nmat2* and *nmat2/pmh2* plants, while reduced levels of complex IV were apparent in the both single and double mutant-lines (Figure 6B). To examine if the reduced complexes I and IV levels affect respiratory functions, we measured the O_2-uptake rates of Arabidopsis wild-type and mutant plants, using a Clark-type electrode. No significant differences in the O_2-uptake rates were observed between the mutants and wild type plants. However, inhibition of mitochondrial respiratory chain complex I by rotenone had a stronger effect on wild-type and *pmh2* plants than on *nmat2* and *nmat2/pmh2* plants (Figure 5), which are affected in complex I biogenesis (Figure 6). We further analyzed the respiratory activities in the presence of KCN, a complex IV-specific inhibitor. As shown in Figure 5, the presence of cyanide had a stronger effect on wild-type plants (about 70% decrease), whereas the respiration rates in the mutant lines decreased by only about 50% in the presence of the inhibitor. These results are correlated with the splicing defects observed in the mutants (Figures 2 and 3). Similar respiration rates seen in wild-type and mutant plants in the absence of inhibitors of the respiratory system (Figure 5), may correspond to the induction of alternative pathways of electron transport, via alternative oxidases (AOX) (Figure 6A) and/or type II NAD(P)H dehydrogenases (ND) [3,5,78–82]. Similarly, notable increases in the expression of various nucleus-encoded alternative oxidase genes, including AOXs and NDs, are seen in many other plants that are affected in mitochondrial RNA metabolism and the biogenesis of the respiratory system in plants (see, e.g., [34,35,40,43,50,64,74,90–92]).

3.5. Morphology of nMAT2 and PMH2 Mutant Alleles

The results in Figure 1 and published data [33,36,37] indicate that the phenotypes of *nmat2* and *pmh2* mutants are comparable with those of wild-type plants. The single and the double mutants are all able to grow, flower and set viable seeds. In light of the expected significance of mitochondria functions to plant physiology, why does the loss of *nmat2* or *pmh2* cause only minor (or no obvious) phenotypic effects? Currently, we cannot provide a definitive explanation, but we speculate that the lack of strong phenotypes may be consequences of the multiplicity of splicing cofactors that are known to exist in plant mitochondria [14,15]. This assumption is supported by the fact that the mutations did not completely abolish the maturation of the pre-RNAs in *nmat2* and *pmh2* mutants, and the protein profiles and O_2-uptake measurements which indicate that the respiratory activity is only partially affected in the mutants (Figures 5 and 6). The splicing defects seem more noticeable in the double mutant line (Figures 2 and 3), which also exhibit more notable growth and developmental defect phenotypes (i.e., shorter roots, altered leaf and flower morphologies, and leaf redness under higher light conditions) (Figure 1). However, despite the maturation defects, the corresponding mRNAs are still accumulating in the double mutant line. We therefore assume that in addition to nMAT2 and PMH2 other factors are likely involved in the splicing of each of the RNA targets.

3.6. Maturases and RNA Helicases as Putative Proto-Spliceosomal Factors in Plant Mitochondria

As indicated above, maturases and RNA helicases serve as key factors in the splicing of group II introns in different organisms [14,15,21,57]. Canonical maturases encoded within group II introns bind with high affinities and specificities to their own host pre-RNAs, and thereby stabilize the catalytically active structure of the intron core [24,93], although in some rare cases these proteins may act on several closely related group II intron targets [94]. More notably are the functions of the organellar maturases in plants, which expanded their splicing functions to multiple intron targets (reviewed by [14,15,21]). Angiosperms contain six maturase-related proteins: MatK encoded by the *trnk* intron in the chloroplasts, the MatR ORF found within *nad1* i4 in the mitochondria, and four nuclear-encoded maturases (nMATs 1 to 4) that exist in the nucleus as self-standing ORFs, out of the context of their cognate introns, and are imported post-translationally into the mitochondria [14,15,31]. Both MatK and MatR function in the splicing of different organellar group II introns, and their functions seem essential in plants [60,65,95]. Likewise, the nuclear-encoded nMAT1 and nMAT4 also promote the splicing of different subsets of group II introns in the mitochondria [32,34,35,60]. Here, we show that nMAT2 acts as a more general splicing cofactor that influences the splicing of numerous introns in Arabidopsis mitochondria (Figures 2 and 3, and Table 1). This evolutionary transition, where the intron-specific factors degenerated and evolved to regulate the splicing of multiple intron targets in plant mitochondria (all of which reside in protein-coding genes), may arose in plants as means to regulate organellar gene-expression in concert with cellular and environmental signals [15,21,57].

In addition to maturases, the splicing of organellar group II introns also depend upon the activities of other proteinaceous cofactors, which may assist in the folding of the large group II intron RNAs. In particular, DEAD-box RNA helicase seem to play central roles in introns splicing and RNA maturation (see, e.g., [36,37,52]). RNA helicases belong to a large group of enzymes that have multiple roles in RNA metabolism and gene-expression (reviewed by [96]). In yeast, *mss116p* mutants are defective in the splicing of the four mitochondrial group II introns, but are also seem to be affected in the splicing of group I introns and in the translation of various mRNAs in the mitochondria [52]. Our data show that, in Arabidopsis mitochondria, PMH2 is involved in the maturation of numerous group II introns-containing pre-RNAs (Figures 2 and 3, and Table 1). However, based on the data in Köhler et al. [37] and our own results (Figures 2 and 3), it is difficult to support any additional roles for PMH2 in mtRNA metabolism other than group II intron splicing.

As with group II introns, the catalytic activities in the spliceosomes are also mediated by the RNA components, while the spliceosomal proteins, including the core pre-mRNA-processing 8 (PRP8) and different RNA helicases, are postulated to play roles in regulating RNA folding and conformational changes needed during the splicing reactions (reviewed by e.g., [97]). The assembly of the spliceosomes involves sequential rearrangements in protein-RNA and RNA-RNA interactions, which are driven by RNA helicases, such as the PRP5 and PRP28 proteins (see, e.g., [98]). PRP5 appears to mediate RNA-protein interactions during early pre-spliceosome formation (i.e., complex A). In a subsequent reaction, involving PRP28, the U4, U5 and U6 snRNPs are recruited for the formation of the pre-catalytic spliceosome (complex B) that catalyzes the first reaction leading to the release of exon 1 and the formation of an intron-exon 2 lariat intermediate. The second catalytic step is mediated by PRP8 and leads to mRNA formation and the release of the intron lariat. Intriguingly, the topology of PRP8 resembles that of maturases [99–101], which also facilitate RNA folding and ribozyme catalysis (see e.g., [15,21]). It was therefore hypothesized that PRP8 has evolved from a maturase-related RNA chaperon, and that during eukaryotic evolution has acquired additional domains and protein cofactors, such as RNA helicases, to facilitate spliceosome assembly and introns splicing [15,97]. The involvement of maturases and DEAD-box proteins in group II intron splicing in plants and yeast mitochondria may be homologous to the spliceosome, where PRP8 and various RNA helicases function at multiple steps to facilitate pre-RNA folding and intron splicing.

4. Materials and Methods

4.1. Plant Material and Growth Conditions

Arabidopsis thaliana (ecotype *Columbia*) was used in all experiments. Wild-type (Col-0) and *nmat2* mutant lines (SALK-064659) were obtained from the Arabidopsis Biological Resource Center (ABRC) at Ohio State University (Columbus, OH, USA). Homozygous *pmh2* mutant line

(SAIL-628C06) was generously donated from the Binder's laboratory (Ulm University, Ulm, Germany). Prior to germination, seeds of wild-type and mutant lines were surface-sterilized with bleach (sodium hypochlorite) solution and sown on MS-agar plates containing 1% (w/v) sucrose. The plates were kept in the dark for 5 days at 4 °C and then grown under long day condition (LD, 16:8-h) in a controlled temperature and light growth chamber (Percival Scientific, Perry, IA, USA) at 22 °C and light intensity of 300 µE m^{-2} s^{-1}. After two weeks, the germinated seedlings were transferred to soil and cultivated in the growth chamber, under similar growth conditions (i.e., 22 °C, 60% RH and light intensity of 300 µE m^{-2} s^{-1}) in either short (SD 8:16-h) or long (LD 16:8-h) day conditions. PCR was used to screen the plant collection and check the insertion integrity of each individual line (specific oligonucleotides are listed in Table S4). Sequencing of specific PCR products was used to analyze the precise insertion site in the T-DNA lines.

4.2. Establishment of a Homozygous Double nmat2/pmh2 Mutant Line

Double *nmat2/pmh2* mutants were generated from single homozygous *nmat2* and *pmh2* lines, SALK-064659 and SAIL-628C06, respectively by genetic crossing. The resultant F1 seeds were sterilized, grown on kanamycin-containing MS medium and their seedlings were screened for double mutants (i.e., *nmat2/pmh2*). The self-fertilized F2 seedlings were further confirmed for the presence of *nmat2* and *pmh2* alleles by genomic PCR, and the double *nmat2/pmh2* mutant was then used for the analyses of its associated growth and developmental phenotypes and organellar activities.

4.3. Microscopic Analyses of Arabidopsis Wild-Type and Mutant Plants

For the analysis of plant morphology, whole plants and different organs obtained from wild-type and homozygous lines were examined under a Stereoscopic (dissecting) microscope.

4.4. Respiration Activity

Oxygen consumption (i.e., O_2 uptake) measurements were performed with a Clarke-type oxygen electrode, and the data feed was collected by Oxygraph-Plus version 1.01 software (Hansatech Instruments, King's Lynn, Norfolk, UK), as described previously [102]. The electrode was calibrated with oxygen-saturated water and by the addition of excess sodium dithionite for complete depletion of the oxygen in water housed in the electrode chamber. Equal weights (100 mg) of 2-week-old seedlings were immersed in water and incubated in the dark for a period of 30 min. Total respiration was measured at 25 °C in the dark following the addition of the seedlings to 2.5 mL of water.

4.5. RNA Extraction and Analysis

RNA extraction and analysis was performed essentially as described previously [35,40,60,103]. In brief, RNA was prepared following standard TRIzol Reagent protocols (Thermo Fisher Scientific, Waltham, MA, USA) with additional phenol/chloroform extraction. The RNA was treated with DNase I (RNase-free) (Ambion, Thermo Fisher Scientific) prior to its use in the assays. RT-qPCR was performed with specific oligonucleotides designed to intron-exon regions (pre-mRNAs) and exon-exon (mRNAs) regions corresponding to the 23 intron-containing mitochondrial transcripts in (Table S5). Reverse transcription was carried out with the Superscript III reverse transcriptase (Invitrogen, Carlsbad, CA, USA), using 1–2 µg of total RNA and 100 ng of a mixture of random hexanucleotides (Promega) and incubated for 50 min at 50 °C. Reactions were stopped by 15 min incubation at 70 °C and the RT samples served directly for real-time PCR. Quantitative PCR (qPCR) reactions were run on a LightCycler 480 (Roche-Diagnostics, Basel, Switzerland), using 2.5 µL of LightCycler 480 SYBR Green I Master mix (Roche-Diagnostics, Basel, Switzerland) and 2.5 µM forward and reverse primers in a final volume of 5 µL. Reactions were performed in triplicate in the following conditions: pre-heating at 95 °C for 10 min, followed by 40 cycles of 10 s at 95 °C, 10 s at 58 °C and 10 s at 72 °C. The nucleus-encoded 18S rRNA (At3g41768) and the mitochondrial 26S ribosomal rRNA subunit (ArthMr001) were used as reference genes in the qPCR analyses.

4.6. Total Protein Extraction and Analysis

Protein analysis was performed essentially as described in [33,60]. Total protein was extracted from three-week-old Arabidopsis leaves or isolated mitochondria by the borate/ammonium acetate method [104]. For this purpose, frozen plant tissue was homogenized in the presence of polyvinylpolypyrrolidone (PVPP) (1:1 w/w ratio). The homogenate was added to microfuge tubes containing 400 µL ice-cold protein extraction buffer (50 mM Na-borate, 50 mM ascorbic acid, 1.25% (w/v) sodium dodecyl sulfate (SDS), 12.5 mM β-mercaptoethanol, pH 9.0) and the protease inhibitor cocktail "complete Mini" from Roche Diagnostics GmbH (Mannheim, Germany). Proteins were recovered by centrifugation (25,000× g) in the presence of three volumes of ice-cold 0.1 M ammonium acetate in methanol buffer (NH_4-OAc-MeOH), generally as described in [104]. Protein concentration was determined according to the Bradford method (BioRad, Hercules, CA, USA), with bovine serum albumin used as a standard. Approximately 20 µg total protein was mixed with an equal volume of 3× protein sample buffer [105], supplemented with 50 mM β-mercaptoethanol, and subjected to 12% SDS-PAGE (at a constant 100 V). Following electrophoresis, the proteins were transferred to a PVDF membrane (BioRad; Hercules, CA, USA) and blotted overnight at 4 °C with specific primary antibodies. Detection was carried out by chemiluminescence assays after incubation with an appropriate horseradish peroxidase (HRP)-conjugated secondary antibody.

4.7. Preparation of Mitochondria from MS-Grown Arabidopsis Seedlings

Crude mitochondria extracts were prepared essentially as described in [83]. For the preparation of organellar extracts from *A. thaliana*, 200 mg of 14-days-old seedlings were harvested and homogenized in 2 mL of 75 mM MOPS-KOH, pH 7.6, 0.6 M sucrose, 4 mM EDTA, 0.2% polyvinylpyrrolidone-40, 8 mM L-cysteine, 0.2% bovine serum albumin and protease inhibitor cocktail "complete Mini" from Roche Diagnostics GmbH (Mannheim, Germany). The lysate was filtrated through one layer of miracloth and centrifuged at 1300× g for 4 min at 4 °C (to remove cell debris). The supernatant was then centrifuged at 22,000× g for 10 min at 4 °C. The resultant pellet, containing thylakoid and mitochondrial membranes, was washed twice with 1 mL of wash buffer 37.5 mM MOPS-KOH, 0.3 M sucrose and 2 mM EDTA, pH 7.6. Protein concentration was determined by the Bradford method (BioRad; Hercules, CA, USA) according to the manufacturer's protocol, with bovine serum albumin (BSA) used as a calibrator. For immunoassays, crude mitochondria fractions were suspended in sample loading buffer [105] and subjected to SDS-PAGE (at a constant 100 V). Following electrophoresis, the proteins were transferred to a PVDF membrane (BioRad; Hercules, CA, USA) and incubated overnight (at 4 °C) with various primary antibodies (Table S4). Detection was carried out by chemiluminescence assay after incubation with an appropriate horseradish peroxidase (HRP)-conjugated secondary antibody.

4.8. Blue Native (BN) Electrophoresis for Isolation of Native Organellar Complexes

Blue native (BN)-PAGE of crude mitochondria fractions was performed generally according to the method described by Pineau et al. [83]. An aliquot equivalent to 40 mg of crude Arabidopsis mitochondria extracts, obtained from wild-type *nmat2*, *pmh2* and *nmat2/pmh2* plants (see above; Protein extraction and analysis) was solubilized with n-dodecyl-β-maltoside [DDM; 1.5% (w/v)] in ACA buffer (750 mM amino-caproic acid, 0.5 mM EDTA, and 50 mM Tris-HCl, pH 7.0), and then incubated on ice for 30 min. The samples were centrifuged 8 min at 20,000× g to pellet any insoluble and Serva Blue G (0.2% (v/v)) was added to the supernatant. The samples were then loaded onto a native 4 to 16% linear gradient gel. For "non-denaturing-PAGE" Western blotting, the gel was transferred to a PVDF membrane (BioRad; Hercules, CA, USA) in Cathode buffer (50 mM Tricine and 15 mM Bis-Tris-HCl, pH 7.0) for 16 h at 4 °C at constant current of 40 mA. The mitochondria were then incubated with antibodies against various organellar proteins and detection was carried out by chemiluminescence assay after incubation with an appropriate horseradish peroxidase (HRP)-conjugated secondary antibody. In-gel complex I activity assays were performed essentially as described previously [92].

4.9. Co-Immunoprecipitations

Mitochondria extracts from nmat2/nMAT2-HA plants were solubilized with 1% NP-40 (v/v) in assay buffer (150 mM NaCl, 10 mM sodium phosphate buffer, pH 7.2) at a protein concentration of about 1 mg/mL. After 30 min incubation on ice, the organellar extract was centrifuged for 10 min at $21,000\times g$. The clear supernatant was then incubated with anti-HA antibodies conjugated to protein A/G sepharose beads, with gentle mixing at 4 °C for 16 h. The beads were collected by a brief centrifugation (1 min at $1000\times g$, at 4 °C) and washed three times with 0.6 M NaCl, 0.5% (v/v) NP-40, 50 mM Tris-HCl (pH 8.3), followed by a single wash with $1\times$ PBS buffer. The identity of the proteins in the co-IPs was established by LC-MS/MS analysis (The Smoler Proteomics Center, Technion, Haifa, Israel) (see also Table S1). Commercial antibodies to SHMT were used as controls for the integrity of the co-IP method. Total mtRNA that co-precipitated with the anti-HA antibodies was DNase digested and then reverse transcribed, using the Superscript III reverse transcriptase and a random hexanucleotide mixture. The identity of the co-purified mtRNAs was established by PCR of the cDNA library with specific oligonucleotides designed to organellar pre-RNAs (see Table S6) and sequencing (The Center for Genomic Technologies, The Hebrew University of Jerusalem, Israel).

Supplementary Materials: The following are available online at www.mdpi.com/1422-0067/18/11/2428/s1.

Acknowledgments: We thank the Arabidopsis biological resource center for providing the T-DNA mutant seeds. This work was supported by grants to O.O.B from the "Israeli Science Foundation" (ISF grant No. 741/15) and in part from the German–Israeli Foundation (GIF 1213/2012) and the "US–Israel Binational Agricultural Research and Development Fund" (BARD grant No. IS-4921-16 F). We thank Stefan Binder (Ulm University, Germany) for his generous donation of *pmh2* mutant lines and antibodies raised against the PMH2 protein. We would also like to thank, in particular, Maya Shimoni and Sam Aldrin for carefully reading this manuscript and making valuable corrections and helpful suggestions regarding the manuscript text and files.

Author Contributions: Michal Zmudjak: Plant growth and analysis, establishment of double mutant lines, analysis of gene expression, BN-PAGE, mtRNA isolation, DNA and RNA sequencing, analysis of the transcriptome profiles of Arabidopsis mitochondria by RT-qPCR and respiration analyses. Sofia Shevtsov: Plant growth and analysis, assistance in BN-PAGE and protein analysis, and RNA extraction and analysis. Laure D. Sultan: Assistance in BN-PAGE and protein analysis. Ido Keren: co-IP experiments. Oren Ostersetzer-Biran: principle investigator, experimental design of this study and manuscript preparation and writing. All authors read and revised the manuscript critically.

Conflicts of Interest: The authors declare no conflicts of interest.

References

1. Gualberto, J.M.; Kuhn, K. DNA-binding proteins in plant mitochondria: Implications for transcription. *Mitochondrion* **2014**, *19*, 323–328. [CrossRef] [PubMed]
2. Gualberto, J.M.; Newton, K.J. Plant Mitochondrial Genomes: Dynamics and Mechanisms of Mutation. *Annu. Rev. Plant Biol.* **2017**, *68*, 225–252. [CrossRef] [PubMed]
3. Millar, A.H.; Whelan, J.; Soole, K.L.; Day, D.A. Organization and regulation of mitochondrial respiration in plants. *Ann. Rev. Plant Biol.* **2011**, *62*, 79–104. [CrossRef] [PubMed]
4. Lee, C.P.; Taylor, N.L.; Millar, A.H. Recent advances in the composition and heterogeneity of the Arabidopsis mitochondrial proteome. *Front. Plant Sci.* **2013**, *4*, 4. [CrossRef] [PubMed]
5. Jacoby, R.P.; Li, L.; Huang, S.; Pong Lee, C.; Millar, A.H.; Taylor, N.L. Mitochondrial composition, function and stress response in plants. *J. Integr. Plant. Biol.* **2012**, *54*, 887–906. [CrossRef] [PubMed]
6. Huang, S.; Shingaki-Wells, R.N.; Taylor, N.L.; Millar, A.H. The rice mitochondria proteome and its response during development and to the environment. *Front. Plant Sci.* **2013**, *4*, 16. [CrossRef] [PubMed]
7. Schertl, P.; Braun, H.P. Respiratory electron transfer pathways in plant mitochondria. *Front. Plant Sci.* **2014**, *5*, 163. [CrossRef] [PubMed]
8. Huang, S.; Jacoby, R.P.; Millar, A.H.; Taylor, N.L. Plant mitochondrial proteomics. *Methods Mol. Biol.* **2014**, *1072*, 499–525. [PubMed]
9. Woodson, J.D.; Chory, J. Coordination of gene expression between organellar and nuclear genomes. *Nat. Rev. Genet.* **2008**, *9*, 383–395. [CrossRef] [PubMed]

10. Liere, K.; Weihe, A.; Börner, T. The transcription machineries of plant mitochondria and chloroplasts: Composition, function, and regulation. *J. Plant Physiol.* **2011**, *168*, 1345–1360. [CrossRef] [PubMed]
11. Small, I. Mitochondrial genomes as living "fossils". *BMC Biol.* **2013**, *11*, 30. [CrossRef] [PubMed]
12. Hammani, K.; Giege, P. RNA metabolism in plant mitochondria. *Trends Plant Sci.* **2014**, *19*, 380–389. [CrossRef] [PubMed]
13. Small, I.D.; Rackham, O.; Filipovska, A. Organelle transcriptomes: Products of a deconstructed genome. *Curr. Opin. Microbiol.* **2013**, *16*, 652–658. [CrossRef] [PubMed]
14. Brown, G.G.; Colas des Francs-Small, C.; Ostersetzer-Biran, O. Group II intron splicing factors in plant mitochondria. *Front. Plant Sci.* **2014**, *5*, 35. [CrossRef] [PubMed]
15. Schmitz-Linneweber, C.; Lampe, M.K.; Sultan, L.D.; Ostersetzer-Biran, O. Organellar maturases: A window into the evolution of the spliceosome. *Biochim. Biophys. Acta* **2015**, *1847*, 798–808. [CrossRef] [PubMed]
16. Liere, K.; Börner, T. Transcription in Plant Mitochondria. In *Plant Mitochondria*; Springer: New York, NY, USA, 2011; pp. 85–105.
17. Kühn, K.; Weihe, A.; Borner, T. Multiple promoters are a common feature of mitochondrial genes in Arabidopsis. *Nucleic Acids Res.* **2005**, *33*, 337–346. [CrossRef] [PubMed]
18. Kühn, K.; Bohne, A.V.; Liere, K.; Weihe, A.; Börner, T. Arabidopsis phage-type RNA polymerases: Accurate in vitro transcription of organellar genes. *Plant Cell* **2007**, *19*, 959–971. [CrossRef] [PubMed]
19. Bonen, L. Cis- and trans-splicing of group II introns in plant mitochondria. *Mitochondrion* **2008**, *8*, 26–34. [CrossRef] [PubMed]
20. Bonen, L.; Vogel, J. The ins and outs of group II introns. *Trends Genet.* **2001**, *17*, 322–331. [CrossRef]
21. Lambowitz, A.M.; Zimmerly, S. Group II introns: Mobile ribozymes that invade DNA. *Cold Spring Harb. Perspect. Biol.* **2011**, *3*, 1–19. [CrossRef] [PubMed]
22. Unseld, M.; Marienfeld, J.R.; Brandt, P.; Brennicke, A. The mitochondrial genome of *Arabidopsis thaliana* contains 57 genes in 366,924 nucleotides. *Nat. Genet.* **1997**, *15*, 57–61. [CrossRef] [PubMed]
23. Cech, T.R. The generality of self-splicing RNA: Relationship to nuclear messenger-RNA splicing. *Cell* **1986**, *44*, 207–210. [CrossRef]
24. Matsuura, M.; Noah, J.W.; Lambowitz, A.M. Mechanism of maturase-promoted group II intron splicing. *EMBO J.* **2001**, *20*, 7259–7270. [CrossRef] [PubMed]
25. Zimmerly, S.; Hausner, G.; Wu, X.C. Phylogenetic relationships among group II intron ORFs. *Nucleic Acids Res.* **2001**, *29*, 1238–1250. [CrossRef] [PubMed]
26. Qu, G.; Kaushal, P.S.; Wang, J.; Shigematsu, H.; Piazza, C.L.; Agrawal, R.K.; Belfort, M.; Wang, H.W. Structure of a group II intron in complex with its reverse transcriptase. *Nat. Struct. Mol. Biol.* **2016**, *23*, 549–557. [CrossRef] [PubMed]
27. Zhao, C.; Pyle, A.M. Crystal structures of a group II intron maturase reveal a missing link in spliceosome evolution. *Nat. Struct. Mol. Biol.* **2016**, *23*, 558–565. [CrossRef] [PubMed]
28. Zimmerly, S.; Semper, C. Evolution of group II introns. *Mob. DNA* **2015**, *6*, 1–19. [CrossRef] [PubMed]
29. Hanley, B.A.; Schuler, M.A. Plant intron sequences: Evidence for distinct groups of introns. *Nucleic Acids Res.* **1988**, *16*, 7159–7175. [CrossRef] [PubMed]
30. Colas des Francs-Small, C.; Small, I. Surrogate mutants for studying mitochondrially encoded functions. *Biochimie* **2014**, *100*, 234–242. [CrossRef] [PubMed]
31. Mohr, G.; Lambowitz, A.M. Putative proteins related to group II intron reverse transcriptase/maturases are encoded by nuclear genes in higher plants. *Nucleic Acids. Res.* **2003**, *31*, 647–652. [CrossRef] [PubMed]
32. Nakagawa, N.; Sakurai, N. A mutation in At-nMat1a, which encodes a nuclear gene having high similarity to group II Intron maturase, causes impaired splicing of mitochondrial nad4 transcript and altered carbon metabolism in Arabidopsis thaliana. *Plant Cell Physiol.* **2006**, *47*, 772–783. [CrossRef] [PubMed]
33. Keren, I.; Bezawork-Geleta, A.; Kolton, M.; Maayan, I.; Belausov, E.; Levy, M.; Mett, A.; Gidoni, D.; Shaya, F.; Ostersetzer-Biran, O. AtnMat2, a nuclear-encoded maturase required for splicing of group-II introns in Arabidopsis mitochondria. *RNA* **2009**, *15*, 2299–2311. [CrossRef] [PubMed]
34. Keren, I.; Tal, L.; Colas des Francs-Small, C.; Araújo, W.L.; Shevtsov, S.; Shaya, F.; Fernie, A.R.; Small, I.; Ostersetzer-Biran, O. nMAT1, a nuclear-encoded maturase involved in the trans-splicing of nad1 intron 1, is essential for mitochondrial complex I assembly and function. *Plant J.* **2012**, *71*, 413–426. [CrossRef] [PubMed]

35. Cohen, S.; Zmudjak, M.; Colas des Francs-Small, C.; Malik, S.; Shaya, F.; Keren, I.; Belausov, E.; Many, Y.; Brown, G.G.; Small, I.; et al. nMAT4, a maturase factor required for nad1 pre-mRNA processing and maturation, is essential for holocomplex I biogenesis in Arabidopsis mitochondria. *Plant J.* **2014**, *78*, 253–268. [CrossRef] [PubMed]
36. Matthes, A.; Schmidt-Gattung, S.; Köhler, D.; Forner, J.; Wildum, S.; Raabe, M.; Urlaub, H.; Binder, S. Two DEAD-Box Proteins May Be Part of RNA-Dependent High-Molecular-Mass Protein Complexes in Arabidopsis Mitochondria. *Plant Physiol.* **2007**, *145*, 1637–1646. [CrossRef] [PubMed]
37. Köhler, D.; Schmidt-Gattung, S.; Binder, S. The DEAD-box protein PMH2 is required for efficient group II intron splicing in mitochondria of Arabidopsis thaliana. *Plant Mol. Biol.* **2010**, *72*, 459–467. [CrossRef] [PubMed]
38. Putnam, A.A.; Jankowsky, E. DEAD-box helicases as integrators of RNA, nucleotide and protein binding. *BBA-Gene Regul. Mech.* **2013**, *1829*, 884–893. [CrossRef] [PubMed]
39. He, J.; Duan, Y.; Hua, D.; Fan, G.; Wang, L.; Liu, Y.; Chen, Z.; Han, L.; Qu, L.J.; Gong, Z. DEXH Box RNA Helicase–Mediated Mitochondrial Reactive Oxygen Species Production in Arabidopsis Mediates Crosstalk between Abscisic Acid and Auxin Signaling. *Plant Cell* **2012**, *24*, 1815–1833. [CrossRef] [PubMed]
40. Zmudjak, M.; Colas des Francs-Small, C.; Keren, I.; Shaya, F.; Belausov, E.; Small, I.; Ostersetzer-Biran, O. mCSF1, a nucleus-encoded CRM protein required for the processing of many mitochondrial introns, is involved in the biogenesis of respiratory complexes I and IV in Arabidopsis. *New Phytol.* **2013**, *199*, 379–394. [CrossRef] [PubMed]
41. Weissenberger, S.; Soll, J.; Carrie, C. The PPR protein SLOW GROWTH 4 is involved in editing of *nad4* and affects the splicing of *nad2* intron 1. *Plant Mol. Biol.* **2017**, *93*, 355–368. [CrossRef] [PubMed]
42. Su, C.; Zhao, H.; Zhao, Y.; Ji, H.; Wang, Y.; Zhi, L.; Li, X. RUG3 and ATM synergistically regulate the alternative splicing of mitochondrial *nad2* and the DNA damage response in *Arabidopsis thaliana*. *Sci. Rep.* **2017**, *7*, 43897. [CrossRef] [PubMed]
43. Kühn, K.; Carrie, C.; Giraud, E.; Wang, Y.; Meyer, E.H.; Narsai, R.; Colas des Francs-Small, C.; Zhang, B.; Murcha, M.W.; Whelan, J. The RCC1 family protein RUG3 is required for splicing of *nad2* and complex I biogenesis in mitochondria of *Arabidopsis thaliana*. *Plant J.* **2011**, *67*, 1067–1080. [CrossRef] [PubMed]
44. Koprivova, A.; Colas des Francs-Small, C.; Calder, G.; Mugford, S.T.; Tanz, S.; Lee, B.R.; Zechmann, B.; Small, I.; Kopriva, S. Identification of a pentatricopeptide repeat protein implicated in splicing of intron 1 of mitochondrial nad7 transcripts. *J. Biol. Chem.* **2010**, *285*, 32192–32199. [CrossRef] [PubMed]
45. Liu, Y.; He, J.; Chen, Z.; Ren, X.; Hong, X.; Gong, Z. ABA overly-sensitive 5 (ABO5), encoding a pentatricopeptide repeat protein required for cis-splicing of mitochondrial nad2 intron 3, is involved in the abscisic acid response in Arabidopsis. *Plant J.* **2010**, *63*, 749–765. [CrossRef] [PubMed]
46. Colas des Francs-Small, C.; Falcon de Longevialle, A.; Li, Y.; Lowe, E.; Tanz, S.; Smith, C.; Bevan, M.; Small, I. The PPR proteins TANG2 and OTP439 are involved in the splicing of the multipartite nad5 transcript encoding a subunit of mitochondrial complex I. *Plant Physiol.* **2014**, *114*, 244616.
47. Xiu, Z.; Sun, F.; Shen, Y.; Zhang, X.; Jiang, R.; Bonnard, G.; Zhang, J.; Tan, B.C. EMPTY PERICARP16 is required for mitochondrial nad2 Intron 4 cis-splicing, complex I assembly and seed development in maize. *Plant J.* **2016**, *85*, 507–519. [CrossRef] [PubMed]
48. Qi, W.; Yang, Y.; Feng, X.; Zhang, M.; Song, R. Mitochondrial Function and Maize Kernel Development Requires Dek2, a Pentatricopeptide Repeat Protein Involved in nad1 mRNA Splicing. *Genetics* **2017**, *205*, 239–249. [CrossRef] [PubMed]
49. Falcon de Longevialle, A.; Meyer, E.H.; Andres, C.; Taylor, N.L.; Lurin, C.; Millar, A.H.; Small, I.D. The pentatricopeptide repeat gene OTP43 is required for trans-splicing of the mitochondrial nad1 intron 1 in Arabidopsis thaliana. *Plant Cell* **2007**, *19*, 3256–3265. [CrossRef] [PubMed]
50. Hsu, Y.W.; Wang, H.J.; Hsieh, M.H.; Hsieh, H.L.; Jauh, G.Y. *Arabidopsis* mTERF15 Is Required for Mitochondrial nad2 Intron 3 Splicing and Functional Complex I Activity. *PLoS ONE* **2014**, *9*, 112360. [CrossRef] [PubMed]
51. Shi, X.; Castandet, B.; Germain, A.; Hanson, M.R.; Bentolila, S. ORRM5, an RNA recognition motif-containing protein, has a unique effect on mitochondrial RNA editing. *J. Exp. Bot.* **2017**, *68*, 2833–2847. [CrossRef] [PubMed]

52. Huang, H.R.; Rowe, C.E.; Mohr, S.; Jiang, Y.; Lambowitz, A.M.; Perlman, P.S. The splicing of yeast mitochondrial group I and group II introns requires a DEAD-box protein with RNA chaperone function. *Proc. Natl. Acad. Sci. USA* **2005**, *102*, 163–168. [CrossRef] [PubMed]
53. Mohr, S.; Matsuura, M.; Perlman, P.S.; Lambowitz, A.M. A DEAD-box protein alone promotes group II intron splicing and reverse splicing by acting as an RNA chaperone. *Proc. Natl. Acad. Sci. USA* **2006**, *103*, 3569–3574. [CrossRef] [PubMed]
54. Fedorova, O.; Solem, A.; Pyle, A.M. Protein-facilitated folding of group II intron ribozymes. *J. Mol. Biol.* **2010**, *397*, 799–813. [CrossRef] [PubMed]
55. Barkan, A. Intron Splicing in Plant Organelles. In *Molecular Biology and Biotechnology of Plant Organelles*; Daniell, H., Chase, C., Eds.; Kluwer Academic: Dordrecht, The Netherlands, 2004; pp. 281–308.
56. Lambowitz, A.M.; Caprara, M.G.; Zimmerly, S.; Perlman, P. Group I and Group II Ribozymes as RNPs: Clues to the Past and Guides to the Future. *Cold Spring Harb. Monogr. Ser.* **1999**, *37*, 451–485.
57. Lambowitz, A.M.; Belfort, M. Mobile Bacterial Group II Introns at the Crux of Eukaryotic Evolution. *Microbiol. Spectr.* **2015**, *3*, 1–26. [CrossRef] [PubMed]
58. Letunic, I.; Doerks, T.; Bork, P. SMART 7: Recent updates to the protein domain annotation resource. *Nucleic Acids Res.* **2012**, *40*, 302–305. [CrossRef] [PubMed]
59. Marchler-Bauer, A.; Anderson, J.B.; DeWeese-Scott, C.; Fedorova, N.D.; Geer, L.Y.; He, S.; Hurwitz, D.I.; Jackson, J.D.; Jacobs, A.R.; Lanczycki, C.J.; et al. CDD: A curated Entrez database of conserved domain alignments. *Nucleic Acids Res.* **2003**, *31*, 383–387. [CrossRef] [PubMed]
60. Sultan, L.D.; Mileshina, D.; Grewe, F.; Rolle, K.; Abudraham, S.; Głodowicz, P.; Khan Niazi, A.; Keren, I.; Shevtsov, S.; Klipcan, L.; et al. The reverse-transcriptase/RNA-maturase protein MatR is required for the splicing of various group II introns in Brassicaceae mitochondria. *Plant Cell* **2016**, *28*, 2805–2829. [CrossRef] [PubMed]
61. Kelley, L.A.; Sternberg, M.J. Protein structure prediction on the Web: A case study using the Phyre server. *Nat. Protoc.* **2009**, *4*, 363–371. [CrossRef] [PubMed]
62. Kim, D.E.; Chivian, D.; Baker, D. Protein structure prediction and analysis using the Robetta server. *Nucleic Acids Res.* **2004**, *32*, 526–531. [CrossRef] [PubMed]
63. Liu, Y.C.; Cheng, S.C. Functional roles of DExD/H-box RNA helicases in Pre-mRNA splicing. *J. Biomed. Sci.* **2015**, *22*, 54. [CrossRef] [PubMed]
64. Colas des Francs-Small, C.; Kroeger, T.; Zmudjak, M.; Ostersetzer-Biran, O.; Rahimi, N.; Small, I.; Barkan, A. A PORR domain protein required for rpl2 and ccmFc intron splicing and for the biogenesis of c-type cytochromes in Arabidopsis mitochondria. *Plant J.* **2012**, *69*, 996–1005. [CrossRef] [PubMed]
65. Zoschke, R.; Nakamura, M.; Liere, K.; Sugiura, M.; Börner, T.; Schmitz-Linneweber, C. An organellar maturase associates with multiple group II introns. *Proc. Natl. Acad. Sci. USA* **2010**, *107*, 3245–3250. [CrossRef] [PubMed]
66. Mohr, S.; Stryker, J.M.; Lambowitz, A.M. A DEAD-box protein functions as an ATP-dependent RNA chaperone in group I intron splicing. *Cell* **2002**, *109*, 769–779. [CrossRef]
67. Van Aken, O.; Whelan, J.; Van Breusegem, F. Prohibitins: Mitochondrial partners in development and stress response. *Trends Plant Sci.* **2010**, *15*, 275–282. [CrossRef] [PubMed]
68. Ostheimer, G.J.; Williams-Carrier, R.; Belcher, S.; Osborne, E.; Gierke, J.; Barkan, A. Group II intron splicing factors derived by diversification of an ancient RNA-binding domain. *EMBO J.* **2003**, *22*, 3919–3929. [CrossRef] [PubMed]
69. Schmitz-Linneweber, C.; Williams-Carrier, R.; Barkan, A. RNA immunoprecipitation and microarray analysis show a chloroplast pentatricopeptide repeat protein to be associated with the 5' region of mRNAs whose translation it activates. *Plant Cell* **2005**, *17*, 2791–2804. [CrossRef] [PubMed]
70. Schmitz-Linneweber, C.; Williams-Carrier, R.E.; Williams-Voelker, P.M.; Kroeger, T.S.; Vichas, A.; Barkan, A. A pentatricopeptide repeat protein facilitates the trans-splicing of the maize chloroplast rps12 Pre-mRNA. *Plant Cell* **2006**, *18*, 2650–2663. [CrossRef] [PubMed]
71. Asakura, Y.; Barkan, A. A CRM domain protein functions dually in group I and group II intron splicing in land plant chloroplasts. *Plant Cell* **2007**, *19*, 3864–3875. [CrossRef] [PubMed]
72. Watkins, K.P.; Kroeger, T.S.; Cooke, A.M.; Williams-Carrier, R.E.; Friso, G.; Belcher, S.E.; van Wijk, K.J.; Barkan, A. A ribonuclease III domain protein functions in group II intron splicing in maize chloroplasts. *Plant Cell* **2007**, *19*, 2606–2623. [CrossRef] [PubMed]

73. Ostersetzer-Biran, O. Respiratory complex I and embryo development. *J. Exp. Bot.* **2016**, *67*, 1205–1207. [CrossRef] [PubMed]
74. Kuhn, K.; Obata, T.; Feher, K.; Bock, R.; Fernie, A.R.; Meyer, E.H. Complete Mitochondrial Complex I Deficiency Induces an Up-Regulation of Respiratory Fluxes That Is Abolished by Traces of Functional Complex I. *Plant Physiol.* **2015**, *168*, 1537–1549. [CrossRef] [PubMed]
75. Fromm, S.; Senkler, J.; Eubel, H.; Peterhansel, C.; Braun, H.P. Life without complex I: Proteome analyses of an Arabidopsis mutant lacking the mitochondrial NADH dehydrogenase complex. *J. Exp. Bot.* **2016**, *67*, 3079–3093. [CrossRef] [PubMed]
76. Chotewutmontri, P.; Barkan, A. Dynamics of Chloroplast Translation during Chloroplast Differentiation in Maize. *PLoS Genet.* **2016**, *12*, e1006106. [CrossRef] [PubMed]
77. Jensen, E.C. Quantitative analysis of histological staining and fluorescence using ImageJ. *Anat. Rec.* **2013**, *296*, 378–381. [CrossRef] [PubMed]
78. Yoshida, K.; Watanabe, C.; Kato, Y.; Sakamoto, W.; Noguchi, K. Influence of chloroplastic photo-oxidative stress on mitochondrial alternative oxidase capacity and respiratory properties: A case study with Arabidopsis yellow variegated 2. *Plant Cell Physiol.* **2008**, *49*, 592–603. [CrossRef] [PubMed]
79. Van Aken, O.; Giraud, E.; Clifton, R.; Whelan, J. Alternative oxidase: A target and regulator of stress responses. *Physiol. Plant* **2009**, *137*, 354–361. [CrossRef] [PubMed]
80. Araújo, W.L.; Nunes-Nesi, A.; Nikoloski, Z.; Sweetlove, L.J.; Fernie, A.R. Metabolic control and regulation of the tricarboxylic acid cycle in photosynthetic and heterotrophic plant tissues. *Plant Cell Environ.* **2012**, *35*, 1–21. [CrossRef] [PubMed]
81. Juszczuk, I.M.; Szal, B.; Rychter, A.M. Oxidation–reduction and reactive oxygen species homeostasis in mutant plants with respiratory chain complex I dysfunction. *Plant Cell Environ.* **2012**, *35*, 296–307. [CrossRef] [PubMed]
82. Rasmusson, A.G.; Geisler, D.A.; Møller, I.M. The multiplicity of dehydrogenases in the electron transport chain of plant mitochondria. *Mitochondrion* **2008**, *8*, 47–60. [CrossRef] [PubMed]
83. Pineau, B.; Layoune, O.; Danon, A.; De Paepe, R. L-galactono-1,4-lactone dehydrogenase is required for the accumulation of plant respiratory complex I. *J. Biol. Chem.* **2008**, *283*, 32500–32505. [CrossRef] [PubMed]
84. Irimia, M.; Roy, S.W. Origin of Spliceosomal Introns and Alternative Splicing. *CSH Perspect. Biol.* **2014**, *6*, a016071. [CrossRef] [PubMed]
85. Barkan, A.; Klipcan, L.; Ostersetzer, O.; Kawamura, T.; Asakura, Y.; Watkins, K.P. The CRM domain: An RNA binding module derived from an ancient ribosome-associated protein. *RNA* **2007**, *13*, 55–64. [CrossRef] [PubMed]
86. Keren, I.; Klipcan, L.; Bezawork-Geleta, A.; Kolton, M.; Shaya, F.; Ostersetzer-Biran, O. Characterization of the molecular basis of group II intron RNA recognition by CRS1-CRM domains. *J. Biol. Chem.* **2008**, *283*, 23333–23342. [CrossRef] [PubMed]
87. Semrad, K. Proteins with RNA Chaperone Activity: A World of Diverse Proteins with a Common Task-Impediment of RNA Misfolding. *Biochem. Res. Int.* **2011**, *2011*, 11. [CrossRef] [PubMed]
88. Tovchigrechko, A.; Vakser, I.A. GRAMM-X public web server for protein-protein docking. *Nucleic Acids Res.* **2006**, *34*, 310–314. [CrossRef] [PubMed]
89. DeLano, W.L.; Lam, J.W. PyMOL: A communications tool for computational models. *Abstr. Pap. Am. Chem. Soc.* **2005**, *230*, U1371–U1372.
90. Lee, K.; Han, J.H.; Park, Y.I.; Colas des Francs-Small, C.; Small, I.; Kang, H. The mitochondrial pentatricopeptide repeat protein PPR19 is involved in the stabilization of NADH dehydrogenase 1 transcripts and is crucial for mitochondrial function and Arabidopsis thaliana development. *New Phytol.* **2017**, *215*, 202–216. [CrossRef] [PubMed]
91. Zhu, Q.; Dugardeyn, J.; Zhang, C.; Takenaka, M.; Kühn, K.; Craddock, C.; Smalle, J.; Karampelias, M.; Denecke, J.; Peters, J.; et al. SLO2, a mitochondrial PPR protein affecting several RNA editing sites, is required for energy metabolism. *Plant J.* **2012**, *71*, 836–849. [CrossRef] [PubMed]
92. Meyer, E.H.; Tomaz, T.; Carroll, A.J.; Estavillo, G.; Delannoy, E.; Tanz, S.K.; Small, I.D.; Pogson, B.J.; Millar, A.H. Remodeled respiration in ndufs4 with low phosphorylation efficiency suppresses Arabidopsis germination and growth and alters control of metabolism at night. *Plant Physiol.* **2009**, *151*, 603–619. [CrossRef] [PubMed]

93. Noah, J.W.; Lambowitz, A.M. Effects of maturase binding and Mg2+ concentration on group II intron RNA folding investigated by UV cross-linking. *Biochemistry* **2003**, *42*, 12466–12480. [CrossRef] [PubMed]
94. Meng, Q.; Wang, Y.; Liu, X.Q. An intron-encoded protein assists RNA splicing of multiple similar introns of different bacterial genes. *J. Biol. Chem.* **2005**, *280*, 35085–35088. [CrossRef] [PubMed]
95. Zoschke, R.; Ostersetzer-Biran, O.; Börner, T.; Schmitz-Linneweber, C. Analysis of the regulation of MatK gene expression. *Endocytobiosis Cell Res.* **2009**, *19*, 127–135.
96. Bourgeois, C.F.; Mortreux, F.; Auboeuf, D. The multiple functions of RNA helicases as drivers and regulators of gene expression. *Nat. Rev. Mol. Cell. Biol.* **2016**, *17*, 426–438. [CrossRef] [PubMed]
97. Papasaikas, P.; Valcarcel, J. The Spliceosome: The Ultimate RNA Chaperone and Sculptor. *Trends Biochem. Sci.* **2016**, *41*, 33–45. [CrossRef] [PubMed]
98. Matera, A.G.; Wang, Z. A day in the life of the spliceosome. *Nat. Rev. Mol. Cell Biol.* **2014**, *15*, 108–121. [CrossRef] [PubMed]
99. Dlakic, M.; Mushegian, A. Prp8, the pivotal protein of the spliceosomal catalytic center, evolved from a retroelement-encoded reverse transcriptase. *RNA* **2011**, *17*, 799–808. [CrossRef] [PubMed]
100. Galej, W.P.; Oubridge, C.; Newman, A.J.; Nagai, K. Crystal structure of Prp8 reveals active site cavity of the spliceosome. *Nature* **2013**, *493*, 638–643. [CrossRef] [PubMed]
101. Yan, C.; Hang, J.; Wan, R.; Huang, M.; Wong, C.C.; Shi, Y. Structure of a yeast spliceosome at 3.6-angstrom resolution. *Science* **2015**, *349*, 1182–1191. [CrossRef] [PubMed]
102. Tomaz, T.; Bagard, M.; Pracharoenwattana, I.; Lindén, P.; Lee, C.P.; Carroll, A.J.; Ströher, E.; Smith, S.M.; Gardeström, P.; Millar, A.H. Mitochondrial malate dehydrogenase lowers leaf respiration and alters photorespiration and plant growth in Arabidopsis. *Plant Physiol.* **2010**, *154*, 1143–1157. [CrossRef] [PubMed]
103. Keren, I.; Shaya, F.; Ostersetzer-Biran, O. An optimized method for the analysis of plant mitochondria RNAs by northern-blotting. *Endocy Cell Res.* **2011**, *1*, 34–42.
104. Maayan, I.; Shaya, F.; Ratner, K.; Mani, Y.; Lavee, S.; Avidan, B.; Shahak, Y.; Ostersetzer-Biran, O. Photosynthetic activity during olive (*Olea europaea*) leaf development correlates with plastid biogenesis and Rubisco levels. *Physiol. Plant* **2008**, *134*, 547–558. [CrossRef] [PubMed]
105. Laemmli, U.K. Cleavage of structural proteins during the assembly of the head of bacteriophage T4. *Nature* **1970**, *227*, 680–685. [CrossRef] [PubMed]

© 2017 by the authors. Licensee MDPI, Basel, Switzerland. This article is an open access article distributed under the terms and conditions of the Creative Commons Attribution (CC BY) license (http://creativecommons.org/licenses/by/4.0/).

Article

Nitric Oxide Regulates Seedling Growth and Mitochondrial Responses in Aged Oat Seeds

Chunli Mao [1,†], Yanqiao Zhu [1,†], Hang Cheng [1], Huifang Yan [1], Liyuan Zhao [1], Jia Tang [1], Xiqing Ma [1,2] and Peisheng Mao [1,2,*]

[1] College of Animal Science and Technology, China Agricultural University, Beijing 100193, China; maochunlim@163.com (C.M.); yanq_zhu@cau.edu.cn (Y.Z.); chenghang0522@126.com (H.C.); yanhui_fang@126.com (H.Y.); yuanlizhao1994@126.com (L.Z.); candytangjia@126.com (J.T.); ma2016@cau.edu.cn (X.M.)
[2] Key Laboratory of Pratacultural Science, Beijing Municipality, China Agricultural University, No. 2 Yuanmingyuan West Road, Haidian District, Beijing 100193, China
* Correspondence: maopeisheng@hotmail.com; Tel.: +86-010-62733311
† These authors contributed equally to this work.

Received: 10 February 2018; Accepted: 26 March 2018; Published: 2 April 2018

Abstract: Mitochondria are the source of reactive oxygen species (ROS) in plant cells and play a central role in the mitochondrial electron transport chain (ETC) and tricarboxylic acid cycle (TCA) cycles; however, ROS production and regulation for seed germination, seedling growth, as well as mitochondrial responses to abiotic stress, are not clear. This study was conducted to obtain basic information on seed germination, embryo mitochondrial antioxidant responses, and protein profile changes in artificial aging in oat seeds (*Avena sativa* L.) exposed to exogenous nitric oxide (NO) treatment. The results showed that the accumulation of H_2O_2 in mitochondria increased significantly in aged seeds. Artificial aging can lead to a loss of seed vigor, which was shown by a decline in seed germination and the extension of mean germination time (MGT). Seedling growth was also inhibited. Some enzymes, including catalase (CAT), glutathione reductase (GR), dehydroascorbate reductase (DHAR), and monodehydroascorbate reductase (MDHAR), maintained a lower level in the ascorbate-glutathione (AsA-GSH) scavenging system. Proteomic analysis revealed that the expression of some proteins related to the TCA cycle were down-regulated and several enzymes related to mitochondrial ETC were up-regulated. With the application of 0.05 mM NO in aged oat seeds, a protective effect was observed, demonstrated by an improvement in seed vigor and increased H_2O_2 scavenging ability in mitochondria. There were also higher activities of CAT, GR, MDHAR, and DHAR in the AsA-GSH scavenging system, enhanced TCA cycle-related enzymes (malate dehydrogenase, succinate-CoA ligase, fumarate hydratase), and activated alternative pathways, as the cytochrome pathway was inhibited. Therefore, our results indicated that seedling growth and seed germinability could retain a certain level in aged oat seeds, predominantly depending on the lower NO regulation of the TCA cycle and AsA-GSH. Thus, it could be concluded that the application of 0.05 mM NO in aged oat seeds improved seed vigor by enhancing the mitochondrial TCA cycle and activating alternative pathways for improvement.

Keywords: nitric oxide; ROS; mitochondria; proteins; alternative pathway; antioxidant enzymes

1. Introduction

High-quality seeds are extremely important to agricultural production, productivity, and germplasm conservation. However, seed deterioration occurs during storage, even under optimal storage conditions, which causes the loss of seed vigor. Various biochemical and metabolic alterations take place during seed aging, including electrolyte leakage, the loss of cell membrane integrity, DNA alteration, and damage of mitochondrial structure and function [1–3]. Although the mechanisms of seed aging are still being researched, reactive oxygen species (ROS) are considered the main factor contributing to seed aging and leading to the damage of lipids, DNA, and proteins [4,5]. ROS, including hydrogen peroxide (H_2O_2), hydroxyl radical ($\cdot OH$), and superoxide radical (O_2^-), accumulate in the aged seeds of sunflowers (*Helianthus annuus* L.) [6], oats [2], and elm (*Ulmus pumila* L.) [7]. Therefore, it is necessary to explore the aging mechanisms of the detrimental role of ROS in deteriorated seeds.

Mitochondrion can provide energy for cell metabolism and transport by respiration and is the main site for the generation and scavenging of ROS [8–10]. Plant mitochondria have two different pathways for electron transport at the ubiquinone pool, the cyanide-sensitive cytochrome pathway and the cyanide-resistant alternative pathway. The cytochrome pathway, consisting of complex I (NADH dehydrogenase), complex II (succinate dehydrogenase), complex III, and finally complex IV (cytochrome oxidase), catalyze the four-electron reduction of O_2 to H_2O [11]. However, complex I and complex III are considered as the main source of ROS [8,12]. There are two terminal oxidases in the plant mitochondrial electron transport chain (ETC). In addition to the cytochrome pathway, alternative oxidase (AOX) can be used as terminal oxidase to reduce O_2 to H_2O in the alternative respiratory pathway descried in plant mitochondria and could produce a branch in the ETC. Then, electrons in ubiquinone are divided between the cytochrome pathway (complex III and complex IV) and AOX [11,13]. It has been reported that severe drought stress induces the accumulation of ROS in wheat (*Triticum aestivum* L.) seedlings; however, alternative pathways could improve drought-resistance by removing ROS [14]. Over-expression of the *AOX* gene reduces the level of ROS in *Arabidopsis* under chilling stress, while suppressing *AOX* induces higher levels of ROS [15]. Thus, studies have shown that AOX could play important role in balancing ROS during plant oxidative stress. However, the relationship between the alternative pathway and ROS accumulation in the mitochondria of aged seeds has not yet been thoroughly studied.

Mitochondria are important sites for the scavenging of ROS, consisting of the enzymatic antioxidant systems and non-enzymatic antioxidant systems, such as superoxide dismutase (SOD), catalase (CAT), and ascorbate-glutathione (AsA-GSH) cycles [16–18]. It has been shown that the activities of antioxidant enzymes decrease as a result of seed aging [2,6]. However, there is no further information on the role of different antioxidant enzymes on ROS scavenging in mitochondria.

Nitric oxide (NO) is a gaseous signaling transduction molecule and plays an important role in responding to diverse stressors in plants. It has been suggested that NO is a regulator of germination as well as H_2O_2 [5,19]. Some research has provided evidence that many of the crucial physiological processes of plants are related with NO, including germination, respiration, stress response, and regulating ROS balance. Exogenous NO could significantly enhance the germination rate of wheat seeds and decreased the content of H_2O_2 and O_2^- in the mitochondria under salt stress [20]. NO treatment has been shown to improve the activities of CAT, SOD, and APX in cucumber (*Cucumis sativus* L.) under salt stress [21] and wheat seed under copper stress [22]. Moreover, NO could inhibit the cytochrome pathway, while inducing the alternative pathway [23]. Royo et al. [24] showed that NO is essential for the induction of the AOX pathway under phosphate-limiting conditions in *Arabidopsis*. However, there is little known about the role of NO in regulating mitochondrial ROS and antioxidant mechanisms in aged seeds.

Recently, proteomics has become an important method to study the mechanisms of aged seeds and detect changes in various cellular processes. Li et al. [7] identified 48 mitochondrial proteins that changed in abundance on aged elm seeds and found that the alteration of voltage-dependent anion channels (VDAC), tricarboxylic acid cycle (TCA), and mitochondrial ETC were related to seed aging. Yin et al. [25] revealed the mitochondrial metabolite of aged rice (*Oryza sativa* L.) seeds under the critical node and found that induction of the alternative pathway led to a decrease in cytochrome c and the accumulation of ROS. However, the way in which events are regulated in the mitochondria of aged seeds should be further studied.

Oat, a low-carbon and eco-friendly crop, can be planted in regions experiencing a variety of environmental stresses, including infertility, salinity, drought, and cold. This study was designed to determine changes in mitochondria as a result of exogenously applied NO, including seedling growth, ROS accumulation, antioxidant enzyme improvement in the AsA-GSH cycle, and proteomics information in the embryo of oat seeds after aging, and to understand the response of mitochondria to seed deterioration.

2. Results

2.1. Changes in Seed Germination and Seedling Growth in Aged Oat Seeds under Nitric Oxide Treatments

The germination percentage and seedling length of aged oat seeds (A0) was significantly ($p < 0.05$) lower than non-aged seeds (CK), and improved significantly ($p < 0.05$) after NO treatment (A1) in all areas except for seed germination (Table 1). On the other hand, there were severe inhibiting roles of higher NO contents (A3 and A4, Figure 1) on root and shoot length. Mean germination time (MGT) of aged oat seeds (A0) was significantly higher ($p < 0.05$) than CK and could also be significantly ($p < 0.05$) reduced with NO treatment (A1). However, there were no effect of higher NO content (A2 and A3) on MGT (Table 1).

Table 1. The germination percentage, mean germination time, and seedling length of aged oat seeds with different NO concentrations. Data are means ± standard deviation (SD) from four replications for each treatment. Different letters indicate significant differences among NO treatments ($p < 0.05$).

Treatment	Germination Percentage (%)	Mean Germination Time (d)	Root Length (cm)	Shoot Length (cm)
CK	99 ± 1.2 [a]	1.9 ± 0.04 [d]	6.2 ± 0.30 [a]	4.7 ± 0.38 [a]
A0	68 ± 5.9 [b,c]	3.4 ± 0.14 [b]	3.9 ± 0.50 [c]	1.9 ± 0.23 [c]
A1	78 ± 4.3 [b]	3.0 ± 0.04 [c]	5.1 ± 0.54 [b]	2.4 ± 0.46 [b]
A2	75 ± 8.3 [b,c]	3.3 ± 0.20 [b]	4.1 ± 0.40 [c]	2.0 ± 0.27 [b,c]
A3	70 ± 8.2 [b,c]	3.4 ± 0.29 [b]	2.0 ± 0.39 [d]	0.7 ± 0.08 [d]
A4	65 ± 6.2 [c]	3.9 ± 0.19 [a]	1.7 ± 0.05 [d]	0.6 ± 0.08 [d]

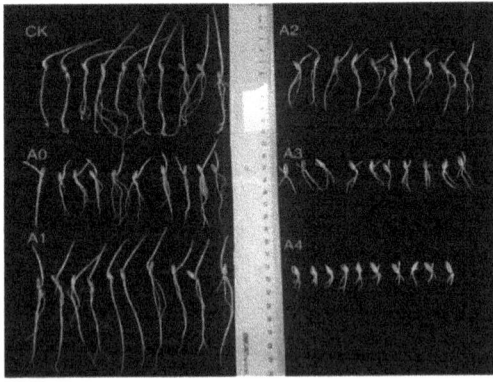

Figure 1. Patterns of seedling length in aged oat seeds with different NO treatments.

2.2. Changes in Mitochondrial H_2O_2 Content, Malate Dehydrogenase (NAD-MDH), and O_2^- Content in Aged Oat Seeds under NO Treatments

The content of mitochondrial H_2O_2 in aged seeds (A0) was significantly higher ($p < 0.05$) compared to CK, and NO application significantly ($p < 0.05$) decreased the content of H_2O_2 in aged seeds

(Figure 2A). However, there were no significant differences ($p > 0.05$) in H_2O_2 content among the treatments of A1, A2, and A3, and H_2O_2 content reached the lowest point at A4.

There was no significant difference ($p > 0.05$) in terms of the activity of mitochondrial NAD-MDH between A0 and CK, but activities of NAD-MDH were significantly improved ($p < 0.05$) with lower level exogenous NO treatments (A1 and A2), compared to A0 (Figure 2B). Furthermore, activity decreased significantly ($p < 0.05$) as NO content increased from A1 to A4 and reached the lowest level at A3. There was no significant difference ($p > 0.05$) between A3 and A4, which were similar to A0 (Figure 2B).

The mitochondrial O_2^- content in aged seeds (A0) increased compared to CK, and NO application decreased the content of O_2^- in aged seeds (Figure 2C). However, there were no significant differences ($p > 0.05$) of O_2^- content among the treatments of CK, A0, A1, A2, A3, and A4.

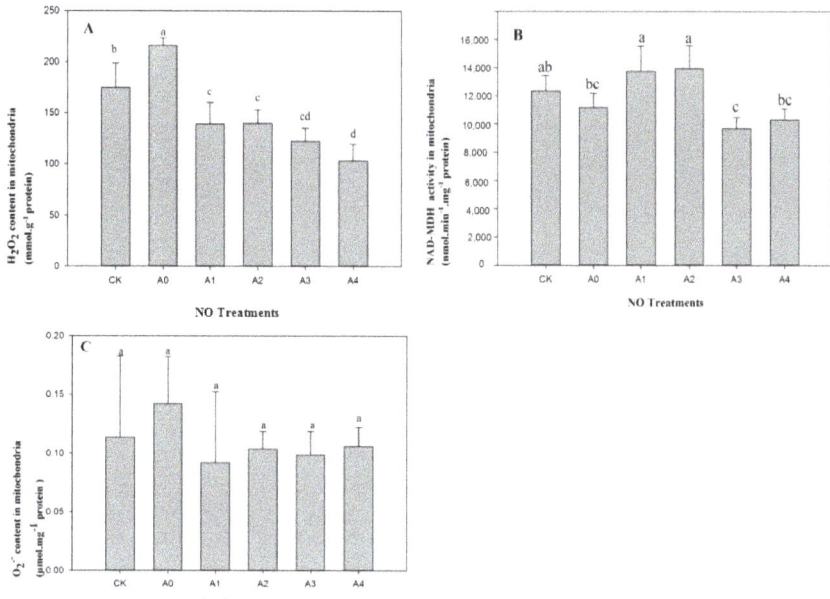

Figure 2. Changes in mitochondrial H_2O_2 content and NAD-MDH activity in aged oat seeds with different NO treatments. (**A**) H_2O_2 content; (**B**) NAD-MDH activities; (**C**) O_2^- content. Data are means ± SD from three replications for each treatment. Different letters indicate significant differences among NO treatments ($p < 0.05$).

2.3. Changes in Mitochondrial Antioxidant Enzymes in Aged Oat Seeds under Nitric Oxide Treatments

There was a significant difference ($p < 0.05$) in the activity of mitochondrial glutathione reductase (GR) between A0 and CK, but no differences were found in terms of CAT, monodehydroascorbate reductase (MDHAR), and dehydroascorbate reductase (DHAR) (Figure 3). Furthermore, activities of GR, CAT, MDHAR, and DHAR were all significantly improved ($p < 0.05$) with lower level of exogenous NO treatment (A1), compared to A0. However, enzyme activities displayed different tendencies as NO application increased from A1 to A4. The activities of GR, MDHAR, and DHAR presented no significant increase among NO treatments from A1 to A4 (Figure 3A,C,D). For CAT, activity decreased significantly ($p < 0.05$) with increasing exogenous NO content and reached the lowest level at A4 treatment, which showed no significant difference to CK or A0 (Figure 3B).

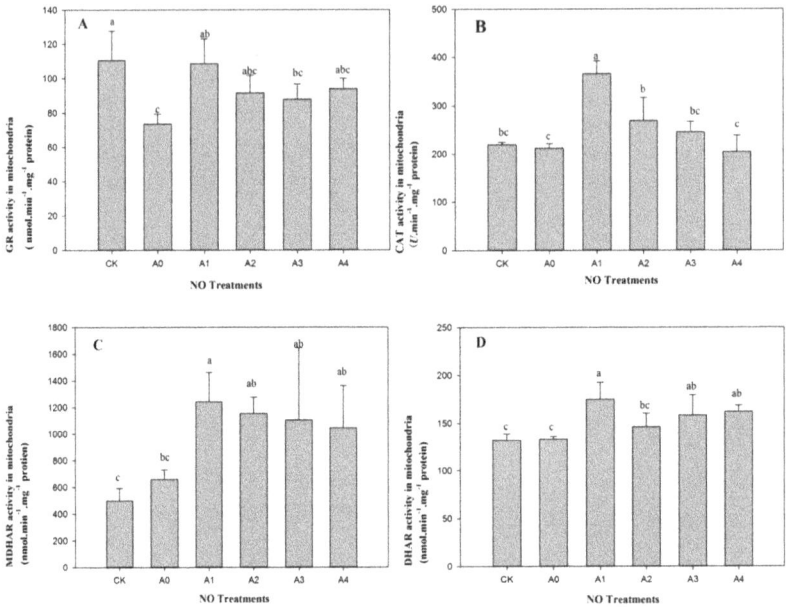

Figure 3. Changes in mitochondrial antioxidant enzyme activities in aged oat seeds with different NO treatments. (**A**) Glutathione reductase (GR) activities; (**B**) Catalase (CAT) activities; (**C**) Monodehydroascorbate reductase (MDHAR) activities; (**D**) Dehydroascorbate reductase (DHAR) activities. Data are means ± SD from three replications for each treatment Different letters indicate significant differences among NO treatments ($p < 0.05$).

2.4. Changes in Mitochondrial Complex IV in Aged Oat Seeds under Nitric Oxide Treatments

The activity of mitochondrial complex IV in aged oat seeds (A0) decreased significantly ($p < 0.05$) compared to un-aged seeds (CK). There were different effects for different NO contents on the activity of mitochondrial complex IV in aged seeds. The activity of mitochondrial complex IV in the aged seeds significant decreased ($p < 0.05$) at the lower NO treatment level (A1), then activity began to increase with NO treatments from A2 to A4. The level of activity of mitochondrial complex IV under A3 and A4 treatments attained the highest level and returned to a similar level to CK (Figure 4).

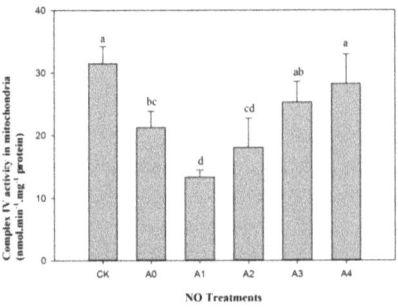

Figure 4. Changes in mitochondrial complex IV activity in aged oat seeds with different NO treatments. Data are means ± SD from three replications for each treatment. Different letters indicate significant differences among NO treatments ($p < 0.05$).

2.5. Changes in Mitochondrial Proteins in Aged Oat Seeds under Nitric Oxide Treatments

According to the results of H_2O_2 and antioxidant enzymes in aged oat seeds treated with NO, we found that A1 treatment was the most sensitive in terms of mitochondrial physiology. Therefore, seed samples from CK, A0, and A1 were selected for the proteomics analysis.

A total of 3874 proteins were found in the quantitative analysis of proteomics. Of the proteins screened with mitochondrion markers, 103 mitochondrial related proteins were identified. At the same time, the results of Gene Ontology (GO) annotation showed that 37 proteins were enriched in mitochondrial related pathways, and there were four overlapping proteins found between mitochondrial markers and GO annotation. Therefore, a total of 136 mitochondrial proteins could be identified. In order to screen the differentially expressed proteins depending on the fold changes, we divided the results of the proteomic analysis into three comparison groups, including A1/A0 (0.05 mM sodium nitroprusside (SNP) treatment compared to 0 mM SNP in aged seeds), A0/CK (aged seeds compared to unaged seeds), and A1/CK (0.05 mM sodium nitroprusside treatment in aged seeds compared to unaged seeds).

In general, only 52 differentially expressed proteins were determined for their name and functions according to the MapMan (Table 2). For A1/A0, 11 differentially expressed proteins were identified, including eight up-regulated and three down-regulated proteins. There were 11 up-regulated proteins and 29 down-regulated proteins in A0/CK, and 11 up-regulated and 24 down-regulated proteins in A1/CK. The number of up-regulated proteins was very similar, and there were much more down-regulated proteins in the group of A0/CK and A1/CK. It could be found that the number of up-regulated proteins were much more than that of down-regulated proteins in aged seeds with NO treatment (A1/A0). However, there were much more down-regulated proteins than up-regulated proteins with aging treatments in groups of A0/CK and A1/CK. According to the MapMan analysis, these differentially expressed proteins were classified into 12 functional categories, including TCA cycle, mitochondrial ETC, protein synthesis and elongation, signaling, RNA transcription, heat stress, cell division, and transport metabolism. Among these proteins, large numbers of differentially expressed proteins belonged to the functional categories of the TCA cycle and protein synthesis and elongation, which, as a proportion of total proteins, attained the highest level of 19.23%. The mitochondrial ETC followed this, at 15.38% (Figure 5).

Furthermore, from the overview of a hierarchical clustering analysis of the differentially expressed proteins in the three comparison groups, the differentially expressed proteins of A1/A0, A0/CK, and A1/CK showed different expression patterns (Figure 6A). Although there were three differentially expressed proteins identified in all three groups, two proteins existed in the groups of A1/A0 and A1/CK, the same 21 proteins in A1/CK and A0/CK, and the same five proteins were found in A1/A0 and A0/CK. Then, the remaining one, nine and 11 differentially expressed proteins only existed in the A1/A0, A1/CK, and A0/CK groups, respectively (Figure 6B). In particular, this occurred during the physiological processes, including the mitochondrial TCA cycle, ETC, and protein synthesis and elongation. Some proteins located in the TCA cycle were down-regulated in the aged seeds (A0/CK), such as the subunit of Succinate-CoA ligase (I1LYN0, K3ZV34, B4FRH5 and W5C4B7), fumarate hydratase (R4X771), malic enzyme (K3ZRI5), aldehyde dehydrogenase (K3YRJ0), and a subunit of Succinate-CoA dehydrogenase (M3AS20). However, the proteins of the subunit of Succinate-CoA ligase (I1LYN0 and K3ZV34) were only up-regulated in the aged seeds with NO treatment (A1/A0). Particularly for the Fumarate hydratase 1(Q10LR5), up-regulation appeared in seeds with NO treatment (A1/A0 and A1/CK). For the mitochondrial ETC proteins, the subunit of ATP synthase (W5BEP1 and A0A0K9R2N3) and external alternative NAD(P)H-ubiquinone oxidoreductase (Q9SKT7) were both up-regulated in the aged seeds (A0/CK) and with NO treatment (A1/CK); while, the subunit of ATP synthase (A0A200QHI3) was down-regulated in the aged seeds with NO application (A1/A0). For protein synthesis and elongation, aging treatment (A0/CK) could lead the proteins to become down-regulated, for example, the factor of elongation (A0A0D3H000, F2DG12, and F2EDF6), lon protease homolog (A0A0D3GV84 and W5C618); however, the factor of elongation (A0A0D3H000 and F2DG12) would be up-regulation with NO treatment (A1/A0). In the group of A1/CK, most proteins related to protein synthesis and elongation were down-regulated, except for the Chaperon in CPN60-like (Q8L7B5). Therefore, the proteins of the mitochondrial TCA cycle, mitochondrial ETC, and protein synthesis and elongation had different patterns across different groups (A1/A0, A0/CK, and A1/CK).

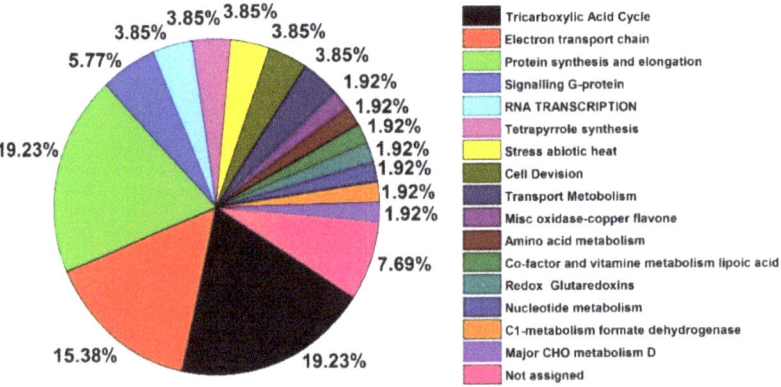

Figure 5. The functional category distribution of the 52 differently expressed mitochondrial proteins in aged oat seeds. Note: Functional classification was based on the MapMan.

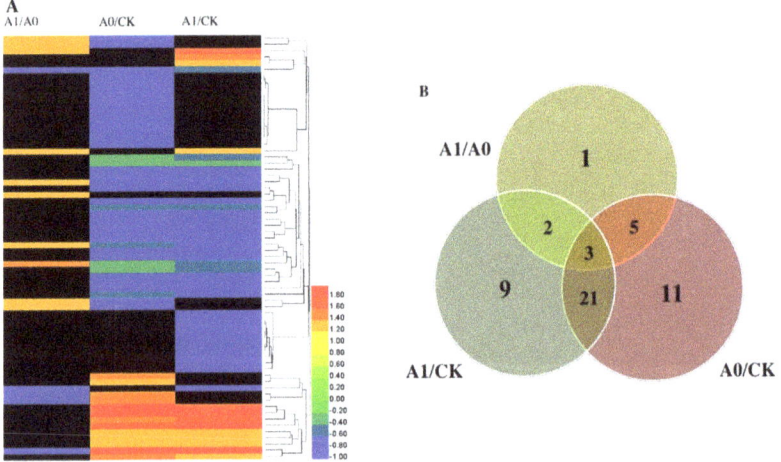

Figure 6. (**A**) Clustering analysis of differentially expressed proteins and (**B**) The number of differentially expressed proteins as a result of the exogenous application of NO across different groups (A1/A0, A0/CK, A1/CK) of aged oat seeds. Note: The color scale bar in the right of the hierarchical clustering analysis indicates the up-regulated (ratio > 0.00) and the down-regulated (ratio < 0.00) proteins. In the Venn diagram, the overlapping regions of cycles indicate proteins that were regulated in both or all treatments, whereas non-overlapping circles indicate proteins regulated in the only treatment.

Table 2. Total of 52 differentially expressed proteins across three comparison groups in aged oat seeds in response to exogenous NO treatment.

Hit Number	Accession No.	Protein Name (Species)	Fold Change		
			A1/A0	A0/CK	A1/CK
		TCA cycle			
I1LYN0	AT2G20420	Succinate-CoA ligase [ADP-forming] subunit beta, mitochondrial *Glycine max*	1.25	0.67	ns
K3ZV34	AT5G23250	Succinate-CoA ligase [ADP-forming] subunit alpha, mitochondrial *Setaria italica*	1.33	0.76	ns
Q10LR5	AT5G50950	Fumarate hydratase 1, mitochondrial, putative, expressed *Oryza sativa* subsp. *japonica*	1.25	ns	2.56
R4X771	AT5G50950	Fumarate hydratase, mitochondrial *Taphrina deformans*	ns	0.40	0.49
K3ZRI5	AT2G13560	Malic enzyme *Setaria italica*	ns	0.70	0.71
K3YRJ0	AT1G79440	Aldehyde dehydrogenase *Setaria italica*	ns	0.77	0.73
B4FRH5	AT2G20420	Succinate-CoA ligase [ADP-forming] subunit beta, mitochondrial *Zea mays*	ns	0.79	ns
W5C4B7	AT5G23250	Succinate-CoA ligase [ADP-forming] subunit alpha, mitochondrial *Triticum aestivum*	ns	0.67	ns
V4KMJ1	AT5G66760	Succinate dehydrogenase [ubiquinone] flavoprotein subunit, mitochondrial *Eutrema salsugineum*	ns	1.52	ns
M3AS20	AT5G40650	Succinate dehydrogenase [ubiquinone] iron-sulfur subunit, mitochondrial *Pseudocercosporafijiensis*	ns	0.74	ns
		Mitochondrial electron transport			
A0A200QHI3	AT3G52300	ATP synthase subunit d, mitochondrial *Macleaya cordata*	0.80	ns	0.67
W5BEP1	AT3G52300	ATP synthase subunit d, mitochondrial *Triticum aestivum*	ns	2.20	2.37
A0A0K9R2N3	AT3G52300	ATP synthase subunit d, mitochondrial *Spinacia oleracea*	ns	1.24	ns
V4LNR7	AT3G52300	ATP synthase subunit d, mitochondrial *Eutrema salsugineum*	ns	ns	0.70
A0A1J3KAQ0	AT5G08530	NADH dehydrogenase [ubiquinone] flavoprotein 1, mitochondrial *Noccaea caerulescens*	ns	0.72	0.72
W5I0L9	AT3G03070	NADH dehydrogenase [ubiquinone] iron-sulfur protein 6, mitochondrial *Triticum aestivum*	ns	ns	0.68
Q9SKT7	AT5G13430	External alternative NAD(P)H-ubiquinone oxidoreductase B4, mitochondrial *Arabidopsis thaliana*	ns	3.93	4.23
K3Y9D3	AT5G13430	Cytochrome b-c1 complex subunit Rieske, mitochondrial *Setaria italica*	ns	ns	0.78
		Protein synthesis and elongation			
A0A0D3H000	AT4G11120	Elongation factor Ts, mitochondrial *Oryza barthii*	1.35	0.65	ns
B6T7S2	AT5G47320	40S ribosomal protein S19 mitochondrial *Zea mays*	ns	0.59	0.58
F2DG12	AT2G45030	Elongation factor G, mitochondrial *Hordeum vulgare* subsp. *vulgare*	1.33	0.46	0.66
Q5JNL6	AT1G51980	Mitochondrial processing peptidase *Oryza sativa* subsp. *japonica*	ns	ns	0.68
B6U5I0	AT2G29530	Mitochondrial import inner membrane translocase subunit Tim10 *Zea mays*	ns	0.72	0.62
Q6EN45	AT5G53140	Probable protein phosphatase 2C member 13, mitochondrial *Oryza sativa* subsp. *japonica*	ns	0.42	0.62
A0A0D3GV84	AT5G26860	Lon protease homolog, mitochondrial *Oryza barthii*	ns	0.66	0.75
Q8L7B5	AT2G33210	Chaperonin CPN60-like 1, mitochondrial *Arabidopsis thaliana*	ns	ns	1.49
F2EDF6	AT4G11120	Elongation factor Ts, mitochondrial *Hordeum vulgare* subsp. *vulgare*	ns	0.74	ns
W5C618	AT5G26860	Lon protease homolog, mitochondrial *Triticum aestivum*	ns	0.76	ns
		Signalling, G-protein			
F2CSX0	AT5G27540	Mitochondrial Rho GTPase *Hordeum vulgare* subsp. *vulgare*	ns	ns	0.75
A0A1Q3B5G5	AT5G27540	Mitochondrial Rho GTPase *Cephalotus follicularis*	ns	0.40	0.42
F2E3Y6	AT5G39900	Translation factor GUF1 homolog, mitochondrial *Hordeum vulgare* subsp. *vulgare*	ns	0.76	ns

Table 2. Cont.

Hit Number	Accession No.	Protein Name (Species)	Fold Change		
			A1/A0	A0/CK	A1/CK
		RNA transcription			
Q7X745-2	AT5G39840	Isoform 2 of ATP-dependent RNA helicase SUV3L, mitochondrial *Oryza sativa* subsp. *japonica*	1.33	0.62	ns
Q6K7E2	AT2G44020	Mitochondrial transcription termination factor-like *Oryza sativa* subsp. *japonica*	ns	0.73	ns
		Tetrapyrrole synthesis, protoporphyrin IX oxidase			
W5GSR7	AT1G48520	Glutamyl-tRNA(Gln) amidotransferase subunit B, chloroplastic/mitochondrial *Triticum aestivum*	ns	0.77	0.77
K3Y6C6	AT5G14220	Protoporphyrinogen oxidase *Setaria italica*	0.75	2.43	1.83
		Stress, abiotic, heat			
G2X6B5	AT5G22060	Mitochondrial protein import protein MAS5 *Verticillium dahliae*	ns	1.56	1.66
A0A1J3J8H0	AT5G22060	Heat shock 70 kDa protein 10, mitochondrial (Fragment) *Noccaea caerulescens*	ns	0.23	0.28
		Cell devision			
F2DZF0	AT3G57090	Mitochondrial fission 1 protein *Hordeum vulgare* subsp. *vulgare*	ns	1.30	1.25
W4ZR59	AT3G57090	Mitochondrial fission 1 protein *Triticum aestivum*	ns	0.73	0.68
		Transport metobolism			
Q5NAJ0	AT1G14560	Graves disease mitochondrial solute carrier protein-like *Oryza sativa* subsp. *japonica*	ns	1.30	1.25
Q10QM8	AT5G64970	Mitochondrial carrier protein, expressed *Oryza sativa* subsp. *japonica*	ns	ns	0.68
		Misc, oxidase-copper, flavone			
B4G146	AT5G06580	D-lactate dehydrogenase [cytochrome] mitochondrial *Zea mays*	0.69	1.58	ns
		Amino acid metabolism			
B6SWZ4	AT4G34030	Methylcrotonoyl-CoA carboxylase beta chain mitochondrial *Zea mays*	ns	ns	1.21
		Co-factor and vitamine metabolism, lipoic acid			
U5H066	AT5G08415	Lipoyl synthase, mitochondrial *Microbotryum lychnidis-dioicae*	ns	0.73	0.72
		Redox, Glutaredoxins			
Q0JQ97	AT3G15660	Monothiol glutaredoxin-S1, mitochondrial *Oryza sativa* subsp. *japonica*	ns	0.78	ns
		Nucleotide metabolism			
K3ZE81	AT5G23300	Dihydroorotate dehydrogenase (quinone), mitochondrial *Setaria italica*	ns	ns	0.75
		C1-metabolism formate dehydrogenase			
A0A0D3GGT7	AT5G14780	Formate dehydrogenase, mitochondrial *Oryza barthii*	ns	0.72	0.67
		Major CHO metabolism, Degradation, sucrose, Invertases, nautral			
Q10MC0	AT1G56560	Neutral/alkaline invertase 1, mitochondrial *Oryza sativa* subsp. *japonica*	ns	0.76	ns
		Not assigned			
W5BQ98	AT4G22310	Mitochondrial pyruvate carrier *Triticum aestivum*	1.31	ns	ns
Q6ZGV8	AT3G52140	Clustered mitochondria protein homolog *Oryza sativa* subsp. *japonica*	ns	1.68	1.91
Q0JDA2	AT1G47420	Succinate dehydrogenase subunit 5, mitochondrial *Oryza sativa* subsp. *japonica*	ns	1.32	1.23
B6SPH3	AT5G08040	Mitochondrial import receptor subunit TOM5-like protein *Zea mays*	1.45	0.33	0.45

Different regulated proteins regarded as the abundance was equal to or more than 1.2-fold or less than 0.8-fold ($p \leq 0.05$). The "ns" indicates no significant difference.

3. Discussion

Seed aging and deterioration during storage could induce the loss of seed vigor, and seed germination, especially MGT, is usually used to reflect the level of seed vigor. It has been reported that seed germination decreases in aged seeds, such as maize [26], oat [2], and elm [7]. In our study, seed germination percentage decreased significantly after 26 days of aging (Table 1). Exogenous NO promotes the germination of seeds under stress or no-stress conditions [27]. For example, NO has been shown to stimulate seed germination under severe salt stress in wheat [20], and NO pretreatment significantly improves seed germination under copper stress, also in wheat [21]. Accordingly, a lower level of NO treatment (i.e., group A1) could improve the seed germination percentage in aged oat seeds. However, the differences among A1, A2, and A3 were not significant, which meant that the improvement in seed germination was sensitive to lower contents of NO. Excessive content or no NO could depress seed germination. MGT, an important index of seed vigor, exhibited a strong vigor with a shorter time [28]. Meanwhile, MGT was dramatically shorter for the A1 compared to other treatments in the aged oat seeds, and the roots and seedlings were significantly elongated. The changes in MGT and seedling length illustrate that the aged oat seed reached a higher level of vigor with the application of 0.05 mM NO. Aging leads to a change in MGT during germination. It has been shown that the transcripts encoding the proteins are associated with protein synthesis and impart changes during germination [29]. Based on the influence of imbibing time on protein abundance, different imbibing time was selected for uniform size aged seeds and non-aged seeds to ensure that they were at the same physiological point.

Mitochondrion is important organelle for respiration and metabolism, including mitochondrial ETC, and the TCA cycle. The ETC, composed of a series of electron carriers on the inner membrane of the mitochondrion, is called a respiratory chain. Furthermore, the mitochondrial ETC is a main site for ROS production, especially complex I and complex III [8,12]. It was found that the content of H_2O_2 significantly increased in aged seeds, but O_2^- content did not differ significantly, most likely because of the speed at which it was produced was not conductive to monitoring. Mitochondrial respiration might be disturbed and cause the over accumulation of ROS during seed storage. In this study, we found the mitochondrial NADH dehydrogenase [ubiquinone] flavoprotein 1 (A0A1J3KAQ0) in aged oat seeds (A0 and A1) was significantly lower than that in unaged seeds (CK). As a subunit of the mitochondrial membrane, respiratory chain NADH dehydrogenase (complex I), NADH dehydrogenase flavoprotein 1 (A0A1J3KAQ0), is believed to belong to the minimal assembly required for catalysis and complex I functions in the electron transfer from NADH to respiratory chain. The activity reduction of complex I limits the transmission of the mitochondrial respiratory chain. Jardim-Messeder et al. [30] illustrated that succinate dehydrogenase was an important site of ROS production in plant mitochondria, in addition to complex I and complex III, for enhancing H_2O_2 release, and could also be a limiting factor in plant growth through mitochondrial ROS generation. Li et al. [7] reported that the succinate dehydrogenase [ubiquinone] flavoprotein subunit 1 in mitochondrial ETC decreased in aged elm seeds. Furthermore, complex IV (Cytochrome c oxidase) was significantly lower in aged oat seeds compared to CK (Figure 5). Also, it was shown that NO could inhibit complex IV [31,32]. In our results, the activity of complex IV was significantly reduced only with a lower content of exogenous NO (A1), which reflected that the cytochrome respiratory chain was inhibited. ATP synthase (complex V) is responsible for ATP production in mitochondria. Different responses were found in terms of ATP synthase in oat seeds; the subunits of ATP synthase (W5BEP1 and A0A0K9R2N3) were up-regulated in the aged seeds (A0/CK), but no significant differences were observed for the NO treatment (A1/A0). The other two subunits (A0A200QHI3 and V4LNR7) were down-regulated with the NO treatment (A1/A0 and A1/CK). Plant mitochondria have two different pathways of electron transport: the cyanide-sensitive cytochrome pathway and the cyanide-resistant alternative pathway. With the administration of NO, the cytochrome pathway was inhibited and the alternative pathway might have changed. It has been reported that NO inhibited the cytochrome pathway, while the alternative pathway was not affected for soybean mitochondria [23].

Succinate dehydrogenase plays a central role in mitochondrion for linking the TCA cycle and ETC by catalyzing the succinate to fumarate. In our proteome analysis results, the subunit of Succinate-CoA ligase [ADP-forming] (I1LYN0 and K3ZV34) in aged seeds was significantly lower compared to CK. Succinate-CoA ligase can catalyze a reversible reaction, transforming succinyl-CoA and ADP or GDP to succinate and ATP or GTP [33]. Supplying 0.05 mM of NO (A1) significantly increased the Succinate-CoA ligase [ADP-forming] subunit (I1LYN0 and K3ZV34) in aged seeds. This protein is involved in the subpathway that synthesizes succinate from succinyl-CoA. Subsequently, succinate dehydrogenase catalyzes succinate to fumarate [30]. Then, the fumarase (fumarate hydratase; E.C. 4.2.1.20) catalyzes the reversible hydration of fumarate to malate, which is a part of TCA. The results showed that the protein abundance of fumarate hydratase (Q10LR5) in the A1 treatment was significantly higher than that of the aged seeds (A0). Meanwhile, NAD-MDH activity also increased dramatically with the supplying of 0.05 mM NO in aged seeds. MDH can catalyze the interconversion of malate and oxaloacetate, combined with the reduction or oxidation of the NAD pool [33]. NAD-MDH catalyzes NADH to reduce oxaloacetic acid and produce malate. The increase in fumarate hydratase and NAD-MDH suggested that there was an accumulation of malate. Malate can be oxidized by malic enzymes to yield pyruvate and CO_2 through oxidative decarboxylation [34]. The mitochondrial pyruvate carrier (W5BQ98) protein was observably expressed in aged seeds with application of 0.05 mM NO (A1). It has been reported that pyruvate acts on AOX to stimulate its activity in mitochondria isolated from the roots of soybean seedlings [35]. Pyruvate could activate AOX in the mitochondria of soybean cotyledons [36] and tobacco leaf [37]. It has been shown that external administration of pyruvate to isolated mitochondria results in activation of AOX pathways in wild type and pyruvate kinase transgenic lines, and transgenic lines decrease pyruvate [38]. The results demonstrated that the AOX pathway was enhanced following the NO treatment.

In general, the cytochrome pathway and alternative pathway branches point to ubiquinone. Exogenous NO inhibits the cyanide-sensitive cytochrome pathway and induces the cyanide-resistant alternative pathway. This suggests complex III is a major site of ROS production in the mitochondrial electron transport chain [39]. Alternative pathways did not pass through complex III, which could decrease the production of ROS in aged seeds with the exogenous NO. On the other hand, there was evidence that AOX played a particularly important role in regulating the balance of ROS. The AOX pathway could decrease the accumulation of ROS [14,40], and there was a higher accumulation of ROS when AOX was suppressed [15]. NO could reduce the production of ROS by inhibiting the cytochrome pathway and inducing the alternative pathway, and the alternative pathway was activated by pyruvate. It has been showed that the presence of exogenous NO induces AOX [11,41]. This was consistent with our results, where NO activated AOX pathways and thus reduced the generation of ROS.

Based on the results obtained in this study, we proposed a possible schematic pathway which might be operating during artificial aging (Figure 7). During oat seed aging, the expression of some proteins related to the TCA cycle were down-regulated and several enzymes related to mitochondrial ETC were up-regulated. Additionally, H_2O_2 of ROS accumulated dramatically, and some enzymes, including CAT, GR, MDHAR, and DHAR maintained the lower level in the AsA-GSH scavenging system. Finally, seed germination and seedling growth were limited as oat seeds aged. This indicated that seed aging led to a decrease in the activities of GR, DHAR, and MDHAR, accompanied by a gradual reduction in the mitochondrial inner membrane [2]. Changes in mitochondrial structure are posited to be responsible for the decrease in antioxidant enzymes in aged seeds, however this is worth further study. In view of the results for NO treatment in aged seeds, some enzyme activities located in the TCA cycle (such as succinate-CoA ligase and fumarate hydratase) and AsA-GSH (such as CAT, GR, MDHAR, and DHAR) were enhanced and H_2O_2 content declined at lower level, before alternative pathways were activated as the cytochrome pathway was inhibited. Seedling growth of aged oat seeds could regain normality and seed germinability could be retained to a certain level. Taken together, our results clearly demonstrate that exogenous NO in aged oat seeds can enhance seed vigor by improving enzyme activities in the AsA-GSH and decreasing the accumulation of H_2O_2. Furthermore, the enhancement of the TCA cycle and activated alternative pathway are beneficial for seedling growth in aged oat seeds with the application of 0.05 mM NO.

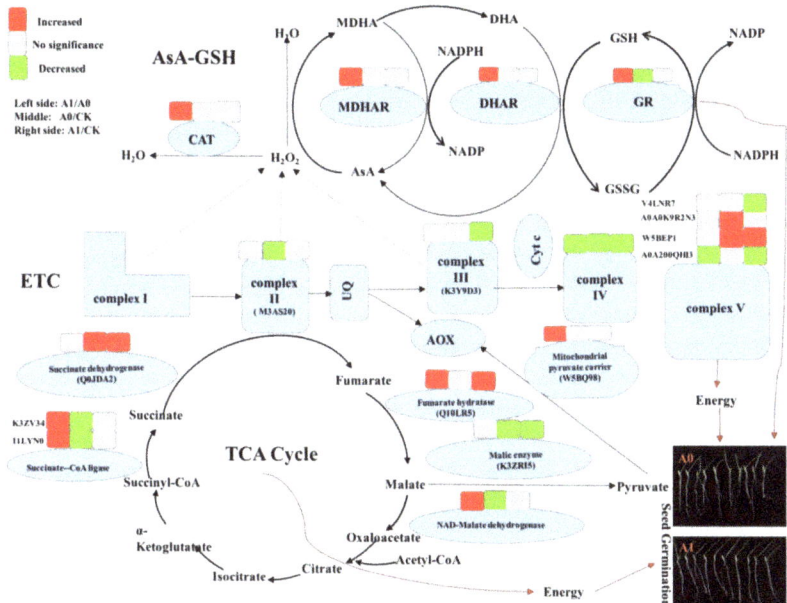

Figure 7. Schematic representation of protein abundance, ROS scavenging, and major biological pathways related to mitochondrial energy synthetic metabolism under NO treatment in aged oat seeds (different groups A1/A0, A0/CK, and A1/CK).

4. Materials and Methods

4.1. Seed Materials

Oat seeds (variety: Haywire) were bought from Beijing Clover Seed & Turf Co. Ltd. (Beijing, China). The germination percentage of the seeds was 99% and original moisture content was 9.2%.

4.2. Determination of Seed Moisture Content

The seed moisture content was determined in accordance with ISTA (2015) [42]. Approximately 4.5 g of seeds were placed in a sample container and weighed, and then they were oven-dried at 130~133 °C for 2 h (two replicates). After cooling for 30 min in a desiccator, the seeds were weighed again and the moisture content was calculated.

4.3. Adjusting the Seed Moisture Content

The moisture content of seed samples was regulated to 10%. Approximately 160 g of seeds were placed in a desiccator with saturated potassium chloride solution and weighed continually. When the seed weight was required to reach the corresponding moisture content, they were immediately placed into an aluminum foil bag and sealed, then incubated at 4 °C for 1 day, at least.

4.4. Seed Aging Treatments

After regulation of seed moisture content, the seeds in the aluminum foil bags were aged in a constant temperature water bath at 45 °C, and then aged 26 days seeds were used for experimental samples. The aged seed germination percentage was 68%.

4.5. NO Treatment and Germination Test

Uniform sized seeds were used in all treatments. The aged seeds were treated with 0 (A0), 0.05 (A1), 0.1 (A2), 0.5 (A3), and 1.0 (A4) mM sodium nitroprusside (SNP) for 8 h in the germination incubator (GXZ-380B, Ningbo, China), and the non-aged seeds (CK) were treated with distilled water for 8 h in the germination incubator (GXZ-380B, Ningbo, China). Then, all treatments were rinsed three times with distilled water. The SNP acted as a donor of NO. The seed samples after NO treatment were used in the germination trials. Germination was assayed according to the rules of ISTA (2015) [42]. Four replicates of 100 seeds were germinated in Petri dishes on filter paper with distilled water. The test was conducted in the germination incubator (GXZ-380B, Ningbo, China) at 20 °C with 8 h light and 16 h dark for 10 days. On the 5th day, the length of the root and shoot were measured, and on the 10th day the number of normal seedlings was counted. When radicles emerged from the seed coat, the whole embryo and radicle were taken as subsequent experimental materials.

Mean germination time (MGT) was calculated according to Ellis and Roberts [43]. Mean germination time (days) = $\Sigma(nd)/\Sigma n$, where n was number of germinated seeds (2 mm radicle growth through seed coat) in day, d, of counting seed germination, and Σn was total germinated seeds.

4.6. Isolation and Purification of Mitochondria

Mitochondria were extracted with 80 embryos collected from seeds with radicles protruding 5 mm after imbibition for 66 h (aged seeds) and 42 h (non-aged seeds). All extraction procedures were carried out at 0~4 °C. The embryos were ground with a mortar and pestle, using a grinding medium. The grinding medium was composed of 50 mM phosphate buffer (pH 7.5), 0.3 M mannitol, 0.5% (w/v) bovine serum albumin (BSA), 0.5% (w/v) polyvinylpyrrolidone-40, 0.2 mM EDTA-2Na, and 20 mM cysteine. The homogenate was centrifuged at 2000× g for 10 min approximately twice. The supernatant was centrifuged at 12,000× g for 15 min. The precipitate was suspended in a wash medium buffer containing 0.3 M mannitol, 0.1% (w/v) BSA, and 10 mM N-[Tris (hydroxymethyl) metyl]-2-aminopropanesulfonic acid (TES) (pH 7.5), and centrifuged again at 12,000× g for 15 min. The final precipitate was washed once with a wash medium and suspended in a small volume of the medium (mitochondrial fraction). The crude mitochondria extract was used for enzyme activity determination.

For pure mitochondria, the suspension was loaded onto a Percoll step gradient consisting of 1:4:2 ratios, bottom to top, of 40% Percoll: 21% Percoll: 16% in a mannitol wash buffer. The mixture was centrifuged for 1 h at 40,000× g, and the mitochondria presented as an opaque band at the 21/40% and 16/21% interface. The mitochondrial band was collected and washed three times by centrifugation at 20,000× g for 15 min in a wash buffer containing 0.3 M mannitol, 0.1% (w/v) BSA, and 10 mM TES (pH 7.5), and the last time without BSA in the wash buffer.

4.7. Determination of H_2O_2 Content

The H_2O_2 content in the mitochondria of the embryo was carried out using a commercial chemical assay kit (Nanjing Jianchen Bioengineering Institute, Nanjing, China) according to the manufacturer's instruction.

4.8. Enzyme Assays

Determination of the protein content in the mitochondria of the embryo was carried out using a commercial chemical assay kit (Suzhou Comin Biotechnology Institute, Suzhou, China) according to the manufacturer's instruction. The assay was based on the Coomassie brilliant blue G-250 bound to protein to form a blue complex in the acidic solution, which shows a maximum absorption peak at 595 nm.

Catalase (CAT) (EC 1.11.1.6) activity in the mitochondria of the embryo was measured by the dynamic change in absorbance at 240 nm after 1 min, due to the decline of extinction of H_2O_2. A total of 20 µL of the supernatant was mixed with 800 µL phosphate buffer (25 mM, pH 7.0, mixed with 0.1 mM EDTA) and 200 µL 100 mM H_2O_2.

GR (EC 1.6.4.2) in the mitochondria of the embryo was carried out using a commercial chemical assay kit (Suzhou Comin Biotechnology Institute, Suzhou, China) according to the manufacturer's instruction. One GR activity unit was defined as the decreasing rate of absorbance of 1 nmol NADPH per min at 340 nm.

DHAR (EC 1.8.5.1) in the mitochondria of the embryo was carried out using a commercial chemical assay kit (Suzhou Comin Biotechnology Institute, Suzhou, China) according to the manufacturer's instruction. One DHAR activity unit was defined as the increase in absorbance of 1 nmol AsA per min at 265 nm.

MDHAR (EC1.6.5.4) in the mitochondria of the embryo was carried out using a commercial chemical assay kit (Suzhou Comin Biotechnology Institute, Suzhou, China) according to the manufacturer's instruction. One MDHAR activity unit was defined as the oxidation of 1 nmol NADPH per min at 340 nm.

Complex IV in the mitochondria of the embryo was carried out using a commercial chemical assay kit (Suzhou Comin Biotechnology Institute, Suzhou, China) according to the manufacturer's instruction. A complex IV activity unit was defined as the catalytic degradation of 1 nmol of reductive cytochrome c per min at 550 nm.

NAD-MDH in the mitochondria of the embryo was carried out using a commercial chemical assay kit (Suzhou Comin Biotechnology Institute, Suzhou, China) according to the manufacturer's instruction. One NAD-MDH activity unit was defined as the consumption of 1 nmol NADH per min at 340 nm.

4.9. Protein Quantification and Digestion

Pure mitochondrial protein concentration was measured using the Bradford method. The appropriate amount of protein sample was mixed with 8 M urea solution to be quantified. According to the quantitative results above, the protein sample (500 µg each sample) was reduced by 10 mM DTT at 37 °C for 2.5 h and IAA was added to the final concentration of 10mM at room temperature for 40 °C min in darkness. All samples were transferred in the filter of a centrifuge tube (MW cutoff was 10 kDa). After reduction with DTT and alkylation with iodoacetamide, the proteins on the filter were washed three times using a lysis buffer and ABC solution (0.05 M NH4HCO3 in water), respectively. Then, the samples were digested by trypsin (1 µg trypsin for 100 µg protein) and incubated at 37 °C for 16 h. Subsequently, the peptide samples were used for LC-MS/MS analysis.

4.10. Mass Spectrometry Method and Data Analysis

4.10.1. DDA Sample Acquisition

In order to generate the spectral library, peptides from each sample were mixed and acquired twice with a data dependent acquisition (DDA) mode using Q Exactive HF (Bremen, Germany, Thermo Fisher). The peptide mixture was separated using an EasyNano LC1000 system (San Jose, Thermo Fisher) with a home-made C18 column (3 µm, 75 µm × 15 cm) at a flow rate of 450 nL/min. A 120-min linear gradient was set as follows: 3% B(0.1% FA in H_2O)/97% A(0.1% FA in H_2O) to 6% B in 12 min; 6% B to 22% B in 75 min; 22% B to 35% B in 20 min; 35% B to 100% B in 6 min; and 7 min for 100% B. For the data acquisition, a top 20 scan mode with MS1 scan range m/z 400–1200 was used, and other parameters were set as below: MS1 and MS2 resolution was set to 120 K and 30 K; AGC for MS1 and MS2 was 3×10^6 and 1×10^6; isolation window was 2.0 Th; NCE was 27; and dynamic exclusion time was 20 s.

4.10.2. Spectral Library Generation

DDA raw files were searched against a Uniprot protein database containing all plant proteins (downloaded on 2017.10.12, 2,304,711 entries) using Proteome Discoverer 2.1 (San Jose). The protein sequence was appended with the iRT fusion protein sequence (Biognosys, Schlieren, Switzerland). A search engine of SequestHT was used with the following searching parameters: enzyme of trypsin with maximum number of two missed cleavages; precursor and fragment ion mass tolerance was set to 10 ppm and 0.02 Da; variable modification was set to Oxidation of M, deamination of N, Q, Acetylation of Protein N terminus and fixed modification was set to carbamidomethylation of C; an algorithm of Percolator [44] was used to keep peptide FDR less than 1% and the q-value used for protein identification was 0.01. The search results of data-dependent acquisition using Proteome Discoverer 2.1 was transferred into a spectral library using Spectronaut 10 (Biognosys). Only a high confidence of peptide was used for the generation of the spectral library. Fragment ions within the mass range of m/z 300–1800 were kept and peptides less than three fragment ions were removed.

4.10.3. DIA Sample Acquisition

Each sample, with the addition of the same amount of iRT, was analyzed using a data independent acquisition (DIA) method. This method consisted of one full MS1 scan, with resolution set at 60 K using AGC of 3×10^6 and a maximum injection time of 20 ms. Sequential 29 isolation mass windows were set as follows: for m/z 400 to 800, the mass isolation window was set to 20 Th; for m/z 800–1000, the mass isolation window was set to 40 Th; and for m/z 1000–1200, the mass isolation window was set to 50 Th. Each DIA MS2 spectrum was acquired using a resolution of 30 K and AGC was set to 1×10^6, maximum injection time was 45 ms and collision energy was set to NCE 30. All the LC conditions were exactly the same as the DDA sample acquisition listed above.

4.10.4. DIA Data Analysis

DIA raw data were processed using Spectronaut 10. Default settings were used for protein identification and quantitation. Peak detection, dynamic iRT, correction factor 1, interference correction, and cross run normalization, were enabled. All peptides were filtered using a Q value ≤ 0.01. The average quantity of fragment ion areas from the top three peptides were used to compare protein abundance between samples.

4.10.5. Identification of Mitochondrial Proteins

Mitochondrial protein analysis was conducted when the quantitative proteins were screened for mitochondrion markers. At the same time, all proteins were submitted to DAVID (https://david.ncifcrf.gov/) and to GO annotation. Mitochondrial proteins were enriched in the mitochondria related pathways by analyzing the Cellular Component. In the study, proteins with a fold change of >1.2 or <0.8 (p-value ≤ 0.05) were regarded as differentially expressed proteins. The functions of these differential proteins were used as Basic Local Alignment Search Tools (BLAST) to find regions of local similarity between sequences in The *Arabidopsis* Information Resource (TAIR). Then, MapMan (MapManlnst-3.5.1 R2) was used to classify the functions of those mitochondrial proteins.

4.11. Statistical Analyses

The mean of three or four replicates was analyzed using analysis of variance (ANOVA), which was performed using SPSS 23.0 software. Duncan's multiple range test was used to compare the treatment means of germination, physiological indicators and enzyme activities. Data was presented as means ± SD from three or four replications for each treatment. Different letters indicated significant differences among NO treatments ($p < 0.05$).

Acknowledgments: This research was financially supported by funding from the National Natural Science Foundation (31572454), China Agriculture Research System (CARS-34), and Beijing Common Construction Project.

Author Contributions: Chunli Mao and Yanqiao Zhu designed and performed the experiments; Chunli Mao analyzed the data and wrote the manuscript; Peisheng Mao designed and supervised the research and edited the manuscript; and Huifang Yan, Hang Cheng, Liyuan Zhao, Jia Tang and Xiqing Ma were involved in performing the experiments.

Conflicts of Interest: The authors declare no conflict of interest.

Abbreviations

CAT	Catalase
DHAR	Dehydroascorbate reductase
ETC	Electron transport chain
GR	Glutathione reductase
H_2O_2	Hydrogen peroxide
MGT	Mean germination time
MDHAR	Monodehydroascorbate reductase
NAD-MDH	NAD-malate dehydrogenase
TCA	Tricarboxylic acid

References

1. El-Maarouf-Bouteau, H.; Mazuy, C.; Corbineau, F.; Bailly, C. DNA alteration and programmed cell death during ageing of sunflower seed. *J. Exp. Bot.* **2011**, *62*, 5003–5011. [CrossRef] [PubMed]
2. Xia, F.S.; Wang, M.Y.; Li, M.L.; Mao, P.S. Mitochondrial structural and antioxidant system responses to aging in oat (*Avena sativa* L.) seeds with different moisture contents. *Plant Physiol. Biochem.* **2015**, *94*, 122–129. [CrossRef] [PubMed]
3. Yan, H.F.; Mao, P.S.; Sun, Y.; Li, M.L. Impacts of ascorbic acid on germination, antioxidant enzymes and ultrastructure of embryo cells of aged elymus sibiricus seeds with different moisture contents. *Int. J. Agric. Biol.* **2016**, *18*, 176–183. [CrossRef]
4. McDonald, M.B. Seed deterioration: Physiology, repair and assessment. *Seed Sci. Technol.* **1999**, *27*, 177–237.
5. Baily, C.; El-maarouf-bouteau, H.; Corbineau, F. From intracellular signaling networks to cell death: The dual of reactive oxygen species in seed physiology. *CR Biol.* **2008**, *331*, 806–814. [CrossRef] [PubMed]
6. Kibinza, S.; Vinel, D.; Côme, D.; Bailly, C.; Corbineau, F. Sunflower seed deterioration as related to moisture content during aging, energy metabolism and active oxygen species scavenging. *Physiol. Plant.* **2006**, *128*, 496–506. [CrossRef]
7. Li, Y.; Wang, Y.; Xue, H.; Pritchard, H.W.; Wang, X.F. Changes in the mitochondrial protein profile due to ROS eruption during ageing of elm (ulmus pumila L.) seeds. *Plant Physiol. Biochem.* **2017**, *114*, 72–87. [CrossRef] [PubMed]
8. Møller, I.M. Plant mitochondria and oxidative stress: Electron transport, NADPH turnover, and metabolism of reactive oxygen species. *Annu. Rev. Plant Physiol. Plant Mol. Biol.* **2001**, *52*, 561–591. [CrossRef] [PubMed]
9. Fleury, C.; Mignotte, B.; Vayssière, J.L. Mitochondrial reactive oxygen species in cell death signaling. *Biochimie* **2002**, *84*, 131–141. [CrossRef]
10. Rhoads, D.M.; Umbach, A.L.; Subbaiah, C.C.; Siedow, J.N. Mitochondrial reactive oxygen species, contribution to oxidative stress and interorganellar signaling. *Plant Physiol.* **2006**, *141*, 357–366. [CrossRef] [PubMed]
11. Vanlerberghe, G.C. Alternative oxidase: A mitochondrial respiratory pathway to maintain metabolic and signaling homeostasis during abiotic and biotic stress in plants. *Int. J. Mol. Sci.* **2013**, *14*, 6805–6847. [CrossRef] [PubMed]
12. Navrot, N.; Rouhier, N.; Gelhaye, E.; Jacquot, J.P. Reactive pxygen species generation and antioxidant systems in plant mitochondria. *Physiol. Plant.* **2007**, *129*, 185–195. [CrossRef]
13. Affourtit, C.; Krab, K.; Moore, A.L. Control of plant mitochondrial respiration. *Biochim. Biophys. Acta* **2001**, *1504*, 58–69. [CrossRef]

14. Wu, Q.; Feng, H.Q.; Li, H.Y.; Wang, D.S.; Liang, H.G. Effects of drought stress on cyanide-resistance respiration and metabolism of reactive oxygen in wheat seedling. *J. Plant Physiol. Mol. Biol.* **2006**, *32*, 217–224.
15. Fiorani, F.; Umbach, A.L.; Siedow, J.N. The alternative oxidase of plant mitochondria is involved in the acclimation of shoot growth at low temperature. A study of Arabidopsis AOX1a transgenic plants. *Plant Physiol.* **2005**, *139*, 1790–1805. [CrossRef] [PubMed]
16. Bailly, C. Active oxygen species and antioxidants in seed biology. *Seed Sci. Res.* **2004**, *14*, 93–107. [CrossRef]
17. Xin, X.; Tian, Q.; Yin, G.K.; Chen, X.L.; Zhang, J.M.; Sophia, N.G.; Lu, X.X. Reduced mitochondrial and ascorbate-glutathione activity after artificial ageing in soybean seed. *J. Plant Physiol.* **2014**, *177*, 140–147. [CrossRef] [PubMed]
18. Das, K.; Roychoudhury, A. Reactive oxygen species (ROS) and response of antioxidants as ROS-scavengers during environment stress in plants. *Front. Environ. Sci.* **2014**, *2*, 1–13. [CrossRef]
19. Krasuska, U.; Gniazdowska, A. Nitric oxide and hydrogen cyanide as regulating factors of enzymatic antioxidant system in germination apple embryos. *Acta Physiol. Plant.* **2012**, *34*, 683–692. [CrossRef]
20. Zheng, C.F.; Jiang, D.; Liu, F.L.; Dai, T.B.; Liu, W.C.; Jing, Q.; Cao, W.X. Exogenous nitric oxide improves seed germination in wheat against mitochondrial oxidative damage induced by high salinity. *Environ. Exp. Bot.* **2009**, *67*, 222–227. [CrossRef]
21. Lin, Y.; Liu, Z.Z.; Shi, Q.; Wei, M.; Yang, F.J. Exogenous nitric oxide (NO) increased antioxidant capativity of cucumber hypocotyl and radicle under salt stress. *Sci. Hortic.* **2012**, *142*, 118–127. [CrossRef]
22. Hu, K.D.; Hu, L.Y.; Li, Y.H.; Zhang, F.Q.; Zhang, H. Protective roles of nitric oxide on germination and antioxidant metabolism in wheat seeds under copper stress. *Plant Growth Regul.* **2007**, *53*, 173–183. [CrossRef]
23. Millar, A.H.; Day, D.A. Nitric oxide inhibits the cytochrome oxidase but not the alternative oxidase of plant mitochondria. *FEBS Lett.* **1996**, *398*, 155–158. [CrossRef]
24. Royo, B.; Moran, J.F.; Ratcliffe, R.G.; Gupta, K.J. Nitric oxide induces the alternative oxidase pathway in *Arabidopsis* seedlings deprived of inorganic phosphate. *J. Exp. Bot.* **2015**, *60*, 6273–6280. [CrossRef] [PubMed]
25. Yin, G.K.; Whelan, J.; Wu, S.H.; Zhou, J.; Chen, B.; Chen, X.L.; Zhang, J.M.; He, J.J.; Xin, X.; Lu, X.X. Comprehensive mitochondrial metabolic shift during the critical node of seed ageing in rice. *PLoS ONE* **2016**, *11*, 1–19. [CrossRef] [PubMed]
26. Xin, X.; Lin, X.H.; Zhou, Y.C.; Chen, X.L.; Liu, X.; Lu, X.X. Proteome analysis of maize seeds: The effect of artificial ageing. *Physiol. Plant.* **2011**, *143*, 126–138. [CrossRef] [PubMed]
27. Belibni, M.V.; Lamattina, L. Nitric oxide stimulates seed germination and de-etiolation and inhibits hypocotyl elongation, three light-inducible responses in plants. *Planta* **2000**, *210*, 215–221. [CrossRef] [PubMed]
28. Mattews, S.; Noli, E.; Demir, I.; Khajeh-Hosseini, M.; Wagner, M.H. Evaluation of seed quality: From physiology to international standardization. *Seed Sci. Res.* **2012**, *22*, S69–S73. [CrossRef]
29. Law, S.R.; Narsai, R.; Whelan, J. Mitochondrial biogenesis in plants during seed germination. *Mitochondrion* **2014**, *19*, 214–221. [CrossRef] [PubMed]
30. Jardim-messeder, D.; Caverzan, A.; Rauber, R.; Ferreira, E.S.; Margis-Pinheiro, M.; Galina, A. Succinate dehydrogenase (mitochondrial complex II) is a source of reactive oxygen species in plants and regulates sevelopment and stress responnses. *New Phytol.* **2015**, *208*, 776–789. [CrossRef] [PubMed]
31. Brown, G.C. Nitric oxide regulates mitochondrial respiration and cell function by inhibiting cytochrome oxidase. *FEBS Lett.* **1995**, *369*, 136–139. [CrossRef]
32. Brown, G.C. Regulation of mitochondrial respiration by nitric oxide inhibition of cytochrome c oxidase. *Biochim. Biophys. Acta* **2001**, *1504*, 46–57. [CrossRef]
33. Johnson, J.D.; Mehus, J.G.; Tews, K.; Milavetz, B.L.; Lambeth, D.O. Genetic evidence for the expression of ATP- and GTP-specific Succinyl-CoA synthetases in multicellular eucaryotes. *J. Biochem.* **1998**, *273*, 27580–27586. [CrossRef]
34. Maurino, V.G.; Gerrard Wheeler, M.C.; Andreo, C.S.; Drincovich, M.F. Redoundancy is sometimes seen only by the uncritical: Does Arabidopsis need six malic enzyme isoform. *Plant Sci.* **2009**, *176*, 715–721. [CrossRef]
35. Millar, A.H.; Wiskich, J.T.; Whelan, J.; Day, D.A. Organic acid activation of the alternative oxidase of plant mitochondria. *FEBS Lett.* **1993**, *329*, 259–262. [CrossRef]
36. Day, A.A.; Millar, A.H.; Wiskich, J.T.; Whelan, J. Regulation of alternative oxidase by pyruvate in soybean mitochondria. *Plant Physiol.* **1994**, *106*, 1421–1427. [CrossRef] [PubMed]

37. Vanlerberghe, G.C.; Day, D.A.; Wiskich, J.T.; Vanlerberghe, A.E.; McIntosh, L. Alternative oxidase activity in tobacco leaf mitochondria (Dependence on trucarboxylic acid cycle-mediated redox regulation and pyruvate activation). *Plant Physiol.* **1995**, *109*, 353–361. [CrossRef] [PubMed]
38. Oliver, S.N.; Lunn, J.E.; Urbanzcyk-Wochniak, E.; Lytovchenko, A.; Van Dongen, J.T.; Faix, B.; Schmalzlin, E.; Fernie, A.R.; Geigenberger, P. Decreased expression of cytosolic pyruvate kinase in potato tubers lead to a decline in pyruvate resulting in an in vivo repression of the alternative oxidase. *Plant Physiol.* **2008**, *148*, 1640–1654. [CrossRef] [PubMed]
39. Raha, S.; Robinson, B.H. Mitochondria, oxygen free radicals, disease and ageing. *Trends Biochem. Sci.* **2000**, *25*, 502–508. [CrossRef]
40. Maxwell, D.P.; Wang, Y.; Mclotosh, L. The alternative oxidase lowers mitochondrial reactive oxygen production in plant cells. *Proc. Natl. Acad. Sci. USA* **1999**, *96*, 8271–8276. [CrossRef] [PubMed]
41. Huang, X.; Rad, U.V.; Durner, J. Nitric oxide induces transcriptional activation of the nitric oxide-tolerant alternative oxidase in Arabidopsis suspension cells. *Planta* **2002**, *215*, 914–923. [CrossRef] [PubMed]
42. ISTA. *International Rules for Seed Testing*; International Seed Testing Association: Basserdorf, Awitzerland, 2015.
43. Ellis, R.A.; Roberts, E.H. The quantification of ageing and survival in orthodox seeds. *Seed Sci. Res.* **1981**, *9*, 373–409.
44. Spivak, M.; Weston, J.; Bottou, L.; Kall, L.; Noble, W.S. Improvements to the percolator algorithm for peptide identification from shotgun proteomics data sets. *J. Proteom. Res.* **2009**, *8*, 3737–3745. [CrossRef] [PubMed]

© 2018 by the authors. Licensee MDPI, Basel, Switzerland. This article is an open access article distributed under the terms and conditions of the Creative Commons Attribution (CC BY) license (http://creativecommons.org/licenses/by/4.0/).

MDPI
St. Alban-Anlage 66
4052 Basel
Switzerland
Tel. +41 61 683 77 34
Fax +41 61 302 89 18
www.mdpi.com

International Journal of Molecular Sciences Editorial Office
E-mail: ijms@mdpi.com
www.mdpi.com/journal/ijms

www.ingramcontent.com/pod-product-compliance
Lightning Source LLC
LaVergne TN
LVHW071936080526
838202LV00064B/6617